Introduction to Software for Chemical Engineers

Introduction to Software for Chemical Engineers

Second Edition

Edited by
Mariano Martín Martín

CRC Press
Taylor & Francis Group
Boca Raton London New York

CRC Press is an imprint of the
Taylor & Francis Group, an **informa** business

CRC Press
Taylor & Francis Group
6000 Broken Sound Parkway NW, Suite 300
Boca Raton, FL 33487-2742

CRC Press is an imprint of Taylor & Francis Group, an Informa business

Printed on acid-free paper

International Standard Book Number-13 978-1-138-32422-0 (Hardback); 978-1-138-32421-3 (Paperback)

Library of Congress Cataloging-in-Publication Data

Names: Martín, Mariano Martín, editor.
Title: Introduction to software for chemical engineers / [edited by] Mariano Martín Martín.
Description: Second edition. | Boca Raton, FL: CRC Press/Taylor & Francis Group, 2018. | Includes bibliographical references and index.
Identifiers: LCCN 2018061322| ISBN 9781138324220 (hardback: acid-free paper) | ISBN 9781138324213 (pbk.: acid-free paper) | ISBN 9780429451010 (ebook)
Subjects: LCSH: Chemical engineering–Computer programs. | Chemical engineering—Data processing.
Classification: LCC TP184 .I595 2018 | DDC 660.0285—dc23
LC record available at https://lccn.loc.gov/2018061322

Visit the Taylor & Francis Web site at
http://www.taylorandfrancis.com

and the CRC Press Web site at
http://www.crcpress.com

Contents

SECTION III Detailed Equipment Design and Analysis

SECTION IV Process Simulation

SECTION V Process Design & Optimization

Preface

The field of Chemical Engineering and its link to Computer Science through the use of software packages is in constant evolution. New engineers have at their disposal a large number of tools to tackle their everyday problems. In the second edition of this book we have included several additional packages to the already wide spectrum in order to provide a better overview of the software tools at hand. This book is inspired by the difficulties in introducing during class hours the use of different software and the need for a textbook that is more user friendly than instruction manuals while at the same time able to address typical examples within the subjects of the Chemical Engineering curriculum. Therefore, our intention is to provide an introduction that serves as a quick reference guide to the use of different computer software for Chemical Engineering applications so that the students become familiar with the capabilities. The book covers a wide range of computer packages from general purpose ones such as MS EXCEL®; common mathematical packages such as MATLAB®, Python, R® or MATHCAD®; process simulators such as CHEMCAD®, ASPEN PLUS®, or ASPEN-HYSYS®; equation-based modeling tools such as gPROMS® and Engineering Equation Solver (ESS)®; specialized software including computational fluid dynamics (CFD), COMSOL, and ANSYS Fluent) or discrete element methods (DEM), codes, and optimization software such as GAMS®, AIMMS®, LINGO®, Julia, or SolverStudio. The different packages are introduced and applied to solve typical problems in fluid mechanics, heat and mass transfer, mass and energy balances, unit operations, reactor engineering, and process and equipment design and control, whose solution by hand is complex and most of the times untractable. We do this also to show that real world problems require the use of simulation and optimization tools. At this point, we would like to highlight that the capabilities of the packages are only introduced; and thus more complex problems are left for advanced users.

I would like to take the opportunity to thank the companies of the software mentioned above, for their kind support in the elaboration of this book; the access to trial versions, such as Chemstations for CHEMCAD® and ANSYS Fluent; and suggestions and corrections to the manuscript, such as Mathworks for MATLAB® or Process Systems Enterprise for gPROMS®. For the images used in Chapter 8 we would like to acknowledge COMSOL Multiphysics® from the COMSOL® Model Gallery without any representation or warranties of any kind including, but not limited to any implied warranties of merchantability, fitness for a particular purpose or non-infringement. Furthermore, some images in Chapter 8 are made using COMSOL Multiphysics® and are provided courtesy of COMSOL.® It is also important to acknowledge those who introduced us in the use of computers for chemical process calculations and analysis. Some of the problems in the book are inspired by their teachings. Furthermore, I would like to thank the contributors since, without their expertise and effort, the book could not have covered such a wide spectrum of packages and topics. We also appreciate the work of colleagues for their comments, and those who provided comments to the first edition: Dr. Hossein Amadian,

Prof. Lam Hon Loong, Dr. BoonHo NG, and Dr. Ben Weinstein. We also thank our previous students at the University of Salamanca (Laura Bueno Romo, William Davis Fernández, Estela Peral Elena, Mónica Rodríguez Martín, and Marta Vidal Otero), as well as previous students at Carnegie Mellon University (Kristen A. Severson) and the University of Birmingham (Dr. Emilia Nowak) for their valuable comments and suggestions. We are also grateful at the University of Salamanca—Jose A. Luceño, Jose Enrique Roldán, Paula Manteca, Judit Redondo and Arantza Criado, Manuel Taifouris—who helped correct the current edition. It is not expected that all students will love process modeling and simulation. However, this is a world open to them. We hope that this book can help them take their first steps.

Mariano Martín Martín
Salamanca, November 2018

Editor Biography

Dr. Mariano Martín Martín is Associate Professor of Chemical Engineering at the University of Salamanca and Head of the Master Studies. He graduated cum laude in an integrated program (BSc+MEng) (2003) and holds a PhD in Chemical Engineering from the University of Salamanca (2008). Dr. Martín joined the Modeling and Simulation Department at Procter and Gamble. He left Prector and Gamble for a Fulbright Postdoctoral Fellowship at Carnegie Mellon University. He received the Accesit Mapfre Award in 2003, the Extraordinary PhD Award from the University of Salamanca in 2008, and the P&G Award for his contributions to modeling and simulation in 2009. He was named Rising Star in Chemical Engineering by Imperial College London in 2016 and has been nominated four times to the ENI Awards. Dr. Martín has published over 100 peer-reviewed papers and authored/edited four books for major editorials. http://diarium.usal.es/marianom3/

Contributors

EDITOR

Mariano Martín Martín
Associate Professor
Department of Chemical Engineering
University of Salamanca
Salamanca, Spain

AUTHORS

Mohammadreza Alizadeh Behjani
Research Fellow
School of Chemical and Process Engineering
University of Leeds
Leeds, United Kingdom

Carlos Amador Zamarreño
Principal Engineer Formula Design
Modelling and Simulation, Procter and Gamble
Newcastle Innovative Centre
Newcastle upon Tyne, United Kingdom

Lorenz T. Biegler
Bayer Professor Chemical Engineering
Carnegie Mellon University
Pittsburgh Pennsylvania

Laura Bueno Romo
PhD Student
School of Chemical Engineering
The University of Birmingham
Birmingham, United Kingdom
and
Department of Chemical Engineering
University of Salamanca
Salamanca, Spain

José A. Caballero Suárez
Professor
Department Chemical Engineering
Universidad de Alicante
Alicante, Spain

Francesco Coletti
Associate Professor
Department of Chemical Engineering
College of Engineering, Design and Physical Sciences
Brunel University London
and
Hexxcell Ltd
London, United Kingdom

María Soledad Díaz
Associate Professor
Planta Piloto de Ingeniería Química (PLAPIQUI)
Universidad Nacional Del Sur
Bahía, Blanca, Argentina

Emilio Díaz-Bejarano
Senior Consultant
Hexxcell Ltd
London, United Kingdom

Victor Francia
Assistant Professor
Institute of Mechanical, Process and Energy Engineering
School of Engineering and Physical Sciences
Heriot-Watt University
Edinburgh, United Kingdom

Ignacio E. Grossmann
Dean University Professor
Department of Chemical Engineering
Carnegie Mellon University
Pittsburgh, Pennsylvania

Gonzalo Guillén-Gosálbez
Professor
Department of Chemistry and Applied Biosciences
ETH
Zurith, Switzerland

Bostjan Hari
Project Engineer
WMG, International Manufacturing Centre
University of Warwick
Coventry, United Kingdom

Iiro Harjunkoski
Corporate Research Fellow
ABB Corporate Research Center Germany
Ladenburg, Germany

Ali Hassanpour
Associate Professor
School of Chemical and Process Engineering
University of Leeds,
Leeds, United Kingdom

Borja Hernández Blázquez
Department of Chemical Engineering
University of Salamanca
Salamanca, Spain

Luis G. Hernández-Pérez
Department of Chemical Engineering
Universidad Michoacana de San Nicolás de Hidalgo
Guanajuato, Mexico

Jordan Jalving
PhD Student
Department of Chemical and Biological Engineering
University of Wisconsin-Madison
Madison, Wisconsin

Arturo Jiménez Gutiérrez
Departamento de Ingeniería Química
Instituto Tecnológico de Celaya
Celaya, Guanajuato, México

Ricardo Manuel Pinto de Lima
Research Scientist
Computer, Electrical and Mathematical Sciences and Engineering Division
King Abdullah University of Science and Technology
Thuwal, Saudi Arabia

Luis Martín de Juan
Principal Engineer
Procter and Gamble
Newcastle Innovative Centre
Newcastle upon Tyne, United Kingdom

Mariano Martín Martín
Associate Professor
Department of Chemical Engineering
University of Salamanca
Salamanca, Spain

Edgar Martín Hernández
Department of Chemical Engineering
University of Salamanca
Salamanca, Spain

Alberto Martínez Bécares
Senior Engineer
Procter and Gamble
Brussels Innovative Center
Strombeek-Bever, Belgium

W. Alex Marvin
Engineer
BASF Corporation
Florham Park
New Jersey, USA

Massih Pasha
The Chemours Company
Wilmington, Delaware

Salvador I. Pérez-Uresti
PhD Student
Departamento de Ingeniería Química
Instituto Tecnológico de Celaya
Celaya
Guanajuato, México

José María Ponce Ortega
Professor
Department of Chemical Engineering
Universidad Michoacana de San Nicolás de Hidalgo
Guanajuato, Mexico

César Ramírez-Márquez
Department of Chemical Engineering
University of Guanajuato
Guanajuato, Mexico

Rubén Ruiz-Femenía
Associate Professor
Department Chemical Engineering
Universidad de Alicante
Alicante, Spain

Antonio Sánchez García
PhD Student
Department of Chemical Engineering
University of Salamanca
Salamanca, Spain

Lidia Sánchez Guerras
PhD Student
Department of Chemical Engineering
University of Salamanca
Salamanca, Spain

Juan Gabriel Segovia
Professor
Department of Chemical Engineering
University of Guanajuato
Guanajuato, Mexico

Kristen A. Severson
PhD Student
Department of Chemical Engineering
Massachusetts Institute of Technology
Cambridge, Massachusetts

Martijn A.H. van Elzakker
Edwin Zondervan Laboratory of Process System Engineering
Technische Universiteit Eindhoven
Eindhoven, Netherlands

Ángel L. Villanueva Perales
Associate Professor
Department of Chemical and Environmental Engineering
School of Engineering
University of Sevilla
Sevilla, Spain

Fengqi You
Roxanne E. and Michael J. Zak Professor and David Croll Sesquicentennial Faculty Fellow
Department of Chemical and Biomolecular Engineering
Cornell University
Ithaca, New York

Victor M. Zavala
Baldovin-DaPra Associate Professor
Department of Chemical and Biological Engineering
University of Wisconsin-Madison
Madison, Wisconsin

Qi Zhang
Assistant Professor
Department of Chemical Engineering and Materials Science
University of Minnesota
Minneapolis, Minnesota

Edwin Zondervan
Professor
Assistant Professor
Laboratory of Process Systems Engineering
University of Bremen
Bremen, Germany

Section I

Modeling and Simulation in the Chemical Engineering CV and its Application to Industry

1 Modeling, Simulation, and Optimization in the Chemical Engineering Curriculum

Mariano Martín, Ignacio E. Grossmann

CONTENTS

1.1 INTRODUCTION

It is not possible to describe the history of the Chemical Engineering curriculum without highlighting a number of important events from the development of Chemical Engineering as a field and discipline. The Chemical Heritage Foundation has published a list of major events in the history of Chemical Engineering "The First Century of Chemical Engineering, A Timeline of Discoveries and Achievements" [1]. Apart from this list, Hougen in 1977 [2] and a few years later Freshwater and Yates in 1985 [3] proposed a number of stages that somehow define the evolution of Chemical Engineering as a field. More recently, CEP [4] published the list of the 30 authors and books that have influenced the most the Chemical Engineering education. Based on these and other reviews [5,6] we can describe how Chemical Engineering curriculum has evolved. Thus, we extend the stages presented by Hougen, Freshwarer, & Yates to start a few years before and finish today. We present here 7 stages where the main events and the books launched have been responsible for the current chemical engineering curriculum.

1.1.1 Establishment Stage

We can trace back this era to 1882 when a course in "Chemical Technology" was offered at University College, London. A few years later, in 1885, Henry E. Armstrong offered a course in "Chemical Engineering" at Central College also in London (later Imperial College, London). Apart from London, in 1887 George E. Davis, a former industrial inspector in England, offered lectures on "chemical operations" at the Manchester Technical School. The main feature of those lectures was that he presented the existence of a number of relative limited "common operations" that were applied to different chemical processes. Based on those lectures he published *Handbook of Chemical Engineering* (1901 and 1904). However, even though it was in the UK where the first ideas were put into practice, the first Chemical Engineering curriculum was established in 1888 when Lewis M. Norton proposed Course X: Chemical Engineering at the Massachusetts Institute of Technology (MIT). This course's idea was to provide the students with a background on mechanical engineering together with chemistry basics and the engineering applications that were used in the production of chemical products. The first graduates as bachelors of science in Chemical Engineering were awarded in 1891. Just a year later the University of Pennsylvania started her bachelor's program in Chemical Engineering. The first PhD in Chemical Engineering was awarded in 1905 by the University of Wisconsin. Over these first years, the teaching methodology on Chemical Engineering was mostly descriptive and most people at this point consider Chemical Engineering as an art rather than a science. We end this stage in 1908, when the American Institute of Chemical Engineers (AIChE) is founded in Philadelphia.

1.1.2 Unit Operations Stage

One of the key founding contributions to Chemical Engineering was provided in 1915 when Arthur D. Little stated that many operations typically used in the chemical industry were common in different processes. This was not exactly new, he based his statement on Davis' "Chemical Operations" report to the MIT Corporation (Board of Trustees); Little and Prof. William Walker restructured the MIT Chemical Engineering curriculum around the "Unit Operations" concept providing Chemical Engineering with identity enabling the first stage in the transformation of Chemical Engineering from a descriptive discipline into a science. By 1922, the Institution of Chemical Engineers is founded in the UK just after the first Chemical Engineering department was established at MIT. Furthermore the first textbook in Chemical Engineering appears in 1923 *Principles of Chemical Engineering,* by W.H. Walker, W.K. Lewis and W.H. Mc Adams and soon later *Industrial Stoichiometry* by W.K. Lewis which can be considered as the starting point for the next stage. In 1925 the AIChE begins accreditation of Chemical Engineering programs in the U.S., awarding it to 15 institutions. This year is also remembered by the presentation of the McCabe and Thiele's graphical method for computing the number of equilibrium plates required in a fractionating column for binary mixtures. A method that is still in use in class rooms worldwide.

1.1.3 Mass and Energy Balances & Thermodynamics

This third stage is based on the look for better understanding on the fundamentals of the processes based on the application of thermodynamic principles to them. The pioneers in introducing the importance of thermodynamics in Chemical Engineering Education were Hougen & Watson, at the University of Wisconsin. Thus, in 1931 they published *Industrial Chemical Calculations* (later entitled *Chemical Process Principles*), which constitutes a comprehensive treatment of material and energy balances applied to process analysis replacing the descriptive nature of Chemical Engineering used until then. During this period at Michigan, Katz, Brown, White, Kurata, Standing, & Sliepcevich help establish some foundations in phase equilibria, heat, momentum and mass transfer. Following this trend it also began the systematic analysis of chemical reactors with important names in the Chemical Engineering field devoted to that subject; Damkohler in Germany, Van Heerden in Holland, and Danckwerts and Denbigh in England. They explored mass transfer, temperature variations, flow patterns, and multiple steady states.

In 1934 another famous book was published. Perry's first edition of the *Chemical Engineers Handbook*. Process scale up was based on dimensionless studies and a number of dimensionless numbers appear to design different chemical equipment.

By the end of this period the first books on thermodynamics applied to Chemical Engineering appeared starting with *Chemical Engineering Thermodynamics* by Bennet F. Dodge in 1944.

1.1.4 Applied Kinetics and Process Design

Only three years later, 1947, Hougen and Watson published *Kinetics and Catalysis* (*Chemical Process Principles* Part 3) which represents the foundation for modern Chemical Reaction Engineering. This same year W. Robert Marshall and Robert L. Pigford published *Applications of Differential Equations to Chemical Engineering Problems*, illustrating applications of mathematics to many areas of Chemical Engineering. Up to a certain point it constituted the first step towards the systematic process design.

1.1.5 Transport Phenomena: Science in Chemical Engineering

Mathematics and Chemical Engineering became partners and allies in this journey to establish the core of Chemical Engineering. Professors Amundson & Aris highlighted the importance in mathematical modeling in chemical reactor engineering. This work is key in the history of Chemical Engineering curriculum since from now on more competence in mathematics is required. It was in 1950 when Brown et al published *Unit Operations* where solid handling, fluid, mass and energy transport were classified for the first time. Furthermore, professors Bird, Steward and Lightfoot, from the University of Wisconsin, released in 1956 the unifying concepts of mass, momentum, and energy transport. Their textbook, *Transport Phenomena*, provides a highly mathematical treatment of heat, mass, and momentum transfer. It was quickly adopted in graduate Chemical Engineering education, but it did not get the same attention at the undergraduate level or in Chemical Engineering practice. The book provided the scientific basis not only for unit operations but also for chemical reaction engineering. As a result, from that moment on, Chemical Engineering became a science rather than an art. In 1959 Stanley Wallas published *Kinetics for Chemical Engineers* and a few years later, in 1962, Octave Levenspiel published *Chemical Reactor Engineering*, which is seen as the last contribution of the first collection of Chemical Engineering books.

1.1.6 Process System Engineering: Computer Science

Up to the decade of the 1960's chemical engineers lack the tools to design a complete process. However, by the end of the decade of 1950's computer science was rising as a promising field providing new tools that lead to the new concept. Furthermore, control was becoming an important discipline for Chemical Engineering. A few years later, 1964, Roger Sargent, the father of Process System Engineering released "SPEEDUP" (Simulation Program for the Economic Evaluation and Design of Unsteady-State Processes). Some books were also published by the 60's and 70's on numerical analysis, process design and flowsheeting [7]. By this time CACHE organization was created. CACHE stands for Computer Aids for Chemical Engineering. Its main aim was to develop new methods for introducing computers into chemical engineering teaching. In 1969 the CACHE committee was formed initially sponsored by the National Academy of Engineering and the National Science Foundation. In 1985 CACHE came up with a report suggesting the capabilities than by that time a chemical engineer must achieve from word processing and graphics to, scientific programming language competence [7]. With all of this into place, even though the presence of the computers can be traced back to 1959, it was not until 1981 that different specialized packages became available for the public such as DESIGN II, ASPEN, SIMSCI (PROII), HYSYS & CHEMCAD which start appearing on engineering desktops. The development of computers allowed better analysis and understanding of the processes. However, one of the reasons why the Chemical Engineering curriculum did not utilize computer based methods was the fact that popular textbooks were not computer oriented [7] until very recently. The development of computers over the last decades has allowed the solution to more complex problems and/or to provide a different approach. This can be easily seen by direct comparison between traditional books in the field of fluid mechanics where the equations were solved for particular asymptotic cases where certain simplifications hold, compared to the use of Computational Fluid Dynamics (CFD) codes to solve the momentum equations. This fact has been reflected in the text books for Chemical Engineering at bachelor level. A well-known example is the comparison between Octave Levenspiel's book on reactor engineering and Fogler's book, where computer software is used in order to present the student with some basics on reactor design solving simultaneously mass, energy balances and pressure drop. Another comparison is the one between Rudd et al [8], Douglas' 1988 [9] book on process design, based on heuristics, and the use of mathematical programming techniques for process design by Biegler et al 1997 [10] just to mention a couple of examples.

1.1.7 Extension from Process to Product Design

Wintermantel [11], an engineer at BASF, published a paper on process and product design in 1999 highlighting the importance of considering the properties of the product into the production process. This results in the need for better control of the production process as well as the need for better knowledge and tools in modeling, simulation and optimization for process scale up and operation. In the early 2000, Cussler [12] published his book on product and process design highlighting the need to consider both in order to be able to produce the right product for the customer. Lately, a number of institutions have included either in the degree or at Master level modules on product design.

1.2 CHEMICAL ENGINEERING CURRICULUM

One interesting and particular feature of Chemical Engineering, shared with other major technical degrees, is that the curriculum across the globe is very similar because the basic principles of nature are common. In Figure 1.1 we present, from the out to the core, the main subjects that are currently covered in the Chemical Engineering curriculum.

On the one hand there are the basis on Mathematics including Algebra and Calculus, Statistics and Numerical analysis. We also have Physics (Mechanics and Electricity) and, as we go to the next level inside, the different fields within Chemistry

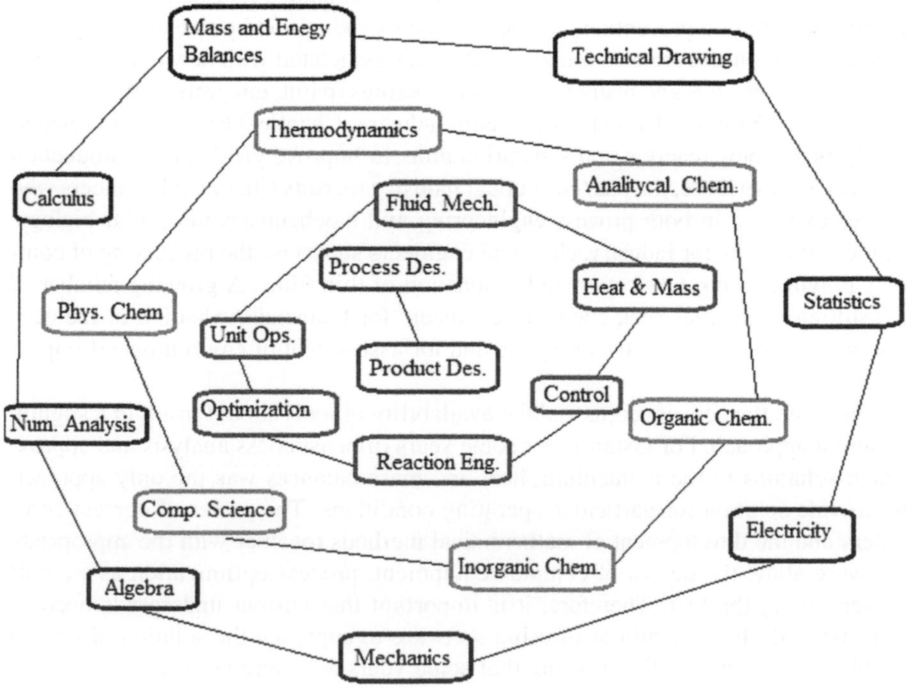

FIGURE 1.1 Typical subjects of the ChemE curriculum.

such as Organic, Inorganic, Analytics, Physical Chemistry, including Computer Science. In particular, as we have seen in the previous section, thermodynamics and applied thermodynamics for Chemical Engineers is a core subject to provide strong foundations for undergrads. Biology, Biochemistry or Biotechnology courses are also being incorporated in some universities, above all in those whose department names include the word “bio”. Double majors in Chemical and Biochemical engineering are also awarded. Building on all these, stands the core subjects of Chemical Engineering such as, transport phenomena, fluid mechanics, heat and mass transfer, unit operations, reactor engineering ending with process and product design, control and optimization. In most of the programs in the U.S. and the UK technical drawing, mechanics, inorganic chemistry and electricity are no longer offered. The development in computer science has had an impact in everyday life at all levels, and in particular, for the way chemical engineering is applied. The basics remain the same, however, the approach to the solution of the problems changes.

Furthermore, the chemical engineering profession offers careers in a wide range of topics from the process industries, high-technology areas, the petroleum and energy industries, environmental technologies. The future of Chemical engineering depends indeed on the needs of the society [13]. In order to be flexible, the core must be maintained. The reason is simple, part of the great success of Chemical Engineering is due to the relationship between basic science, process design and manufacture providing a strong scientific and technical background that enables the chemical engineer into different industries. Some of the main tasks of Chemical engineers are to supervise the operation of chemical plants, redesign chemical processes for pollution prevention, and develop new products and processes. Moreover, Chemical Engineers are also found in industries associated with new materials such as polymers (plastics and resins), fibres, and coatings (paint, integrated circuits, magnetic tapes). Traditionally, in the petroleum industry, Chemical Engineers developed catalysts and new reaction and separation units to improve yields in the production of fuels. Furthermore, the pharmaceutical industry recruits Chemical Engineers who posses expertise in both process engineering and biochemistry/molecular biology. In the semiconductor industry, chemical engineers supervise the processing of complex polymers, chip fabrication and production of thin films. A growing number of consulting companies seek chemical engineers for financial evaluation of the economic feasibility of industrial projects and for assessment of environmental impact of projects.

However, the new concerns and the availability of tools should result in a slightly different approach. For instance, for some years dimensionless analysis and approximate solutions to the momentum, heat and mass balances was the only approach to provide solution for particular operating conditions. The power of current computers and the development of mathematical methods together with the appropriate software allow the design of complex equipment, process optimization as we will present along the text. Therefore, it is important that current undergrads become familiar with the possibilities of using software to approach the solution of typical problems in Chemical Engineering that some years ago were not trackable, solved iteratively or by approximation. Therefore, there is a need to introduce the student to the use of software to help solve the common problems as well as how they can take advantage of computer science for their jobs, from evaluating scenarios and alternatives to providing early estimations. However, getting started in the use of any software is challenging since the students need to get used to the interface, the typical mistakes and, the capabilities and put them into perspective so that they can evaluate which are the opportunities within the software available in the market and also at his workplace to address that problem.

1.3 BOOK ORGANIZATION

In this book we present the use of different software to address problems in a broad spectrum of the bachelor curriculum of chemical engineering. In particular, we present examples based on problems whose solution by hand represent a tedious process. We start with general software (EXCEL®, MATLAB®, MATHCAD®, Python®, EES®, R®) going to process simulators (CHEMCAD®; ASPEN PLUS and HYSYS®) and the specialized software such as Computational Fluid Dynamics, CFD (COMSOL® and FLUENT®), Discrete Element Method, DEM (EDEM®), gPROMS®, Julia®, AIMMS®, LINGO® and GAMS®. We use the capabilities of such packages to solve mass and energy balances, fluid flow, heat and mass transfer. We continue to process analysis and synthesis and equipment design and, finally, we include basic notions of optimization from process operation and synthesis, to plant location and operation. Table 1.1 presents a summary of the examples studied in this book and the software used. We must highlight that, in general, most software is capable of solving most of the problems in one way or another. The book aim is not

TABLE 1.1
Summary of examples and subfields

Subject	Software	Example
Fluid Mechanics	CFD (COMSOL®, FLUENT®)	Pipe flow
	EXCEL®	Cyclone
	CHEMCAD®	Piping
	MATLAB®	Fluid flow
Heat Transfer	CFD (COMSOL®)	2D Conduction
	MATLAB®	
Mass Transfer	MATLAB®	2D diffusion
	gPROMS®	Mass and momentum transfer
	CFD(COMSOL®)	
Mass and Energy Balances	EXCEL®	Ammonia production
	MATHCAD®	NH_3 oxidation
	gPROMS®	$CaCO_3$ furnace
	Python®	Synthesis gas
	R®	CSTR in steady state
	EES®	Adiabatic reactor/Furnace
	Julia	Methanol synthesis

(Continued)

TABLE 1.1 (*Continued*)
Summary of examples and subfields

Subject	Software	Example
Unit Operations	MATLAB®	Distillation column
	EXCEL®	(Continuum and Discontinuum)
	Python®	Cooling tower
	GAMS®	CO_2 Capture
	CHEMCAD®	Reactive distillation
	gPROMS®	
Unit Operations (particles)	EDEM®	Rotary drum
	ANSYS- FLUENT®	Cyclone
	Python®	Fluidization
Reactor Engineering	MATLAB®	SO_2 to SO_3 (Pressure drop)
	gPROMS®	Bioethanol 2nd Gen. (Batch)
	MATHCAD®	Simultaneous reactions
	GAMS®	Polymers reactor (Semibatch)
	Python®	Multibed reactor for SO_3
	EES®	Semibatch reactor
Control & Dynamics	MATLAB®/Simulink	System response
	Julia	Control loop
		Controller Tuning
Equipment Design	EXCEL®/MATHCAD®	Evaporators
	MATLAB®	Distillation columns
	CHEMCAD®	Heat exchangers
	ASPEN-PLUS®	Flash
	gPROMS®	
	EES®	
	GAMS	
Process Analysis and Design	EXCEL®	HNO_3
	ASPEN-HYSYS®	Methanol production
	ASPEN-PLUS®	Ammonia synthesis
	CHEMCAD®	Utility plant
	GAMS®	Water treatment
	gPROMS®	Steam reforming
	Julia	Flowsheet design
Product Design	LINGO®	Metabolic network
		Beer and refinery operation
Plant Allocation	GAMS®	Supply Chain design
Plant Operation	AIMMS®	Scheduling
	SolverStudio®	Lot sizing problem
Data Management	R®	

to provide a through manual for the use of each of the packages, and more complete books and user guides that are referenced can be used for a deeper study of the software. The idea is to introduce the main features of the software to undergrads in the Chemical Engineering field to identify the capabilities and help them get started in the use of such packages to solve problems. In this sense it is expected to serve as an initialization to independent usage of the software based on typical examples in that are familiar to the Chemical Engineer undergraduate student.

REFERENCES

1. *The First Century of Chemical Engineering: A Timeline of Discoveries and Achievements, Chemical Heritage Foundation*, Philadelphia, pamphlet ca. 1990.
2. Hougen, O.A. (1977) Seven Decades of Chemical Engineering, *Chem. Eng. Prog.*, 73, 89–104.
3. Freshwater, D.C. and Yates, B. (1989) The Development of Process Engineering at the United Kingdom, Working Party on Chemical Engineering Education, EFCE, Loughborough.
4. CEP 30 Authors and their groundbreaking chemical engineering books. CEL, August, 2008.
5. Peppas, N.A. (Editor) (1999) *One Hundred Years of Chemical Engineering*, Kluwer Academic Publishers, Dordrecht/Boston/London.
6. Furter, W.F. (Editor) (1980) *History of Chemical Engineering*, Advances in Chemistry Series 190, American Chemical Society, Washington, DC.
7. Kantor, J.C. and Edgar, T.F. (1996) Computer skills in the chemical engineering curriculum. http://cache.org/site/super_store/Mono_File_no._2.pdf
8. Rudd, D., Powers, G., Siirola, J. (1973) *Process Synthesis*. Prentice Hall, Englewood Cliffs, NJ.
9. Douglas, J.M. (1988) *Conceptual Design of Chemical Processes*. McGraw-Hill, New York.
10. Biegler, L.T., Grossmann, I.E., Westerberg, A.W. (1997) *Systematic Methods of Chemical Process Design*, Prentice Hall, Englewood Cliffs, NJ.
11. Wintermantel, K. (1999) Process and Product Engineering: Achievements, Present and Future Changes, *Trans. IChemE A.*, 77, 175–188.
12. Cussler, E.L., Wagner, Q., Maarchal-Heusler, L. (2010) Designing Chemical Products Requires More Knowledge of Perception. *AICHE J.*, 56(2), 283–288.
13. Caruana, C.M. and Amundson, N. (1987) Assesses a Changing Profession, *Chem. Eng. Prog.*, 12, 76–79.

2 Modeling, Simulation, and Optimization in the Process and Commodities Industries

Iiro Harjunkoski, Mariano Martín

CONTENTS

2.1 INTRODUCTION

With the recent rapid development of information technologies and the increasing level of automation in production systems it is clear that more and more tasks and responsibilities will be taken over by software solutions. A model is a mathematical representation of a system that is useful in a number of levels. Apart from the fact that it can help clarify the behavior of the system, models are currently used to reduce production and design costs, reduce time to market, evaluate alternatives or train operators, among the most common [1–4]. The final aim of any simulation is to improve the operation and eventually, to determine the optimal solution. Optimization is a field where even a very simple PC can easily outperform any personnel—not owing to the knowledge it possesses—but owing to the capability of systematically and reliably testing thousands of alternatives within seconds.

One of the main questions is how to use this opportunity to create added value. Global competition has become very present all over the world and laying off people or moving a production site to another country with more advantageous cost structure is common under the pressure to produce cheaper, faster and more flexibly than ever. In many industries the amount of various product alternatives has exploded through individually tailored products and shorter lead times, which also emphasizes the cost and risk of higher inventories. The consequence of this is that smaller batch sizes and shorter campaigns call for more agile solutions able to adapt to frequently changing situations. As a response to these challenges new business strategies, requirements and models have appeared, driven by the multi-level needs of a processing plant also taking into account economic, environmental and legislative factors, which makes traditional approaches and production philosophies not anymore fully functional.

In managing and optimization of a production process, energy also plays a more important role than ever, partly because of the increasing and volatile costs triggered by more uncertain availability of electricity e.g., due to the increasing share of renewable energy sources (spot market electricity prices can vary by a factor of more than 20 depending on the point in time). Also, the less predictable prices of conventional fuels, e.g., crude-oil and the overall larger focus on total costs and emissions motivate efforts towards better energy-efficiency. Thus, there is a large demand and opportunity to improve optimization of industrial processes. Before going into details it is worth putting some attention into the landscape that constitutes a production environment.

2.2 INDUSTRIAL PRODUCTION SYSTEM LANDSCAPE

One of the main issues that must be considered in any industrial application of simulation and optimization technologies is the system architecture. Figure 2.1 shows a layout of a typical processing landscape from the process layer including controllers, sensors and actuators (hardware) up to the business layer that mainly comprises software solutions managing and integrating the overall process. One of the most important practical challenges is how to realize an efficient information exchange—both between internal and external system components. The main challenge is not to be able to transmit the data as bytes across the system (in intelligent systems almost any device can be connected with each other) but to exchange information in a correct format.

Why should the integration aspect in first place be mentioned within the context of simulation and optimization? There are of course several possible answers but the most generic is: Even the most brilliant optimization solution can never solve an industrial problem in an industrial environment, unless it has been built into as an integrated part of a production system. Some of the most critical aspects are:

- Data input and output should be automatic and not require any manual support.
- Current production situation should always be considered–every optimization builds on top of a starting situation.

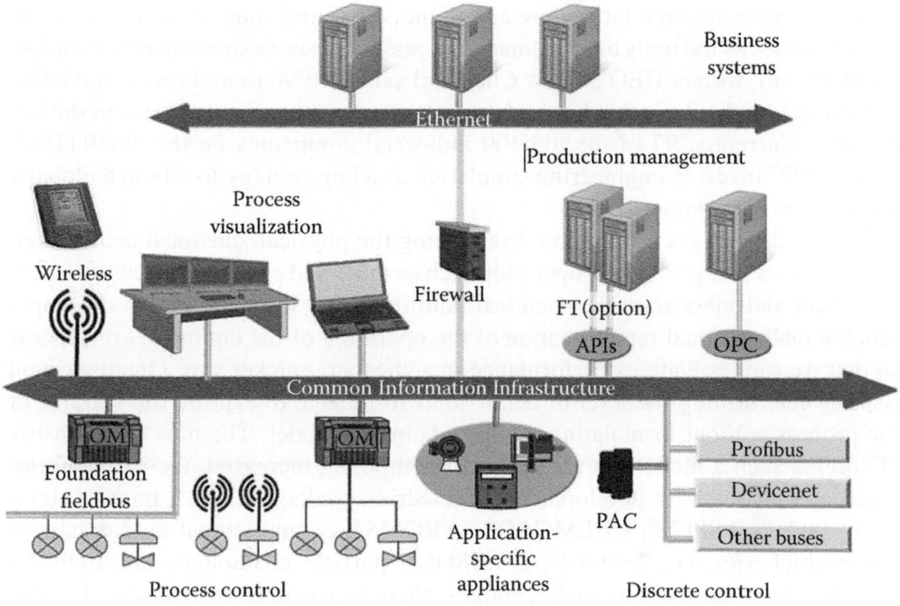

FIGURE 2.1 Logical view of a production system (Source: [5], with permission).

- Solution algorithms should be configurable in order to take into account normal daily and possibly frequently changing requirements.
- Optimization results should be returned sufficiently quickly (milliseconds to minutes).
- Results from the optimization should be allowed to be manually tuned and adapted by the operator, if feasible.
- Easy-of-use is critical for the acceptance of any solution.

Some of these aspects are easily lost if the professional communities creating the industrial production systems and those working on optimization algorithms are not connected and their main interests are not well aligned. Therefore, the more individuals there are with insights to both "worlds" the higher success probability there is to achieve enhanced and increased amount of productized optimization solutions. There are industrial standards for data exchange, e.g., ISA-88 [6] and ISA-95 [7], that define how to structure the necessary data for exchange, which may be of significant help in projects related to optimization as they allow to focus the existing resources on more challenging topics. More information about industry-specific requirements for manufacturing execution systems can be found in [8].

2.3 MODELING AND SIMULATION IN INDUSTRY

For many years, developments and improvements in the chemical industry relied on a trial and error approach. However, the expensive experimental trials at pilot and industrial scale in terms of time and money and the current simulation capacities have placed a lot of pressure on modeling and simulation of equipment and processes as the tools for systematic process analysis, design and optimization. Frank Popoff, former CEO at Dow Chemical said in 1996 that "Process modeling is the single technology that has had the biggest impact on our business in the last decade." Currently, 97 of the top 100 industrial companies on the "FORTUNE Global 500" invest in engineering simulation as a key strategy to win in a globally competitive environment [9].

The modeling task is based on determining the physical, chemical and biological principles that govern any operation such as mass and energy balances, momentum, heat and mass transfer, chemical equilibria and kinetics, etc. to develop a reliable mathematical representation of the operation of the equipment or process so that we can evaluate its performance in a cheaper, quicker way. One important issue is determining the level of detail so as to be able to capture the features of the process without formulating an overwhelming model. The next step consists of solving such a model. As the power of computers increased, their capabilities together with software development (engineering packages, EES®, process simulators such as ASPEN®, CHEMCAD®, gPROMS®, computational fluid dynamics and multiphysics, i.e., ANSYS®, COMSOL®, particle technology, i.e., EDEM®) have provided the tools to solve complex phenomena more realistically. Finally, and before we can use the model as a decision making tool, validation is required. At this point we can use the model to evaluate designs and operating conditions, reduce design time and production costs, improve productivity and efficiency, evaluate risks, in essence, making informed decisions and train personnel [3, 4]. As a note, BASF believes that their net benefits from the broad use of process simulation, in a comprehensive way, have been between 10 and 30% of installed capital cost of projects [3].

2.4 OPTIMIZATION PROBLEMS IN INDUSTRY

An industrial production process is always a chain of tasks and processes through which raw- and intermediate material is fed and which finally results in various end products. A successful production process requires thus a seamless collaboration between many various components, of which many in fact comprise some optimization capability. Figure 2.2 (based on [10]) shows the decision layers of a typical batch process. Optimization is critical for each of these layers and an important aspect is that they should not work against each other in a competitive manner such that the total production targets are met.

According to [10], the planning layer sets the production targets. The scheduling layer transforms this plan into batches, assigns them to equipment and sequences the batches. The recipe control system governs the execution of the batches making use of production recipes and the recipe execution triggers the phases of the recipes and provides the set-points for the process parameters in the phases. The continuous optimization layer optimizes the trajectories during the phases of the batch (RTO for continuous processes). The advanced control layer implements the optimal trajectory, typically applying linear model-predictive control (MPC), and provides reference values to the low-level controllers. Any

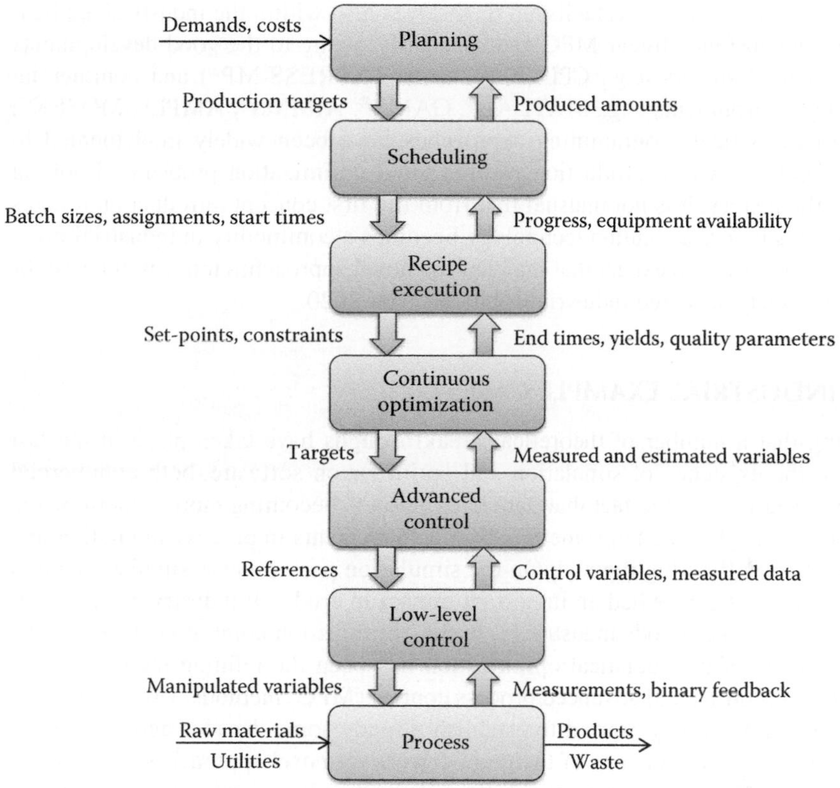

FIGURE 2.2 Various decision layers of batch production.

changes or disturbances in the process or any of these optimization layers need to be communicated properly across the other layers.

Nevertheless, the above example illustrates that optimization is present at all layers of a production process, be it manual or automated. It should also emphasize the fact that a single unconnected optimization function is incapable of contributing to a running production process. In [11], many aspects of collaborative process automation systems are described in more detail. Apart from the process-centric optimization problems, recent focus has been put into supply-chain optimization and enterprise-wide optimization (EWO) [12] problems, where the optimization focus is clearly expanded to larger logistic problems, which makes it possible to ensure that the overall strategy of an enterprise with possibly several production units is also considered in the short-term decision making. Another current trend is to identify optimal ways to deal with uncertainties both in the control (e.g., process disturbances) and scheduling (e.g., uncertainties in production orders) levels. In the scheduling literature, approaches vary from robust [13] and stochastic optimization [14] and multi-parametric programming [15] to re-scheduling concepts, i.e., initiate a new schedule as soon as something changes. However, these more recent research activities still need some years to be established as an integral part of industrial solutions.

Technologies that have established themselves also within the industrial applications are for instance linear MPC and especially owing to the good developments of commercial solvers (e.g., CPLEX®, Gurobi®, XPRESS-MP®) and commercial modeling environments (e.g., MATLAB®, GAMS®, AIMMS®, AMPL®, MOSEK®) many mathematical programming approaches have been widely implemented to tackle problems within production planning and optimization problems. Looking back in the history, it is not unusual that from the first concept introduction it takes 20–40 years before a solution technology becomes a commodity in industrial practice. Thus it is valid to expect that many of the novel approaches today will be established as widely accepted industrial solutions after 2020.

2.5 INDUSTRIAL EXAMPLE CASES

Knowing that a number of theoretical breakthroughs have taken place in the last decades, the existence of simulation and optimization software–both commercial and open-source, and the fact that data integration is becoming more smooth owing to standards, in this section some practical achievements in process simulation and optimization will be highlighted. On the simulation part, process simulators, CFD or DEM modeling have had an important impact in crude oil industry, pharmaceuticals and consumer goods industry. From the optimization point of view, one of the early adopters of mathematical optimization has been the refining industry. Here, linear programming and advanced process control (MPC) methods with strong optimization emphasis have started to establish already some decades ago. The oil & gas industry is still today one of the main drivers for novel approaches, e.g., supply chain logistics for transportation and optimization under uncertainty. Examples of other industries where the optimization potential has been early identified are the so-called heavy industries (integrated pulp & paper mills, metals plants) and power generation and distribution. In pulp & paper apart from process control optimization applications emphasis has been put on production planning and scheduling and solving the cutting stock problem in order to minimize material losses. Within the metals industry, especially in the 80s several expert systems were generated to improve the optimality of earlier manual decision making. The demand for a better optimization support is growing due to more complex problem instances that are for instance triggered by significant changes in the energy markets.

However, in general, it is difficult to prove the goodness of an optimization solution, which also decreases the openness of companies to do significant investments into optimization solutions. Some of these complicating factors follow:

- The objective function (i.e., the target of the optimization activity) is always a simplification of reality.
- Optimization problem constraints cannot comprise all relevant production-related aspects.
- A direct comparison between an optimized and non-optimized production situation is difficult due to the unforeseeable dynamics of production.

- Most processes aim at continuous improvement which makes it more difficult to separate the optimization benefits from other independent process improvements.
- In most processes, many decisions are still made manually and an optimization solution may only serve as a guideline.

There are of course a vast number of academic contributions to the topic of optimization as well as industrial sales-oriented documents aiming at promoting the commercial application of optimization technologies but a relatively small amount of publications that also highlight the industrial needs in a more analytical and technical manner. Recent industrial perspectives in the area of process control are given by [16] and in the area of production scheduling a more industrially oriented review is provided by [17].

The following subsections discuss some short examples of reported success stories within the field of optimization and production manufacturing systems.

2.5.1 Dairy Industry

Dairy industry is a typical batch process with critical time constraints, due to the fact that the shelf life of intermediates is limited. Also, cleaning policies must be fully respected due to the regulations. A benchmarking study on scheduling of a dairy industry process has been reported in [18], which discusses a more complex ice cream production problem (one processing lines feeding 8 packaging lines). The logistics of this problem is challenging due to the fact that the optimal sequence based on sequence-dependent setup-times is the opposite for the process and the packaging lines. The study tests a number of scheduling software packages and using the most successful one resulted in a throughput increase of around 30%.

2.5.2 Long-term Planning and Scheduling of a Refinery

The example of a refinery is discussed in [19] and also highlights the value of integration. In this case the manufacturing execution system plays an important role as multiple products are produced and material planning modules must be integrated by product line changes. The model-based solution will analyze jointly the feasibility and economics of alternative product strategies performing the monthly planning, which also sets targets for the optimized production and manages inventory levels and product distribution. The refinery's LP model is used for scheduling and blending optimization and to calculate the APC parameters. The comprehensive decision support results in estimated savings of 3 million euro on raw material costs. Apart from this, the blending units can be utilized optimally and the computer-based workflows are precise and simplify the handling of the model structures.

2.5.3 P&P Industry

Also in the paper industry example the collaboration of various optimization components is of key [8]. Most production facilities in the paper industry produce paper and board from large quantities of ground wood, pulp, and recycled paper together with water, chemicals and additives. There are also multiple products ranging from lightweight tissue to heavy cardboard, each of which have high quality requirements for instance on basis weight, caliper and brightness. After the paper production the paper is further processed in a converting mill and therefore the controlling and visualization of material flows as well as continuous and comprehensive quality monitoring and traceability present crucial challenges. In order to further improve production efficiency and transparency, the individual systems require apart from optimization also better functional interconnection. The objective is to increase the total plant efficiency by reducing inventory levels of raw material, intermediate and finished goods and by optimized use of energy. A manufacturing execution system (MES) can, among others, contribute to smooth coordination of production and material flows and provide decision support. The introduction of an MES resulted in improved profitability as the production efficiency increased by 1.2% and 4%, largely because of reduction of trim loss and improved production quality. At one major paper company the estimated savings were 10–20 million €/year, mainly owing to a holistic planning and scheduling optimization approach, corresponding to a 2% improvement of the overall efficiency of the plant.

2.5.4 Crude-oil Blend Scheduling Optimization

In crude-oil blend scheduling optimization, even small improvements can have a huge impact due to the large production volumes [19]. The complexity of operations is high due to the mixing and splitting processes, nonlinear blending models, storage management, and pipeline availability. Therefore, detailed modeling is required for the scheduling of a refinery's crude oil feed stocks from receipt to charging of the pipe stills. There are numerous crude oils that can finally be processed into different products, which makes the problem even further complex. Crude oils are often planned and purchased long before they arrive at the refinery but several practical aspects on off-loading, storing, blending and charging to meet pipe still feed quantity and quality specifications need to be based on current information, i.e., short-term requirements. An optimized crude-oil blend shop scheduling can generate several feasible and consistent schedules, react and adapt to spot-market opportunities, reduce penalties and working capital. According to [19], approximate annual savings from optimization are 2.85 million USD/year. Due to the many couplings between sub-problems, it is impossible to achieve this with manual solutions.

2.5.5 Equipment Design and Operation

Maybe the most exciting examples for the use of Computational Fluid Dynamics (CFD) in industry do not come from the chemical sector. Airplane and F1 cars design have benefited from the use of computational fluid dynamics to reduce tests and improve the performance [20]. In terms of process industry, mixing of viscous fluids is an important topic in customer goods industries. The design of static and dynamic mixers for high density viscous fluids such as tooth paste or predicting nozzle flow benefits from CFD solutions. Another example is the design of the bottles to be friendly to handle and to pour the viscous typically non-Newtonian fluid out of them also benefit from multiphysics [21, 22].

In pharmaceuticals and customer goods, particle technology is one of the most common operations. Discrete element methods (DEM) provide the tools for evaluating the behavior of such systems optimizing the geometry of the equipment and reducing the time to market [21].

Solar incidence varies along the year and also during a single day. In order to extend the use of solar energy along the day molten salt technology is being implemented in several plants across the U.S., Italy and Spain. The key to this technology is to keep the salts liquid to store the energy and use them as heat transfer fluid. COMSOL® was used to help design the storage tanks evaluating the heat losses [23].

Another example of interest in the chemical industry is the design of heat exchangers since they are one of the most abundant equipment. The design is based on the selection of the proper configuration of the tubes in order to provide the contact area for heat transfer while maintaining acceptable pressure drops. This problem has attracted the attention of process simulators which has resulted in the development of heat exchanger design tools packages to evaluate fooling problems, heat exchange limits, exchanger configurations etc. As an example, Dow Chemical reported $65 million savings by using the Aspen Shell & Tube Exchanger tools [24].

Finally, BP claimed that the use of gPROMS® saved them $1.5 million in the design of a depressurization vessel for an African offshore oilfield based on the fact that only a part of the vessel required an expensive alloy [25].

2.5.6 Process Design

The need to evaluate beforehand the outcome of complete processes has driven the development of process simulators (i.e., CHEMCAD®. ASPEN®) to improve their operation, see Chapter 12. We can find several examples on the technical and economical advantages of the use of process simulators in industry. For instance, Kuwait oil recently reported a 60% savings in process design by using ASPEN-HYSYS® and its economic analyzer to screen alternatives [26]. BASF reported 30% batch time savings in the production of expandable polystyrene as a result of changing the recipe while Shell presented that the redesign of an azeotropic distillation unit allowed $0.5 million savings a year using gPROMS® [25]. Another example, REPSOL reported that the design of a new propylene oxide process based on the results of a simulation in gPROMS® where simultaneous optimization of the reactor and the separation sections allowed 5M €/yr of savings [27]. Petrochemical Plant Troubleshoots with Aspen Plus® and saves $2.4M USD per year [28] and LG Chem reports capacity increases by 15% and saves energy using Aspen Plus® [29].

2.6 CONCLUSIONS

As has been seen in the above examples the use of simulation and optimization provides industrial advantage. Furthermore the optimization of industrial processes is almost never a standalone solution but heavily affected by neighboring components and their decision making. Due to the fact that in normal production there are multiple optimization solutions, each of which follows its own objectives and dynamics (runs at its own pace), the integration challenge is present even without taking into account normal disturbances and external factors that influence the production landscape. As many of the existing optimization solutions may even drive the process into contradictory directions, one of the greatest challenges is how to deal with conflicting objectives and the fact that individual optimization problems only have limited view of the reality. Defining an overall objective and linking these seamlessly into the various optimizers, measuring the true "goodness" of a solution are very difficult tasks that require a holistic knowledge of the process and the economic factors affecting the profitability of a production facility.

For industrial optimization solutions the modeling should be simplified, the algorithms should be solved in reasonable computing times and the resulting solutions must be robust and feasible. Also the usability of any industrial solution must be simple enough in order to gain the acceptance from the end user. Systems aspects and challenges again come from the integration of different software systems and solutions, which calls for standardized data structures. Also, in a system with multiple roles it must be clear how to deal with conflicting situations, which components should be the driver in which situations, how to balance the various targets of the individual optimization components. The challenges to the engineers designing and implementing integrated advanced solutions are enormous as deep knowledge is required both in the application domain (chemical batch production, metals, food, etc.), in the specific constraints and targets of the application (DCS, MES, communication standards, security mechanisms, etc.) as well as in the state of the art in control and scheduling. Normally, this can only be achieved by well-established project teams of experts who also can communicate across the borders of their expertise.

Once the above challenges have been solved, many industrial optimization applications can be produced and the main focus can be put on emerging challenges, one of the biggest being the periodically changing requirements due to e.g., pressure towards cost-efficiency, changing regulations, appearing new technologies and standards, as well as changing competitive situation. Other interesting questions are how to deal with energy (flexible markets and pricing), raw-material (availability and volatile pricing), larger problem instances where an entire enterprise and its internal logistics transportation needs should be considered, new processes that also poses design challenges, to name a few. Whatever the answer to these challenges are, it is certain that optimization will play a central role in shaping the future industrial solutions.

REFERENCES

1. Eykhoff, P. (1974) *System Identification.* John Wiley & Sons, London, U.K.
2. Hangos, K., Cameron, I. (2003) *Process Modeling and Model Analysis.* Academic Press, Cambridge, MA.

3. Polt, A. (Oct 2004) Collaborative conceptual engineering at BASF. *AspenWorld 2004*. Orlando, FL.
4. http://www.idac.co.uk/enews/articles/wp-cfd-business-benefits.pdf.
5. ARC Advisory Group. (2010) *The Collaborative Process Automation System for the 21st Century—CPAS 2.0*.
6. ANSI/ISA-88.00.01-2010. (2010) *Batch Control Part 1: Models and Terminology*.
7. ANSI/ISA-95.00.03-2005. (2005) *Enterprise-Control System Integration. Part 3: Activity Models of Manufacturing Operations Management*, ISBN: 1-55617-955-3.
8. Adams, M., Bangemann, T., Fittler, H., Friedl, C., Harjunkoski, I., Hochfellner, G., Kara, E., et al. (2011) Manufacturing Execution Systems—Industry-specific Requirements Solutions, ZVEI—Zentralverband Elektrotechnik und Elektronikindustrie e.V. Fachverband Automation ZVEI AG MES Brochure. ISBN: 978-3-939265-23-8.
9. ANSYS Inc. (2009) *Engineering Simulation Solutions for the Chemical Industry*. http://www.cadit.com.sg/imagestore/userfiles/image/industry/Materials_and_Chemical_Processing/chemical-industry.pdf.
10. Engell, S., Harjunkoski, I. (2012) Optimal Operation: Scheduling, Advanced Control and Their Integration. *Comput. Chem. Eng.*, 47, 121–133.
11. Hollender, M. (ed.) (2009) *Collaborative Process Automation Systems*. ISA, 2009. [ISBN 978-1936007103].
12. Grossmann, I.E. (2005) Enterprise-Wide Optimization: A New Frontier in Process System Engineering. *AIChE J.*, 51(7), 1846–1857.
13. Janak, S.L., Lin, X., Floudas, C.A. (2007) A New Robust Optimization Approach for Scheduling under Uncertainty: II. Uncertainty with Known Probability Distribution. *Comput. Chem. Eng.*, 31, 171–195.
14. Birge, J.R., Louveaux, F. (1997) *Introduction to Stochastic Programming*. Springer, Berlin, Germany.
15. Ryu, J.-H., Dua, V., Pistikopoulos, E.N. (2007) Proactive Scheduling under Uncertainty: A Parametric Optimization Approach. *Ind Eng Chem Res*, 46, 8044–8049.
16. Sand, G., Terwiesch P. (2013) Closing the Loops: An Industrial Perspective on the Present and Future Impact of Control. *Eur J Control*. DOI: 10.1016/j.ejcon.2013.05.020.
17. Harjunkoski, I., Maravelias, C., Bongers, P., Castro, P., Engell S, Grossmann I.E., Hooker J., et al. (2013) Scope for Industrial Applications of Production Scheduling Models and Solution Methods. *Comput. Chem. Eng.*
18. Bongers, P.M.M., Bakker, B.H. (2008) Validation of an ice cream factory operations model, *European Symposium on Computer Aided Process Engineering 18*, proceedings.
19. Kelly, J.D., Mann J.L. (2003). Crude oil blend scheduling optimization: an application with multimillion dollar benefits. *Hydrocarbon Processing*, June 2003, 47–53.
20. Slagter, W. (2011) Cutting design costs: How industry leaders benefit from fast and reliable CFD. ANSYS White paper.
21. http://computing.ornl.gov/workshops/FallCreek12/presentations/Lange-ComplexEngineeredSystems-SMC12.pdf.
22. http://www.comsol.com/story/download/158423/ENEL_CN2011.pdf.
23. Elabbasi, N., Liu, X., Brown, S., Vidal, M., Pappalardo, M. Modeling of Laminar Flow Static Mixers. http://www.comsol.com/story/download/157395/Veryst_Nordson_CN12.pdf.
24. Kolesar, D., The Dow Chemical Company. (2010) Aspen EDR helps troubleshoot thermosyphon problems. *AspenTech Global Conference*. Boston, MA, May 2010.
25. http://www.psenterprise.com/concepts/examples.html.
26. Kapavarapu, V. (2011) Kuwait Oil Company, "Project Optimization at a Conceptual Level by using Aspen HYSYS and Aspen Integrated Economics", *OPTIMIZE 2011 AspenTech Global Conference*. Washington DC, May 2011.

27. Martín-Rodríguez, H., Cano, A., Marzopoulos, M. (2010) Improve engineering via whole-plant design optimization. *Hydrocarbon Processing*. December, 43–49.
28. AspenTech. (2015) Case Study. Petrochemicals. Last accessed March 2018. https://www.aspentech.com/en/-/media/aspentech/home/resources/case-study/pdfs/11-7756-cs-reliance.pdf.
29. AspenTech. (2015) Case Study. Chemicals. Last accessed March 2018. https://www.aspentech.com/en/resources/case-studies/lg-chem-significantly-increases-plant-capacity-and-reduces-energy-usage.

Section II

General Tools

3 EXCEL® for Chemical Engineering

Mariano Martín, Luis Martín de Juan

CONTENTS

3.1 INTRODUCTION

Microsoft EXCEL® forms part of the package Microsoft Office. It is a spreadsheet application that features calculation, graphing tools, tables and a macro programming language, Visual Basic. The main advantage of EXCEL® is that it has spread and is widely used in industry and academia which makes it a perfect tool or interface not only to perform calculations, but also to connect different software so that the end user can interact with EXCEL® and run "behind the scenes" other software such as CHEMCAD®, MATLAB®, MATHCAD® or GAMS® and report the results back to EXCEL®. Therefore, we consider that this book should start by presenting the main features of this software and its application to Chemical Engineering. Note, we use different font to denote the code we input in the cells. The entire models can be found on the editorial web page.

3.2 EXCEL BASICS

The first image that the user sees when opening EXCEL® is a large table, with letters to define columns and numbers to denote rows. The key element is the cells, which as a result, are denoted by a letter and a number, and where we define equations and variables. Any cell reference may be modified from its coordinate to any arbitrary name the user may want to define by replacing the actual coordinate name in the left area in the formula bar by the desired name. Cells naming is especially useful when dealing with large and multiple calculations as the equations become more intuitive than a large string of coordinates. Cells, rows, columns, and small matrixes are named by selecting them and then typing on formula bar the desired name (Figure 3.1). Alternatively, the user may click *Define name* in the *Formulas* menu tab (Figure 3.1). We cannot use names that are existing functions, (e.g., PI, sin, … or references to other cells such as, T1, P3…).

3.2.1 Built-in Functions

There are a number of functions defined by default, such as trigonometric ones (SIN; COS; TAN; ACOS…), statistical (MEAN, AVERAGE, t-test), algebraic (SUM, ROUND, LOG, LOG10), logical (IF, FALSE,…), reference, database, information,… In order to find the desired function and learn how to use it, the user should click on the *fx* button on the formula bar or alternatively navigate through the *Formulas* menu tab. The user may manually type the function on the cell or click into the function in the menu to access the help menus with examples and use the function step-by-step. (See Figure 3.2).

3.2.2 Operations with Columns and Rows

a. If an equation is defined in a cell, we can replicate dynamically the same operation across rows and/or columns just by clicking on the bottom right corner of the cell and dragging the box across the rows or cells. When a formula is introduced in EXCEL® it is interpreted as a set of calculations of different cells versus the reference cell where calculations are written, therefore by copying the formula in other cells it's just changing the system of coordinates.

$$A3 = B1 + B2$$

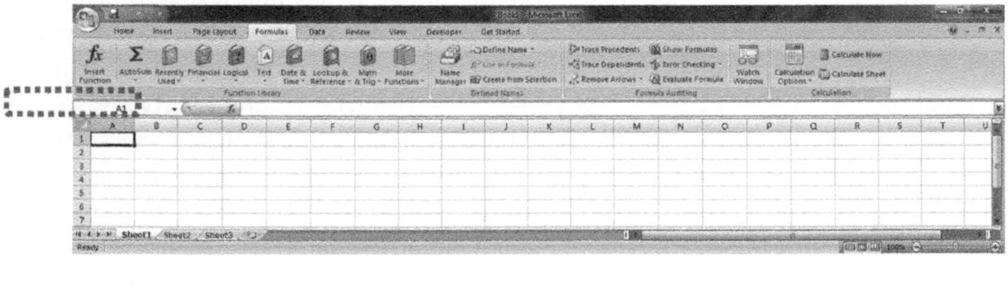

FIGURE 3.1 General window. Naming groups of cells.

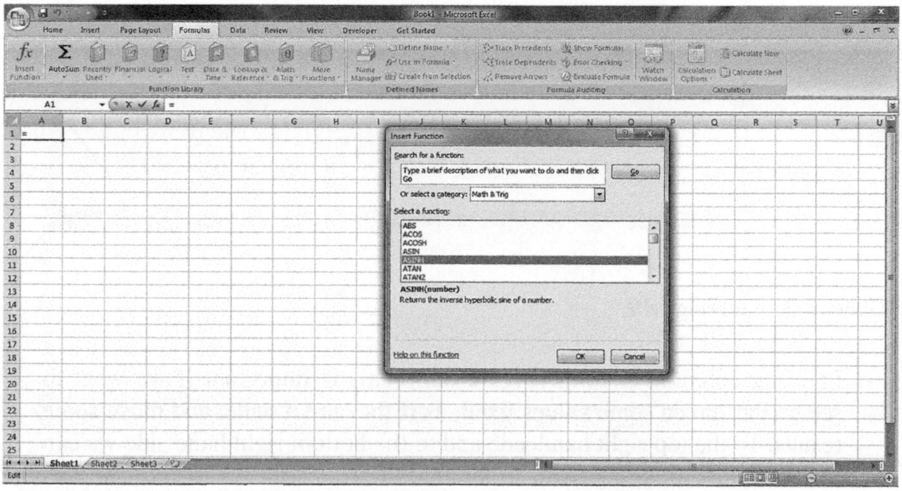

FIGURE 3.2 Use of formulas.

if we drag the box to B3 the operation will be:

$$B3 = C1 + C2$$

if we drag the box to A4 the operation in A4 will be:

$$A4 = B2 + B3$$

Sometimes a calculation within a cell contains a parameter that is constant and therefore the reference to this cell is not desired to move dynamically as we copy the formula across multiple cells. Then, in order to fix that cell we use "$". We use the "$" before any letter or number in a reference to a cell to ensure that the character before it does not move dynamically as we copy the formula in different cells. So, if the operation is A3 = B1 + B2 and B1 is a constant parameter we must define:

$$A3 = \$B\$1 + B2.$$

We can use F4 also to fix a cell. Alternatively, we could name the cell B1 with any desired name like "constant" and define the calculation as:

$$A3 = \text{constant} + B2$$

If we want to fix a column or a row we use $ before the letter or the number respectively. Alternatively, we can name the column or row with any descriptive name and include it in the formula. In this way, the formula will always refer to the same array of cells.

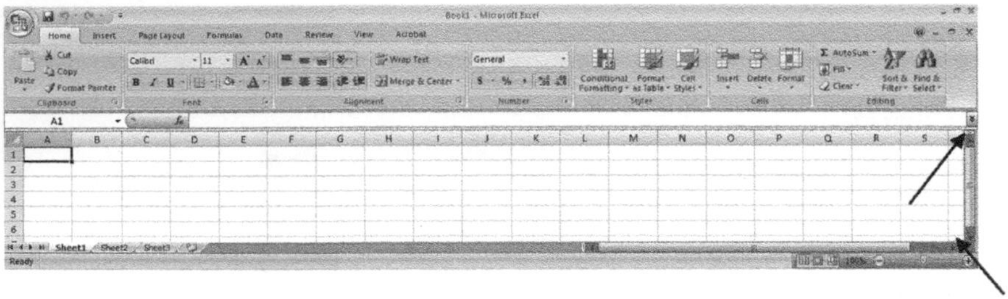

FIGURE 3.3 Freezing cells.

b. Freezing columns or rows (Figure 3.3): This is important if we want to see on the screen some values fixed, as if they are a scale, and move some results. To the top right hand side and to the bottom right hand side we can find this. Just dragging them to the middle in a vertical direction we can split our spreadsheet into 2 × 2. This is especially useful when dealing with a large set of data and the user wants to compare visually columns or rows that are separated in such a way they cannot be seen on the same screen.
c. Inserting rows or columns: Using the left button of the mouse we get a dialogue box that allows including rows and columns and allows displacing the content of the previous rows or columns. (See Figure 3.4).
d. Format: This is particularly interesting in order to be able to provide with a constant number of significant digits, use scientific notations for the numbers, percentages, font size and type.
e. Sort: If we have data in a column, we can organize them from lower to higher, alphabetically/numerically just using the tool under the data tab. Please ensure when you sort data within a table where data is correlated across multiple

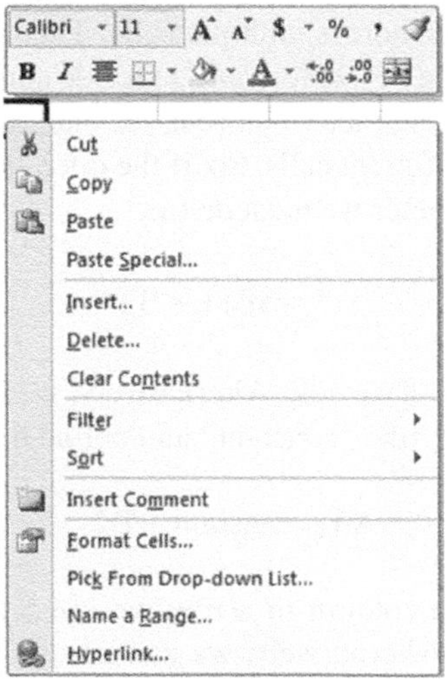

FIGURE 3.4 General options.

columns or rows that you select the full table to be sorted instead of one of the columns or rows. Otherwise you will lose the correlation as the selected columns will reorganize independently of the ones that have not been selected.

3.2.3 Plotting

Plots are another major capability of EXCEL®. We have a large number of options depending on the needs. We have columns and bars, typically to represent results on a process; sectors, area, bubbles, that are useful to present share of different variables to a total value; dispersions and lines, to show the experimental points or profiles of dependent variables versus independent ones.

To plot you just have to select the cells that you desire to plot. Next, go to the Ribbon tab *Insert* and select within the submenu *Charts,* the type of graph to be used. Once you've done this, you will get a figure on the spreadsheet. In order to edit it, just click with the right button of the mouse. You can click on different parts of the figure.

Axis: To edit them and determine the units, the lower and upper limits give format to the axis, provide the format of the numbers and select the type of scale (linear and logarithmic). To add a grid to the figure:

Area: The background.

Legend: The location of the legend.

However, in order to include labels for the axis once you've selected the figure, you need to go to the Ribbon tabs at the top of the EXCEL® window and select the *Presentation* tab, where you can look for the axis labels and legend text additions as well as adding the data on the points or bars of the figure. Double x and y axis is a nice feature for advanced users.

3.2.4 Solver

This is a particularly interesting feature that is widely used along the examples. It is the way we can use EXCEL® to numerically solve a set of equations, for problem optimization, including fitting a set of data to a given linear and non-linear equation and more. *Solver* is an add-in that needs to be activated to be used. We need to Click on *Office Button* (top left corner) then *EXCEL® Options*, in the Tab *Adds-Ins.* Click on bottom *Go* (Manage EXCEL® Adds-Ins) then look for *Solver* and enable *Solver* to activate the function. Once activated, *Solver* function can be found on the Ribbon tab *Data* sub menu *Analysis* on the right side of the bar.

The objective function must be defined in the *Target Cell.* Remember that the target cell must contain an equation with reference to one or more cells. The decision variables or parameters to be estimated are denoted as *Changing Cells.* The constraints are introduced under the *Subject to* dialogue box which is also used to define the variables as integer or binaries in optimization problems. *Solver* will find the value of the *Changing Cells* that maximize, minimize, or get closer to the objective value proposed. Solver allows the solution to nonlinear optimization using the GRG2 method; a reduced gradient method, while uses the simplex method for linear programming problems and branch- and bound-type of algorithms for mixed integer linear programming problems. It is capable of solving up to 200 variables and 100 constraints. These basic properties are enhanced for premium versions. The user can control several options and tolerances used by the optimizers through the *Solver Options* button.

- The *Max Time* and the *Iterations* edit boxes control the Solver's running time. The *Show Iteration Results* check box instructs the Solver to pause after each major iteration and displays the current "trial solution" on the spreadsheet. Instead of these options, however, the user can simply press the ESC key at any time to interrupt the Solver, inspect the current iterate, and decide whether to continue or to stop.
- The *Assume Linear Model* check box determines whether the simplex method or the GRG2 nonlinear programming algorithm will be used to solve the problem. The *Use Automatic Scaling* check box causes the model to be rescaled internally before solution. The *Assume Non-Negative* check box places lower bounds of zero on any decision variables that do not have explicit bounds in the Constraints list box.
- The Precision edit box is used by all of the optimizers and indicates the tolerance within which constraints are considered binding and variables are considered integral in mixed integer programming (MIP) problems. The *Tolerance* edit box is the integer optimality or MIP-gap tolerance used in the branch and bound method. The GRG2 algorithm uses the Convergence edit box and Estimates, Derivatives, and Search option button groups.

3.2.5 Building Functions in VBA

EXCEL® has built in capability to generate customized functions using VBA. This is a powerful tool that can save time without becoming an expert on programming as it opens the possibilities to run loops and conditionals in the background. This capability also allows the user building relatively large equations that are used in several areas of the worksheet (e.g., a polynomial for the estimation of specific heat of components) and that allows to read the calculations easily when looking at the formulas in the cells. In order to be able to build functions in EXCEL® *Developer* tab in the Ribbon needs to be activated. To do so, click on *Office Button* (top left corner) then *EXCEL® Options* and in the *Popular* tab enable *Show Developer Tab* in the Ribbon.

In order to build a new function, the user will have to go to *Developer* tab, then click on *Visual Basic*. It will prompt a new window with VBA environment. To write the new function we will have to *Insert* a new *Module* where we will write the new function.

Syntax for the function is very simple. First, we indicate that it is a function just by including the word "function" before the name we want to assign to it. Then, in brackets we indicate what the variables of the function are. Finally, at the end of the string we will add *As Double* to indicate that the function will return a numerical variable.

Once we have defined the function we can just introduce any equation, conditional or loop, that help with our calculations. A simple example in Figure 3.5 shows how to develop a function to estimate vapor pressure of water from Antoine Equation:

$$Ln(Pv_{sat}(mmHg)) = 18.3036 - \frac{3816.44}{227.02 + T(C)} \tag{3.1}$$

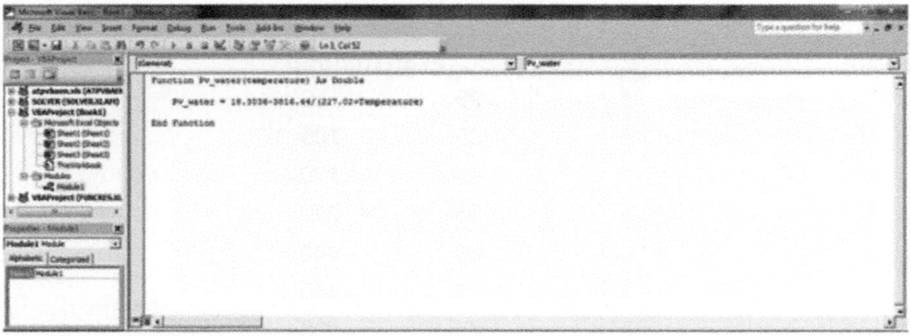

FIGURE 3.5 Developer screen.

Once the module is saved, whenever we introduce this function in a cell in EXCEL® it will return the result of the calculation.

$$= \text{Pv\_Water(A1)}$$

Functions can be much complicated and be composed of several equations as we will see in the examples. Whenever we introduce new variables in the function we will always have to define them as numerical by indicating the *Set Variable as Double.*

3.3 EXAMPLES

3.3.1 Fitting, Plotting, and Solving

Typically in the lab we get profiles of experimental data versus a few independent variables. We are interested in finding a function that can allow us to predict the operation of our equipment or operation. Let's start with the basics. Imagine we get the table of results in the EXCEL® file shown in Figure 3.6 corresponding to the vapor pressure of ethanol.

The problem at hand is to calculate the boiling point of a water/ethanol mixture 5, 10, 25, and 35% molar in ethanol at the ambient pressure of 760 mmHg.

For this example we use some of the basic operations that are available in EXCEL®. First, we plot the data and prepare a figure. Next, we fit the data. At this point it is interesting to present two fittings. The first one is based on the use of the built-in models in EXCEL®. The second is based on the common Antoine correlations. We use *Solver* to determine the parameters A, B, and C of those correlations by solving a small parameter estimation problem. Finally we use *Solver* to determine the boiling point of the water-ethanol mixture.

For plotting we just select both columns clicking on cell A2 and dragging the mouse to B9. We go to the Ribbon tab *Insert* and look in the submenu *Charts* for *Scatter.* With a simple intro we obtain a raw figure, see Figure 3.7, with a generic legend.

We include the *x* and *y* labels by clicking on the presentation tab, axis labels, and selecting title for horizontal and vertical axis. A generic title appears that we can modify. With regards to the legend we can overlap it with the figure by either using the *legend tab* in the *label tab* and selecting *legend format* or by using the right mouse button on the legend and dragging it across the figure. (See Figure 3.8).

	A	B
1	T(ºC)	Pvap (mmHg)
2	25	59
3	35	105
4	45	170
5	55	279
6	65	440
7	75	660
8	85	990
9	95	1420

FIGURE 3.6 List of data.

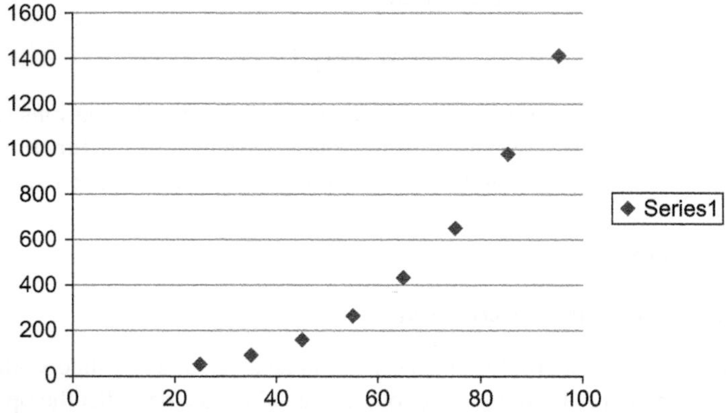

FIGURE 3.7 Plotting data.

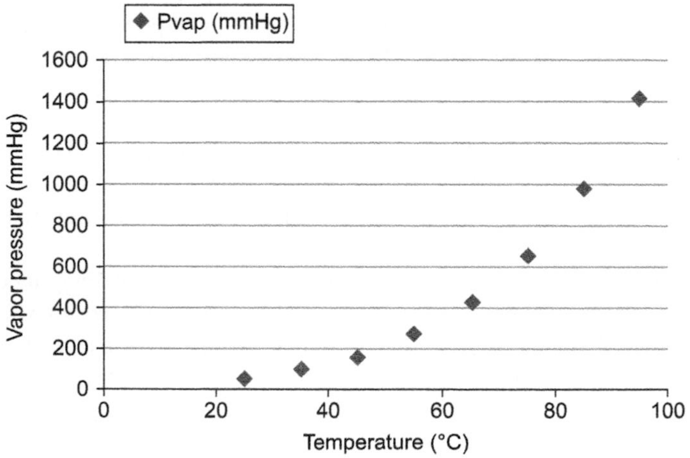

FIGURE 3.8 Plotting data.

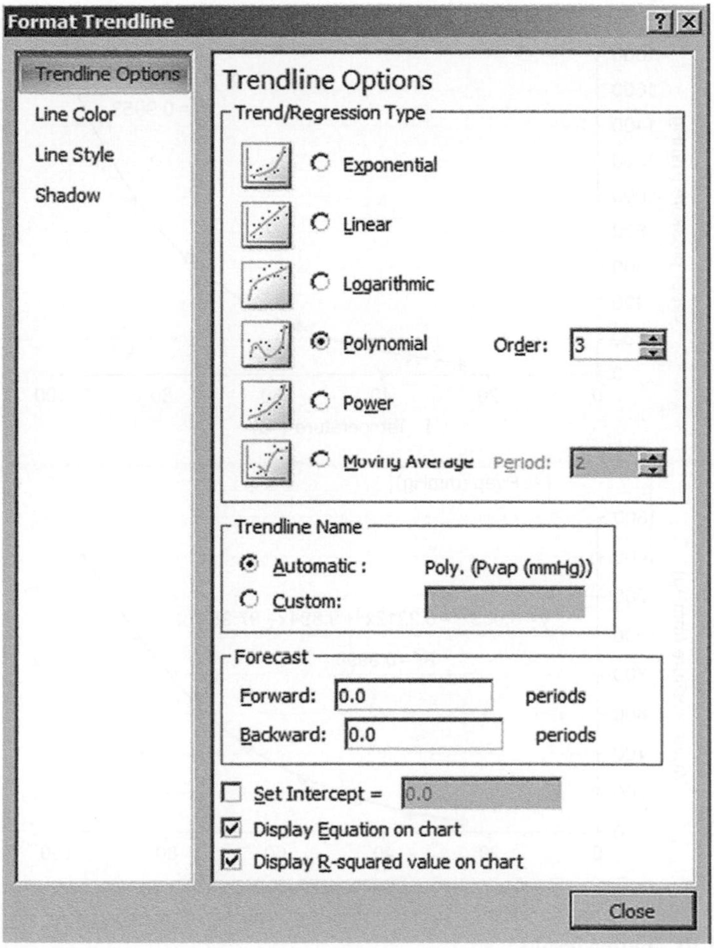

FIGURE 3.9 Adding trend lines.

Now that we have represented the data set we fit it to an equation. By the shape of the profile we can foresee that a polynomial or exponential function can fit our data. We click on to the data with the right button of the mouse. A menu appears and we select *add trend line*. (See Figure 3.9).

We can try with exponential or polynomial types. We also enable *Display Equation on Chart* and *Display R-squared Value on Chart* to show us the equations and the coefficient of variation. For the polynomial fitting we can select the degree, in this case a third degree polynomial is selected. We see that we get better fitting with the polynomial, see Figure 3.10.

An important idea to point out is the fact that higher degree polynomials do not mean better fitting and it is also good to remember that the maximum degree of a polynomial is the number of data points we have minus 1.

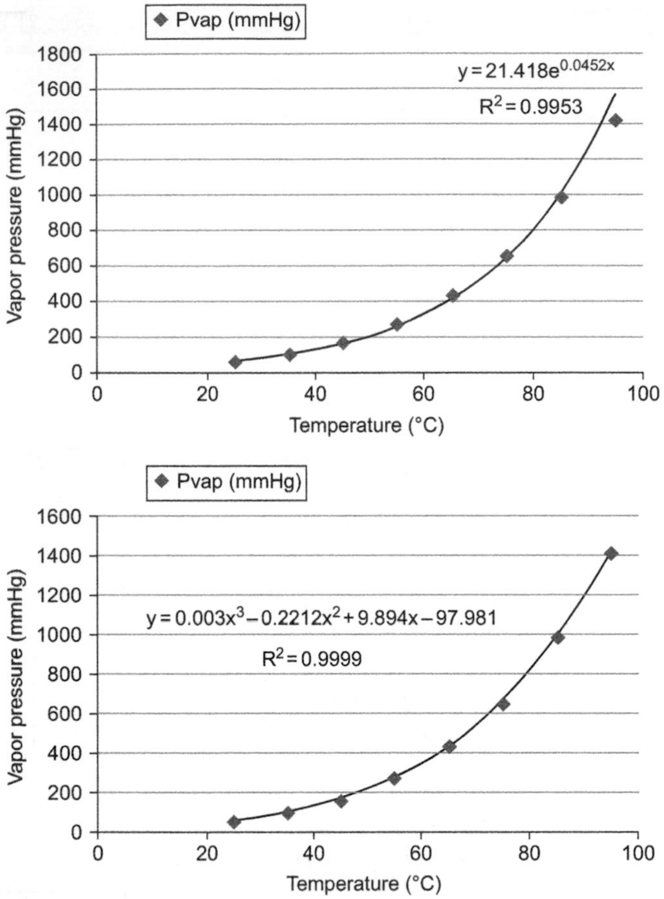

FIGURE 3.10 Fitting results.

Finally, we would like to solve a small parameter estimation problem to determine the coefficients of an Antoine type correlations where:

$$Ln(Pv(mmHg)) = A + \frac{B}{C + T(°C)} \tag{3.2}$$

For parameter estimation of a given equation from experimental data a common procedure is minimum square error. Using this solver we will be looking to minimize the square of the difference between the actual experimental results and the results obtained by the proposed equation that is based on the parameters that need to be estimated. Figure 3.10 contains all the cells, we have included in bold the names we have given to the correspondent columns and cells as it will be indicated in order to follow the calculations.

We define the name of the cells that contain temperatures as **T** and the cells containing vapor pressure as **Pvw.** Once the initial set of data is named, we define a column, C in our spreadsheet as **Pvant** where we include Antoine equation as a function of the parameters we want to estimate (A, B, and C) that are included in cells, G2, G3, and G4. We name each of the cells as **ant_a**, **ant_b,** and **ant_c** thus the equation in C2 (**Pvant**) of our spreadsheet becomes:

```
Pvant=EXP(ant_a+ant_b/(ant_c+T))
```

We just click on the right bottom corner of C2 and drag to fill in up to C9. We next calculate the square error (**S_Error**) between the calculated vapor pressure (**Pvant**) and the experimental one (**Pvw**). So we call cells in column D as **S_Error** and the formula will be:

```
=(1-Pvant/Pvexp)^2
```

It is highly recommended to use (1-Pvant/Pvexp)^2 vs (Pvant-Pvexp)^2 unless one of the values in Pvexp is 0. By doing so we avoid situations where the value of the variables change significantly along the set of data to be fitted (e.g., viscosity vs. shear rate) and the fitting will be conditioned by large value of the variables vs. the lower value.

We define the sum of square error as **S_S_Error**, that is the sum of the column **S_Error** eq (3.3) and using EXCEL we use the function SUM:

```
= SUM(S_Error)
```

$$TotalE = \sum_i Error_i^2 \tag{3.3}$$

We now use *Solver* to calculate ant_A, ant_B, and ant_C by minimizing the error. The target cell is the sum of square errors **S_S_Error**, cell G6. The changing cells are **ant_a**, **ant_b,** and **ant_c** and we set the calculation to minimizing. The results can be seen in Figure 3.11. Please note that as any numerical solution, results may depend on the initial conditions. In this case it is interesting to provide initial values for the cells **ant_a**, **ant_b**, **ant_c**. If the range of values is known, it is recommended to introduce initial values in the cells close to this range and add as constraints the maximum and minimum value of the cells.

Finally we present an example to compute the boiling point of the water-ethanol mixture at 25% molar. From the literature we find the coefficients from the Antoine correlation for water to be A = 18.3036 B = 3816.44 C = 227.02 [1]. The boiling point of any ideal mixture is calculated as:

$$P_T = \sum_i x_i Pvap_i \tag{3.4}$$

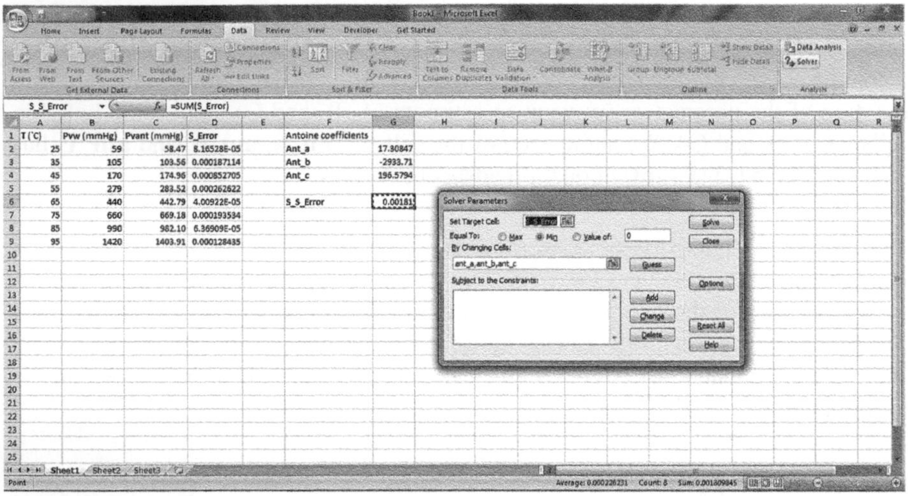

FIGURE 3.11 Solver use for fitting.

For our case, assuming ideal behavior in the range of values, we need to solve equation (3.5):

$$P_T = x_{ethanol} \cdot \exp\left(18.9 + \frac{-3800}{231.3 + T(°C)}\right) + x_{water} \cdot \exp\left(18.3036 - \frac{3816.44}{227.02 + T(°C)}\right) \quad (3.5)$$

The equation has one variable only, temperature, that we introduce in cell B1 and defined as **Temperature**. We introduce the equation 3.5 in cell B2 and we name it as **Pressure**. We will use *Solver* function to find the value of the cell Temperature that provides the value of the cell Pressure as atmospheric pressure that is 760 mmHg.

If we attempt to estimate a boiling point for different mixtures, we should have assigned the concentration of ethanol to a different cell and make the reference in the equation for Pressure instead of the value of 0.25 and concentration of water as (1-concentration of alcohol) instead of the constant value of 0.75 on the equation. The target cell (equation) will be set to take values of 0. When *Solver* is clicked, we get 93°C (see Figure 3.12).

As in previous cases, an initial estimation in the cell Temperature may condition the results depending on the mathematical form of the function to be solved (multiple solutions), therefore it is recommended to add an initial value to the Temperature close to the expected solution and add constraints to the values that Temperature can achieve.

The particularities in the thermodynamics of the systems may result in problems with multiple solutions. Since *Solver* uses a Newton-based method it is difficult to identify the presence of a second solution. Let's consider the mixture of

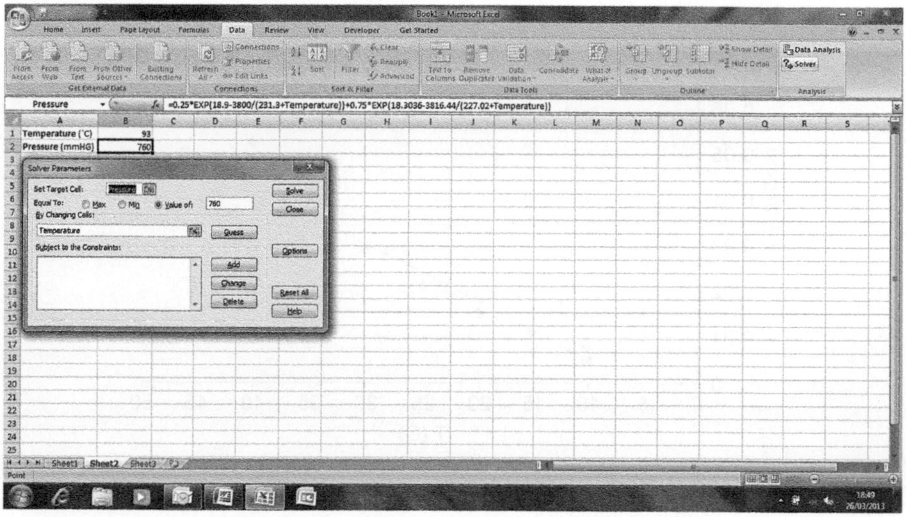

FIGURE 3.12 Solving equations.

two streams of air at different conditions at the pressure of Salamanca (Spain), around 700 mmHg. The first stream, A, is at 10°C and a relative humidity (RH) of 60% and the second stream, B, is at 46°C and a relative humidity of 70%. The problem would be to determine on what proportion these two streams can lead to condensation. We plot the psychometric diagram for the total pressure of our example as:

$$y = 0.62\frac{P_v}{P_T - P_v} = 0.62\frac{\varphi P_v^{sat}}{P_T - \varphi P_v^{sat}} \tag{3.6}$$

Where:

$$Ln(Pv_{sat}(mmHg)) = 18.3036 - \frac{3816.44}{227.02 + T(°C)} \tag{3.7}$$

We build a table on EXCEL® for different values of temperature (e.g., 5, 10, 15…, 45°C) in column E and we determine the saturation humidity according to Eq. 3.7 in column F. Next, as in previous cases, we plot the two columns resulting in Figure 3.13, diamonds. Now, with the characteristics of the two streams we locate them in the chart, squares. To do that we need to create two pair of data (e.g., B2, C2 and B3, C3) containing the conditions indicated in the problem. We will indicate the name of the variables as **Temp_A**, **RH_A**, **Temp_B,** and **RH_B** (combination of bold letters in column and rows). Then, we click on the figure with the right bottom of the mouse, it will pop up a menu, click on *Select data*, then *Add* and select the *x* and *y* values of the 2 streams.

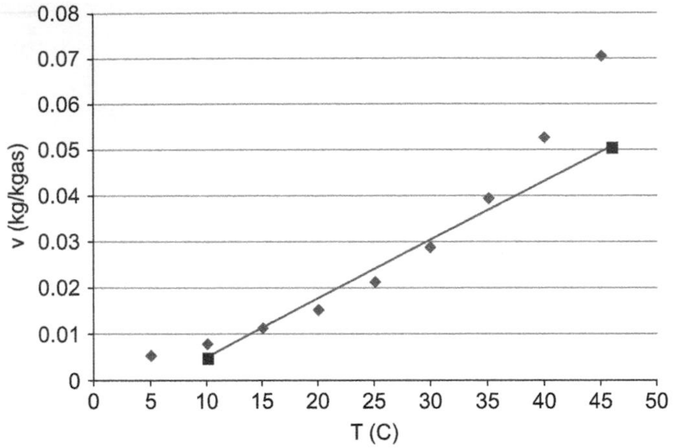

FIGURE 3.13 Mixture of two air stream.

We now formulate the problem as a mass and an energy balance as Eq. (3.8):

$$\begin{aligned}
&m_1 \cdot y_1 + m_2 \cdot y_2 = y_{sat}(m_1 + m_2) \\
&m_1 \cdot H_1 + m_2 \cdot H_2 = (m_1 + m_2) H_{m=sat} \\
&H_{sat} = (0.24 + 0.46 \cdot y_{sat}) \cdot T_{sat} + 597.2 \cdot y_{sat} \\
&m_1 = 100 \\
&m_2 = \text{variable}
\end{aligned} \tag{3.8}$$

Effectively we have 2 equations, mass and energy balances, and 2 variables temperature of the mix (that we will define as **Temp_Sat**) and mass of the second stream **m_2** since we look for the saturation point. In order to solve them we will use *Solver*. The *Target cell* is the mass balance and the energy balance is added as a constraint. The changing variables or parameters to compute are **m_2** and **Temp_Sat.**

$$\frac{100 \cdot y_1 - 100 y_{sat}}{y_{sat} - y_2} = m_2 = \frac{100 \cdot H_1 - 100((0.24 + 0.46 \cdot y_{sat}) \cdot T_{sat} + 597.2 \cdot y_{sat})}{((0.24 + 0.46 \cdot y_{sat}) \cdot T_{sat} + 597.2 \cdot y_{sat}) - H_2} \tag{3.9}$$

In order to solve the exercise, we could define functions to estimate the absolute humidity at saturation point and the enthalpy of the humid air as per equations (3.6, 3.7 and 3.8). As indicated before we go to the developer tab and click on *VBA*, *Insert* and then *Module*, see Figure 3.14.

Please note that Function y has included 2 calculations: estimation of vapor pressure and then estimation of absolute humidity y. As a result we have to define a new variable within the function called *vapour_pressure* with the corresponding equation. In order to be able to use the variable vapour_pressure we have to include the sentence *Dim vapour_pressure as Double* so the program is able to understand that it is a numerical variable.

```
Function H(temperature, y) As Double
    H = (0.24 + 0.46 * y) * temperature
End Function
```

```
Function y(temperature, RH, Pressure) As Double
    Dim vapour_pressure As Double
    vapour_pressure = Exp(18.3036 - 3816.44 / (227.02 + temperature))
    y = 0.62 * RH / 100 * vapour_pressure / (Pressure - RH / 100 * vapour_pressure)
End Function
```

FIGURE 3.14 Developer function definition.

With these functions to obtain the enthalpy of a stream we just need to write in a cell:

= H(temperature, absolute humidity), where temperature and absolute humidity can be actual values or references to other cells. To obtain the absolute humidity we just need to write

= y(temperature, relative humidity, pressure), where temperature, relative humidity and pressure can be actual values or references to other cells.

Note that in the equations we have divided RH by 100 as relative humidity in those equations is typically used as a fraction. If we had defined the format of the cell as percentage we would have not needed to divide between 100.

Finally, we need to define the stream that is the result of the mixture it will have a temperature that we will define in cell D2 as **Temp_sat**, and D3 as **RH_sat** that is 100%.

Therefore we can now define a new cell that contains the full mass balance and that we will define as **Mass_bal:**

```
=100*y(Temp_1,RH_1,Total_Pressure)+M_2*y(Temp_2,RH_2,Total_
Pressure)-(100+M_2)*y(Temp_Sat,RH_Sat,Total_Pressure)
```

and another that contains the energy balance that we will call **Energy_bal.**

```
=100*H(Temp_1,y(Temp_1,RH_1,Total_Pressure))+M_2*H
(Temp_2,y(Temp_2,RH_2,Total_Pressure))-(100+M_2)*H
(Temp_Sat,y(Temp_Sat,RH_Sat,Total_Pressure))
```

If we consider Figure 3.13, we can see that there are two points of saturation. Be careful, the straight line does not mean that we can use the lever rule. It is only a guide. In order to find both we need to help *Solver* by adding a range of temperatures where it will look for the solution. We add two constraints to limit the range of temperature for each of the condensation points following the figure. For the first solution from 10°C to 20°C and we obtain m_2 = 20 kg and T_{sat} = 16.5°C, see Figure 3.15.

And by looking for the solution in a different range, from 25°C to 50°C, we obtain m_2 = 115 kg and T_{sat} = 30°C

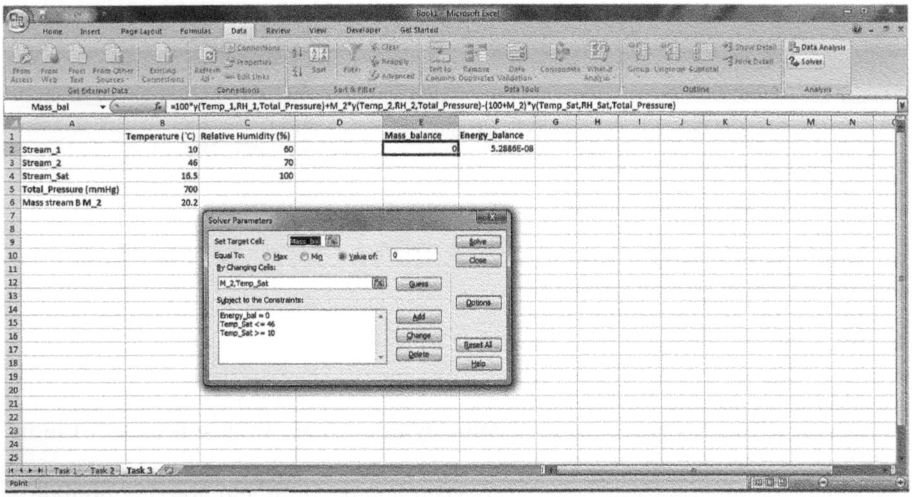

FIGURE 3.15 Solver model for air mixing.

3.3.2 Fluid Mechanics: Piping Design

Fluid flow calculation in pipes presents the difficulty of computing the energy losses through the pipes. These losses are related to the fanning factor. While the fanning factor has an algebraic exact solution as function of the velocity in laminar regime [2] the turbulent regime is more challenging. The friction loss in the pipes is a function of the roughness factor, related to the pipe material, and the flow rate in the pipe. Several correlations can be found in the literature. Typically we need to transfer a certain flow rate of liquid but do not know the velocity. Thus, we cannot estimate the energy losses either and an iteration procedure is presented to solve such a problem, so that the energy balance in the form of Bernouilli equation and the mass balance holds [3–4]. Here we present one of those problems.

We need to feed tanks T2 and T3 from the storage tank T1. The total flow rate from Tank 1 is 0.042 m^3/s. The characteristics of the pipes, diameter and equivalent length are given in Table 3.1. We need to determine the power of the pump to be bought for this operation and the flow rates that reach each of the tanks. We assume

TABLE 3.1
Data for the example

	Pipe 1	Pipe 2	Pipe 3
h (m)	8	35	50
d (m)	0.1524	0.0625	0.1016
L (m)	600	200	100
ε	0.000005	0.000005	0.000005

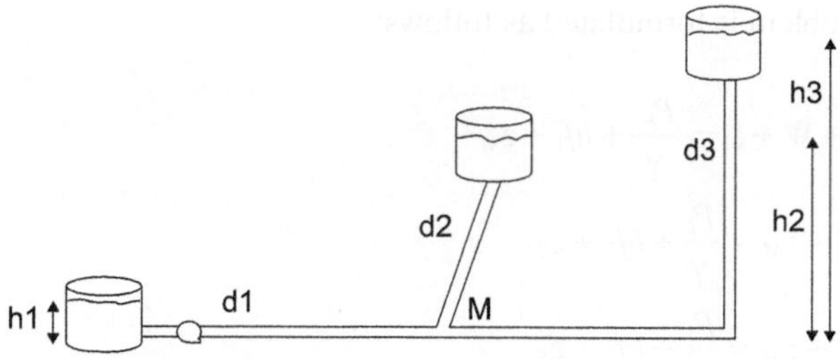

FIGURE 3.16 Tank piping example.

that the liquid that is transported has a density of 1000 kg/m^3 and the tanks are open to atmosphere. The layout of the tanks can be seen in Figure 3.16.

The problem formulation consists of applying Bernoulli equation to the three different pipes, with diameters d1, d2, and d3, and a global mass balance at the splitter. The fanning factor for the friction loss is calculated using the Colebrook correlation [5]:

$$\begin{aligned}
\frac{P_1}{\gamma}+\frac{v_1^2}{2g}+W+z_1&=\frac{P_M}{\gamma}+\frac{v_1^2}{2g}+hf_1+z_M\\
\frac{P_M}{\gamma}+\frac{v_2^2}{2g}+z_M&=\frac{P_2}{\gamma}+\frac{v_2^2}{2g}+hf_2+z_2\\
\frac{P_M}{\gamma}+\frac{v_2^2}{2g}+z_M&=\frac{P_3}{\gamma}+\frac{v_3^2}{2g}+hf_3+z_3\\
hf_i&=f_i\frac{L_i}{d_i}\frac{v_i^2}{2g}
\end{aligned} \tag{3.10}$$

where:

$$\begin{aligned}
\frac{1}{\sqrt{f_i}}&=-4.0\log_{10}\left(\frac{\varepsilon/d_i}{3.7}+\frac{1.256}{\mathrm{Re}\sqrt{f_i}}\right)\\
\mathrm{Re}&=\frac{\rho v_i d_i}{\mu}
\end{aligned} \tag{3.11}$$

We simplify the formulation since in a pipe the velocity is constant based on a mass balance for incompressible fluids

$$\begin{aligned}
Q_1\rho&=Q_2\rho+Q_3\rho\\
Q_i&=\pi\left(\frac{d_i}{2}\right)^2\cdot v_i
\end{aligned} \tag{3.12}$$

Thus the problem is formulated as follows:

$$\frac{P_1}{\gamma}+W+z_1=\frac{P_M}{\gamma}+hf_1+z_M$$
$$\frac{P_M}{\gamma}+z_M=\frac{P_2}{\gamma}+hf_2+z_2$$
$$\frac{P_M}{\gamma}+z_M=\frac{P_3}{\gamma}+hf_3+z_3 \tag{3.13}$$
$$hf_i=f_i\frac{L_i}{d_i}\frac{v_i^2}{2g};\frac{1}{\sqrt{f_i}}=-4.0\log_{10}\left(\frac{\varepsilon/d_i}{3.7}+\frac{1.256}{\mathrm{Re}\sqrt{f_i}}\right);\mathrm{Re}=\frac{\rho v_i d_i}{\mu}$$
$$d_1^2\cdot v_1=d_2^2\cdot v_2+d_3^2\cdot v_3$$

Now, we present the equations that are going to be solved after a few transformations of the general formulation of the problem:

Momentum balance

$$\frac{P_1}{\gamma}+W+z_1=\frac{P_M}{\gamma}+hf_1+z_M$$
$$\frac{P_M}{\gamma}+z_M=\frac{P_2}{\gamma}+hf_2+z_2$$
$$\frac{P_M}{\gamma}+z_M=\frac{P_3}{\gamma}+hf_3+z_3 \tag{3.14}$$

becoming

$$W+z_1=hf_2+z_2+hf_1$$
$$W+z_1=hf_3+z_3+hf_1$$

Mass balance

$$d_{1-M}^2 v_{1-M}=d_{M-2}^2 v_{M-2}+d_{M-3}^2 v_{M-3} \tag{3.15}$$

Friction losses

$$hf_1=f_{1-M}\frac{L_{1-M}}{d_{1-M}}\frac{v_{1-M}^2}{2g}$$
$$hf_2=f_{M-2}\frac{L_{M-2}}{d_{M-2}}\frac{v_{M-2}^2}{2g} \tag{3.16}$$
$$hf_3=f_{M-3}\frac{L_{M-3}}{d_{M-3}}\frac{v_{M-3}^2}{2g}$$

And Colebrook

$$0 = \frac{1}{\sqrt{f_i}} + 4.0log_{10}\left(\frac{\varepsilon / d_i}{3.7} + \frac{1.256}{Re\sqrt{f_i}}\right) \quad \forall i \in \{1 - M, M - 2, M - 3\} \quad (3.17)$$

where

$$\frac{1}{\sqrt{f_i}} = -4.0\log_{10}\left(\frac{\varepsilon / d_i}{3.7} + \frac{1.256}{\mathrm{Re}\sqrt{f_i}}\right)$$
$$\mathrm{Re} = \frac{\rho v_i d_i}{\mu} \quad (3.18)$$

Where M is the split point and i-M denotes pipes and velocities between any of the deposits and M.

Problem is reduced to 6 equations (2 energy balances, 1 mass balance, 3 Colebrook equations) and 6 variables (3 friction coefficients, the power required and the velocity in each of the two splits.

Therefore we define the data given in Table 3.1.

- Potential energy z_i for each of the deposits will be included in cells B3, C3, D3 and we name the 3 cells as **z_1, z_2,** and **z_3**.
- Diameter of each of the pipes is included in cells B6, C6, D6. The selection of the three cells will be named as **d_1_m, d_2_m,** and **d_3_m**.
- Equivalent length of each of the pipes is included in cells B7, C7, D7. The selection of the three cells will be named as **L_1_m, L_2_m,** and **L_3_m**.
- Velocity in each of the pipes is included in cells B8, C8, D8. We will name the cells as **v_1_m, v_2_m,** and **v_3_m**. **v_1_m** can be calculated as volumetric flow rate divided by area of pipe. = 0.042/(PI()*(d/2)^2). The other will be variables to be calculated.
- Pipe roughness is given in B10, C10, D10. We name the cells as **rough**.
- The Reynolds number is calculated in cells B11, C11, and D11 as:

```
=1000*v_1_m*d_1_m/0.001.
=1000*v_2_m*d_2_m/0.001.
=1000*v_3_m*d_3_m/0.001.
```

We name the cells as **Re_1_m, Re_2_m, Re_3_m:**

- Colebrook equation is written as:

```
=1/sqrt(f_1_m)+4*Log((rough/d_1_m)/3.7+1.256/
(Re_1_m*sqrt(f_1_m)))
=1/sqrt(f_2_m)+4*Log((rough/d_2_m)/3.7+1.256/
(Re_2_m*sqrt(f_2_m)))
=1/sqrt(f_3_m)+4*Log((rough/d_3_m)/3.7+1.256/
(Re_3_m*sqrt(f_3_m)))
```

```
Function hf(f, d, L, v) As Double

    hf = f * d / L * v ^ 2 / (2 * 9.81)

End Function
```

FIGURE 3.17 Developer for function definition.

Please note that "sqrt" is the function for square root and Log for logarithm. All the cells will be named as **Ceq_1_m, Ceq_2_m, Ceq_3_m.**

Finally the power required by the pump, W will be included in the cell C13 and named as **W**.

Calculation of friction losses may be done in a different cell or as a function that we can define in VBA as in previous exercise. To do that, we go to the ribbon tab *Developer* then we click on *Visual Basic*, then *Insert* new *module*. As in previous situations, we include the formula for the friction losses (Figure 3.17).

Once the variables have been defined we just need to write the momentum and mass balances.

The mass balance is included in cell E14 and we name it as **Mass_bal** this will be:

```
=v_1_m*d_1_m^2-v_2_m*d_2_m^2-v_3_m*d_3_m^2
```

The momentum balances are included in cells C16 and D15 and cells will be named as **Mom_bal_1_2:**

```
=W+z_1-hf(f_1_m,d_1_m,L_1_m, v_1_m)-hf(f_2_m,d_2_m,L_2_m,
v_2_m)-z_2
```

and **Mom_bal_1_3:**

```
=W+z_1-hf(f_1_m,d_1_m,L_1_m, v_1_m)-hf(f_3_m,d_3_m,L_3_m,
v_3_m)-z_3
```

As in previous situations we use *Solver* to solve the system of equations. The mass balance is considered as the target cell while the momentum and Colebrook equations will be used as constrains.

This is the problem we solve using EXCEL®. We use a line per energy balance, for the sake of simplicity, and another line for the mass balance to the splitter. Basically, we have three equations and three variables, W, v_2 and v_3 or equivalently four equations where P_M and Z_M are together the fourth variable. We consider that Z_M is 0, but we will modify the example later to see the effect of the friction losses on the flow rates sent to each of the tanks, Figure 3.18.

The solution yields a power of the pump of 47.7 m of column of water and 40% of the flow goes to Tank 3 while 60% goes to Tank 2.

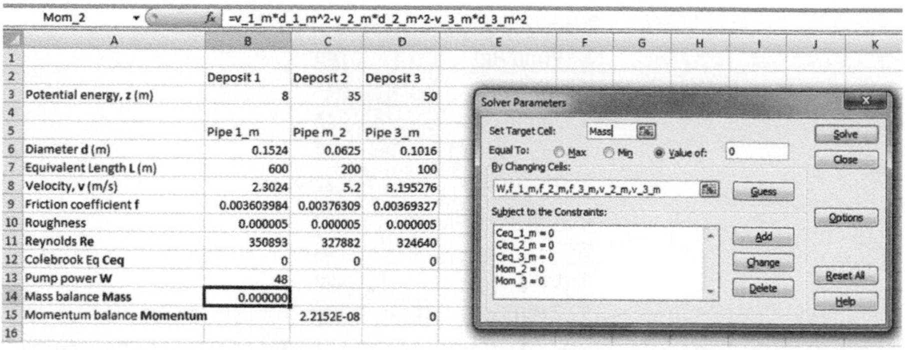
Mom_2
=v_1_m*d_1_m^2-v_2_m*d_2_m^2-v_3_m*d_3_m^2
Deposit 1 Deposit 2 Deposit 3
Potential energy, z (m) 8 35 50
Pipe 1_m Pipe m_2 Pipe 3_m
Diameter d (m) 0.1524 0.0625 0.1016
Equivalent Length L (m) 600 200 100
Velocity, v (m/s) 2.3024 5.2 3.195276
Friction coefficient f 0.003603984 0.00376309 0.00369327
Roughness 0.000005 0.000005 0.000005
Reynolds Re 350893 327882 324640
Colebrook Eq Ceq 0 0 0
Pump power W 48
Mass balance Mass 0.000000
Momentum balance Momentum 2.2152E-08 0
Solver Parameters
Set Target Cell: Mass
Equal To: Max Min Value of: 0
By Changing Cells:
W,f_1_m,f_2_m,f_3_m,v_2_m,v_3_m
Subject to the Constraints:
Ceq_1_m = 0
Ceq_2_m = 0
Ceq_3_m = 0
Mom_2 = 0
Mom_3 = 0
Solve
Close
Guess
Options
Add
Change
Delete
Reset All
Help

FIGURE 3.18 Solving the piping problem.

QUESTIONS

As exercises to extend this problem we leave to the student the following questions:

a. How does the pipe diameter affect the power?
b. What if we need to fill in a fourth or a fifth tank?
c. What would have been the energy consumption if the splitter would have been located at 5 m from the ground?
d. Suppose that we now have a pump with 35 m of power. How much flow rate can you send?

3.3.3 Unit Operations: Cooling Tower

Cooling towers are typical equipment in the power industry where they are used in order to have a closed cycle for cooling water. Their operation is based on the evaporation of part of the water so that the energy involved in the mass transfer from the liquid phase to the gas phase, the air, cools down the mass of water. As a result, the operation of cooling towers depends on the atmospheric conditions, the air temperature and humidity so that the cooling is season dependent.

The problem at hand is to determine the mass transfer coefficient and the air flow needed in an industrial cooling tower that is meant to refrigerate 2000 kg/s of water that enters at 27°C and is cooled down to 20°C to be reused as cooling agent. The ambient air is at 21°C with a relative humidity of 60% and leaves the column with 90% humidity at 22°C. The specific contact area, a, is 250 m^{-1} and the cross sectional area of the column, S, is 25m^2. (See Figure 3.19). The packing height, Z, is 5 m.

For this example we need to remember basic definitions on humidification:

$$\text{Relative humidity } \varphi = \frac{P_v}{P_v^{sat}} \tag{3.19}$$

$$\text{Absolute humidity } y = \frac{M_{cond}}{M_{dry\ air}} \frac{P_v}{P_T - P_v} = \frac{M_{cond}}{M_{dry\ air}} \frac{\varphi P_v^{sat}}{P_T - \varphi P_v^{sat}} \tag{3.20}$$

	A	B
1	P(mmHg)	760
2		
3	Tair in (ºC)	21
4	Tair out (ºC)	22
5	Tliq in (ºC)	27
6	Tliq out (ºC)	20
7		
8	L (Kg/s)	2000
9	Z (m)	5

FIGURE 3.19 Operating data.

$$\text{Vapor pressure } Ln(Pv(mmHg)) = 18.3036 - \frac{3816.44}{227.02 + T(°C)} \tag{3.21}$$

$$\text{Air enthalpy (reference temperature 0°C)} = \mathrm{c_h \cdot T_{dry}} + 597.2 \cdot \mathrm{Y_s} \tag{3.22}$$

Where c_h (kcal/kg) is given by $(0.24 + 0.46 \cdot Y_s)$

$$\text{The specific volume, } v_B = \frac{RT}{P_T}\left(\frac{1}{M_{air}} + \frac{Y_s}{M_{cond}}\right) \tag{3.23}$$

Based on these definitions and the data provided we compute the enthalpy of the air at the inlet and outlet of the column, Figure 3.20. Alternatively, we can use the functions defined in the exercise for mixtures of air streams to estimate absolute humidity and enthalpy.

Cooling towers are designed based on the mass and energy transfer required for cooling the water flow. Thus we perform an energy balance to a small element of volume along the vertical axis. Figure 3.21 shows a scheme of the cooling tower. We name 1 to the bottom of the column where the air enters and the water leaves and 2 the top of the column, where the air leaves the column and we feed the water. The

	C	D	E	F	G
1					
2	Pvap (mmHg)	Sat	Y (kg/kg)	Vb (m3/kgas)	H (kcal/kg)
3	18.46671588	0.6	0.0091727	0.843595648	10.6065448
4	19.64381868	0.9	0.0147662	0.853982058	14.2478061

FIGURE 3.20 Air streams information.

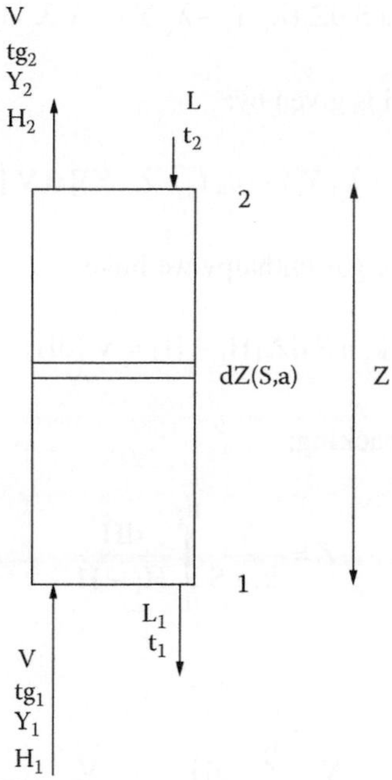

FIGURE 3.21 Cooling tower.

energy balance is defined from the interphase of the air to the bulk of the air. We consider two terms, latent and sensible heat. We denote with "t" the temperature, "h" is the heat transfer coefficient where sub-indexes and L are used to denote air and liquid respectively. k_y is the mass transfer coefficient g [6]:

Sensible heat from interphase to air bulk:

$$h_g \, a \, S \, dZ \, (t_i - t_g) = V \, c_h \, dt_g \tag{3.24}$$

where:

$$c_h = h_g / k_y \tag{3.25}$$

$$k_y \, c_h \, a \, S \, dz \, (t_i - t_g) = V \, c_h dt_g \tag{3.26}$$

$$k_y \, a \, S \, dZ \, (c_h t_i - c_h t_g) = V \, c_h dt_g \tag{3.27}$$

Latent heat from interphase to air bulk:

$$k_y \, a \, S \, dZ \, (Y_i - Y) \, \lambda_o = V \, \lambda_o \, dy \tag{3.28}$$

$$k_y\, a\, S\, dZ\, (\lambda_o\, Y_i - \lambda_o\, Y) = V\, \lambda_o\, dy \tag{3.29}$$

The total heat transferred is given by:

$$k_y\, a\, S\, dZ\, [(c_h\, t_i + \lambda_o\, Y_i) - (c_h\, t_g - \lambda_o\, Y)] = V\, [c_h\, dt_g + \lambda_o dy] \tag{3.30}$$

based on the definition for gas enthalpy we have:

$$k_y\, a\, S\, dZ\, [H_i - H] = V\, [dH] \tag{3.31}$$

so that the height of the packing:

$$Z = \frac{V}{k_y a\, S} \int_{H_1}^{H_2} \frac{dH}{H_i - H} \tag{3.32}$$

Thus,

$$k_y = \frac{V}{Z\, a\, S} \int_{H_1}^{H_2} \frac{dH}{H_i - H} = \frac{V}{Z\, a\, S} NUT \tag{3.33}$$

On the one hand we need to calculate the number of transfer units, NUT, and the air flow rate we need to use. The problem is typically solved by using the Mickley method that involves iterating the slope between the operating line and the equilibrium line. However, this is tedious and time consuming.

We define two sheets in EXCEL®. The first one is used to determine the equilibrium curve, the enthalpy of saturated air. The second performs the calculations of the operation of the column.

First, we determine the equilibrium line, the enthalpy of saturated air, based on the atmospheric conditions, assumed 760 mmHg of pressure (in EXCEL® file A2 = 760 mmHg) is given by eq. (3.34). The key point in here is that we need an equation that predicts the equilibrium line (eq. 3.20 & 3.21) so that we can compute the interphase points. Thus, for a certain temperature and the corresponding vapor pressure, we compute the saturating moisture and the enthalpy of the humid air:

$$T_i \rightarrow P_{v,\, sat} \rightarrow Y_{i,sat} \rightarrow H_i(T_i, Y_i) \tag{3.34}$$

In the example (3.3.1) for the mixture of two streams we showed how to create a function to return the enthalpy of a gas for using a function. We can use the function generated before considering a relative humidity of 100% or alternatively generate a new function that simply estimates the enthalpy for saturated air, Figure 3.22.

In this case the function has three equations so we need to define two additional variables: *vapour_pressure* and *ysat*. Now when we use the function Hsat and

```
Function Hsat(temperature, Pressure) As Double

    Dim vapour_pressure As Double
    Dim ysat As Double

    vapour_pressure = Exp(18.3036 - 3816.44 / (227.02 + temperature))

    ysat = 0.62 * vapour_pressure / (Pressure - vapour_pressure)

    Hsat = (0.24 + 0.46 * ysat) * temperature

End Function
```

FIGURE 3.22 Function definition for the air enthalpy.

provide the temperature and the pressure via a number or a reference to another cell the function will return the enthalpy of saturation at this temperature.

Thus, we could follow the following procedure. In EXCEL® we build a table:

```
B3= 1
```

For the rest of the values, we do:

Bi = Bi − 1 + 1. Click on the bottom right corner of the cell and drag down. This allows generating a series of data for temperature with an increment of 1 degree.

In column C, we define the saturated vapor pressure given by Antoine Correlation (eq. 3.21) for all the temperature points. In column D we calculate the absolute humidity (eq. 3.20) and finally in column E the enthalpy of saturated air (eq. 3.22). Now, as in a previous example, we define an equation that may fit our data. In this case a polynomial will do the work. In cells G1 to K1 we define the coefficients of the polynomial.

$$H_{sat} \approx G1{\cdot}x^4 + H1{\cdot}x^3 + I1{\cdot}x^2 + J1{\cdot}x + K1 \tag{3.35}$$

To fit G1:K1 we define an error between the calculated H_{sat} at column E and the one computed by this polynomial, where x corresponds to the air temperature:

$$\texttt{G3 = 1 - F3/E3} \tag{3.36}$$

Just as in the previous example we define a total error as:

$$\sum_{i=3}^{n} 1 - Fi/Ei \tag{3.37}$$

In the solver we define the target cell to be this total error, the variable cells will be the coefficients G1:K1 and we minimize the error. We can also plot this curve.

Now we start a second sheet for the operation of the column where we actually input the data in Figures 3.19 and 3.20. We define **PT**, the total pressure (cell B1), **Tairin**, **Tairout**, **Tliqin**, **Tliqout** are the inlet and outlet temperatures for the air and liquid in cells B3 to B6. We also define as Liquid the cell B8 containing the flow of water and B9 is defined as **Z**. Finally, we define B11, as **slope**, it will be our main variable.

With the vapor pressure and the relative humidities, C3:C4 and D3:D4, we compute the absolute humidity (E3:E4).

```
E3(yini)=0.62*Phiin*Pvapin/(PT-Phiin*Pvapin)
```

the specific volume (F3:F4)

```
F3(vb)= (1/29+E3/18)*(0.082*(Tairin+273))/(PT/760)
```

and the enthalpy (G3:G4)

```
G3(enthalpy) =(0.24+0.46*E3)*Tairin+597.2*E3
```

For the inlet properties and we do the same for the outlet in E4, F4 and G4 respectively, see Figure 3.20. Next, we need to determine the operating line given by the flows that go through the column. The operating line is calculated using an energy balance between the air and the water flows.

$$\mathrm{V\,dH = L\,dh = L\,c_L\,dt_L = L\,dt_L} \tag{3.38}$$

We use $c_L = 1$ kcal/(kg°C) for simplicity in the equations:

$$\frac{\mathrm{H_2 - H_1}}{\mathrm{t_{L2} - t_{L1}}} = \frac{\mathrm{L}}{\mathrm{V}} \tag{3.39}$$

```
F10(Slope L/V)=(G4-G3)/(Tliqin-Tliqout)
```

The air inlet and outlet enthalpies are calculated using the experimental temperatures and humidities. In this equation only V is unknown:

$$\mathrm{V} = \frac{\mathrm{L(t_{L2} - t_{L1})}}{\mathrm{H_2 - H_1}} \tag{3.40}$$

```
B15(V)=Liquid*(Tliqin-Tliqout)/(G4-G3)
```

Furthermore, we can plot this line as the line with a slope L/V that starts at t_{L1} and H_1 or else a line from t_{L1}, H_1 to t_{L2}, H_2:

$$H - H_1 = \frac{L}{V}(t_L - t_{L1}) \tag{3.41}$$

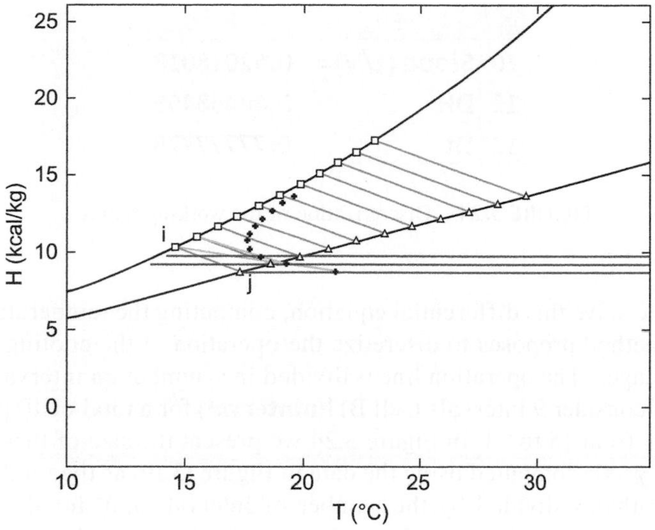

FIGURE 3.23 Enthalpy - temperature diagram. **(See color insert.)**

Finally, the temperature of the air describes a profile along the column. In order to determine this profile of temperature along the cooling tower, in black dots in Figure 3.23, the Mickley method is considered [6]. The Lewis relationship relates the operation line with the equilibrium line and results from the resistance to mass and energy transfer between the gas and liquid phases. We compute this using a global energy balance.

$$h_l\, a\, S\, dZ\, (t_L - t_i) = Ldh = VdH = k_y\, a\, S\, dZ\, (H_i - H) \tag{3.42}$$

$$\frac{H_i - H}{t_i - t_L} = -\frac{h_L}{k_y} \tag{3.43}$$

This balance links the equilibrium and the operating lines from i to j in Figure 3.21. It is a straight line that has as slope $\frac{H_i - H}{t_i - t_L}$ and starts at the operation line at t_{L1}, H_1. Typically, the slope ($-h_L/k_y$) is between −3 and −10, from i to j in Figure 3.23. The profile of the air temperature corresponds with the evolution of the air temperature with the change in its enthalpy. Thus,

$$\frac{k_y\, a\, S\, dZ\, (H_i - H)}{k_y\, c_h\, a\, S\, dZ\, (t_i - t_g)} = \frac{VdH}{Vc_h dt_g} \tag{3.44}$$

$$\frac{(H_i - H)}{(t_i - t_g)} = \frac{dH}{dt_g} \tag{3.45}$$

	E	F
10	Slope (L/V)=	0.52018018
11	DH	0.40458459
12	Dt	0.77777778

FIGURE 3.24 Discretization of the working range.

In order to solve this differential equation, computing the temperature profile, Mickley's method proposes to discretize the operation of the cooling tower in a number of stages. The operation line is divided in a number on intervals. For this example we consider 9 intervals (cell B11:**ninterval**) for a total of 10 points, column E, cells from 15 to 24. In Figure 3.24 we present the size of these intervals in the x and y axis computed using the data in Figure (3.19) as the total increment in the air enthalpy divided by the number of intervals and, for the increment of temperature, the total range of liquid temperature divided by the number of intervals:

```
F11 =(G4-G3)/ninterval
F12 =(Tliqin-Tliqout)/ninterval
```

We define F11 = **DH** and F12 as **Dt**

For the operating line, in EXCEL® we start at $t_{L1} = t_j = 20°C$ (G15) and we calculate 9 division points until $t_{L2} = 27°C$ (G24) as:

```
Gj+1 = Gj + (Dt)*(j-1) →G16 =$G$15+Dt*E16
```

Where E16 represents the point j of the discretization, F12 is the increment in the temperature along the x axis, Dt, and G15 the initial temperature. And we do the same for the enthalpies starting from H_{L1}, H_j (Column F15 = G3 to F24) where F11 is the increment in the enthalpy per interval, DH, and F11 is the enthalpy of the air fed to the column.

```
Fj+1 = Fj + (DH)*(j-1) →F16 =$F$15+DH*E16
```

For each point in the operation line we have a liquid temperature, $t_{L,j}$, and an air enthalpy H_j. This line, as presented before, is related to the equilibrium curve by a line that has a slope, $(-h_L/k_y)$, which is not known. Thus, for each point j we define a line that goes from j to a point in the equilibrium line i, see Figure 3.23.

$$H_i - H_j = slope(t_i - t_j) \tag{3.46}$$

We write this set of equations in column J (J15 to J24). *Slope* will be our main variable in the problem and we locate it in a cell, for example B13. Thus we can build straight lines from each j point as:

```
J15=F15+Slope*(I15-G15)
```

These lines intersect the equilibrium curve at points i which correspond to the saturation enthalpy of the air. Each of these points has an x coordinate of t_i (Column I, I15 to I24) and a y coordinate of H_i (Column H, H15 to H24). Furthermore, H_i depends on the temperature as calculated in the first sheet of EXCEL®.

$$H15 = H_i \approx G1{\cdot}t_i^4 + H1{\cdot}t_i^3 + I1{\cdot}t_i^2 + J1{\cdot}t_i + K1 \tag{3.47}$$

Where t_i in equation (3.47) is given by column I, I15 for our example in eq. (3.47). Finally,

```
H15 = J15 ; H16=J16.... ; H24=J24
```

The air temperature at point j + 1, $t_{g,j+1}$, is calculated by solving the system between the line (t_{gj}, H_j) and (t_i, H_i), from the horizontal line that represents the enthalpy of the humid air to the enthalpy curve in Figure 3.23:

```
K16=K15+((I15-K15)/(H15-F15))*(F16-F15)
```

This calculation is repeated at each of the intervals until the enthalpy of the exiting air is reached:

```
K24=K23+((I23-K23)/(H23-F23))*(F24-F23)
```

Resulting in the black dot profile shown in Figure 3.23. At this point, the air exit, the enthalpy, moisture and temperature of the air must be consistent to each other. In other words, K24 must be equal to the air outlet temperature given in B4. We define an error function:

```
K26 = K24-B4
```

And set out target cell to K26. We minimize the error or just look for a value of zero, and our variables will be the slope, B13, and t_i, in I15:I24.

Subject to $H_k = J_k$, as presented before (see Figure 3.25).

We thus obtain a series of points, A-A′,A″, ... A^n that constitutes the temperature profile of the air along the column, see Figure 3.24 where we plot the columns F15:F24 vs. G15:G24, squares for the operating line, together with H15:H24 vs. I15:I24, diamonds for the interphase points, and F15:F24 vs. K15:K24, triangles, for the air temperature profile, see Figure 3.26.

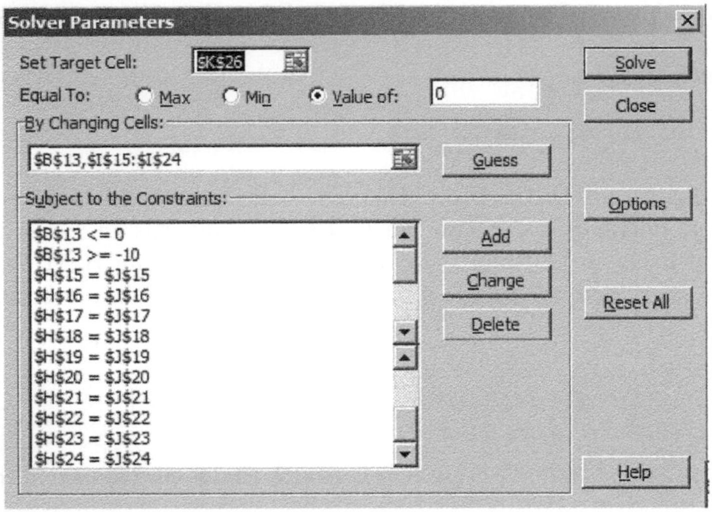

FIGURE 3.25 Solving the cooling tower design.

In order to determine the mass transfer coefficient, we compute $1/(H_i - H)$ and evaluate the integral. There are a number of options to evaluate an integral numerically (Figure 3.27). Maybe the easiest one is by adding the areas below the curve. Thus we have:

```
M16=1/(H16-F16)
```

The mean point is the height of the rectangle:

```
N16 = 0.5*(M15+M16)
```

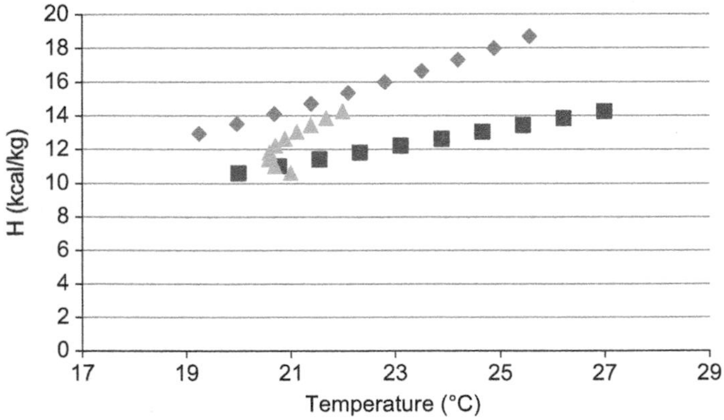

FIGURE 3.26 Cooling tower operation.

	A	B	C	D	E	F	G	H	I	J	K	L	M	N	O	P
1	P(mmHg)	760							Poly							
2			Pvap (mmHg)	Sat	Y (kg/kg)	Vb (m3/kgas)	H (kcal/kg)		-7.13E-07	2.99E-04	1.73E-03	4.09E-01	2.38E+00			
3	Tair in (ºC)	21	18.46671588	0.6	0.0091727	0.843595648	10.6065448									
4	Tair out (ºC)	22	19.64381868	0.9	0.0147662	0.853982058	14.2478061									
5	Tliq in (ºC)	27														
6	Tliq out (ºC)	20														
7																
8	L (Kg/s)	2000														
9	Z (m)	5														
10					Slope (L/V)=	0.520180181										
11	n	9			DH	0.404584585										
12					Dt	0.777777778										
13	Beta	-3.1008533														
14						Hj	tj	Hi	ti	Hi recta	Tg		1/(Hi-Hj)	D (1/(Hi-H))	DH	NUT
15	V(kg/s)	3844.82161			0	10.6065448	20	1.29E+01	19.2493689	12.9341416	21		0.42962767			1.15145704
16	a(m-1)	250			1	11.01112938	20.7777778	13.5120823	19.9712408	13.5120823	20.695704		0.39984759	0.41473763	0.40458459	0.16779645
17	S(m2)	25			2	11.41571397	21.5555556	14.1047381	20.6883671	14.1047381	20.578506		0.37188213	0.38586486	0.40458459	0.15611497
18					3	11.82029855	22.3333333	14.7122356	21.4007071	14.7122356	20.5950355		0.345789	0.35883556	0.40458459	0.14517934
19		Ky			4	12.22488314	23.1111111	15.3346802	22.1082268	15.3346802	20.7077497		0.32156439	0.33367669	0.40458459	0.13500045
20		0.1416687			5	12.62946773	23.8888889	15.9721571	22.8108987	15.9721571	20.8899517		0.29916031	0.31036235	0.40458459	0.12556782
21					6	13.03405231	24.6666667	16.6247323	23.5087015	16.6247323	21.1224548		0.27849878	0.28882955	0.40458459	0.11685598
22					7	13.4386369	25.4444444	17.2924537	24.2016198	17.2924537	21.3913283		0.25948301	0.2689909	0.40458459	0.10882957
23					8	13.84322148	26.2222222	17.9753524	24.8896436	17.9753524	21.6863606		0.24200588	0.25074445	0.40458459	0.10144734
24					9	14.24780607	27	18.6734431	25.5727679	18.6734434	22		0.22595617	0.23398103	0.40458459	0.09466512
25																
26									Error		7.1054E-14					

FIGURE 3.27 Results of the cooling tower example.

While the base of the rectangle is DH = F11
The area is given by:
P16 = N16*O16
And we just add:
NUT = SUM(P16:P24)
Going back to eq. 3.33 now we have:

$$k_y = \frac{V}{Z\,a\,S}\int_{H_1}^{H_2}\frac{dH}{H_i - H} = \frac{V}{Z\,a\,S}NUT = 0.14\ \text{kg/m}^2\text{s} \qquad (3.33^*)$$

This procedure for integration is useful when we have discrete data. In situations where we have an equation that correlates *x* and *y*, we could develop a function that estimates the integral also numerically within any interval of x values.

Let's consider also previous an example where we would like to evaluate the integral from the calculation. We have shown before how we could define a function to estimate the enthalpy of saturated air in the interphase that we called Hsat(temp).

In this case we allow the variables to change dynamically within the loop to estimate at each of the integral steps the value of the temperature of the interphase and the enthalpy of the gas. Then we estimate the enthalpy of the interphase using the function we created before called *Hsat* and finally we estimate the value the integral in each of the elements. As in previous cases we go to *Developer*, *Visual Basic*, *Insert*, *Module* and we write the following code, see Figure 3.28.

The data we enter in the function is in the form of arrays. We need to declare the variables *x* and *y* as Range instead of as values. Also we also define two new variables within the function that will allow us to make the needed calculations:

"inter": we use this value to calculate the number of intervals for the integration. It is equal to the number of points we have minus one. Therefore we use the formula:

```
Function Integral(x As Range, y As Range) As Double

    Dim inter As Double
    Dim ent As Double

    inter = x.Rows.Count - 1
    ent = 0

    For n = 1 To inter

        ent = ent + (y(n) + y(n + 1)) / 2 * (x(n + 1) - x(n))

    Next n

    Integral = ent

End Function
```

FIGURE 3.28 Functions for cooling tower solution.

inter = x. Rows. Count –1, where. Rows Count returns the number of data we have in the array x.

"ent": is a variable that accumulates the sum of all the areas of the different intervals. Its initial value must be 0.

Finally to make the integration we need to introduce the concept of a loop. In this case the loop is indicating that the instructions that are recorded in the rows between *For* and *Then* are repeated till the value of variable *n* achieve the value we have defined for *inter.*

Value of n is incremented by 1 at each step on the loop.

Within the loop we have introduced the simple calculation to estimate the value of the area within an interval n to n + 1 plus the accumulated value of previous intervals that is collected in "*ent*".

Finally we will indicate that the value that function needs to return is the value of "*ent*" once the loop has finished. In order to use the function we only have to introduce the following command in any cell in EXCEL®. = Integral (x data, y data).

3.3.4 Process Analysis

3.3.4.1 Ammonia Production: Reactor Operation and Recycle

A number of typical processes in the chemical industry consist of a reactor, which is fed by a stream that contains inerts, a separation stage to recover the unconverted raw materials a recycle and a purge to avoid building up of the inert. Examples like methanol or ammonia production are among the most typical. In this example we are going to evaluate a system that involves these units including the reactor conversion as function of the feed composition and operating conditions. We need to analyze the process presented in Figure 3.29:

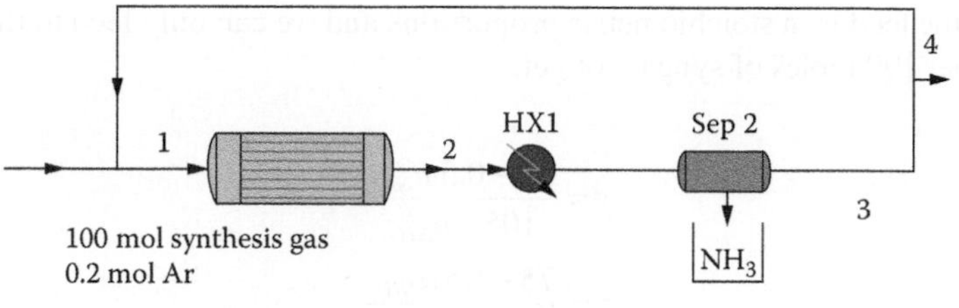

FIGURE 3.29 Flowsheet for the production of ammonia.

It is a simplification of the synthesis loop for the production of ammonia from synthesis gas consisting of hydrogen and nitrogen. The fresh stream to the process contains 0.2 mol of Argon, an inert gas, per 100 mol of mixture $N_2 - H_2$ in stoichiometric proportions. The fresh feed is mixed with the recycle and fed to the reactor. The limit of Ar into the reactor is 5 mol per 100 mol of synthesis gas. The reactor operates at 200 atm and 740 K isothermally:

$$0.5\,N_2 + 1.5\,H_2 \longleftrightarrow NH_3 \tag{3.48}$$

The equilibrium constant (K_p) for the reaction as function of the temperature is given by eq. (3.49) [7]

$$\log(K_p) = \frac{2250.322}{T} - 0.85340 - 1.51049\log(T) - 2.58987\cdot10^{-4}T + 1.48961\cdot10^{-7}T^2 \tag{3.49}$$

$$T[=]\,K,\ K_p[=]\,atm^{-1}$$

The ammonia is assumed to be completely condensed and the unconverted gases recycled. We need to determine the conversion of the reactor and the purge.

Solution:

The problem consists of solving the equilibrium at the reactor together with the mass balances to the system.

Balance to the reactor

$$0.5\,N_2 + 1.5\,H_2 \longleftrightarrow NH_3 \tag{3.50}$$

We first compute the conversion of the reaction. We call x to the ammonia molar fraction in the equilibrium, y the molar fraction of nitrogen and z the hydrogen molar fraction. k represents the molar fraction of impurity, Ar. Thus:

$$n_T = n_{H_2} + n_{N_2} + n_{NH_3} + n_{Ar} \tag{3.51}$$

Since the feed is in stoichiometric proportions and we can only feed to the reactor 5 moles per 100 moles of syngas we get:

$$y = \frac{25 - 0.5 \cdot n_{NH_3}}{105 - n_{NH_3}}$$
$$z = \frac{75 - 1.5 \cdot n_{NH_3}}{105 - n_{NH_3}} \quad (3.52)$$
$$x = \frac{n_{NH_3}}{105 - n_{NH_3}}$$

The partial pressures of each of the components are given by eq. (3.53):

$$P_{NH_3} = xP_T ; P_{N_2} = yP_T ; P_{H_2} = zP_T ; \quad (3.53)$$

And the equilibrium constant given by:

$$K_p = \frac{P_{NH_3}}{P_{N_2}^{0.5} \cdot P_{H_2}^{1.5}} = \frac{x \cdot P_T}{(y \cdot P_T)^{0.5} (z \cdot P_T)^{1.5}} = \frac{1}{P_T} \frac{x}{y^{0.5} \cdot z^{1.5}} \quad (3.54)$$

$$K_p = \frac{\frac{n_{NH_3}}{105 - n_{NH_3}}}{P_T \left(\frac{25 - 0.5 \cdot n_{NH_3}}{105 - n_{NH_3}} \right)^{0.5} \left(\frac{75 - 1.5 \cdot n_{NH_3}}{105 - n_{NH_3}} \right)^{1.5}}$$
$$= \frac{(105 - n_{NH_3}) n_{NH_3}}{P_T (25 - 0.5 \cdot n_{NH_3})^{0.5} (75 - 1.5 \cdot n_{NH_3})^{1.5}} \quad (3.55)$$

We can compute K_p independently since we have the total pressure and temperature. To estimate x we have to solve a non linear equation. We could calculate the conversion of the reaction X already as

$$X = \frac{1.5 \cdot n_{NH_3}}{n_{H_{2,ini}}} \quad (3.56)$$

Per 4 moles of reactants we obtain 2 moles of ammonia

	Inlet	Reacts	Outlet
$N_2 + 3H_2$	100 + R	−(100 + R)·X	(100 + R)·(1 − X)
NH_3		(2/4) (100 + R)·X	(100 + R)·0.5·X
Ar	(100 + R)·0.05		(100 + R)·0.05

Mass balance to the separator

	Inlet	Recovery	Outlet
$N_2 + 3H_2$	$(100 + R)\cdot(1 - X)$		$(100 + R)\cdot(1 - X)$
NH_3	$(100 + R)\cdot 0.5\cdot X$	$(100 + R)\cdot 0.5\cdot X$	
Ar	$(100 + R)\cdot 0.05$		$(100 + R)\cdot 0.05$

Mass balance to the recycle

Is a simple splitter, we call α the split fraction:

	Inlet	Purge	Recycle
$N_2 + 3H_2$	$(100 + R)\cdot(1 - X)$	$(100 + R)\cdot(1 - X)\cdot\alpha$	$(100 + R)\cdot(1 - X)\cdot(1 - \alpha)$
NH_3			
Ar	$(100 + R)\cdot 0.05$	$(100 + R)\cdot 0.05\cdot\alpha$	$(100 + R)\cdot 0.05\cdot(1 - \alpha)$

Thus, in steady state the argon entering with the fresh raw material must leave through the purge. Furthermore, the mass balance must hold for the recycle

$$\begin{aligned} 0.2 &= (100+R)\cdot 0.05\alpha \\ R &= (1-\alpha)\cdot(100+R)\cdot 0.75 \end{aligned} \tag{3.57}$$

Thus, we have a total of three equations (3.55–3.57) and three variables, the conversion, the recycle and the split fraction.

Now we go to EXCEL®:

We write the operating conditions of the reactor (see Figure 3.30)

We define the K_p, where C3 is the **temperature** and C2 as **Pt**. Both directly include the named variable in the equation.

```
Kp=10^(2250.322/Temperature-0.8534-1.51049*LOG10(Temperature)
-2.58987*10^-4*Temperature+1.48961*10^(-7)*Temperature^2)
```

Using the mass balance in (3.52) we write the initial and final molar fractions of the species, see C9:C12 and D9:D12 respectively in Figure 3.31.

Based on these molar fractions K_p is defined using eq. (3.55) as:

```
Kp2 = (D9+D10+D11+D12)*(D11)/((D9^0.5*D10^1.5)*(Pt))
```

	A	B	C
1			
2		P(atm)	200
3		T(K)	740

FIGURE 3.30 Initial data.

	A	B	C	D
1				
2		P(atm)	200	
3		T(K)	740	
4				
5		Kp=	0.00554052	
6		Kp=	0.00554052	
7				
8			ini	fin
9		N2	25	16.28301406
10		H2	75	48.84904217
11		NH3	0	17.43397188
12		Ar	5	5
13				
14		Conversion	0.348679438	

FIGURE 3.31 Results of the reactor.

Where C2 is the pressure, from D9 to D12 are the moles in the equilibrium. We have that a maximum of 5 moles per 100 moles of synthesis gas can be fed to the reactor. Thus, we can calculate the conversion based on the reaction. We have an equation ($K_{p_1} = K_{p_2}$) with one variable, the production of ammonia (D11). The conversion is written in C14 following the definition in eq. (3.56) as follows:

$$C14(\text{Conversion}) = (-D10 + C10)/C10$$

Now we write the mass balances presented above in the cells of the spread sheet. In yellow we present variables and in green the input data. We define C14 as **Conversion**, F15 as **Alpha** and H17 as **Recycle:**

Under React in (see Figure (3.27)):

```
C17=B17+H17
C19 =(B17+G17)*0.05
```

Reactor Outlet:

```
D17 =(1-Conversion)*C17
D18 =Conversion*0.5*C17
D19=C19
```

Separation (NH_3 recovery)

```
E18=D18
```

Purge

```
F17=C17*(1-Conversion)*alpha
F19=alpha*0.05*C17
```

Recycle

```
G17=D17-F17
```

The moment we include this equation, EXCEL® send a message indicating that there is a cycle. Therefore, we need to habilitate the iterative calculation. We go to the main menu → EXCEL® OPTIONS →Formula→ Tick "Enable iterative calculation" to allow it.

```
G19=D19-F19
```

Our two equations of the mass balance are:

The balance to the argon → D22 = F19-B19

The Recycle → Recycle = G17

Changing cells to be estimated are the split fraction, F15, and the recycle, H17, and the concentration of ammonia D11.

Thus we write the three equations in *Solver*. In this case we select as target cell the balance to Ar while equations for the equilibrium constant and the recycle are included as constraints, Figure 3.32.

Now we can solve obtaining the results presented in Figure 3.33. We could have selected any of the three equations as target cell.

QUESTIONS

- Now that the example has been formulated and solved we can go forward and perform some sensitivity studies.
- What is the effect of the P and T on the conversion?

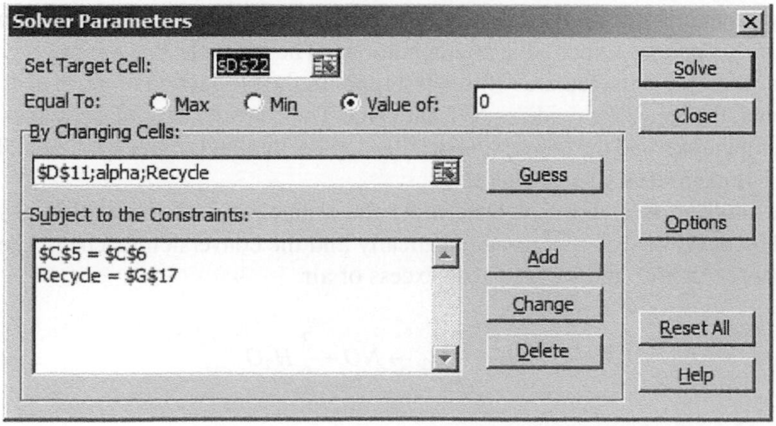

FIGURE 3.32 Solver model for the system.

	A	B	C	D	E	F	G	H
1								
2		P(atm)	200					
3		T(K)	740					
4								
5		Kp=	0.00554052					
6		Kp=	0.00554052					
7								
8			ini	fin				
9		N2	25	16.28301406				
10		H2	75	48.84904217				
11		NH3	0	17.43397188				
12		Ar	5	5				
13								
14		Conversion	0.348679438			alpha		
15						0.01432026		
16		Feed	React In	React Out	Rec NH3	Purge	Recir	Recycle
17	Feed	100	279.3245234	181.9298057		2.60528256	179.324523	179.324523
18	NH3			48.69735888	48.6973589			
19	Ar	0.2	13.96622616	13.96622616		0.20000002	13.7662261	
20								
21								
22		Argon balance		2.38664E-08				

FIGURE 3.33 Results of the ammonia reaction and recycle example.

- How does the operating pressure and temperature affect the purge?
- What should the operating condition be to obtain 25% of conversion?

3.3.4.2 Ammonia Oxidation: Reactor Operation, Mass and Energy Balances

Many times the conversion of a reactor depends on the operating temperature. Even if isothermal operation is recommended, sometimes it is not easy to remove the energy generated during the reaction or because it can be an asset for the process, the reactor operates under adiabatic conditions. The example that we are trying to solve now presents the features of this last type. In the production of nitric acid from ammonia, the converter oxidizes NH_3 into NO. The conversion of that reactor varies with temperature and the conversion profile may be obtained from experimental data in the Ullmann's Encyclopaedia [8].

The problem is as follows. Determine the temperature of the exiting gases of the NH_3 converter that operates adiabatically and the conversion reached. The feed stream is at 65.5°C, and we feed 10% excess of air.

$$NH_3 + \frac{5}{4}O_2 \rightarrow NO + \frac{3}{2}H_2O \tag{3.58}$$

We assume that no secondary reactions take place and we use dry air. Using the information from [8], see Figure 3.34, we plot Figure 3.35.

	A	B	C
1	T(K)	T(ºC) Read	X
2	1023	750	0.875
3	1073	800	0.91
4	1123	850	0.93
5	1173	900	0.945
6	1223	950	0.925
7	1273	1000	0.9

FIGURE 3.34 Data read for Figure 3.35.

Data for the heat capacity and the enthalpy of formation of the species involved can be found elsewhere [1,8].

The formulation of the problem is straightforward consisting of an energy balance to the reactor.

$$\sum_{prod}\Delta H_{f,prod} - \sum_{react}\Delta H_{f,react} = 0;$$

$$\sum m_i \int_{T_{ref}}^{T_{entrada}} Cp_i dT + \left(\sum_{prod}\Delta H_{f,prod} - \sum_{react}\Delta H_{f,react}\right)_{T=tref} = \sum m_i \int_{T_{ref}}^{T_{salida}} Cp_i dT \quad (3.59)$$

However, there are two problems to solve the previous equation by hand. First, the heat capacity of the outlet stream is a function of the temperature, which we intend to calculate. Furthermore, the conversion is also a function of the temperature and

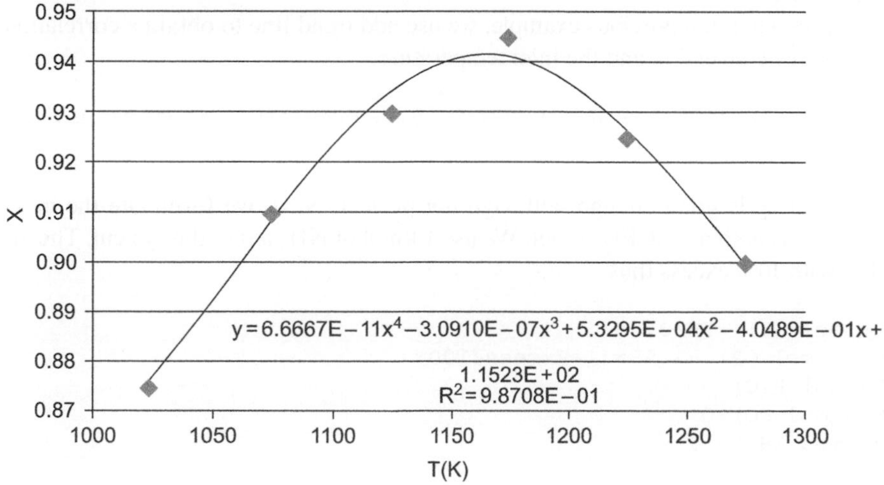

FIGURE 3.35 Fitting of the effect of the temperature on the conversion.

it is what gives us the composition of the outlet gas. Let's expand the right hand side term of the previous equation:

$$\sum n_i \int_{T_{ref}}^{T_i} Cp_i \, dT + Q_{generated} = n_{N_2,out} \int_{T_{ref}}^{T_{out}} Cp_{N_2} \, dT + n_{NH_3,out} \int_{T_{ref}}^{T_{out}} Cp_{NH_3} \, dT$$
$$+ n_{O_2,out} \int_{T_{ref}}^{T_{out}} Cp_{O_2} \, dT + n_{H_2O,out} \int_{T_{ref}}^{T_{out}} Cp_{H_2O} \, dT + n_{NO,out} \int_{T_{ref}}^{T_{out}} Cp_{NO} \, dT \quad (3.60)$$

We denote X as the conversion

$$\begin{aligned} n_{N_2,out} &= n_{N_2,in} \\ n_{O_2,out} &= n_{O_2,in} - \frac{5}{4} n_{NH_3,in} \cdot X \\ n_{H_2O,out} &= \frac{3}{2} n_{NH_3,in} \cdot X \\ n_{NO,out} &= n_{NH_3,in} \cdot X \\ n_{NH_3,out} &= n_{NH_3,in} \cdot (1 - X) \end{aligned} \quad (3.61)$$

The solution of the problem by hand would be to choose a temperature, read the conversion in the figure, compute the mass balance and see if the energy balance holds. If not, go back and select another temperature.

Now we are going to formulate the problem in EXCEL®. First, for a number of temperature points, we read the conversion and plot a figure. Since the energy balance must be in K, we prepare another column with the corresponding temperatures in K, Figure 3.34.

As presented in a previous example, we use add trend line to obtain a correlation between the conversion and the inlet temperature.

```
X = 6.6667E-11T⁴ - 3.0910E-07T³ + 5.3295E-04T² - 4.0489E-01T +
1.1523E+02
```

The fitting is good enough, although not perfect. Next, we formulate the mass balances in another EXCEL® sheet. We use 1 kmol of NH_3 fed to the system. The air is fed with 10% excess thus

```
F5=(mol N2) =F6*3.76
F6 = (mol O2) =1.25*(1+Excess/100)
F7=(mol H2O) =0
F8 =(mol NO)=0
F9=(mol NH3)= 1
```

F1 is used to determine the excess of oxygen. We define Cell F1 as **Excess**.

F2 (defined as Conversion) is where we write the conversion of the reactor as function of temperature (Tout = I10) as:

```
Conversion=0.000000000066667* Tout ^4 - 0.0000003091* Tout ^3
+ 0.00053295* Tout ^2 - 0.40489* Tout + 115.23
```

The outlet stream is defined using the mass balances given by eq. (3.61) as follows

```
I5=(mol N2) =F5
I6 =(mol O2) =F6-F9*(5/4)*( Conversion)
I7=(mol H2O) =1.5*F9* Conversion
I8 =(mol NO) =F9* Conversion
I9=(mol NH3) =F9*(1- Conversion)
```

Next we compute the energy generated in the reaction. So, in column L we use i to denote the species from N_2 to NH_3 in the table and define as formation enthalpy (ΔH_f) the column C5:C9

$$Li = (Ii-Fi) \cdot \Delta H_f \tag{3.62}$$

Moreover, we need to calculate the energy that enters and leaves the system with the flow (the inlet temperature, Tin = F10). For the feed we have a flow enthalpy of:

```
Hi =Gi*(a*( Tin -298)+(b/2) *( Tin ^2-298^2)+(c/3)
(Tin ^3-298^3)-(d/4)*( Tin ^4-298^4))
```

Where a, b, c and d denote the coefficients of the heat capacity of the each of the species in gas phase. We store these ones as well as other data such as molecular weights and enthalpies of formation in a section of the spreadsheet. Note that we are integrating the heat capacity already in this expression. Moreover, we have to be careful with the units since we are performing the balance in kcal/kg. Gi corresponds to the mass of species i (N_2 to NH_3) that is computed using the data for the molecular weights stored in column B that we define as **MW.**

```
Gi =Fi*MW
```

The advantage is that the temperature of this stream is known. Alternatively we can define modules for each of the functions defining the enthalpy of flow. We leave it to the reader as an exercise.

We do the same for the exiting stream in column K where our variable, the outlet temperature, **Tout** = I10 and **Ji** is the mass flow:

```
Ji=Ii*MW
Ki =Ji*(a*(Tout -298)+(b/2) *(Tout ^2-298^2)+(c/3)*
(Tout ^3-298^3)-(d/4)*(Tout ^4-298^4))
```

The total enthalpy of the flows is computed as

```
H11= SUM(H5:H9)
```

And

```
K11=SUM(K5:K9)
```

And the energy generated in the reaction is calculated based on the difference in the moles of each species.

```
Li=(Ii-Fi)*Ci
```

Where Ci is the normal enthalpy of formation at 298 K and Ii and Fi are the molar flows of each species. The total energy generated in the reaction is computed in L11 as:

```
L11=SUM(L5:L9)
```

Therefore, the energy balance is our target cell. We write the LHS of the energy balance in H13 as

```
H13=H11-L11
```

and the RHS in I13. The target cell is defined in H14 as

```
H14=H13-I13
```

The solver model uses cell H14 as objective and Tout as variable. We just solve the equation to obtain the results presented in Figure 3.36. In order to fix the number of decimals we click with the right button of the mouse, select format cell, and click in Number and to the right we select 2 decimals. Thus the conversion is 93%.

QUESTIONS

Now that the problem is formulated, we can evaluate different variables

a. Write the problem again defining the heat capacities as modules.
b. What is the effect of the excess of oxygen on the outlet temperature?
c. Determine the excess of air temperature to get a conversion of 90%. This is tricky.

3.3.4.3 HNO_3 Production: Process Mass and Energy Balances

The idea of using EXCEL® to perform the mass and energy balances of a production facility is interesting from the user point of view. Mainly, the presentation of the results is easy to understand if they are organized properly and thus it helps operators and engineers to see the actual values and the effect of changes in the

	E	F	G	H	I	J	K	L
1	Excess air	10.00						
2	Conver=	0.93						
3		In			Out			
4		kmol	kg	Q(kcal), flow	kmol	kg	Q(kcal), flow	Q(kcal)react
5	N2	5.17	144.76	1464.85	5.17	144.76	32994.52	
6	O2	1.38	44.00	403.61	0.21	6.72	1897.93	
7	H2O	0.00	0.00		1.40	25.16	10924.45	-80799.38
8	NO	0.00	0.00		0.93	27.96	6178.44	20130.65
9	NH3	1.00	17.00	352.08	0.07	1.16	679.50	10214.44
10	Temp	338.60			1157.32			
11	Q			2220.55			52674.84	-50454.30
12			LHS		RHS			
13				52674.84	52674.84			
14			Error	0.00				

FIGURE 3.36 Results of the ammonia reactor example.

process conditions. We understand that in solving the models EXCEL® may not be the best option when we need rigorous thermodynamics but even though we can use EXCEL® to show the results of different software such as gPROMS®, MATLAB®, GAMS® just to mention a few of the packages covered in this text. For this example we are going to perform the mass and energy balances for the production of nitric acid from ammonia. We are going to use EXCEL® for all the calculations but bear in mind that could link EXCEL® to other software if needed. We build on the previous examples in order to reduce the explanation for the sake of the length of the text.

A basic problem for the production of nitric acid from ammonia can be found in [1,7]. However typically dual processes have become preferred in order to take advantage of the physic-chemical principles so that it is possible to enhance the conversion of ammonia oxidation at low pressure and improve the gases absorption downstream at a higher pressure. The problem that we address is presented as follows. The production of HNO_3 from ammonia starts with atmospheric air. The ratio between oxygen and ammonia to be fed to the converter must be 2.11, and the ratio between air and added oxygen 12 due to security issues [8], see Table 3.2. Consider 1 kmol of atmospheric air as basis and liquid ammonia as raw materials.

TABLE 3.2
Initial conditions of the nitric acid production problem

T air (°C)	26
Ptotal (mmHg)	810
Relative humidity	0.26
Rel Air/$O_{2\ added}$	12
Rel O_2 total/NH_3	2.11

TABLE 3.3
Operation data for the absorption tower

T in tower (°C)	30
Pt (mmHg)	7600
conc liq abs (%)	2.5
Conc HNO_3 prod (%)	55
L/V	0.25

The mixture enters the converter at 26°C and 810 mmHg. We assume a conversion in the reactor equal to 1. The gases are cooled down to 25°C and compressed to 10 atm. During the cooling, NO is oxidized with a conversion of 15% and HNO_3 is formed and dissolved in water with a composition of 3.5%. The gas phase is sent to an oxidation tower where the rest of NO is oxidized completely at 7600 mmHg.

The gases are cooled down to 30°C at constant pressure and are fed to an absorption – oxidation tower that operates isothermically. We feed an HNO_3 solution 2.5% w/w with a ratio liquid to vapor of 0.25. The acid produced is expected to reach 55% concentration. A summary of the operating conditions of the tower is given in Table 3.3. Determine the flowrate of acid solution used, flows and temperatures at the different units.

3.3.4.3.1 Data and information

We can find the data needed for this model elsewhere [1,7,8,9]

$$Pv_{Agua}(\text{mmHg}) = \text{EXP}\left(18.3036 - \frac{3816.44}{227.02 + T(C)}\right) \tag{3.63}$$

$$2 \cdot NO_2 \leftrightarrow N_2O_4 \qquad K_p = \frac{P_{N_2O_4}}{P_{NO_2}^2} \tag{3.64}$$

$$\log_{10}\left(\frac{1}{K_p}\right) = -\frac{2692}{T} + 1.75\log_{10} T + 0.00483 \cdot T - 7.144 \cdot 10^{-6} T^2 + 3.062 \tag{3.65}$$

$$T\,[=]\text{K},\ K_p[=]\text{atm}^{-1}$$

Dilution heat for HNO_3 is taken from the figures in [7]

Solution:

We divide the process into five zones. Each one will be defined in a different sheet within the EXCEL® file that will be linked to each other so that the result of the previous sheet is the inlet of the following one. We organize the solution of the problem as follows, starting with feedstock, converter, cooling down, oxidation tower and finally, absorption-oxidation tower. The process operates at dual pressure to improve the performance. Thus the converter operates at low pressure, 1.05 atm, to increase the conversion, while the last stages of oxidation and absorption will operate at a pressure of 10 atm. For the sake of the length of the explanation we build on the previous examples.

3.3.4.3.2 1st Excel Sheet: Feedstock Preparation

As it is indicated in the description of the exercise, we have air at 26°C and 26% relative humidity at 810 mmHg. We also have liquid ammonia at the same temperature and pressure. We store the data provided in the example, Table 3.4, as:

We first compute the water contained in the air

```
B4(H2O in air)=B9*B11/(B10-B9*B11) where
B10(Ptotal)=EXP(18.3036- 3816.44/( 227.02+B8))
```

We have to take into account that our basis is 1 kmol of air (including water). To verify that the pressure of O_2 (cell B6) and N_2 (cell B5) is proportional to the number of moles and is a fraction of the total atmospheric pressure we use molar humidity thus our air composition as function of the moles of oxygen is given by:

$$1kmol_{aire} = n_{O_2} + 3.76n_{O_2} + 4.76n_{O_2}\left(\frac{P_v}{P_T - P_v}\right) \tag{3.66}$$

TABLE 3.4
Table in EXCEL format

	A	B
1		
2	Composicion del aire	
3		
4	H2O	0.00809918
5	N2	0.78357048
6	O2 (air atm)	0.2083964
7	O2 (added)	0.08333884
8	T (ºC)	26
9	Pv (mmHg)	25.0293539
10	Pt (mmHg)	810
11	Humidity	0.26
12	Rel Air/O2 added	12
13	Rel O2total/NH3	2.11

To compute the moles of oxygen we perform a balance to the dry air:

```
B14(moles of dry air by pressure)=(B10-(B9*B11))/B10
C14(dry air moles)=B6+B5
B5(N2 in air) =B6*3.76
```

We have one variable, B6 (moles of oxygen in air), that we solve with *Solver* defining as target cell:

```
D14=B14-C14
```

Once that we have the air composition (Table 3.3), we use the data provided to calculate the oxygen added and the ammonia fed to the converter:

$$\frac{n_{aire}}{12} = n_{O_2,added} \tag{3.67}$$

$$\frac{n_{O_2,total}}{2.11} = n_{NH_3} \tag{3.68}$$

```
B7(Oxygen added) =(1/O2added)*(B4+B5+B6)
B16( NH3 fed) =(1/RelO2total)*(B6+B7)
```

For the ammonia we can compute the energy needed to evaporate it as:

$$Q(\text{vap}) = m(\lambda\ m_{NH3} + cp_{liq}(T - T_{ref})) \tag{3.69}$$

```
I18 (Q to evaporate)=G16*17*(1186/4.18+(4.5/4.18)*(26-25))
```

We organize the results in the EXCEL® sheet so that it is easy to follow, Figure 3.37. Thus we define the different streams that are mixed. Air, pure oxygen, and ammonia.

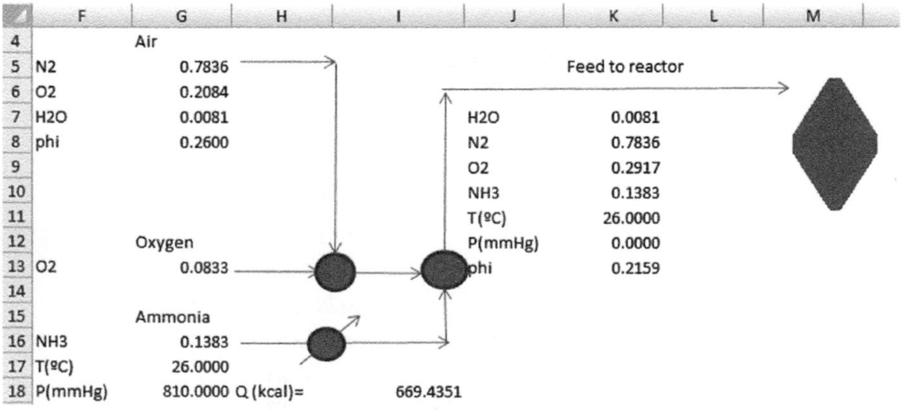

FIGURE 3.37 Reactants preparation. Flows in kmol.

Since it is a simple addition we just define the stream that enters the reactor component by component as the sum of the components from the previous one.

3.3.4.3.3 2nd Excel Sheet: Converter

Next the ammonia is oxidized. Thus the EXCEL® sheet is built as the previous example. We assume conversion of the oxidation equal to 1.

$$NH_3 + \frac{5}{4}O_2 \rightarrow NO + \frac{3}{2}H_2O \tag{3.70}$$

The energy balance is as follows:

$$\begin{aligned} \sum n_i \int_{T_{ref}}^{T_i} Cp_i\, dT + Q_{generated} &= n_{N_2,out} \int_{T_{ref}}^{T_{out}} Cp_{N_2}\, dT + n_{NH_3,out} \int_{T_{ref}}^{T_{out}} Cp_{NH_3}\, dT \\ &+ n_{O_2,out} \int_{T_{ref}}^{T_{out}} Cp_{O_2}\, dT + n_{H_2O,out} \int_{T_{ref}}^{T_{out}} Cp_{H_2O}\, dT + n_{NO,out} \int_{T_{ref}}^{T_{out}} Cp_{NO}\, dT \end{aligned} \tag{3.71}$$

In EXCEL® we organize the list of components in column A and define each one related to the previous sheet as:

```
Bi=Feedstock!Bi
```

Where i represents a component rather than a row number. From now on we have in each of the sheets information on the molecular weights and formation heat, Figure 3.38. Figure 3.39 summarizes the results of this stage including a flowsheet scheme of the results that is appropriate to present the calculation, Figure 3.40.

3.3.4.3.4 3rd Excel Sheet: Compression and Cooling

This stage could have also been divided in two, the compression and the cooling. First the gases are compressed. Since the ratio between the final pressure and the

	MW(kg/kmol)	DHf (kcal/kmol)
N2	28	0
O2	32	0
H2O	18	-57798
NO	30	21600
NO2	46	8091
NH3	17	-10960
HNO3	63	-15224.8804
N2O4	92	-13712.9187

FIGURE 3.38 Data within the spread sheet.

	A	B	C	D	E	F	G	H
1								
2								
3								
4	Conver=	1						
5		Inlet			Out			
6		kmol	kg	Q(Flow)(kcal)	kmol	kg	Q(Flow) kcal	Reaction
7	N2	0.78357048	21.93997349	5.469337413	0.78357048	21.9399735	4354.57311	
8	O2	0.29173524	9.335527803	2.089874894	0.11890631	3.80500185	885.357874	
9	H2O	0.00809918	0.1457853		0.21549391	3.87889032	1456.37017	-11987.0002
10	NO	0	0		0.13826315	4.14789446	797.588841	2986.48401
11	NH3	0.13826315	2.350473529	1.178699829	0	0	0	1515.36411
12	Temp (K)	299			1053.20197			
13	Q			8.737912137			7493.89	-7485.15208
14			Q in + Q generated		Qout			
15				7493.889996	7493.89			
16			Error	3.00133E-10				

FIGURE 3.39 Results of the reactor operation.

initial one is larger than 4 we decide to consider a two-stage compression with cooling between stages. We assume ideal polytropic compression and the ratio between the heat capacity at constant pressure and heat capacity at constant volume, k, equal to 1.4. In between the compressors, stream is cooled down to the initial temperature of 780°C. To calculate the final temperature, compressor power and the cooling needs for the heat exchanger the equations are summarized in eq. (3.72). Assume that the heat capacity for the N_2O_4 is the same as for the NO_2.

NH3 + (5/4) O2 --> NO + (3/2) H2O

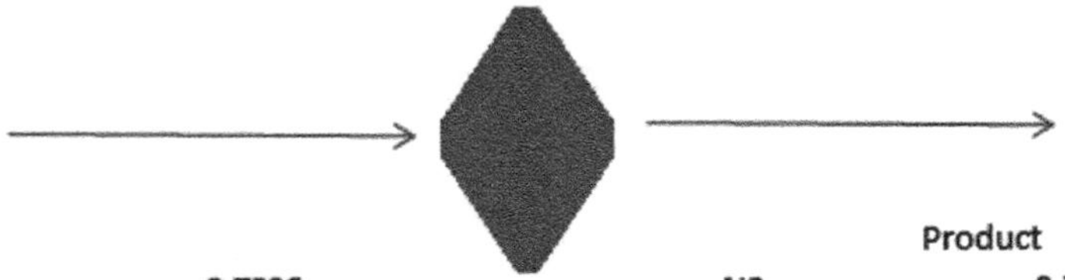

	J	K	L	M	N	O
10	Feed					Product
11	N2	0.7836			N2	0.7836
12	O2	0.2917			O2	0.1189
13	H2O	0.0081			H2O	0.2155
14	NO	0.0000			NO	0.1383
15	NH3	0.1383			NH3	0.0000
16	T(ºC)	26.0000			T(ºC)	780.2020

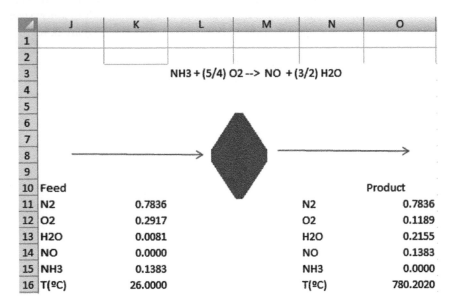

FIGURE 3.40 Summary of the results of the reactor operation. Flows in kmol.

$$T_2 = T_1\left(\frac{P_2}{P_1}\right)^m$$
$$m = 0.5\frac{1.4-1}{1.4}$$
$$W = 2T_1R\frac{1.4}{1.4-1}\left[\left(\frac{P_2}{P_1}\right)^m - 1\right] \quad (3.72)$$
$$Q = -\sum_i m_i \int_{Ti}^{Tout} Cpdt$$

In EXCEL® we organize the list of components in column A and define each one related to the previous sheet as:

```
Bi=Converter!Oi
```

As before, i actually represents the component and not the number of the row. There is no reaction during the compression and thus:

```
Ei = Bi; Fi = Ci
```

We compute the final temperature of the stream as:

```
F17=(B17+273)*(F18/B18)^((1/2)*(1.4-1)/1.4)
```

Be careful since these calculations need the use of temperatures in column K.

```
W(kcal) =0.24*2*(B17+273)*8.314*((1.4-1)/1.4)*((0.5*(F18/
B18))^(0.4/1.4)-1)
```

We use the flow enthalpies computed in column G for the exit and D for the inlet. Bear in mind that the intercooling systems works in such a way that the inlet and outlet temperatures for each of the compressors are the same. Figure 3.41 show the mass balances.

```
Q(intercooling)= C27= SUM(G11:G14)-SUM (D11:D14)
```

Next step is gases cooling down to 20.6°C at 10 bar. During this process there is condensation of the water to produce nitric acid 3.5%. 15% of the NO is oxidized to NO_2 and HNO_3 is produced in a small amount:

$$NO + \frac{1}{2}O_2 \rightarrow NO_2 \quad (3.73)$$

In EXCEL® this is the easy part. As in previous examples we have a chemical reaction with a conversion 15% (stored in B3). The feed to the reaction comes from

	A	B	C	D	E	F	G
8	Compression						
9	In				Out		
10	GAS	kmol	kg	Q (kcal)flow	kmol	kg	Q(kcal)flow
11	N2	0.783570482	21.939973	4354.5731	0.7835705	21.9399735	6811.47738
12	O2	0.118906308	3.8050019	885.35787	0.1189063	3.80500185	1755.90444
13	H2O	0.215493906	3.8788903	1456.3702	0.2154939	3.87889032	2358.79697
14	NO	0.138263149	4.1478945	797.58884	0.1382631	4.14789446	1259.15036
15	NO2	0	0		0	0	
16	N2O4	0	0		0	0	
17	T (ºC)	780.201975	1053.202		1177.1594	1450.15942	
18	P (mmHg)	810				7600	
26	W (kcal)		568.58795				
27	Q (kcal)		4691.4392				

FIGURE 3.41 Compressor results.

the compressor. Thus, we compute the formation of NO_2 and the consumption of NO and O_2 based on the stoichiometry:

```
L12(O2 out) =H12-0.5*L15
L14(NO out) =H14*(100-Conversion)*0.01
L15(NO2 out) = Conversion*0.01*E14
```

We also compute the mass flow in column M as

$$\mathrm{Mi} = \mathrm{Li} \cdot \text{Molar weight} \tag{3.74}$$

Again the index i represents the component rather than the row number. We also compute the energy generated in this particular reaction

```
Qrec1=SUM(N14:N15)
```

$$\mathrm{Ni} = (\mathrm{Li} - \mathrm{Hi})\Delta \mathrm{H_f} \tag{3.75}$$

The results of this stage are the initial point for the next one. Now, the main challenge here is the equilibrium between the NO_2 and the N_2O_4. We consider that the equilibrium is established once the nitric acid is generated since while the acid is formed, the equilibrium will move to produce NO_2 that is disappearing from the system.

$$2NO_2 \leftrightarrow N_2O_4 \tag{3.76}$$

$$2NO_2 + H_2O + \frac{1}{2}O_2 \rightarrow 2 \cdot HNO_3 \tag{3.77}$$

Thus in the heat exchanger water condenses

$$Q_{cond} = -m_{H_2O}\lambda \tag{3.78}$$

The amount of condensed water is responsible for the concentration of acid that is given by the exercise data. The non-condensed water will saturate the gas:

$$\%_{acido} = \frac{n_{HNO_3} \cdot M_{HNO_3}}{n_{HNO_3} \cdot M_{HNO_3} + M_{H_2O}\left(n_{H_2O} - y_{sat} \cdot GAS - \frac{1}{2} n_{HNO_3}\right)} \tag{3.79}$$

$$GAS = n_{NO} + n_{NO_2,ini} - n_{HNO_3} + n_{N_2} + n_{O_2} - \frac{1}{4} n_{HNO_3}$$

$$m_{H_2O} = \left(\frac{100 - \%_{acido}}{\%_{acido}}\right) m_{HNO_3} \tag{3.80}$$

We relate the results of this stage to the nitric acid produced. We have two phases, liquid phase comprised of water and nitric acid and gas phase whose components are N_2, NO, NO_2, O_2 that are saturated with water.

Thus the oxygen remaining is defined as

```
P12 =L12-P25/4
```

The water in the gas phase

```
P13=L13-P24-0.5*(P25)
```

The NO_2

```
P15=L15-P25
```

The nitric acid formed

```
P25=0.01*B4*(B36*L13-P20*B36*(L11+L12+L14+L15))/
(B40-0.01*B4*(B40-B36*0.5+B36*P20*1.25))
```

Thus the water that generates the solution

```
Q24=Q25*(1-0.01*B4)/(0.01*B4)
```

And the energy involved in this process is due to the formation of nitric acid liquid and that due to the condensation of water. It is an exothermic process

```
R26 (Q condensation) =-Q24*597
R27 (Q HNO3 formation and dilution) =P25*(C40-7500)
R28 (Sensible heat)= 1*(Q24+Q25)*(P17-25)
```

With the gas remaining the equilibrium is established

$$2 \cdot NO_2 \leftrightarrow N_2O_4 \qquad K_p = \frac{P_{N_2O_4}}{P_{NO_2}^2} \tag{3.81}$$

Where

$$\log_{10}\left(\frac{1}{K_p}\right) = -\frac{2692}{T} + 1.75\log_{10} T + 0.00483 \cdot T - 7.144 \cdot 10^{-6} T^2 + 3.062 \quad (3.82)$$

We denote y as the moles of N_2O_4 and x the NO_2 moles in the equilibrium so that the equilibrium constant becomes:

$$Kp = \frac{\frac{y}{n_t}}{P_T\left(\frac{x}{n_T}\right)^2} = \frac{y \cdot n_t}{P_T x^2} \quad (3.83)$$

Furthermore the total moles of NO_2, denoted as equivalent, are constant and can be calculated upfront

$$2y + x = NO_{2,eq} = cte \quad (3.84)$$

Thus we can actually solve x since the equation if is of second degree

$$Kp\left(P_{T,atm}\right)x^2 = \left(\frac{NO_{2,eq} - x}{2}\right)\left(n_{H_2} + n_{O_2} + n_{N_2} + n_{H_2O} + x + \left(\frac{NO_{2,eq} - x}{2}\right)\right)$$

$$n_{H_2} + n_{O_2} + n_{N_2} + n_{H_2O} = N_{ini} \quad (3.85)$$

$$x = \frac{-0.5N_{ini} + \sqrt{(0.5N_{ini})^2 - 4Kp\left(P_{T,atm} + 0.25\right)\left(0.25NO_{2,eq}{}^2 + 0.5NO_{2,eq}N_{ini}\right)}}{2Kp\left(P_{T,atm} + 0.25\right)}$$

```
S19 (Kp) =(10^(-2692/(S17+273)+1.75*LOG10(S17+273)+0.00483*
(S17+273)-7.144*10^(-6)*(S17+273)^2+3.062))^(-1)
```

Thus the NO remaining

```
S15 (NO2 out) =
((-0.5*(S11+S12+S13+S14))+((0.,5*(S11+S12+S13+S14))^2+
4*(S19*(B2/760)+ 0.25)*(0.25*P21^2+0.5*P21*(S11+S12+S13+S14)))
^0.5)/(2*(S19*(B2/760)+0.25))
```

Where the number of initial moles is S11+S12+S13+S14

And the N_2O_4 out:

```
S16 (N2O4) =(P15-S15)/2
```

This last stage has also energy involved due to the reaction.

```
Q(react) = SUM(V11:V16)
```

	A	H	I	J	K	L	M	N	O	P	Q	R	S	T	U	V
7																
8	Compression															
9	In	In				Oxidation NO - NO2		Q reacc	NO2 - HNO3			Q reacc	Out		NO2- N2O4	Reacc
10	GAS	kmol	kg	Q(kcal)flow					GAS	kmol	kg		kmol	kg	Q(kcal)flow	
11	N2	0.783570482	21.93997349	6811.477378		0.78357048	21.94	0		0.78357048	21.9399735		0.783570482	21.9399735	-24.0667514	0
12	O2	0.118906308	3.805001853	1755.904443		0.10853657	3.47317	0		0.10798747	3.45559896		0.107987467	3.45559896	-3.79057164	0
13	H2O	0.215493906	3.878890317	2358.796974		0.21549391	3.87889	0		0.00244144	0.04394591		0.00244144	0.04394591	-0.04820586	0
14	NO	0.138263149	4.147894462	1259.150359		0.11752368	3.52571	-447.973		0.11752368	3.52571029		0.117523676	3.52571029	-3.8674793	0
15	NO2	0	0	0		0.02073947	0.95402	167.803		0.01854306	0.85298054		0.007606099	0.34988055	-0.38379664	-88.4909131
16	N2O4	0	0	0			0				0		0.005468478	0.09296413	-0.10197572	-74.9887958
17	T (ºC)	1177.159416	1450.159416							20.6	293.6		20.6	293.6		
18	P (mmHg)	0	7600							7600			7600			
19	Kp		532527.2008							9.68491689			9.684916891			
20	y sat									0.00237581			0.002375809			
21	NO2 eq									0.01854306						
22																
23	LIQ								LIQ	kmol	kg					
24	H2O									0.21195426	3.81517665					
25	HNO3									0.00219642	0.13837428					
26	W (kcal)										Q cond	-2277.660459				
27	Q (kcal)		Qin(flow)	12185.32915			Q re1	-280.17			Qrec2	-49.91331778		Qout (Flow)	-32.2587806	
28											Qout	-17.3956241			Q react=	-163.479709
29																
30								Q (cooldown)		-15006.2066						

FIGURE 3.42 Results of the condenser.

Where:

$$Vi = (Si - Pi) * \Delta H_f$$

Finally, once that we have the final composition we compute the energy that exits the system with the liquid and the gas flows as before, see Figure 3.42. As a summary of results we can organize the streams as in Figure 3.43.

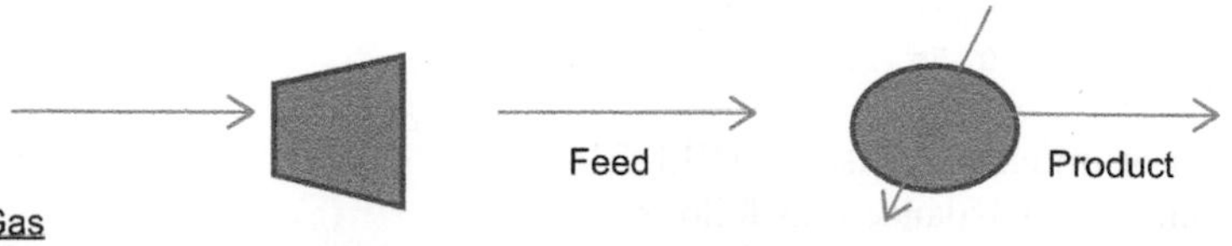

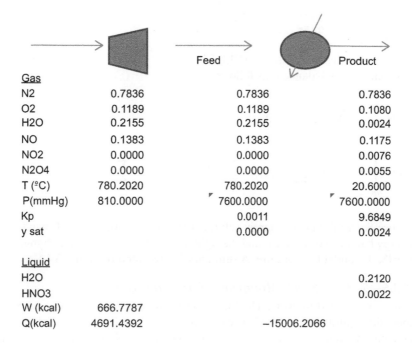

Gas				
N2	0.7836	0.7836		0.7836
O2	0.1189	0.1189		0.1080
H2O	0.2155	0.2155		0.0024
NO	0.1383	0.1383		0.1175
NO2	0.0000	0.0000		0.0076
N2O4	0.0000	0.0000		0.0055
T (ºC)	780.2020	780.2020		20.6000
P(mmHg)	810.0000	7600.0000		7600.0000
Kp		0.0011		9.6849
y sat		0.0000		0.0024
Liquid				
H2O				0.2120
HNO3				0.0022
W (kcal)	666.7787			
Q(kcal)	4691.4392		–15006.2066	

FIGURE 3.43 Summary of results.

3.3.4.3.5 4th Excel Sheet: Oxidation Tower

The gases coming from the condenser are fed to the oxidation tower so that we obtain NO_2. We assume conversion of 100%. The tower is supposed to operate adiabatically. This example involves the energy balance and the fact that the temperature determined the equilibrium between NO_2 and N_2O_4 that must be solved simultaneously. We can take a look at the previous example were the conversion of the NH_3 was a function of the temperature to see that both examples share similarities. See the material in the web for the full table.

The mass balance takes into account the oxidation of the NO to NO_2 and the equilibrium on NO_2 and N_2O_4. The equations that describe this equilibrium are the same as the ones presented above. We here only present the balance to the species whose composition changes. B3 is the conversion, equal to 1.

The NO_2 equivalent is given by:

```
B21(NO2 eq)=B16+B15+2*B17
```

The kp is calculated at the final temperature (E18) and the pressure (10 atm)

```
E20(kp out) = (10^(-2692/(E18)+1.75*LOG10(E18)
+0.00483*(E18)-7.144*10^(-6)*(E18)^2+3.062))^(-1)
E13(O2 out)=B13-B3*0.5*B15
E15(NO out) = B15*(1-B3)
E16 (NO2 out)=
((-0.5*(E12+E13+E14+E15))+((0.5*(E12+E13+E14+E15))^2+
4*(E20*(B2/760)+ 0.25)*(0.,25*B21^2+0.5*B21*(E12+E13+E14+E15)))
^0.5)/(2*(E20*(B2/760)+0.25))
E17(N2O4 out)= 0.5*(B21-E16)
```

Number of initial moles is: E12+E13+E14+E15

Next, the energy balance is as follows:

$$\sum_{prod}\Delta H_{f,prod} - \sum_{react}\Delta H_{f,react} = 0$$

$$\sum m_i \int_{T_{ref}}^{T_{entrada}} Cp_i dT + \left(\sum_{prod}\Delta H_{f,prod} - \sum_{react}\Delta H_{f,react}\right)_{T=tref} = \sum m_i \int_{T_{ref}}^{T_{salida}} Cp_i dT \tag{3.86}$$

We use *Solver* defining as target cell the difference between the left hand side of the energy balance above, D24, and the right hand side (E24), and the change variable is E18, the outlet temperature. A summary is presented in Figure 3.44.

3.3.4.3.6 5th Excel Sheet: Absorption – Oxidation Tower

The gases are cooled to 30°C. The important thing here is that, again, we need to compute the equilibrium at 30°C. The cooling that is required involves the sensible heat together with the energy of the reaction from NO_2 to N_2O_4. Since it is similar as the previous case, here we only show the results, see Figure 3.45.

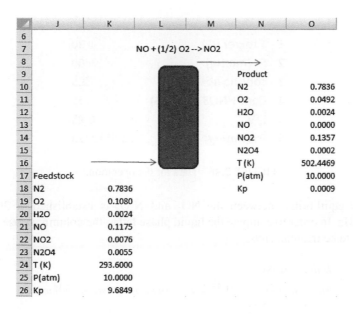

	J	K	L	M	N	O
6						
7			NO + (1/2) O2 --> NO2			
8						
9					Product	
10					N2	0.7836
11					O2	0.0492
12					H2O	0.0024
13					NO	0.0000
14					NO2	0.1357
15					N2O4	0.0002
16					T (K)	502.4469
17	Feedstock				P(atm)	10.0000
18	N2	0.7836			Kp	0.0009
19	O2	0.1080				
20	H2O	0.0024				
21	NO	0.1175				
22	NO2	0.0076				
23	N2O4	0.0055				
24	T (K)	293.6000				
25	P(atm)	10.0000				
26	Kp	9.6849				

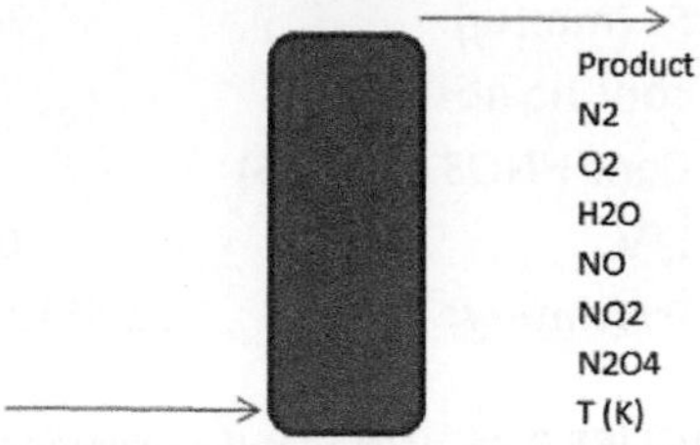

FIGURE 3.44 Summay of results of the oxidation tower.

```
B22 (NO2 eq) =B16+B17
B16 (NO2 out) =((-0.5*(E12+E13+E14+E15))+((0.5*(E12+E13+E14
+E15))^2 +4*(E21*(B2/760)+0.25)*(0.25*B22^2+0.5*B22*(E12+E13+
E14+E15)))^0.5)/(2* (E21*(B2/760)+0.25))
B17(N2O4 out) =0.5*(B22-E16)
```

And the energy balance is

```
D31 (Q ref) =SUM(G12:G17)-SUM(D12:D17)+SUM(H12:H17)
```

This gas phase is fed to the tower. The data related to the liquid phase added is presented in the following Figure 3.46.

	A	B	C	D	E	F	G	H	I	J	K
7					Absorption tower						
8											
9		Cooling									
10		In			Cooling and tower input				Out		
11	GAS	kmol	kg	Q(kcal)	kmol	kg	Q(kcal)	Q(kcal) reac	kmol	kg	Q(kcal) Reac
12	N2	0.78357048	21.9399735	1123.19207	0.783570482	21.93997349	27.345476	0	0.783570482	21.9399735	27.345476
13	O2	0.04922563	1.57522014	76.2969795	0.049225629	1.575220136	1.76473568	0	0.019190017	0.61408054	0.68796089
14	H2O	0.00244144	0.04394591	4.11911515	0.00244144	0.043945913	0.09834864	0	0.003435318	0.06183572	0.13838508
15	NO	0	0	0	0	0	0	0	0	0	0
16	NO2	0.135732	6.24367219	266.886941	0.031882113	1.46657718	1.40698738	-840.249472	0.009122864	0.41965175	0.40260051
17	N2O4	0.00016736	0.01539746	0.65816747	0.052008628	4.78479374	4.59037854	-710.895035	0.004772628	0.08113467	0.07783802
18	Total	0.97113692	29.8182092	1471.15328	0.919128291	29.81051046	35.2059262	-1551.14451	0.820091308	23.1166762	28.6522605
19	T (K)	502.446889			303				303		
20	P(atm)	10			10				10		
21	Kp	0.00088222			4.702809047				4.702809047		
22	NO2 eq	0.13589937							0.01866812		
31	Q (kcal)			-2987.09186							-2671.14406

FIGURE 3.45 Results of the fed to the absorption tower.

	A	B
1	T tower (C)	30
2	Pt (mmHg)	7600
3	conc liq abs (%)	2.5
4	Conc HNO3 prod (%)	55
5	L/V	0.25
6	Pv (mmHg)	31.6517823

FIGURE 3.46 Data for the operation.

A new equilibrium between the NO_2 and N_2O_4 is established at 30°C and 7600 mmHg. In order to compute the liquid phase fed to the column we use the total gas phase to be treated. Thus,

$$\begin{aligned}
&\textit{Balance}\ \mathrm{HNO_3}\\
&n_{HNO_3,total}\cdot M_{HNO_3} = 0.55\cdot Liq = (n_{HNO_3,in} + n_{HNO_3,prod})M_{HNO_3}\\
&n_{HNO_3,prod} = \frac{0.55\cdot Liq}{M_{HNO_3}} - n_{HNO_3,in}\\
&\textit{Balance}\ \text{water}\\
&(n_{H_2O,vap} + n_{H_2O,liq})_{in} = 0.45\cdot Liq + n_{H_2O,vap} + 0.5\cdot n_{HNO_3,prod}\\
&n_{H_2O,vap} = (n_{N_2} + n_{O_2} + n_{NO_2,eq})\frac{P_v}{P_T - P_v}\\
&n_{NO_2,sale} = n_{NO_2,ini} - n_{HNO_3,prod}\\
&n_{O_2,sale} = n_{O_2,in} - \frac{1}{4}n_{HNO_3,prod}\\
&n_{N_2,sale} = n_{N_2,in}\\
&Liq = 0.25V_{ini}
\end{aligned} \tag{3.87}$$

From the balance to the water we obtain the flow of liquid, Liq, and with it the rest of the components. So, the liquid inlet is:

```
F28 (Total liquid ) =C18*B5
F27(HNO3 formed) =F28*0.01*B3
F26(H2O liquid) =F28-F27
```

The gas outlet, for the species whose amount change:

```
I13 (O2 out) =B13-(0.01*B4*J28/B41-E27/B41)*0.25
I14(H2O out gas) =(I12+I13+I22)*B6/(B2-B6)
I16 (NO2 out ) = ((-0.5*(I12+I13+I14+I15))+((0.5*(I12+I13+I14
+I15))^2+ 4*(I21*(B2/760)+0.25)*(0.25*I22^2+0.5*I22*(I12+I13+
I14+I15)))^0.5)/(2*(I21*(B2/760)+0.25))
I17 (N2O4 out)=(I22-I16)/2
```

```
I21 (kp out) =(10^(-2692/(I19)+1.75*LOG10(I19)
+0.00483*(I19)-7.144*10^(-6)*(I19)^2+3.062))^(-1)
I22 (NO2 eq) =B22-(0.01*B4*J28/B41-E27)
```

We use *Solver* so that the water balance holds:

```
Water in (J29) =E26+B14 = Water out =J26/B37+I14+0.5*(I27-E27)
```

Where the changing cell is J28. With this value we compute:

```
J26 (Water liquid out) =0.01*(100-B4)*J28
J27 (HNO3out) =J28-J26
```

We need to cool the tower since the formation and dilution of HNO_3 is exothermic. We can find the data elsewhere [1,7]. As previously, we compute the energy of the flows that enter and leave the system. Furthermore we have to take into account the new equilibrium established and the formation and dilution of HNO_3:

```
K24(Formation of HNO3) =(I27-E27)*C41
```

$$Q_{dis} = -5600 kcal/kg \quad (55\%); Q_{dis} = -7500 kcal/kg \quad (2.5\%) \quad [7]$$
$$cp = 0.67 kJ/kg(45\%); 0.64 kJ/kg(60\%)$$

```
K30(dilution)
=-5600*I27+E28*1*(I19-298)-((-7500)*E27+0.63*(E29-298))
K31(Energy to cool) =K30+K24+SUM(K12:K18)-G28-SUM(G12:G18)
```

Figure 3.47 the summary of the results, the details can be found in the excel file.

QUESTION

- Evaluate the effect of the pressure on the production of nitric acid.

3.3.5 Heat Exchange and Equipment Design: Multi-effect Evaporators

Na_2CO_3 industry is one of the most important ones. The history of industrial chemistry and Chemical Engineering is directly related to Leblanc and Solvay processes. The first one was the reason why nowadays we pay so much attention to products while Solvay process is somehow the starting point of process engineering. In the production of NaOH from Na_2CO_3 we typically deal with dilute solutions that must be concentrated.

The example we study here is a three effect evaporator to concentrate a solution 15000 kg/h of NaOH from 15% to 45%. The feedstock is fed at 25°C in parallel and we have saturated steam at 4 atm. The absolute pressure of the last effect is 0.1 atm and the heat transfer coefficients for the three effects are 1500, 1250 and 1000 kcal/m^2 h C. We would like to calculate the energy needed for the process

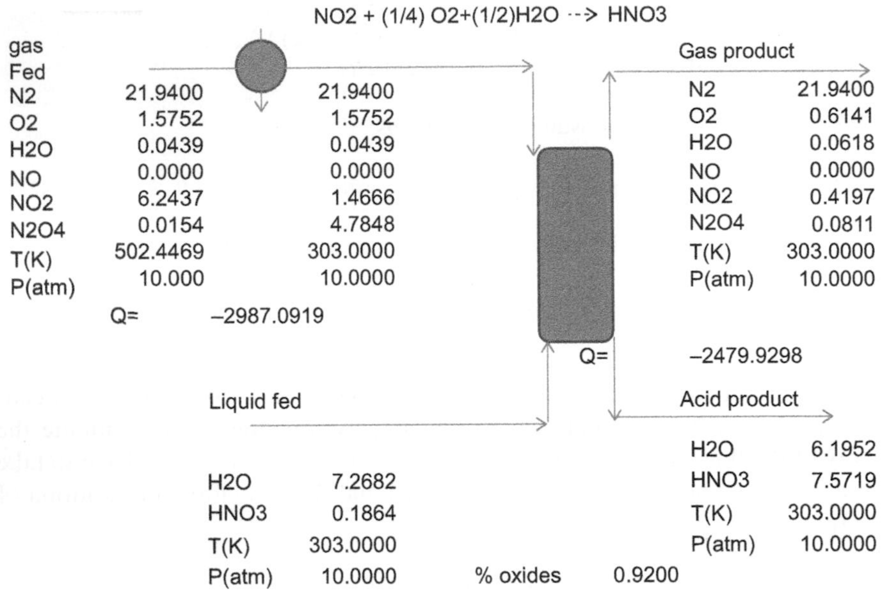

FIGURE 3.47 Summary of results of the operation of the absorption tower (Q(kcal)).

and the cost of the evaporator system, assuming that all the effects are similar, see Figure 3.48 for a scheme.

Solution

The design of such systems was traditionally done by trial and error. The procedure can be found elsewhere [10]. In short:

- Perform a global mass balance
- Assume boiling points of the intermediate effects and equal area for all of them.
- Distribute the total ΔT into the number of effects inversely proportional to the global heat transfer coefficient.

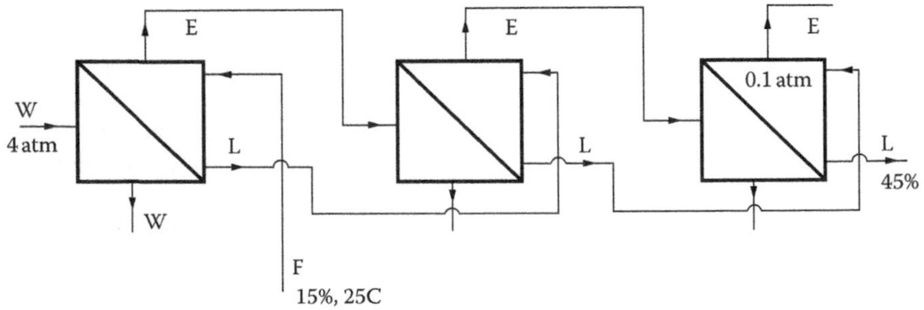

FIGURE 3.48 Multieffect evaporator system.

- Perform energy balances to all the effects
- Solve the system and check if the areas calculated for each of the effects match. If not, redistribute the useful increment of temperature for each effect.

The procedure is easy but tedious. The main challenge is the thermodynamics of the solution. The enthalpy of the solution and the boiling point depend on the concentration and thus the mass and energy balances. Thus we need to correlate the effect of the concentration of NaOH with the boiling point of the solution and the enthalpy and, to a lesser point, the steam saturated enthalpy and latent heat. Once we have developed the correlations to predict the properties of the streams, we can proceed to propose the mass and energy balances to the system of evaporators.

1. *Parameter estimation problem:*

We need to develop correlations for the heat of vaporization on water as well as the enthalpy of saturated vapor [4]. From the table, using "add trend" line in EXCEL® we can easily correlate these two as:

$$\lambda = -4.3722E\text{-}06t^3 + 4.3484E\text{-}04t^2 - 5.8433E\text{-}01t + 5.9748E{+}02$$
$$H_{Vap} = -3.0903E\text{-}06t^3 + 2.2613E\text{-}04t^2 + 4.2436E\text{-}01t + 5.9742E{+}02$$

The solutions of NaOH are more difficult. However, there are a lot of data in the literature. We focus on the effect of the NaOH on the boiling point, Dühring diagram, and on the enthalpy, enthalpy-concentration diagram. We can find these diagrams elsewhere [4, 10-13]. The first one represents the difference between the boiling point of water, which is easy to calculate using Antoine correlation, and that of the solution of NaOH of different concentrations of NaOH by weight. Basically we have straight lines relating $T_{eb}(NaOH, \%_i)$ and T_{eb} of water.

First, for a particular composition, we correlate:
$T_{eb}(NaOH, \%_i) - T_{eb}$ (water) vs T_{eb} (water)
We get $a_i T_{eb}(water) + b_i$
Next, we look for the relationship between a_i and $\%_i$ and between b_i and $\%_i$
Thus we obtain a correlation:

$$T_{eb}(NaOH, \%_i) = (-0.0000030233 * (\%_i)^3 + 0.00021232 * (\%_i)^2 - 0.0042815 * (\%_i) + 1.0459) * T_{eb}(water) + (0.00024541 * (\%_i)^3 + 0.0060805 * (\%_i)^2 - 0.009887 * (\%_i) + 0.17863)$$

For the enthalpy of the solution the figure is more complex. However, we follow the same procedure. We see that the figure is composed by a number of curves that look like parables or easy polynomial for each temperature as function of the concentration. Thus, for a particular temperature, we look for correlation in the form of

$$H(NaOH,\%,T = cte) = a_i\,\%_i^3 + b_i\,\%_i^2 + c_i\,\%_i + d_i$$

we do it for a number of temperatures. Next, we look for the effect of the temperature on a_i, b_i, c_i, d_i

we obtain another correlation, a little more complex:

```
hL(NaOH,%,T)=( -0.0413* T ^2 - 0.8245* T + 816.925)* (%i)^3+
( 0.0273* T ^2 + 1.2025* T - 176.35)* (%i)^2+(-0.0048* T ^2
- 0.723* T - 12.433)* (%i)+(1.0675* T - 1.975)
```

With these ones and Antoine correlation, eq. (3.88),

$$\text{Ln}(\text{Pv}_{\text{Agua}}(\text{mmHg})) = \left(18.3036 - \frac{3816.44}{227.02 + \text{T}(^\circ\text{C})}\right) \tag{3.88}$$

we are ready to proceed.

2. *Mass and energy balances:*

We present the mass and energy balance to the three effects where the enthalpies of the vapor corresponded to superheated vapor at the boiling point of the solution and that of the liquid phase is given by the correlations obtained in the previous section as function of the composition, eq. (3.89).

Now we write them in EXCEL®. For each of the effects we compute the liquid feed, the liquid out, the steam used to provide the energy for the concentration of the liquid and the evaporated stream. Another important thing to point out is the fact that the steam leaving each effect is superheated as presented in the problem formulation.

$$\begin{aligned}
&F - L_n = \sum_n E_i \\
&W \cdot \lambda_w + F \cdot h_f = E_1 \cdot H_{e1} + L_1 \cdot h_{L1} \\
&E_1 \cdot (\lambda_{e1} + 0.46\Delta e_1) + L_1 \cdot h_{L1} = E_2 \cdot H_{e2} + L_2 \cdot h_{L2} \\
&E_2 \cdot (\lambda_{e2} + 0.46\Delta e_2) + L_2 \cdot h_{L2} = E_3 \cdot H_{e3} + L_3 \cdot h_{L3} \\
&E = E_1 + E_2 + E_3 \\
&H_{ei} = H_{sat,eb(water)} + cp\Delta e_i \\
&h_{L,i} = h_{solution}(\%_i, t_{eb,water} + \Delta e_i)
\end{aligned} \tag{3.89}$$

$$Q = UA\Delta T \rightarrow \frac{Q_1}{U_1(T_w - (T_{1,ebwater} + \Delta e_1))} = A_1 = A_2 = \frac{Q_2}{U_2(T_{1,ebwater} - (\Delta e_2 + T_{2,ebwater}))} = A_3$$

F stands for the feed to each effect, L the concentrated solution produced and E the vapor generated. The temperatures of the liquid and vapor are the same. We use one row for the temperature of the streams. We have as data the feed temperature **FeedT**, 25°C.

```
F10(FeedT)=25
```

Just below we allow a row for the concentration of the solutions. We have two data here, the concentration of the feed, 15%, and the concentration of the product, 45%.

```
E11:Concf=0.15;
F11:Conc1, F11=I11;
J11:Conc2; J11=M11;
N11:Conc3=0.45
We define Conceni=100·Conci
```

The next row is used to calculate the enthalpy of the streams. Thus for the liquid streams (F and L) we use the correlations we developed in the previous section of the problem, h_L(NaOH,%,T). For the steam generated, **E**, the enthalpy is calculated as a superheated steam. Thus we calculate the enthalpy of the saturated steam at the boiling point of water due to the operating pressure at the evaporator plus the ebullioscopy increment due to the presence of NaOH in the solution calculated using the first of the equations developed in the previous section, T_{eb}(NaOH, $\%_i$):

```
H20(Teb(NaOH):Teffecf1)=(-0.0000030233*Concen1^3 +
0.00021232*Concen1^2 - 0.0042815*Concen1 + 1.0459)*Tebull1 +
(0.00024541*Concen1^3 + 0.0060805*Concen1^2 - 0.009887*Concen1
+ 0.17863)
G12(HEi)= (-0.0000030903*Tebull1^3 + 0.00022613*Tebull1^2 +
0.42436*Tebull1 + 597.42)+0.46*(Teffec1-Tebull1)
```

So the variables are the boiling temperature of water (F20)

```
F20 (boiling temperature water)= Tebull1=VARIABLE
```

And the concentration of the NaOH in the liquid, calculated through the mass balance to the first effect. Thus the real variable here is the amount that is evaporated:

```
F11(Concentration of the liquid) =E13*E11/F13=Feed*Concf/
Liquid1
```

The flows of the streams are written in row 13

```
E13(Feed)=15000
F13(Liquid)=E13-G13=Feed-Vapor
G13(Vapor)= VARIABLE
```

While the operating pressure at the evaporation chamber is given by Antoine correlation of the boiling temperature of water (F20)

```
E21(Operating pressure)= EXP(18.3036-3816.44/
(227.02+Tebull1))
```

The energy balance is established as

```
E26 = E27
E26(Energy in)=E7*E25+E13*E12=lambda*Steam+Feed*Hf
```

Where E25 is the flow rate of the steam used to provide energy to the system and is another variable

```
E27(Energy out)= F13*F12+G13*G12= Liquid1*Hliq1+Vapor1*Hvap1
```

The area of the effect is calculated with an energy balance. The energy input is that given by the steam

```
E28=E25*E7 = lambda*Steam
```

Where the latent heat (E7) is calculated at the boiling point of the stream as follows:

```
E7 =-0.0000043722*E6^3 + 0.00043484*E6^2 - 0.58433*E6 + 597.48
E7=-0.0000043722*Tvapor^3 + 0.00043484*Tvapor^2 -
0.58433*Tvapor + 597.48
F5=E5 =EXP(18.3036-3816.44/(227.02+ Tvapor))
E5=4*760
```

And that energy is to be transferred to the solution with a DT equal to the difference between the temperature of the saturated steam (E6) and the temperature of the streams leaving the evaporator:

```
E29=E14*E15*(E6-G10)
E15(Area) =VARIABLE
```

We repeat this analysis for the following effect so that either the liquid phase or the vapor generated is the variable, as well as the boiling point and operating pressures. The last effect has the particularity that using a global mass balance we determine the liquid phase and we have as data the operating pressure at the chamber. Furthermore the energy provided to the effects from the second onwards is given as

$$E_1 \cdot (\lambda_{e1} + 0.46\Delta e_1)$$

Thus as an example, for the second effect we have

```
J27(Energy provided)=G13*(J7+E23*0.46)
J7 =-0.0000043722*J6^3 + 0.00043484*J6^2 - 0.58433*J6 + 597.48
J6= Boiling point at the previous effect=Tvapor2
```

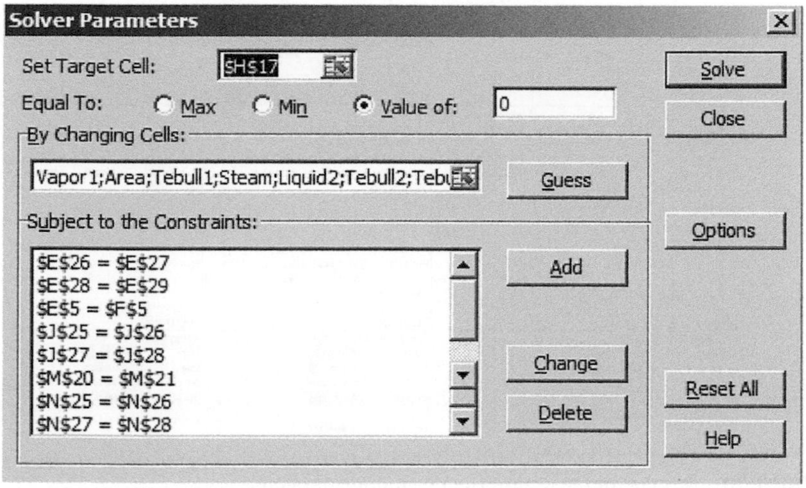

FIGURE 3.49 Model of a multi-effect evaporator.

```
E23=DT useful=E6-G10
```

Thus we present here the *Solver* parameters

Target cell H17 (The area of the different effects is the same)

```
H17=E15-I15
```

Variable cells: Vapor1;Area;Tebull1;Steam;Liquid2;Tebull2;Tebull3;Tvapor

Subject to the mass and energy balances, see Figure 3.49 for the details.

We solve it and we get the results presented in Figure 3.50. The economy of the system, the vapor generated per kg of saturated steam used is:

$$Economy = \frac{\sum_i E}{W} = 1.67 \tag{3.90}$$

Finally to determine the price of the evaporators we use www.matche.com [14]. We select falling film evaporator and we obtain a correlation for the cost of the evaporators as a function of the heat exchange areas, assuming stainless steel 316. We fit the data using add trend line and we obtain:

$$\text{Cost (\$)} = -2.249(\text{Area m}^2)^2 + 1601.3(\text{Area m}^2) + 55866 \tag{3.91}$$

For our three evaporators, with individual areas of 116 m^2 we have \$634000.

Now that we have formulated the problem the questions to be answered:

	B	C	D	E	F	G	H	I	J	K	L	M	N	O	P
3			1st effect					2nd Effect				3rd Effect			
4			Vapor												
5			P(mmHg)	3040	3040				P(mmHg)	1488.55325			P(mmHg)	684.923663	
6			T vapor(ºC)	144.08509				T vapor(ºC)	119.990921			T vapor(ºC)	97.113273		
7			λ (kcal/kg)	509.235785				λ (kcal/kg)	526.073007			λ (kcal/kg)	540.830387		
8			Solution												
9				F	L	E (vapor rec)		F	L	E		F	L	E	
10			Tini (ºC)	25	126.298625	126.298625		126.298625	108.432824	108.432824		108.432824	80.8761746	80.8761746	
11			C(%w/w)	0.15	0.18969545			0.18969545	0.26624544			0.26624544	0.45		
12			Hdis(kcal/kg)	19.379175	113.802197	649.157844		113.802197	98.6295363	643.140286		98.6295363	102.216171	633.157425	
13			Flow(kg/h)	15000	11861.1174	3138.88265		11861.1174	8450.849	3410.26835		8450.849	5000	3450.849	
14			U(kcal/hm2ºC)	1500				1250				1000			
15			A(m2)	116.071922				116.071922				116.071922			
16															
17							0								
18															
19				P(mmHg)	T(eb water)	C(%)	T +Δe	P(mmHg)	T(eb water)	C(%)	T +Δe	P(mmHg)	T(eb water)	C(%)	T +Δe
20			Salida		119.990921	18.9695451	126.299		97.113273	26.6245439	108.4328	76	46.1122135	45	80.8761746
21				1488.55325				684.923663				76			
22															
23			Dtutil	17.7864657				11.5580963				16.2370984			
24			Energy balance 1st effect					Energy balance 2nd effect				Energy balance 3rd effect			
25			W=	6081.19848				Entra	3026784.28			Entra	2696011.53		
26			Entra	3387451.51				Sale	3026784.28			Sale	2696011.53		
27			Sale	3387451.51				Q (Wλ)	1676963.06			Q (Wλ)	1884671.22		
28			Q (Wλ)	3096763.88				Q(UADT)	1676963.06			Q(UADT)	1884671.22		
29			Q(UADT)	3096763.88											
30								Economy	1.6444127						

FIGURE 3.50 Results for the multi-effect evaporator system.

QUESTION

- What is the effect of the pressure on the last effect on the performance of the system and what about that of the pressure of the steam?

3.3.6 Link EXCEL with Other Software

Excel has become a standard package so that a number of other specialized software use it as a source of information, to report data since it is more user friendly. Therefore we can use the information in EXCEL® to be loaded in MATLAB®, MATHCAD®, gPROMS®, GAMS® or CHEMCAD®, or transferred back to EXCEL® just to mention some of the ones presented in the book. Simple data can be easily transferred while advanced uses are left to the user to explore.

REFERENCES

1. Sinnot, R.K. (1999) *Coulson and Richardson. Design*, Vol 6. Singapore.
2. Bird, R.B., Steward, W.E., Lightfoot, E.N. (2007) *Transport Phenomena*. 2nd edn. John Wiley & Sons, Hoboken, NJ.
3. Streeter, V.L., Wylie, E.B., Bedford, K.W. (1998) *Fluid Mechanics*. 9th edn. McGraw-Hill, New York.
4. Ocon, J, Tojo, G. (1968). *Problemas de de Ingeniería Química*. Tomo 1. Aguilar Madrid, Spain.
5. Colebrook, C.F. and White, C.M. (1937) Experiments with Fluid Friction-Roughened Pipes, *Proc. R. Soc. A.*, 161.
6. Geankoplis, C.J. (1993) *Transport Processes and Unit Operations*, 3rd edn. Prentice Hall, Upper Saddle River, NJ.
7. Hougen, O.A., Watson, K.M., Ragatz, R.A. (1954) *Chemical Process & Principles*. John Wiley & Sons, Hoboken, NJ.
8. Vian Ortuño, A. (1999) *Introducción a la Química Industrial*. Ed. Reverté, Barcelona.

9. *Ullman's Encyclopedia of Industrial Chemistry*, Ed. Wiley-VCH, (1998). Nitric Acid, Nitrous Acid, and Nitrogen Oxides.
10. McCabe, W.L., Smith, J.C. (1968) *Unit Operations*. McGraw-Hill, New York.
11. McCabe, W.L. (1935) The enthalpy-concentration chart: A useful device for chemical engineering calculations. *Trans. Am. Inst. Chem. Eng.*, 31, 129-169.
12. Coulson, J.M., Richardson, J.F. (1991) *Particle Technology and Separation Process*, Vol 2. Singapore
13. Perry, R.H., Green, D.W. (1997) *Perry's Chemical Engineer's Handbook*, McGraw-Hill, New York.
14. MATCHE. (2003) http://www.matche.com/EquipCost/Exchanger.htm. (Last accessed March 2013).

4 MATLAB® for Chemical Engineering

Mariano Martín, Luis Martín de Juan

CONTENTS

4.1 INTRODUCTION

MATLAB® stands for Matrix Laboratory, which is a good indication of the kind of operations it typically performs. MATLAB® is one of the most used software in Engineering, in general, and also in Chemical Engineering. Much has been written about this popular software, more than 1,500 books serving more than 1 million users. Therefore, we will focus on the use for Chemical Engineering applications, presenting a number of examples. However, we see the need for providing a bit of background to get started. Note that we use different font to denote MATLAB® code. The models can be found on the editorial web page.

4.2 MATLAB® BASICS

The layout of MATLAB® can be seen in Figure 4.1. There are a number of different layouts that the user can define, see Layout Tab. We select one basic layout that consists of the workspace, in the figure to the left top corner, where the variables are stored, a command history, to the left bottom corner, where a log of operations is stored and the command window, where calculations can be executed.

4.2.1 Basic Functions

Needless to say that within MATLAB® we have all the algebraic, statistical functions predefined along with plotting capabilities. Before briefly commenting on the basics, it is interesting to point out the use of help within the command window. We use the "help" function followed by any built-in function and MATLAB® returns a small tutorial on the use of that function, the arguments and the options (i.e., try "help ode45").

Before starting any new calculation we can use the command "clear all" in order to erase the workspace.

MATLAB® operations rely on matrices, thus the first topic is how to input vectors and matrices. A row vector is written using "a = [a11 a12 a13]" with the elements separated by spaces or commas. Instead, for entering a column vector we can either use the transpose function applied to a row vector "a'" or directly with a = [a11;a21;a31] separating the rows with semicolons.

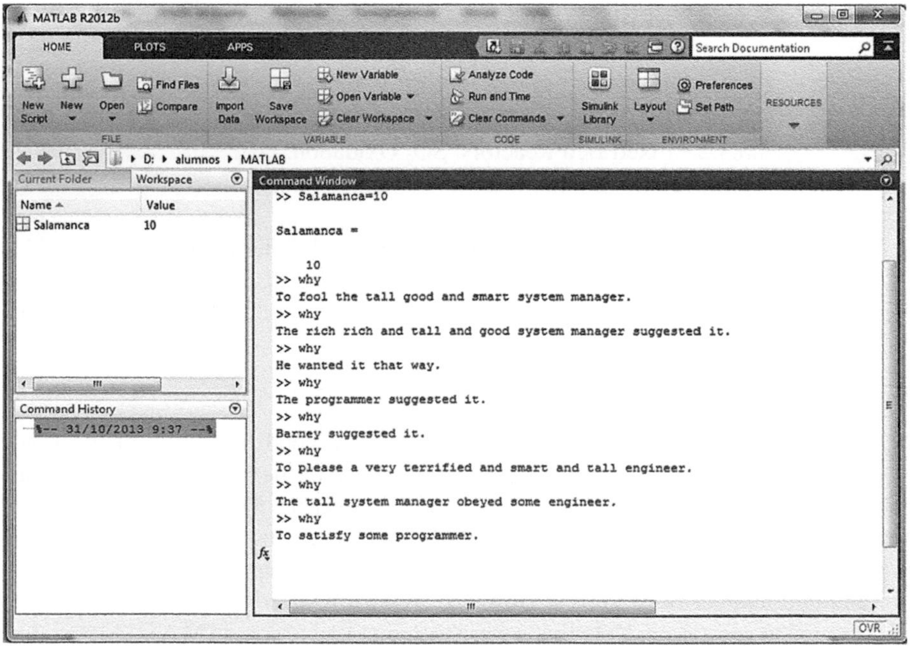

FIGURE 4.1 General framework.

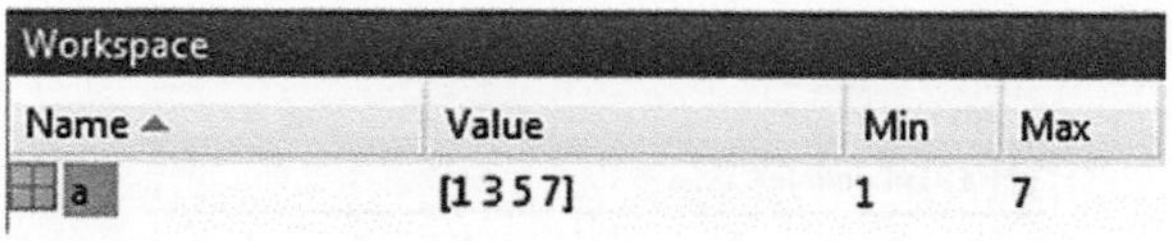

FIGURE 4.2 Workspace.

It is also possible to create a vector out of a series of numbers, which is particularly useful to provide a list of values for independent variables.

To use this function, we determine the initial element, the step (it can also be negative) and the final one:

```
a=10:-2:2
a=[10 8 6 4 2]
```

To input matrices the syntaxes are similar:

```
A=[[a];[b];[c]]
Or
A=[a;b;c]
```

Where a, b, and c are row vectors.

Vectors and matrices (it is possible to store up to 3D matrices) can be used to store information, experimental data and results of calculations. Whenever a vector is entered, a variable is generated in the workspace where the name, values, and class are presented, see Figure 4.2:

We can save the workspace as a file (.mat) that can be loaded anytime. This is particularly interesting to save the results of a calculation. We can save it directly by clicking on the disk symbol over the workspace or using the command:
"Save FILENAME"

Alternatively we can save only some variables by adding their names after FILENAME.

Later, we use:
"Load FILENAME"

To load the information previously stored or we can open it using the open symbol over the workspace.

The variables stored in the workspace can be opened or edited by clicking twice on it in the workspace and a new window, Figure 4.3, over the command window is opened, array editor, presenting the values. We can input data directly here.

Furthermore, we can use EXCEL® to store the information and load it. We use "xlsread" to import data from the excel file to the workspace. The general syntaxis is:

Matrix_name = xlsread(filename, sheet,xlRange,'basic')
There is no need to use all the parameters.
Filename: is the name of the file you use as source.
Sheet: is the number of the sheet where the data are located.

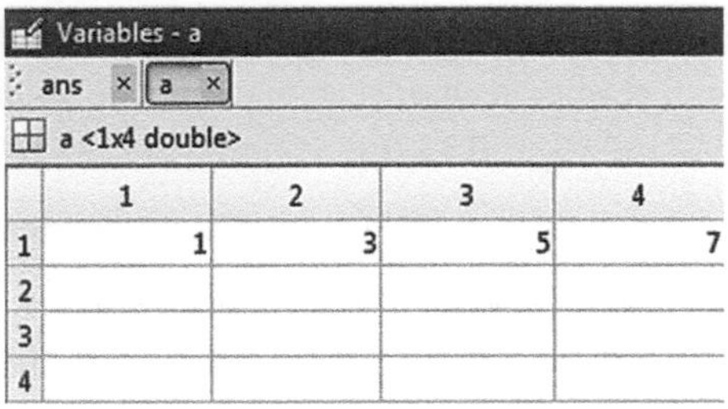

FIGURE 4.3 Array editor.

xlRange: is the range of cells you want to import.
basic: represents the format.

You can go the opposite direction from MATLAB® to EXCEL® using the command "xlswrite" to export a matrix. The general formulation is:

Xlswrite(filename,A,sheet,xlRange)
Where:
Filename: is the name of the file where you send the date.
A: The matrix.
Sheet: The sheet of excel where you write.
xlRange: Is the range of cells you export.
For advanced users we leave the spreadsheet link feature.

The information stored in the form of vectors or matrices can be called element by element, for example, to retrieve the element of the first row and of the first column we have:

```
b=a(1,1)
```

or we can call an entire row or one column respectively as:

```
c(i,:) or c(:,i)
```

There are some interesting functions when dealing with matrices:

Size: Returns a vector with the number of columns and rows of a matrix or vector.
Length: Returns the number of elements of a vector.

Many times it is interesting to store this piece of information to be used in our calculations.
Furthermore the common matrix operations are also built-in functions such as:

INV(A): Returns the inverse of matrix A.
Det(A): Returns the determinant of matrix A.
Rank(A): Returns the rank of matrix A.
Eye(n): Identity matrix of size [n,n].

Ones(m,n): A matrix of size [m,n] full of ones.
Zeros(m,n): A matrix of size [m,n] full of zeros.
[L,U] = LU(A): LU decomposition of matrix A. Typically used within MATLAB® to solve systems of linear equations. You need to provide [L,U] so that both matrices are stored.
[Q,R] = QR(A): QR decomposition of matrix A stored in matrices Q and R.
[] = **find**(A == value): Returns the position of value in matrix A. Alternatively, it is possible to find the values larger than value.

To add, or multiply two matrices, we can use the + and * operators. We can use the "^" symbol to compute the power of a matrix or exp(A) for the exponential.

On the other hand, we may need to multiply/divide or calculate the power element by element. In this case we use a dot after the vector or matrix name and then the operator, a.*.

4.2.2 Fitting and Plotting

Many problems in engineering and science require the use of graphic tools to fit experimental data and plot them. If we start with plotting, the basic function is "plot". We show different plotting routines in the examples. We first define the vectors:

```
t=[1:1:50]
c1=25*exp(-0.05*t)
c2=0.5*t
```

Hint: We use ";" after the line if we do not want to see on the command window the result of the computation. Finally the use of the function plot requires the vectors and we can add as options the symbols and line types for each of the series between '':

```
plot(t,c1,'k-', t,c2,'k--')
```

Alternatively we can use:

```
plot(t, [25*exp(-0.05*t)],'k-', t,0.5*t,'k--')
```

A brief summary of the plotting options can be seen in the command window by using:
"help plot"

We can add labels, legend and title directly in the command window (or in a script):

```
xlabel('Time (s)')
ylabel('Concentration (M)')
legend('A','B')
```

The resulting plot can be seen in Figure 4.4.

A hint: In the command window we can use the arrows up to bring back the previous operations and reuse the information so that we do not have to retype them.

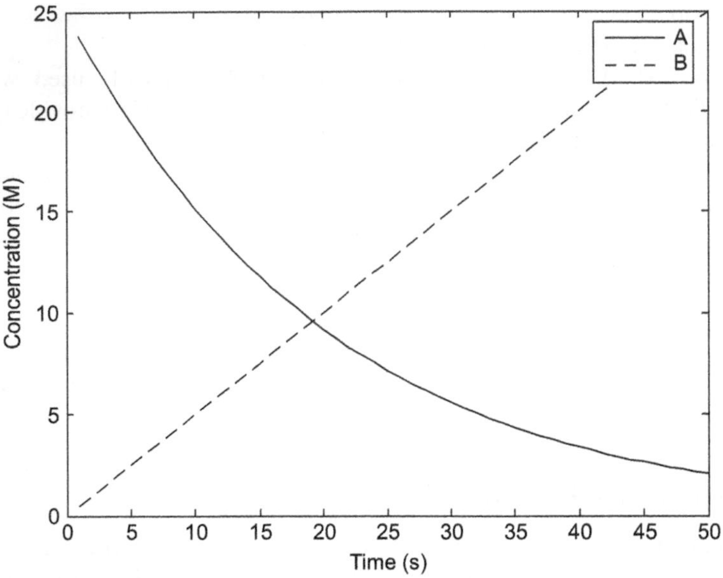

FIGURE 4.4 Simple plotting example.

However, once the Figure is plotted, we can edit it to modify the type of line as well as the labels or the location of the legend, see Figure 4.5.

One important feature is the possibility of saving the Figure as .tiff or .eps which are generic formats for other image processing software.

To fit data we can use "polyfit". The function requires the vectors and the degree of the polynomial as inputs. Using the same vectors as before we have:

```
Polyfit(t,c1,3)
```

The function returns the coefficients of the polynomial:

```
polyfit(t,c1,3)
ans =
    -0.0002     0.0219    -1.1541    24.7519
```

We can also store the coefficients in a proper vector instead of the default vector:

```
ans
coeff=ans
```

To compute values using the fitted polynomial we can use "polyval":

```
C125=polyval(coeff, 12.5)
13.4352
```

We can also use least-squares model fitting such as "lsqcurvefit":

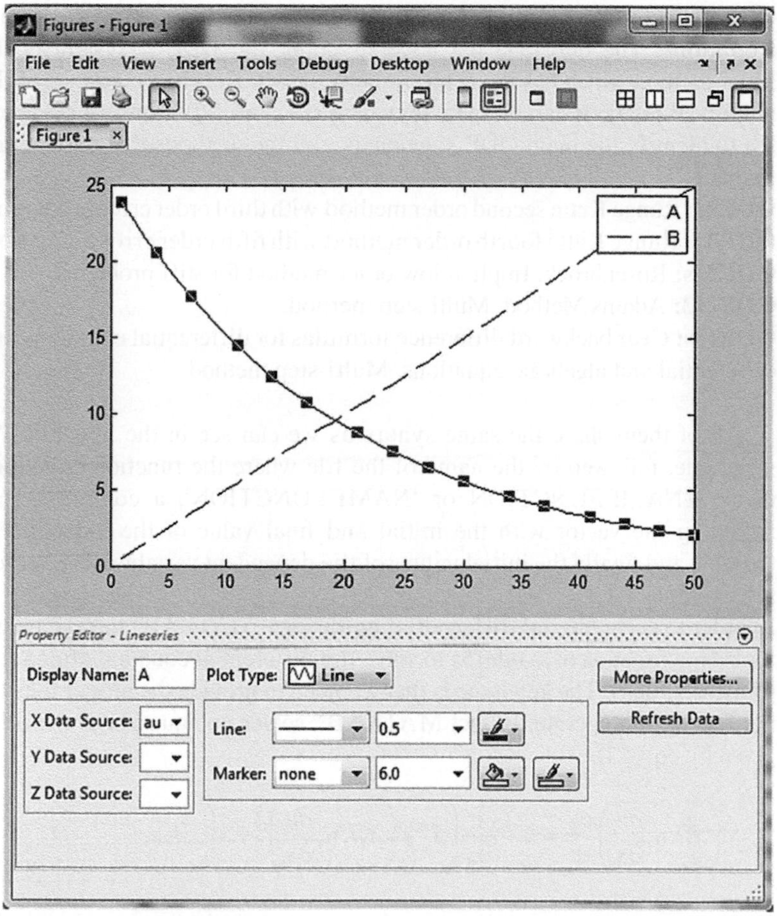

FIGURE 4.5 Editing plots.

4.2.3 Using Built-in Functions

MATLAB® has a number of functions that allow solving linear and non-linear equations (fzero: for one variable alone, fsolve), optimizing a function (fmincon: constrained optimization, linprog: linear programming; fminin or fminsearch: unconstrained optimization; bintprog: binary and integer optimization), solving differential equations (ode__) or partial differential equations (pdepe).

a. "fsolve" is a built-in function to solve mainly **non-linear equations**. The algorithm behind "fsolve" uses a Newton or quasi-Newton method that implements three different routines to control the step size and direction. The default one is the 'trust-region-dogleg' while as options we can select the 'trust-region-reflective' or the 'levenberg-marquardt' option. Those interested in the details can refer to ref. [1].

b. ODE__ comprises a family of methods for solving **Ordinary Differential Equations**. The only one not covered under the ODE__ system is the Euler method which has its own routine "eulers". The characteristics of the numerical methods can be seen anywhere (i.e., ref. [2]). Here we present a list to identify the main ODE statements and the methods that correspond to them.
ODE23: Runge Kutta second order method with third order error prediction.
ODE45: Runge Kutta fourth order method with fifth order error prediction.
ODE23s: Rosenbrock. Implicit low order method for stiff problems.
ODE113: Adams Method. Multi-step method.
ODE15s: Gear backward difference formulas for differential equations and differential and algebraic equations. Multi-step method.

All of them share the same syntax as we can see in the first file. We write ode, followed by the name of the file where the function is defined using @NAMEFUNCTION or 'NAMEFUNCTION', a comma before including the vector with the initial and final value of the independent variable and finally the initial values of the dependent variables as a vector, see example 4.3.3.

c. In order to **solve partial differential equations**, MATLAB® has the pdepe tool. This function uses ode15s to solve the differential equation after a few transformations. The key issue is that we need to provide the proper files for the function to operate. In fact MATLAB® solves an equation of the form given by eq. (4.1).

$$c\left(x,t,u,\frac{\partial u}{\partial x}\right)\frac{\partial u}{\partial t}=x^{-m}\frac{\partial}{\partial x}\left(x^{m}f\left(x,t,u,\frac{\partial u}{\partial x}\right)\right)+s\left(x,t,u,\frac{\partial u}{\partial x}\right)=0 \tag{4.1}$$

d. Typically we can use two functions for simple **optimization problems** with MATLAB®. The first one is the function "fminsearch". This function finds the minimum of an unconstrained multivariable function using the Nelder_Mead simplex algorithm, a derivative-free method. Alternatively we can use "fmincon". We can select among four optimization algorithms: active-set, interior-point, sqp or trust-region-reflective. To do so, we write the algorithm at the command line with "optimoptions()".

```
options = optimoptions('fmincon','Algorithm','active-set');
```

The default algorithm is 'trust-region-reflective'. It is not the aim of this book to comment on the solution methods but for the interested reader the details can be found in specialized literature [3,4].

In general these functions require the definition of a number of files to include the equations that we are going to process and later another one where we call the built-in function. We will see the use of these in the examples.

4.2.4 Programming Language

Apart from performing operations, MATLAB® is a programming language. Its operation is based on the use of .m files that can be divided in two classes, scripts and functions. A script is basically a number of operations that we want to perform in a certain sequence. It is required that they are in order so that the variables and parameters are already defined. We can use them as executables and within them we can also call other subrogate scripts that contain other operations or functions. In fact we can write any of the examples above as a script to be executed altogether (i.e., the plot of a couple of vectors providing labels for the axis and legend).

Functions are a particular type of scripts that must begin with the word "function" at the top of them. Functions can be user defined or typical operations such as equation or solving differential equations. The name of the function should be the same as that of the file .m. One particular and interesting thing is the fact that the variables within the functions are local; this means that we are not going to be able to use them outside the function unless we declare them as "global". To do so we need to start by defining global X where X will be the variable that is global along the complete study.

The typical commands of programming can be used such as:

FOR:

```
For  i=1:10
       a(i) = 2*i
end
```

IF

```
If x>3
       M=1
Elseif x<1
       M=2
Else
       M=5
End
```

We use ==, <, >, <=, >=, or ~= to impose the conditions.

WHILE

```
While x > 3
M=2
A=3+B
x=M+0.5
End /Break
```

We use the same symbols as before.

Be aware that for executing a .m file you have to define the path where it is stored. Look for "Current Folder" in the MATLAB® main tab.

4.3 EXAMPLES

It is important to notice that there is no unique way of solving the following problems but the syntaxes presented allow us to show the capabilities of the software and the use of different approaches.

4.3.1 Momentum, Mass and Energy Transfer

4.3.1.1 Heat and Mass Transfer in 2D

Let's consider the mass transfer in 2D by diffusion. Imagine a piece of solid in contact with atmosphere and the moisture goes into the solid, see Figures 4.6 and 4.7. The problem is modeled using Laplace equation.

$$\frac{\partial^2 C_A}{\partial x^2}+\frac{\partial^2 C_A}{\partial y^2}=0 \tag{4.2}$$

The equation has an analytic solution given as:

$$C_A = X(x)Y(y)$$
$$C_A = C_A(x)=\sum_{n=1}^{\infty} A_n \sin\left(\frac{n\pi x}{w}\right)\sinh\left(\frac{n\pi L}{w}\right) \tag{4.3}$$

The numerical solution is computed by discretizing the region and the differential equations as finite differences:

$$\frac{\partial^2 C_A}{\partial x^2}=\frac{\partial\left(\partial C_A/x\right)}{\partial x}=\frac{\frac{C_{A,i+1,j}-C_{A,i,j}}{\Delta x}+\frac{C_{A,i-1,j}-C_{A,i,j}}{\Delta x}}{\Delta x}=\frac{C_{A,i+1,j}-2C_{A,i,j}+C_{A,i-1,j}}{\Delta x^2}$$
$$\frac{\partial^2 C_A}{\partial y^2}=\frac{C_{A,i,j+1}-2C_{A,i,j}+C_{A,i,j-1}}{\Delta y^2} \tag{4.4}$$

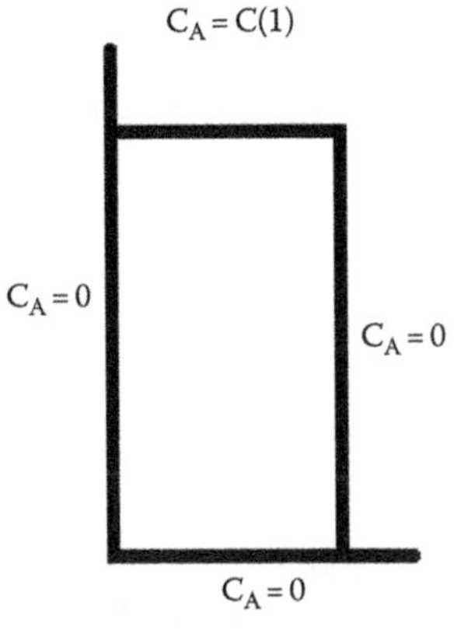

FIGURE 4.6 Scheme for Example 4.3.1.1.

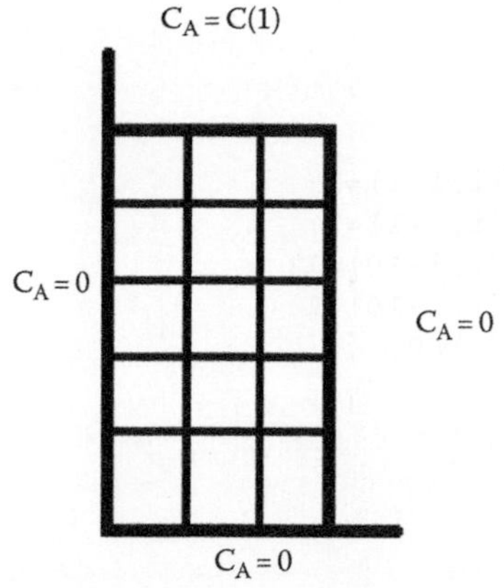

FIGURE 4.7 Space discretization.

For each of the internal nodes we have:

$$C_{A,i+1,j} + C_{A,i-1,j} + C_{A,i,j+1} + C_{A,i,j-1} - 4C_{A,i,j} = 0 \tag{4.5}$$

Thus, we have a matrix of the form:
$A \cdot C_{A(i,j)} = B$

$$\begin{bmatrix} -4 & 1 & \dots & & 0 \\ 1 & -4 & & & . \\ . & & \dots & & . \\ . & & & -4 & 1 \\ 0 & & \dots & 1 & -4 \end{bmatrix} \begin{bmatrix} C_{A(1,1)} \\ C_{A(2,1)} \\ \dots \\ C_{A(2,1)} \\ C_{A(10,10)} \end{bmatrix} = \begin{bmatrix} C_{A0} \\ C_{A0} \\ \dots \\ 0 \\ 0 \end{bmatrix} \tag{4.6}$$

The boundary conditions are $C_A = 0$, but at the top of the region $C_A = 25$. With these data we need to build our system of equations by filling the matrix. Let's consider a mesh of 10 × 10 elements. As we can see from the previous equation, we have a system of linear equations where all the coefficients of the nodes surrounding the point i, j have a coefficient equal to 1. We use a counter "a" to fill the ones in the diagonals and a counter "q" so that all the internal lines of the matrix are filled. Next, we fill in the boundary conditions of the geometry at the maximum of the *x* and *y* axis. Finally we take care of the vertex of the geometry and the diagonal. Thus our model is as follows:

```
CAxmin=0;
CAxmax=0;
CAymax=25;
CAymin=0;
```

```
a=0;
%Internals
for q=1:8;
for i=12+a:19+a
                   C(i,i-1)=1;
                   C(i,i+1)=1;
                   C(i,i+10)=1;
                   C(i,i-10)=1;
                   B(i)=0;
end
a=a+10;
end
%Boundary
for i=2:9
    C(i,i-1)=1;
    C(i,i+1)=1;
    C(i,i+10)=1;
    B(i)=-CAymin;
end
for i=92:99
    C(i,i-1)=1;
    C(i,i+1)=1;
    C(i,i-10)=1;
    B(i)=-CAymax;
end
for i=11:10:81
    C(i,i+10)=1;
    C(i,i-10)=1;
    B(i)=-CAxmin;
end

for i=20:10:90
    C(i,i+10)=1;
    C(i,i-10)=1;
    B(i)=-CAxmax;
end
%Vertex
for i=[1,10,91,100]
    if i==1
         C(i,i+1)=1;
         C(i,i+10)=1;
         B(i)=-CAxmin-CAymin;
     elseif i==10
         C(i,i-1)=1;
         C(i,i+10)=1;
         B(i)=-CAxmax-CAymin;
         elseif i==91
         C(i,i+1)=1;
         C(i,i-10)=1;
         B(i)=-CAxmin-CAymax;
```

```
        elseif i==100
        C(i,i-1)=1;
        C(i,i-10)=1;
        B(i)=-CAxmax-CAymax;
    end
 end
%Diagonal
   for i=1:100
   for j=1:100
              if i==j
              C(i,j)=-4;
              end
   end
   end
```

We use "%" to include a comment within the code. We solve the linear equation. MATLAB® typically uses LU decomposition for this:

```
Conc=C^-1*B';
```

Or

```
Conc=C\B';
```

In order to represent the results we use a **contour plot** where we use the data in the vector Conc to prepare the matrix profile as:

```
Profile=[Conc(1:10)';Conc(11:20)';Conc(21:30)';Conc(31:40)';
Conc(41:50)';Conc(51:60)';Conc(61:70)';Conc(71:80)';
Conc(81:90)';Conc(91:100)'];
colormap([gray])
contourf([1:10],[1:10],Profile)
colorbar;
```

We can write all this code in the same script to execute it to obtain Figure 4.8:

QUESTION

- Do the same for the heat conduction transfer problem in 2D.

4.3.1.2 Simultaneous or Unsteady Momentum/Heat/Mass Transfer

There are a number of examples in the transport phenomena field [5] that cover different topics but that can be mathematically described using a partial differential equation. In this section we present three apparently different problems, two as exercises, that are solved using the "pdepe" toolbox.

Unsteady flow in a pipe: One typical example that can be addressed in this way is the start-up of a flow through a pipe. We can consider a one-dimensional problem for

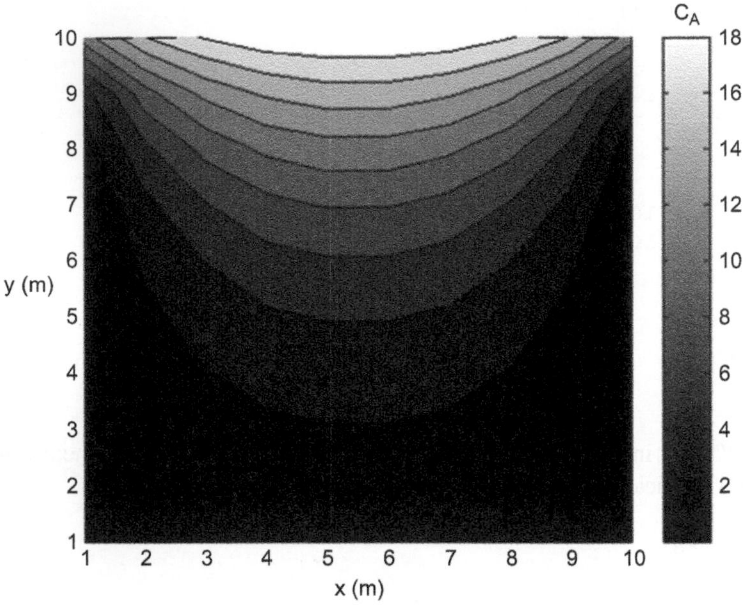

FIGURE 4.8 Results of the concentration profile.

the profile in the radial direction. We evaluate now the unsteady fluid flow through a long pipe with constant fluid properties (ρ, μ). The momentum balance is given by eq. (4.7) [5]:

$$\rho\frac{\partial v_z}{\partial \tau} = \frac{P_o - P_L}{L} + \mu\frac{1}{r}\frac{\partial}{\partial r}\left(r\frac{\partial v_z}{\partial r}\right) \tag{4.7}$$

As boundary conditions we have:

$$\begin{aligned} &t = 0 \ v_z = 0 \\ &r = 0 \ v_z \text{ is finite} \\ &r = R \ v_z = 0 \end{aligned} \tag{4.8}$$

We prepare the equation in dimension-less variables by using:

$$\phi = \frac{v_z}{(p_o - p_L)R^2/(4\mu L)} = \frac{v_z}{v_{max}}, \xi = \frac{r}{R}, \tau = \frac{vt}{R^2} \tag{4.9}$$

So that we get:

$$\frac{\partial \phi}{\partial \tau} = 4 + \frac{1}{\xi}\frac{\partial}{\partial \xi}\left(\xi\frac{\partial \phi}{\partial \xi}\right) \tag{4.10}$$

The initial conditions become:

$$\phi(0,\xi) = 0 \tag{4.11}$$

And the boundary conditions:

$$\phi(\tau,1) = 0 \tag{4.12}$$

$$\frac{\partial\phi(\tau,0)}{\partial\xi} = 0 \tag{4.13}$$

We need to make use of the form provided by MATLAB® to input the functions that describe the model. We start by eq. (4.1).

$$c\left(x,t,u,\frac{\partial u}{\partial x}\right)\frac{\partial u}{\partial t} = x^{-m}\frac{\partial}{\partial x}\left(x^m f\left(x,t,u,\frac{\partial u}{\partial x}\right)\right) + s\left(x,t,u,\frac{\partial u}{\partial x}\right) = 0 \tag{4.1}$$

The specific syntaxis imposed by MATLAB® is as follows:

```
sol=pdepe(m,@ecuation,@initialcond,@boundary,x,t)
```

By direct comparison between our equation and MATLAB® generic equation for pdepe we have:

$$\mathrm{m} = 1$$

$$c\left(x,t,u,\frac{\partial u}{\partial x}\right) = 1 \tag{4.14}$$

$$s\left(x,t,u,\frac{\partial u}{\partial x}\right) = 4 \tag{4.15}$$

$$f\left(x,t,u,\frac{\partial u}{\partial x}\right) = \frac{\partial\phi}{\partial\xi} \tag{4.16}$$

Now we need to be able to write our initial conditions in the form:

$$u(x,t_o) = u_o(x) \tag{4.17}$$

While for the boundary conditions we need to use as template:

$$p(x,t,u) + q(x,t)f\left(x,t,u,\frac{\partial u}{\partial x}\right) = 0 \tag{4.18}$$

Therefore, for the initial condition given by eq. (4.11) we have:

$$u_o(x) = 0 \tag{4.11a}$$

For the boundary conditions:

$$p(x,t,u)+q(x,t)f\left(x,t,u,\frac{\partial u}{\partial x}\right)=0 \quad \text{vs} \quad \phi(\tau,1)=0 \rightarrow \mathrm{p}=\mathrm{ur},\ \mathrm{q}=0 \quad (4.18a)$$

$$p(x,t,u)+q(x,t)f\left(x,t,u,\frac{\partial u}{\partial x}\right)=0 \quad \text{vs} \quad \frac{\partial\phi(\tau,0)}{\partial\xi}=0 \rightarrow \mathrm{q}=1,\ \mathrm{p}=0 \quad (4.18b)$$

We start a script where we define *m*, the range of values for the independent variables *x* (in our case *x* will be *x*) and *t* (in our case the coordinate ξ). We can use any of the vector defining syntaxes but in this case we use a particular one, "linespace". As arguments we provide the starting point, the step, and the number of points to evaluate. After that we can call "pdepe". However, before running our script we need to define three function scripts.

Furthermore, we are using a **surface plot** to represent the results obtained for ϕ versus the two coordinates. For that we need to provide a vector for each of the coordinates, which in fact are already defined since it determines the range of the variables, and a matrix *u* that will store the solution. We use "colormap" gray for the sake of the book but we can use any other. Finally we also plot a parametric Figure with the results and for that we actually plot over the same Figure a number of plots. We can use the command plot and the particular feature is the use of "hold on" so that the new plot is actually plotted on the same Figure. A hint, if we want to write greek letters in the label of the axis we use "\" followed by the name. The script looks like this:

```
clear all;
x=linspace(0,1,50);%We use 50 values from 0 to 1
t=linspace(0,1,60);%the time span in 60 s
m=1;

sol=pdepe(m,@ecuation,@initialcond,@boundary,x,t);

u=sol(:,:,1);

% A surface plot i
surf(x,t,u)
colormap([gray])
xlabel('\xi')
ylabel('\tau')
zlabel('\phi')
shading interp
hold on
surf(-x,t,u)
colormap([gray])
xlabel('\xi')
ylabel('\tau')
zlabel('\phi')
shading interp
```

```
Figure
for j=1:length(t)
    plot(x,u(j,:),'k')
    xlabel('\xi')
    ylabel('\tau')
    hold on
end

hold on
for j=1:length(t)
    plot(-x,u(j,:),'k')
    xlabel('\xi')
    ylabel('\tau')
    hold on
end
```

Now we write the three functions:

```
@ecuation
function [c,f,s]=ecuation(x,t,u,DuDx)
c=(1);
f=DuDx;
s=4;

@initialcond
function u0=initialcond(x)
u0=0;

@boundary
function [pl,ql,pr,qr]=boundary(xl,ul,xr,ur,t)
%First boundary
pl=0;
ql=1;
%%Second boundary
pr=ur;
qr=0;
```

We now can run the script to obtain the results presented in Figures 4.9 and 4.10.

EXERCISES AND QUESTIONS

a. Evaluate the Simultaneous mass and momentum transfer in an absorption column.
b. Plot the operation of a vertical pipe in laminar flow while it exchanges heat [5].

   The details of the code for these two cases can be found on the web at CRC press.
c. Determine the unsteady state heat transfer from a sphere.

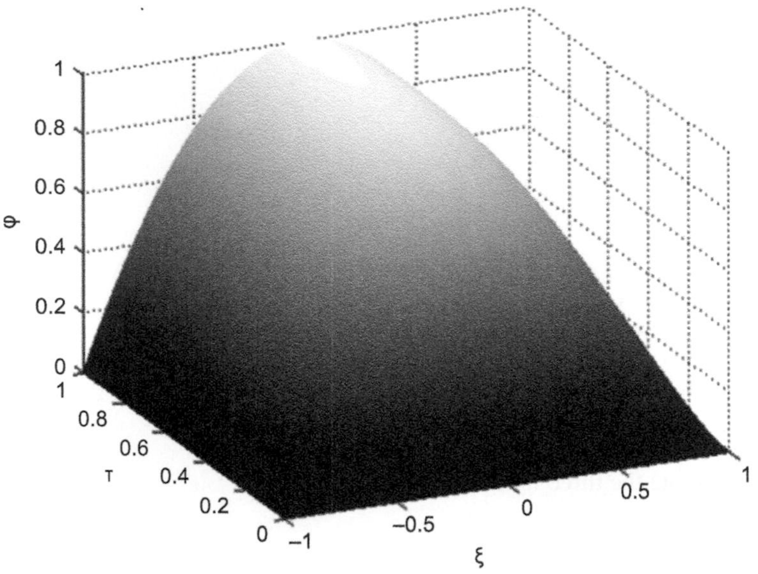

FIGURE 4.9 Axysimetric results.

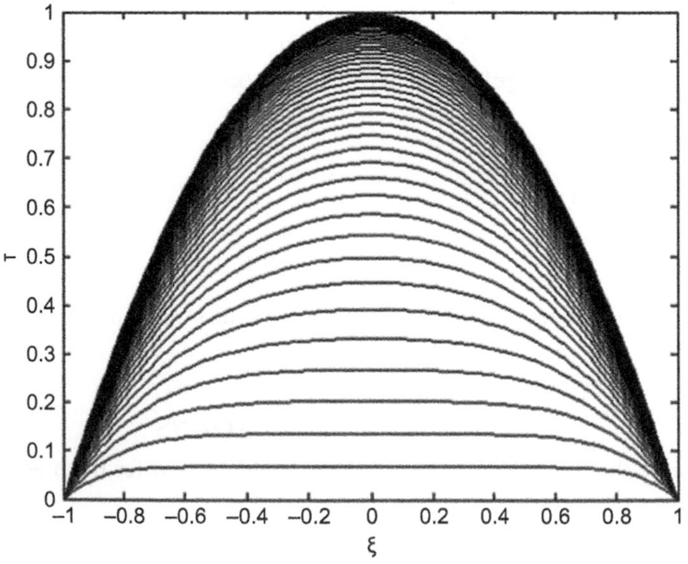

FIGURE 4.10 Parametric plot of the results for the flow in a pipe.

4.3.2 Distillation Column Operation: McCabe Method

McCabe method for distillation column design is a typical shortcut approach for the initial conceptual estimation of the operation of binary distillation columns released in 1925 [6]. It is a graphical method based on mass balances to the two sections of a distillation column, the rectifying and stripping sections. As such, it can be easily transformed into an iterative algorithm for evaluating the performance of distillation columns for binary mixtures in a simple way. In this section we are presenting the general modeling approach for distillation columns including the equilibrium curve, the operations lines (for the rectifying and the stripping sections of the column) and the q line. Next we apply this analysis for the study of two cases. The first one deals with the operation of the Linde double column for separation of O_2 and N_2. In this case we evaluate the number of trays needed for the separation at different pressures. The second example corresponds to a discontinuous distillation of an ethanol – water mixture where the main point is to determine the operation of the column for obtaining a certain purity of the distillate fraction. The basic modeling for both examples is as follows:

$$\text{Global balance to the column: } F = D + W \tag{4.19}$$

$$\text{Balance to the most volatile component: } F \cdot x_f = D \cdot x_D + W \cdot x_w \tag{4.20}$$

The operating line in the rectifying section is calculated by performing a mass balance to the upper part of the column from a tray n, just above the feed, including the condenser:

$$V_n = L_{n-1} + D \tag{4.21}$$

$$V_n\, y_n = L_{n-1}\, x_{n-1} + D \cdot x_D -> y_n = \frac{L_{n-1}}{V_n} x_{n.-1} + \frac{D}{V_n} x_D \tag{4.22}$$

Similarly we obtain the operating line for the stripping section by means of a mass balance from tray m, just below the feed, involving the reboiler.

$$V_m = L_{m-1} - W \tag{4.23}$$

$$V_m y_m = L_{m-1} x_{m-1} + W x_w -> y_m = \frac{L_{m-1}}{V_m} x_{m-1} + \frac{W}{V_m} x_W \tag{4.24}$$

It is assumed that the mass transfer from the liquid phase to the vapor phase is small in each of the sections so that the liquid and vapor streams are constant across them.

$$y_n = \frac{L}{V} x_{n.-1} + \frac{D}{V} x_D \tag{4.25}$$

$$y_m = \frac{L'}{V'} x_{m-1} + \frac{W}{V'} x_W \tag{4.26}$$

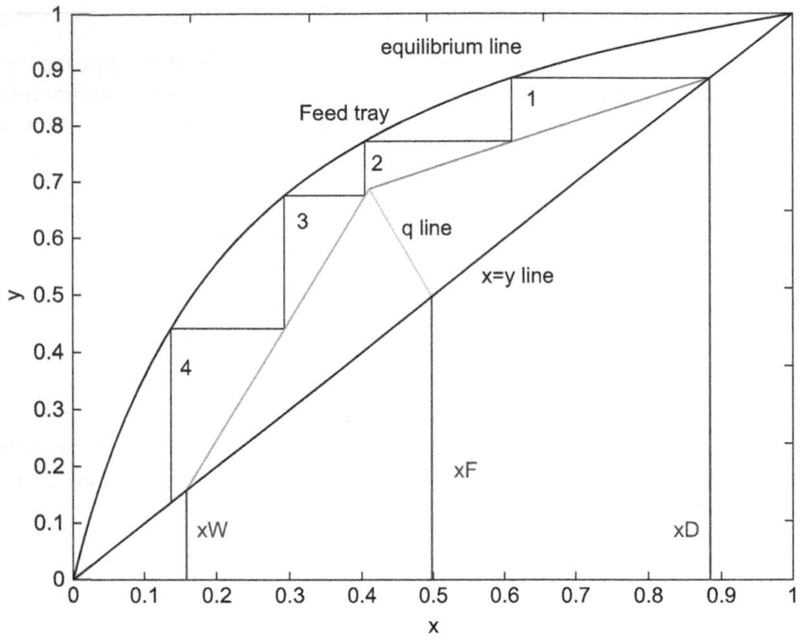

FIGURE 4.11 McCabe Thiele method for binary distillations. **(See color insert.)**

The feed can be in different aggregation states as compressed liquid, saturated liquid, a vapor liquid mixture, saturated vapor and superheated vapor. We define the liquid and vapor fraction as Φ_L and Φ_V respectively:

$$\Phi_L = \frac{L' - L}{F} \tag{4.27}$$

$$\Phi_V = \frac{V - V'}{F} \tag{4.28}$$

The q line, that defines the feed, is given by:

$$q = \frac{\Phi_V - 1}{\Phi_V} x_{n.} + \frac{1}{\Phi_V} x_f \tag{4.29}$$

In Figure 4.11, we present in green the operating line in the rectifying section, in pink, the operating line in the stripping section and, in blue, the q line. The steps representing the theoretical trays are plotted in black.

4.3.2.1 Linde Double Column N_2-O_2 Distillation

The double column of Linde can be defined as a multi-effect column where the reboiler of the section of the column operating at low pressure is used as a condenser for the section operating at higher pressure. The typical operation starts with cool, compressed air. The air is used to provide energy to the reboiler of the high pressure section, later it is expanded to be liquefied and fed to the high pressure column. The distillate of the low pressure column after expansion, mainly oxygen, is fed to the top as a reflux, while the bottom product is also expanded and fed to the low pressure column.

The final products from the low pressure column are nitrogen from the top and oxygen from the bottoms. Figure 4.12 shows the scheme of the operation. We use subindex 1 for the high pressure column and 2 for the low pressure column.

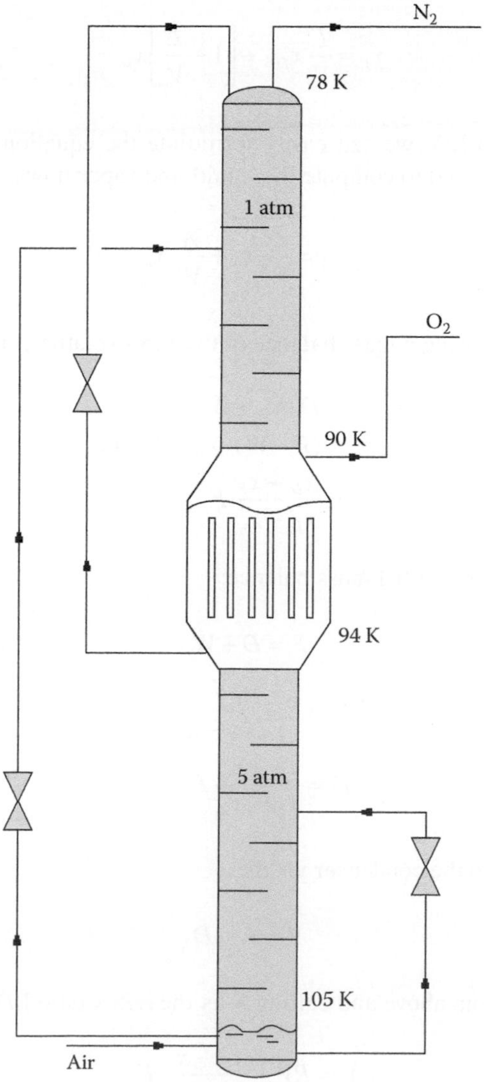

FIGURE 4.12 Linde double column.

The feed to the high pressure column (5 atm) is air, thus the composition is $x_{F,1}$ = 0.79 and we assume that the fraction of liquid is 0.5. The distillate is mainly nitrogen.

Let's assume $x_{D,1}$ = 0.99 since its fed to the top of the low pressure column. The residue is assumed to have a composition in nitrogen of $x_{W,1}$ = 0.63. This residue, assumed as liquid, Φ_L = 1; in Figure 4.12 is fed to the medium part of the low pressure column (1 atm). The distillate, mainly nitrogen, is assumed to have a composition of $x_{D,2}$ = 0.996 in nitrogen while the residue of the low pressure column has a composition of $x_{w,2}$ = 0.008. The problem is to compute the number of trays needed for such a separation.

For the rectifying section we have an operating line as:

$$y_f = \frac{L}{V} x_{f-1} + \left[1 - \frac{L}{V}\right] x_D \tag{4.30}$$

Thus, for a given L/V we can easily formulate the equation. However, for the stripping section we need to compute the liquid and vapor flows:

$$y_{n+1} = \frac{L'}{V'} x_n + \frac{W}{V'} x_w \tag{4.31}$$

We start with *W*. Using a mass balance to the most volatile component we have:

$$\begin{aligned} F \cdot x_F &= D \cdot x_D + W \cdot x_W \\ F \cdot x_F &= (F - W) \cdot x_D + W \cdot x_W \\ W &= \frac{x_D - x_F}{x_D - x_w} F \end{aligned} \tag{4.32}$$

Substituting into the global mass balance:

$$F = D + W \tag{4.33}$$

$$\begin{aligned} D &= F - \frac{x_D - x_F}{x_D - x_w} F \\ D &= \frac{x_F - x_w}{x_D - x_w} F \end{aligned} \tag{4.34}$$

A mass balance to the condenser yields:

$$V = L + D \tag{4.35}$$

Substituting *D* from above and calling *R* as the reflux ratio L/V we have:

$$V = RV + \frac{x_F - x_w}{x_D - x_w} F \tag{4.36}$$

Thus we can solve for V,

$$V(1-R)=F\frac{x_F-x_w}{x_D-x_w}$$
$$V=\frac{F}{(1-R)}\frac{x_F-x_w}{x_D-x_w} \tag{4.37}$$

Using the definition above for R, we can compute L,

$$L=RV=R\frac{F}{(1-R)}\frac{x_F-x_w}{x_D-x_w}=\frac{\left(L/V\right)\cdot F\cdot\frac{x_F-x_w}{x_D-x_w}}{1-\left(L/V\right)} \tag{4.38}$$

To compute L and V, we use the liquid fraction of the feed:

$$\phi_V+\phi_L=1 \tag{4.39}$$

By definition we have:

$$\phi_L=\frac{L'-L}{F}\Rightarrow L'=\phi_L F+L \tag{4.40}$$

Substituting L from the development above we have L as:

$$L'=\phi_L F+\frac{\left(L/V\right)\cdot F\cdot\frac{x_F-x_w}{x_D-x_w}}{1-\left(L/V\right)}=F\frac{\phi_L+\left(L/V\right)\cdot\left[\frac{x_F-x_w}{x_D-x_w}-\phi_L\right]}{1-\left(L/V\right)} \tag{4.41}$$

Similarly, using the vapor fraction we compute V from V:

$$\phi_V=\frac{V-V'}{F}\Rightarrow V'=V-\phi_V F$$
$$\phi_V+\phi_L=1\Rightarrow\phi_V=(1-\phi_L) \tag{4.42}$$
$$V'=V-(1-\phi_L)F$$

Using the equation for V obtained above we have:

$$V'=\frac{F}{(1-L/V)}\frac{x_F-x_w}{x_D-x_w}-(1-\phi_L)F=F\frac{\phi_L-1+L/V(1-\phi_L)+\frac{x_F-x_w}{x_D-x_w}}{(1-L/V)} \tag{4.43}$$

We can now compute *L*/*V* by dividing:

$$\frac{L'}{V'}=\frac{\phi_L+\frac{L}{V}\left[\left(\frac{x_F-x_W}{x_F-x_W}\right)-\phi_L\right]}{\phi_L-1+\frac{L}{V}(1-\phi_L)\left(\frac{x_F-x_W}{x_D-x_W}\right)} \tag{4.44}$$

Thus the operating line in the stripping section becomes:

$$y_{n+1}=\frac{L'}{V'}x_n-\frac{\left(1-\frac{L}{V}\right)\left[\left(\frac{x_D-x_F}{x_D-x_w}\right)\right]}{\phi_L-1+\frac{L}{V}(1-\phi_L)\left(\frac{x_f-x_w}{x_d-x_w}\right)}x_w \tag{4.45}$$

In order to compute the vapor-liquid equilibrium we have the Antoine correlations for the vapor pressure [7]. For Nitrogen:

$$\ln(P_v)=\left(14.9542-\frac{588.72}{(T(^\circ C)+273.15-6.6)}\right) \tag{4.46}$$

For Oxygen:

$$\ln(P_v)=\left(15.4075-\frac{734.55}{(T(^\circ C)+273.15-6.45)}\right) \tag{4.47}$$

Assuming ideal behavior of the air for T from the boiling point of N_2 to the boiling point of O_2 we compute the *XY* diagram as:

$$\begin{aligned} x&=\frac{P_{Total}-P_{Vap,N_2}}{P_{Vap,O_2}-P_{Vap,N_2}}\\ y&=\frac{P_{Vap,N_2}x}{P_{Total}} \end{aligned} \tag{4.48}$$

Next, we need an equation to predict the equilibrium line. We can easily fit these data to a polynomial. We use two fittings, one for P = 1 atm and another one for P = 5 atm for the low and high pressure columns, respectively. We prepare a script that will perform the operation and four functions. The script will handle the iteration to compute the number of trays while the functions are used to define the equilibrium line, the operation lines above and below the feed and finally another function to compute the cut of the operating lines, defining the feed tray. In order to follow the iteration we make use of the global command so declare "S" as global. Variable "S" represents the value for the composition of the vapor in the equilibrium, y_{eq}, in the *xy* diagram.

In the script we start by declaring "S" as global. Next, we input the data for each column L/V reflux ratio, x_D, Φ_L, x_F, xw. We initialize *k*, a counter that computes the number of trays.

Next, we need to define the feed tray since from that point we need to change the operation line from that of the rectifying section to that of the stripping section. We call it inter. We compute the transition from the rectifying section to the stripping section defining a function as the cut between the two lines "cut" function. Alternatively we can solve the two lines ourselves. For an example like this we present the next case of study.

Subsequently, we define the steps by the cut between the equilibrium line and the value of the vapor composition at the equilibrium given by "S". See function equilibrium. S is updated at each tray, corresponding with each iteration *k*.

Next, we need to declare as functions the two operating lines. The defined functions "linerect" and "linestrip" using the data provided and the equations developed in the previous section.

Going back to the script we include the fact that, if the *x* at the equilibrium line is higher than "inter", the *x* calculated at the crossing of the operating lines, then we use the operating line for the rectifying section. Otherwise we use the line for the stripping section. The iteration continues while Xeq, the *x* at the equilibrium line, is larger than the residue concentration. We also plot the functions we have created.

Thus we have four functions and a script. At the end of the script we plot the equilibrium, the $x = y$, the operating lines and the data representing the concentrations at the trays, the equilibrium concentrations.

```
% N2 O2 distillation
global S
%Data
LtoV=0.5;
xd=0.99;
phiL=0.5;
xf=0.79;
xw=0.63;
k=1;
S=xd;
xeq(k)=xd;
yeq(k)=xd;

%Tray calculation

inter=fsolve(@corte,0.5);

while xeq(k)-xw>0;

   xeq(k+1)=fsolve(@equilibrium,0.5);
   yeq(k+1)=S;

    if    (xeq(k+1)-inter)>0
        S=  linerect(xeq(k+1));
    else
```

```
          S= linestrip(xeq(k+1));
     end

k=k+1;
end;

%Results plotting
xx=[0:0.05:1];
dim=length(xx);

for j=1:dim;
yy(j)=equilibriocut(xx(j));
yydest(j)= linerect (xx(j));
yyagot(j)=linestrip(xx(j));
yyxx(j)=xx(j);
end;
plot(xx,yy,'k-',xx,yydest,'r-',xx,yyagot,'b-',xx,yyxx,
'k-',xeq,yeq,'ro')
```

The four functions are as follows:

```
function F = equilibrium(x)
 global S
 %a 1 atm
% F=2.7231*x^5 - 8.9493*x^4 + 11.918*x^3 - 8.5302*x^2 +
3.8378*x + 0.0016 - S;
  % a 5 atm
  F= -0.4773*x^4 + 1.7554*x^3 - 2.6667*x^2 + 2.3819*x + 0.0031 -S;
% equilibrium data of the O2-N2 system

function ydest = linerect(x)
LtoV=0.5;
xd=0.99;
phiL=0.5;
xf=0.79;
xw=0.63;
ydest=LtoV*x+xd*(1-LtoV);

function yagot = linestrip(x)
LtoV=0.5;
xd=0.99;
phiL=0.5;
xf=0.79;
xw=0.63;
LtoVp=(phiL+LtoV*(((xf-xw)/(xd-xw))-phiL))/(phiL-1+LtoV*
(1-phiL)+((xf-xw)/(xd-xw)));
yagot=LtoVp*x-xw*(LtoVp-1);

% Cut between the operation lines
function cut = corte(x)

cut = linerect(x) - linestrip(x);
```

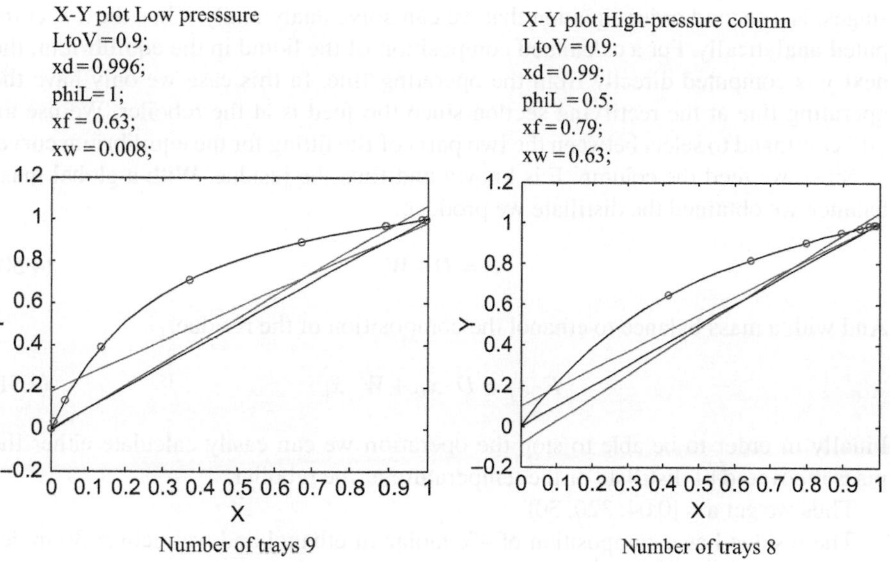

FIGURE 4.13 Result of the double column of Linde.

Figure 4.13 shows the tray number calculation for an example:

4.3.2.2 Distillation of Water: Ethanol Mixture in a Discontinuous Column

The problem is formulated as follows. We have a batch distillation with 11 trays. We would like to process a water–ethanol mixture under constant reflux conditions to obtain a distillate of 83% molar in ethanol. We would like to determine the residue composition and the amount of distillate and residue if we feed 350 kmol of mixture 11% molar in ethanol.

The operation of such a column is based on the fact that the composition of the residue changes with time during the operation so that the following equation holds [6].

$$\ln\frac{F}{W} = \int_{Xw}^{X_f} \frac{dx_w}{x_D - x_w} \tag{4.49}$$

We need to evaluate the integral to compute the ratio between the feed and the residue. In order to do so, for a given distillate composition, a fixed reflux ratio and a fixed number of trays, we calculate the corresponding composition of the residue. We have to do that for a number of feasible x_D's so that we can have a number of values that are in the range of our x_f and x_w compositions. The equilibrium line for the ethanol water system is complex due to the azeotrope, thus we fit it in two parts. We use two second order polynomials for fitting it. See coea (1 or 2) in the script below. Since the polynomials are second order, the cut between the horizontal lines, the composition of the distillate and the vapour composition at each of the following stages, is a second order equation that we can solve analytically. Thus the x is computed analytically. For a calculated composition of the liquid in the equilibrium, the next y is computed directly from the operating line. In this case we only have the operating line at the rectifying section since the feed is at the reboiler. We use an "if" command to select between the two parts of the fitting for the equilibrium curve.

Since we feed the column, F is known and thus the residue. With a global mass balance we obtained the distillate we produce.

$$F = D + W \tag{4.50}$$

And with a mass balance to ethanol the composition of the residue.

$$F \cdot x_f = D \cdot x_D + W \cdot x_w \tag{4.51}$$

Finally in order to be able to stop the operation we can easily calculate either the mass recovered of distillate or the temperature at the reboiler.

Thus we get a = [0.04, 320, 30]

The residue has a composition of 4% molar in ethanol and we recover 30 moles of distillate at 83%.

```
%Unsteady state column

a=fsolve(@rectdis,[0.05,300,50])

function  resid=rectdis(var);

%Variable specification
xres=var(1);
Residue=var(2);
Distillate=var(3);
%tempfin=var(4);

%Data input
plates=11;

LtoD=3;
Pend=(LtoD)/(LtoD+1);

%List of Xd to solve the integral
ydest=[0.75:0.01:0.85];
yd=0.83;
xf=0.11;
Feed=350;

a=size(ydest);
long=a(1,2);

%Polynomial coefficients for vapor liquid equilibrium
coea2=-16.78846;
```

```
coeb2=5.35043771;
coec2=0.07820301;

coea1=0.2774380;
coeb1=0.21006435;
coec1=0.4766639;
%Antoine coefficents
Awa=18.3036;
Bwa=3816.44;
Cwa=227.02;

Aet=18.9119;
Bet=3803.98;
Cet=231.47;

%itereation for computing the integral
for i=1:1:long

    yd=ydest(i);
    yeq(i,1)=yd;

    for j=1:1:plates
%Algebraic solution for tray calculation

          if yeq(i,j) > 0.5044
              xeq(i,j)=(-coeb1+(coeb1^2-4*(coea1*
              (coec1-yeq(i,j))))^0.5)/(2*coea1);
          else
              xeq(i,j)=(-coeb2+(coeb2^2-4*(coea2*
              (coec2-yeq(i,j))))^0.5)/(2*coea2);
          end
          if j<11
          yeq(i,j+1)=Pend*xeq(i,j)+(1-Pend)*yeq(i,1);
          end

    end
    xw(i)=xeq(i,11);
    inv(i)= 1/(ydest(i)-xw(i));
end
%Integral computation
for i=1:1:10
    difer(i) =inv(i)*(xw(i+1)-xw(i));
end
coefint=polyfit(xw,inv,2);
Integral=(coefint(1)/3)*(xf^3-xres.^3)+(coefint(2)/2)*
(xf^2-xres.^2)+(coefint(3))*(xf-xres);

%List of equations to be solved
resid(1)=Residue*exp(Integral)-Feed;
resid(2)=Feed-Distillate-Residue;
resid(3)=Feed*xf-Distillate*yd-Residue*xres;
```

4.3.3 Reactor Design

In this section we comment on the use of MATLAB® to model chemical reactors. We focus on two examples, a semi-batch reactor and a fixed bed reactor. Both represent common cases of study; such as the production of polymethyl methacrylate (PMMA) in suspension and the oxidation of SO_2 to SO_3 that is one of the steps in the production of sulfuric acid via heterogeneous method. The models are based on explicit algebraic equations and differential equations. Thus we use "odexx" function in MATLAB® to solve the concentration, temperature, and/or pressure profiles along the operation of such equipment.

4.3.3.1 Batch Reactors: Polymerization Reactor

The example we consider is the suspension polymerization of methyl metacrylate (MMA) for the production of polymethyl methacrylate (PMMA). The basic principle for polymerization reactions in suspension is the fact that we disperse the monomer in water in the form of drops using the proper agitation in a jacketed stirred tank reactor. Different surfactants can be added to maintain that suspension. The modeling of the operation of the reactor is based on the assumption that each of the drops behaves as a batch reactor operating in bulk [8]. For this particular example we can find the kinetic data and basic model in the papers by Seth [9] and Ghosh [10], even though this last one uses it for bulk polymerization, where for the sake of the example we neglect the evaporation rate assuming that the dispersion generated in water isolates the monomer and the only losses may be due to the solubility of the monomer in water and thus both are the moisture that is dragged by the nitrogen used to maintain the pressure within the reactor.

The polymerization of MMA is radical based polymerization, similar to the production of PVC or PE. It requires the use of initializers. We assume that the small amount needed to initialize the reaction does not affect the volume of the liquid phase. The reaction mechanism proposed is as follows:

Initialization:	$I \xrightarrow{k_d} 2R + CO_2$
Propagation:	$R + M \xrightarrow{k_i} P_1$
Propagation:	$P_n + M \xrightarrow{k_p} P_{n+1}$
Termination by combination	$P_n + P_m \xrightarrow{k_{tc}} D_{n+m}$
Termination by disproportion	$P_n + P_m \xrightarrow{k_{td}} D_n + D_m$
Transfer to monomer	$P_n + M \xrightarrow{k_f} P_1 + D_n$

Where I is the initiator, R represents the active radicals, M is the monomer, P is the polymer in formation and D is the dead polymer. The subindexes "n" and "m" correspond to the number of monomeric units. The different constants and definitions are taken from the papers mentioned above:

k_d: Initiator decomposition rate constant (s^{-1}).

$$k_d = 1.69 \cdot 10^{-14} \cdot Exp\left[\frac{-125400}{8.314 \cdot T}\right] \tag{4.52}$$

k_i: Initial propagation rate constant ($m^3 \cdot mol^{-1} \cdot s^{-1}$).

$$k_i = k_p \tag{4.53}$$

k_{po}: Propagation rate constant in absence of gel, cage and crystal effects, ($m^3 \cdot mol^{-1} \cdot s^{-1}$)

$$k_{po} = 4.917 \cdot 10^2 \cdot Exp\left[\frac{-18220}{8.314 \cdot T}\right] \tag{4.54}$$

Based on the literature we assume that the propagation rate constants are independent of the chain length, $k_i = k_p$[9]. It is also considered that the transfer to the monomer is small $k_f \approx 0$ [9].

k_f: Monomer transfer rate constant ($m^3 \cdot mol^{-1} \cdot s^{-1}$).

$$k_f = 0.0 \tag{4.55}$$

In the case of the PMMA the termination of the polymerization is due to disproportion and thus, $k_{tc} \approx 0$ [9].

k_{tc}: Termination rate constant by combination ($m^3 \cdot mol^{-1} \cdot s^{-1}$).

$$k_{tc} = 0.0 \tag{4.56}$$

$k_{td,o}$: Disproportion constant rate in absence of gel, cage and crystal effects ($m^3 \cdot mol^{-1}\ s^{-1}$).

$$k_{to} = k_{tdo} = 9.8 \cdot 10^4 \cdot Exp\left[\frac{-2937}{8.314 \cdot T}\right] \tag{4.57}$$

ψ is used to represent the volumetric relationships:

$$\psi = \frac{\gamma\left(\dfrac{\rho_m \cdot \varphi_m \cdot V_m^*}{\xi_{13}} + \rho_p \cdot \varphi_p \cdot V_p^*\right)}{\rho_m \cdot \varphi_m \cdot V_m^* \cdot V_{fm} + \rho_p \cdot \varphi_p \cdot V_p^* \cdot V_{fp}} \tag{4.58}$$

Where the monomer density with temperature, ($kg \cdot m^{-3}$), is given by:

$$\rho_m = 966{,}5 - 1{,}1 \cdot (T - 273{,}1) \tag{4.59}$$

And polymer density ($kg \cdot m^{-3}$) is taken to be:

$\rho_p = 1200$

V_{fm} is the volume of gel, cage and crystal effects for the monomer ($m^3 \cdot kg^{-1}$) and V_{fp}: is the corresponding one for the polymer, ($m^3 \cdot kg^{-1}$).

$$V_{fm} = 0.149 + 2.9 \cdot 10^{-4} [T - 273.1] \quad (4.60)$$

$$V_{fp} = 0.0194 + 1.3 \cdot 10^{-4} [T - 273.1 - 105] \text{ for } T < (105 + 273.1) \quad (4.61)$$

γ is equal to 1.

ϕ_m and ϕ_p are the volumetric fraction of the monomer with respect to the reaction volume V and is the corresponding one for the polymer with respect to V respectively:

$$\phi_m = \frac{\dfrac{M \cdot (MW_m)}{\rho_m}}{\dfrac{M \cdot (MW_m)}{\rho_m} + \dfrac{(M_o - M) \cdot (MW_m)}{\rho_p}} \quad (4.62)$$

$$\phi_p = 1 - \phi_m \quad (4.63)$$

ξ_{i3} are the ratio of critical volumes of the monomer, i = 1, and the initiator, i = I; over the polymer:

$$\xi_{13} = \frac{V_m^* \cdot (MW_m)}{V_p^* \cdot M_{jp}} \quad (4.64)$$

$$\xi_{I3} = \frac{V_I^* \cdot (MW_I)}{V_p^* \cdot M_{jp}} \quad (4.65)$$

Where the different critical volumes are V_I^*, for the initiator, $8.25 \cdot 10^{-4}$ ($m^3 \cdot kg^{-1}$), V_m^*: for the monomer, $8.22 \cdot 10^{-4}$ ($m^3 \cdot kg^{-1}$) and V_p^* for the polymer $7.70 \cdot 10^{-4}$ ($m^3 \cdot kg^{-1}$).

The molecular weights used are, for the initiator, M_{jp}: 0.18781, ($kg \cdot mol^{-1}$), for the monomer, MW_m, for the monomer 0.10013 ($kg \cdot mol^{-1}$), for the initiator (BPO), MW_I, 0.077 $kg \cdot mol^{-1}$. While we define the average molecular weight for the polymer by number as M_n ($kg\ mol^{-1}$). And by weight M_w ($kg\ mol^{-1}$).

The initiator efficiency is given as:

$$f = \frac{f_o}{1 + \theta_f(T) \cdot \dfrac{M}{V} \cdot \dfrac{1}{\text{Exp}\left[\xi_{I3}\left\{-\psi + \psi_{ref}\right\}\right]}} \quad (4.66)$$

Where

$$\psi_{ref} = \frac{\gamma}{V_{fp}} \quad (4.67)$$

f_o is the initial efficiency of the BPO, equal to 1.

k_t: Termination constant rate ($m^3 \cdot mol^{-1} \cdot s^{-1}$).

$$k_t = \frac{1}{\frac{1}{k_{to}} + \theta_t(T) + \mu_n^2 \cdot \frac{\lambda_o}{V} \cdot \frac{1}{Exp\left[-\psi + \psi_{ref}\right]}} \tag{4.68}$$

k_p: Propagation rate constant ($m^3 \cdot mol^{-1} \cdot s^{-1}$).

$$k_p = \frac{1}{\frac{1}{k_{po}} + \theta_p(T) \cdot \frac{\lambda_o}{V} \cdot \frac{1}{Exp\left[\xi_{13}\left\{-\psi + \psi_{ref}\right\}\right]}} \tag{4.69}$$

Where θ_f, ($m^3 \cdot mol^{-1}$), θ_p, (s), and θ_t, (s) are adjustable parameters for the initiator efficiency, the propagation constant rate and for the disproportion termination constant rate respectively and are given by:

$$Log_{10}\left[\theta_t(T), s\right] = 1.241 \cdot 10^2 - 1.0314 \cdot 10^5 \cdot \frac{1}{T} + 2.2735 \cdot 10^7 \cdot \frac{1}{T^2} \tag{4.70}$$

$$Log_{10}\left[\theta_p(T), s\right] = 8.03 \cdot 10^1 - 7.50 \cdot 10^4 \cdot \frac{1}{T} + 1.765 \cdot 10^7 \cdot \frac{1}{T^2} \tag{4.71}$$

$$Log_{10}\left[10^3 \cdot \theta_f(T), m^3 \cdot mol^{-1}\right] = -40.86951 + 1.7179 \cdot 10^4 \cdot \frac{1}{T} \tag{4.72}$$

The volume of the mixture, V (m^3) is calculated as:

$$V = \frac{M \cdot (MW_m)}{\rho_m} + \frac{(M_o - M) \cdot MW_m}{\rho_p} \tag{4.73}$$

Based on the mechanism proposed above the differential equations describing the polymerization of MMA to PMMA are as follows:

$$\begin{aligned}
\frac{dI}{dt} &= -k_d \cdot I + \text{feedrate} \\
\frac{dM}{dt} &= -(k_p + k_f) \cdot \frac{\lambda_o \cdot M}{V} - k_i \cdot \frac{R \cdot M}{V} \\
\frac{dR}{dt} &= 2 \cdot f \cdot k_d \cdot I - k_i \cdot \frac{R \cdot M}{V} \\
\frac{d\lambda_o}{dt} &= k_i \cdot \frac{R \cdot M}{V} - k_t \cdot \frac{\lambda_o^2}{V} \\
\frac{d\lambda_1}{dt} &= k_i \cdot \frac{R \cdot M}{V} + k_p \cdot M \cdot \frac{\lambda_o}{V} - k_t \frac{\lambda_o \cdot \lambda_1}{V} + k_f \cdot M \cdot \frac{(\lambda_o - \lambda_1)}{V} \\
\frac{d\lambda_2}{dt} &= k_i \cdot \frac{R \cdot M}{V} + k_p \cdot M \cdot \frac{\lambda_o + 2 \cdot \lambda_1}{V} - k_t \cdot \frac{\lambda_o \cdot \lambda_2}{V} + k_f \cdot \frac{(\lambda_o - \lambda_2)}{V}
\end{aligned} \tag{4.74}$$

$$\frac{d\mu_o}{dt} = k_f \cdot M \cdot \frac{\lambda_o}{V} + \left(k_{td} + \frac{1}{2} \cdot k_{tc}\right)\frac{\lambda_o^2}{V}$$
$$\frac{d\mu_1}{dt} = k_f \cdot M \cdot \frac{\lambda_1}{V} + k_t \cdot \frac{\lambda_o \cdot \lambda_1}{V}$$
$$\frac{d\mu_2}{dt} = k_f \cdot M \cdot \frac{\lambda_2}{V} + k_t \cdot \frac{\lambda_o \cdot \lambda_2}{V} + k_{tc} \cdot \frac{\lambda_1^2}{V}$$
$$\frac{dQ}{dt} = -57700\left(-(k_p + k_f) \cdot \frac{\lambda_o \cdot M}{V} - k_i \cdot \frac{R \cdot M}{V}\right)$$

Where I represents the BPO moles (mol), M, the monomer moles (mol), M_o, the initial moles of monomer, Q the energy (J), R the moles of growing polymer chains (mol), t is time (s) and T is temperature (K). Furthermore, λ_i is the momentum i of the radical species.

$$\lambda_i = \sum_{n=1}^{\infty} n^i \cdot R_n, \text{ (mol)}. \tag{4.75}$$

μ_i is the momentum i of the dead polymeric species given by:

$$\mu_i = \sum_{n=1}^{\infty} n^i \cdot P_n, \text{ (mol)}. \tag{4.76}$$

And μ_n is the length of the chain at time t. After solving we can use the result to compute some characteristics:

$$M_n = MW_m \cdot \left[\frac{\lambda_1 + \mu_1}{\lambda_o + \mu_o}\right] \qquad M_w = MW_m \cdot \left[\frac{\lambda_2 + \mu_2}{\lambda_1 + \mu_1}\right] \tag{4.77}$$

We use Ross [11] definition to compute the average molecular weight by number and by weight of the polymer respectively. We compute the conversion of the reaction as follows:

$$X = \frac{M_o - M}{M_o} \tag{4.78}$$

We assume that we feed the reactor with 1500 kg of monomer and fix the temperature to 350 K. In order to start the reaction the BPO is added. We assume that we feed the mass of it, 1% of the monomer, in 1 min at a constant rate. After that the addition rate becomes 0. The reaction progresses over two hours. For further details on the complete production process and economic evaluation we refer to Martín, M. [13].

The conversion, the energy generation profile and the monomer and initiator moles profiles, as well as the polymer averages molecular weights, are computed and plotted. We produce here two files: a script that is used to provide the data for the ode statement and a function that will contain the differential equations and the definitions of the different parameters needed.

In terms of the script we use the statement "global" so that we can easily test the effect of the temperature on the conversion. We use the order "input" so that when we run the script MATLAB® will ask us for the temperature in K. For temperature to be used within the function file "reac" we declare it as global. We can define multiple variables as global writing them after the statement "global" separated by spaces. We also input the initial number of moles.

The differential equations are solved using the ordinary differential equation order ode## as presented above.

We store the results of the ode function in a matrix [t,b] that is used later to plot the results. In order to plot all of the Figures one after the other in different frames we use the statement "Figure" before the statement "plot". In case we want to over impose one plot on the other we can use the statement "hold on". Apart from the general plot function presented before we can use **subplot** (*x,y,n*) before creating the plot. The first argument, *x*, is the number of plots in a row and y, the number of plots in a column. Finally, *n* corresponds to the location of the plot within the combined framework.

```
Mmma=15000;
global Temperature Qgen
Temperature=input('Temperatura de reacción en K\n');
[a,b]=ode15s('reac',[0 7200],[0,Mmma,0,0,0,0,0,0,0,0]);
conversion=(-b(:,2)+Mmma)/Mmma
subplot(3,2,1)
plot(a,conversion)
xlabel('Time (s)')
ylabel('Conversion')
subplot(3,2,2) %figure
iniciador=(b(:,1))
plot(a,iniciador)
xlabel('Time (s)')
ylabel('Moles Initiator')
subplot(3,2,3) %figure
monomer=(b(:,2))
plot(a,monomer)
xlabel('Time (s)')
ylabel('Moles Monomer')
subplot(3,2,4) %figure
energy=(b(:,10))
plot(a,energy)
xlabel('Time (s)')
ylabel('Energy (kJ)')
Momentos0=b(:,4)+b(:,7);
Momentos1=b(:,5)+b(:,8);
Momentos2=b(:,6)+b(:,9);
Longitud=size(Momentos0)
Long=Longitud(1,1);
```

```
for j=2:1:Long;
   Mn(j,1)=Momentos1(j-1,1)/Momentos0(j-1,1);
end
for h=2:1:Long;
   Mw(h,1)=Momentos2(h-1,1)/Momentos1(h-1,1);
end
for k=1:1:Long;
     PD(k,1)=Mw(k,1)/Mn(k,1);
end
PMn=100.13*Mn;
PMw=100.13*Mw;
%Ross, R.T., 1976
subplot(3,2,5)
plot(a,PMn)
xlabel('Time (s)')
ylabel('Molecular weight PMn')
subplot(3,2,6)
plot(a,PMw)
xlabel('Time (s)')
ylabel('Molecular weight PMw')
```

Within the function file we define it as such in the first line. The set of ordinary equations "Reactor" is written as a function of the independent variable, *t*, and a vector of dependent variables x. In order to use variables names closer to the real ones we can define them as columns of the dependent variables vector (i.e., i = x(1)). Next, in the second line we write the global statement so that these variables can be used within the script. MATLAB® needs that all the parameters, constants or equations are explicit of the variables and defined before the system of differential equations. Thus we write the rate constants and other parameters defined above. This reactor is a semi-batch in the sense that we add the initiator during the first minute of the reaction at a constant rate. Thus we define a rate that is constant if the time is lower or equal to 60s and zero otherwise using the if-then-end syntax. Finally we write the differential equations. The right hand side of the differential equations must be the same word as the one we use in the title of the function "Reactor". Each equation will be Reactor (n,1). We present the results in Figure 4.14.

```
function Reactor=Reac(t,x)
global Temperature Qgen
%Units, kg, mol,m^3 y s
%Lambda==M; Mu==s
I=x(1);
M=x(2);
R=x(3);
Mo=x(4);
M1=x(5);
M2=x(6);
so=x(7);
s1=x(8);
s2=x(9);
T=Temperature;
Nmma=15000;
```

```
rhom=966.5-1.1*(T-273.1);
rhop=1200;
kd=1.69*10^14*exp(-125400/(8.314*T));
kpo=4.917*10^2*exp(-18220/(8.314*T));
ktdo=9.8*10^4*exp(-2937/(8.314*T));
%Ray A.B., 1995
MWm=0.10013;
Mjp=0.18781;
gamma=1;
Vmm=8.22*10^-4;
Vpp=7.70*10^-4;
Vm=0.149+2.9*10^-4*(T-273.1);
Vp=0.0194+1.3*10^-4*(T-273.1-105);
%Ray A.B., 1995
phit=10^(1.241*10^2-1.0314*10^5*(1/T)+2.2735*10^7*(1/T^2));
phip=10^(8.03*10-7.50*10^4*(1/T)+1.765*10^7*(1/T^2));
phif=0.001*10^(-40.86951+1.7179*10^4*(1/T));
%Ghosh, P., 1998
effo=1;
%Seth, V., 1995
VII=8.25*10^-4;
MWI=0.077;
%Seth, V., 1995
gref=gamma/Vp;
sigma13=Vmm*MWm/(Vpp*Mjp);
sigmaI3=VII*MWI/(Vpp*Mjp);
%Ghosh, P., 1998
V=0.10013*M/rhom+(Nmma-M)*0.10013/rhop;
%Seth, V., 1995
thetam=(M*0.10013/rhom)/V;
thetap=1-thetam;
%Seth, V., 1995
if Mo+so==0;
   sn=0;
else
   sn=(M1+s1)/(Mo+so);
end
g=gamma*(rhom*thetam*Vmm/sigma13+rhop*thetap*Vpp)/
(rhom*thetam*Vmm*Vm+rhop*thetap*Vpp*Vp);
%Ray A.B., 1995

eff=effo/(1+phif*(M/V)*(1/(exp(sigmaI3*(-g+gref)))));
ktd=1/(1/ktdo+phit*sn^2*(Mo/V)*(1/exp(-g+gref)));
kp=1/(1/kpo+phip*(Mo/V)*(1/exp(sigma13*(-g+gref))));
%Ghosh, P., 1998
ki=kp;
kf=0;
%Seth, V., 1995
if t<60;
   feedrate=Nmma*0.01*100/(242*60);
   else
   feedrate=0;
end
```

```
Reactor(1,1)=-kd*I+feedrate;
Reactor(2,1)=-kp*Mo*M/V-ki*R*M/V;
Reactor(3,1)=2*eff*kd*I-ki*R*M/V;
Reactor(4,1)=ki*R*M/V-(ktd*Mo^2)/V;
Reactor(5,1)=ki*R*M/V+kp*M*Mo/V-ktd*Mo*M1/V;
Reactor(6,1)=ki*R*M/V+kp*M*(Mo+2*M1)/V-ktd*Mo*M2/V;
Reactor(7,1)=(ktd*Mo^2)/V;
Reactor(8,1)=ktd*Mo*M1/V;
Reactor(9,1)=ktd*Mo*M2/V;
Reactor(10,1)=-57700*(-kp*Mo*M/V-ki*R*M/V);

%Seth, V., 1995
```

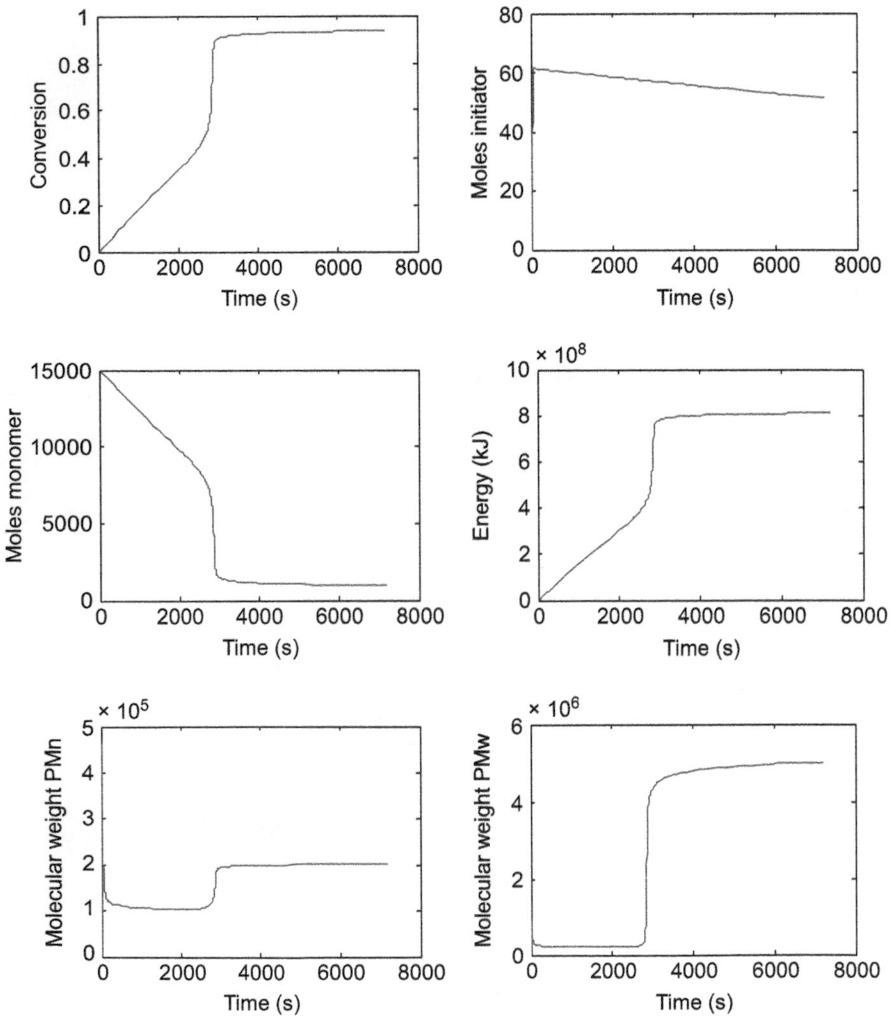

FIGURE 4.14 Profiles of the reactor performance.

QUESTIONS

- Compute the time for conversion equal to 0.9.
- Evaluate other types of initiator injection such as initial value.

4.3.3.2 Fixed Bed Reactors: SO_2 Oxidation

Sulfuric acid production was for decades the industrialization index of any country due to its use for the production of a number of chemicals from dyes to explosives. The first method available was the lead chambers by Rosenbrock in 1746. The need for more concentrated sulfuric acid improved the process with the addition of the Gay-Lussac Tower in 1835 and the Glover tower in 1859. However, the concentration of sulfuric acid, around 60%, was not enough for the growing chemical industry. Thus, the use of heterogeneous catalysis was in place to develop the contact method. The method consists of obtaining SO_2 from sulfur or pyrite by burning them. The next stage involves the equilibrium:

$$S + SO_2 \longleftrightarrow SO_3 \tag{4.79}$$

Finally, the SO_3 is absorbed in water or in a solution of H_2SO_4.

In this problem we evaluate the performance of the reactor that produces SO_3 from SO_2. It is a complex equilibrium. The reaction being exothermic does not allow high conversions. There are two typical designs for the reactor. A multiple bed reactor, an example of which we present in Chapter 14, and a fixed bed reactor. For the sake of the example we focus on this last one, even though it is not the most common it allows for developing a model based on ordinary differential equations that considers mass balance, including equilibrium, energy balance and pressure drop along the bed. We consider just one of the packed bed tubes of the reactor using (V_2O_5).

The problem statement is as follows. We use sulfur as raw material for the production of sulfuric acid. After burning the sulfur, the composition that we feed to the convertor can be seen in Table 4.1:

Where a = 0.1 to match the example presented in Chapter 14.

Assume that we have a packed bed reactor using V_2O_5 in a pipe of 2 in 12 BWG (internal diameter is 0.0453m). The volumetric flow rate is assumed to be

TABLE 4.1
Mass Balance for SO_3 Production from Sulfur

	$S + O_2 \rightarrow SO_2$		$SO_2 + (1/2)\ O_2 \longleftrightarrow SO_3$
	Initial	Final	Initial
SO_2	—	a	a
N_2	0.79	0.79	0.79
O_2	0.21	0.21 – a	0.21 – a
SO_3	—		

30.48 m^3/min·m^2 [14] and the operation is recommended from 400°C to 600°C so that the reaction occurs at a reasonable rate without energy transfer problems. The converters operate close to atmosphere; we assume that we feed the gas at 202 kPa. The initial flow rate is 0.0215 mol/s of the mixture. Determine the length of the tube (the mass of catalyst whose density is 542 kg/m^3) to obtain a conversion of SO_2 of 85%.

We can proceed in two ways. The iterative one, the easiest to start with, and a more complex one based on the solution of a differential algebraic system using an error function. In this example, we focus on the first one and leave the second as an exercise to the reader. We start with modeling the reactor following the principles in Fogler [15] where a similar problem but for the design of a complete multi-tubular reactor can be found using polymath.

Kinetics of the reactor:

$$F_{Ao}\frac{dX}{dW} = -r_A{}' \tag{4.80}$$

If the conversion is below 5%, the reaction rate does not depend on the conversion, thus and it is given as:

$$\begin{aligned} &x \le 0.05 \\ &-r_{SO_2} = -r_{SO_2}(X = 0.05) \end{aligned} \tag{4.81}$$

Where the reaction rate given by [16]:

$$-r'_{SO_2} = k\sqrt{\frac{P_{SO_2}}{P_{SO_3}}}\left[P_{O_2} - \left(\frac{P_{SO_3}}{K_p P_{SO_2}}\right)^2\right] \tag{4.82}$$

The equilibrium constant is given as:

$$\begin{aligned} &K_p = 0.0031415\exp\left(\frac{42{,}311}{1{,}987(1{,}8{\cdot}T - 273{,}15) + 491{,}67} - 11.24\right) \\ &\mathrm{K_p\,[=]\,Pa^{-0.5},\ T\,[=]\,K} \end{aligned} \tag{4.83}$$

For the rake constant, based on Eklund results [16] we have:

$$\begin{aligned} k = 9.86{\cdot}10^{-6}\exp\Bigg[&\frac{-176008}{((1.8*\mathrm{T} - 273.15) + 491.67)} - 110.1\ln(((1.8*\mathrm{T} - 273.15) \\ &+ 491.67)) + 912.8\Bigg] \end{aligned} \tag{4.84}$$

k is in mol of SO_2 /kg cat S·Pa while T is in K.

The stoichiometrics of the reaction is:

$$
\begin{array}{llll}
SO_2 + & \frac{1}{2}O_2 + & N_2 \leftrightarrow SO_3 + N_2 & \\
F_{Ao} & F_{Bo} & F_{Co} & \\
F_{Ao}(1-x) & F_{Bo} - \frac{1}{2}F_{Ao}x & F_{Co} & F_{Ao}x
\end{array}
$$

$$\Theta_i = \frac{F_{io}}{F_{Ao}} \tag{4.85}$$

$$F_T = F_{Ao}\left[(1-x)+\Theta_{O_2} - \frac{1}{2}x + \Theta_{N_2} + \Theta_{SO_3} + x\right]$$

$$= F_{Ao}(1+\Theta_{O_2}+\Theta_{N_2}+\Theta_{SO_3}) + F_{Ao}\delta x$$

$$\delta = -\frac{1}{2}$$

$$F_T = F_{To} + F_{Ao}\delta x$$

Assuming ideal gases:

$$C_T = \frac{F_T}{v} = \frac{P}{RT}; \qquad C_{To} = \frac{F_{To}}{v_o} = \frac{P_o}{RT_o} \tag{4.86}$$

Dividing both:

$$v = v_o \frac{P_o}{P}\frac{T}{T_o}\frac{F_T}{F_{To}} = v_o \frac{P_o}{P}\frac{T}{T_o}\frac{F_{To}+F_{Ao}\delta x}{F_{To}} = v_o \frac{P_o}{P}\frac{T}{T_o}\left(1+\frac{F_{Ao}}{F_{To}}\delta x\right) \tag{4.87}$$

Then, the concentration of the species i is as follows:

$$
\begin{aligned}
C_i &= \frac{F_i}{v} = \frac{F_i}{v_o \frac{P_o}{P}\frac{T}{T_o}\frac{F_T}{F_{To}}} = C_{To}\frac{F_i}{F_T}\frac{P}{P_o}\frac{T_o}{T} \\
C_i &= C_{To}\frac{F_i}{F_{To}+F_{Ao}\delta x}\frac{P}{P_o}\frac{T_o}{T} = C_{To}\frac{F_{Ao}(\Theta_i + v_i x)}{F_{To}+F_{Ao}\delta x}\frac{P}{P_o}\frac{T_o}{T}
\end{aligned}
\tag{4.88}
$$

Dividing both:

$$C_i = C_{To}\frac{F_{Ao}}{F_{To}}\frac{(\Theta_i + v_i x)}{1+(F_{Ao}/F_{To})\delta x}\frac{P}{P_o}\frac{T_o}{T} = C_{Ao}\frac{(\Theta_i + v_i x)}{1+\varepsilon x}\frac{P}{P_o}\frac{T_o}{T} \tag{4.89}$$

$$\varepsilon = y_{Ao}\delta = (F_{Ao}/F_{To})\delta$$

Thus the partial pressure of component i:

$$P_i = C_i RT = C_{Ao}\frac{T_o}{P_o}\frac{(\Theta_i + v_i X)}{1+\varepsilon X}RT \cdot P = P_{Ao}\frac{(\Theta_i + v_i X)}{1+\varepsilon X}\frac{P}{P_o} \tag{4.90}$$

$$P_{Ao} = C_{Ao}RT_o$$

Therefore the kinetics of the reaction becomes:

$$-r'_{SO_2} = k\sqrt{\frac{P_{SO_2}}{P_{SO_3}}}\left[P_{O_2} - \left(\frac{P_{SO_3}}{K_p P_{SO_2}}\right)^2\right] \tag{4.91}$$

$$\frac{dx}{dW} = \frac{-r'_{SO_2}}{F_{Ao}} = \frac{k}{F_{Ao}}\sqrt{\frac{P_{SO_2,o}\left(\frac{1-x}{1+\varepsilon x}\right)\frac{P}{P_o}}{P_{SO_2,o}\left(\frac{0+x}{1+\varepsilon x}\right)\frac{P}{P_o}}}$$

$$\left[P_{SO_2,o}\left(\frac{\Theta_{O_2} - \frac{1}{2}x}{1+\varepsilon x}\right)\frac{P}{P_o} - \left(\frac{P_{SO_2,o}\left(\frac{0+x}{1+\varepsilon x}\right)\frac{P}{P_o}}{K_p P_{SO_2,o}\left(\frac{1-x}{1+\varepsilon x}\right)\frac{P}{P_o}}\right)^2\right] \tag{4.92}$$

$$= \frac{k}{F_{Ao}}\sqrt{\frac{1-x}{(x)}}\left[P_{SO_2,o}\left(\frac{\Theta_{O_2} - \frac{1}{2}x}{1+\varepsilon x}\right)\frac{P}{P_o} - \left(\frac{x}{K_p(1-x)}\right)^2\right]$$

$$\varepsilon = \delta y_{Ao} = -\frac{1}{2}\frac{0.10}{0.1+0.11+0.79} = -0.05$$

$$P_{SO_2,o} = P_T y_{SO_2,o} = 202650 \cdot 0.1 = 20265 Pa$$

$$\Theta_{O_2} = \frac{0.11}{0.1} = 1.1$$

$$\Theta_{N_2} = \frac{0.79}{0.1} = 7.9$$

$$F_{Ao} = 0.00237 \text{ mol/s}$$

$$\frac{dx}{dW} = \frac{-r'_{SO_2}}{F_{Ao}} = \frac{k}{0.00237}\sqrt{\frac{1-x}{(x)}}\left[20265\left(\frac{1.1-\frac{1}{2}x}{1-0.05x}\right)\frac{P}{P_o} - \left(\frac{x}{K_p(1-x)}\right)^2\right] \tag{4.93}$$

The energy balance is as follows:

$$\dot{Q} - W_s - F_{Ao}\sum_{i=1}^{n}\int_{T_{io}}^{T}\Theta_i Cp_i dT - \left[\Delta H_R(T_R) + \int_{T_R}^{T}\Delta Cp dT\right]F_{Ao}x = 0$$

$$\frac{d\dot{Q}}{dV} - F_{Ao}\left(\sum_{i=1}^{n}\Theta_i Cp_i + x\Delta Cp\right)\frac{dT}{dV} - \left[\Delta H_R(T_R) + \int_{T_R}^{T}\Delta Cp dT\right]F_{Ao}\frac{dx}{dV} = 0 \tag{4.94}$$

$$-r_A = F_{Ao}\frac{dx}{dV}$$

$$\frac{dT}{dV}=\frac{UA(T_a-T)+(-r_A)\left[\Delta H_R(T_R)+\int_{T_R}^{T}\Delta CpdT\right]}{F_{Ao}\left(\sum_{i=1}^{n}\Theta_iCp_i+x\Delta Cp\right)}$$

$$\frac{dT}{dW}=\frac{\frac{4\cdot U}{\rho_bD}(T_a-T)+(-r_A)\left[\Delta H_R(T_R)+\int_{T_R}^{T}\Delta CpdT\right]}{F_{Ao}\left(\sum_{i=1}^{n}\Theta_iCp_i+x\Delta Cp\right)} \tag{4.95}$$

Thus the heat of reaction is computed as follows:

$$\begin{aligned}
&\Delta H_R(298K)=-98480J\,/\,molSO_2\\
&C_{p,SO_2}=23.852+66.989\cdot10^{-3}T-4.961\cdot10^{-5}T^2+13.281\cdot10^{-9}T^3\\
&C_{p,O_2}=28.106-3.680\cdot10^{-6}T+17.459\cdot10^{-6}T^2-1.065\cdot10^{-8}T^3\\
&C_{p,SO_3}=16.370+14.591\cdot10^{-2}T-1.120\cdot10^{-4}T^2+32.324\cdot10^{-9}T^3\\
&C_{p,N_2}=31.150-1.357\cdot10^{-2}T+26.796\cdot10^{-6}T^2-1.168\cdot10^{-8}T^3
\end{aligned} \tag{4.96}$$

Where Cp is in J/mol K and temperature, T, in K [7]:

$$\begin{aligned}
&\Delta H_R=\Delta H_R(T_R)+\int_{T_R}^{T}\Delta CpdT=\Delta H_R(T_R)+\Delta\alpha(T-T_R)+\frac{\Delta\beta}{2}(T^2-T_R^2)\\
&\quad+\frac{\Delta\gamma}{3}(T^3-T_R^3)+\frac{\Delta\xi}{4}(T^4-T_R^4)\\
&\Delta\alpha=\alpha_{SO_3}-\frac{1}{2}\alpha_{O_2}-\alpha_{SO_2}=16.370-0.5(28.106)-23.852=-21.535\\
&\Delta\beta=0.07892\\
&\Delta\gamma=-7{,}112\cdot10^{-5}\\
&\Delta\xi=2{,}4467\cdot10^{-8}\\
&\Delta H_R=-98480-21.535(T-298)+0.0395(T^2-298^2)-2.371\cdot10^{-5}(T^3-298^3)\\
&\quad+6.11675\cdot10^{-9}(T^4-298^4)\\
&\Delta Cp=\Delta\alpha+\Delta\beta T+\Delta\gamma T^2=-21.535+0.0789T-7.112\cdot10^{-5}T^2+2.447\cdot10^{-8}T^3\\
&\sum\Theta_iCp_i=300.85-0.0402T+0.00018T^2-9.071\cdot10^{-8}T^3
\end{aligned} \tag{4.97}$$

$$\frac{U\pi D}{\rho_bA_c}=\frac{4U}{\rho_bD}=\frac{4\cdot17\frac{J}{s\cdot m^2K}}{542\frac{kg}{m^3}0.045m}=2.87\frac{J}{s\cdot kgK} \tag{4.98}$$

We use the following information for the energy balance:

$$\begin{aligned} &U = 17J/(s\cdot m^2K) \\ &A_c = 0.0016m^2 \\ &T_o = 750K \\ &\rho_b = 542kg/m^3 \end{aligned} \tag{4.99}$$

$$\frac{dT}{dW} = \frac{9.3(T_a - T) + (-r_A)\left[\Delta H_R(T_R) + \int_{T_R}^{T} \Delta Cp dT\right]}{F_{Ao}\left(\sum_{i=1}^{n} \Theta_i Cp_i + x\Delta Cp\right)} \tag{4.100}$$

The pressure drop along the tube is given by Ergun's equation:

$$\frac{dP}{dz} = -\frac{G(1-\phi)}{\rho D_p \phi^3}\left[\frac{150(1-\phi)\mu}{D_p} + 1.752G\right] \tag{4.101}$$

G is the superficial mass velocity (kg/m^2 s), ϕ is the porosity of the bed, D_p the particle diameter (m), μ the viscosity of the gas (Pa·s), ρ is the gas density (kg/m^3) that can be found in Fogler [15].

$$\begin{aligned} &\phi = 0.45 \\ &\rho_o = 0.866\ kg/m^3 \\ &P_o = 202650\ Pa \\ &D_p = 0.00457\ m \\ &\mu = 3.72\cdot 10^{-5}\ Pa\cdot s \end{aligned} \tag{4.102}$$

$$\begin{aligned} &\frac{dP}{dz} = -\frac{G(1-\phi)}{\rho_o g_c D_p \phi^3}\left[\frac{150(1-\phi)\mu}{D_p} + 1.752G\right]\frac{P_o}{P}\frac{T}{T_o}\frac{F_T}{F_{To}} \\ &G = \frac{\sum_i F_{io} M_i}{A_c} = 0.433\frac{kg}{m^2 s} \\ &A_c = \pi D^2 4 \\ &W = \rho_b A_c z \\ &\frac{dP}{dW} = -\frac{G(1-\phi)(1+\varepsilon x)}{\rho_b A_c \rho_o g_c D_p \phi^3}\left[\frac{150(1-\phi)\mu}{D_p} + 1.752G\right]\frac{P_o}{P}\frac{T}{T_o} \\ &\mu \approx cte \end{aligned} \tag{4.103}$$

Now that we have the model developed we implement it in MATLAB®. We create a script to write the ode statement. In this script we also include the statement for plotting the results. The ode statement includes the function React, the range of the independent variable, in this case the weight of catalyst, and finally the initial values for the variables involved, the concentration of SO_2, the temperature and the pressure. In the function file we write the model developed above including an if-then-else-end statement in order to account for the fact that the kinetics of the reaction only depends on the conversion if the conversion is higher than 0.05. We start by relating the generic dependent variable x with the actual variables so that our equations look closer to the ones developed above. Next, we just type the different parameters of the model as we have defined above. Finally, we write the three differential equations which must start with the same word as the one we have used to produce our function, in this case:

```
Reactor=Reac(w,x)
```

Thus we use Reactor (i:1) for the differential equation i.

In order to plot the results we use subplot. This command allows us to present all the Figures in one by deciding the location of the different plots.

Subplot (*n*,*m*,*x*).

Where n is the number of rows of Figures, m the number of columns of Figures and x is the location. Bear in mind that the order is sequential, not like in a matrix. Thus, for a subplot 2,2 in the Figure we present the x location. Figure 4.15 shows the results of the problem.

1	2
3	4

M file: ReactorSO$_2$

```
[a,b]=ode15s('Reac',[0 20],[0,750,202650]);

plot(a,b(:,1))
xlabel('W (kg)')
ylabel('X')
Figure
plot(a,b(:,2))
xlabel('W')
ylabel('T (K)')
Figure
plot(a,b(:,3))
xlabel('W')
ylabel('P (Pa)')
```

M file Reac

```
function Reactor=Reac(w,x)
X=x(1);
T=x(2);
Presion=x(3);
```

```
k=9.8692e-3*exp(-176008/(1.8*(T-273.15)+491.67) -110.1*log
((1.8*(T-273.15)+491.67))+912.8);
Kp=0.0031415*exp(42311/(1.987*(1.8*(T-273.15)+491.67) )-11.24);
Ta=700;
Pto=202650;
ySO2o=0.10;
yO2o=0.11;
yN2o=0.79;
PhiO2=yO2o/ySO2o;
PSO2o=Pto*ySO2o;
Fto=0.02153;
Fao=Fto*ySO2o;
epsilon=-0.05;
G=0.433;
ph=0.45;
rhoo=0.866;
Dp=0.00457;
visc=3.72e-5;
Ac=0.0016;
rhob=542;
To=770;
sumCp=300.85-0.0402*T+0.00018*T^2-9.071e-8*T^3;
dCp=-21.535+0.0789*T-7.112*10^(-5)*T^2+2.447e-8*T^3;
deltaHr=-98480-21.535*(T-298)+0.0395*(T^2-298^2)-2.371*10^
(-5)*(T^3-298^3)+6.117*10^(-9)*(T^4-298^4);
Ucorreg=9.3;
if X<0.05;
   ra=-k*((1-0.05)/0.05)^(0.5)*(PSO2o*((PhiO2-0.5*X)/
   (1+epsilon*0.05))*(Presion/Pto)-(0.05/(Kp*(1-0.05)))^2);
   else
   ra=-k*((1-X)/X)^(0.5)*(PSO2o*((PhiO2-0.5*X)/
   (1+epsilon*X))*(Presion/Pto)-(X/(Kp*(1-X)))^2);
end
Reactor(1,1)=-ra/Fao;
Reactor(2,1)=(Ucorreg*(Ta-T)+(-ra)*(-deltaHr))/
(Fao*(sumCp+X*dCp));
Reactor(3,1)=-G*(1-ph)*(1+epsilon*X)*Pto*T*(150*(1-ph)*visc/
Dp+1.75*G)/(rhob*Ac*rhoo*Dp*ph^3*To*Presion);
```

We need 5kg of catalyst to obtain 85% conversion in the reaction. Comparing this reactor with the one in Chapter 11, we can see that the refrigeration allows higher conversion compared to adiabatic reactors where multi-bed designs are needed to drive the equilibrium to SO_3.

QUESTIONS

- Evaluate the effect of the initial concentration of SO_2 on the operation of the reactor.
- Use an error function to compute the catalyst weight as a variable.

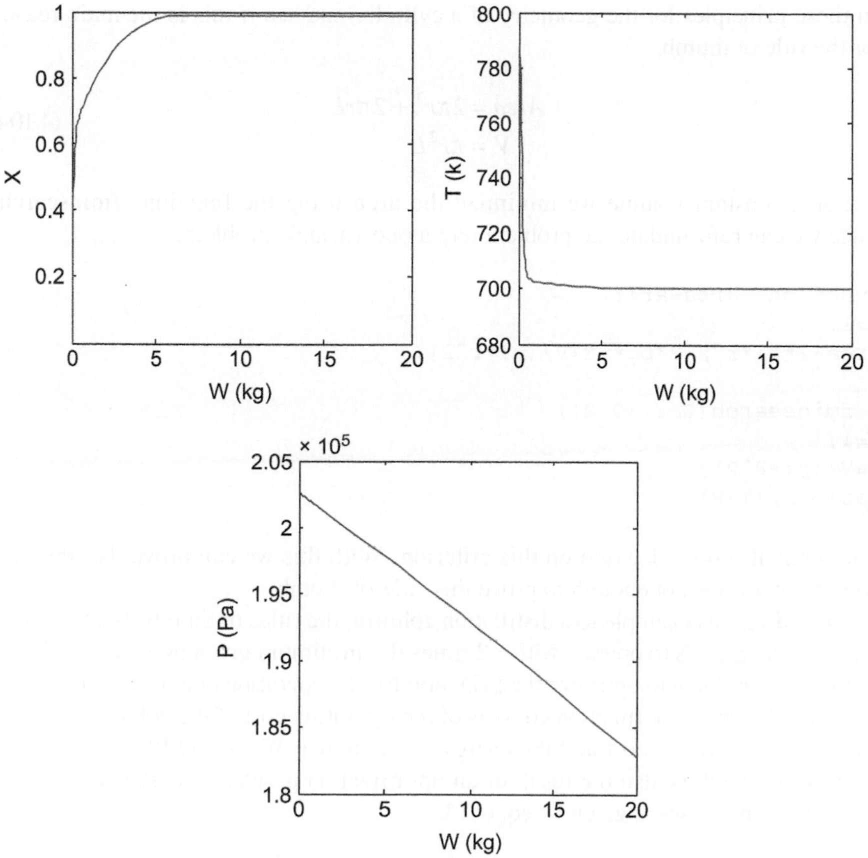

FIGURE 4.15 Simulation of a fixed bed pipe reactor for the oxidation of SO_2.

- Determine the effect of the heat transfer coefficient on the temperature profile.
- Evaluate the effect of the operating pressure on the equilibrium and the final conversion.

4.3.4 Optimization

We are using here the optimization routines in order to validate and discuss a couple of typical rules of thumb. The first one dealing with equipment cost, the second one with the minimum reflux ratio in a binary distillation column.

Rules of thumb for tank sizes dictate that typically L/D ranges from 2 to 4 [17]. The cost of a vessel is based on the cost of the metal that is proportional to the superficial area. In this small example we are going to determine the optimal ratio based on these principles for the geometry of a cylinder and see if this is the main reason for the rule of thumb.

$$\begin{aligned} Area &= 2\pi r^2 + 2\pi r L \\ V &= \pi r^2 L \end{aligned} \tag{4.104}$$

For a constant volume we minimize the area using the function "fminsearch" since we can reformulate the problem into a one variable problem.

```
function  area=ar(r)
V=1;
area=2*pi*r^2+2*pi*r*(V/(pi*r^2));

R=fminsearch(@ar,[0.2])
V=1;
L=V/(pi*R^2);
ratio=L/(2*R);
```

The optimal ratio is 1 based on this criterion. With this we can prove that the economic criterion is not enough to prove this rule of thumb.

Extending this example to a distillation column, the rules of thumb about the minimum reflux ratio is to operate with 1.2 times the minimum given by Fenske [17]. We formulate a problem to optimize the L/D ratio for the operation of a binary distillation column. The objective function consists of the operating cost of the column involving the annualized vessel cost and the energy in the boiler. We would like to prove that the rule of thumb is also the result of an optimization problem. Fenske equation for comparison purposes is given by eq. (4.132):

$$\left.\frac{L}{D}\right|_{\min} = \frac{1}{\alpha - 1}\left[\frac{x_d}{x_f} - \frac{\alpha(1-x_d)}{(1-x_f)}\right] \tag{4.105}$$

We consider the operation of a distillation column for 100 kmol/s of feed, where $x_d = 0.85$, $x_w = 0.05$ and $x_f = 0.4$. The feed to the column is saturated liquid. We assume constant relative volatility equal to 2.3. The mean operating temperature is assumed to be 70°C and 1 atm of pressure. The liquid in the boiler has a $\lambda = 800$ kJ/kmol.

Based on the same equations for the distillation columns design presented in Section 4.3.2 we formulate the problem to determine the number of trays as function of the L/D ratio. We use McCabe – Thiele method to compute the number of theoretical trays, where the equilibrium composition is given by eq. (4.106):

$$y_{eq} = \frac{\alpha x_{eq}}{1 + (\alpha - 1)x_{eq}} \tag{4.106}$$

And the operating lines for the rectifying and striping sections as given by eq. (4.30) (4.31).

For the sake of the length we do not write the calculation of W and D as function of their composition again, the reader can find the definition in Section 4.3.2 of this chapter.

The cost of the column is computed as the cost of the steel as follows:

$$Investment = \left(Area + Ntrays \cdot \pi \cdot \left(\frac{Diameter}{2} \right)^2 \right) \cdot width \cdot density \cdot csteel \tag{4.107}$$

We calculate the area of a vessel, assuming spherical ends. The diameter of the tower is calculated as eq. (4.108)

$$\begin{aligned} Diameter &= \sqrt{\frac{4 \cdot V \cdot 22.4 \cdot (273.15 + Temp) \cdot 760}{\pi \cdot u \cdot Pressure \cdot 273}} \\ u &= K \sqrt{\frac{\rho_L - \rho_G}{\rho_L}} \\ K &= 0.050\ m/s \\ \rho_L &= 850\ kg/m^3 \\ \rho_G &= 2\ kg/m^3 \end{aligned} \tag{4.108}$$

Where we can compute V from the equations in section 3.2 as:

$$\mathrm{V} = \left(1 + \frac{L}{D}\right) \cdot \mathrm{F} \cdot \left(\frac{(\mathrm{x_f} - \mathrm{x_w})}{(\mathrm{x_d} - \mathrm{x_w})} \right); \tag{4.109}$$

Where the steel density is assumed to be 8000 kg/m³, the cost of the steel 4 €/kg and the area is calculated as eq. (4.110) where the height of the column is given by the number of trays, considered separated by 0.6 m (2 ft). Furthermore, 1m below and 1m above the last and first trays are also added [17]. The efficiency of the theoretical trays, eff, is assumed to be 0.75.

$$\begin{aligned} Area &= 4\pi \left(\frac{Diameter}{2} \right)^2 + \pi Diameter \cdot height; \\ height &= \left(\frac{Ntrays - 1}{eff} + 1 \right) \cdot 0.6 + 2; \end{aligned} \tag{4.110}$$

The width of the steel is computed as eq. (4.111):

$$width(m) = \frac{P(psi) \cdot radius(m)}{S(psi) \cdot E - 0.6P(psi)} + 0.0032 \tag{4.111}$$

Where P is the operating pressure (psi), E the welding efficiency (0.8), and S is the tensor stress for the material, 12000 psi. We add 3.2 mm for corrosion allowance. The energy in the boiler is given by:

$$\mathrm{Energy} = \lambda * \left(1 + \frac{L}{D}\right) \cdot \mathrm{F} \cdot \left(\frac{(\mathrm{x_f} - \mathrm{x_w})}{(\mathrm{x_d} - \mathrm{x_w})} \right) \cdot \frac{1}{\lambda_{steam}} \cos t_{steam} \tag{4.112}$$

With $cost_{steam} = 0.02$ €/kg and $\lambda_{steam} = 1800$ kJ/kg
We minimize the cost of the operation as:

$$Cost = \frac{1}{3} Investment + Energy \tag{4.113}$$

Where our only variable is L/D. We define a script to write the "fminsearch" statement:

```
CalcLtoD=fminsearch(@Mincolumn,[1.5]);
alpha=2.3;
xd=0.85;
xw=0.05;
xf=0.4;
LtoDmin=(1/(alpha-1))*(xd/xf-alpha*(1-xd)/(1-xf));
LtoDaprox=1.2*LtoDmin
CalcLtoD
```

And we define a function to optimize including the McCabe Thiele method for the number of trays assuming constant relative volatility. We compute the operating lines as function of the compositions of the feed, the distillate and the residue. Next, we use a while statement to compute the number of theoretical trays. Finally we compute the cost of the vessel and that of the energy required at the boiler.

```
      function cost=Mincolum(LtoD)
alpha=2.3;
LtoV=LtoD/(LtoD+1);
DtoV=1/(1+LtoD);
F=100;
Temp=70;
Pressure=760;
Lambda=800;
eff=0.75;
xd=0.85;
xw=0.05;
xf=0.4;

LptoVp=(LtoD+((xd-xw)/(xf-xw)))/(1+LtoD);
WtoVp=( ((xd-xw)/(xf-xw))-1 )/(1+LtoD);
Np=1;
yeq(1,1)=xd;
xeq(1,1)=yeq(1,1)/(alpha*(1-yeq(1,1))+yeq(1,1));

while xeq(Np)>xw
             if xeq(1,Np) > xf
               yeq(1,Np+1)=(LtoV)*xeq(1,Np)+DtoV*xd;
           else
               yeq(1,Np+1)=LptoVp*xeq(1,Np)-WtoVp*xw;
           end
```

```
        xeq(1,Np+1)=yeq(1,Np+1)/(alpha*(1-yeq(1,Np+1))+yeq
        (1,Np+1));
        Np=Np+1;

end

V=(1+LtoD)*F*((xf-xw)/(xd-xw));
vel=0.05*((850-2)/850)^0.5;
Diameter=(4*V*22.4*(273+Temp)*760/(pi*vel*Pressure*273))
^0.5;
Energy=Lambda*(1+LtoD)*F*((xf-xw)/(xd-xw));
height=((Np-1)/eff+1)*0.6+2;
Area=4*pi*(Diameter/2)^2+pi*Diameter*height;
csteel=4;
density=8000;
width=(Pressure/760)*14.7*(Diameter*0.5)/
(12000*0.8-0.6*(Pressure/760)*14.7)+0.0032;
lambdawa=1800;
coststeam=0.02;

cost=(1/3)*(Area+Np*pi*(Diameter/2)^2)*width*density*csteel
+(Energy/lambdawa)*coststeam;
```

The L/D_{min} calculated from the optimization results in 1.55 and the estimated is 1.43 thus we can prove this rule of thumb based on the optimization of the operation of the distillation column.

4.3.5 Simulink

Simulink (Simulation and Link) is a software add-on to MATLAB® based on the concept of block diagrams that are common in the control engineering areas. It is an environment for dynamic simulation and process control.

We can start Simulink either by clicking on the "Simulink library" icon on the menu bar of MATLAB® or just typing:

>>simulink

in the command window, see Figure 4.16.

We have three sections. The explanatory window is at the top where the description of the block can be read. The Browser shows to the left the groups of blocks available within Simulink. For each group, we can click and the different blocks appear in the main window. Of particular interest are the ones classified as "Commonly used Blocks". Among them we can find the sources of signals "Sources", the sinks "Sinks" the operators.

Sources: Define the kind of source of signal to the system, i.e. Ramp, step, ground, pulse, random number, sine wave, clock, constant, from file, from workspace, signal generator, etc.

Sinks: Results from the simulation, display, floating scope, scope, stop simulation, to file, to workspace, XY graph (allows plotting graphs from the results of the calculation.), out 1, terminator.

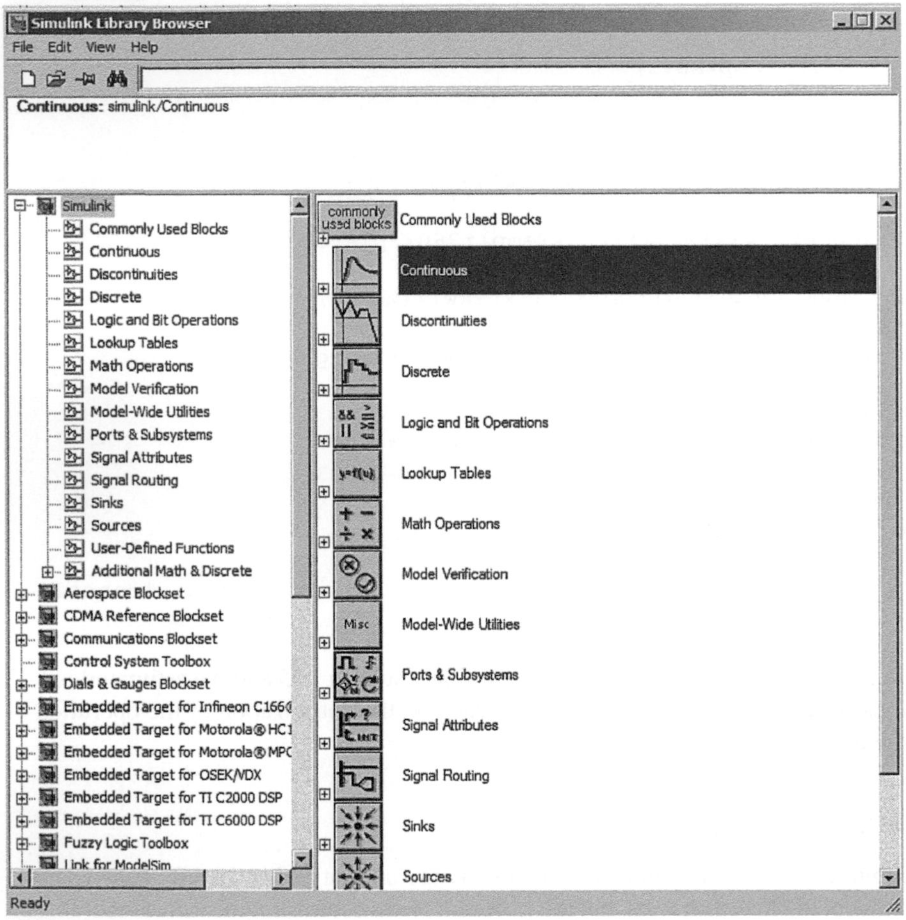

FIGURE 4.16 Simulink library.

Continuous: including derivative, integrator, state-space, transfer function, transport delay, variable transport delay, zero-pole.

A particular feature is the fact that you can define the parameters of the transfer function as variables and input those values in the command window before running the simulation in Simulink. In that sense we save clicking on the block anytime we want to change them and evaluate the effect of the new parameters.

Operators: To build our block diagram including, add, divide, gain, sign, math function, product, sum, subtract, sum of elements, product of elements, minmax, assignment, algebraic constraint, etc.

User defined functions: So that we can write our models in MATLAB® and use them within a control structure for example.

Each of the blocks can contain a subsystem inside which is helpful for big problems. We only need to select a number of blocks and with the right button of the mouse, click and select create subsystem.

In order to start we go to New→ Model and a new window appears. To the bottom right hand corner we can read ode45. It is the default ode solver that Simulink is going to use. You may want to change it depending on the problem. For doing so we go to:

Simulation→ Configuration Parameters.

And to the left-hand side, Solver.

4.3.5.1 Getting Started: First and Second Order Response

A typical example in process control is a tank that holds a liquid and whose exiting flow rate may be due to gravity or because of a pump. This kind of system is modeled based on a mass balance in unsteady state:

$$A\rho\frac{dh}{dt} = F_{in} - F_{out} = F_{in} - kh(t) \tag{4.114}$$

The transfer function of such a system, G, can be calculated using Laplace transforms to become:

$$G = \frac{K}{\tau s + 1} \tag{4.115}$$

Where "K" is the gain and "τ" the time constant. Let us simulate the response to such a tank to a step input with a delay.

To draw the block diagram just select the block in the menu and drag it to the model window. We have a number of options to organize blocks whose dialogue box can be opened when clicking with the right button as can be seen in Figure 4.17. From rotating or flipping the block to hide the name, etc.

To link the blocks just go with the cursor to the ">" sign of each block, click on it, and move the cursor to the > symbol of the next block. There are alternative wiring possibilities, the so call automatic connection; we can link two blocks by selecting the source, holding down the "Ctrl" key while we select the destination block with the length button of the mouse. Finally to connect from a wire to a block we need to hold "Ctrl" while we select the wire with the left button of the mouse and then do the same for the input of the destination block.

Once the diagram is complete we input the parameters of the different blocks by clicking on them, see Figure 4.18.

Scope block will report the results of the simulation. Furthermore, they will be stored in the workspace. We select a system with a step source, a time delay, a model of the system given by a simple transfer function and a scope as final monitoring block. In order to run we can click on the arrow, see Figure 4.19, and we obtain the results shown in Figure 4.20.

QUESTION

- Evaluate the effect of the gain and the time constant on the response of the system.

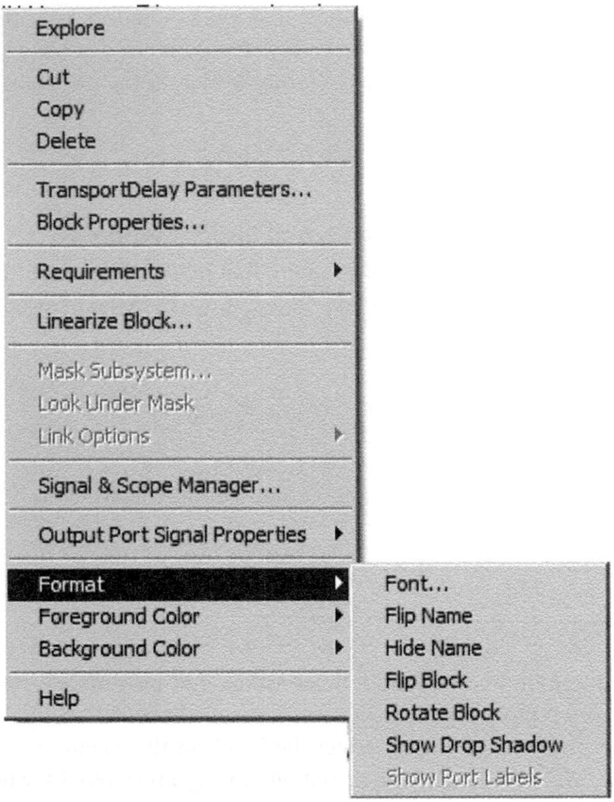

FIGURE 4.17 General options for a block.

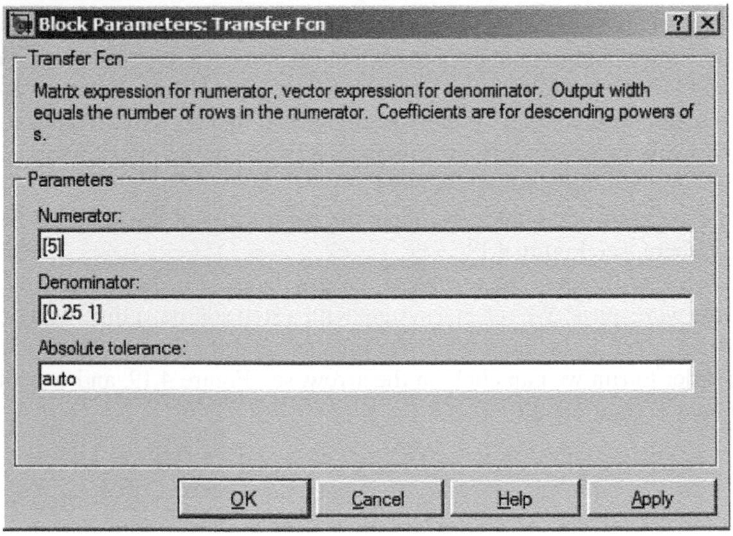

FIGURE 4.18 Dialogue box for inputting parameters.

FIGURE 4.19 Running a simulation with Simulink.

When instead of a tank we have a number of them in series or if together with our system we have a controller or finally a system that itself is a second order we find the phenomena of oscillations in the response. Depending on the damped factor and the frequency we can have a system that behaves like underdamped, over damped or critically damped. The transfer function of this kind of systems is as follows:

$$G = \frac{K}{\tau s^2 + 2\delta\tau s + 1} \tag{4.116}$$

Let us evaluate how it behaves under a step input signal. We go to Simulation → Configuration Parameters and select 60 as final calculation time or directly in the simulation window, change 10 in Figure 4.20 by 60. In Figure 4.21 we present a small sensitivity analysis of the effect of the parameters on the response of the second order system. As τ increases the system perturbation extends in time and eventually it becomes unstable.

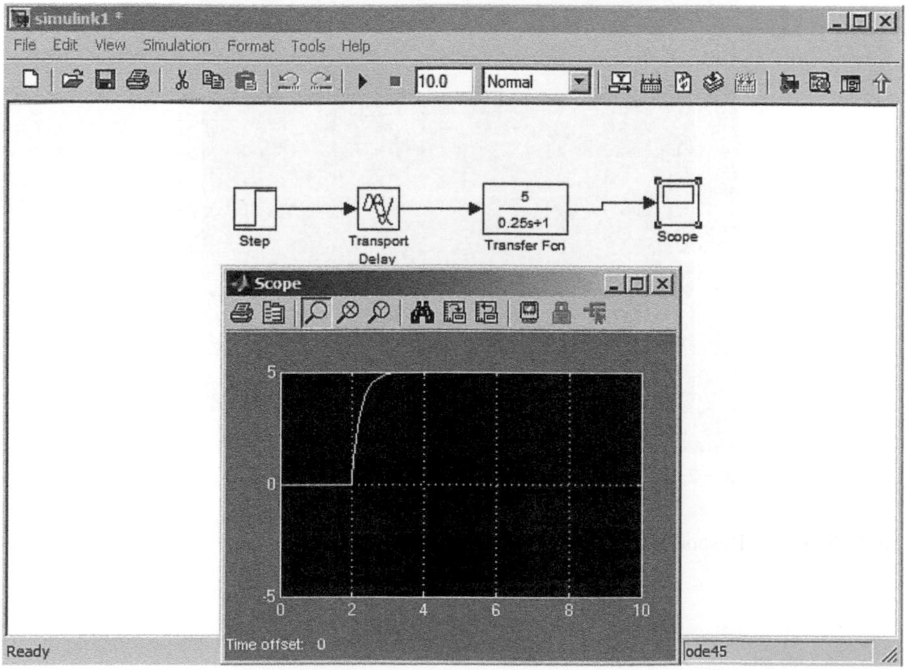

FIGURE 4.20 Results of the step input to the system.

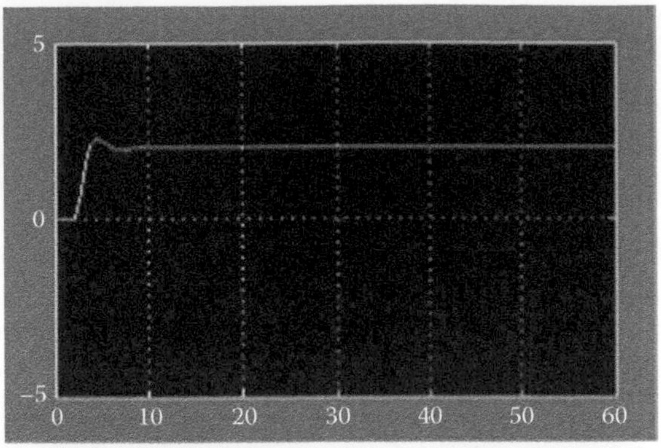

K = 2; τ = 0.5; 2τδ = 0.75

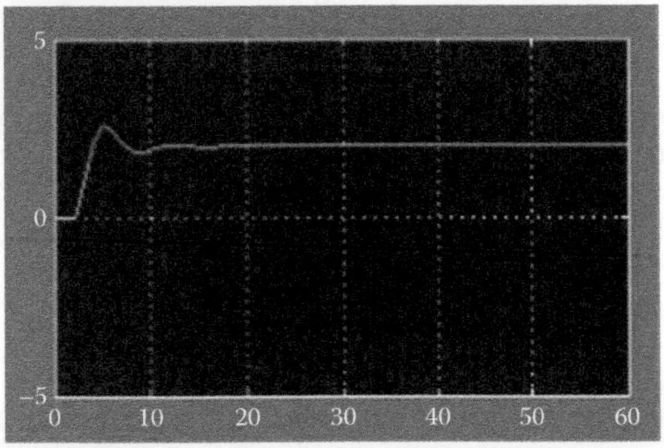

K = 2; τ = 1; 2τδ = 0.75

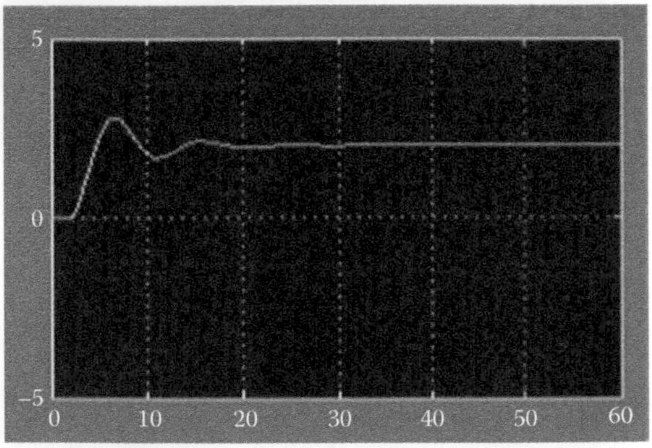

K = 2; τ = 2; 2τδ = 0.75

FIGURE 4.21 Response of a second order system.

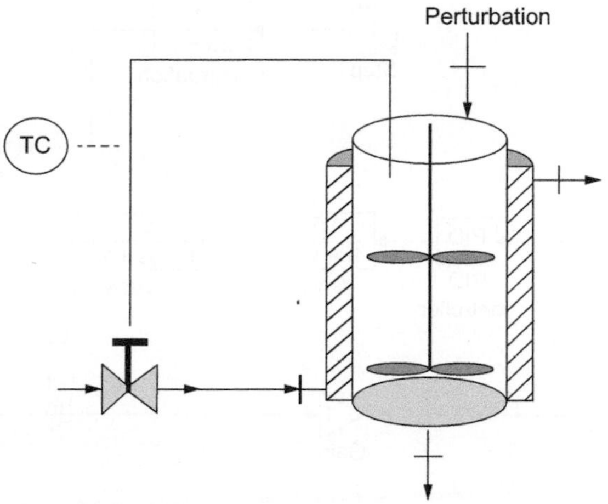

FIGURE 4.22 Control loop for a jacketed reactor.

4.3.5.2 Control Loops

Many chemical reactions are non-isothermal. In order to carry out a particular reaction we must maintain the temperature of the reaction volume within some limits to avoid species decomposition. In particular, exothermic reactions are problematic. We need to be able to eliminate the generated energy so that we can maintain the reaction under control. Let's consider a system of a jacketed stirred tank reactor where we need to control the temperature by means of the flowrate through the jacket and that there is a perturbation in the flow that feeds the reactor. In our system we have a controller, a valve, the actuator, the system, our reactor, the perturbation and a sensor, see Figure 4.22. We assume that the control valve has a transfer function of the form [18] given by:

$$G_{valve} = \frac{K_{valve}}{\tau_{valve}s + 1} \tag{4.117}$$

The process and the perturbation:

$$G_{Process} = \frac{K_{proc}}{\tau_{proc}s + 1} \tag{4.118}$$

$$G_{Per} = \frac{K_{per}}{\tau_{per}s + 1} \tag{4.119}$$

The sensor transfer function is a gain and the controller, that has no derivative contribution, is a PI. The block diagram looks like the following Figure 4.23 where we can see the transfer functions and their values. We run the model and the results can be seen in Figure 4.24. We see that we can control the perturbation.

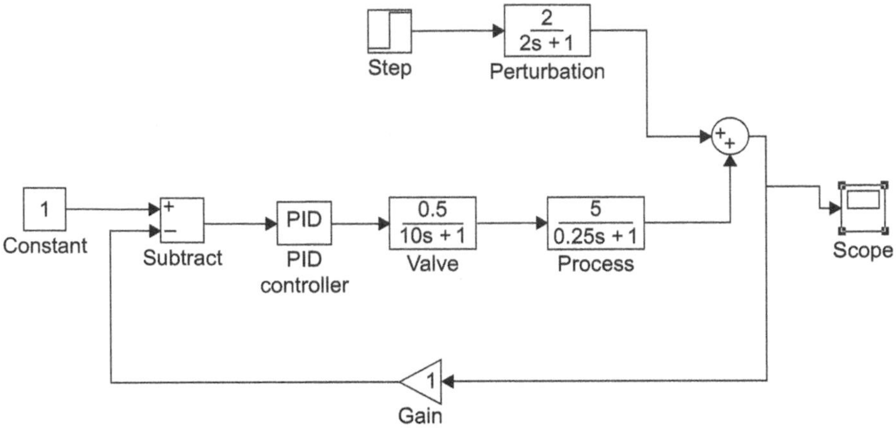

FIGURE 4.23 Control loop for a jacketed stirred tank.

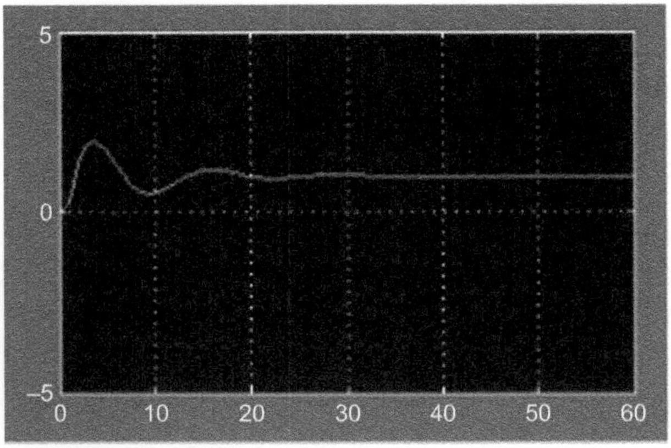

FIGURE 4.24 Perturbation damping in a closed loop control system.

QUESTIONS

- Evaluate the effect of the derivative control.
- Study the effect of the time constant in the valve. Can the system become unstable?

REFERENCES

1. Biegler, L.T. (2010) *Nonlinear Programming: Concepts, Algorithms, and Application to Chemical Engineering Processes*, SIAM, Philadelphia, PA.
2. Kincaid, D., Cheaney, W. (1991) *Numerical Analysis*, 3rd edn. Addison-Wesley, New York.
3. Nocedal, J., Wright., S.J. (1999) *Numerical Optimization*. Springer-Verlag, New York.

4. Biegler, L.T., Grossmann, I.E. (2004) Retrospective on Optimization. *Comput. Chem. Eng.*, 28(8), 1169–1192.
5. Bird, R.B., Stewart, W.E., Lightfoot, E.N. (2002) *Transport Phenomena*, 2nd ed. John Wiley & Sons, New York.
6. McCabe, W.L., Smith, J.C., Harriot, P. (1999) *Unit Operations of Chemical Engineering*, 7th edn. McGraw-Hill, New York.
7. Sinnott, R.K. (1999) *Coulson and Richardson: Chemical Engineering*, 3rd edn. Butterworth-Heinemann, Singapore.
8. Stickler, M., Panke, D., Hamielec, A.E. (1984) Polymerization of methyl methacrylate up to high degrees of conversion: Experimental investigation of the diffusion-controlled polymerization, *J Polym Sci: Polym Chem Ed*, 22, 2243–2253.
9. Seth, V., Gupta, S.K. (1995) Free radical polymerizations associated with the Trommsdorff effect under semi-batch reactor conditions: An improvement model. *J Polym Eng*, 15(3–4), 283–323.
10. Ghosh, P., Gupta, S.K., Saraf, D.N., (1998) An experimental study on bulk and solution polymerization of methyl methacrylate with responses to step changes in temperature. *Chem Eng J*, 70, 25–35.
11. Ross, R.T., Laurence, R.L. (1976) Gel effect and free volume in the bulk polymerization of methyl methacrylate. *AIChE Symp Ser*, 72, 74–79.
12. Ray, A.B., Saraf, D.N., Gupta, S.K. (1995) Free radical polymerizations associated with the Trommsdorff effect under semi-batch reactor conditions. I: Modelling. *Polym Eng Sci*, 35, 1290–1299.
13. Martín, M. (2003) *Production Plant of PMMA Beads* (in Spanish). Final degree project. University of Salamanca, Salamanca, Spain.
14. Harrer, T.S. (1969) *Kirk Othmer Encyclopedia of Chemical Technology*, 2nd edn. Vol. 9, p. 470. Wiley-Interscience, New York.
15. Fogler, S.H. (2000) *Elements of Chemical Reaction Engineering*, 3rd edn. Prentice-Hall International Series, Upper Saddle River, NJ.
16. Eklund, R. (1956) *The Rate of Oxidation of Sulfur Dioxide with a Commercial Vanadium Catalyst*. Almqvist & Wiksells Boktr, Uppsala, Sweden.
17. Walas. S.M. (1990) *Chemical Process Equipment: Selection and Design*, Butterworth-Heinemann Series in Chemical Engineering, Newton, MA.
18. Stephanopoulos, G. (1984) *Chemical Process Control: An Introduction to Theory and Practice*. Prentice Hall, Englewood Cliffs, NJ.

5 PTC Mathcad® for Chemical Engineering

Kristen A. Severson, Alberto Martínez, and Mariano Martín

CONTENTS

5.1 INTRODUCTION

Mathcad® is a symbolic software tool for engineering applications. It has a "what you see is what you get" interface which allows the user to work in Mathcad more naturally. It is important for an engineer or scientist to be able to find solutions as well as to be able to communicate the results. Mathcad's strengths, among others, include a readable format for displaying equations, built-in unit functionality and the ability to deal with symbolic calculations.

This text demonstrates some of Mathcad's capabilities applied to solving basic problems in chemical engineering. Please note that throughout this chapter, regions of Mathcad code are presented with a different font mainly within figures.

5.2 MATHCAD FUNDAMENTALS

5.2.1 Mathcad Workspace

The Mathcad workspace allows the user to write equations anywhere in the worksheet. The only restriction is that equations will be executed from left to right and top to bottom (with a few noted exceptions). Each block of code (e.g., a variable or equation definition) can be dragged and dropped anywhere on the screen, or the space bar can be used to add space between blocks of code, see Figure 5.1.

Mathcad recognizes math regions with two blue editing lines. When entering comments, these blue editing lines will disappear as the user continues to type text, as shown below:

$$x := y + z$$

Above I define the variable x which is equal to y + z|

All equation blocks are evaluated continuously; there is no need for the user to run the code. Additionally, any error messages will appear immediately when the line is written, denoted by red text. If the user hovers over the red text, a description of the error will appear. A sample error message, for a variable that is not defined, is shown below.

$$\mathrm{Temp} := 300\mathrm{K} \quad \mathrm{Vol} := 1\mathrm{L} \quad R_{\mathrm{gasconst}} := 8.214\frac{\mathrm{J}}{\mathrm{mol}\cdot\mathrm{K}}$$

$$n := \frac{P \cdot V}{R}$$

This variable is undefined.

Mathcad has a variety of toolbars such as: standard, math, calculator, graph and matrix. These toolbars can be docked on the top ribbon or floating in the workspace. The user is free to customize the view. To see a list of toolbars, choose View > Toolbars.

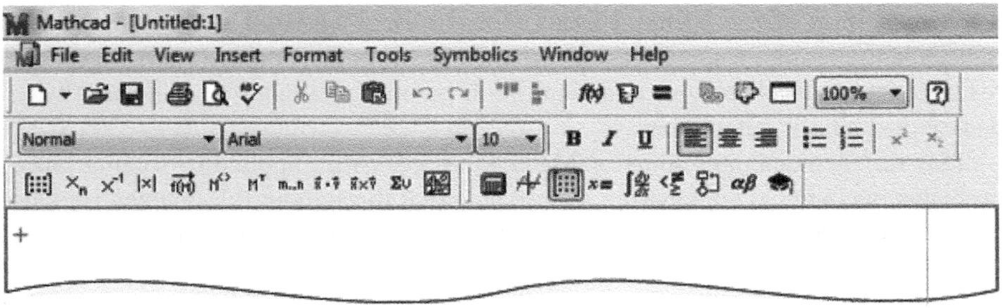

FIGURE 5.1 Mathcad® workspace.

5.2.2 Equal Signs & Defining Equations

Mathcad has four types of equals signs, each of which has a different function. The first is to define a variable or equation. For instance, to define a density ρ, insert the Greek character rho and the equals key. Alternatively, you can enter rho then type ":"

$$\rho := 1.5 \frac{kg}{L}$$

The initial definition of a variable will always appear with this type of equals sign (:=). This can also be used to define units that are not built in (see Section 5.2.3 for more detail). For instance,

$$mm\_Hg := \frac{in\_Hg}{25.4}$$

The second type of equals sign is also inputted using the equals key; however, it evaluates a variable or equation. Continuing with the example above, if we again type ρ=, the following appears:

$$\rho = 1.5 \times 10^3 \frac{kg}{m^3}$$

The third type of equals sign is used to symbolically define a variable. It is entered by typing CRTL + =. This format is used in solve blocks (See Section 5.2.5). Typically Mathcad will show an error when using an undefined variable; this is not the case in symbolic equations, however.

$$P \cdot V = n \cdot R \cdot T$$

The final type of equals sign is for a global assignment of a variable. This is similar to the first type, but the expression will be evaluated first regardless of where it is in the worksheet. This type of equals sign can be entered by typing ~.

$$P \equiv 1atm$$

The two types of assignment equals signs and the evaluation equals sign are all available in the Evaluation toolbar. The symbolic equals sign is available under the Boolean toolbar. Please see Figure 5.2.

To modify an equation after it has been entered, select it and make the desired changes. To change the definition of an equation within a workspace, re-write it at another point in the workspace with an assignment equals sign and it will overwrite the previous definition for the remainder of the workspace.

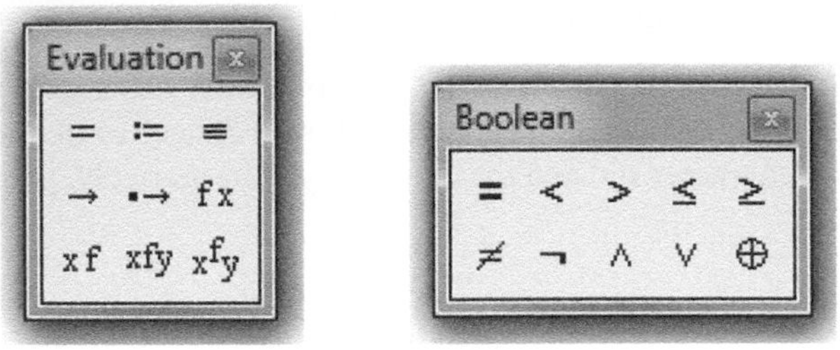

FIGURE 5.2 Toolbars.

5.2.3 Units

As noted above, Mathcad offers built-in definitions for various engineering units. The user can search for available units by clicking Insert > Unit or using Ctrl + U, see Figure 5.3.

A menu with unit choices will appear and the user can make a selection, see Figure 5.4.

In the event a unit is not available, the user can create units using the predefined ones, as shown with mmHg in Section 5.2.2. Mathcad will automatically convert units to SI; however, if another unit system is desired, the user can change the units that are displayed by clicking on the answer and typing the desired unit in the black box on the right-hand side as shown in Figure 5.5.

To change the units more generally, select Tools > Worksheet Options and alter the unit system as shown in Figure 5.6.

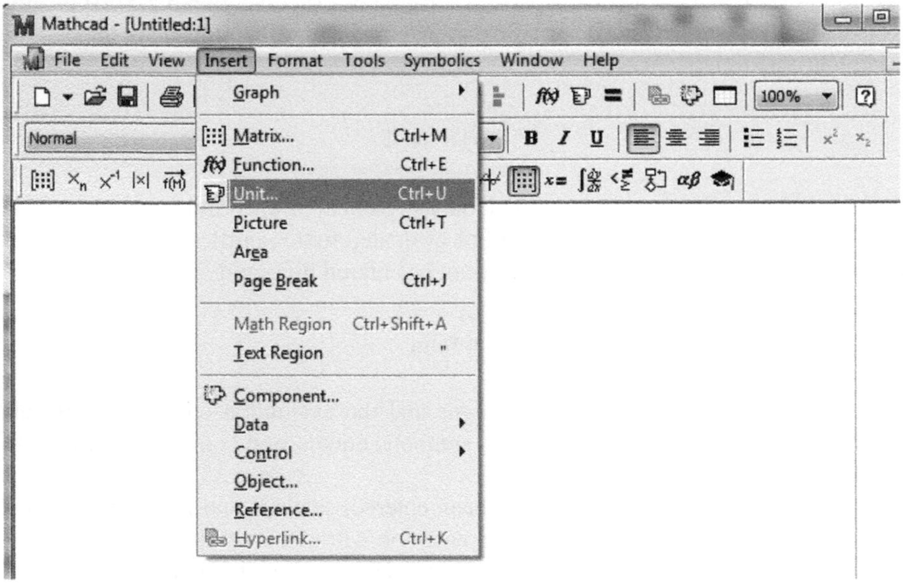

FIGURE 5.3 Units menu.

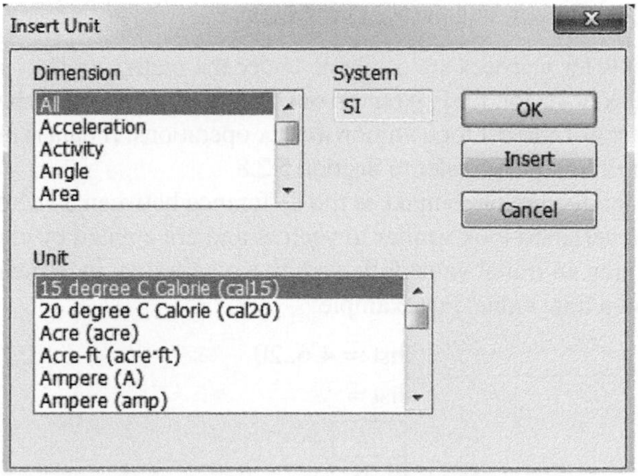

FIGURE 5.4 Units menu (II).

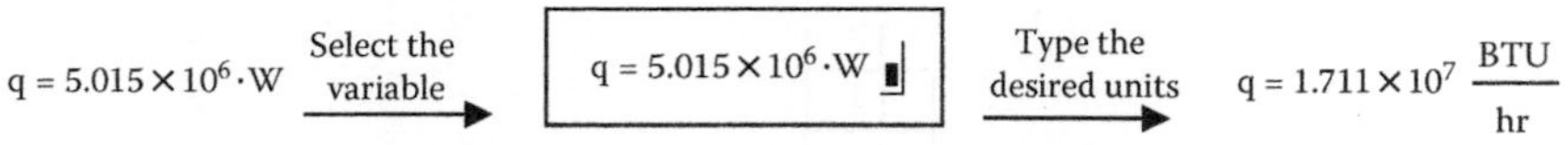

FIGURE 5.5 Altering displayed units.

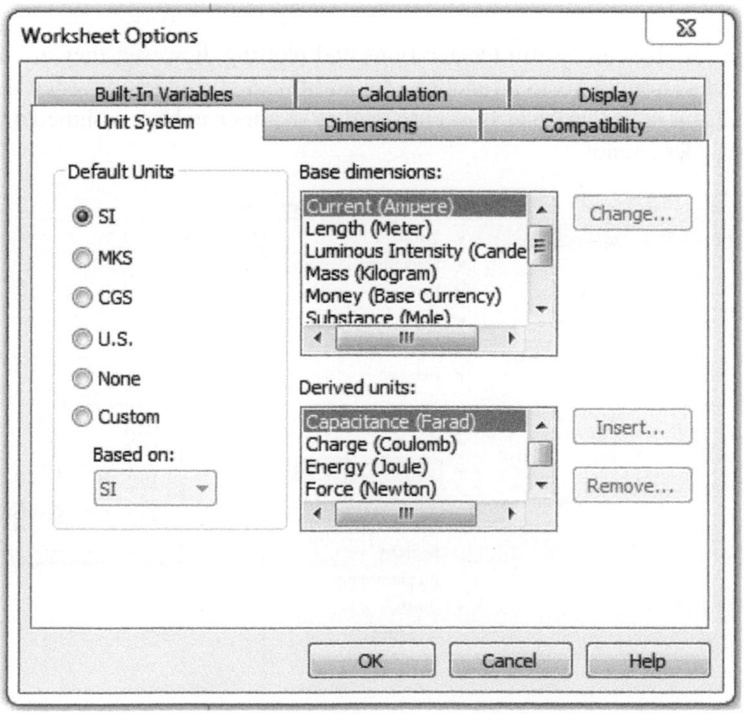

FIGURE 5.6 Worksheet Options menu.

5.2.4 Vectors, Range Variables, and Matrices

Many useful tools for matrices are available under the matrix toolbar, see Figure 5.7. To create a matrix or vector, the [:::] button or Ctrl + M is used and each value is populated. Please refer to Table 5.1 for common matrix operations. If data is being imported into a matrix or vector, please refer to Section 5.2.8.

One important distinction to make is the difference between a vector and a range variable. Range variables look similar to vectors and are created by choosing a variable then assigning an initial value followed by a comma, an increment followed by a semicolon and a final value. For example:

list := 4,6..20

list =

4
6
8
10
12
14
16
18
20

Range variables are useful for iterations and plotting; however, they are not recommended as arguments to a function. An additional consideration when deciding between using range variables or vectors is that vectors can have assigned units and range variables cannot.

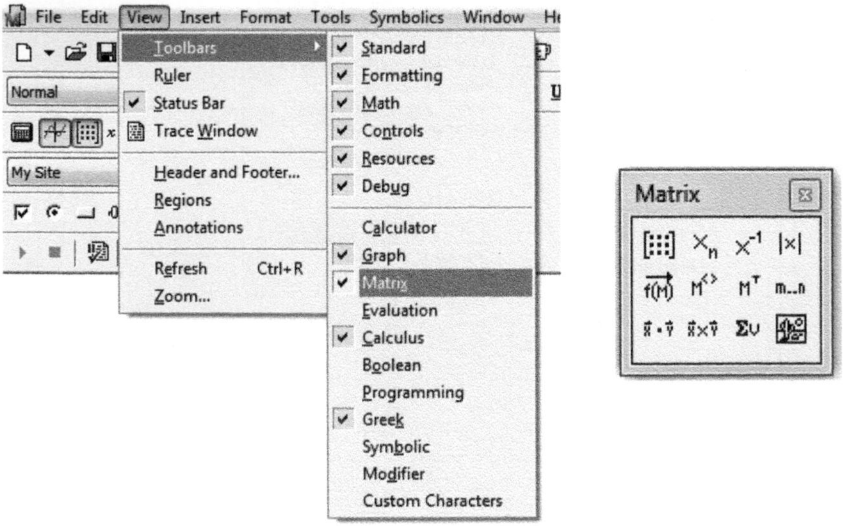

FIGURE 5.7 Matrix menu and tool bar.

TABLE 5.1
Mathcad Functions for Common Matrix Operations

Button or Function	Description
X_n	Enter subscripts.
$M^{<>}$	Refer to a specific column.
X^{-1}	Calculate the inverse.
M^T	Calculate the transpose.
length(**v**)	Returns the number of elements in vector **v**.
max(**A**)	Returns the largest element in array **A**.
min(**A**)	Returns the smallest element in array **A**.
cols(**A**)	Returns the number of columns in array **A**. If **A** is a scalar, returns 0.
rows(**A**)	Returns the number of rows in array **A**. If **A** is a scalar, returns 0.
identity(n)	Returns a matrix of 0's with 1's on the diagonal.
eigenvals(**A**)	Returns a vector containing the eigenvalues of the matrix **A**.
lsolve(**A**, **v**)	Returns a solution vector **x** such that **A**·**x** = **v**
lu(**A**)	Returns a single matrix containing the three square matrices **P**, **L**, and **U**, all having the same size as **A** and joined together side by side, in that order. These three matrices satisfy the equation **P**·**A** = **L**·**U**, where **L** and **U** are lower and upper triangular respectively.
qr(**A**)	Returns a matrix whose first *n* columns contain the square, orthonormal matrix **Q**, and whose remaining columns contain the upper triangular matrix, **R**.
cholesky(**A**)	Returns a lower triangular matrix **L** such that **L L'** = **A**

5.2.5 Equation Solving

Mathcad solves equations and systems of equations both symbolically and numerically. To solve equations in Mathcad, the user types expressions very similar to what he would write on a sheet of paper. As a simple example, if the user wanted to calculate the integral of $f(x) = x^2$ from 0 to 6, he would use the toolbars and type exactly that expression, see Figure 5.8.

Once these variables are defined, they can continue to be used throughout the workspace. The user interface is largely intuitive, as indicated by the "what you see is what you get" design. Typical trigonometric, statistic and algebraic functions are built-in within Mathcad. Furthermore, Table 5.2 lists some other common engineering math functions and the Mathcad syntax.

Two particularly useful functions are those for finding roots: *root* (f(x),x) and *polyroots*(v). *Root* will find one root of a function if one exists but requires an initial guess of x as well as the function as inputs. *Polyroots* will find all of the roots of a polynomial using the coefficients of the polynomial in increasing order of x. Please see Figure 5.9 for an example.

Many calculations can be done using these simple configurations; however, more complex problems can also be solved using solve blocks.

$f(x) := x^2$

$$\text{area} := \int_0^6 f(x)\,dx$$

area = 72

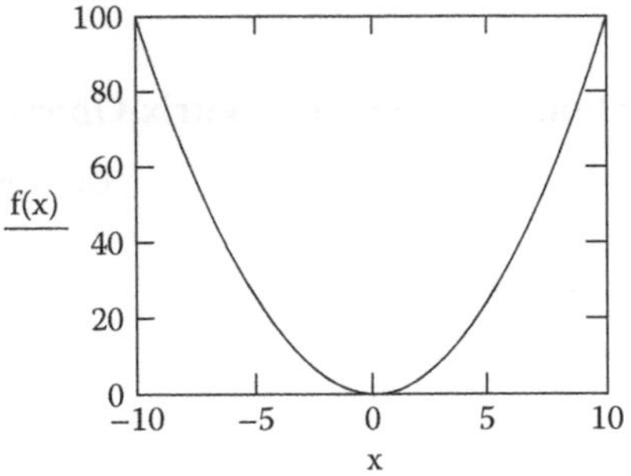

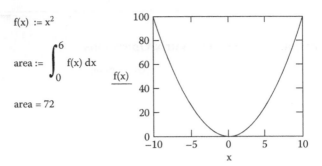

FIGURE 5.8 Solving and plotting in Mathcad.®

TABLE 5.2
Mathcad® Functions for Common Math Operations

Function	Description
Jacob(f(x),x,k)	Returns the Jacobian of the vector function f(x) with respect to x. k is an optional argument that specifies the number of variables for the Jacobian.
J0(x), J1(x)	Returns the value of the zeroth-order and first-order Bessel function of the first kind, respectively.
K0(x), K1(x)	Returns the value of the zeroth-order and first-order modified Bessel function of the second kind, respectively.
line(vx,vy)	Returns a vector containing the coefficients a and b for a line of the form a + bx that best approximates the data in vectors vx and vy in the least-squares sense.
linterp(vx,vy,x)	Returns a linearly interpolated value at x for data vectors xv and vy.
log(x, [b])	Returns the base *b* log of x. If b is omitted, base 10 is assumed.

$f(x) := x^3 + 7x^2 + x - 8 \qquad g(x) := 0$

$x := -6$

$\text{root}(f(x),x) = -6.67$

$$v := \begin{pmatrix} -8 \\ 1 \\ 7 \\ 1 \end{pmatrix}$$

$$\text{polyroots}(v) = \begin{pmatrix} -6.67 \\ -1.272 \\ 0.943 \end{pmatrix}$$

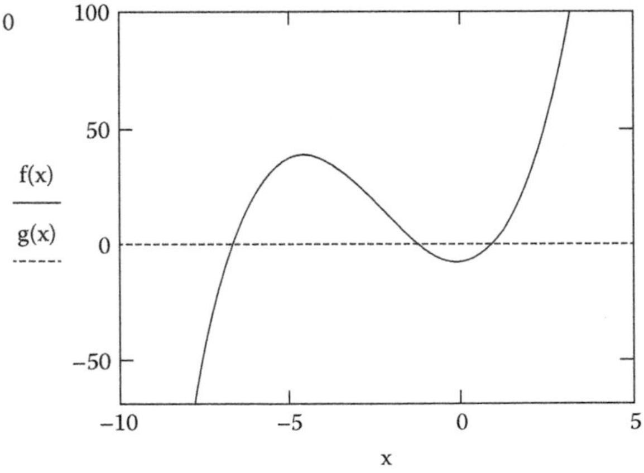

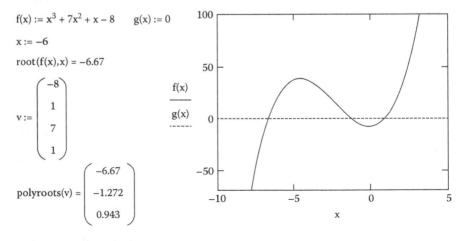

FIGURE 5.9 Finding roots in Mathcad®.

In general, to create a solve block, first type an initial guess for each of the unknowns. Below, type the word *Given*. Note that the blue bars indicating a math region are around *Given*, indicating that it is not a comment. Type all of the constraining equations using Boolean equals signs. Call the function that is being used to solve the problem and evaluate the answer. Examples of using solve blocks to solve differential equations, non-linear equations, and optimization problems are provided here and in Section 5.3.

To solve a **differential equation**, as outlined generally, start by typing *Given*, the differential equation and the initial condition(s) using the symbolic equals sign from the Boolean toolbar. Assign the variable of interest to *Odesolve*([vector],x,b,[intvls]) where x is the variable of integration and b is the endpoint of the integration. Assign x to a range variable, starting at the initial point and ending at b. vector and `intvls` are optional arguments. vector is used for a system of ODEs and is a vector of function names without any variable names; for example f for f(x). intvls is the number of integration steps; the default value is 1000. Mathcad is able to solve ordinary differential equations of higher order and there is no need to transform the problem to a first order ODE prior to coding it in Mathcad. The default methods used by *Odesolve* are Adams-Bashforth unless the solver discovers that the system is stiff in which case backward differentiation formulae are used (BDF). By right-clicking on *Odesolve* in the worksheet, the user can choose another solver. The options include: Fixed, which uses a fixed-step Runge-Kutta method, Adaptive, which uses an adaptive step Runge-Kutta method or Radau, which uses a Radau method. Radau is the only solver that will solve systems with algebraic constraints. If the system contains algebraic constraints, Radau will always be used even if another solver is selected. Please see Figure 5.10 for the complete sample code of a second order ODE.

To solve an ODE without a solve block, use an ODE solver. Mathcad offers ODE solvers for stiff and non-stiff systems of first order ODEs, as outlined in Table 5.3.

Each ODE solver function has the same first five inputs: init, x1, x2, intvls, and D. inits is the vector of initial conditions, x1 is the starting point of the integration, x2 is the final point of the integration, intvls is the number of integration steps and D is the vector function specifying the right-hand side of the system. Some solvers have additional required or optional arguments. tol is an optional tolerance or vector of tolerances for each independent variable in the system. tol is not available when using *odesolve*. J is a function in the form J(x,y) that returns the Jacobian matrix of D. M is a matrix representing any coupling of the variables in the form $\mathbf{M}\frac{dy}{dt} = \mathbf{f}(t, \mathbf{y})$. AJ is a function in the form AJ(x,y) that returns the augmented Jacobian. For further reference on differential equation solving, please see Biegler or Asher and Petzold [1,2]. Please refer to Example 5.3.3 for an example of the implementation of one of these functions.

To solve **non-linear equations** using a solve block, the user makes an initial guess of the variable of interest, types *Given* and the constraints using the symbolic equals sign. Below the equation definition, the user types the variable of interest again equals, using an assignment equals sign, *Find*(variable). The solution can then be displayed using an evaluation equals sign. For non-linear equations, Mathcad will automatically choose among three different methods for

Given

$$\frac{d^2}{dx^2}y(x) + y(x) = 0$$

$y(0) = 3 \qquad y'(0) = -0.5$

$y := \text{Odesolve}(x, 10)$

$x := 0, .1 .. 10$

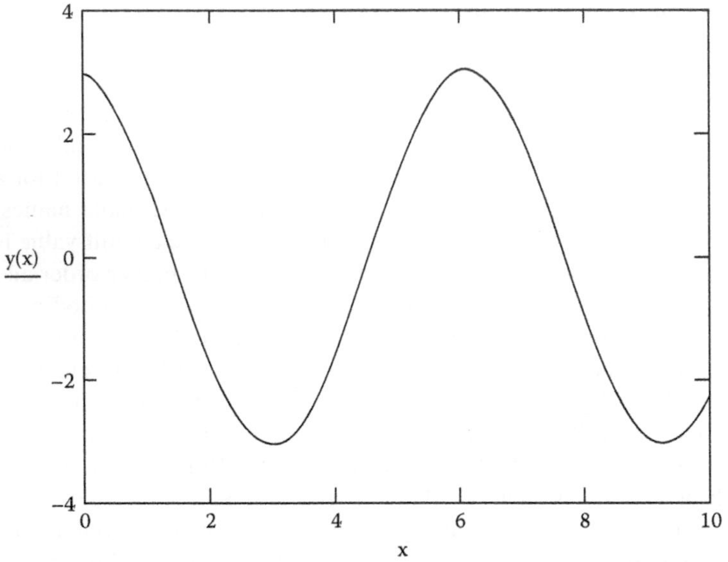

y(x) =

3
2.935
2.841
2.718
2.568
2.393
2.194
1.972
...

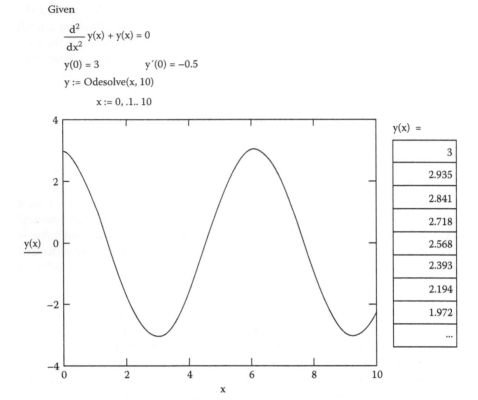

FIGURE 5.10 Solving differential equations.

TABLE 5.3
Mathcad® Functions for Solving Systems of Differential Equations

Function Name	Description
Adams(init,x1,x2,intvls,D,[tol])	Adams method.
AdamsBDF(init,x1,x2,intvls,D,[J],[tol])	Adams or BDF depending on the stiffness of the system. Use either J or tol or omit both.
BDF(init,x1,x2,intvls,D,[J], [tol])	Backward differentiation formula methods. Use either J or tol or omit both.
Bulstoer(init,x1,x2,intvls,D)	Bulrish-Stoer method (for smoothly varying systems).
Radau(init,x1,x2,intvls,D,[J],[M],[tol])	Radau5 method. Use any combination of J, M and tol.
rkfixed(init,x1,x2,intvls,D)	Fourth order Runge-Kutta method with fixed step.
Rkadapt(init,x1,x2,intvls,D)	Fourth order Runge-Kutta method with an adaptive step.
Stiffb(init,x1,x2,intvls,D,AJ)	Bulirsch-Stoer method (for stiff systems).
Stiffr(init,x1,x2,intvls,D,AJ)	Rosenbrock method.

TABLE 5.4
Mathcad® Functions for Solving Systems of Differential Equations

Function Name	Description
Maximize / Minimize(f, var1, var2…)	Returns values for the variables of the function f that meet the constraints so that f is maximized or minimized. KNITRO solver is used.
Find(var1, var1) or Minerr(var1, var2)	Returns the value of the variables so that the equations (equalities and inequalities) are met. N equations must be provided for n variables.

the solver: conjugate gradient, Levenberg-Marquardt or quasi Newton. To check which solver was used or to change the selection manually, right-click on *Find*. For further reference on the methods see Biegler, Grossmann & Westerberg [3]. Please see Example 5.3.2 for the implementation of a solve block for a non-linear equation.

The last example of solve blocks that is demonstrated here is **optimization.** Mathcad has four primary functions that are used in the context of optimization. The general format follows that of a solve block. Specifics for the function syntax are outlined in Table 5.4. Please note that the difference between *Find* and *Minerr* is that *Find* selects a suitable algorithm for the problem and *Minerr* returns an error message if the system does not converge. By default, both use Levenberg-Marquardt but KNITRO is also available. Please see Figure 5.11 for an example of the implementation of *Find*.

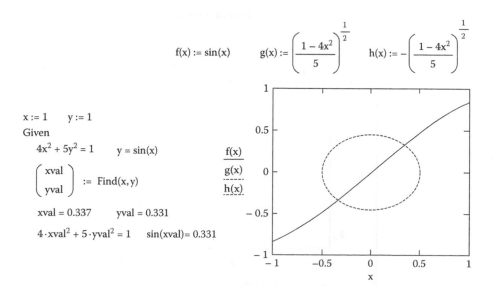

$$f(x) := \sin(x) \qquad g(x) := \left(\frac{1-4x^2}{5}\right)^{\frac{1}{2}} \qquad h(x) := -\left(\frac{1-4x^2}{5}\right)^{\frac{1}{2}}$$

$x := 1 \quad y := 1$

Given

$$4x^2 + 5y^2 = 1 \qquad y = \sin(x)$$

$$\begin{pmatrix} xval \\ yval \end{pmatrix} := \text{Find}(x, y)$$

$$xval = 0.337 \qquad yval = 0.331$$

$$4 \cdot xval^2 + 5 \cdot yval^2 = 1 \qquad \sin(xval) = 0.331$$

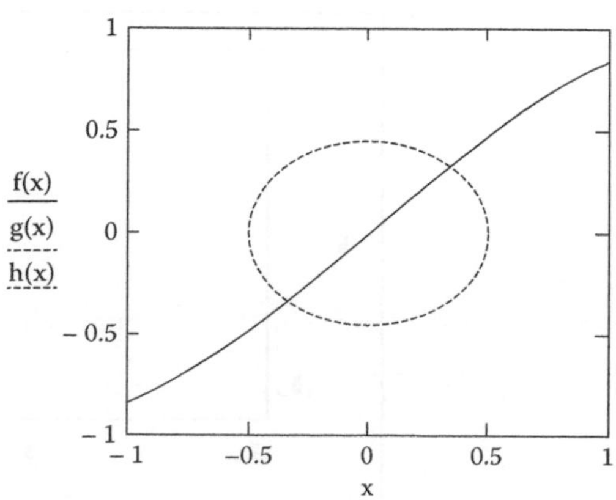

FIGURE 5.11 Using *Find* in Mathcad®.

5.2.6 Plotting and Graphing

To add a graph in Mathcad, choose Insert, Graph, and the type of plot. Alternatively, the user may also select the plotting wizard from the Insert > Graph menu. Mathcad offers X-Y plots, polar graphs, surface plots, contour plots, 3D scatter plots, 3D bar plots and vector field plots. Once a type of plot has been chosen, a blank graph will appear with placeholders for the axis and the limits, please see Figure 5.12.

Click on each of these areas to add the dependent and independent variable as well as the limits of each axis. To add an additional dependent variable type "," and another input placeholder will appear. Double click on the plot to view additional options for formatting and labeling, as shown in Figure 5.13.

5.2.7 Programming Functions

It is also possible to create functions in Mathcad. The authors leave it up to the reader to explore function options more fully but review control structures for functions here, including *for* loops, *while* loops and *if* statements. The general format for control statements is:

```
function(input) := | "function statement"
                   | "control statement"
                   | for i ∈ 0..10
                   | "condition statement"
                   | "output statement"
```

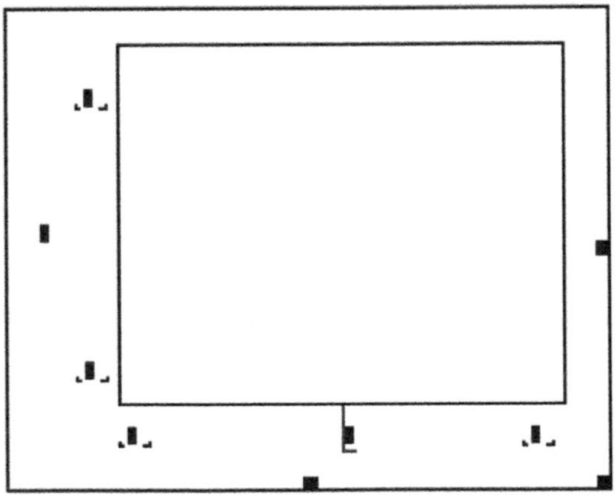

FIGURE 5.12 Initial plot.

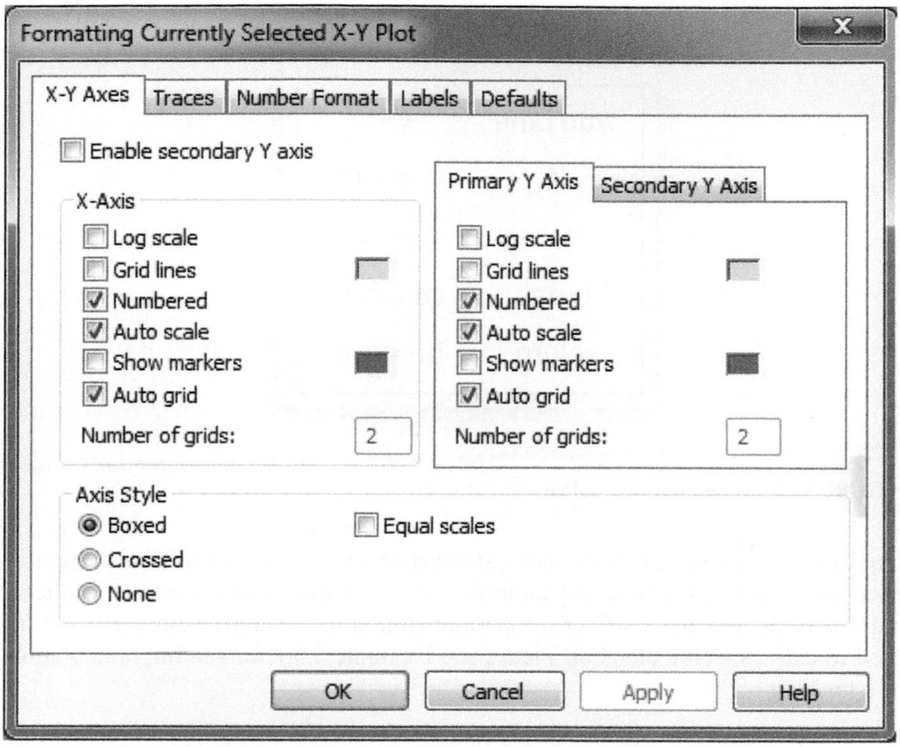

FIGURE 5.13 Formatting plots.

Each type of control structure varies slightly in its construction however all operators can be found in the programming toolbar, as shown in Figure 5.14.

To insert a *for* loop, type the name of the variable, use a definition equals sign, then select *Add Line* on the Programming toolbar and choose *for*. For loops will continue to execute a line of code for a given range variable therefore as a minimum, a range variable and a loop statement are required. A simple example is shown below:

$$\text{sum}(x) := \left| \begin{array}{l} s \leftarrow 0 \\ \text{for } i \in 0..x \\ \quad s \leftarrow s + i \\ s \end{array} \right.$$

$$\text{sum}(5) = 15$$

While loops are entered in a similar way to *for* loops however *while* is chosen. The condition statements will continue to be executed so long as the *while* criteria is met. *If* loops are also from the same toolbar, choosing *if*. *If* condition statements are executed only if the condition statement is evaluated as the Boolean true. Control statements can be made increasingly complex by adding lines. It is

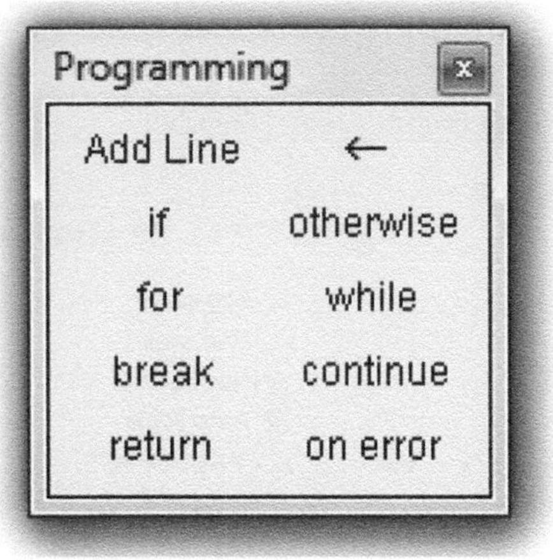

FIGURE 5.14 Programming toolbar.

important to remember that the user cannot type *while*, *if* or *for* but must select the operator form the programming toolbox. It is also important to note that Mathcad indexes from zero and a subscript operator (found in the matrix toolbar) must be used to call a specific element. Please see Example 5.3.1 for the implementation of a for loop.

5.2.8 Interactions - Mathcad® with Other Software

A useful aspect of many software products is the ability to share information among programs. In this section we will illustrate how Mathcad can exchange information with other software, focusing on Microsoft Excel and Word. While Microsoft Office is not designed as a mathematical tool, it is used commonly to write reports. The user may receive information in Word and want to transfer it into Mathcad or vice versa. Please note for Mathcad 14, the user will need to save Excel files in Excel 2003 (.xls). Mathcad 15 supports the new Excel file format (.xlsx).

5.2.8.1 Mathcad® and Microsoft Word®

To copy information from Mathcad into Word, click outside of the math and comment regions, then drag the mouse over the regions to be selected. Next select Edit/Copy or (Ctrl+C) to copy the selected regions into the windows clipboard. Navigate to the Word document and choose Paste special > Pasting as rich text region or (Ctrl+V). This is the simplest way of copying Mathcad regions into Word. Additional options are Windows Metafile, which preserves large operators and Mathcad graphs, and Enhanced Metafile, which creates a scalable, resizable graphic image but does not correctly display graphs. It is also possible to create a Mathcad Document Object which adds the possibility of creating Mathcad regions within Word. The user can edit the object and make changes to equations after it has been inserted into the word document by pasting as bitmap. A bitmap is an image file, typically with a very large resolution image file. This can be used when the rest of the options fail.

5.2.8.2 Mathcad® and Microsoft Excel®

There are several different methods to import data from Excel to Mathcad or vice versa. Before attempting to import information from Excel, always save the Excel file to ensure the imported data are current. The main difference among the methods is if they are a one-time type of transfer or if the information will be updated automatically.

If only one-time transfer of information is required, the authors recommend a simple copy paste method. However data transferred in this manner will not be automatically updated. To copy data from Mathcad into Excel, start by selecting the array or matrix (not the entire equation section). Then copy (Ctrl+C) the data and paste (Ctrl+V) into the desired starting cell in Excel (Figure 5.15). Similarly to go from Excel to Mathcad, create a variable in Mathcad and paste data from Excel into the placeholder (Figure 5.16).

Another option to use the Insert Data Menu. Here, three options are illustrated: Data Table, File Input and Data Import Wizard. Please see Figure 5.17.

Loading a **Data Table** is also a one-time transfer of information from Excel to Mathcad. The data table is never automatically updated and must be manually updated by transferring the information.

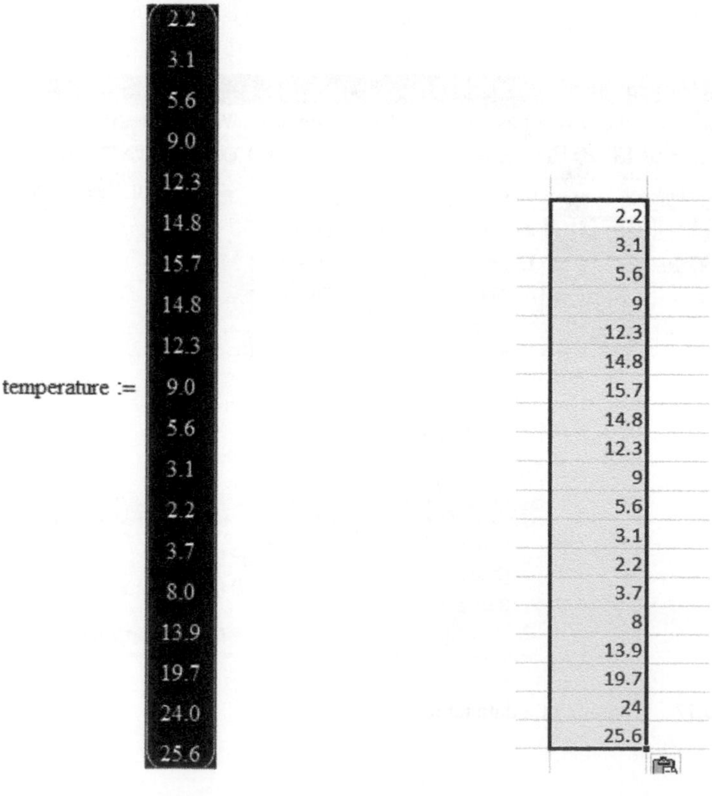

FIGURE 5.15 Copy/paste results from Mathcad® into Excel®.

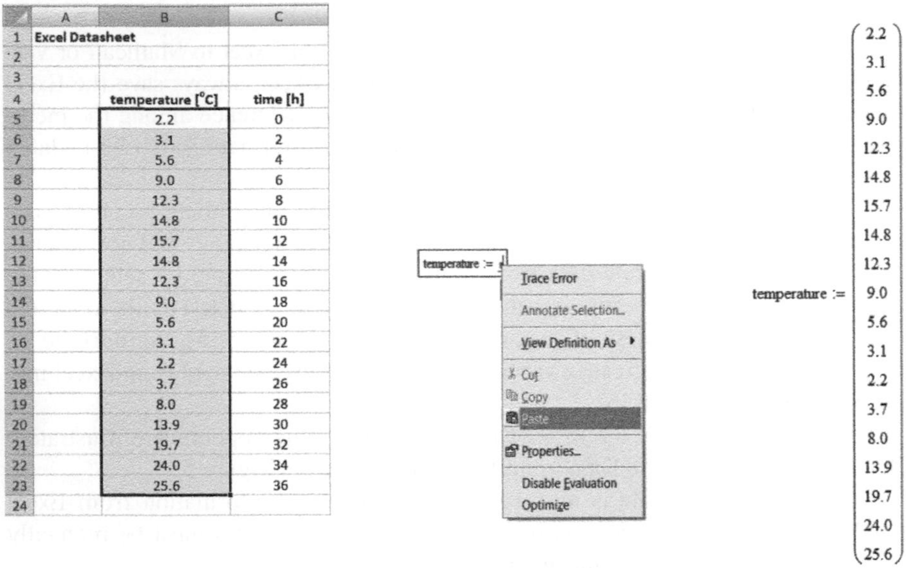

FIGURE 5.16 Copy/paste data from Excel® into Mathcad®.

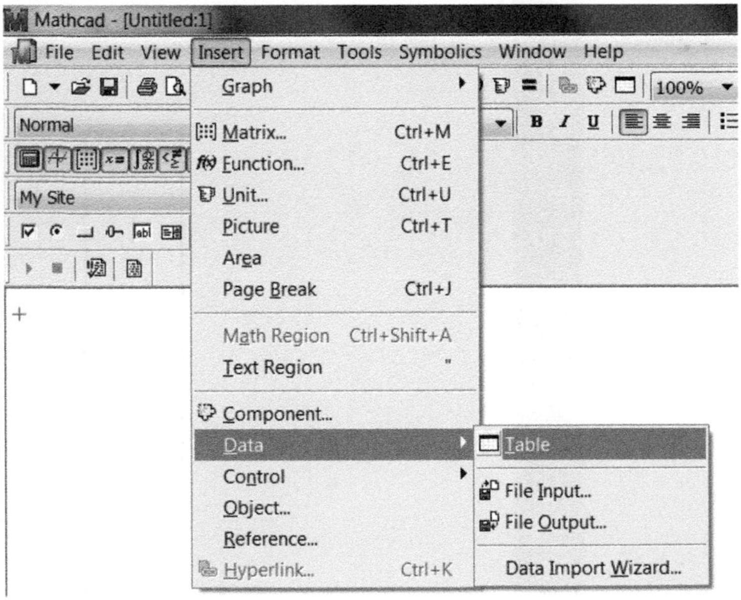

FIGURE 5.17 Creation of a data table.

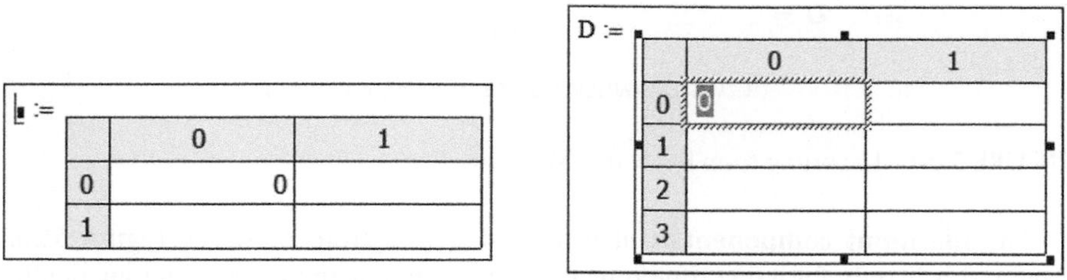

FIGURE 5.18 Filling-in the data table.

When the table is inserted, there is a placeholder for the name of the variable. By default the table is 2 × 2 but by dragging the handles (small black squares on the edges) the table can be resized. Figures 5.17 and 5.18 illustrate the steps.

The table can be filled in by typing manually the values in the cells or by using the command Copy in Excel (or Ctrl+C) and Paste (or Ctrl+V) in Mathcad. Another option is to import the data from Excel. First click on any cell in the data table to select it. Right-click on the cell to open a pop-up menu as shown in Figure 5.19.

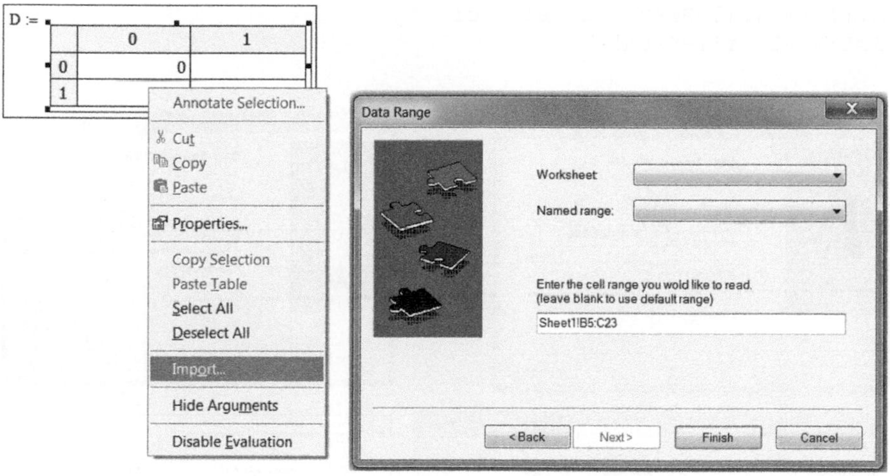

D :=

	0	1
0	2.234	0
1	3.134	2
2	5.593	4
3	8.952	6
4	12.311	8
5	14.771	10
6	15.671	12
7	14.771	14
8	12.311	16
9	8.952	...

FIGURE 5.19 Importing from Excel® into Mathcad®.

FIGURE 5.20 Importing from Excel into Mathcad.

The **file input component** is able to read values from files, see Figure 5.20. Again, this is only for a one time transfer of information and is never automatically updated. To insert a File input component go to **Insert/Data/ File input.** This option is similar to importing data in a data table (see Figure 5.17).

The **data import wizard**, unlike the Data Input content, allows the user to see the Excel file content and select the desired values. If the values in Excel are updated, the Mathcad worksheet is updated next time it recalculates. Figure 5.21 shows how to do this.

The READFILE function option offers same options as Data Import Wizard but without the dialog boxes. The READFILE function is executed each time Mathcad recalculates the worksheet so it will update the information everytime the Excel file changes.

```
READFILE (filePath, "Excel") or
READEXCEL (filePath)
```

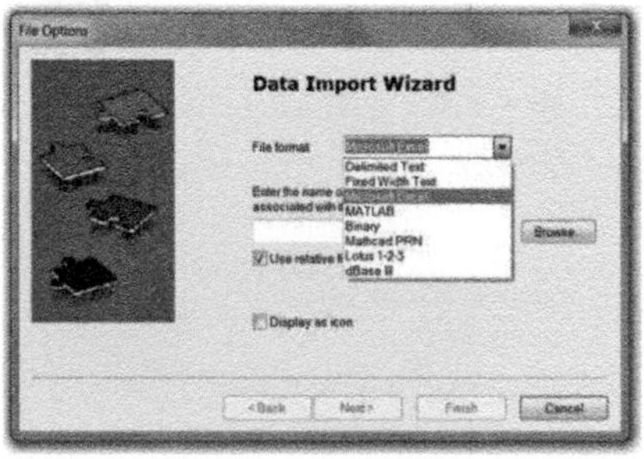

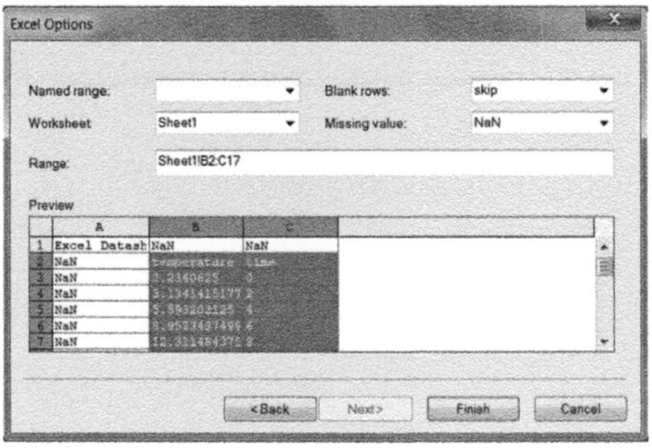

:=

	0	1
0	"temperature"	"time"
1	2.234	0
2	3.134	2
3	5.593	4
4	8.952	6
5	12.311	8
6	14.771	10
7	15.671	12
8	14.771	14
9	12.311	16
10	8.952	18
11	5.593	20
12	3.134	22
13	2.154	...

FIGURE 5.21 Importing from Excel® into Mathcad® using the Import Wizard.

filePath := "example Mathcad-Excel transfer.xlsx"

excelfile_path := concat(CWD,filePath)

D := READFILE(excelfile_path,"Excel")

D =

	1	2	3
1	"Excel Datasheet"	NaN	NaN
2	NaN	NaN	NaN
3	NaN	NaN	NaN
4	NaN	"temperature"	"time"
5	NaN	2.234	0
6	NaN	3.134	2
7	NaN	5.593	4
8	NaN	8.952	6
9	NaN	12.311	8
10	NaN	14.771	10
11	NaN	15.671	12
12	NaN	14.771	14
13	NaN	12.311	16
14	NaN	8.952	...

FIGURE 5.22 Example on how to use READFILE.

This option reads all contiguous data in the excel file. Sometimes this will require some data clean-up. Figure 5.22 illustrates these steps.

```
READFILE (filePath, "Excel", startRow, startCol) or
READEXCEL (filePath, "B2:B4")
```

Mathcad will read continuous values starting from startRow and startCol.

IB := READEXCEL (excelfile_path,"B5:C10")

$$IB = \begin{pmatrix} 2.234 & 0 \\ 3.134 & 2 \\ 5.593 & 4 \\ 8.952 & 6 \\ 12.311 & 8 \\ 14.771 & 10 \end{pmatrix}$$

READFILE (filePath, "Excel", vRow, vCol)

vRow and vCol will be vectors indicating the starting and stopping columns.

$$\text{vRows} := \begin{pmatrix} 5 \\ 10 \end{pmatrix} \qquad \text{vCols} := \begin{pmatrix} 2 \\ 3 \end{pmatrix}$$

AF := READFILE(excelfile_path,"Excel",vRows,vCols)

$$AF = \begin{pmatrix} 2.234 & 0 \\ 3.134 & 2 \\ 5.593 & 4 \\ 8.952 & 6 \\ 12.311 & 8 \\ 14.771 & 10 \end{pmatrix}$$

Mathcad additionally has an option to install an add-in for Excel. This add-in inserts a Mathcad worksheet directly into Excel. It is used when the user wants to access Mathcad features from Excel. The authors encourage the reader to investigate this option further depending on his utilization of Excel.

5.3 EXAMPLES

This section highlights Mathcad functionalities for chemical engineering applications.

5.3.1 Functions, Matrices, Graphing, and Fitting

Let's consider the operation of a vapor-recompression evaporator [4]. Imagine we are trying to minimize operating costs by determining the optimum operating pressure. We are provided the following information about this system, see Figure 5.23:

Additionally, we are given that the heat transfer coefficient is 1500 W/m^2 K and an Excel spreadsheet with properties of saturated steam and water. The following data is provided [5]:

Row 1: Temperature (°C)
Row 2: Pressure (kPa)
Row 3: Specific Volume of Liquid (m^3/kg)
Row 4: Specific Volume of Saturated Vapor (m^3/kg)
Row 5: Enthalpy of Liquid (kJ/kg)
Row 6: Enthalpy of Saturated Vapor
Row 7: Entropy of Liquid (kJ/kg K)
Row 8: Entropy of Saturated Vapor (kJ/kg K)

Lastly, we are provided the following economic data: steam costs \$18/1000 kg, electricity costs \$0.04/kW-hr and the heater costs \$700/m^2. The overall compressor efficiency is 70%. Assume the lifetime of the heat exchanger is 4 years.

The authors acknowledge that there are many ways to formulate and solve this problem however we have chosen the following method to highlight Mathcad functionalities. Therefore, we have chosen to use a vector of pressure guesses and find the

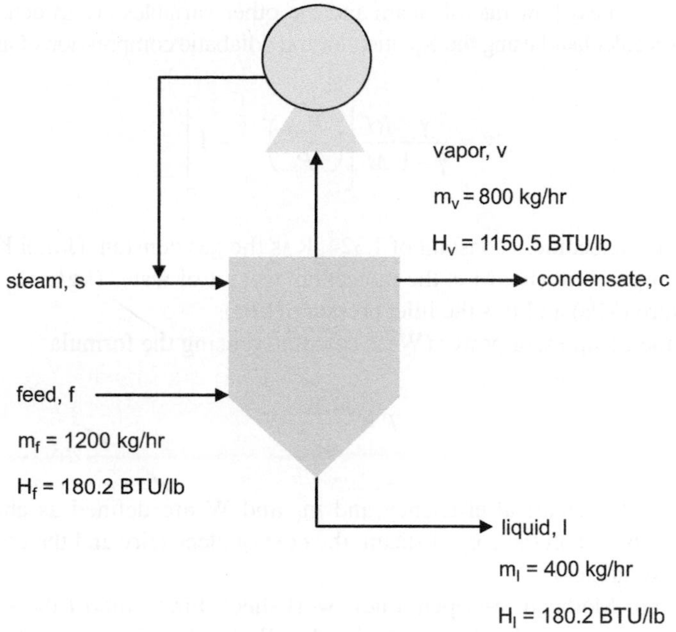

FIGURE 5.23 Scheme of the evaporator system.

annualized cost for each pressure guess. Another viable option would have been to perform an iterative calculation, changing the guess for the optimal operating pressure.

We choose to start the problem by finding the heat duty by using the general equation for an evaporator:

$$q = UA\Delta T = S\lambda \tag{5.1}$$

where q is the rate of heat transfer (W), U is the overall heat transfer coefficient (W/m^2 K), A is area (m^2), ΔT is change in temperature (K), S is the mass flow rate of steam entering the system (kg/h) and λ is the latent heat of steam (J/kg). At this point in the problem we are interested in the ratio of heat transfer to area. To do this, we calculate the temperatures of saturated steam at our pressure guess vector values via linear interpolation of the steam table data and calculate a vector of fluxes for each temperature.

Next we use an energy balance to calculate q directly using:

$$q = m_v H_v + m_l H_l - m_f H_f \tag{5.2}$$

where q is again the rate of heat transfer (W), m is the mass flow rate of the vapor, liquid and feed respectively (kg/h) and H is the enthalpy of the vapor, liquid and feed respectively (kJ/kg). Combining these two equations for q, we can calculate vectors of area, mass flow rate of steam and work, all depending on the operating pressure. Mass flow rate of steam is calculated using the energy balance equation

$$m_s = \frac{q}{\lambda_s} - m_v \tag{5.3}$$

where m_s is the mass flow rate of steam and the other variables are as defined above. Work (J/kg) is calculated using the equation for the adiabatic compression of an ideal gas

$$W = \frac{\gamma}{\gamma - 1}\frac{RT}{M}\left[\left(\frac{P_{out}}{P}\right)^{1-\frac{1}{\gamma}} - 1\right] \tag{5.4}$$

where γ is a dimensionless constant of 1.324, R is the gas constant (J/mol K), T is the operating temperature (K), M is the molecular weight of water (kg/mol), P_{out} is the outlet pressure (kPa) and P is the inlet pressure (kPa).

Finally, the compressor power (W) is calculated using the formula

$$P = \frac{m_v W}{\eta} \tag{5.5}$$

where η is the mechanical efficiency and m_v and W are defined as above. Total annual cost is based on the cost of steam, the cost of electricity and the cost per area of the heat exchanger.

Moving into Mathcad, we open a new worksheet. First, import the steam table data. Because the data table is not large and will not change, copy and paste from Excel. First we define the variable that we want to house the steam data. We use a definition equations sign (:=) and paste our copied data over the red marker.

steam := ■

steam :=

0.01	3	6	9	12	15	18	21	...
0.6113	0.7577	0.9349	1.1477	1.4022	1.7051	2.064	2.487	...
0.0010002	0.0010001	0.0010001	0.0010003	0.0010005	0.0010009	0.001004	0.001002	...
206.136	168.132	137.734	113.386	93.784	77.926	65.038	54.514	...
0	12.57	25.2	37.8	50.41	62.99	75.58	88.14	...
2501.4	2506.9	2512.4	2517.9	2523.4	2528.9	2534.4	2539.9	...
0	0.0457	0.0912	0.1362	0.1806	0.2245	0.2679	0.3109	...
9.1562	9.0773	9.0003	8.9253	8.8524	8.7814	8.7123	8.645	...

The imported data is arranged in rows however, it is more convenient to work in columns therefore we use the **transpose function** to create variable *steam2*. Next we can define the units of our data within the matrix *steam2*. To simplify this process, we also create a new unit kJ = 1000J.

$\text{steam2} := \text{steam}^T$ $\text{kJ} := 1000\text{J}$

$\text{Tcol} := \text{steam2}^{\langle 0 \rangle}$ $\text{Pcol} := \text{steam2}^{\langle 1 \rangle} \cdot \text{kPa}$

$\text{Hxcol} := \text{steam2}^{\langle 4 \rangle} \cdot \frac{\text{kJ}}{\text{kg}}$ $\text{Hycol} := \text{steam2}^{\langle 5 \rangle} \cdot \frac{\text{kJ}}{\text{kg}}$

$\lambda\text{col} := \left(\text{steam2}^{\langle 5 \rangle} - \text{steam2}^{\langle 4 \rangle}\right)\frac{\text{kJ}}{\text{kg}}$

We can confirm that steam2 is the transpose of steam by typing steam = (note not :=)

steam2 =

	0	1	2	3	4	5	6	7
0	0.01	0.611	$1\cdot10^{-3}$	206.136	0	$2.501\cdot10^{3}$	0	9.156
1	3	0.758	$1\cdot10^{-3}$	168.132	12.57	$2.507\cdot10^{3}$	0.046	9.077
2	6	0.935	$1\cdot10^{-3}$	137.734	25.2	$2.512\cdot10^{3}$	0.091	9
3	9	1.148	$1\cdot10^{-3}$	113.386	37.8	$2.518\cdot10^{3}$	0.136	8.925
4	12	1.402	$1.001\cdot10^{-3}$	93.784	50.41	$2.523\cdot10^{3}$	0.181	8.852
5	15	1.705	$1.001\cdot10^{-3}$	77.926	62.99	$2.529\cdot10^{3}$	0.225	8.781
6	18	2.064	$1.001\cdot10^{-3}$	65.038	75.58	$2.534\cdot10^{3}$	0.268	8.712
7	21	2.487	$1.002\cdot10^{-3}$	54.514	88.14	$2.54\cdot10^{3}$	0.311	8.645
8	24	2.985	$1.003\cdot10^{-3}$	45.883	100.7	$2.545\cdot10^{3}$	0.353	8.579
9	25	3.169	$1.003\cdot10^{-3}$	43.36	104.89	$2.547\cdot10^{3}$	0.367	8.558
10	27	3.567	$1.004\cdot10^{-3}$	38.774	113.25	$2.551\cdot10^{3}$	0.395	8.516
11	30	4.246	$1.004\cdot10^{-3}$	32.894	125.79	$2.556\cdot10^{3}$	0.437	8.453
12	33	5.034	$1.005\cdot10^{-3}$	28.011	138.33	$2.562\cdot10^{3}$	0.478	8.393
13	36	5.947	$1.006\cdot10^{-3}$	23.94	150.86	$2.567\cdot10^{3}$	0.519	8.334
14	40	7.384	$1.008\cdot10^{-3}$	19.523	167.57	$2.574\cdot10^{3}$	0.572	8.257
15	45	9.593	$1.01\cdot10^{-3}$	15.258	188.45	$2.583\cdot10^{3}$	0.639	...

Now that we have successfully imported and organized our data, we will define the parameters and their units from the problem statement.

$$mf := 12000\cdot\frac{kg}{hr} \qquad mv := 8000\cdot\frac{kg}{hr} \qquad mliq := 4000\cdot\frac{kg}{hr}$$

$$U := 1500\frac{W}{m^2K}$$

$$Pguess := \begin{pmatrix} 105 \\ 110 \\ 115 \\ 120 \\ 125 \\ 130 \\ 135 \\ 140 \\ 145 \\ 150 \\ 155 \\ 160 \\ 165 \\ 170 \\ 175 \\ 180 \\ 185 \\ 190 \\ 195 \\ 200 \end{pmatrix}\cdot kPa$$

Our first calculation is to find flux, which we do by creating a vector of possible temperatures based on the interpolation of the given data set for steam.

$$\text{Tsteam} := \text{linterp(Pcol, Tcol, Pguess)}$$
$$\text{deltaT} := \text{Tsteam} - 100$$
$$\text{qoverA} := \text{deltaT} \cdot \text{K} \cdot \text{U}$$

Next we perform an energy balance and find the area. Notice that the units of enthalpy are converted automatically.

$$\text{Hv} := 1150.5 \cdot \frac{\text{BTU}}{\text{lb}} \qquad \text{Hl} := 180.2 \cdot \frac{\text{BTU}}{\text{lb}} \qquad \text{Hf} := \text{Hl}$$
$$\text{q} := \text{mv} \cdot \text{Hv} + \text{mliq} \cdot \text{Hl} - \text{mf} \cdot \text{Hf}$$
$$\text{q} = 5.015 \times 10^6\,\text{W}$$
$$\text{Area} := \frac{\text{q}}{\text{qoverA}}$$

Next we find the mass of steam (*msteam*) and work as functions of our pressure vector (***Pguess***). Here we use two for loops so that the flow rate of steam and work are only calculated for a single pressure and temperature pair.

$$\eta := 0.7 \quad \gamma := 1.324 \quad \text{P} := 1\text{atm} = 1.013 \times 10^5\text{Pa}$$

$$\text{msteam(Pguess)} := \text{for } i \in 0..19$$
$$\left| \begin{array}{l} \lambda s_i \leftarrow \text{linterp(Pcol, } \lambda\text{col, Pguess}_i) \\ \text{msteam}_i \leftarrow \dfrac{\text{q}}{\lambda s_i} - \text{mv} \\ \text{msteam} \end{array} \right.$$

$$\text{Work(Pguess)} := \text{for } i \in 0..19$$
$$\left| \begin{array}{l} \text{temp}_i \leftarrow \text{linterp(Pcol, Tcol, Pguess}_i) \\ \text{Work}_i \leftarrow \dfrac{\gamma}{\gamma - 1} \cdot 8.314 \dfrac{\text{J}}{\text{mol} \cdot \text{K}} \cdot \dfrac{\text{temp}_i \cdot \text{K}}{0.018020 \cdot \dfrac{\text{Kg}}{\text{mol}}} \cdot \left[\left(\dfrac{\text{Pguess}_i}{\text{P}} \right)^{1 - \frac{1}{\gamma}} - 1 \right] \\ \text{Work} \end{array} \right.$$

We correct the work term for efficiency and calculate the total annual cost as functions of pressure.

$$\text{Pp(Pguess)} := \frac{\text{mv} \cdot \text{Work(Pguess)}}{\eta}$$

$$\text{totalcost(Pguess)}:$$
$$= \left(\frac{18}{1000 \cdot \text{kg}} \cdot \text{msteam(Pguess)} + \frac{0.03}{\text{kW hr}} \cdot \text{Pp(Pguess)} + \frac{0.25}{\text{yr}} \cdot \frac{700}{\text{m}^2} \text{Area} \right) \cdot \text{yr}$$

Plotting the result, yields the following graph (see Figure 5.24):

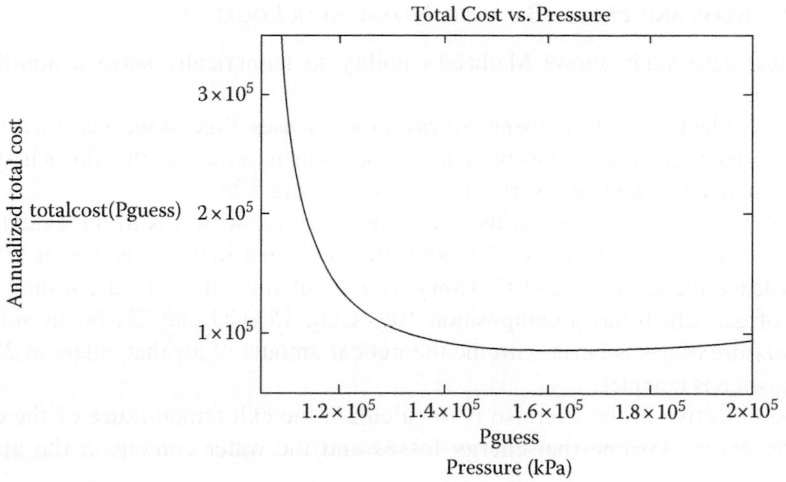

FIGURE 5.24 Annualized cost for the evaporator system.

To determine the minimum pressure, change the axis to zoom in on the minimum (see Figure 5.25)

We find that the optimum operating pressure for the compressor is 160 kPa.

QUESTIONS

- Evaluate the effect of the cost of steam and electricity on the optimal cost and area.
- How would the optimal pressure change if the cost function was given by the more complex formula? $Cost\,(\$) = -2.249\,(Area\,m^2) + 1601.3\,(Area\,m^2) + 55866$ [6]

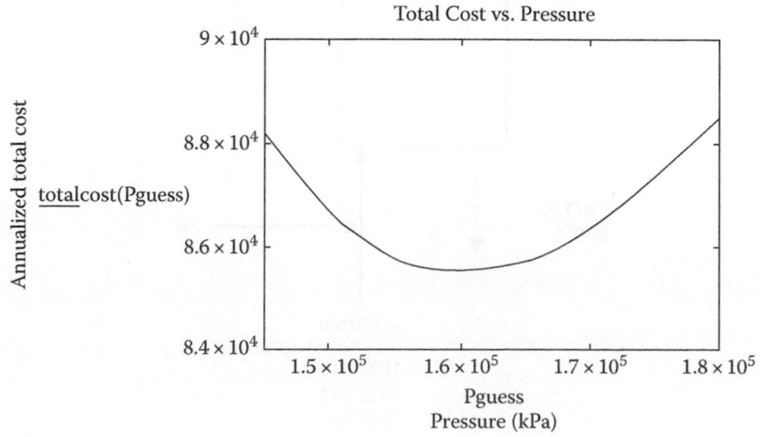

FIGURE 5.25 Profile of the effect of the pressure on the annualized cost.

5.3.2 Mass and Energy Balance: Non-Linear Equations

The next case study shows Mathcad's ability to numerically solve a non-linear equation.

The production of CO_2 using Solvay process uses lime stone (assume 100% $CaCO_3$), that is calcinated in a continuous process using a flue gas that flows in direct contact with the $CaCO_3$ in a vertical tower, see Figure 5.26.

The gases move upwards as the lime stones goes downwards in the calcination process. Lime stone enters at 25°C and the calcinated material leaves at 900°C. Gases leave the tower at 250°C. Every 100 kg of lime stone feed consumes 4.7 kmol of gas which has a composition 10% CO_2, 15%CO and 75%N_2 in volume. The mixture of gas is burnt with the theoretical amount of air that enters at 25° C. Combustion is complete.

The objective of the exercise is to calculate the exit temperature of the combustion gases. Assume that energy losses and the water content in the air are negligible.

Mass Balance

We write the equations as if the problem is formulated by hand.

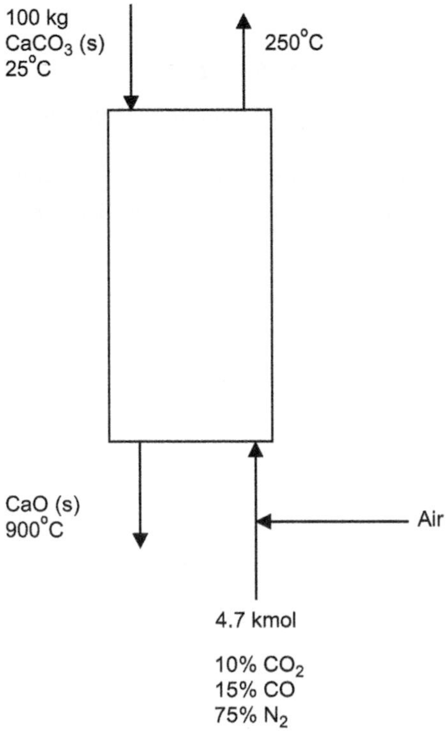

FIGURE 5.26 Scheme for the furnace operation.

Inlet streams

Flue gas

$$n_{flue\ gas} := 4.7$$

$$n_{N2} := 0.75 \cdot n_{flue\ gas} = 3.525$$

$$n_{CO} := 0.15 n_{flue\ gas} = 0.705$$

$$n_{CO2} := 0.10 \cdot n_{flue\ gas} = 0.47$$

Combustion air

$$n_{O2air} := 0.5 \cdot n_{CO} = 0.353$$

$$n_{N2air} := (79 \div 21) \cdot n_{O2air} = 1.326$$

$CaCO_3$

$$n_{Ca\,CO3} := (100 \div 100) = 1$$

Outlet streams

Exhaust gas

$$n_{N2exhaust} := n_{N2} + n_{N2air} = 4.851$$

$$n_{CO2exhaust} := n_{CO2} + 0.5 \cdot n_{O2air} + n_{CaCO3} = 1.646$$

CaO

$$n_{CaO} := n_{CaCO3} = 1$$

Energy Balance

***Heat** capacity (Kcal/kmol K)*

$$C_{pCO2}(T) := \int_{298}^{T} 6.339 + 10.14 \cdot 10^{-3} \cdot T - 3.415 \cdot 10^{-6} \cdot T^2 \, dT$$

$$C_{pCO}(T) := \int_{298}^{T} 6.350 + 1.811 \cdot 10^{-3} \cdot T - 0.2675 \cdot 10^{-6} \cdot T^2 \, dT$$

$$C_{pN2}(T) := \int_{298}^{T} 6.457 + 1.389 \cdot 10^{-3} \cdot T - 0.069 \cdot 10^{-6} \cdot T^2 \, dT$$

$$C_{pO2}(T) := \int_{298}^{T} 6.117 + 3.167 \cdot 10^{-3} \cdot T - 1.005 \cdot 10^{-6} \cdot T^2 \, dT$$

$$C_{p298CaCO3} := 0.18 \cdot 100 = 18$$

$$C_{p900CaO} := 0.23 \cdot 56.07 = 12.896$$

Standard formation enthalpy at 298 K (Kcal/kmol)

$$H_{fCaCO3} := -289540$$

$$H_{fCaO} := -151700$$

$$H_{fCO2} := -94052$$

$$H_{fCO} := -26416$$

Standard reaction enthalpy at 298 K

$$H_{R298CaCO3} := H_{fCO2} + H_{fCaO} - H_{fCaCO3} = 4.379 \times 10^4$$

$$H_{R298CO} := H_{fCO2} - H_{fCO} = -6.764 \times 10^4$$

Enthalpy (Inlet)

Flue gas inlet temperature (T) will be left as variable that is needed to solve the non-linear equation.

$$H_{air\_in}(T) := n_{O2air} \cdot C_{pO2}(T) + n_{N2air} \cdot C_{pN2}(T)$$

$$H_{flue\ gas\_in}(T) := n_{N2} \cdot C_{pN2}(T) + n_{CO} \cdot C_{pCO}(T) + n_{CO2} \cdot C_{pCO2}(T)$$

$$H_{CaCO3\_in} := n_{CaCO3} \cdot C_{p298CaCO3} \cdot (298 - 298) = 0$$

Reaction Enthalpy (Kcal)

$$H_{R298} := n_{CO} \cdot H_{R298CO} + n_{CaCO3} \cdot H_{R298CaCO3} = -3.895 \times 10^3$$

Enthalpy (Output)

$$H_{CaO\_Out} := n_{CaO} \cdot C_{P900CaO} \cdot (900 + 273 - 298) = 1.128 \times 10^4$$

$$H_{N2exhaust} := n_{N2exhaust} \cdot C_{pN2}(250 + 273) = 7.657 \times 10^3$$

$$H_{CO2exhaust} := n_{CO2exhaust} \cdot C_{pCO2}(250 + 273) = 3.671 \times 10^3$$

$$H_{Total\_out} := H_{CaO\_out} + H_{N2exhaust} + H_{CO2exhaust} = 2.261 \times 10^4$$

Now that the mass and energy balances have been formulated, we will solve the non-linear equation using a solve block, as described in Section 5.2.5.

$$T := 300$$

Given

$$H_{Total\_out} = n_{O2air} \cdot C_{pO2}(T) + n_{N2air} \cdot C_{pN2}(T) + ((n_{N2} \cdot C_{pN2}(T) + n_{CO} \cdot C_{pCO}(T) + n_{CO2} \cdot C_{pCO2}(T)) + H_{CaCO3\_in} + H_{R298}$$

$$T := \mathrm{Find}(T)$$

$$T = 849.891$$

The temperature of the flue gas stream is around 850 K.

5.3.3 Reaction Engineering: Differential Equations

Let's consider a plug flow reactor with the flowing reactions:

$$A \xrightarrow{k_1} B; A \xrightarrow{k_3} D; B \xrightarrow{k_2} C$$

Where

$$k_1 = 0.1 \, m^3/kg\,s\,;\; k_2 = 0.15 \, m^3/kg\,s\,;\; k_3 = 0.1 \, m^3/kg\,s$$

The reactor operates isothermically. The feed to the reactor is pure A at 1 atm and 1000 K at 0.15 m^3/s. The molar flow rate of A is 18.29 mol/s. The reactor contains 15 kg of catalyst and α, the characteristic value for a simplified Ergun equation, is equal to 0.01 kg^{-1}. Compute the profile of the species along the reactor.

We first need to develop the model for the reactor. We can follow the example for the SO_2 oxidation in the MATLAB® Chapter 4 for reference or for further details of reactor design and modeling see Fogler [7]. Here is the system of equations:

$$-\frac{dF_a}{dW} = (k_1 + k_3)\cdot\frac{F_a \cdot P}{v_o \cdot P_o} \tag{5.6}$$

$$\frac{dF_b}{dW} = (k_1 \cdot F_a - k_2 \cdot F_b)\cdot\frac{P}{v_o \cdot P_o} \tag{5.7}$$

$$\frac{dF_c}{dW} = k_2 \cdot F_b \cdot \frac{P}{v_o \cdot P_o} \tag{5.8}$$

$$\frac{dF_d}{dW} = k_3 \cdot F_a \cdot \frac{P}{v_o \cdot P_o} \tag{5.9}$$

The final equation for the change in pressure is found by simplifying the Ergun equation

$$\frac{dP}{dW} = -\frac{\alpha}{2}\cdot\frac{P_o^2}{P} \tag{5.10}$$

We will solve for each of the species flow rates and the pressure profile. First, we define all of our constants in Mathcad. We include one additional value, "Num", which is the number of integrating steps between the start and end points. Please note, the authors do not recommend using the built in units when working with ode solvers. Next, we create a matrix of initial values for each of our equations and a matrix of the differential equations. Mathcad indexes from zero and therefore our first equation is f_0. To solve the system we choose rkfixed and assign the result to fn. Fn is now a matrix of our solution set which we can plot. Alternatively, we can see our matrix result by typing fn=. The final workspace is shown below.

$$k_1 := 0.1 \qquad k_3 := 0.01 \qquad v_0 := 0.15 \qquad Num := 100$$

$$k_2 := 0.15 \qquad \alpha := 0.01 \qquad T_0 := 1000 \qquad P_0 := 1$$

$$ini := \begin{pmatrix} 18.29 \\ 0 \\ 0 \\ 0 \\ 1 \end{pmatrix} \qquad function(x, f) := \begin{bmatrix} -k_1 + k_3 \cdot \dfrac{f_0 \cdot f_4}{v_0 \cdot P_0} \\ (k_1 \cdot f_0 - k_2 \cdot f_1) \cdot \dfrac{f_4}{v_0 \cdot P_0} \\ k_2 \cdot f_1 \cdot \dfrac{f_4}{v_0 \cdot P_0} \\ k_3 \cdot f_0 \cdot \dfrac{f_4}{v_0 \cdot P_0} \\ \dfrac{-\alpha}{2} \cdot \dfrac{P_0^2}{f_4} \end{bmatrix}$$

$$fn := rkfixed(ini, 0, 15, Num, function)$$

$$x := fn^{\langle 0 \rangle} \qquad fB := fn^{\langle 2 \rangle} \qquad fD := fn^{\langle 4 \rangle}$$

$$fA := fn^{\langle 1 \rangle} \qquad fC := fn^{\langle 3 \rangle} \qquad Pressure := fn^{\langle 5 \rangle}$$

To plot our result, assign a column of data to each of our variables using the $M^{\langle\rangle}$ button in the matrix toolbar. Finally, insert an X-Y plot by choosing insert and the plotting suggestions from Section 5.2.5, see Figures 5.27 and 5.28.

fn =

	0	1	2	3	4	5
0	0	18.29	0	0	0	1
1	0.15	16.385	1.606	0.126	0.173	0.999
2	0.3	14.68	2.82	0.462	0.328	0.998
3	0.45	13.154	3.715	0.954	0.467	0.998
4	0.6	11.787	4.351	1.561	0.591	0.997
5	0.75	10.563	4.779	2.245	0.702	0.996
6	0.9	9.467	5.04	2.981	0.802	0.995
7	1.05	8.486	5.169	3.744	0.891	0.995
8	1.2	7.606	5.194	4.518	0.971	0.994
9	1.35	6.819	5.139	5.289	1.043	0.993
10	1.5	6.113	5.023	6.046	1.107	0.992
11	1.65	5.481	4.862	6.782	1.164	0.992
12	1.8	4.915	4.668	7.491	1.216	...

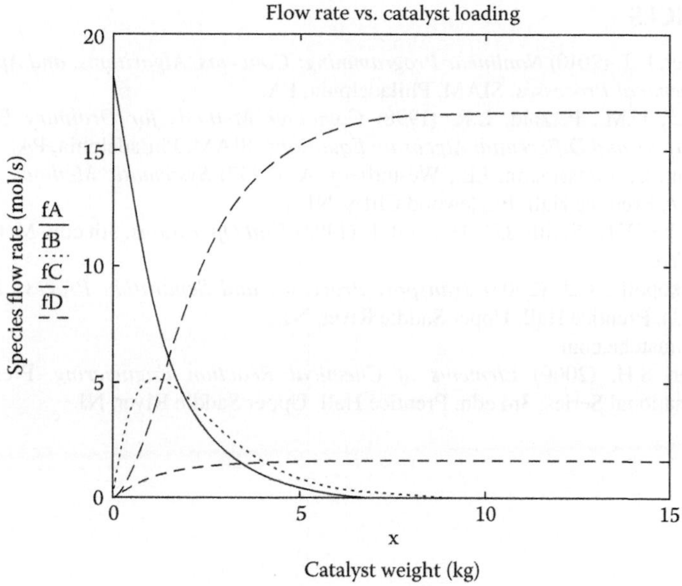

FIGURE 5.27 Profiles of the species along the reactor.

QUESTIONS

- How do your results change if you also have a flow of B to the reactor of 10 mol/s?
- Evaluate the effect of a in the solution of the problem.
- Compute the weight of catalyst for a conversion of 90%.

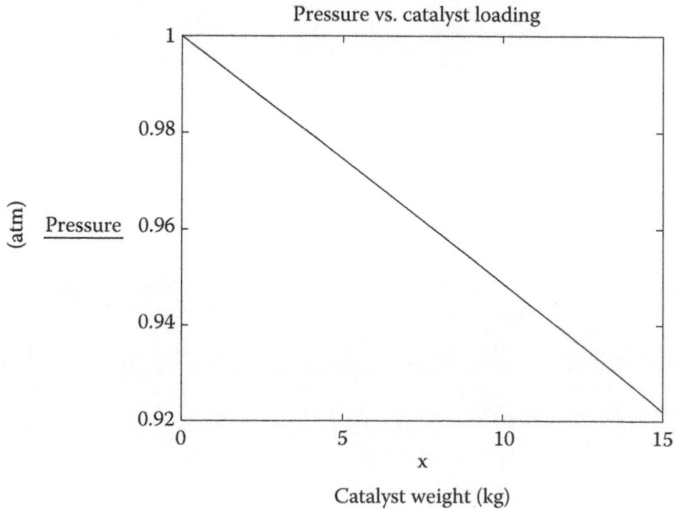

FIGURE 5.28 Pressure change along the reactor.

REFERENCES

1. Biegler, L.T. (2010) *Nonlinear Programming: Concepts, Algorithms, and Applications to Chemical Processes*. SIAM, Philadelphia, PA.
2. Ascher, U.M., Petzold, L.R. (1998) *Computer Methods for Ordinary Differential Equations and Differential Algebraic Equations*, SIAM, Philadelphia, PA.
3. Biegler, L., Grossmann, I.E., Westerberg, A. (1997) *Systematic Methods of Process Design*. Prentice Hall, Englewood Cliffs, NJ.
4. McCabe, W.L., Smith, J.C., Harriott. P. (1993) *Unit Operations*, 5th edn. McGraw-Hill, New York.
5. Geankopolis, C.J. (2003) *Transport Processes and Separation Process Principles*, 4th edn. Prentice Hall, Upper Saddle River, NJ.
6. www.matche.com
7. Fogler, S.H. (2000) *Elements of Chemical Reaction Engineering*. Prentice Hall International Series, 3rd edn. Prentice Hall, Upper Saddle River, NJ.

6 Python for Chemical Engineering

Edgar Martín Hernández, Mariano Martín

CONTENTS

6.1 INTRODUCTION

Python is an open source programming language created by Guido van Rossum in 1991 [1]. As MATLAB®, it is an interpreted high-level language, which enhances the readability of the code. This allows the user to create programs in an easier way and producing shorter codes than other programming languages. Python has a large user's community that provides support and creates packages, libraries, and modules which extends the capabilities and applications of Python to areas such as scientific and engineering simulations, machine learning, computational chemistry, data visualization, or its implementation in embedded systems, among others. It is available for the major operating systems, including Linux based OS, Windows, and Mac OS.

There exists two versions of Python, Python 2.x and Python 3.x, with minor differences in their syntaxes. Python 2.x is the classical Python programming language, whereas Python 3.x provides some new features and is the current version of Python. In this chapter Python 3.6 it has been used [2].

6.2 PYTHON BASICS

Python, along with other languages such as MATLAB® or C, is a high-level language. This means that Python language is relatively close to human language, which eases reading and writing programs in Python. However, computers only understand machine language, composed by a combination of zeros and ones, so it is necessary to translate the high-level language code into machine language. To carry out this conversion, there exist two strategies:

- Interpreted languages: the interpreter reads the code written in the high-level language, analyzes it, and interprets the instructions contained in the code directly. Python is an interpreted language.
- Compiled languages: the compiler needs to manage the high-level code to create a file which contains the machine code from the original source code. This new source code file generated will be executed when the user wants to execute the program.

Python can be used through a text editor, such as Notepad++, Atom or Gedit, to create the script and then executed using a terminal to run the program. However, there exists several Integrated Development Environments (IDE's) which can help in the creation of our projects. Some IDEs used for programming in Python are Spyder, PyDev, PyCharm, and Geany.

We will focus on one of them, Spyder, which is very useful to develop scientific scripts and, as its framework is similar to MATLAB's interface, its interface is very friendly. Spyder's layout, which can be seen in Figure 6.1, is divided into three interchangeable windows. To the left it is the editor, where the code is developed. It has multiple functions such as code completion, horizontal and vertical splitting, and code recognition among others. To the top right there is a window which shares three functions: *variable explorer*, where the variables can be read, the *file explorer*, to manage files and *Spyder's help*. Finally, to the bottom right there is another window with three functions: the *Python standard console*, the *history log*, where all executed commands are saved, and the *IPython console*, which is an interactive computational environment, in which it is possible to combine code execution, plots, mathematics, and other media.

6.2.1 Packages

The core provides the basic capabilities: data types, data structures, functions definition, control structures as conditionals sentences, and loops. However, the specific

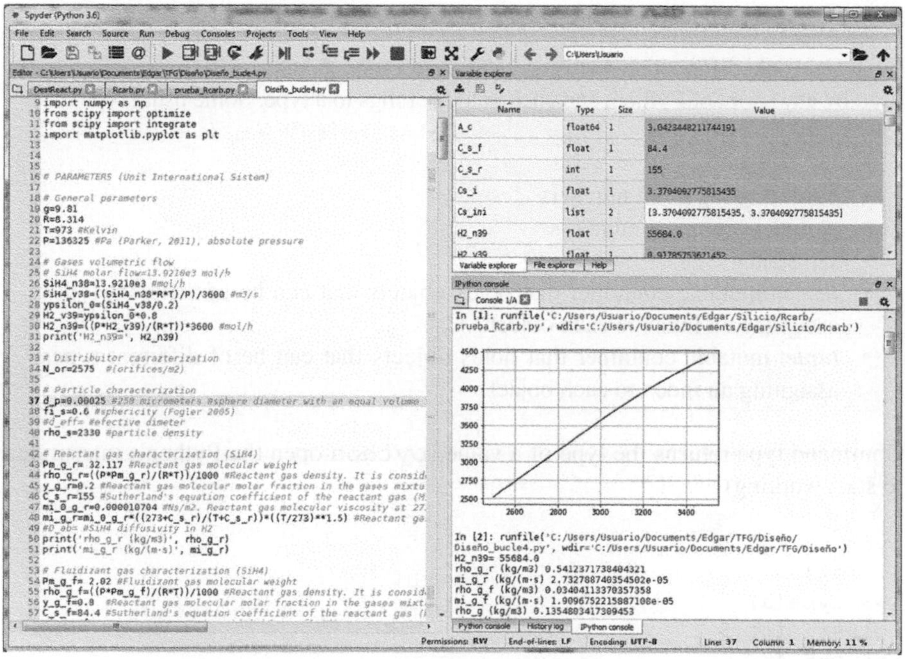

FIGURE 6.1 Spyder interface.

functions for scientific computing are supplied by packages. The main modules used are the following:

- NumPy: provides support to work with arrays and matrices of numeric data, linear algebra and other number capabilities [3].
- SciPy: this module complements NumPy and provides algorithms for linear algebra, integration, interpolation, linear algebra and optimization, among others [4].
- Matplotlib: it is a 2D and 3D plotting library which, in combination with NumPy, it is capable to produce high quality figures in different formats [5].

Nevertheless, there other packages for other specific tasks, as symbolic mathematics (SymPy), visualization of 3D graphics (MayaVi), etc.

6.2.2 Python Distributions

Both IDE's and packages can be installed separately from Python through command lines in Python terminal. However, there are Python distributions for scientific computing which contains Python and the main modules and IDE's, including Spyder, Jupyter, NumPy, ScyPy and Matplotlib. Some Python distributions are Anaconda, Python-XY and PyCharm. To develop this chapter, Anaconda has been used (more details about Anaconda can be found in the developer website, www.anaconda.com/distribution/ [6]).

6.2.3 Datatype, Operators, and Comments

First, we would differentiate between values and types. A value is the basic information used by any script of Python, and each value belongs to a type. Some usual types are:

- *int*: integers
- *float*: floating point numbers
- *complex*: complex numbers
- *str*: string
- *list*: immutable container that holds objects that can be of different types assigning an index to each object
- *tuple*: mutable container that holds objects that can be of different types assigning an index to each object

Command type returns the type of a value (`python` open the Python's interpreter to start working)

```
python
>>> a=4
>>> type(a)
<class 'int'>
>>> b=5.45
>>> type(a)
<class 'float'>
>>> c='Chemical Engineering'
>>> type(c)
<class 'str'>
```

Data types can be changed between some types. For instance, an '*int*' can be converted into '*float*', a '*float*' into '*int*' removing the decimal component, a '*string*' formed by a number can be converted to '*int*' or '*float*', but a string formed by text cannot be converted into a number type.

```
>>> float(a)
6.0
>>> int(b)
5
>>> int(c)
Traceback (most recent call last):
  File "<stdin>", line 1, in <module>
ValueError: invalid literal for int() with base 10: 'Chemical
Engineering'>>> type(c)
>>> d='3'
>>> int(d)
3
```

Furthermore, each value can be referred to a named variable. We have used it in previous examples, assign the word 'a' to the value '4' or 'c' to the string 'Chemical Engineering'. If we call 'a' or 'c', the referred value is returned.

```
>>> a=4
>>> a
4
>>> c='Chemical Engineering'
>>> c
'Chemical Engineering'
```

Number variables can be used for mathematical computations using the basic Python operators. These operators are +, -, *, /, and ** for addition, subtraction, multiplication, division, and exponentiation operations, respectively. There is also an operator, '%', that returns the remainder of a division. These operators have the usual mathematical preference order, which can be modified and specified using parentheses.

Python can also work with Boolean operators. Note that the command '=' is used to assign a value to a variable, while the operator equal is represented by '==', and not equal is symbolized by '!='. Other Boolean operators are '>' and '<' for greater than and less than, '>=' and '<=' for equal or greater than and equal or less than, and 'is' and 'is not' to represent if one value is the same or is not the same than other, respectively. Boolean operators are very used in conditional sentences. As it will be presented later, NumPy provides additional functions for scientific computing, as logarithms or trigonometric functions among other elements.

In Python, comments are inserted preceded by a '#' symbol in each line. If there are various lines in the code that to comment, they can be commented writing three quotes, """", at both the start and the end of the commented lines.

6.2.4 Data Structures

Data can be arranged in various structures to be managed later. The main structures to arrange data are lists, tuples, sets, and dictionaries.

Both lists and tuples are sequences of data arranged with a specific order with an assigned index for each object. *Lists* are mutable and are defined by brackets, '[]', with the different elements separated by commas. The elements of a list can be of different types, including other nested lists. The index for data structures in Python starts with 0, and the elements of a list can be also indexed from the end with negative indexes, starting in –1. This index can be used to return an element from the list through the variable assigned to the list and the index between brackets. Also can be returned a list section specifying the indexes of the first element index and the final element, which is not included in the output, separated by ':'.

```
>>> a=[1, 2.5, 'A', 8.6]
>>> a
[1, 2.5, 'A', 8.6]
>>> a[0]
1
>>> a[2]
'A'
>>> a[-1]
```

```
8.6
>>> a[0:3]
[1, 2.5, 'A']
```

For each element in a list, a variable can be assigned, and this can be returned and managed separately from the other elements.

```
>>> a=[1, 2.5, 'A', 8.6]
>>> n1, n2, n3, n4=a
>>> n1
1
>>> n3
'A'
>>> n4+2
10.6
```

Lists can be modified in various ways. The value of an element can be changed selecting the element list index and equaling it to another value. Additionally, *lists* can be extended at the last position adding an '*x*' new element using the list built-in function *append(x)*. The build-in functions are used calling to the list variable followed by a dot and the built-in function name. If the new item needs to be added to a specific '*i*' position, the *insert(I, x)* function is used. List '*b*' can be added to another list using the *extend(b)* build-in function. *remove(x)* removes the first list element which value match with '*x*', and *pop(i)* remove and return the element with the '*i*' index in the list. The index of an element '*x*' can be returned with *index(x)* function, and the number of the elements which value is the same than *x* is obtained using *count(x)*. Finally, if a list it is formed only by numbers (integer and float type can be mixed), or by strings, their elements can be sorted using *sort()*, and they can be reversed with *reverse()*.

```
>>> a=[1, 2.5, 'A', 8.6]
>>> a.append(3)
>>> a
[1, 2.5, 'A', 8.6, 3]
>>> b=['C', 85, 6.9]
>>> a.extend(b)
>>> a
[1, 2.5, 'A', 8.6, 3, 'C', 85, 6.9]
>>> a.pop(2)
'A'
>>> a
[1, 2.5, 8.6, 3, 'C', 85, 6.9]
>>> c=[2.6,0.2,0.33,8.9]
>>> c.sort()
>>> c
[0.2, 0.33, 2.6, 8.9]
>>> d=['Pluto', 'Charon', 'Eris', 'Sedna', 'Hydra']
>>> d.sort()
>>> d
['Charon', 'Eris', 'Hydra', 'Pluto', 'Sedna']
```

It is possible to define an empty list to be filled later, where each element is represented by a pair of parenthesis, (), instead of a number or a string.

Tuples are elements similar to lists, with indexation of their elements, but they are immutable. Tuples are defined separating their elements by commas, and can be contained within parenthesis, (), or without them. As in the case of lists, the elements can belong to different types. If one try to change an element or modify the tuple, Python returns an error.

```
>>> f=(1,  2.89, 'Ceres', 963, 'Vesta')
>>> f
(1, 2.89, 'Ceres', 963, 'Vesta')
>>> f=1,  2.89, 'Ceres', 963, 'Vesta'
>>> f
(1, 2.89, 'Ceres', 963, 'Vesta')
>>> f[3]=7
Traceback (most recent call last):
  File "<stdin>", line 1, in <module>
TypeError: 'tuple' object does not support item assignment
>>> f.append(3)
Traceback (most recent call last):
  File "<stdin>", line 1, in <module>
AttributeError: 'tuple' object has no attribute 'append'
['Charon', 'Eris', 'Hydra', 'Pluto', 'Sedna']
```

As lists, tuple elements can be returned using its index and each one can be assigned to a variable to be returned and managed separately from the other elements.

Sets are sequences of data arranged without a specific order, and cannot contain repeated data. Therefore, sets do not support indexing. These data elements can be of several types, similar to lists and tuples. If within a set there are repeated data, when the command is executed these repeated data are eliminated. Sets are defined using the function *set()*, where data elements are introduced between brackets and separated by commas. If the input is a string, the set elements will be formed by each string character separately.

```
>>> g=set([2, "Gliese", 5.6, "Kepler"])
>>> g
{2, 'Gliese', 'Kepler', 5.6}
>>> h=set("And yet it moves")
>>> h
{'n', 'i', 'e', 't', 'y', 'o', 's', 'A', 'd', 'v', ' ', 'm'}
>>> g[0]
Traceback (most recent call last):
  File "<stdin>", line 1, in <module>
TypeError: 'set' object does not support indexing
```

In a similar way than sets theory in mathematics, sets in Python can be object of operations of subtraction using command '-', union through command '|', intersection using '&', and symmetric difference (elements which are not in both sets) using the command '^'. As lists and tuples, each element can be assigned to a variable to be returned and managed separately from the other elements. However, if the element is a string, the number of individual elements which form the string is retuned.

```
>>> i=set(["Gliese", 8, 5.6, 5.5])
>>> g-i
{2, 'Kepler'}
>>> g|i
{2, 5.6, 5.5, 8, 'Kepler', 'Gliese'}
>>> g&i
{'Gliese', 5.6}
>>> g^i
{2, 5.5, 8, 'Kepler'}
>>> y1,y2,y3,y4=i
>>> y1
8
>>> y2
5.5
```

Finally, dictionaries are collections of pairs *key:value*. *key* represents a keyword, and *value* the value associated to the keyword. The *key* can be used to call the assigned *value* and to work with it as if it would be a variable. *Values* must be defined through any immutable type, so tuples can be used for it. Dictionaries can be defined in various ways. They can be constructed writing between curly braces the pair *key:value* using ':', separating each pair with a comma, using the built-in function *dict()*, using a tuple where each pair *key:value,* which is assigned using a comma, is contained in a parenthesis, or using the symbol equal, '=', to assign the pairs. Also, an empty dictionary can be created and assign the items later.

```
>>> j={'one':1, 'two':2, 'three':3}
>>> j
{'one': 1, 'two': 2, 'three': 3}
>>> j=dict([('one',1),('two',2),('three',3)])
>>> j
{'one': 1, 'two': 2, 'three': 3}
>>> j=dict(one=1, two=2, three=3)
>>> j
{'one': 1, 'two': 2, 'three': 3}
>>> j=dict()
>>> j['one']=1
>>> j['two']=2
>>> j['three']=3
>>> j
{'one': 1, 'two': 2, 'three': 3}
>>> j['two']
2
>>> j['three']+3
6
```

If we need to return the *keys* of a dictionary, we can use the built-in function *keys()*. If it is necessary return the *values*, the corresponding built in function is *values()*.

Other useful functions for dictionaries are *in*, which check if a parameter appears as *key*, or *len()*, which returns the number of *key:value* pairs contained in a dictionary. *len()* can be also used for lists.

```
>>> j.keys()
dict_keys(['one', 'two', 'three'])
>>> j.values()
dict_values([1, 2, 3])
>>> len(j)
3
>>> 'two' in j
True
>>> 2 in j
False
```

6.2.5 Conditionals and Loops

Python allows us to write programs with conditional expressions whose verification results in a decision. To achieve it, Boolean expressions are used in combination with the conditional statement *if*. *If* statement checks the condition, and if it is true, the following statement is executed, if it is false, it is skipped.

The structure of a conditional statement is formed by the *if* followed by the Boolean condition, ended by a colon character, ':'. After that, the statements which will be executed if the condition is verified are indented. This indentation is the way to indicate in Python that this code belongs to the conditional statement. If the condition is not verified, another execution can be specified using the command *else*, which have the same structure than *if*, followed by and indented block which contains the alternative statement. Finally, different conditions can be specified in the same conditional expression using *elif*, which is a contraction of *else if*, containing a new condition with its respectively execution statement. *elif* and *else* commands can be combined in the same expression.

```
>>> a=3
>>> if a!=4:
        print ('a is not equal than 4')
a is not equal than 4

>>> b=4
>>> if b<1:
        print ('b is less than 1')
    elif b>10:
        print ('b is greater than 10')
    else:
        print('b is between 1 and 10')
b is between 1 and 10.
```

Different conditionals statements can be nested within one another. This allows to make statements with different alternatives, simultaneous conditions, and so on.

```
>>> c=-20
>>> if c%2==0:
        print('c is an even number')
    if c<0:
        print('c is an even and negative number')
c is an even number
c is an even and negative number
```

In Python, as in other programming languages, there exist loops that allow to execute multiple times code fragments.

There are two kinds of loops: *for* and *while* loops. *for* loops executes a sequence of code multiple times given by a sequence. Therefore *for* loops are definite loops. The sequence can be formed by a list of words, a list of numbers, etc. The structure of a *for* loop is composed by the *for* statement declaration, an iteration variable, which is the variable which changes its value as the loop moves forward, the reserved word *in,* and the variable which contains the elements of the sequence:

```
>>> Fibonacci=[1, 1, 2, 3, 5, 8, 13, 21, 34, 55, 89, 144, 233,
377]
>>> for number in Fibonacci:
        print(number)
```

where *number* is the iteration variable.

On the other hand, *while* loops are indefinite loops, because there is not a predefined number of iterations. *while* loops repeat the execution of a portion of code as long as a Boolean condition is verified. An infinite loop can be constructed using *while* assigning as Boolean condition the value *True.* Its structure is similar to *if* statement.

```
>>> count=0
>>> while count<7:
       count=count+1
       print(count)
```

There exist two control statements which can be used for both *for* and *while* loops. *break* can be used to exit from a loop before than the condition changes, while the *continue* statement can be used to skip the current block and return to the *for* or *while* statement, skipping to the next iteration. In addition, the statement *else* can be used to execute an alternative code if the condition is not verified, in a similar way as conditional execution.

```
>>> count=3
>>> while True:
            count = count-1
    if count==0:
        continue
            print (count)
    if count==-2:
        break
```

```
>>>   count=0
>>> while count<3:
        count=count+1
        print(count)
     else:
        print('end')
end
```

Note that, for loops, as for *if* statements, the loop code must be indented for Python to recognize it as part of the loop.

6.2.6 Functions

A function is a block of code in a program which performs a specific task. To specify a function, a name for it and the statements which will be contained in the function must be defined. Later, the function could be called using the assigned name. Functions can take variables previously defined in the main program, which are called function arguments, and returns values which can be used in the main program or by other functions. To call a function, simply write the function's name followed by the arguments between parentheses.

Functions can also be used to divide the code in more legible blocks, and many times can be reusable from one program to other.

To define a new function, the *def* command is used, followed by the function name and the arguments between parenthesis and ended by a colon character, ':'.

```
>>>def a_power_b (a,b):
      return a**b
>>> x=a_power_b(2,3)
>>> print(x)
```

6.2.7 Scientific Computing with Python

6.2.7.1 Import Packages

The first step to use Python in scientific computing is to import the packages which allow us to use additional mathematical expressions and functions. As we have seen before, the main packages for scientific computing are NumPy, SciPy and Matplotlib. Packages are imported using the command *import* followed by the name of the package. Packages can be imported complete, or just a specific module of the package. Import the whole package is useful when multiple of its features will be used. For instance, Numpy contains several elements commonly used, such as trigonometric functions, logarithms, exponential, etc. However, if a few functions of a package are used, it could be more interesting to import only the modules which contain these functions instead of the complete package to make the program lighter. This partial importation is carried out using the command *from,* specifying the name of the main package followed by the command *import,* and the name of the module which we want to import. Another way to import a module of a package is using the command *import,* specifying the name of the main package followed by a dot and the name of the module which we want to import. In both cases, whole package importation and partial importation, an alias can be assigned to rename the imported package and make easier calling it within the program through the command *as* and the name of the alias after specifying the imported package. For instance, we are going to import the whole NumPy package, renaming it as *np*, the modules *optimize* and *integrate* belonging to the SciPy package without alias, and the module *pyplot* of Matplotlib package with the alias *plt*.

```
>>> import numpy as np
>>> from scipy import optimize
>>> from scipy import integrate
>>> import matplotlib.pyplot as plt
```

Finally, it should be noted that the command *help()* returns Python's help about the package or sub-package called as argument.

6.2.7.2 Mathematical Functions

Numpy contains common mathematical functions such as trigonometric functions, exponents and logarithms. It also allows the use of some constants as π or *e* numbers.

Constants: Constants from NumPy can be used by calling them as *np.constant* (or numpy instead *np* if the alias is no defined).

```
>>> np.pi
3.141592653589793
>>> np.e
2.718281828459045
```

Trigonometric functions: Sine, cosine, tangent, their inverse ones, and the conversions from radians to degrees and from degrees to radians are also implemented in NumPy. They are called in a similar way than constants using their respective functions: *sin*, *cos*, *tan*, *arcsin*, *arccos*, *arctan*, *deg2rad,* and *rad2deg,* respectively. Hyperbolic functions are also defined through the following NumPy functions: *sinh*, *cosh*, *tanh*, *arcsinh*, *arccosh*, *arctanh*. By default, these functions work with angles measured in radians. The argument can be a single number, or an array of numbers. In the last case, the function calculates the result for each number in the array.

```
>>> np.sin(np.pi/2)
1.0
>>> np.tan(np.pi)
-1.2246467991473532e-16
>>> np.rad2deg(2*np.pi)
360.0
```

Exponents and logarithms: Natural logarithm is calculated using the function *log,* whereas *log10* calculates the 10-base logarithm. The exponential of a number is defined by *exp*. In the last case, the function calculates the result for each number in the array.

```
>>> np.log10(100)
2.0
>>> np.exp([1,2])
array([ 2.71828183,  7.3890561 ])
```

Arrays: Arrays can be created using the NumPy function *array(),* defining the array elements between brackets and separated by commas. In addition, a list or a tuple can be created with various elements, and converted into an array using the same function.

Arrays can also be created using other special built-in functions as *zeros*, which creates an array filled with 0. This function can only have one argument, which represents the number of elements in a one-dimensional array, or two elements, defined between brackets and separated by commas, where the first represents the number of rows and the second one the number of column. Another function is *arange*, which creates arrays with regular distribution. If only one number is used as argument, this is the number of natural numbers, starting at 0, which will be contained in the array. If there are two numbers as arguments, they represent the initial and final number which will be contained in the array, and if there is a third argument, this number represents the incrementing value between elements, whose default value is 1. If the objective is to create a space equally separated between the elements, *linspace* is the function of choice, where its three arguments are the initial value, the final value and number of elements in the array.

```
>>> np.array([4,6.5,6,9.9])
array([ 6. ,  6.5,  6. ,  9.9])
>>> x=[3,8.1,21]
>>> np.array(x)
array([  3. ,   8.1,  21. ])
>>> np.zeros([2,4])
array([[ 0.,  0.,  0.,  0.],
       [ 0.,  0.,  0.,  0.]])
>>> np.arange(2,4,0.2)
array([ 2. ,  2.2,  2.4,  2.6,  2.8,  3. ,  3.2,  3.4,  3.6,
3.8])
>>> np.linspace(0,1,6)
array([ 0. ,  0.2,  0.4,  0.6,  0.8,  1. ])
```

Two dimensional arrays, in the form of a matrix, can be created defining the elements of each row as it is described above, and separating different rows by commas, adding additional brackets which enclose all rows. There exists a command to return the dimensions of an array, *shape*, and another command to return the total number of the elements of the array, *size*.

```
>>> a=np.array([[1,2,3,4], np.linspace(1,2,4)])
>>> a
array([[ 1.        ,  2.        ,  3.        ,  6.        ],
       [ 1.        ,  1.33333333,  1.66666667,  2.        ]])
>>> a.shape
(2, 4)
>>> a.size
8
```

6.2.7.3 Solving Nonlinear Algebraic Equations

It is possible to solve non-linear equations in Python using a function provided by SciPy. This is the *root* function, which is contained in the module *optimize*. The basic syntax is as follows:

root(fun, x0, args = (), method = 'hybr')

In the called to the *root* function, the first argument is the name of the function where the equation is defined. In the case that various equations will be solved, as a system of equations, they are defined within the function in the return statement, separated by commas and with brackets involving the whole system. The second argument of *root* is the initial guess, and the third argument is the variables of the equations that are being solved. The arguments are declared as *args=(),* and between the parenthesis the parameters are inserted separated by commas. It is possible to select different solution methods. We can add them as a fourth argument of the function root. Currently, methods such as Powel hybrid (hybr), least squares based on Levenberg-Marquardt (lm), derivative free (df-sane), and different inexact Newton methods including Broyden's method (broyden1), among others, are available

If we call to the calculated variable, SciPy will return a report about the convergence of the process, number of iterations required, etc. If the calculated variable is not called, this report will be not returned.

An aspect that must be highlighted is that the returned variable type is a *root* SciPy specific type. Therefore, to manage it later, the returned variable must convert to float (or integer) variable.

6.2.7.4 Integrating with Python

In Python it is possible to integrate using the *integrate* module of SciPy. *Integrate* contains functions to integrate expressions, as well as to solve ordinary differential equations (ODEs).

Regarding integration, the module contains several integration algorithms, some of them are classical numerical integration algorithms as the trapezoidal rule, which built-in function is *trapz,* the composite trapezoidal rule through *cumtrapz,* the Simpson's rule defined in the function *simps,* or the Gauss's quadrature method, which is implemented in *fixed_quad.* The integration computing is carried out using a *n* fixed order, or *quadrature* if the integration computing uses a fixed tolerance to perform the integration. However, integrate also contains the function *quad,* which is used for a general purpose integration, and it has two variants, *dblquad* and *tplquad,* for double and triple integrations respectively.

The syntax of the *quad* function has four principal arguments. First, the users defined function, which contains the equation or equations to be integrated, similar to function *root,* explained previously. Second, the lower limit of integration, where *–numpy.inf* (or *–np.inf* if we are working with the alias *np*) can be used for negative infinity. The third argument is the upper integration limit, which as the lower level *numpy.inf* (or *np.inf*) can be used for positive infinity, and the fourth is the argument *args=(),* which contains the equation defined function parameters called from the main program, inserted separated by commas.

The user defined function which contains the equation to be integrated has the same structure as the *root* function: the equation or equations (equaled to zero) to be integrated are defined in the return statement, the first function parameter must be the integration variable, and the parameters for the equations, which can be defined in the main program and be called for by the equation defined function as arguments behind unknown variable argument, or be defined within the self-function. *quad* function returns two parameters, the first one is the result of the integration between the defined limits, while the second is the estimated absolute error in the result. The returned values are tuple type, so in contrast with *root* function, no type conversion is required to work with them.

```
>>> import numpy as np
>>> from scipy import integrate
>>> def  int_func(x, a, b):
        return  a/(b+x**2)
>>> a=2
>>> b=1
>>> result=integrate.quad(int_func, -2, np.inf, args=(a,b))
>>> result
(5.355890089177968, 6.084862969713607e-10)
>>> type(result)
tuple
```

6.2.7.5 Solving Differential Equations

As we have pointed out previously, the sub-package integrate contains built-in functions to solve ordinary differential equations (ODEs). In particular, these functions are *ode* and *odeint*. We will focus on the *odeint* function, which is a general purpose function to solve an ordinary differential equation or ODEs systems.

To work with *odeint,* it is necessary to define a function which contains the differential equation, or equations, where the first argument is the dependent variable, and the second arguent is the independent variable(s). After these arguments, other parameters can be defined in the main program. These additional arguments provide the parameters used by the function where the differential equation is defined, and define additional features for the solvers. If there are several equations, they can be defined using an array, where the first element corresponds with the first equation, the second element with the second equation, etc., or assigning different variables for each equation. The *return* argument should contain the variable that represents the derivatives.

After defining the function, the initial conditions of the variables should be defined, as well as the independent variable interval to evaluate the differential equations. Usually it can be defined using the *arrange* or *linspace* functions.

Finally, the *odeint* function is called, where the first function argument is the function name that includes the equations to be solved, the second argument is the initial conditions value, and the third argument is the value of the independent equations over the differential equations will be evaluated. *odeint* returns an array which contains the value of each defendant variable evaluated over each independent variable value.

It uses the *lsoda* from the FORTRAN library odepack [3], that automatically switches between implicit Adams method (for non-stiff problems), and a method based on backward differentiation formulas (BDF) (for stiff problems).

6.2.7.6 Plotting

Matplotlib Python's package allows producing two and three dimensional plots from data contained in lists or arrays. Matplotlib contains the *pyplot* module, which provides an interface similar to MATLAB® with the aim of ease the plots generating process. Additionally, Matplotlib provides functions to save plots in several image formats, including .jpeg, .pdf, .png, .ps. raw, .svg and .tiff.

Matplotlib package provides a large number of options. However, we will focus in two ways to use this package, the sub-package *pyplot,* and the object-oriented interface.

pyplot is a set of functions which allows the user prepare graphics easily, with a similar structure to MATLAB®, see Chapter 4. The main basic *pyplot* function is *plot*, whose arguments include the elements to be represented, defined as lists, arrays, o tuples. The order of the arguments is, first, the values for the x axis, and in second place, the values for the y axis. Finally, it is possible to used additional arguments to modify the figure shape. By default, the plot will be a marker linked line graphic without grid, but this presentation can be modified trough additional arguments. The color can be modified using an argument which specifies the color between quotes. Color can be specified following different strategies: writing full names as *'yellow'*, abbreviated names as *'y'*, hexadecimal color code strings as *'#FFFF00',* or RGB or RGBA codes as tuples (255, 255, 0, 1). If the color is defined as an abbreviation, the shape of the line style or markers also can be modified together with the color parameter. To do this, it is necessary to write a string character which represents a specific line style or marker within the same quotes than color abbreviation, resulting in a two-element string, one of them for color and another for the line style or marker. Finally, two shape control characters can be added, one for line style and other for markers. The abbreviated names supported and the control strings for line style and markers are collected in Table 6.1.

Labels can be added using the *pyplot* functions *xlabel* and *ylabel* respectively. The title of the label is added between parenthesis and with quotes if the label is a

TABLE 6.1
Abbreviated Names Supported and Control Strings for Line Style and Markers (Matplotlib documentation, http://matplotlib.org/api/pyplot_api.html#matplotlib.pyplot.plot)

Character	Line style or marker	Character	Line style or marker	Character	Color
'-'	solid line style	'3'	tri_left marker	'b'	blue
'--'	dashed line style	'4'	tri_right marker	'g'	green
'-.'	dash-dot line style	's'	square marker	'r'	red
':'	dotted line style	'p'	pentagon marker	'c'	cyan
'.'	point marker	'*'	star marker	'm'	magenta
','	pixel marker	'h'	hexagon1 marker	'y'	yellow
'o'	circle marker	'H'	hexagon2 marker	'k'	black
'v'	triangle_down marker	'+'	plus marker	'w'	white
'^'	triangle_up marker	'x'	x marker		
'<'	triangle_left marker	'D'	diamond marker		
'>'	triangle_right marker	'd'	thin_diamond marker		
'1'	tri_down marker	'|'	vline marker		
'2'	tri_up marker	'_'	hline marker		

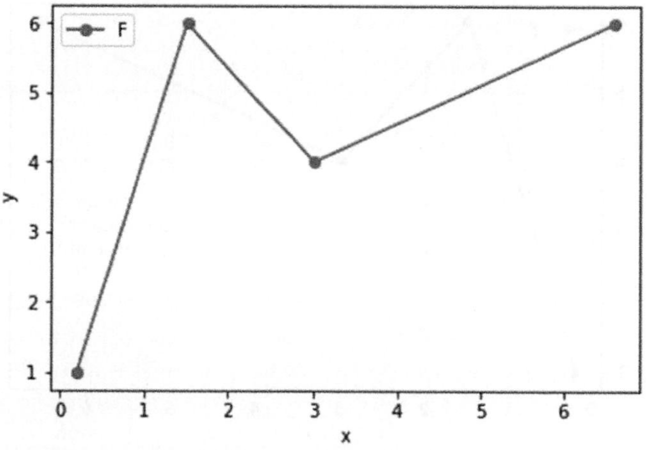

FIGURE 6.2 Example of pyplot default style plot.

string. These arguments must be added following the *plot* function to be recognized as the labels of the figure. The function *legend* prepares the legend for the figure. As in the labels case, the legend title should be defined between parenthesis, and with quotes if it is a string. A grid can also be added through the function *grid*, which argument can be a Boolean value *True* or *False*, or a string value, '*on*' and '*off*', to insert or not the grid. Figure 6.2 shows the figures generated with the example below:

```
import numpy as np
from scipy import integrate
import matplotlib.pyplot as plt
a=[0.2, 1.5 ,3, 6.6]
b=[1,6,4, 6]
plt.plot(a,b, 'ro-')
plt.xlabel('x')
plt.ylabel('y')
plt.legend('F')
```

However, *plot* can be used with a non-abbreviate structure which enables preparing more detailed figures. This structure is based on keywords arguments trough which different figure elements can be controlled: the line width, color, line style, marker shape, marker color, and marker size. Furthermore, it can also create an auto legend, specifying in the *plot* function the label and later calling it to the *label* function, change the colormap used in the figure, etc. All arguments which can be added are collected in the Matplotlib site, in the *pyplot* documentation section (https://matplotlib.org/api/pyplot_api), see Figure 6.3.

```
plt.plot(a, b, color='r', linestyle='dashed', marker='o',
     markerfacecolor='k', markersize=6, label='F')
plt.legend()
plt.grid(True)
```

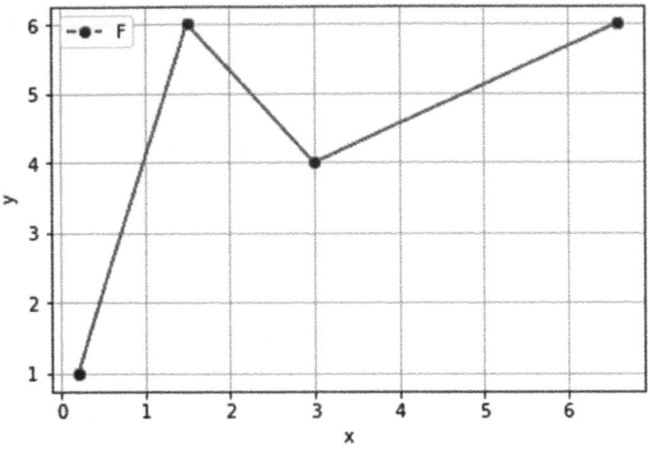

FIGURE 6.3 Example of a figure created with a non-abbreviated structure.

The object-oriented interface allow us even greater control over the figure creation process than the function *pyplot*. The object-oriented interface is used through the *subplots* function, creating two objects, *fig* and *axes* (we are going to use these nomenclature, but this is arbitrary and any other can be used), which enables controlling the figure and the axes system independently. When *fig* and *axes* are created, all the actions to build the figure will be performed through them. We work with the object *axes* to build the figure, and this is composed in the object *fig*. The object *axes* can use the *plot* function, and accepts as arguments the elements to be represented, defined in first place the vales for the x axis, and in second place the values for the y axis, and the control shape arguments as abbreviate way or with the keywords, in a similar way than the previous case. To represent different data sets in the same figure, *axes.plot* declarations will be built for each representation.

There are other additional functions that can be used with the *axes* object, as *grid,* to add or remove the grid, *set_xlabel* and *set_ylabel* to insert the axes labels, *legend* to insert a legend in the figure, etc. To know more about the different functions that can be used with the object axes, we recommend see the Matplotlib documentation. Let us remember that these actions can be also carried out using the function *pyplot*. The axes limits can be specified manually using the function *set_xlim* and *set_ylim* within the object *figure,* entering the low limit and the upper limit as the first and the second arguments of the function respectively.

Other useful function is *savefig*, which save the figure created in the same directory where the main program is located. This function is used with the object *fig,* and can accept several extensions, among them the ones enumerated at the beginning of the section, specifying the archive name and its format extension as an argument of the function, write within quotes.

As an example, we will represent the differential equation system solution solved in the Section 6.3.3.

There are other kinds of figures that can be made with Matplotlib and *pyplot*. For example, a scattering representation can be obtained with the *pyplot* command *scatter*, and a 3D figure can be represented though its level curves through the *contour* function. These functions accept as arguments the elements to be represented. In the first place, the x values are provided. In a second place, the y values, and in the case of a *countour* function, the z values in third position. These functions accept their own shape control abbreviate arguments and keywords. These functions also can be used in both sub-package *pyplot* and object oriented interface through the object *axes.*

6.3 EXAMPLES

Once we know how work with the basic Python functions and with the scientific computing libraries, some common examples applied to Chemical Engineering will be shown. In particular, we will show how we can model fluidized bed reactors, chemical reactions in a CSTR and in a reactive distillation process, and a cooling tower, with the aim to present the different features described along the chapter.

6.3.1 FLUIDIZATION

As an example, we will solve the terminal velocity equation, u_t, provided by Kunii and Levenspiel [7] in their book "Fluidization Engineering", Eq. 6.1, using the arbitrary set of parameters described below.

$$u_t^* = u_t \left(\frac{\rho_g^2}{\mu_g (\rho_s - \rho_g) g} \right)^{1/3} \tag{6.1}$$

$$u_t^* = 0.5879 \ (dimensionless)$$
$$\rho_g = 0.1355 \ kg / m^3$$
$$\mu_g = 2.6394 \cdot 10^{-5} \ kg / (m \cdot s)$$
$$\rho_s = 2330 \ kg / m^3$$
$$g = 9.81 \ m / s^2$$

```
>>> import numpy as np
>>> from scipy import optimize
>>> g=9.81
>>> fi_s=0.6 #sphericity
>>> rho_s=2330 #particle density (kg/m3)
>>> rho_g=0.1355 #gas density (kg/m3)
>>> mi_g=2.6394E-5 #gas viscosity (kg/m·s)
>>> u_t_star=0.5879
#Solving non-linear algebraic equations
>>> def  f_u_t(x_u_t, rho_g, mi_g, rho_s, u_t_star, g):
        return x_u_t*((rho_g**2)/(mi_g*(rho_s-rho_g)*g))
        **(1/3)-u_t_star
>>> u_t=optimize.root(f_u_t, 1, args=(rho_g, mi_g, rho_s,
u_t_star,g))
```

```
>>> u_t
    fjac: array([[-1.]])
     fun: array([ 0.])
 message: 'The solution converged.'
    nfev: 4
     qtf: array([ -1.33503164e-09])
       r: array([-0.31221797])
  status: 1
 success: True
       x: array([ 1.88297935])
>>> type (u_t)
scipy.optimize.optimize.OptimizeResult
>>> u_t=np.float_(u_t.x)
>>> print ('u_t (m/s) =',u_t)
u_t (m/s) = 1.88297935311
```

6.3.2 Chemical Reaction in a CSTR

In this example, we are going to consider two consecutive reactions, where a product from first reaction performs the role of reactant in the second one [8]. The reactions are the following:

$$\begin{aligned} M + H_2 &\rightarrow X + Me \\ X + H_2 &\rightarrow T + Me \end{aligned} \tag{6.2}$$

The kinetics of each reaction referred to the reactants are described below, where r_j is referred to the change of concentrations with time, k_1 and k_2 are the kinetic coefficients for each reaction, which values are temperature dependent, and C_j is the concentration of the component j.

$$\begin{aligned} -r_M &= k_1 C_M C_H^{0.5} \\ r_T &= k_2 C_X C_H^{0.5} \end{aligned} \tag{6.3}$$

These reactions are carried out in a continuous stirred tank reactor (CSTR), operating in continuous mode. The molar balances for the reactor, evaluated for each component, are the following.

$$\begin{aligned} H_2 &: F_{H_0} - F_H = \left((-r_M) + (r_T)\right)V \\ M &: F_{M_0} - F_M = \left((-r_M)\right)V \\ X &: F_X = \left((-r_M) + (r_T)\right)V \\ T &: F_T = \left((r_T)\right)V \\ Me &: F_{Me} = \left((-r_M) + (r_T)\right)V \end{aligned} \tag{6.4}$$

Where the F_{j0} is referred to input molar flow of component j, F_j is referred to exit molar flow of component j, and V is the volume of the reaction media within the reactor.

We can consider that there is no changes in the density of the fluid, and if we also consider that reactor operates in steady state and there is no mass accumulation inside of the reactor, both total mass flow and total volumetric flow are equal at the input and the output reactor streams, with a value which will be designed as v_0. Considering this, component molar flows can be substituted by the product of component concentration and total volumetric flow.

$$\begin{aligned} F_H &= v_o C_H \\ F_M &= v_o C_M \\ F_X &= v_o C_X \\ F_T &= v_o C_T = F_{Mo} - F_M - F_X \\ F_{Me} &= v_o C_{Me} = v_o (C_{Ho} - C_H) \end{aligned} \tag{6.5}$$

Finally, if we combine this set of equations with the component molar balances, we obtain the following expressions, where the V / v_0 results in the residence time, denoted by τ.

$$\begin{aligned} H_2 &: C_{Ho} - C_H = \left(k_1 C_M C_H^{0.5} + k_2 C_X C_H^{0.5}\right)\tau \\ M &: C_{Mo} - C_M = \left(k_1 C_M C_H^{0.5}\right)\tau \\ X &: C_X = \left(k_1 C_M C_H^{0.5} - k_2 C_X C_H^{0.5}\right)\tau \\ T &: C_T = \left(k_2 C_X C_H^{0.5}\right)\tau \\ Me &: C_{Me} = \left(k_1 C_M C_H^{0.5} + k_2 C_X C_H^{0.5}\right)\tau \end{aligned} \tag{6.6}$$

The resolution of this set of equations returns the values of components concentrations in steady state. To solve this model, we have considered the following values for the kinetic coefficients, residence time, and initial concentration:

$$\begin{aligned} &k_1 = 55.2\ min^{-1}; k_2 = 30.2\ min^{-1}; \tau = 0.5\ min; C_{H_0} = 0.006\ mol/l; \\ &C_{M_0} = 0.003\ mol/l; C_{X_0} = 0.005\ mol/l; C_{T_0} = 0\ mol/l; C_{Me_0} = 0\ mol/l \end{aligned} \tag{6.7}$$

Considering the equations and conditions presented, we can solve the CSTR model in Python using the program shown below. Spyder IDE has been used in this model.

```
#Import libraries
import numpy as np
from scipy import optimize
from scipy import integrate
import matplotlib.pyplot as plt

#We create an empty list to store the concentrations data
Conc=[(),(),(),(),()]
```

```
#Parameters
#min-1
k1=55.2
k2=30.2
CH_0=0.006
CM_0=0.003
tau=0.5

#Equations: We define a function which collect the equations
to solve

def  Eq(Conc, CHo, CMo, tau, k1,k2):

    return [CH_0-Conc[0]-tau*(k1*Conc[1]*Conc[0]**0.5+k2*
           Conc[2]*Conc[0]**0.5),
            CM_0-Conc[1]-tau*(k1*Conc[1]*Conc[0]**0.5),
            Conc[2]-tau*(k1*Conc[1]*Conc[0]**0.5-k2*Conc[2]
            *Conc[0]**0.5),
            Conc[3]-tau*(k2*Conc[2]*Conc[0]**0.5),
            Conc[4]-tau*(k1*Conc[1]*Conc[0]**0.5+k2*Conc[2]*
            Conc[0]**0.5)]

#We collect the results in a list calles Res
Res=optimize.root(Eq, [0,0,0,0,0], args=(CH_0, CM_0, tau,
k1,k2))
#Change the data type to float to manage the results
Res=np.float_(Res.x)
#Print the results list
print ('Res',Res)

#Store each concentration result in a new variable
CH=Res[0]
CM=Res[1]
CX=Res[2]
CT=Res[3]
CMe=Res[4]

#Print the new variables to show the concentration of each
component separately
print("CH=", CH, "\n" "CM=", CM, "\n" "CX=", CX, "\n" "CT=",
CT, "\n" "CMe=", CMe)
```

6.3.3 Chemical Reaction in a Reactive Distillation System

We consider the reaction of acetic acid with methanol to obtain methyl acetate based on [8]. This process, which is a chemical equilibrium, can be carried out through a reactive distillation, so that methyl acetate will be eliminated from the reactor and recovered as gas, as reaction moves forward and the equilibrium is displaced to the products formation. It can be achieved because the methyl acetate boiling point is lower than the reactants, enabling the elimination of the methyl acetate from the reactor while the reaction takes place. The reaction occurs as follows:

$$CH_3 - COOH + CH_3OH \leftrightarrow CH_3 - COOCH_3 \uparrow + H_2O \tag{6.8}$$

We consider that the operation is carried out in a vessel where a certain amount of acetic acid is deposited, while methanol will be added at a constant rate, with a given volumetric flow v_0, and mass flow. Additional acetic acid will not be added, so only the initial acetic acid contained in the vessel is involved in the reaction. Methyl acetate will be produced, and a fraction of methyl acetate is eliminated by evaporation, as we have indicated above, with an evaporation flow F_{Acme}, while water will be retained within the vessel. Total volume will increases as reaction take place until reaches a certain volume. Then the reactor is emptied, and the process starts again.

Equilibrium is given by the equilibrium coefficient, which is function of the temperature, and it is evaluated through the next correlation:

$$K_{eq} = \frac{[CH_3 - COOCH_3]\cdot[H_2O]}{[CH_3 - COOH]\cdot[CH_3OH]} = 5.2\cdot e^{\left(-\left(\frac{8000}{1.978}\right)\cdot\left(\frac{1}{298}-\frac{1}{T}\right)\right)} \tag{6.9}$$

The kinetic coefficient, which is function of temperature, is evaluated using the following expression:

$$k = \left(8.8\cdot 10^8\right)\cdot e^{\left(\frac{-7032.1}{T}\right)} \tag{6.10}$$

Different parameters are collected in the next expressions:

$$MW_{acetic\ acid} = 60; \rho_{acetic\ acid} = 104.9\ g/cm^3; MW_{methyl\ acetate} = 74; \rho_{methyl\ acetate} = 943.4\ g/cm^3$$

Finally, we have the following initial values:

$$T = 300\ K;\ P = 101\ kPa;\ N_{Ace_0} = 300\ kmol;\ [MetOH]_0 = 5\ kmol/cm^3; v_0 = 1\ cm^3$$

To compute the initial volume within the reactor, we calculate the volume of the initial acetic acid:

$$V_0 = N_{Ace0}\cdot\frac{MW_{acetic\ acid}}{\rho_{acetic\ acid}} \tag{6.11}$$

Mass balances for each component should consider for the mass flow of the components at the inlet and outlet streams, and for the disappearance and appearance of components due to chemical reaction. Volume variation along time will be defined by the difference between the inlet and outlet volumetric flow:

$$\begin{aligned}
-\frac{dN_{Ace}}{dt} &= k\left(C_{Ace}C_{Met} - \frac{C_{acme}C_{H2O}}{K_{eq}}\right)V \\
-\frac{dN_{MetOH}}{dt} &= k\left(C_{Ace}C_{Met} - \frac{C_{acme}C_{H2O}}{K_{eq}}\right)V - F_{MetOH} \\
\frac{dN_{H2O}}{dt} &= k\left(C_{Ace}C_{Met} - \frac{C_{acme}C_{H2O}}{K_{eq}}\right)V \\
\frac{dN_{Acme}}{dt} &= k\left(C_{Ace}C_{Met} - \frac{C_{acme}C_{H2O}}{K_{eq}}\right)V - F_{Acme} \\
\frac{dV}{dt} &= vo - F_{Acme}\frac{M_{Acme}}{\rho_{Acme}} \\
F_{Acme} &= x_{Acme}\frac{Pv_{Acmet}}{P_{total}}
\end{aligned} \tag{6.12}$$

The evaporated flow of methyl acetate is given by the product of the methyl acetate molar fraction by the methyl acetate vapor pressure and total pressure. Methyl acetate vapor pressure is given by the next correlation:

$$P_{vap} = 4750 \cdot e^{\left(10.703 - \left(\frac{11.0088}{T_r}\right) - 5.431 \cdot log(T_r) + 0.3058 \cdot T_r^6\right)} \tag{6.13}$$

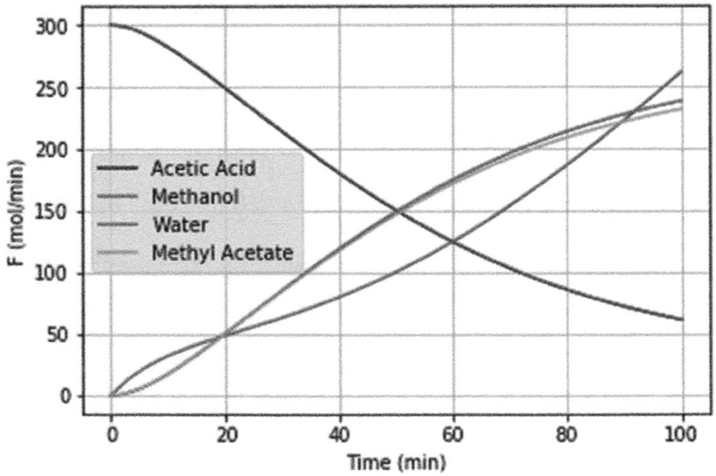

FIGURE 6.4 Results of example 6.3.3. **(See color insert.)**

Where T_r is the reduced temperature of methyl acetate given by:

$$T_r = T / 506.5 \tag{6.14}$$

Considering the equations and conditions exposed above, we can solve the reactive distillation model in Python using the program shown below. Spyder IDE's have been used in this model. The result is plotted in Figure 6.4.

```
#Packages import
import numpy as np
from scipy import optimize
from scipy import integrate
import matplotlib.pyplot as plt

#Initial conditions and parameter definition
N_AA_0=300 #intial number of kmol of AA

MW_AA=60 #AA molecular weight
rho_AA=106.9 #AA density (g/cm3)
MW_AAMe=74 #AAMe molecular weight
rho_AAMe=943.4 #AA_Me density (g/cm3)
T=300 #Temperature
T_r=T/506.5 #Reduced temperature
P_tot=101 #Pressure (kPa)
V_0=N_AA_0*MW_AA/rho_AA #Initial volumen wtihin the reactor

#Reaction parameters
k=(8.8E8)*np.exp(-7032.1/T)
K_eq=5.2*np.exp(-(8000/1.978)*(1/298-1/T))
P_vap=4750*np.exp(10.703-(11.0088/T_r)-5.431*np.
log(T_r)+0.3058*T_r**6)

#The empty list N is created to save the number of moles
variation of each
#component in the first four places and the volume variation
in the last position
N=[(),(),(),(),()]

#The empty list Conc is created to save theconcetration
variation
Conc=[(),(),(),()]

#A function is defined to collect the equations to be solved
def kinetics (N,t):
# Concentration calculation of each comonent
        Conc[0]=N[0]/N[4]
        Conc[1]=N[1]/N[4]
        Conc[2]=N[2]/N[4]
        Conc[3]=N[3]/N[4]
#Acetic acid reaction rate
        r_AA=k*(Conc[1]*Conc[0]-Conc[3]*Conc[2]/K_eq)
```

```
#Additional equations
        x_AAMe=N[3]/(N[0]+N[1]+N[2]+N[3]) #Molar fraction of
        methyl acetate
        F_AAMe=x_AAMe*P_vap/P_tot #Methyl acetate removed flow
        v_0=1 # inlet volumetric flow
        C_MetOH_0=5 #inital MetOH concentration
        F_MetOH_0=v_0*C_MetOH_0 #inlet MetOH molar flow
        alpha=MW_AAMe/rho_AAMe #alpha parameter
#An empty array is created to save define the equations
        dNdt=np.zeros(5)
        dNdt[0] = -r_AA*N[4]
        dNdt[1] = -r_AA*N[4]+F_MetOH_0
        dNdt[2] = r_AA*N[4]
        dNdt[3] = r_AA*N[4]-F_AAMe
        dNdt[4] = v_0-F_AAMe*alpha
        return dNdt
#Creation of the time array over the equations will be evaluated
t=np.linspace(0,100,500)

#Initial conditions
N_ini=[N_AA_0, 0, 0, 0, V_0]

#ODE system solver
Res=integrate.odeint(kinetics,N_ini,t)

#Print the results
print ('Res',Res)

#Create plots:
#Number of moles variation
fig,axes=plt.subplots()
axes.plot(t,Res[:,0],'b', label='Acetic Acid')
axes.plot(t,Res[:,1],'g', label='Methanol')
axes.plot(t,Res[:,2],'r', label='Water')
axes.plot(t,Res[:,3],'c', label='Methyl Acetate')
axes.set_xlabel('Time (min)')
axes.set_ylabel('F (mol/min)')
axes.grid('on') #Insert grid
legend = axes.legend(loc='best', fontsize='medium') #Legend
location and font size
legend.get_frame().set_facecolor('#FFD700') #Nicer background
color on the legend.

#Volumen variation
fig,axes=plt.subplots()
axes.plot(t,Res[:,4],'k', label='Volume')
axes.set_xlabel('Time (min)')
axes.set_ylabel('V (m3)')
axes.grid('on')
legend = axes.legend(loc='best', fontsize='medium')
legend.get_frame().set_facecolor('#FFD700')
```

6.3.4 Cooling Tower Modeling

Cooling towers are used to dissipate heat from water streams. Their performance is based on the partial evaporation of the water inside the tower, where water and air flow counter currently. Therefore, the latent heat involved in the phase change cools down the liquid water. The tower is packed to distribute water and provide large contact surface between water and air, enhancing the heat transfer. It must be noted that, as the operation is carried out with atmospheric air, the operation of the cooling towers depends on the atmospheric conditions, in particular, the air humidity and temperature.

As there is an evaporation process in the contact between water and air, there are water mass losses in the process. However, these loses are small with respect to the total water mass flow. Therefore, the total water flow is assumed as constant for the calculations.

The water cooling limit is determined by the wet bulb temperature because the water evaporation driving force is approximately equal to the real water vapor pressure minus the water vapor pressure at the wet bulb temperature. The gas phase resistances control the process.

The designing of cooling towers is based on the mass and energy exchanges that take place inside the tower between the water stream and the air flow. Water is evaporated in the interphase between liquid and gas phase due to a heat transfer process where energy is taken from liquid warm water. Water vapor diffuses to the air bulk due to enthalpy gradient, which is the driving force of the diffusion process. The driving force for heat transfer is the temperature gradient between the liquid warm water and the interphase, and between the interphase and the cool air. Sensible heat flows in both liquid and gas phase, while latent heat flows from the interphase to the bulk air. The sensible heat, which flows from the bulk liquid to the interphase, is equal to the sum of the latent heat plus the sensible heat flowing from the interphase to the bulk gas. The detail mathematical analysis can be seen in Chapter 3.

The model is initialized importing the necessary packages and defining the tower inlet and outlet streams temperatures, total pressure, and other initial parameters as follows:

```
import numpy as np
import matplotlib.pyplot as plt
from scipy import optimize

#~ Parameters

Ptotal=698                #Total pressure (mmHg)
T1=21.7                   #Air Inlet Dry Temperature point 1 (°C)
T2=16.5                   #Air Inlet Wet Temperature point 2 (°C)
T3=19                     #Air Outlet Wet Temperature point 3 (°C)
T4=19.4                   #Air Outlet Dry Temperature point 4 (°C)
T5=21.9                   #Water Inlet  Temperature point 5 (°C)
T6=19.2                   #Water Outlet Temperature point 6 (°C)

orif=25                   #Orifice diferential(mmH2O)
Magua=40                  #Water mass
```

Regarding energy balance, two terms must be considered: latent and sensible heat, due to there are an energy exchange between both phases, and therefore, their temperature change as well. The energy balance will be evaluated in the interphase between water and air. Henceforth, subindexes are used for each phase, *g* for the gas phase and *l* for the liquid phase. It is assumed that the process is carried out adiabatically. The reference conditions considered are 0 °C and atmospheric pressure.

The interphase surface is unknown. Therefore, a specific exchange area is set, defined as interphase area per volume of packed section (m^2/m^3).

On the other hand, the enthalpy of the air-water vapor mixture, H_y, is calculated adding the dry air and vapor water enthalpies.

To calculate air and vapor water enthalpies, it is necessary to compute water vapor pressure through the following correlation:

$$\begin{aligned} P_v(\text{mmHg}) = & -2.73933{\cdot}10^{-11}\left(T(^{\circ}C)\right)^6 + 2.85333{\cdot}10^{-8}\left(T(^{\circ}C)\right)^5 \\ & +1.27164{\cdot}10^{-6}\left(T(^{\circ}C)\right)^4 + 2.36198{\cdot}10^{-4}\left(T(^{\circ}C)\right)^3 \\ & +1.00275{\cdot}10^{-2}\left(T(^{\circ}C)\right)^2 + 3.38158{\cdot}10^{-1}T(^{\circ}C) + 4.57466 \end{aligned} \tag{6.15}$$

Using the absolute humidity definition with the appropriate molecular weights relation, we obtain the following relation:

$$Y = 0.62 \cdot \frac{P_V}{P_T - P_V} \tag{6.16}$$

The enthalpy of the air-water vapor mixture is calculated as follows:

$$H_y(\text{kcal/kg}) = (0.24 + 0.46{\cdot}Y){\cdot}T(^{\circ}C) + 597.2{\cdot}Y \tag{6.17}$$

It has been evaluated in the temperature range from 1 to 50, which is the tower expected performance range. The implementation in Python is the following: first, the expressions described above are evaluated along the temperature points, then a fourth grade polynomial is adjusted to the points to obtain the air-water vapor mixture enthalpy curve.

```
#~ Computing H_WetAir(T) curve
T_evaluation=np.arange(1,51,1) #Temperatures to evaluate H

PV=[]
Yvap=[]
H=[]

for j in T_evaluation: #The curve is calculated from 1 to
50 points
      PV_j=-2.73933E-11*j**6+2.85333E-8*j**5+1.27164E-
6*j**4+2.36198E-
4*j**3+1.00275E-2*j**2+3.38158E-1*j+6.57466
      Yvap_j=0.62*PV_j/(Ptotal-PV_j)
```

```
    H_j=(0.24+0.46*Yvap_j)*j+597.2*Yvap_j

    PV.append(PV_j)
    Yvap.append(Yvap_j)
    H.append(H_j)

print('PV=', PV, '\n')
print('Yvap=', Yvap, '\n')
print('H=', H, '\n')

a=np.polyfit(T_evaluation,H,4)
```

The air outlet stream properties are calculated as:

- Wet vapor pressure: the equation 6.15 is used.
- Vapor pressure of the wet bulb: for the air–water system we have the following expression.

$$\frac{P_{Bulb} - P_V}{T_{dry} - T_{wet}} = \frac{h_c}{k_y \cdot MW_{vapor} \cdot \lambda_o} = 0.5 \tag{6.18}$$

- Absolute humidity: the equation 6.16 is used.
- Specific volume of the water vapor - air mixture:

$$v = \frac{RT}{P}\left(\frac{1}{MW_{air}} + \frac{y}{MW_{vapor}}\right) \tag{6.19}$$

- Enthalpy: the equation 6.17 is used.
- Dry air mass.

Where *MW* represent the molecular weight.

```
#~ AIR OUTLET STREAM
#~ Outlet Wet Vapor Pressure (point 3)
PWts=-2.73933e-11*(T3)**6+2.85333e-8*(T3)**5+1.27164e-
6*(T3)**4+2.36198e-4*(T3)**3+1.00275e-2*(T3)**2+3.38158e-1*
(T3)+6.57466
#~ Water Vapor Pressure (mmHg)
PVts=PWts-0.5*(T4-T3)
#~ Outlet humidity (kg/kg_dry_air)
Ys=0.62*PVts/(Ptotal-PVts)
#~ Specific volumen of the water vapor - air mixture which
leave the column in the head stream
VH=(1/29+Ys/18)*(0.082*(273+T4))/(Ptotal/760)
#~ Outlet stream enthalpy (kcal/kg_a_s)
Hs=(0.24+0.46*Ys)*T4+597.2*Ys
#~ Dry air mass (kg/s)
Mas=0.0137*(orif/VH)**0.5
```

The same parameters, except dry mass, are calculated for the inlet stream:

```
#~ AIR INLET STREAM
#~ Inlet Wet Vapor Pressure (point 2)
PWth=-2.73933e-11*(T2)**6+2.85333e-8*(T2)**5+1.27164e-6*
(T2)**4+2.36198e-4*(T2)**3+1.00275e-2*(T2)**2+3.38158e-1*
(T2)+6.57466
#~ Water Vapor Pressure (mmHg)
PVth=PWth-0.5*(T1-T2)
#~ Inlet humidity (kg/kg_dry_air)
Ye=0.62*PVth/(Ptotal-PVth)
#~ Inlet stream enthalpy (kcal/kg_a_s)
He=(0.24+0.46*Ye)*T1+597.2*Ye
```

The operation line of the tower can be derived from an overall enthalpy balance:

$$V\,dH = L\,dh = L\,c_L\ dT_L \tag{6.20}$$

Assuming that, for water, the heat capacity value is equal to 1 kcal/(kg°C), we obtain the following:

$$V\,dH = L\,dT_L \tag{6.21}$$

The operation line slope is obtained from the following relation:

$$\frac{H_2\text{-}H_1}{T_{L2}\text{-}T_{L1}} = \frac{L}{V} \tag{6.22}$$

Finally the operation line of the tower can be computed as:

$$\text{Operation line} = \frac{L}{V}\cdot T + T_{L1} \tag{6.23}$$

This is formulated in Python as follows:

```
#~ L/V slope (Operation line)
LV=(Hs-He)/(T5-T6)
#Operation line: y=LV*T+T6
OpLine=[]
OpLine_Teval=np.arange(T6,T5,0.05)
for i in OpLine_Teval:
    OpLine_i=LV*(i-T6)+He
    OpLine.append(OpLine_i)
```

With the operation line, the air-water vapor mixture enthalpy curve, and the inlet and outlet streams parameters defined, the air temperature and enthalpy evolution along the tower can be calculated. In order to estimate this air profile, the Mickley method is used, which iterates the slope between the operating line and the equilibrium line to calculate the number of transfer units, NUT. For this purpose, the Lewis relationship is applied, which relates the operating line with the equilibrium line, and results from the resistance to mass and energy transfer between the gas and liquid phases. It can be deducted trough an overall energy balance:

$$h_l \; a \; S \; dZ \; (T_L - T_i) = L \; dh = V \; dH = k_y \; a \; S \; dZ \; (H_i - H) \tag{6.24}$$

$$\frac{H_i - H}{T_i - T_L} = -\frac{h_L}{k_y} \tag{6.25}$$

This balance links the equilibrium and the operating line through a straight line that has as slope $\frac{H_i - H}{T_i - T_L} = -\frac{h_L}{k_y}$. Typically, this value of the slope ranges from –3 to –10.

The profile of the air temperature corresponds to the evolution of the air temperature with the change in its enthalpy:

$$\frac{k_y \; a \; S \; dZ \; (H_i - H)}{k_y \; c_h \; a \; S \; dZ \; (T_i - T_g)} = \frac{VdH}{Vc_h dT_g} \tag{6.26}$$

$$\frac{(H_i - H)}{(T_i - T_g)} = \frac{dH}{dt_g} \tag{6.27}$$

In order to solve this differential equation, the operation of the cooling tower is discretized, dividing the operation line in steps. For this example, 10 steps were chosen.

For each point in the operating line, we have a liquid temperature, $T_{L,j}$, and an air enthalpy H_j. This line, as presented before, is related to the equilibrium curve by a line that has a slope $-\frac{h_L}{k_y}$, which value is not known. Therefore, we have the following relation:

$$H_i - H = \text{slope} \; (T_i - T) \tag{6.28}$$

A more detailed explanation about the modeling of a cooling tower can be found in Chapter 3.

The proposed algorithm to solve the model is as follows:

1. Set a slope.
2. Calculate the straight line which links the operating line with the saturated air-water mixture curve for each step, starting from the bottom of the column (water outlet temperature and inlet stream enthalpy).
3. Calculate the temperature corresponding to the intersection of the straight line and the enthalpy-temperature curve.
4. Calculate the operation line temperature for the next step.
5. Calculate the operation line enthalpy for the next step.
6. Calculate the enthalpy in the saturated air-water mixture curve for the next step using the temperature calculated in 3.
7. Calculate the temperature in the saturated air-water mixture curve for the next step using the enthalpy calculated in 5.
8. Compute 2-7 for all steps.

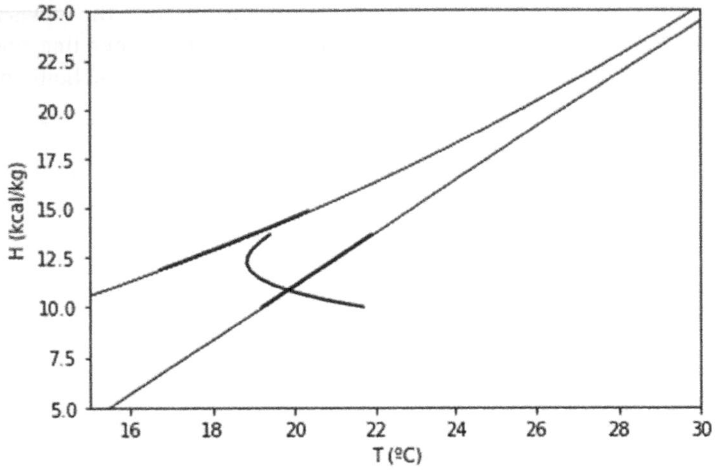

FIGURE 6.5 Operation of the cooling tower.

9. In the last step, the finalization criterion is computed as the difference between the calculated and experimental air outlet temperature, which are placed over the air saturation curve. If the difference value is below a tolerance set by the user, the corresponding saturated air-water mixture curve enthalpy at this point is calculated, and the iterative is finished. If not, the calculation process is restarted at 1, changing the slope.
10. If the iterative process finishes the air enthalpy evolution has been obtained and NUT and k_y can be computed.

The final plot obtained is seen in Figure 6.5, where the operation line, the enthalpy-temperature curve, and the air enthalpy evolution can be observed.

The rest of the code, following the algorithm described above, is attached below:

```
#fig, axes = plt.subplots()
##axes.set_xlim([OpLine_Teval[0]-3, OpLine_Teval[-1]+3])
##axes.set_ylim([0,35])
#axes.plot(OpLine_Teval,OpLine)
#axes.plot(T_evaluation,H)

#~ Coeficiente
Points=10
Coef=(Hs-He)/Points
pm=-5 #Slope
step=0.005
pm=pm-step

ii=np.arange(1,1000,1)
#ii=np.arange(1,100,1)
```

```
#contador=[]
#contador_k=[]

for i in ii:
    pm=pm+step
    Tinterphase=[]
    T_OpLine=[T6]
    T_Satline=[T1]
    #T_OpLine=[]
    He_OpLine=[He]
    Hi_Satline=[]

#     contador.append(i)

    for k in np.arange(0,Points,1):
#         contador_k.append(k)

        for x in np.arange(10,35,0.01): #Temperature range
            Eq1=a[0]*x**4+a[1]*x**3+a[2]*x**2+a[3]*x+a[4]
            Eq2=(pm*(x-T_OpLine[k])+He_OpLine[k])

            if abs(Eq1-Eq2)<0.05:
                Tinterphase.append(x)
                break
        #Operation line temperature for the next step
        T_OpLine_k=(Coef/LV)+T_OpLine[k]
        T_OpLine.append(T_OpLine_k)

        #Operation line enthalpy for the next step
        He_OpLine_k=He_OpLine[k]+Coef
        He_OpLine.append(He_OpLine_k)

        #Saturation line enthalpy for the next step
        Hi_Satline_k=a[0]*Tinterphase[k]
        **4+a[1]*Tinterphase[k]**3+a[2]
        *Tinterphase[k]**2+a[3]*Tinterphase[k]+a[4]
        Hi_Satline.append(Hi_Satline_k)

        #Saturation line temperature for the next step
        T_Satline_k=((He_OpLine[k]+Coef)-Hi_Satline[k])
        *((Tinterphase[k]-
        T_Satline[k])/(Hi_Satline[k]-He_OpLine[k]))+
        Tinterphase[k]

        T_Satline.append(T_Satline_k)

#Computing of the air outlet stream temperature to finish the
iterations
    for x in np.arange(10,35,0.01): #Temperature range
            Eq1=a[0]*x**4+a[1]*x**3+a[2]*x**2+a[3]*x+a[4]
            Eq2=(pm*(x-T_OpLine[k+1])+He_OpLine[k+1])
```

```
            if abs(Eq1-Eq2)<0.05:
                Tinterphase.append(x)
                break
    Hi_Satline_k=a[0]*Tinterphase[-1]**4+a[1]
    *Tinterphase[-1]**3+a[2]
    *Tinterphase[-1]**2+a[3]*Tinterphase[-1]+a[4]
    Hi_Satline.append(Hi_Satline_k)

#Finalization criteria
    if abs(T_Satline[-1]-T4)<0.005:
        print('Error = ', abs(T_Satline[-1]-T4))
        print('h_l/k_y= = ', pm)

        InvHimHe=[]

#Plot the results
        for k in np.arange(0,Points+1,1):
            InvHimHe_k=1/(Hi_Satline[k]-He_OpLine[k])
            InvHimHe.append(InvHimHe_k)

        b=np.polyfit(He_OpLine,InvHimHe,3)

        #Calculate NUT
        NUTb=(1/4)*b[0]*Hs**4+(1/3)*b[1]*Hs**3+0.5*b[2]
        *Hs**2+b[3]*Hs

          NUTa=(1/4)*b[0]*He_OpLine[0]**4+(1/3)
          *b[1]*He_OpLine[0]**3+0.5*b[2]*
          He_OpLine[0]**2+b[3]*He_OpLine[0]
        NUT=NUTb-NUTa
        z=0.48
        #Calculate ky
        ky=NUT*Mas/(110*0.15*0.15*z)
         x=np.arange(10,35,0.01)
        y=a[0]*x**4+a[1]*x**3+a[2]*x**2+a[3]*x+a[4]
        y2=He_OpLine[0]+LV*(x-T_OpLine[0])
        fig2, axes2 = plt.subplots()
        axes2.plot(x,y, 'ko', markersize=0.1)
        axes2.plot(x,y2, 'ko', markersize=0.1)
        axes2.plot(T_Satline,He_OpLine, 'k-')
        axes2.plot(Tinterphase,Hi_Satline, 'k', markersize=5)
        axes2.plot(T_OpLine,He_OpLine, 'k', markersize=5)
        axes2.set_xlim([15, 30])
        axes2.set_ylim([5, 25])
        axes2.set_xlabel('T (°C)')
        axes2.set_ylabel('H (kcal/kg)')

        break
```

REFERENCES

1. van Rossum, G., Drake, F.L. (2001) *Python Reference Language Manual.* PythonLabs.
2. Python documentation. https://www.python.org/doc/.
3. Numpy documentation. https://docs.scipy.org/doc/numpy/.
4. Scipy documentation. https://docs.scipy.org/doc/scipy/reference/.
5. Matplotlib documentation. https://matplotlib.org/contents.html.
6. Anaconda Python distribution. https://www.anaconda.com/download/.
7. Kunii, D., Levenspiel O. (1991) *Fluidization Engineering*, 2nd edn. Butterworth-Heinemann, Boston, MA.
8. Fogler, S. (2006). *Elements of Chemical Reactor Engineering.* 4th edn. McGraw Hill, New York.

7 Data Analysis for Chemical Engineers: Introduction to R

Emilio Díaz-Bejarano, Victor Francia, and Francesco Coletti

CONTENTS

7.1 INTRODUCTION

R is a free, multiplatform, open source software for statistical analysis, programming, data mining and graphics, created by Robert Gentleman and Ross Ihaka in 1993. R has grown in popularity in the past decade becoming one of the most widely used statistical packages, ranked the 6th most popular of all data science languages by the IEEE in 2017 [1].

R is in continuous development, coordinated by the R Core Team Development (R Project) and enjoys a large, active community of users which provides invaluable support. This is particularly important for neophytes, as it enables them to quickly pick up the programming language, but it is also useful for expert users, should they need advice on a specific problem. The Comprehensive R Archive

Network, CRAN (https://cran.r-project.org/), makes available a wide range of packages that provide R with extensive capabilities for computation, statistical analysis, data visualization and cloud deployment. As it will be seen in what follows, R language is very compact and powerful, which simplifies the development of complex statistical models and allows the user to focus on what really matters: the analysis and interpretation of the results. One of the major strengths of R lies in its ecosystem which includes not just numerous free packages and the vibrant community of users mentioned above, but also tools specifically developed to build interactive documents, interfaces and dashboards and further enhanced capabilities. For example:

- Connections with databases (e.g., MySQL), spreadsheets and external objects (e.g., Fortran and C++ programmes) enhance the flexibility of the applications that can be developed.
- A number of editors with Graphical User Interface (GUI) make the development experience easier (e.g., R Studio [2]).
- Documents that integrate text and computation, enabling robust and reproducible research, can be produced with R Markdown [3].
- Interactive web apps can be built and deployed on the cloud via Shiny [4].
- R commander [5] enables users to perform powerful analysis with R without having to learn how to program using intuitive GUIs (see Section 4.3).
- Solution of computationally intensive problems can be significantly sped up with the use of parallelisation algorithms, including the new free, fully compatible distribution of R by Microsoft (Microsoft R Open [6]).

Compared to other software for statistical analysis, R has the advantage of combining the flexibility of a programming language with the convenience of a range of powerful capabilities already built-in or that can be easily added on. For this reason, it is extensively used in statistics, engineering, medicine, biology, finance, marketing and, more recently, management consulting to tackle a variety of complex problems in the real world. Examples of successful applications include genomics, predictive maintenance, quality control and assurance, analysis and prediction of customer behaviour, business analytics, etc. However, R is a tool particularly valuable in education as it provides a free and powerful platform to introduce the students to the world of programming and data science. Simple examples such as those contained in this Chapter can be used to illustrate key concepts in statistical analysis. Assignments on specific topics can help the students to learn and consolidate programming skills while working on topics relevant to Chemical Engineering.

This Chapter aims to provide an overview of the basic use of R and its use for statistical analysis in Chemical Engineering. The next section, Section 2, provides a primer on R and includes some useful examples to get the reader started. Section 3 is dedicated to plotting and data visualisation while Section 4 covers statistical analysis presenting additional examples pertaining to Chemical Engineering. Section 5 provides the references cited.

7.2 R PRIMER

7.2.1 Download, Installation, Development Environment, and Resources

In order to download and install R, the user needs to visit the R homepage http://www.r-project.org, select one of the CRAN sites, and find the right downloadable executable file for the operating system (Windows, MacOS, Linux) and configuration (32 or 64 bit). Once downloaded, the executable can be opened and installed following the onscreen instructions. The CRAN sites provide detailed information on installation and documentation in several languages.

R comes with its own basic development interface, RGui. However, development in R can be performed in a number of other environments with additional capabilities and enhanced ease of use. A widely used open source integrated environment for development is RStudio [2], which includes a console, a syntax highlighting editor, and plotting, history, debugging and workspace management tools. A screenshot of the RStudio interface and indication of key components is shown in Figure 7.1.

Extensive documentation on the use of R and related programming packages are accessible by using the "help" commands in the command window (Figure 7.2) or the Help tab in RStudio. Information on specific functions can also be accessed by typing "?" followed by the function name. This information can also be found in the R homepage [7]. The website also provides a comprehensive list of over 150 edited books related to R, many of them focusing on the application of R to specific scientific or technical fields. R "Cheat sheets", like the ones available in [8], are very useful resources for quick introduction to the basic commands and tools in R. Finally, programming forums, such as Stack Overflow [9], are excellent platforms for exchange of ideas on programming in R and other languages.

R benefits from an extensive archive of programming packages. The full list of packages by subject area can be accessed in http://cran.r-project.org/web/views/. The installation of a new package can be done by using the "Packages" tab in RStudio

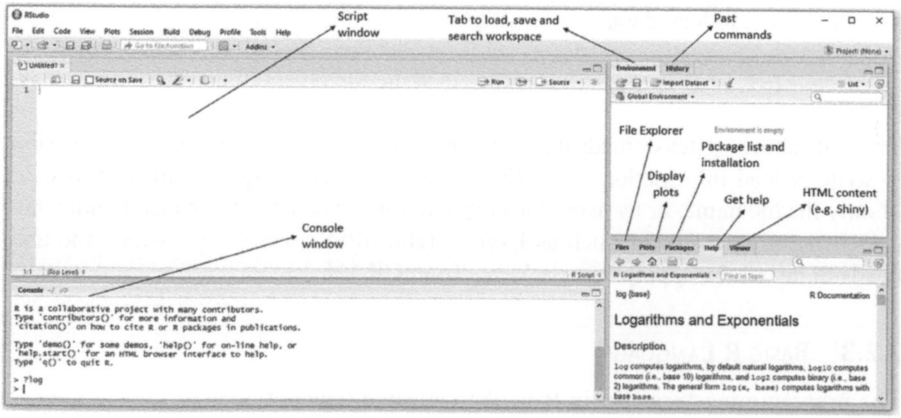

FIGURE 7.1 Screenshot of the RStudio interface.

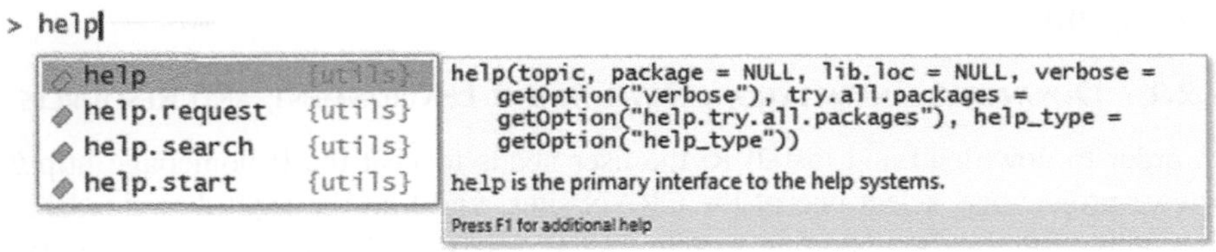
> help
help {utils}
help.request {utils}
help.search {utils}
help.start {utils}
help(topic, package = NULL, lib.loc = NULL, verbose = getOption("verbose"), try.all.packages = getOption("help.try.all.packages"), help_type = getOption("help_type"))
help is the primary interface to the help systems.
Press F1 for additional help

FIGURE 7.2 Help commands.

(Figure 7.1) or by using the command `install.packages("Name")`, which will install the package called "Name". Once installed, the package needs to be loaded into the work environment by using the command `library(Name)`.

7.2.2 Working Directory, Loading, and Saving Data

The working directory is the location where R looks for inputs and sends outputs. The user can access or change the working directory as follows:

- `getwd()`: shows the working directory.
- `setwd("C:/myfolder ")`: selects a new working directory in the path specified. If the command `choose.dir()` is typed instead of the path, the working directory is selected interactively.
- Including `"../"` in front of a subfolder's path selects the upper level folder to the current working directory.

Setting up the right working directory is important when importing and exporting data. R allows reading and writing information in a wide variety of formats, the most common being .csv, .txt and .RData files (the last being the data format specific of R). Some useful commands are:

- `read.table('data.txt')`, `read.csv('data.csv')` and `load('data.RData')`, to load data.
- `write.table(df, 'data.txt')`, `write.csv(df, 'file.csv')`, `save(df, file='file.Rdata')`, to write the object df as a data file.

The code above writes or reads information from the working directory. It is possible to write or load information from other folders by specifying the full path, instead of only the file name, or by using the command file.choose(). R can also handle data files from other software, such as Excel, Stata, SPSS, or SAS, for which the user needs to install the appropriate packages from CRAN.

7.2.3 Basic R Language

The programming language in R works with two types of objects: data structures and functions. Both data and functional objects are created by using an "assignment" operator. Once created, R stores the new object and its value in memory. For example, in order to create a variable called "Emissions" with value "400":

```
> Emissions <- 400     Variable saved in the local environment
                       where it is defined
> Emissions <<- 400    Variable saved in the global environment
> Emissions = 400      Normally used to provide a value to an
                       argument of a function
```

The assignment can also be done left to right, e.g. > 400 -> Emissions. The variable can be called within the environment where it has been saved. If created as a local variable in the main script or as a global variable, it can be called by typing its name in the console and pressing enter:

```
> Emissions
[1] 400
```

It is important to note that the R language is *case sensitive*; so, in the example, R will find variable "Emissions", but not "emissions".

7.2.3.1 Data Objects

Data can be stored and organized in different types of structures: vectors, matrices, arrays, factors, data frames and lists. The simplest object is the *vector* (or atomic vector). Individual values, such as the "Emissions" variable above, are vectors of length one. Vectors are categorized in six classes depending on the type of data that they contain, see Table 7.1.

Vectors are created with the "combine" function **c()**. The class of a vector is consulted with the command "class()". Data can be converted between classes by using the command "as." + class. For example,

```
> a <- c(1,2,3) # anything typed after a '#' is considered a
comment
> class(a)
[1] "numeric"
```

TABLE 7.1
Classes of Atomic Vector

Data class	Description	Example (vector length 3)
Logical	Logical values TRUE or FALSE	TRUE, TRUE, FALSE
Numeric	Numbers	3, 4.21, 16.55
Integer	Integer numbers	1, 2, 2
Complex	Complex numbers	2 + 5i, 3 + 2i, 1 – 1i
Character	Character strings	"Emissions", "CO2", "A"
Raw	Intended to hold raw bytes.	

If the data is to be treated as integer values:

```
> a <- as.integer(c(1,2,3))
> a
[1] 1 2 3
> class(a)
[1] "integer"
```

The other data structure types are comprised of a combination of atomic vectors, i.e., linear vectors of a single primitive type (e.g., logical, integer, double, complex, character). Data objects can be homogeneous, if all the vectors composing them are of the same class; or heterogeneous, if they allow for vectors of different classes. Table 7.2 summarizes the characteristics and basic commands for creating and sub-setting the various data structures. Some functions of interest to manage, view or obtain information on data objects are:

- `str(a)` provides a summary of an object's structure.
- `View(a)` see the full data set in a spreadsheet.
- `ls()` lists all variables in the environment.
- `rm(a)` removes a from the environment.
- `rm(list = ls())` removes all variables from the environment.
- `nrow(), ncol(), dim()` provides the number of rows, columns and rows and columns, respectively, of a matrix or data frame.
- `cbind(a,b)` and `rbind(a,b)` are used to bind columns and rows, respectively (e.g. join to vectors to form a matrix, or join a new vector to an existing matrix).

Some of these actions can also be done by using the environment tab in RStudio.

7.2.3.2 Functions and Operators

Functions are set of instructions programmed to perform a specific task. They are used to avoid repeating lines of code every time the task needs to be done, which significantly reduces the complexity of a piece of code and the chances to make mistakes. A function takes a number of inputs (function arguments), executes the code inside the function (body) and returns an output to the working environment:

```
> function_name(argument1 = input1,argument2 = input2)
[1] Output
```

R includes many built-in functions (some of them already shown above, e.g., set.wd(), read.csv(), class() or factor()). Among them, an important group is composed by mathematical functions or operators. A summary of the most relevant math functions is provided in Table 7.3 (statistical functions are not shown as they will be discussed later). Basic operations, such as sum, subtraction or multiplication, are usually specified with the corresponding symbol (+, -, *). However, all of them are function calls and can also be specified as any other function by coding the

TABLE 7.2
Types of Data Structure. Commands to Create them and Subset Data

		Vector	
Dimensions	1	Data classes	Homogeneous
	Create vectors	Select elements (sub setting)	
Combine elements	`c(1,2,3)`	By position	
Integer sequence	`1:3`	• First element	`a[1]`
		• All but first element	`a[-1]`
		• First and third element	`a[c(1,3)]`
Complex sequence	`seq(1, 3, by = 1)`	By value	
Repeat	`rep(1:3, times = 1)`	• Using logical operators	`a[a<2]`
		• Elements in a set	`a[a %in% c(1,5)]`
		Matrix	
Dimensions	2	Data classes	Homogeneous
	Create matrix from vector "a"	Select elements (sub-setting)	
`> a <- 1:8`		Select first row	
`> M <- matrix(a, nrow = 2,`		Select first column	
`ncol = 4)`		Select element	`M[1,]`
`> M`		Also selection by column	`M[,1]`
`     [,1] [,2] [,3] [,4]`		name, using *%in%*,	`M[1,1]`
`[1,]    1    3    5    7`		logical operators...	
`[2,]    2    4    6    8`			
		Array	
Dimensions	n	Data classes	Homogeneous
	Create array	Select elements (sub-setting)	
`A <- array(a, dim = c(2,2,2))`		As for matrices, but with *n* dimensions.	
		List	
Dimensions	1 (each element may have *n* dimensions)	Data classes	Heterogeneous
	Create List	Select elements (sub-setting)	
`L<- list(a, M, A)`		Element of the list	`L[[1]]`
Where a, M and A are the vector, matrix and array above.		New list with an element	`L[1]`
		Element by name	`L$a`
Data Frames: special type of list with all elements being vectors of the same length			
Dimensions	2	Data classes	Heterogeneous
	Create data frame	Select elements (sub-setting)	
`b <- c('a', 'b', 'c', 'd')`		Can use sub-setting methods of matrices and lists	
`Df <- data.frame(b, a=1:4)`			
Factors: created from a vector and used to identify the levels of a data set			
Dimensions	1	Data classes	Homogeneous (character)
	Create factor		
From vector	`factor(a)`		

TABLE 7.3
Some of the Most Common Mathematical and Logical Operators

Math operator	Description	Logical operator	Description
`x + y` (or `'+' (x,y))`	Sum of x and y	x < y (or `'<' (x,y)`)	Less than
`x - y`	Subtraction	`x <= y`	Less than or equal to
`x * y`	Multiplication	`x > y`	Greater than
`x / y`	Division	`x >= y`	Greater than or equal to
`x ^ y`	Power	`x == y`	Equal to
`x %% y`	Modulus	`x != y`	Not equal to
`x %/% y`	Integer division	`!x`	Not x
`max(x)`	Maximum	`x | y` or `x||y`	Or
`min(x)`	Minimum	`x & y` or `x&&y`	And
`abs(x)`	Absolut value of x	`is.na(x)`	Test if x is missing
`log(x, base=y)`	Logarithm (no base for natural log)	`is.null(x)`	Test if x is null
`exp(x)`	Exponential of x	`isTRUE(x)`	Test if x is true
`sqrt(x)`	Square root of x	`any(expression)`	Test if any element is TRUE
`factorial(x)`	Factorial of x	`all(expression)`	Test if all elements are TRUE
`round(x, n)`	Round x to n decimal places	`is.numeric(x)`	Test if x is of class numeric

operator symbol in between inverted commas, e.g. `'+'`(). For example, the sum of two numbers:

```
> 3+4              > '+'(3,4)
         or
[1] 7              [1] 7
```

Another important group is made up by logical operators, also summarized in Table 7.3. Similarly to math operators, these can be called with the corresponding symbol or with the function syntax.

7.2.4 Programming in R

Loops, conditional statements and functions are used in programming to repeat specific pieces of codes, run a task for different inputs or automatize selection between various options. Three basic programming routines are:

a. For loop: a task is repeated for a number of values of an iteration variable over a sequence:

```
for (i in sequence){ expressions }
```

b. For while: a task is repeated while certain condition is satisfied:

```
while (condition){ expressions }
```

c. If statement: if certain condition is satisfied, the task is executed; otherwise, (optionally) a different task is performed.

```
if (condition1){ expressions1 } else {
         expressions2 # the 'else' statement is not
mandatory}
```

Finally, a sequence of commands or expressions can be wrapped up in the form of a function. The syntax to write a function is:

```
function1 <- (argument1, argument 2){
  expressions
  return(output_variable)}
```

Once the piece of code containing the function has been executed for the first time, the function is saved as an object in the working environment and can be called at any time. Functions return the result of the last expression evaluated (implicit result) or the variable specified in the command `return()`. It is noticed that 'return' works for a single object, although this limitation can be easily overcome by returning a list containing multiple objects.

Functions can be defined in the command window before using them. However, it is convenient to write functions within *scripts* (.r) files. This allows creating a repository of functions that can be loaded at the beginning of a new session as any other R package, by using the following command:

```
source('C:/myfolder/Functions.r'))
```

7.2.5 Examples

Example 7.1: Load a .csv File and Inspect a Dataset

The following example uses data of CO_2 emissions associated to the power generation in Spain in 2017, made publicly available by *Red Electrica de España* through its website [10].

```
> Emissions <- read.csv('C:/myfolder/Emissions_CO2_07_2017.
csv',header= T) # creates an object from the data in the .csv
file
> Emissions # Calls the object to view its contents
  X Jan. Feb. Mar. Apr. May. Jun. Jul. Aug. Sep. Oct.
1 Coal 5117837 3340224 1870687 1986764 3575524 4330564 4098060
3079792 2919966 3902305
2 Fuel + gas  465272  411754  445314  459884  463700  497519
527079  541340  489127  485943
3 Combined cycle 1319176  761479  743942  646712  807893
1347083 1567137 1512389 1379043 1674643
4 Thermal non-renewable/Cogeneration  906493  816592  885116
826308  860774  856645  887294  835192  840641  885325
5 Residues non-revewable   53965   47648   52821   45066
44938   54202   51626   55104   54910   58294
6 Residues renewable   17456   15275   17008   14804   11021
19552   20009   19288   18697   19346
     Nov.    Dec.
1 4586284 4128447
2  443687  465162
3 1933105 1343568
4  903341  919345
5   52294   54972
6   18702   19322
> class(Emissions) # checks the class of the object
[1] "data.frame"
> # Emissions is loaded as a data.frame, with two types of
data vectors:
> class(Emissions$X)
[1] "factor"
> # Characters are transformed to factors by read.csv.
> class(Emissions$Jan.)
[1] "numeric"
> # Numbers are treated as numeric data.
> # We can now subset and operate with the data. For example,
we can obtain the total emissions in January by
> sum(Emissions$Jan.)
[1] 7880199
> # the total emissions per month
> colSums(Emissions[,2:ncol(Emissions)])
   Jan. Feb. Mar. Apr. May. Jun. Jul. Aug. Sep. Oct. Nov. Dec.
   7880199 5392972 4014888 3979538 5763850 7105565 7151205
   6043105 5702384 7025856 7937413 6930816
> # and the total emissions in 2017
> sum(colSums(Emissions[,2:ncol(Emissions)]))
[1] 74927791
> #or, simply
> sum(Emissions[,2:ncol(Emissions)])
[1] 74927791
> # we can also calculate the total emissions in a year for a
given row, e.g. Coal
```

```
> sum(Emissions[which(Emissions$X %in%
'Coal'),2:ncol(Emissions)]) # "which" is used to identify
which element of the vector is equal to 'Coal'
[1] 42936454
```

Example 7.2: Programming a Function with Conditional Statements and Loops

The example considers the calculation of pressure drop due to friction for an incompressible flow of a Newtonian fluid through a horizontal pipe, which is given by the following equation:

$$\Delta P = \frac{2f\rho u^2 L}{D} \tag{7.1}$$

where ΔP is the pressure drop, f is the Fanning friction factor, ρ the density of the fluid, u the linear velocity, L the length of the pipe, g the local acceleration of gravity and D the pipe diameter. The friction factor depends on the flow regime. Below Reynolds number (Re) of ca. 2100, the flow regime is laminar and the friction factor can be simply calculated as:

$$f = \frac{16}{Re} \tag{7.2}$$

While for Re > 2100, the flow transitions towards turbulent regime. A simple expression for the calculation of the friction factor in turbulent flow is given by the Blasius correlation [11]:

$$f = \frac{0.0791}{Re^{0.25}} \tag{7.3}$$

In order to compute the pressure drop, we can program a function that includes an *if* statement to choose the right equation for the friction factor as a function of the Reynolds number.

```
Pressure_drop <- function(fluid_density, viscosity,
                 volumetric_flowrate_m3perh, pipe_length,
                 diameter){
velocity <-   4 * volumetric_flowrate_m3perh / 3600 / (pi *
        diameter^2)
Re       <- fluid_density * velocity * diameter / viscosity
if(Re < 2100) {
  friction_factor <- 16 / Re
}else{
  friction_factor <- 0.0791/(Re ^ 0.25)
}
```

```
Pressure_drop <- 2 * friction_factor * fluid_density *
             (velocity ^ 2) * pipe_length / diameter
return(Pressure_drop)
 }
```

For example, for water at 20°C, in international units, flowing at 25 m^3/h through a pipe of length 100 m and diameter 10 cm, the pressure drop in Pa is:

```
> Pressure_drop(fluid_density=1000, viscosity=1E-3,
volumetric_flowrate_m3perh=25, pipe_length=100,
diameter=0.10)
[1] 7172.405
```

Once the function is ready, it may be useful, for example, to compute the pressure drop for a range of pipe diameters. Two options are shown below that use the *for* and *while* loops for that purpose, respectively.

```
> # A "for" loop can be used over a pre-defined sequence of
diameter values
> diameter_sequence <- seq(0.1,0.5, by=0.1) # we create the
diameter values
> dP_estimated <- rep(0, times = length(diameter_sequence)) #
and a vector of corresponding dP p values that we want to
compute
> for(i in 1:length(diameter_sequence)){
+   dP_estimated[i] <- Pressure_drop(fluid_density=1000,
viscosity=1E-3, volumetric_flowrate_m3perh=25, pipe_
length=100, diameter=diameter_sequence[i])
+ }
> dP_estimated # the for loop computed each value into
dp_estimated
[1] 7172.405346  266.546108   38.845335    9.905579
3.432079
> # or we ca do the same with a while loop for a set of
increasing diameters
> diameter_iterator <- 0.1 # set a diameter increment
> dP_estimated       <- list()# and a list of dP's
> while(diameter_iterator <= 0.5){
+   dP_estimated[[length(dP_estimated) + 1]] <- Pressure_
drop(fluid_density=1000, viscosity=1E-3, volumetric_flowrate_
m3perh=25, pipe_length=100, diameter=diameter_iterator)
+   diameter_iterator <- diameter_iterator + 0.1
+ } # the while loop will keep running this piece of code
computing a new dP every iteration for diameters increasing by
0.1 until the diameter turns larger than 0.5
> unlist(dP_estimated) # once it is done, we can the print
results
[1] 7172.405346  266.546108   38.845335    9.905579
3.432079
```

7.3 DATA VISUALIZATION IN R

R provides a powerful platform for the visualisation of one-dimensional and multi-dimensional datasets. Figure 7.3 shows examples of how a histogram or a box plot can be used to visualise univariate data. In the example, the number of bins used to construct the histogram is set assigning the value of the argument "breaks" in the call to the function "hist". Many other adjustable graphical parameters are available to customise the rendering of a plot. In this case, the filling pattern in the histogram has been defined in the call by setting a given "density" in the call to the function.

Scatter and line plots are given by function "plot" for data in a vectorised format or as a mathematical function. The use of "plot" and many other associated functions allows creating different types of graphs. Modifying the argument "type" in the call to "plot" it is possible to create a scatter graph of points, a set of lines, both or several other different standard graphs (see 'plot' in R help for further details). The format is also very flexible. One may define attributes such as the type and colour of makers or lines, the axes, major and minor ticks, limits, colour pallets, labels and many other with a number of graphical parameters (search for "par" in the R help). Each parameter has a default value, so there is no need to define them in the call unless intended.

In order to include several plots in the same graph an extra attribute can be simply included in the call to the plot function making "add = TRUE". Alternatively, it is possible to use specific functions to add a scatter plot, "points", or a set of lines, "lines". The result of a mathematical expression can be plotted either by using "plot" or "curve". Figure 7.4 shows an example where combining different types of lines and vectorised data and formulas.

```
> windows(4,4)
> hist(Concentration, freq=TRUE, breaks =15, density = 30,
xlim = c(0,600))
> windows(4,4)
> boxplot(Concentrations)
```

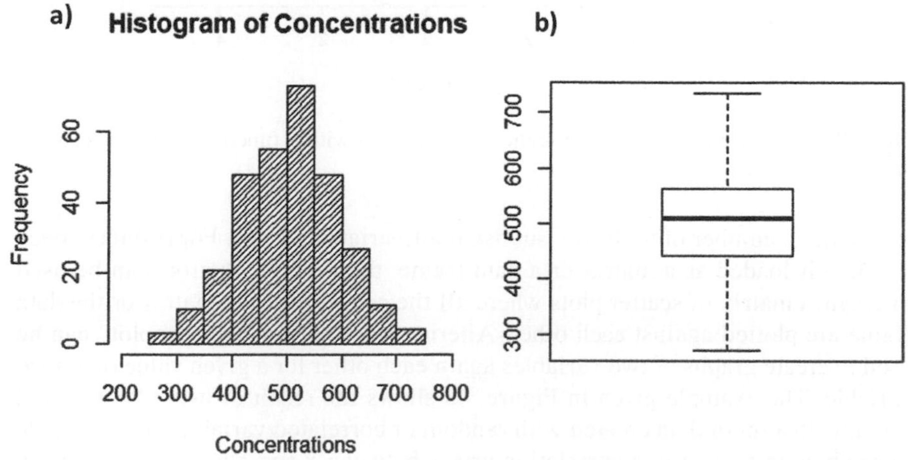

FIGURE 7.3 Examples of histograms (a) and box plots (b) to visualize univariate data sets.

```
> windows(7,7)
> plot(X,Z,lty = 2, col='blue',type="b", ylim = c(0,40), xlim
= c(-5,5), xlab = 'Variable X', ylab = 'Variable Y, Z or B')
> curve(x^2-2,add=TRUE)
> lines(X,B,lt=3, col='red', lwd=3)
> points(rep(0,40),seq(0,40,length.out = 40),
pch=16,col='black')
> title('A series of plots')
> legend(x = -5, y=35, legend = c('Z with plot', 'Y with
curve','A = Y+Z with lines', 'X with points'), lty =
c(2,1,3,0),lwd = c(1,1,3,0), pch= c(NA,NA,NA,16),col =
c('blue','black','red','black'),box.lty = 0)
```

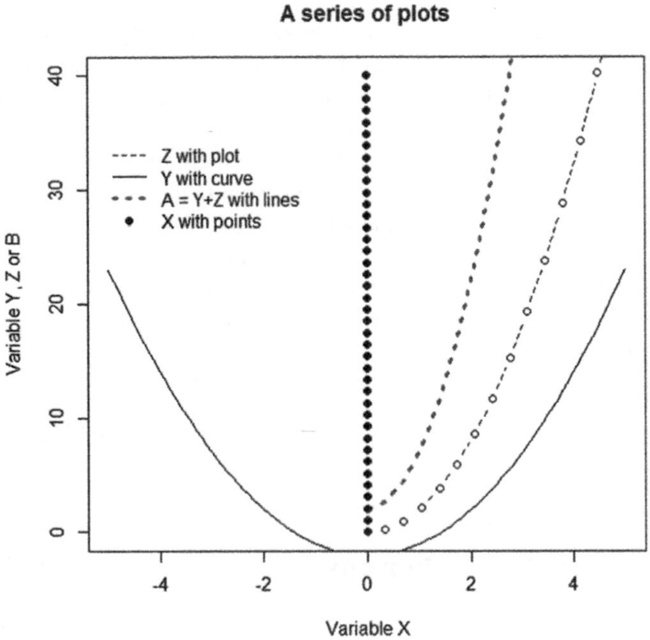

FIGURE 7.4 Different ways to plot scatter and line plots with R functions plot, points, lines and curve.

R also has a number of tools to visualise multivariate datasets. For instance, once the data is loaded as a matrix or a data frame, the function "pairs" can be used to create a matrix of scatter plots where all the columns of the matrix or the data frame are plotted against each other. Alternatively, the function "coplot" can be used to create graphs of two variables again each other for a given value of a third variable. The example given in Figure 7.5 shows the result of using "pairs" and "coplot" in a set of data created with random or correlated variables. It is possible to see how in this case, a correlation arises between x and z because z has been defined as a linear function of x with a certain level of noise. In Figure 7.6 we observe that the same linear correlation between x and z appears independently

```
> windows (5,5)
> pairs(Data)
```

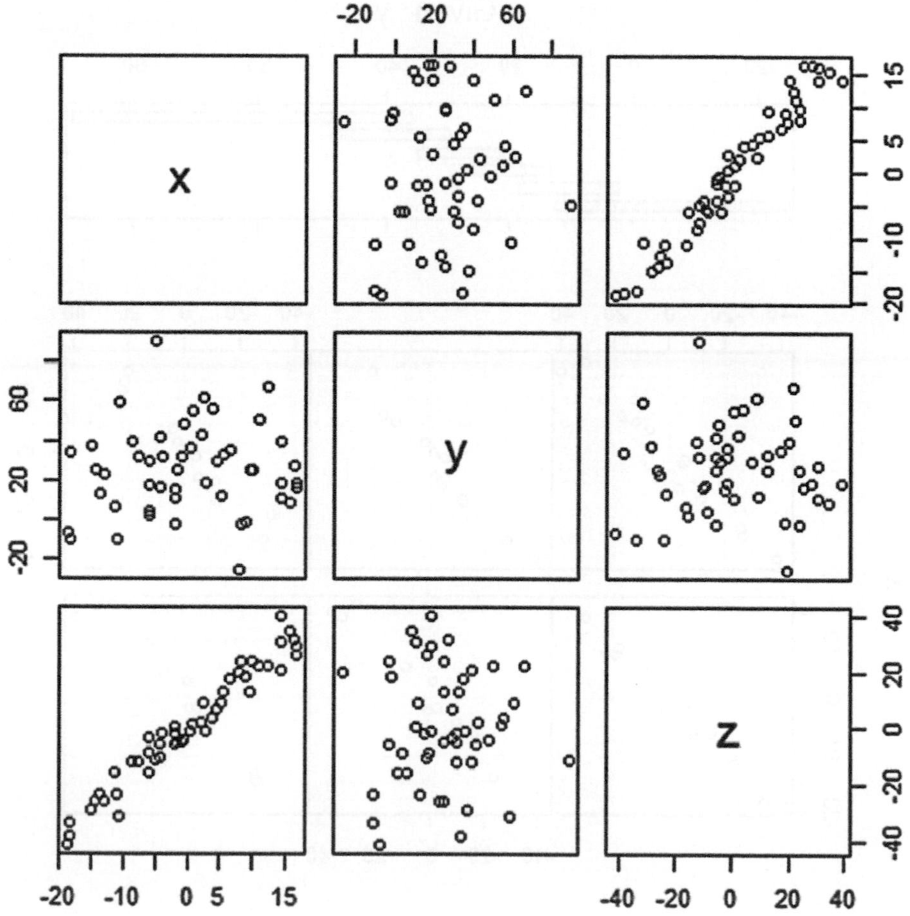

FIGURE 7.5 Pair plot to observe the correlation in a multivariate data set.

of the value of *y*. This is obvious in the example because *x* and *z* were correlated to each other but not to *y*, which was sampled from an independent normal distribution. 3D plots, surfaces and contours are other ways to visualise multivariate datasets. A surface or a contour plot show the variation of a given variable in the space of other two. For example, the results for "contour" in Figure 7.7 and "filled.contour" in Figure 7.8 describe the change of a response, *z*, in a grid (a set of equally spaced values) of *x* and *y*. This is the typical situation when measurements are collected in space (either in a surface with a grid in 2D or in a volume in a grid in 3D).

For instance, *x* and *y* may be the coordinates of the measurements of temperature taken in the cross-section of reactor. Let us consider that we have installed

```
> windows (5,5)
> coplot(x~z|y)
```

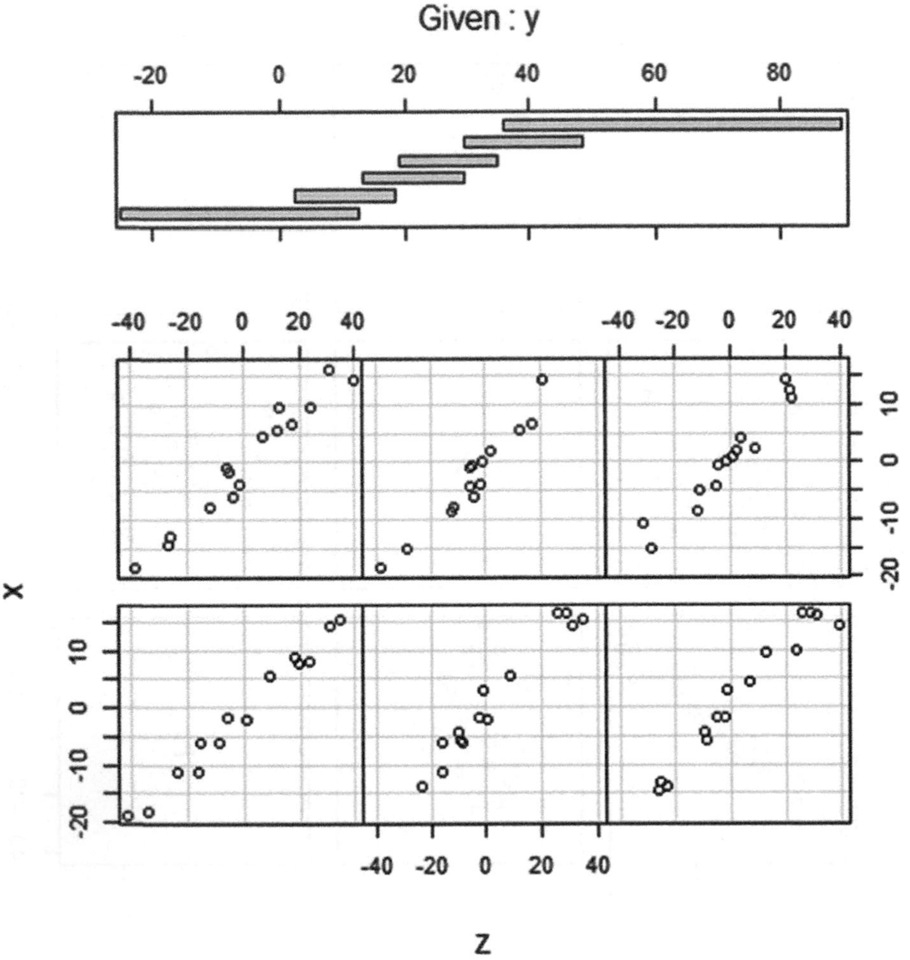

FIGURE 7.6 Coplot to observe the correlation between two variables with a third variables as a factor.

thermocouples at a number of equally spaced positions *xy*, so that each provides us with a measurement of the temperature of the fluid around it, denoted *z*. In the example below, we create the location of these measurements with a homogenous grid of *x* and *y* computing equally spaced locations by using "seq" to form a sequence of numbers. We then assign a value for the temperature. *z* must be defined as a matrix where the value of the element *z*(i,j) refers to the temperature measured at position x_i and y_j in the grid. In the example, *z* is made a function of *x* and *y* such as $z = x^2 + 100\,y$ and it is made to contain a certain degree of noise by adding a random component responsible of the wobbles in observed the contour.

```
> windows(7,7)
> contour(x,y,z,nlevels=15,xlab = 'x', ylab = 'y')
> title('Z contour')
```

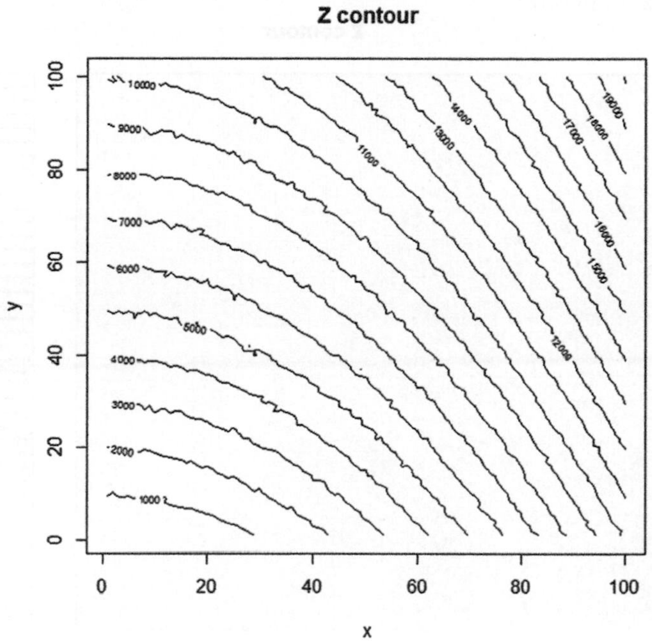

FIGURE 7.7 Contour plot.

```
> x    <- rnorm(50,0,10)
> y    <- rnorm(50,30,20)
> z    <- x*2+rnorm(50,0,5)
> Data <-data.frame(x,y,z)
```

As an open source code, R contains a very wide range of tools for visualization and image analysis, even evolving to include new available packages (i.e., rasterimage). Details of each available function, library and package go beyond the scope of this book chapter, but the interested reader is referred to the manuals and sources of information provided in section 7.5.5

```
> x <- seq(1,100,length.out = 100) # we create an equally
spaced grid
> y <- seq(1,100,length.out = 100) # we create an equally
spaced grid
> z <- matrix(0,100,100) # and a matrix of zero values for the
temperature
>
> for (i in 1:100){
     for (j in 1:100){
          z[i,j] <- x[i]^2+100*y[j]+rnorm(1,50,50)
     }}
```

```
> windows(8,7)
> filled.contour(x,y,z,nlevels=30,xlab = 'x', ylab = 'y',color
= gray.colors)
> title('Z contour')
```

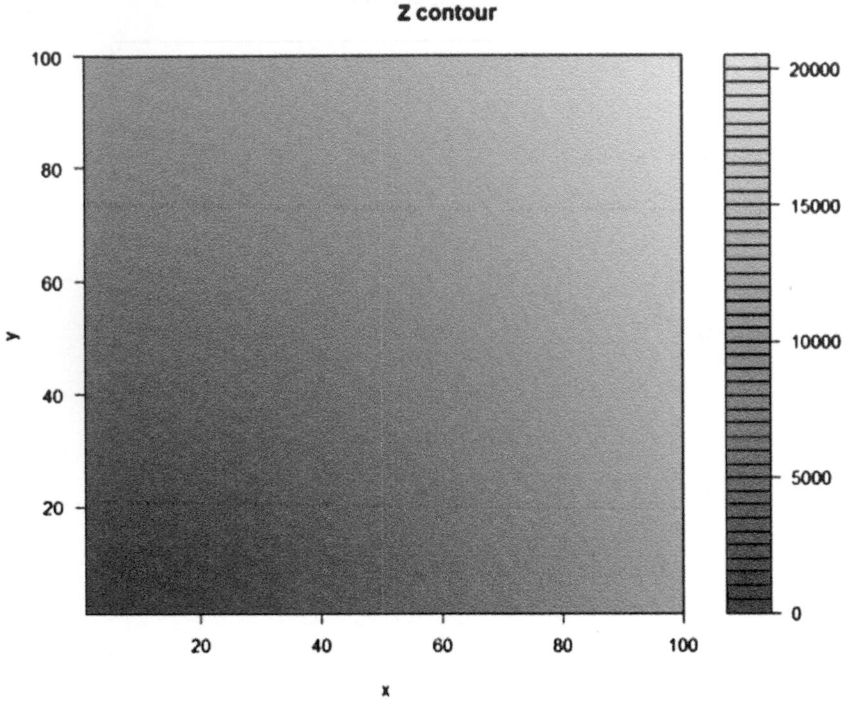

FIGURE 7.8 Filled contour plot. (See color insert.)

```
# the loop assigns a value for temperature at each position
function of
# x and y plus a normal error
```

Apart from the standard plot functions, there are a number of plotting packages that allow creating a variety of customizable plots with a neat appearance. Some of them are *ggplot2* (the most famous), *ggvis* (for interactive web-based graphics), *htmlwidgets* (javascript based) or *rgl.*

7.4 STATISTICAL DATA ANALYSIS IN R

R includes many functions and packages for statistical analysis (descriptive statistics, inference, correlation, regression, etc.). Some basic statistical functions available in the base R installation are summarized in Table 7.4.

7.4.1 Probability and Basic Statistics: Standard Density Functions

R allows to easily operate with standard probability density functions i.e. the function f describing the probability for a measured variable x to lie between a given

Table 7.4 Statistical Functions and Probability Density Functions

Function	Description	Function	Description
	Central tendency and variability		*Distributions*
mean()	Mean	r+name	Random sample
median()	Median	d+name	Density function
var()	Variance	p+name	Cumulative distribution
sd()	Standard deviation	q+name	Quantile
	Relative standing		*names:*
rank()	Ranking of the numbers	norm	Normal distribution
sort()	Sort numbers in increasing order	pois	Poison distribution
quantile()	Quantiles	binom	Binomial distribution
	Correlation and regression	unif	Uniform distribution
cor(x,y)	Correlation	*e.g.* rnorm	Random normal sample.
lm(y ~ x, data = A)	(Multiple) Regression analysis		
	Tests and analysis		*Summary*
aov()	Analysis of variance	summary()	Summary of descriptive statistics
t.test(x, y, mu, alternative)	t-test (two tailed, one tailed, mean comparison, etc.)		

range x_1 to x_2 such that $P = \int_{x1}^{x2} f\,dx$. Other functions derived from f are the cumulative distribution function F, i.e. the function that provides for any given value x_1, the probability $P(x < x_1)$ or the quantile functions i.e. for a given threshold q, the smallest value of x_1 such that the probability $P(x > x_1) > q$. R contains functions to compute f, F and q for a number of standard distributions, i.e. normal, beta, binomial, Cauchy, chi-squared, exponential, F, gamma, geometric, hypergeometric, log-normal, multinomial, negative binomial, Poisson, Student's t, Weibull distribution. Figure 7.9 shows a comparison between different standard distributions.

The internal R functions are based in the name that R allocates to each standard. f, F and q are computed with "dname", "pname" and "qname". For example, the name given by R to a normal distribution is norm (see Table 7.4 and distributions in the R package stats), so f, F and q are computed with "dnorm", "pnorm" and "qnorm" respectively. Using "rname" one can also simulate the distribution sampling values from it.

For example, let us consider that we know that the error of an unbiased NOx detector follows a normal distribution with x_{mean}= 0 ppm and a σ = 150 ppm. We may want to know how often we will incorrectly state that the NOx concentration is below a given safety threshold. Let us consider that we have a measurement of x_s= 200 ppm, and a safe threshold of 300 ppm. Which is the probability that the NOx concentration is actually unsafe? F would tell us the probability that any measurement has an error lower than a given value. In this case, we are interested in knowing how often the error is lower than 100 ppm so that we can ensure that if our measurement is $x_{measured}$ = 200 ppm, then the real concentration is x_{real} < 300 ppm. This is

```
windows(7,7)

curve(dnorm(x,mean=0,sd=2),ylim = c(0,0.4),xlim =
c(-10,20),xlab = 'Size',ylab = 'f', lt=1)

curve(dt(x,df = 5, ncp=1), lt=2, col='red',add=TRUE)

curve(dchisq(x,df = 5, ncp=1),lt=3, col='blue',add=TRUE)

legend(x = 5, y=0.4, legend = c('normal, mean 0, std 2',
't-student df 5, ncp 1','chi-squared df 5, ncp 1'), lty =
c(1,2,3), col = c('black','red','blue'),box.lty = 0)
```

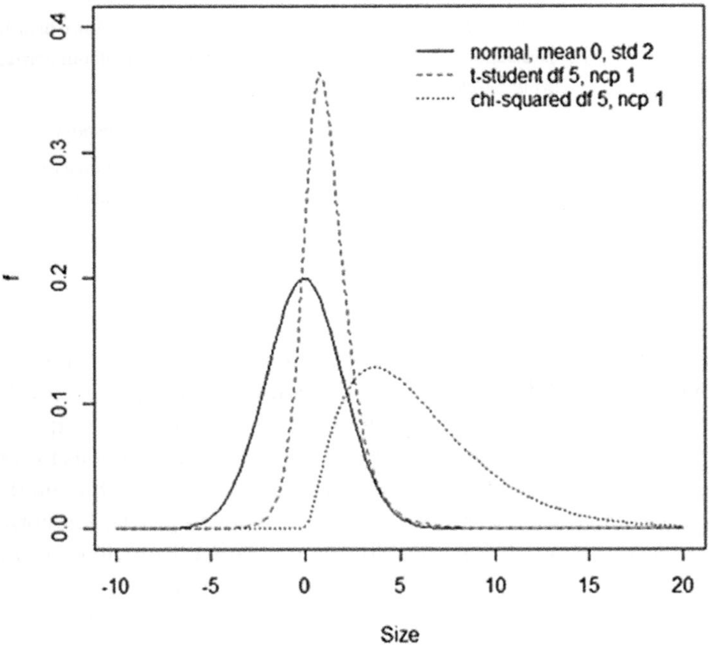

FIGURE 7.9 Comparison of different standard distribution functions, normal, t-Student, Chi-squared.

given by "pnorm" for a value of 100 ppm, which results in 74.75%. In other words, 25.25% of the times that we measure a NOx concentration of 200 ppm, the real one is beyond 300 ppm.

```
> pnorm(100, mean=0, sd=150)    > qnorm(0.7475075, mean=0, sd=150)
[1] 0.7475075                   [1] 100
```

Another way of stating the same would be to say that the 74.75 percentile of the distribution of the errors is 100 ppm. Working with quantiles is useful to compare distributions. For instance, if we look at the 90% percentile of different sensors, we can compare the maximum error that they would comply with 90% of the times. For this sensor the 90 percentile is 192.23 ppm, but in a better one with x_{mean}= 0 ppm and a $\sigma = 60$ ppm it drops to 76.89 ppm. If we expect the NOx concentration to be most of the time around x_s= 200 ppm, we would need to select the better sensor to be sure that more than 90% of the times the sensor is reliable in maintaining a safe NOx concentration. If, for example, we were constructing a stochastic model using measurements from this sensor, we would to associate an error to each measurement. The example below shows how it is possible by using "rnorm" to sample 5 errors from a normal distribution with x_{mean}= 0 ppm and a $\sigma = 150$ ppm.

```
> qnorm(0.90, mean = 0, sd = 150) > qnorm(0.90, mean = 0,
  sd = 60)
[1] 192.2327                          [1] 76.89309
> rnorm(5, mean = 0, sd = 150)
[1]    48.26330    55.98685   -61.99086   -54.85821 -148.69787
```

The statistics of non-standard experimental datasets can also be easily studied. For instance, let us consider that we have used laser diffraction to measure the size of the droplets forming a water spray and we have loaded a variable in R as a vector that gives the size of each droplet in μm (in the example below the data is created sampling measurements from a normal distribution with x_{mean} = 10 μm and a $\sigma = 3$ μm). As described earlier, we can use summary, mean, median, quantile, var and many other statistical functions to analyse the average and the dispersion of the data. We may construct a histogram with the function "hist" creating a histogram object, Size_histo, where the data are arranged in a number of bins, characterised by their limits (breaks) and their middle points (mids). For each of these bins, the histogram object computes a probability density f. Figure 7.10 show the resulting histogram of droplet sizes for 200 measurements sampled from a normal distribution in counts (the absolute frequency of number of droplets in each bin) or the resulting density frequency computed with no smoothing by using "hist", or applying a smoothing kernel by using the function "density".

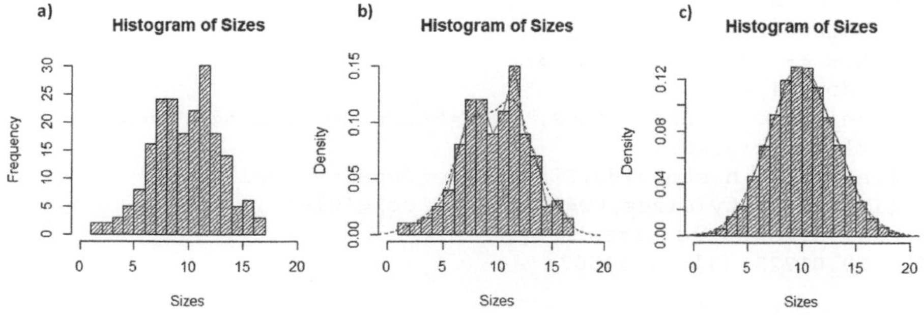

FIGURE 7.10 Histograms, (a) absolute frequency (counts), (b) probability density frequency f. Lines show the function f without (solid red) and with (blue dashed) smoothing kernel. 200 sizes (a and b) or 20000 (c) sampled from a normal distribution.

```
> Sizes <- rnorm(200,10,3)
> summary(Sizes) # 'summary' provides a range of descriptive
statistics at once.
   Min. 1st Qu.  Median    Mean 3rd Qu.    Max.
  1.516   7.511   9.865   9.730  11.860  16.570
> mean(Sizes) > median(Sizes)       > quantile(Sizes)
[1] 9.730266  [1] 9.865394          0%         25%        50%
75%       100%
                                    1.515521  7.511077  9.865394
                                    11.863339 16.572554
> var(Sizes)  > sd(Sizes)   > range(Sizes)
[1] 9.010885  [1] 3.001814  [1]  1.515521 16.572554
> windows(4,4)
  Size_histo <- hist(Sizes, freq=TRUE, breaks =15, density =
  30, xlim = c(0,20))
  Size_histo
  windows(4,4)
  Size_histo <- hist(Sizes, freq=FALSE, breaks =15, density =
  30, xlim = c(0,20))
  lines(Size_histo$mids,Size_histo$density, col='red')
  lines(density(Sizes,bw=1),lt =2, col='blue')
$breaks
 [1]  1  2  3  4  5  6  7  8  9 10 11 12 13 14 15 16 17
$counts
 [1]  2  2  3  5  8 16 24 24 18 22 30 18 14  5  6  3
$density
 [1] 0.010 0.010 0.015 0.025 0.040 0.080 0.120 0.120 0.090
 0.110 0.150 0.090 0.070 0.025 0.030
[16] 0.015
$mids
 [1]  1.5  2.5  3.5  4.5  5.5  6.5  7.5  8.5  9.5 10.5 11.5
 12.5 13.5 14.5 15.5 16.5
$xname
[1] "Sizes"
$equidist
[1] TRUE
attr(,"class")
[1] "histogram"
> Sizes <- rnorm(20000,10,3)
> windows(4,4)
  Size_histo <- hist(Sizes, freq=FALSE, breaks =15, density =
30, xlim = c(0,20))
  lines(Size_histo$mids,Size_histo$density, col='red')
  lines(density(Sizes,bw=1),lt =2, col='blue')
> mean(Sizes) > sd(Sizes)
[1] 10.01725  [1] 3.014567
```

Notice that the shape of the distribution in Figure 7.10(b) is still not Gaussian due to the low number of samples used to construct it (200). For that reason, the statistics of our sample (mean and standard deviation) are still not the same as those of the real population (in this case the normal distribution used to generate the data). Figure 7.10(c) shows how by increasing the number to 20000, the ability of the sample to represent the real population increases: the distribution approaches a Gaussian profile and the statistics converge to those used to create the dataset.

7.4.2 Statistical Tests and Regression

In a real experimental dataset, the distribution of probabilities does not necessarily need to comply with a given standard distribution. In order to assess whether it does, we can compare the shape of the density and cumulative functions f and F for a standard and experimental case and study the differences in parity plots. A more rigorous manner involves performing statistical tests to determine the actual probability that our set of measurements is in fact a sample from a given standard distribution. For example, how likely is it that we have a mean measurement of 100 μm from samples extracted from a distribution with a mean of 200 μm? If our measurements are very different from what one expects from a given distribution, the probability that they were actually extracted from it will be low (but always possible). If the differences were small (a mean close to 200 μm), the more likely the system follows this distribution. A large experimental sample size is needed for any test to be statistically significant. In other words, very small samples of just a few experiments could be extracted from almost any distribution, particularly if they contain a large variability, but it would be unlikely to sample frequent values that deviate from the expected. R contains many of the most common statistical tests in science and engineering. For example, it is possible to perform various normality tests (a test to evaluate whether a dataset can be considered to follow a Gaussian distribution) such as the Shapiro-Wilk test or the Kolmogorov-Smirnov test that can be extended to bivariate systems. In the cases shown below, both tests result in large p-values, which means that they are unable to reject the null hypothesis H_0 (i.e., that the data are normally distributed). Note that this is correct since the data were in fact sampled from a normal distribution.

```
> Sizes <- rnorm(200,10,3)
> shapiro.test(Sizes)
      Shapiro-Wilk normality test
data:  Sizes
W = 0.99467, p-value = 0.701
> ks.test(Sizes, "pnorm", mean = mean(Sizes), sd =
sqrt(var(Sizes)))
      One-sample Kolmogorov-Smirnov test
data:  Sizes
D = 0.041868, p-value = 0.8746
alternative hypothesis: two-sided
```

The R environment, particularly with the help of the package "Stats", provides a very extensive array of statistical tools, including statistical tests for other standard distributions, tests for comparison of the means of two populations, regression tools and the analysis of variance for univariate and multivariate systems. For example, see below a simple case where a linear function is created with $y = mx + n$ using "seq" and adding noise via a normal error in each measurement of *y*. We can use the least mean square regression method "lm" to perform a linear fitting to the data.

```
> x <- seq(1:20)
> y <- 4*x+25
> y <- y + rnorm(20, mean = 0, sd = 3)
> Data <- data.frame(x,y)
> lm(formula = y~x, data = Data)

Call:
lm(formula = y ~ x, data = Data)
Coefficients:
(Intercept)            x
      26.75         3.92
```

After defining the results of such a model as an object, it is possible use the functions "summary" and "anova" to extract the details of the fitting and perform an analysis of the variability in the errors. This allows us to evaluate the statistical significance of each of the parameters fitted and whether the model used is in fact a good representation or not of the data. The summary below shows a coefficient of determination R^2 of 0.9742; each parameter (intercept and slope) is found to be statistically significant by performing a t-test (here the low p-value rejects the null hypothesis that the value of the parameter does not affect the response). The overall p-value shows a high level of confidence in the existence of this correlation (remember that we have actually created "*y*" from a linear correlation with "*x*" adding some random error).

Alternatively, a similar study can be conducted with an analysis of variance "anova" test. In this case, we perform an F-test that evaluates the significance of a combination of parameters instead of an individual t-test for each of them. We achieve the same level of confidence from the F statistic, where the low p-value rejects the null hypothesis that a model with no predictors i.e., a model where each fitted value would simply be the mean of the variable, is as good as the fitted model.

More detailed information about linear regression, non-linear regression in R and the breath of tools available for multivariate analysis can be found in the books and references provided in the R homepage [7]. Some examples of books focusing on the use of R for statistical analysis are [12]–[14].

```
> model<-lm(formula = y~x, data = Data)
> summary(model)
Call:
lm(formula = y ~ x, data = Data)
Residuals:
    Min      1Q  Median      3Q     Max
-6.9588 -2.8060  0.5457  2.5514  8.7982
Coefficients:
            Estimate Std. Error t value Pr(>|t|)
(Intercept)  26.7537     1.7996   14.87 1.50e-11 ***
x             3.9201     0.1502   26.09 9.35e-16 ***
---
Signif. codes:  0 '***' 0.001 '**' 0.01 '*' 0.05 '.' 0.1 ' ' 1
```

```
Residual standard error: 3.874 on 18 degrees of freedom
Multiple R-squared:  0.9742,     Adjusted R-squared:  0.9728
F-statistic: 680.9 on 1 and 18 DF,  p-value: 9.355e-16
> anova(model)
Analysis of Variance Table
Response: y
          Df  Sum Sq Mean Sq F value    Pr(>F)
x          1 10219.4   10219  680.91 9.355e-16 ***
Residuals 18   270.2      15
---
Signif. codes:  0 '***' 0.001 '**' 0.01 '*' 0.05 '.' 0.1 ' ' 1
```

7.4.3 R Commander

R Commander (Rcmdr) [5], [15] is a graphical user interface (GUI) that provides a user-friendly, easy-to-use and free platform to perform statistical analysis on data taking advantage of the R engine. It is aimed at introducing students and researchers to data management and statistical analysis without requiring programming skills. However, it is also compatible with user-defined programs and other packages in R, allowing for a gradual transition from basic data analysis using windows and interfaces to more advanced analysis using the full capabilities of R.

R Commander is installed as any other package in R. After installation, the GUI, shown in Figure 7.11(a), is open by executing the command library(Rcmdr). Data can be input manually or imported from external files of different extensions (same as those discussed earlier for R). The latter is done by clicking on **Data > Import data**. As example, we can load the Emissions dataset discussed earlier in this chapter. The data can be inspected and edited by clicking on "View data set" and "Edit data set", which opens a new window with an editable spreadsheet Figure 7.12. The software allows filtering data, working with multiple datasets, modifying variables or creating new ones.

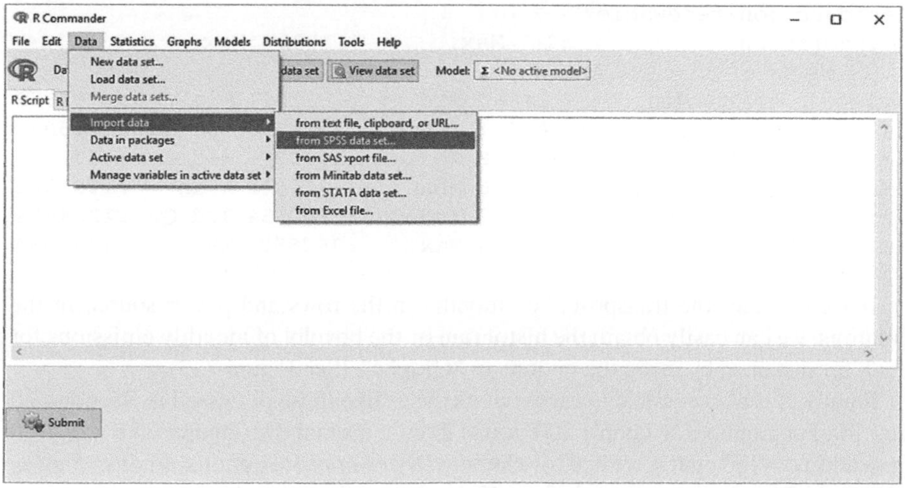

FIGURE 7.11 R Commander GUI.

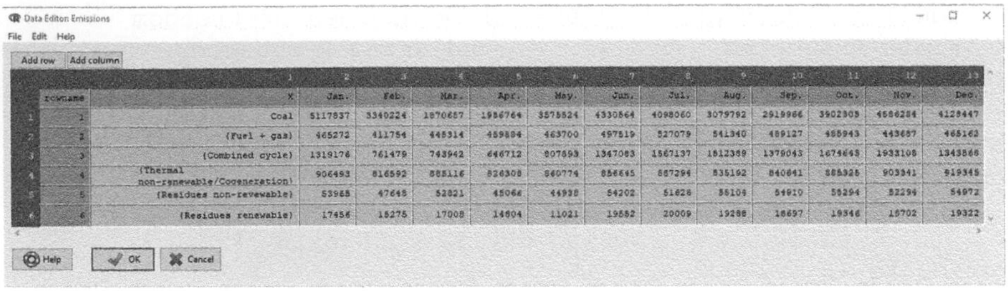

FIGURE 7.12 Inspecting and editing data in R Commander.

With the data loaded and ready for analysis, the user can easily perform basic statistical analysis, dimensional analysis (e.g., Principal Component Analysis, cluster analysis) and fit models (regression) by using the option tree called "Statistics" button; visualise data with different types of plots available in "Graphs" (histogram, boxplot, xy plots, etc.); work with fitted models ("Models" button); and generate distributions by using the "Distribution" button. The actions performed in the R Commander GUI and the results produced are recorded in the console of RStudio.

For example, with the Emissions data selected, clicking on *Statistics > Summaries > Active data sheet* returns the basic statistics (min, max, quantiles, mean and median) for each of the variables:

```
Rcmdr>                     Summary(Emissions_data)
                            x           Jan.               Feb.
Coal                        :1   Min.     :  17456  Min.    : 15275
Combined cycle              :1   1st Qu. :  156792 1st Qu.: 138674
Fuel + gas                  :1   Median  :  685882 Median  : 586616
Residues non-revewable      :1   Mean    : 1313366 Mean    : 898829
Residues renewable          :1   3rd Qu. : 1216005 3rd Qu.: 802814
Residues non-renewable/
Cogeneration                :1   Max.    : 5117837 Max.    :3340224
     Mar.                Apr.                May.               Jun.
Min    :  17008 Min.    :  14804 Min.    :  11021 Min.    :  19552
1st Qu.: 150944 1st Qu.: 148771 1st Qu.: 149628 1st Qu.: 165031
Median : 594628 Median : 553298 Median : 635796 Median : 677082
Mean   : 669148 Mean   : 663256 Mean   : 960642 Mean    :1184261
3rd Qu.: 849823 3rd Qu.: 781409 3rd Qu.: 847554 3rd Qu.:1224474
Max.   :1870687 Max.   :1986764 Max.   :3575524 Max.    :4330564
```

If we consider the transpose, i.e., months in the rows and power source in the columns, we can easily obtain the histogram or the boxplot of monthly emissions for coal, for instance, by using the options in "Graphs": (See Figure 7.13).

Finally, it is also possible to carry out analyses like those discussed in Sections 7.0 and 7.0. For example, a sample 200 points from a normal distribution with $x_{mean} = 0$ ppm and a $\sigma = 150$ ppm is created by selecting *Distributions > Continuous Distribution > Normal Distribution > Sample from Normal Distribution* and input the parameters in the dialog box that opens Figure 7.14(a). This action creates a new data set that

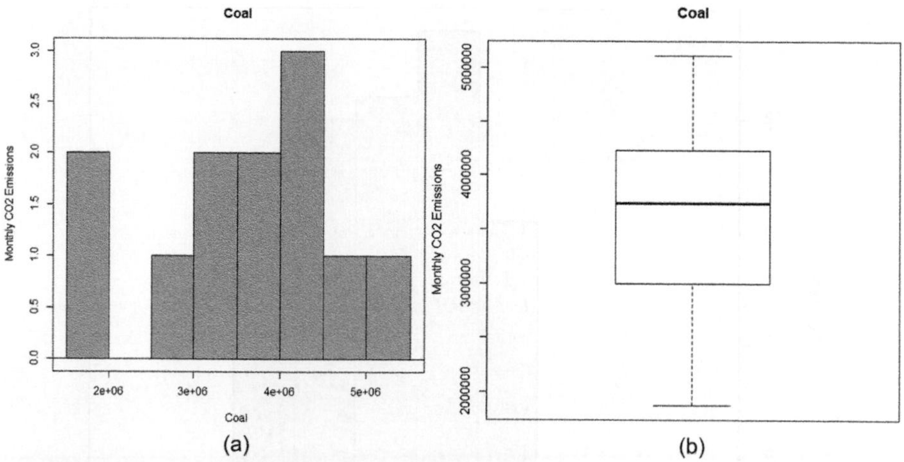

FIGURE 7.13 Monthly CO2 Emissions due to power generation in Spain in 2017: histogram (a) and boxplot (b).

can be used in further visualization, e.g., creating a histogram (Graphs > Histogram; see Figure 7.14b), and statistical analysis, such as the Shapiro-Wilk test (Statistics > Summaries > Shapiro-Wilk test of normality; see Figure 7.14c).

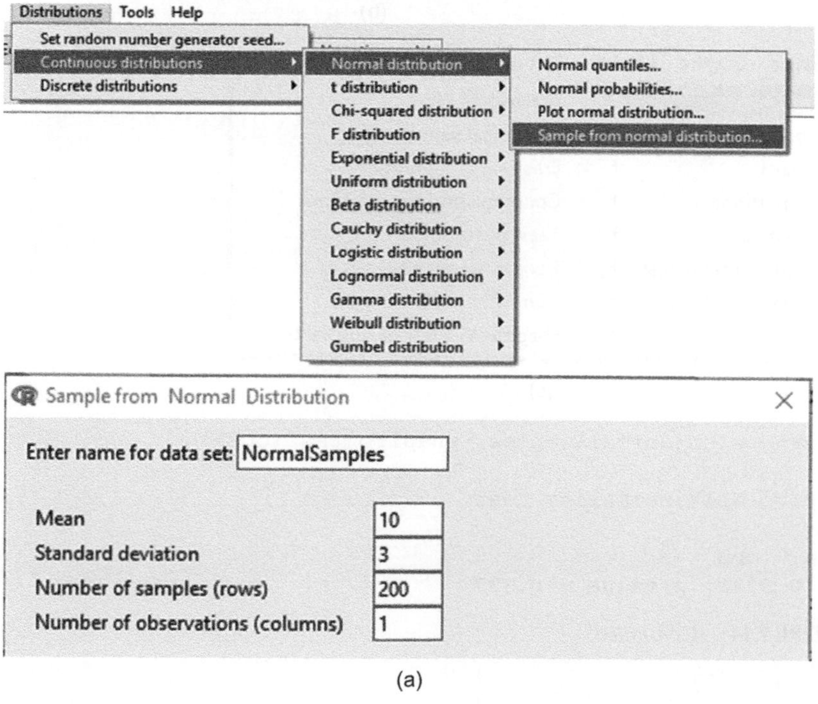

FIGURE 7.14 Example of data sampling from a normal distribution (a), plotting the results in a histogram (b) and Shapiro-Wilk test on the sample.

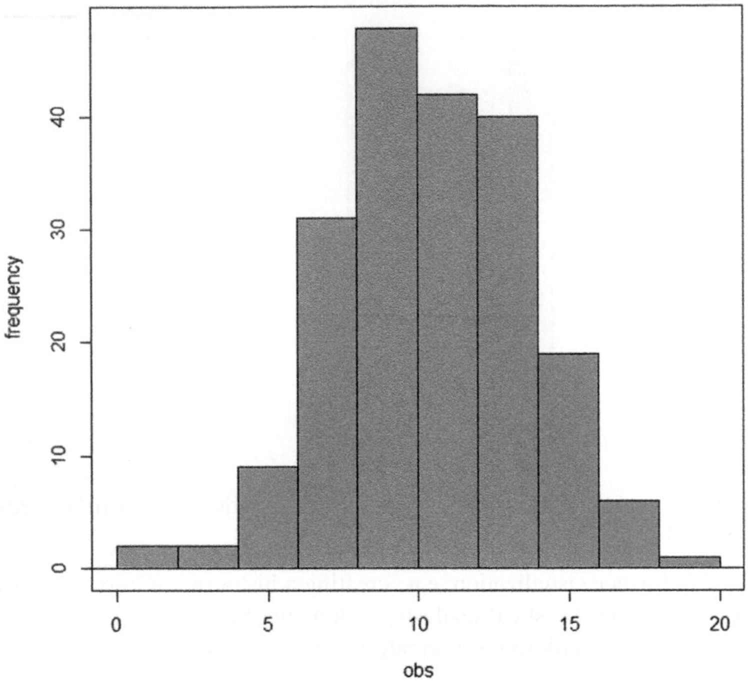

(b)

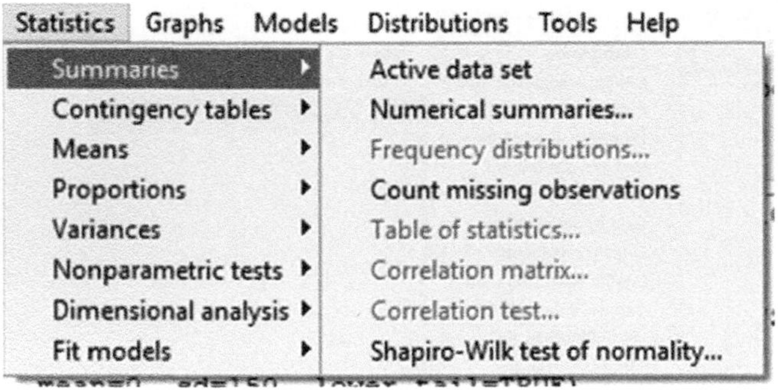

(c)

```
Rcmdr>  with(NormalSamples, shapiro.test(obs))

Shapiro-Wilk normality test

data:  obs
W = 0.9921, p-value = 0.357
```

FIGURE 7.14 *(Continued)*

REFERENCES

1. Cass S., "The 2017 Top Programming Languages," IEEE Spectrum, 2018. [Online]. Available: https://spectrum.ieee.org/computing/software/the-2017-top-programming-languages. [Accessed: 01-Nov-2018].
2. RStudio Inc., "RStudio," 2018. [Online]. Available: https://www.rstudio.com/products/rstudio. [Accessed: 25-Sep-2018].
3. RStudio Inc., "Rmarkdown," 2017. [Online]. Available: https://rmarkdown.rstudio.com. [Accessed: 01-Nov-2018].
4. RStudio Inc., "Shiny," 2017. [Online]. Available: https://shiny.rstudio.com. [Accessed: 01-Nov-2018].
5. Rcommander.com, "Rcommander," 2013. [Online]. Available: https://www.rcommander.com/.
6. Microsoft, "Microsoft R Application Network," 2018. [Online]. Available: https://mran.microsoft.com/open. [Accessed: 02-Nov-2018].
7. The R Foundation, "R," 2018. [Online]. Available: https://www.r-project.org. [Accessed: 25-Sep-2018].
8. RStudio Inc., "R Cheat Sheets," 2017. [Online]. Available: https://www.rstudio.com/resources/cheatsheets/. [Accessed: 25-Sep-2018].
9. Stack Exchanger Inc., "Stack Overflow," 2018. [Online]. Available: https://stackoverflow.com/.
10. Red Electrica de España, "Statistical data - National Statistical Series," 2018. [Online]. Available: https://www.ree.es/en/statistical-data-of-spanish-electrical-system/statistical-series/national-statistical-series.
11. J. Ward-Smith, (2012). *Mechanics of Fluids*, 9th edn. CRC Press, Boca Raton, FL.
12. M. Kohl, (2015). *Introduction to statistical data analysis with R*. Bookboon. ISBN 978-87-403-1123-5.
13. S. Stowell, (2014). *Using R for Statistics*. Apress, Berkely, CA.
14. F. Husson, S. Le, and J. Pages, (2011). *Exploratory Multivariate Analysis by Example Using R*. CRC Press, Boca Raton, FL.
15. J. Fox, (2016). *Using the R Commander: A Point-and-Click Interface for R*. CRC Press, Boca Raton, FL.

Section III

Detailed Equipment Design and Analysis

8 Computational Fluid Dynamics and Multiphysics

Bostjan Hari, Borja Hernández, and Mariano Martín

CONTENTS

8.1 INTRODUCTION

Computational fluid dynamics, known as CFD, is the numerical method of solving mass, momentum, energy and species conservation equations, and related phenomena on computers by using programming languages.

CFD presents an intersection between science and different engineering disciplines, among which mathematics, fluid mechanics and computer science are the most important. Users, who want to work as CFD engineers, need to have a broad knowledge and understanding of physics, especially fluid dynamics, heat and mass transfer apart from their field of expertise. The basics of CFD are partial differential equations and thus knowledge of numerical mathematics is essential to solve them with appropriate numerical technique. Since these so-called conservation equations are designed and solved on computers, knowledge of programming languages, such as FORTRAN, C++, Java® or MATLAB® is equally important.

Scientists have always been interested in describing flows with mathematical equations. The idea comes from French engineer Claude-Louis Navier and British physicist and mathematician George Gabriel Stokes in the middle of the 19th century when they described the motion of fluids based on Newton's Second Law. At that time their equations were not possible to be solved numerically and neither applied to real science and engineering systems. The equations, known as the Navier-Stokes conservation equations are commonly referred to continuity, momentum and energy equations. They represent the basic equations of computational fluid dynamics and can be applied to flows inside pipes, blood vessels, around airplane wings, turbines blades or even to predict weather and ocean currents just to name some applications. These equations are very complex and thus almost impossible to be solved analytically. With the invention of integrated circuits and development of computing hardware a new era has begun in this challenging science and engineering field called computational fluid dynamics.

The emergence of personal computing towards the end of the 1980s marks the beginning of widespread CFD based software modelling tools. Such software tools, popular in scientific and engineering communities are commercial software tools, such as ACE+ Suite, ANSYS® CFX®, ANSYS® Fluent®, ANSYS® Multiphysics™, COMSOL Multiphysics®, FLOW-3D®, STAR-CD® and STAR-CCM+® and an open-source software tool OpenFOAM®. Other CFD based software tools, such as AVL FIRE® or ANSYS® Polyflow® are also available on the market, but they are specialised for particular physical systems, such as internal combustion engines, powertrains, polymers, glass, metals and cement process technologies.

The computational branch of fluid dynamics has expanded beyond simple modelling and simulation of flow. The reason is the advent of powerful and cost-effective desktop computers. They can capture real-world physical phenomena, which are very complex in nature and include combination of different physical laws, such as conservation of mass, momentum, heat, species, electric or magnetic charge. With the availability of desktop computing the term computational fluid dynamics or CFD modelling and simulation is often now referred to as multiphysics modelling and simulation.

Figure 8.1 shows knowledge and skills any CFD engineer needs to possess for successful work with CFD and multiphysics software tools.

Computer based modelling and simulation has become an important and essential part of research and development process in science and engineering departments in academia and industry. Computational models namely represent virtual prototypes that can predict behaviour of physical systems, which might be impossible to observe on real-world prototypes. Visualisation of such models is especially important for

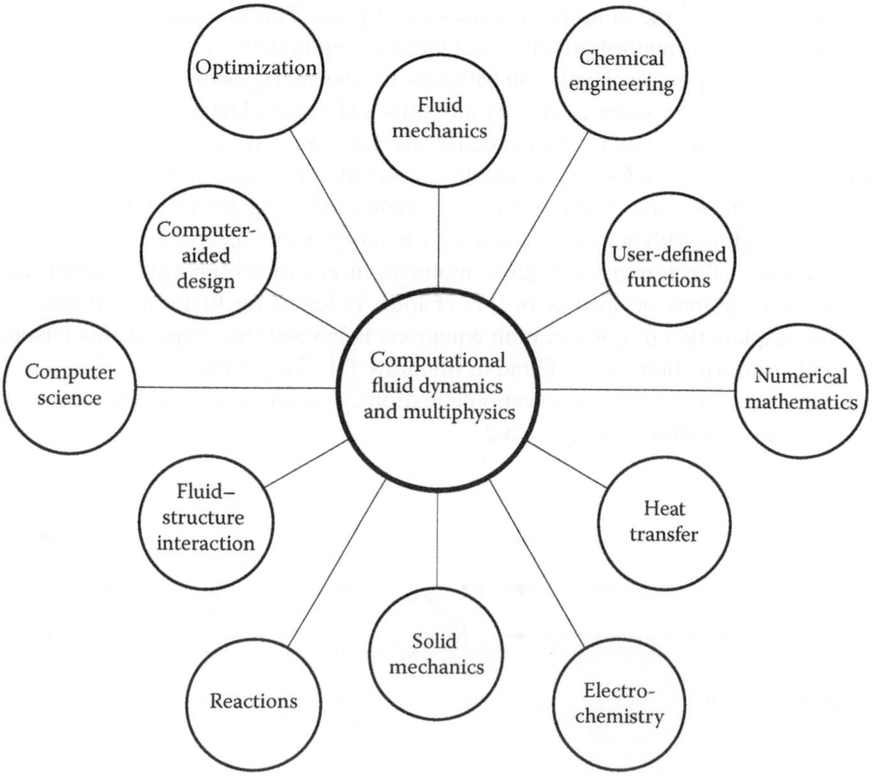

FIGURE 8.1 Knowledge and skills any CFD engineer needs to possess for successful work with CFD and multiphysics software tools.

understanding very complex systems or systems with difficulties to obtain required measurements. CFD based models thus offer relatively easy, reliable and cost-effective option to mimic real-world problems, depending on model assumptions. They are now included in an initial stage of product research and development process prior designing physical prototypes. The more accurate real-world physical systems need to be modelled, the more complex geometry and assumptions with appropriate boundary conditions need to be applied to computational models. This is very challenging task that requires multi core computers called supercomputers. Examples of complex modelling and simulation physical systems include weather forecast, climate research, molecular or multi scale modelling, folding of macromolecules, aerodynamics, detonation of weapons, combustion processes or nuclear fusion.

Further information about computational fluid dynamics and multiphysics can be found in literature [1], [2] and [3].

8.2 CONSERVATION EQUATIONS

Computational fluid dynamics and multiphysics modelling and simulation are based on four conservation equations; conservation of mass, conservation of momentum, conservation of energy and conservation of species. They represent mathematical formulations of the conservation laws of physics. These equations are usually enough to solve the majority of chemical engineering systems, except fuel cells and batteries as examples. Fuel cells and batteries are special types of chemical reactors, which additionally produce electricity and thus additional electric charge equations are needed. Additional conservation equations, equation terms or chemical reaction mechanisms must be added to an existing computational system if more complex physical system is used. Such examples include modelling microfluidic flows and chemical reactors with homogeneous and/or heterogeneous reactions.

Derivation of continuity equation, momentum equations, energy equation and continuity equations of species in this chapter is based on literature [1] and [2]. Detailed explanation of conservation equations is beyond the scope of this chapter but further information can be found in literature [1], [2], [3] and [4].

In order to understand conservation equations, consider a flow field represented by streamlines as shown in Figure 8.2.

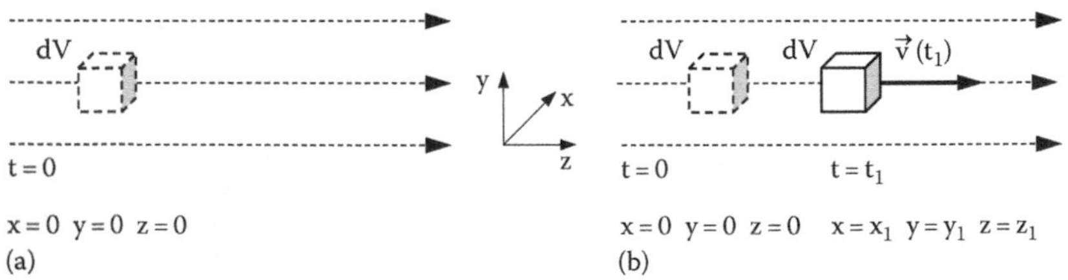

FIGURE 8.2 Fluid flow over the infinitesimal fluid element: (a) infinitesimal fluid element fixed in space to describe conservation form of conservation equations, (b) infinitesimal fluid element moving with the flow to describe non-conservation form of conservation equations.

The flow includes an infinitesimal fluid element with a differential volume dV and molecules inside this volume. The fluid element can be fixed in space with the flow moving through it as shown on the left-hand side of Figure 8.2. The fluid element can also move together with the flow and with the same velocity as shown on the left-hand side of Figure 8.2. Conservation equations derived from this concept are known as partial differential equation form. Partial differential equations obtained from the stationary fluid element, as shown on the left-hand side of Figure 8.2, are called the conservation form of conservation equations. Partial differential equations obtained from the moving fluid element, as shown on the right-hand side of Figure 8.2, are called non-conservation form of conservation equations. For clarification purposes, the difference between the conservation form and the non-conservation form of partial differential equations will be demonstrated on the continuity equation in the next paragraph.

An example of the continuity equation derived from the stationary fluid element and thus written in the conservation form with the vector notation appropriate for any coordinate system is

$$\frac{\partial \rho}{\partial t} + \vec{\nabla} \cdot (\rho \, \vec{v}) = 0 \tag{8.1}$$

where the vector operator nabla in the three-dimensional space with the Cartesian coordinates is defined as

$$\vec{\nabla} \equiv \frac{\partial}{\partial x} \vec{i} + \frac{\partial}{\partial y} \vec{j} + \frac{\partial}{\partial z} \vec{k} \tag{8.2}$$

By applying the definition for the vector operator divergence of the density and velocity product as

$$div(\rho \, \vec{v}) \equiv \vec{\nabla} \cdot (\rho \, \vec{v}) = \rho \, \vec{\nabla} \cdot \vec{v} + \vec{v} \cdot \vec{\nabla} \, \rho \tag{8.3}$$

we can expand the vector form of the continuity equation written in the conservation form appropriate for any coordinate system as

$$\frac{\partial \rho}{\partial t} + \rho \, \vec{\nabla} \cdot \vec{v} + \vec{v} \cdot \vec{\nabla} \, \rho = 0 \tag{8.4}$$

or alternatively in the three-dimensional space of the Cartesian coordinate system for time-dependent, compressible and viscous flow as

$$\frac{\partial \rho}{\partial t} + \rho \left[\frac{\partial v_x}{\partial x} + \frac{\partial v_y}{\partial y} + \frac{\partial v_z}{\partial z} \right] + v_x \frac{\partial \rho}{\partial x} + v_y \frac{\partial \rho}{\partial y} + v_z \frac{\partial \rho}{\partial z} = 0 \tag{8.5}$$

By introducing the substantial derivative written with the vector notation and expanded to the Cartesian coordinate system as

$$\frac{D}{Dt} \equiv \frac{\partial}{\partial t} + \left(\vec{v} \cdot \vec{\nabla} \right) = \frac{\partial}{\partial t} + v_x \frac{\partial}{\partial x} + v_y \frac{\partial}{\partial y} + v_z \frac{\partial}{\partial z} \tag{8.6}$$

equation (8.5) can be written as

$$\frac{D\rho}{Dt} + \rho\, \vec{\nabla} \cdot \vec{v} = 0 \tag{8.7}$$

which is the vector notation of the continuity equation written in the non-conservation form for the three-dimensional space of the Cartesian coordinate system for dependent, compressible and viscous flow representing the fluid element moving with the flow.

It must be emphasised that the equations (8.1) and (8.7) are the same continuity equations but written in different forms. The first equation is the conservation form applied to the fluid element fixed in the flow whereas the second equation is the non-conservation form applied to the fluid element moving with the flow. Understanding the difference between these two forms is essential to understand the conservation equations explained in the next sections in order to eliminate confusion with different notations of the conservation equations written in literature.

Contrary to the partial differential equation form of conservation equations is the integral form of conservation equations. Detailed explanation of conservation equations written in the integral form is beyond the scope of this chapter but further information can be found in literature [1] and [2].

8.2.1 Laminar Flow

A British engineer and scientist Osborne Reynolds demonstrated an experiment in 1883 in which water was introduced through a glass tube. The flow pattern was observed by injected dye at the inlet side of the tube and the flow rate was controlled at the outlet by a valve. Reynolds found out that the dye filament remained intact and parallel with the tube at low flow velocities. This flow is known as laminar flow. When the flow velocity was increased the dye filament initially began to oscillate up and down and finally broke up. The dye particles randomly mixed with water and occupied different positions of the tube. This flow is known as turbulent flow.

Laminar flow is important in engineering systems, such as flow in pipes, fuel cells, chemical reactors, microreactors, microfluidic and nanofluidic devices just to name some applications.

8.2.1.1 Conservation of Mass

The conservation of mass is mathematically described by the continuity equation. Its physical interpretation means that a mass, represented by the total mass flow rate, entering a fluid element must be equal to the total mass flow rate leaving it. In other words, the total mass flow out of a fluid element must be equal to the time rate of total mass decrease in the fluid element as

$$\underbrace{A}_{\text{net mass flow}} = \underbrace{B}_{\text{time rate of mass decrease}} \tag{8.8}$$

The mathematical representation of the mass flow out of the fluid element in x direction of the surface $dy\ dz$ is

$$\underbrace{\left[\rho v_x + \frac{\partial(\rho v_x)}{\partial x} dx\right] dy\ dz}_{\textit{mass outflow in x direction}} - \underbrace{(\rho v_x)\ dy\ dz}_{\textit{mass inflow in x direction}} \tag{8.9}$$

in y direction of the surface $dx\ dz$ is

$$\underbrace{\left[\rho v_y + \frac{\partial(\rho v_y)}{\partial y} dy\right] dx\ dz}_{\textit{mass outflow in y direction}} - \underbrace{(\rho v_y)\ dx\ dz}_{\textit{mass inflow in y direction}} \tag{8.10}$$

and in z direction of the surface $dx\ dy$ is

$$\underbrace{\left[\rho v_z + \frac{\partial(\rho v_z)}{\partial z} dz\right] dx\ dy}_{\textit{mass outflow in z direction}} - \underbrace{(\rho v_z)\ dx\ dy}_{\textit{mass inflow in z direction}} \tag{8.11}$$

The total mass flow out of the fluid element is sum of the mass flow out of the individual fluid element surfaces represented by the equations (8.9), (8.10) and (8.11)

$$A = \left[\frac{\partial(\rho v_x)}{\partial x} + \frac{\partial(\rho v_y)}{\partial y} + \frac{\partial(\rho v_z)}{\partial z}\right] dx\ dy\ dz \tag{8.12}$$

The time rate of total mass decrease in the fluid element is

$$B = \left(\frac{\partial \rho}{\partial t}\right) dx\ dy\ dz \tag{8.13}$$

By equalising the equations (8.12) and (8.13) according to the equation (8.8), we get the conservation form of the continuity equation written in the three-dimensional space of the Cartesian coordinate system for time-dependent, compressible and viscous flow. This can be written as

$$\frac{\partial \rho}{\partial t} + \frac{\partial(\rho v_x)}{\partial x} + \frac{\partial(\rho v_y)}{\partial y} + \frac{\partial(\rho v_z)}{\partial z} = 0 \tag{8.14}$$

which is the same equation as the equation (8.1) written for any coordinate system.

The conservation form of the continuity equation (8.14) written in the three-dimensional space of the Cartesian coordinate system for dependent, compressible and viscous flow can be expanded as

$$\underbrace{\frac{\partial \rho}{\partial t}}_{\textit{acceleration}} + \underbrace{\rho\left[\frac{\partial v_x}{\partial x} + \frac{\partial v_y}{\partial y} + \frac{\partial v_z}{\partial z}\right]}_{\textit{advection}} + \underbrace{v_x \frac{\partial \rho}{\partial x} + v_y \frac{\partial \rho}{\partial y} + v_z \frac{\partial \rho}{\partial z}}_{\textit{compression}} = 0 \tag{8.15}$$

and divided into acceleration, advection and compression terms.

The continuity equation (8.15) written in the three-dimensional space of the Cartesian coordinate system for time-dependent but incompressible and viscous flow simplifies to

$$\frac{\partial \rho}{\partial t}+\rho\left[\frac{\partial v_x}{\partial x}+\frac{\partial v_y}{\partial y}+\frac{\partial v_z}{\partial z}\right]=0 \tag{8.16}$$

because the mass density $\rho\left[kg/m^3\right]$ of the fluid element moving with the flow is constant and thus the compression term in the equation (8.15) is neglected.

If the fluid element in the incompressible flow is stationary and does not change by time, the equation (8.16) can be further simplified to

$$\frac{\partial v_x}{\partial x}+\frac{\partial v_y}{\partial y}+\frac{\partial v_z}{\partial z}=0 \tag{8.17}$$

The equation (8.17) is the simplest form of the continuity equation in the three-dimensional space of the Cartesian coordinate system, written for the steady-state, incompressible and viscous flow.

8.2.1.2 Conservation of Momentum

The conservation of momentum is mathematically described by the momentum equations, also known as the Navier-Stokes equations. Its physical interpretation is based on Newton's second law, meaning that the net force on the infinitesimal fluid element is equal to its mass times the acceleration of the fluid element defined as

$$\vec{F}=m\cdot\vec{a} \tag{8.18}$$

The total force vector $\vec{F}\,[N]$ consists of surface forces (pressure, normal and shear stresses) acting on the fluid element surface and body forces (gravity and electromagnetic force) acting on the volume of the same fluid element. By replacing the x component of the force vector, we get the following notation

$$F_x=\underbrace{\left[-\frac{\partial p}{\partial x}+\frac{\partial \tau_{xx}}{\partial x}+\frac{\partial \tau_{yx}}{\partial y}+\frac{\partial \tau_{zx}}{\partial z}\right]dx\;dy\;dz}_{\textit{surface forces in x direction}}+\underbrace{\rho\;f_x\;dx\;dy\;dz}_{\textit{body forces in x direction}} \tag{8.19}$$

By replacing the moving fluid element mass $m\,[kg]$ with

$$m=\rho\;dx\;dy\;dz \tag{8.20}$$

and the acceleration $a_x\left[m/s^2\right]$ of the same fluid element in the x direction with the substantial derivative

$$a_x=\frac{Dv_x}{Dt} \tag{8.21}$$

we get the x component of the equation (8.18) written as

$$\left[-\frac{\partial p}{\partial x}+\frac{\partial \tau_{xx}}{\partial x}+\frac{\partial \tau_{yx}}{\partial y}+\frac{\partial \tau_{zx}}{\partial z}\right]dx\ dy\ dz+\rho\ f_x\ dx\ dy\ dz=\rho\ dx\ dy\ dz\cdot\frac{Dv_x}{Dt} \quad (8.22)$$

With some mathematical manipulation, the equation (8.22) can be rewritten as

$$\rho\frac{Dv_x}{Dt}=-\frac{\partial p}{\partial x}+\frac{\partial \tau_{xx}}{\partial x}+\frac{\partial \tau_{yx}}{\partial y}+\frac{\partial \tau_{zx}}{\partial z}+\rho f_x \quad (8.23)$$

which is known as the x component of the momentum or Navier-Stokes equations, written in the non-conservation form of the Cartesian coordinate system for time-dependent, compressible and viscous flow.

By introducing the substantial derivative in the equation (8.23) as

$$\rho\frac{Dv_x}{Dt}\equiv\rho\frac{\partial v_x}{\partial t}+\rho\left(\vec{v}\cdot\vec{\nabla}\ v_x\right) \quad (8.24)$$

and expanding the derivative as

$$\rho\frac{\partial v_x}{\partial t}=\frac{\partial(\rho v_x)}{\partial t}-v_x\frac{\partial \rho}{\partial t} \quad (8.25)$$

we can write equation (8.24) as

$$\rho\frac{Dv_x}{Dt}=\frac{\partial(\rho v_x)}{\partial t}-v_x\frac{\partial \rho}{\partial t}+\rho\left(\vec{v}\cdot\vec{\nabla}\ v_x\right) \quad (8.26)$$

By applying the definition for the vector operator divergence as the product of the scalar ρv_x and the vector $\vec{v}$ as

$$div\left(\rho v_x\ \vec{v}\right)\equiv\vec{\nabla}\cdot\left(\rho v_x\ \vec{v}\right)=v_x\vec{\nabla}\cdot\left(\rho\vec{v}\right)+\rho\vec{v}\cdot\vec{\nabla}\ v_x \quad (8.27)$$

and substituting the equation (8.27) into the equation (8.26) and rearranging equations we get

$$\rho\frac{Dv_x}{Dt}=\frac{\partial(\rho v_x)}{\partial t}-v_x\left[\frac{\partial \rho}{\partial t}+\vec{\nabla}\cdot\left(\rho\vec{v}\right)\right]+\vec{\nabla}\cdot\left(\rho v_x\ \vec{v}\right) \quad (8.28)$$

where the term in parentheses is the continuity equation (8.1) which is equal to zero. The final form of the equation (8.28) thus simplifies to

$$\rho\frac{Dv_x}{Dt}=\frac{\partial(\rho v_x)}{\partial t}+\vec{\nabla}\cdot\left(\rho v_x\ \vec{v}\right) \quad (8.29)$$

By substituting the substantial derivative in the equation (8.23) with the equation (8.29) we get the x component of the momentum equations as

$$\frac{\partial(\rho v_x)}{\partial t}+\vec{\nabla}\cdot(\rho v_x\ \vec{v})=-\frac{\partial p}{\partial x}+\frac{\partial \tau_{xx}}{\partial x}+\frac{\partial \tau_{yx}}{\partial y}+\frac{\partial \tau_{zx}}{\partial z}+\rho f_x \tag{8.30}$$

The y and z components of the momentum equations can be derived in similar fashion as the x component. Thus, the y component is written as

$$\frac{\partial(\rho v_y)}{\partial t}+\vec{\nabla}\cdot(\rho v_y\ \vec{v})=-\frac{\partial p}{\partial y}+\frac{\partial \tau_{xy}}{\partial x}+\frac{\partial \tau_{yy}}{\partial y}+\frac{\partial \tau_{zy}}{\partial z}+\rho f_y \tag{8.31}$$

and the z component as

$$\frac{\partial(\rho v_z)}{\partial t}+\vec{\nabla}\cdot(\rho v_z\ \vec{v})=-\frac{\partial p}{\partial z}+\frac{\partial \tau_{xz}}{\partial x}+\frac{\partial \tau_{yz}}{\partial y}+\frac{\partial \tau_{zz}}{\partial z}+\rho f_z \tag{8.32}$$

The equations (8.30), (8.31) and (8.32) represent the complete momentum or Navier-Stokes equations written in the conservation form of the Cartesian coordinate system for time-dependent, compressible and viscous flow.

Normal stresses describe the time rate of the deformation change applied to the fluid element whereas shear stresses describe the time rate of the volume change applied to the fluid element. Normal stresses are smaller compared to shear stresses and are usually neglected. If we assume the fluid element as Newtonian flow, we can write the following definition for normal and shear stresses as

$$\tau_{xx}=\lambda\left(\vec{\nabla}\cdot\vec{v}\right)+2\ \mu\ \frac{\partial v_x}{\partial x} \tag{8.33}$$

$$\tau_{yy}=\lambda\left(\vec{\nabla}\cdot\vec{v}\right)+2\ \mu\ \frac{\partial v_y}{\partial x} \tag{8.34}$$

$$\tau_{zz}=\lambda\left(\vec{\nabla}\cdot\vec{v}\right)+2\ \mu\ \frac{\partial v_z}{\partial x} \tag{8.35}$$

$$\tau_{xy}=\tau_{yx}=\mu\left[\frac{\partial v_y}{\partial x}+\frac{\partial v_x}{\partial y}\right] \tag{8.36}$$

$$\tau_{xz}=\tau_{zx}=\mu\left[\frac{\partial v_x}{\partial z}+\frac{\partial v_z}{\partial x}\right] \tag{8.37}$$

$$\tau_{yz}=\tau_{zy}=\mu\left[\frac{\partial v_z}{\partial y}+\frac{\partial v_y}{\partial z}\right] \tag{8.38}$$

where $\tau_{xx}\ \left[N/m^2\right]$, $\tau_{yy}\ \left[N/m^2\right]$ and $\tau_{yy}\ \left[N/m^2\right]$ denote the normal stress in the x, y and z directions, $\tau_{xy}\ \left[N/m^2\right]$ the shear stress in the y direction applied on a plane perpendicular to the x axis, τ_{yx} $\left[N/m^2\right]$ the shear stress in the x direction applied on a plane perpendicular to the y axis, τ_{xz} $\left[N/m^2\right]$ the shear stress in the z direction applied on a plane perpendicular to the x axis, τ_{zx} $\left[N/m^2\right]$ the shear stress in the x direction applied on a plane perpendicular to the z axis, τ_{yz} $\left[N/m^2\right]$ the shear stress in the z direction applied on a plane perpendicular to the y axis and τ_{zy} $\left[N/m^2\right]$ the shear stress in the y direction applied on a plane perpendicular to the z axis.

The equations (8.33), (8.34) and (8.35) include factor λ, which is only important for compressible fluids. This factor is based on the Stokes hypothesis and is determined as

$$\lambda = -\frac{2}{3}\mu \tag{8.39}$$

where μ $\left[Pa \cdot s\right]$ denotes the dynamic viscosity of the fluid.

If we substitute the equations (8.33) to (8.38) with the equations (8.30), (8.31) and (8.32), we get the complete momentum or Navier-Stokes equations written in the conservation form of the Cartesian coordinate system for time-dependent, compressible and viscous flow in terms of velocities. The x component of the momentum equations is

$$\frac{\partial(\rho v_x)}{\partial t} + \frac{\partial(\rho v_x^2)}{\partial x} + \frac{\partial(\rho v_x v_y)}{\partial y} + \frac{\partial(\rho v_x v_z)}{\partial z}$$
$$= -\frac{\partial p}{\partial x} + \frac{\partial}{\partial x}\left[\lambda \vec{\nabla} \cdot \vec{v} + 2\mu \frac{\partial v_x}{\partial x}\right] + \frac{\partial}{\partial y}\left[\mu\left(\frac{\partial v_y}{\partial x} + \frac{\partial v_x}{\partial y}\right)\right] + \frac{\partial}{\partial z}\left[\mu\left(\frac{\partial v_y}{\partial z} + \frac{\partial v_z}{\partial x}\right)\right] + \rho f_x \tag{8.40}$$

the y component of the momentum equations is

$$\frac{\partial(\rho v_y)}{\partial t} + \frac{\partial(\rho v_x v_y)}{\partial x} + \frac{\partial(\rho v_y^2)}{\partial y} + \frac{\partial(\rho v_y v_z)}{\partial z}$$
$$= -\frac{\partial p}{\partial y} + \frac{\partial}{\partial x}\left[\mu\left(\frac{\partial v_y}{\partial x} + \frac{\partial v_x}{\partial y}\right)\right] + \frac{\partial}{\partial y}\left[\lambda \vec{\nabla} \cdot \vec{v} + 2\mu \frac{\partial v_y}{\partial y}\right] + \frac{\partial}{\partial z}\left[\mu\left(\frac{\partial v_z}{\partial y} + \frac{\partial v_y}{\partial z}\right)\right] + \rho f_y \tag{8.41}$$

and the z component of the momentum equations is

$$\frac{\partial(\rho v_z)}{\partial t} + \frac{\partial(\rho v_x v_z)}{\partial x} + \frac{\partial(\rho v_y v_z)}{\partial y} + \frac{\partial(\rho v_z^2)}{\partial z}$$
$$= -\frac{\partial p}{\partial z} + \frac{\partial}{\partial x}\left[\mu\left(\frac{\partial v_x}{\partial z} + \frac{\partial v_z}{\partial x}\right)\right] + \frac{\partial}{\partial y}\left[\mu\left(\frac{\partial v_z}{\partial y} + \frac{\partial v_y}{\partial z}\right)\right] + \frac{\partial}{\partial z}\left[\lambda \vec{\nabla} \cdot \vec{v} + 2\mu \frac{\partial v_z}{\partial z}\right] + \rho f_z \tag{8.42}$$

If we consider the flow with constant density, we can cancel items from the continuity equation and effect of terms due to constant property and neglect body forces. Thus we can derive the momentum or Navier-Stokes equations written in the conservation form of the Cartesian coordinate system for time-dependent, incompressible and viscous flow in terms of velocities. The x component of the momentum equations is

$$\underbrace{\frac{\partial v_x}{\partial t}}_{acceleration} + \underbrace{v_x \frac{\partial v_x}{\partial x} + v_y \frac{\partial v_x}{\partial y} + v_z \frac{\partial v_x}{\partial z}}_{advection} = \underbrace{-\frac{1}{\rho}\frac{\partial p}{\partial x}}_{pressure} + \nu \left[\underbrace{\frac{\partial^2 v_x}{\partial x^2} + \frac{\partial^2 v_x}{\partial y^2} + \frac{\partial^2 v_x}{\partial z^2}}_{diffusion} \right] \tag{8.43}$$

the y component of the momentum equations is

$$\underbrace{\frac{\partial v_y}{\partial t}}_{acceleration} + \underbrace{v_x \frac{\partial v_y}{\partial x} + v_y \frac{\partial v_y}{\partial y} + v_z \frac{\partial v_y}{\partial z}}_{advection} = \underbrace{-\frac{1}{\rho}\frac{\partial p}{\partial y}}_{pressure} + \nu \left[\underbrace{\frac{\partial^2 v_y}{\partial x^2} + \frac{\partial^2 v_y}{\partial y^2} + \frac{\partial^2 v_y}{\partial z^2}}_{diffusion} \right] \tag{8.44}$$

and the z component of the momentum equations is

$$\underbrace{\frac{\partial v_z}{\partial t}}_{acceleration} + \underbrace{v_x \frac{\partial v_z}{\partial x} + v_y \frac{\partial v_z}{\partial y} + v_z \frac{\partial v_z}{\partial z}}_{advection} = \underbrace{-\frac{1}{\rho}\frac{\partial p}{\partial z}}_{pressure} + \nu \left[\underbrace{\frac{\partial^2 v_z}{\partial x^2} + \frac{\partial^2 v_z}{\partial y^2} + \frac{\partial^2 v_z}{\partial z^2}}_{diffusion} \right] \tag{8.45}$$

8.2.1.3 Conservation of Energy

The continuity equation (8.1) or (8.7) and the momentum equations in x direction (8.40), y direction (8.41) and z direction (8.42) are mechanical equations and are not affected by chemical reactions. The energy equation for chemically non-reacting flows with included heat flux due to thermal conduction and thermal radiation is usually sufficient. But in chemically reacting flows, as common in chemical engineering applications, the energy equation with the heat flux due to additional diffusion of species has to be included.

The conservation of energy is mathematically described by the energy equation. Its physical interpretation is based on First law of thermodynamics, meaning that the rate of change of the total energy A in the fluid element is equal to the sum of the heat flux B into the fluid element and the rate of work C done on the same fluid element due to body and surface forces defined as

$$\underbrace{A}_{rate\ of\ change\ of\ energy} = \underbrace{B}_{heat\ flux} + \underbrace{C}_{rate\ of\ work} \tag{8.46}$$

The rate of change of the total energy A on the moving fluid element consists of the kinetic energy due to its translational motion and of the internal energy due to random molecular motion inside moving fluid element. The random motion of molecules is combination of translational, rotational, vibrational and electronic energies.

The rate of change of the total energy A on the moving fluid element is mathematically defined with the substantial derivative as

$$A = \rho \frac{DE}{Dt} dx\ dy\ dz \tag{8.47}$$

where $E\ [J/kg]$ denotes the total energy of the fluid element defined as

$$E = e + \frac{1}{2}\left(v_x^2 + v_y^2 + v_z^2\right) \tag{8.48}$$

where $e\ [J/kg]$ denotes the internal energy of the fluid element and $\frac{1}{2}\left(v_x^2 + v_y^2 + v_z^2\right)\ [J/kg]$ the kinetic energy of the fluid element.

The net heat flux B for chemically non-reacting flow represents the net flux of heat into the fluid element due to volumetric heating (thermal radiation) and temperature gradients (thermal conduction). It is mathematically defined as

$$B = \underbrace{\rho \dot{q}\ dx\ dy\ dz}_{volumetic\ heating} - \underbrace{\left(\frac{\partial \dot{q}_x}{\partial x} + \frac{\partial \dot{q}_y}{\partial y} + \frac{\partial \dot{q}_z}{\partial z}\right) dx\ dy\ dz}_{temperature\ gradients} \tag{8.49}$$

where $\dot{q}_x\ \left[W/m^2\right]$, $\dot{q}_y\ \left[W/m^2\right]$ and $\dot{q}_z\ \left[W/m^2\right]$ denote the heat flux components into the fluid element in the x, y and z directions.

By replacing the heat flux due to thermal conduction with Fourier's law of heat conduction in the x direction defined as

$$\dot{q}_x = -k \frac{\partial T}{\partial x} \tag{8.50}$$

in the y direction defined as

$$\dot{q}_y = -k \frac{\partial T}{\partial y} \tag{8.51}$$

and in the z direction defined as

$$\dot{q}_z = -k \frac{\partial T}{\partial z} \tag{8.52}$$

we get the final notation for the net heat flux B

$$B = \left[\rho \dot{q} + \frac{\partial}{\partial x}\left(k \frac{\partial T}{\partial x}\right) + \frac{\partial}{\partial y}\left(k \frac{\partial T}{\partial y}\right) + \frac{\partial}{\partial z}\left(k \frac{\partial T}{\partial z}\right)\right] dx\ dy\ dz \tag{8.53}$$

where $\rho\ \left[kg/m^3\right]$ denotes the mass density of the fluid element, $\dot{q}\ \left[W/m^2\right]$ the rate of volumetric heat, $k\ \left[W/(m \cdot K)\right]$ the thermal conductivity of the fluid element and $T\ [K]$ the temperature of the fluid element.

The rate of work C done on the moving fluid element due to body forces (gravity, electromagnetic force) and surface forces (pressure, normal stresses, shear stresses) is mathematically defined as

$$C = \underbrace{\rho\vec{f}\cdot\vec{v}\; dx\; dy\; dz}_{body\ \ forces} + \underbrace{\left[C_x + C_y + C_z\right]}_{surface\ \ forces} \tag{8.54}$$

where the x component of surface forces consisting of the rate of work done by pressure and shear stresses is

$$C_x = \left[-\frac{\partial(v_x p)}{\partial x} + \frac{\partial(v_x \tau_{xx})}{\partial x} + \frac{\partial(v_x \tau_{yx})}{\partial y} + \frac{\partial(v_x \tau_{zx})}{\partial z}\right] dx\; dy\; dz \tag{8.55}$$

the y component of surface forces consisting of the rate of work done by pressure and shear stresses is

$$C_y = \left[-\frac{\partial(v_y p)}{\partial y} + \frac{\partial(v_y \tau_{xy})}{\partial x} + \frac{\partial(v_y \tau_{yy})}{\partial y} + \frac{\partial(v_y \tau_{zy})}{\partial z}\right] dx\; dy\; dz \tag{8.56}$$

and the z component of surface forces consisting of the rate of work done by pressure and shear stresses is

$$C_z = \left[-\frac{\partial(v_z p)}{\partial z} + \frac{\partial(v_z \tau_{xz})}{\partial x} + \frac{\partial(v_z \tau_{yz})}{\partial y} + \frac{\partial(v_z \tau_{zz})}{\partial z}\right] dx\; dy\; dz \tag{8.57}$$

By substituting the equations (8.47), (8.53) and (8.54) with the equation (8.46), we get the non-conservation form of the energy equation for chemically non-reactive flows in terms of total energy in the three-dimensional space of the Cartesian coordinate system for time-dependent, compressible and viscous flow as

$$\begin{aligned}\rho\frac{DE}{Dt} &= \rho\dot{q} + \frac{\partial}{\partial x}\left(k\frac{\partial T}{\partial x}\right) + \frac{\partial}{\partial y}\left(k\frac{\partial T}{\partial y}\right) + \frac{\partial}{\partial z}\left(k\frac{\partial T}{\partial z}\right) \\ &\quad - \frac{\partial(v_x p)}{\partial x} - \frac{\partial(v_y p)}{\partial y} - \frac{\partial(v_z p)}{\partial z} + \Phi + \rho\left(v_x f_x + v_y f_y + v_z f_z\right)\end{aligned} \tag{8.58}$$

where Φ denotes the dissipation function for non-reactive flows due to viscous stresses. It comprises of the rest of the terms defined with the equations (8.55), (8.56) and (8.57) as

$$\begin{aligned}\Phi &= \frac{\partial(v_x \tau_{xx})}{\partial x} + \frac{\partial(v_x \tau_{yx})}{\partial y} + \frac{\partial(v_x \tau_{zx})}{\partial z} + \frac{\partial(v_y \tau_{xy})}{\partial x} + \frac{\partial(v_y \tau_{yy})}{\partial y} \\ &\quad + \frac{\partial(v_y \tau_{zy})}{\partial z} + \frac{\partial(v_z \tau_{xz})}{\partial x} + \frac{\partial(v_z \tau_{yz})}{\partial y} + \frac{\partial(v_z \tau_{zz})}{\partial z}\end{aligned} \tag{8.59}$$

The dissipation function for non-reactive flow represents a source of heat caused by deformation of the fluid element due to work done. It can be neglected in most chemical engineering applications due to small contribution to the increased heat. If we also neglect body forces acting on the fluid element, the equation (8.58) simplifies to

$$\rho\frac{D\left(e+\frac{1}{2}\left(v_x^2+v_y^2+v_z^2\right)\right)}{Dt}=\rho\dot{q}+\frac{\partial}{\partial x}\left(k\frac{\partial T}{\partial x}\right)+\frac{\partial}{\partial y}\left(k\frac{\partial T}{\partial y}\right)+\frac{\partial}{\partial z}\left(k\frac{\partial T}{\partial z}\right)-\frac{\partial(v_x p)}{\partial x}-\frac{\partial(v_y p)}{\partial y}-\frac{\partial(v_z p)}{\partial z} \tag{8.60}$$

It is important to emphasise that the substantial derivative on the left-hand side of the energy equation (8.60) can be also written in terms of internal energy e or enthalpy h. In each case, the right-hand side of the energy equation (8.60) changes.

For the purpose of usability in the CFD and multiphysics chemical engineering applications, it is worth it to write the energy equation (8.60) in terms of enthalpy.

By definition of enthalpy as

$$h=e+\frac{p}{\rho} \tag{8.61}$$

its substantial derivative can be written, based on the differentiation rules, as

$$\frac{De}{Dt}=\frac{Dh}{Dt}-\frac{1}{\rho}\frac{Dp}{Dt}+\frac{p}{\rho^2}\frac{D\rho}{Dt} \tag{8.62}$$

By combining the continuity equation (8.7) in the non-conservation form with the equation (8.62), we get

$$\frac{De}{Dt}=\frac{Dh}{Dt}-\frac{1}{\rho}\frac{Dp}{Dt}+\frac{p}{\rho}\vec{\nabla}\cdot\vec{v}=0 \tag{8.63}$$

or by multiplying the equation (8.63) with the density ρ and introducing the substantial derivative for the pressure p, we get

$$\rho\frac{De}{Dt}=\rho\frac{Dh}{Dt}-\frac{\partial p}{\partial t}-\vec{v}\cdot\vec{\nabla}\,p-p\,\vec{\nabla}\cdot\vec{v}=0 \tag{8.64}$$

or by recalling the definition for the vector operator divergence from the equation (8.3) of the density and velocity product, and rearranging the equation (8.64), we get the substantial derivative for enthalpy as

$$\rho\frac{Dh}{Dt}=\rho\frac{De}{Dt}+\frac{\partial p}{\partial t}+\vec{\nabla}\cdot\left(p\,\vec{v}\right)=0 \tag{8.65}$$

Let multiply the x component of the momentum equation (8.23) with the density ρ and velocity component v_x as

$$\rho\frac{Dv_x^2}{Dt}=-v_x\frac{\partial p}{\partial x}+v_x\frac{\partial\tau_{xx}}{\partial x}+v_x\frac{\partial\tau_{yx}}{\partial y}+v_x\frac{\partial\tau_{zx}}{\partial z}+\rho v_x f_x \tag{8.66}$$

If we use the same analogy for the y component as

$$\rho\frac{Dv_y^2}{Dt}=-v_y\frac{\partial p}{\partial y}+v_y\frac{\partial\tau_{xy}}{\partial x}+v_y\frac{\partial\tau_{yy}}{\partial y}+v_y\frac{\partial\tau_{zy}}{\partial z}+\rho v_y f_y \tag{8.67}$$

and the same analogy for the z component as

$$\rho\frac{Dv_z^2}{Dt}=-v_z\frac{\partial p}{\partial z}+v_z\frac{\partial\tau_{xz}}{\partial x}+v_z\frac{\partial\tau_{yz}}{\partial y}+v_z\frac{\partial\tau_{zz}}{\partial z}+\rho v_z f_z \tag{8.68}$$

we get the substantial derivative of the kinetic energy by summarising the equations (8.66), (8.67) and (8.68) as

$$\rho\frac{D\left(\frac{1}{2}\left(v_x^2+v_y^2+v_z^2\right)\right)}{Dt}=-\vec{v}\cdot\vec{\nabla}p+v_x\left(\frac{\partial\tau_{xx}}{\partial x}+\frac{\partial\tau_{yx}}{\partial y}+\frac{\partial\tau_{zx}}{\partial z}\right)+v_y\left(\frac{\partial\tau_{xy}}{\partial x}+\frac{\partial\tau_{yy}}{\partial y}+\frac{\partial\tau_{zy}}{\partial z}\right)$$
$$+v_z\left(\frac{\partial\tau_{xz}}{\partial x}+\frac{\partial\tau_{yz}}{\partial y}+\frac{\partial\tau_{zz}}{\partial z}\right)+\rho\left(v_x f_x+v_y f_y+v_z f_z\right) \tag{8.69}$$

By splitting the substantial derivative of the total energy E acting on the fluid element into the internal energy part and kinetic energy part as

$$\rho\frac{D\left(e+\frac{1}{2}\left(v_x^2+v_y^2+v_z^2\right)\right)}{Dt}=\rho\frac{De}{Dt}+\rho\frac{D\left(\frac{1}{2}\left(v_x^2+v_y^2+v_z^2\right)\right)}{Dt} \tag{8.70}$$

and replacing the substantial derivative of the internal energy from the equation (8.65) and expressing the equation in terms of substantial derivative of enthalpy, we get

$$\rho\frac{Dh}{Dt}=\rho\frac{D\left(e+\frac{1}{2}\left(v_x^2+v_y^2+v_z^2\right)\right)}{Dt}-\rho\frac{D\left(\frac{1}{2}\left(v_x^2+v_y^2+v_z^2\right)\right)}{Dt}+\frac{\partial p}{\partial t}+\vec{\nabla}\cdot\left(p\,\vec{v}\right) \tag{8.71}$$

Most of chemical engineering flows are chemically reactive. Therefore, the net heat flux equation (8.53) needs to be expanded to include heat fluxes into the fluid element due to thermal conduction, diffusion of all species and thermal radiation. The extended net heat flux equation where heat fluxes in x, y and z

directions due to thermal conduction are replaced by Fourier's law of heat conduction, is written as

$$\vec{q}_{tot} = \underbrace{-k\vec{\nabla}T}_{thermal\ conduction} + \underbrace{\sum_{i=1}^{n} \rho_i\, \vec{u}_i\, h_i}_{diffusion\ of\ species} + \underbrace{\vec{q}_r}_{thermal\ radiation} \tag{8.72}$$

where $\vec{q}_{tot}\,\left[W/m^2\right]$ denotes the net heat flux vector for chemically reactive flow, $\rho_i\,\left[kg/m^3\right]$ the mass concentration of species i, $\vec{u}_i\,[m/s]$ the diffusion velocity vector of species i, $h_i\,[J/kg]$ the enthalpy of species i and $\vec{q}_r\,\left[W/m^2\right]$ the heat flux vector due to thermal radiation.

By replacing the net heat flux equation (8.53) in the equation (8.58) with the equation (8.72) and substituting the substantial derivative of the kinetic energy in the equation (8.71) by the equation (8.69) and substituting the substantial derivative of the total energy in the same equation (8.71) by the equation (8.58) we finally get

$$\rho\frac{Dh}{Dt} = -\vec{\nabla}\cdot\vec{q}_{tot} - \frac{\partial(v_x p)}{\partial x} - \frac{\partial(v_y p)}{\partial y} - \frac{\partial(v_z p)}{\partial z} + \vec{v}\cdot\vec{\nabla}p + \frac{\partial p}{\partial t} + \vec{\nabla}\cdot(p\,\vec{v}) + \Psi \tag{8.73}$$

where Ψ denotes the dissipation function for reacting flows due to viscous stresses as

$$\begin{aligned}\Psi =& \frac{\partial(v_x\tau_{xx})}{\partial x} + \frac{\partial(v_x\tau_{yx})}{\partial y} + \frac{\partial(v_x\tau_{zx})}{\partial z} + \frac{\partial(v_y\tau_{xy})}{\partial x} + \frac{\partial(v_y\tau_{yy})}{\partial y} + \frac{\partial(v_y\tau_{zy})}{\partial z}\\ &+ \frac{\partial(v_z\tau_{xz})}{\partial x} + \frac{\partial(v_z\tau_{yx})}{\partial y} + \frac{\partial(v_z\tau_{zz})}{\partial z} + v_x\left(\frac{\partial\tau_{xx}}{\partial x} + \frac{\partial\tau_{yx}}{\partial y} + \frac{\partial\tau_{zx}}{\partial z}\right)\\ &+ v_y\left(\frac{\partial\tau_{xy}}{\partial x} + \frac{\partial\tau_{yy}}{\partial y} + \frac{\partial\tau_{zx}}{\partial z}\right) + v_z\left(\frac{\partial\tau_{xz}}{\partial x} + \frac{\partial\tau_{yz}}{\partial y} + \frac{\partial\tau_{zz}}{\partial z}\right)\end{aligned} \tag{8.74}$$

By applying the definition for the vector operator gradient of the velocity and normal stresses product for the x component as

$$grad(v_x\tau_{xx}) \equiv \vec{\nabla}(v_x\tau_{xx}) = v_x\vec{\nabla}\tau_{xx} + \tau_{xx}\vec{\nabla}v_x \tag{8.75}$$

and with a similar definition for the rest of the velocity and stresses components, the equation (8.74) further simplifies to

$$\begin{aligned}\Psi =& \tau_{xx}\frac{\partial v_x}{\partial x} + \tau_{yy}\frac{\partial v_y}{\partial y} + \tau_{zz}\frac{\partial v_z}{\partial z} + \tau_{xy}\left(\frac{\partial v_x}{\partial y} + \frac{\partial v_y}{\partial x}\right)\\ &+ \tau_{xz}\left(\frac{\partial v_x}{\partial z} + \frac{\partial v_z}{\partial x}\right) + \tau_{yz}\left(\frac{\partial v_z}{\partial z} + \frac{\partial v_z}{\partial y}\right)\end{aligned} \tag{8.76}$$

The equation (8.73) can also be simplified, by introducing the substantial derivative of the pressure, as

$$\rho \frac{Dh}{Dt} = -\vec{\nabla} \cdot \vec{q}_{tot} + \frac{Dp}{Dt} + \Psi \tag{8.77}$$

Introducing the net heat flux equation (8.72) into the equation (8.77), we get the non-conservation form of the energy equation for chemically reactive flows in terms of enthalpy in the three-dimensional space of the Cartesian coordinate system for time-dependent, compressible and viscous flow as

$$\rho \frac{Dh}{Dt} = \vec{\nabla} \cdot \left(k \vec{\nabla} T\right) - \vec{\nabla} \cdot \sum_{i=1}^{n} \rho_i \, \vec{u}_i \, h_i - \vec{\nabla} \cdot \vec{q}_r + \frac{Dp}{Dt} + \Psi \tag{8.78}$$

The contribution of the dissipation function for reacting flows, pressure gradients and heat flux due to diffusion of all species is relatively low to the increase of the heat compared to the thermal conduction in chemical engineering applications and can be thus neglected. Thermal radiation effects are only significant in high temperature applications, such as a particular type of fuel cells and can be most often neglected for applications working at moderate operating temperatures.

By neglecting these terms, the equation (8.78) simplifies, with only Fourier's law extension of the thermal conduction, to

$$\rho \frac{Dh}{Dt} = \vec{\nabla} \cdot \left(k \vec{\nabla} T\right) \tag{8.79}$$

where the substantial derivative of enthalpy can be written as

$$\frac{Dh}{Dt} \equiv \frac{\partial h}{\partial t} + \left(\vec{v} \cdot \vec{\nabla}\right) h = \frac{\partial h}{\partial t} + v_x \frac{\partial h}{\partial x} + v_y \frac{\partial h}{\partial y} + v_z \frac{\partial h}{\partial z} \tag{8.80}$$

Chemically reacting flows in CFD and multiphysics applications are usually written in terms of the specific heat at the constant pressure c_p and temperature T. By recalling the definition of enthalpy as

$$h = c_p T \tag{8.81}$$

the equation (8.80) can be extended with the following notation

$$\frac{Dh}{Dt} = \frac{\partial (c_p T)}{\partial t} + v_x \frac{\partial (c_p T)}{\partial x} + v_y \frac{\partial (c_p T)}{\partial y} + v_z \frac{\partial (c_p T)}{\partial z} \tag{8.82}$$

The specific heat at the constant pressure c_p is a constant value in most chemical engineering applications and drops out from the derivative. By applying the constant value of the specific heat at the constant pressure c_p in the equation (8.79), we get the final form of the energy equation as known in CFD and multiphysics applications

$$\underbrace{\frac{\partial T}{\partial t}}_{acceleration} + \underbrace{v_x \frac{\partial T}{\partial x} + v_y \frac{\partial T}{\partial y} + v_z \frac{\partial T}{\partial z}}_{advection} = \underbrace{\frac{k}{\rho\, c_p}\left[\frac{\partial^2 T}{\partial x^2} + \frac{\partial^2 T}{\partial y^2} + \frac{\partial^2 T}{\partial z^2}\right]}_{diffusion} \tag{8.83}$$

The equation (8.83) represents the conservation form of the energy equation for chemically non-reactive flow in terms of temperature in the three-dimensional space of the Cartesian coordinate system for time-dependent, incompressible and viscous flow.

8.2.1.4 Conservation of Species

The conservation of species is actually the law of conservation of mass applied to each species in a mixture of various species. The fluid element, as described in Sections 8.2.1.1, 8.2.1.2 and 8.2.1.3, does not comprise of pure fluid with only one species, such as water, but of many species forming a multicomponent mixture. This law is mathematically described by the continuity equation for species, also known as the species equation, advection-diffusion equation or convection-diffusion equation. If the species equation additionally includes a reaction term, it is known as the reaction-diffusion-advection equation.

Similar as for the continuity equation, described in Section 8.2.1.1, the fundamental physical principle of the species equation applied to the fluid element in a mixture of species is based on the law of mass conservation of species. The law states that the net mass flow of species i out of the fluid element A is equal to the sum of the time rate of mass of species i decrease B and rate of mass production of species i due to chemical reactions C inside the fluid element defined as

$$\underbrace{A}_{net\ mass\ flow} = \underbrace{B}_{time\ rate\ of\ mass\ decrease} + \underbrace{C}_{mass\ production} \tag{8.84}$$

and mathematically written for species i in the three-dimensional space as

$$\left[\frac{\partial(\rho_i v_{i,x})}{\partial x} + \frac{\partial(\rho_i v_{i,y})}{\partial y} + \frac{\partial(\rho_i v_{i,z})}{\partial z}\right] dx\ dy\ dz = -\frac{\partial \rho_i}{\partial t} dx\ dy\ dz + r_i\ dx\ dy\ dz \tag{8.85}$$

where ρ_i $\left[kg/m^3\right]$ denotes the mass density of species i in the multicomponent mixture, v_x $[m/s]$, v_y $[m/s]$ and v_z $[m/s]$ velocity components of the mixture in x, y and z directions, t $[s]$ time and r_i $\left[kg/(m^3 s)\right]$ mass rate of production of species i.

By rearranging the equation (8.85), we get the conservation form of the continuity equation of species i written for any coordinate system as

$$\frac{\partial \rho_i}{\partial t} + \vec{\nabla}\cdot\left(\rho_i\ \vec{v}_i\right) = r_i \tag{8.86}$$

A viscous fluid element comprised of the multicomponent mixture of species moves with the flow. The mass velocity of each species i in the mixture $\vec{v}_i$ thus consists of flow velocity of the mixture $\vec{v}$ and diffusion velocity $\vec{u}_i$ of species i defined as

$$\underbrace{\vec{v}_i}_{mass\ velocity} = \underbrace{\vec{v}}_{mixture\ velocity} + \underbrace{\vec{u}_i}_{diffusion\ velocity} \tag{8.87}$$

By replacing the equation (8.86) with the mass velocity of species i defined in the equation (8.87)

$$\frac{\partial \rho_i}{\partial t} + \vec{\nabla} \cdot \left[\rho_i \left(\vec{v} + \vec{u}_i\right)\right] = r_i \tag{8.88}$$

and introducing mass fraction of species i as

$$w_i = \frac{\rho_i}{\rho} \tag{8.89}$$

where $\rho_i \left[kg/m^3\right]$ denotes the mass density or mass concentration of species i and $\rho \left[kg/m^3\right]$ mass density of the mixture, we get

$$\frac{\partial (w_i \rho)}{\partial t} + \vec{\nabla} \cdot \left[w_i\ \rho \left(\vec{v} + \vec{u}_i\right)\right] = r_i \tag{8.90}$$

and by expanding the derivatives as

$$w_i \frac{\partial \rho}{\partial t} + \rho \frac{\partial w_i}{\partial t} + \vec{\nabla} \cdot \left(w_i\ \rho\ \vec{v}\right) + \vec{\nabla} \cdot \left(w_i\ \rho\ \vec{u}_i\right) = r_i \tag{8.91}$$

If we again recall the definition for the vector operator divergence of the mass fraction, density and mass velocity product as

$$div\left(w_i\ \rho\ \vec{v}\right) \equiv \vec{\nabla} \cdot \left(w_i\ \rho\ \vec{v}\right) = w_i\ \vec{\nabla} \cdot \left(\rho\ \vec{v}\right) + \rho\ \vec{v} \cdot \vec{\nabla}\ w_i \tag{8.92}$$

we can expand the equation (8.69) to

$$w_i \left[\frac{\partial \rho}{\partial t} + \vec{\nabla} \cdot \left(\rho\ \vec{v}\right)\right] + \rho \frac{\partial w_i}{\partial t} + \rho\ \vec{v} \cdot \vec{\nabla}\ w_i + \vec{\nabla} \cdot \left(w_i\ \rho\ \vec{u}_i\right) = r_i \tag{8.93}$$

where the term in parenthesis is the continuity equation and thus equalises to zero.

By introducing the substantial derivative of the mass fraction of species i

$$\frac{Dw_i}{Dt} \equiv \frac{\partial w_i}{\partial t} + \left(\vec{v} \cdot \vec{\nabla}\right) w_i = \frac{\partial w_i}{\partial t} + v_x \frac{\partial w_i}{\partial x} + v_y \frac{\partial w_i}{\partial y} + v_z \frac{\partial w_i}{\partial z} \tag{8.94}$$

we get the non-conservation form of the continuity equation of species i written in terms of the mass fraction

$$\rho \frac{Dw_i}{Dt} + \vec{\nabla} \cdot \left(w_i\ \rho\ \vec{u}_i\right) = r_i \tag{8.95}$$

By introducing Fick's law of diffusion for multicomponent mixture

$$\vec{j}_i \equiv \rho_i \, \vec{u}_i = -\rho \, \mathcal{D}_{im} \nabla w_i \tag{8.96}$$

where $\vec{j}_i \left[kg/\left(m^2 \cdot s\right) \right]$ denotes the mass diffusion flux of species i in the mixture and $\mathcal{D}_{im} \left[m^2/s \right]$ the multicomponent diffusion coefficient or diffusivity, we get

$$\rho \frac{Dw_i}{Dt} = \vec{\nabla} \cdot \left(\rho \, \mathcal{D}_{im} \nabla w_i \right) + r_i \tag{8.97}$$

which is the non-conservation form of the continuity equation of species i written in terms of the mass fraction and diffusivity for time-dependent, compressible, viscous and chemically reactive flow. For an inviscid or non-viscous flow, which is assumed as an ideal fluid with no viscosity and thus no diffusion velocity, all species in the mixture thus move with the mixture velocity $\vec{v}$.

Finally, the continuity equation of species i (8.97) can be written in the conservation form in the three-dimensional space of the Cartesian coordinate system for time-dependent, incompressible, viscous and chemically reactive flow. With the notation used in CFD and multiphysics applications by introducing the substantial derivative of the mass fraction of species i in the equation (8.94), we get

$$\underbrace{\rho \frac{\partial w_i}{\partial t}}_{acceleration} + \underbrace{\rho \left[v_x \frac{\partial w_i}{\partial x} + v_y \frac{\partial w_i}{\partial y} + v_z \frac{\partial w_i}{\partial z} \right]}_{advection} = \underbrace{\rho \, \mathcal{D}_{im} \left[\frac{\partial^2 w_i}{\partial x^2} + \frac{\partial^2 w_i}{\partial y^2} + \frac{\partial^2 w_i}{\partial z^2} \right]}_{diffusion} + \underbrace{r_i}_{reaction} \tag{8.98}$$

The multicomponent diffusion coefficient $\mathcal{D}_{im}$ is related to the binary diffusion coefficient $\mathcal{D}_{ij}$ based on the following equation

$$\mathcal{D}_{im} = \frac{1 - X_i}{\sum_j \frac{X_i}{\mathcal{D}_{ij}}} \tag{8.99}$$

where $\mathcal{D}_{im} \left[m^2/s \right]$ denotes the multicomponent diffusion coefficient of species i through the mixture, $\mathcal{D}_{ij} \left[m^2/s \right]$ the binary diffusion coefficient of species i into species j, X_i [−] the mole fraction of species i and X_j [−] the mole fraction of species j.

Binary diffusion coefficients are different for gases, liquids, colloids or polymers. They are based on empirical or semi-empirical equations postulated on different theories depending on different physical quantities. Detailed explanation of binary and multicomponent diffusion coefficients is beyond the scope of this chapter but further information can be found in literature [5], [6] and [7].

8.2.2 Turbulent Flow

Most flows in real-world and engineering systems are turbulent nature. Therefore, modelling and simulation of turbulent flow is equally important as modelling and simulation of laminar flow and will be explained in this section.

A turbulent flow is a random behaviour of a fluid, which increases mixing, as well as heat and species transport. A velocity of the fluid is not constant as it is in a laminar flow regime, but it chaotically changes in space and time. The random velocity in a turbulent flow regime consists of steady-state mean component $\overline{v}$ and time-dependent fluctuating component $v'(t)$. An overall time-averaged velocity is thus written as:

$$\overline{v}'(t) = \overline{v} + v'(t) \tag{8.100}$$

The difference between a laminar flow and turbulent flow is based on dimensionless Reynolds number. The Reynolds number is the ratio between inertial forces and viscous force defined as

$$Re = \frac{\rho \cdot \overline{v}' \cdot L}{\mu} \tag{8.101}$$

where $\rho\ \left[kg/m^3\right]$ denotes the density of the fluid, $\overline{v}'\ [m/s]$ the overall velocity, $L\ [m]$ the characteristic dimension that could be a diameter of a pipe or a length of a plate and $\mu\ \left[Pa \cdot s\right]$ the dynamic viscosity of the fluid.

At low Reynolds numbers inertial forces are small compared to viscous forces and flow remains laminar. At high Reynolds numbers viscous forces become dominant and disturb the flow, which thus becomes turbulent. The transition from a laminar flow regime to a turbulent flow regime never occurs suddenly, there is always a transition flow regime. For example, the laminar flow inside tubular pipes occurs at the Reynolds number around 2300. The turbulent flow in the same pipe occurs at the Reynolds number higher than around 3000. The flow between the Reynolds numbers 2300 and 3000 is thus considered as the transition flow.

Turbulent flow is described by conservation equations of continuity and momentum, known as the Reynolds Averaged Navier-Stokes (RANS) equations. Laminar velocity terms in conservation equations are replaced by the steady-state mean components and time-dependent fluctuating components defined by the equation (8.100).

Detailed explanation of turbulent conservation equations is beyond the scope of this chapter. Further information can be found in literature [3], [4], [8], [9] and [10].

8.2.2.1 Conservation of Mass

The conservation form of the continuity equation describing time-dependent turbulent, incompressible and viscous flow in the three-dimensional space of the Cartesian coordinate system is defined as

$$\frac{\partial \rho}{\partial t} + \rho \left[\frac{\partial \overline{v}_x}{\partial x} + \frac{\partial \overline{v}_y}{\partial y} + \frac{\partial \overline{v}_z}{\partial z} \right] = 0 \tag{8.102}$$

where $\rho\ \left[kg/m^3\right]$ denotes the mass density of the fluid and $\overline{v}_x\ [m/s]$, $\overline{v}_y\ [m/s]$ and $\overline{v}_z\ [m/s]$ the steady-state mean velocity components in x, y and z directions.

The continuity equation (8.102) describing turbulent flow is the same as the continuity equation (8.16) describing laminar flow.

8.2.2.2 Conservation of Momentum

The conservation form of the momentum equations describing time-dependent turbulent, incompressible and viscous flow in the three-dimensional space of the Cartesian coordinate system is defined for the x component as

$$\begin{aligned}&\frac{\partial \overline{v}_x}{\partial t}+\frac{\partial(\overline{v}_x\overline{v}_x)}{\partial x}+\frac{\partial(\overline{v}_y\overline{v}_x)}{\partial y}+\frac{\partial(\overline{v}_z\overline{v}_x)}{\partial z}+\frac{\partial(\overline{v}_x'\overline{v}_x')}{\partial x}+\frac{\partial(\overline{v}_y'\overline{v}_x')}{\partial y}+\frac{\partial(\overline{v}_z'\overline{v}_x')}{\partial z}\\&=-\frac{1}{\rho}\frac{\partial \overline{p}}{\partial x}+\nu\left[\frac{\partial^2\overline{v}_x}{\partial x^2}+\frac{\partial^2\overline{v}_x}{\partial y^2}+\frac{\partial^2\overline{v}_x}{\partial z^2}\right]+\nu\left[\frac{\partial^2\overline{v}_x}{\partial x^2}+\frac{\partial^2\overline{v}_y}{\partial y\partial x}+\frac{\partial^2\overline{v}_z}{\partial z\partial x}\right]\end{aligned} \tag{8.103}$$

for the y component as

$$\begin{aligned}&\frac{\partial \overline{v}_y}{\partial t}+\frac{\partial(\overline{v}_x\overline{v}_y)}{\partial x}+\frac{\partial(\overline{v}_y\overline{v}_y)}{\partial y}+\frac{\partial(\overline{v}_z\overline{v}_y)}{\partial z}+\frac{\partial(\overline{v}_x'\overline{v}_y')}{\partial x}+\frac{\partial(\overline{v}_y'\overline{v}_y')}{\partial y}+\frac{\partial(\overline{v}_z'\overline{v}_y')}{\partial z}\\&=-\frac{1}{\rho}\frac{\partial \overline{p}}{\partial y}+\nu\left[\frac{\partial^2\overline{v}_y}{\partial x^2}+\frac{\partial^2\overline{v}_y}{\partial y^2}+\frac{\partial^2\overline{v}_y}{\partial z^2}\right]+\nu\left[\frac{\partial^2\overline{v}_x}{\partial x\partial y}+\frac{\partial^2\overline{v}_y}{\partial y^2}+\frac{\partial^2\overline{v}_z}{\partial z\partial y}\right]\end{aligned} \tag{8.104}$$

and for the z component as

$$\begin{aligned}&\frac{\partial \overline{v}_z}{\partial t}+\frac{\partial(\overline{v}_x\overline{v}_z)}{\partial x}+\frac{\partial(\overline{v}_y\overline{v}_z)}{\partial y}+\frac{\partial(\overline{v}_z\overline{v}_z)}{\partial z}+\frac{\partial(\overline{v}_x'\overline{v}_z')}{\partial x}+\frac{\partial(\overline{v}_y'\overline{v}_z')}{\partial y}+\frac{\partial(\overline{v}_z'\overline{v}_z')}{\partial z}\\&=-\frac{1}{\rho}\frac{\partial \overline{p}}{\partial z}+\nu\left[\frac{\partial^2\overline{v}_z}{\partial x^2}+\frac{\partial^2\overline{v}_z}{\partial y^2}+\frac{\partial^2\overline{v}_z}{\partial z^2}\right]+\nu\left[\frac{\partial^2\overline{v}_x}{\partial x\partial z}+\frac{\partial^2\overline{v}_y}{\partial y\partial z}+\frac{\partial^2\overline{v}_z}{\partial z^2}\right]\end{aligned} \tag{8.105}$$

where $\overline{v}_x\ [m/s]$, $\overline{v}_y\ [m/s]$ and $\overline{v}_z\ [m/s]$ denote the steady-state mean velocity components in x, y and z directions, $\overline{v}_x'\ [m/s]$, $\overline{v}_y'\ [m/s]$ and $\overline{v}_z'\ [m/s]$ time-averaged velocity components in x, y and z directions, $\overline{p}\ [Pa]$ the steady-state mean pressure, $\rho\ \left[kg/m^3\right]$ the mass density of the fluid and $\nu\ \left[m^2/s\right]$ the kinematic viscosity of the fluid.

The momentum equations (8.103), (8.104) and (8.105) describing turbulent flow are the same as the momentum equations (8.43), (8.44) and (8.45) describing laminar flow with additional terms including steady-state mean velocity components and time-averaged velocity components without body forces.

8.2.2.3 Conservation of Energy

The conservation form of the energy equation describing time-dependent turbulent, incompressible and viscous flow in the three-dimensional space of the Cartesian coordinate system is defined as

$$\frac{\partial \bar{T}}{\partial t}+\frac{\partial(\bar{v}_x\bar{T})}{\partial x}+\frac{\partial(\bar{v}_y\bar{T})}{\partial y}+\frac{\partial(\bar{v}_z\bar{T})}{\partial z}+\frac{\partial(\bar{v}_x'\bar{T}')}{\partial x}+\frac{\partial(\bar{v}_y'\bar{T}')}{\partial y}+\frac{\partial(\bar{v}_z'\bar{T}')}{\partial z}$$
$$=\frac{k}{\rho\, c_p}\left[\frac{\partial^2 \bar{T}}{\partial x^2}+\frac{\partial^2 \bar{T}}{\partial y^2}+\frac{\partial^2 \bar{T}}{\partial z^2}\right] \tag{8.106}$$

where $c_p\left[J/(kg\cdot K)\right]$ the specific heat at the constant pressure of the fluid, $k\left[W/(m\cdot K)\right]$ the thermal conductivity of the fluid, $\bar{T}\ [K]$ the steady-state mean temperature of the fluid and $\bar{T}'\ [K]$ time-averaged temperature of the fluid.

The energy equation (8.106) describing turbulent flow is the same as the energy equation (8.83) describing laminar flow with additional terms including time-averaged velocity components and time-averaged temperature of the fluid.

8.2.2.4 Conservation of Species

The conservation form of the continuity equation of species i describing time-dependent turbulent, incompressible and viscous flow in the three-dimensional space of the Cartesian coordinate system is defined as

$$\rho\left[\frac{\partial \bar{w}_i}{\partial t}+\frac{\partial(\bar{v}_x\bar{w}_i)}{\partial x}+\frac{\partial(\bar{v}_y\bar{w}_i)}{\partial y}+\frac{\partial(\bar{v}_z\bar{w}_i)}{\partial z}+\frac{\partial(\bar{v}_x'\bar{w}_i')}{\partial x}+\frac{\partial(\bar{v}_y'\bar{w}_i')}{\partial y}+\frac{\partial(\bar{v}_z'\bar{w}_i')}{\partial z}\right]$$
$$=\rho\,\mathcal{D}_{im}\left[\frac{\partial^2 \bar{w}_i}{\partial x^2}+\frac{\partial^2 \bar{w}_i}{\partial y^2}+\frac{\partial^2 \bar{w}_i}{\partial z^2}\right]+r_i \tag{8.107}$$

where $\rho\left[kg/m^3\right]$ denotes the mass density of the mixture, $\bar{w}_i\ [-]$ the steady-state mean mass fraction of species i and $\bar{w}_i'\ [-]$ time-averaged mass fraction of species i.

The continuity equation for species i (8.107) describing turbulent flow is the same as the continuity equation for species i (8.98) describing laminar flow with additional terms including time-averaged velocity components and time-averaged mass fraction of species i.

8.2.2.5 Turbulence Models

Turbulence causes swirls or eddies in turbulent flows with different length and time scales. Describing turbulent flows is hence challenging and requires powerful computers. Scientists and engineers have therefore developed different turbulence models, which accurately enough mimic real-world turbulent flows. There are several models such as zero-equation models, one-equation models, two-equation models, second-order closure models, Large Eddy Simulation (LES) models, and Direct Numerical Simulation (DNS) models. This chapter will focus only on extending the Reynolds Averaged Navier-Stokes (RANS) equations with widely used two-equation turbulence models. The most widely used turbulence models are summarised in Table 8.1.

Further information of turbulent flow and turbulence models can be found in literature [3], [4], [8], [9] and [10].

8.3 MODELLING AND SIMULATION

Modelling and simulation of CFD and multiphysics physical systems consists of:

- pre-processing unit,
- processing unit and
- post-processing and visualisation unit.

8.3.1 Pre-Processing Unit

8.3.1.1 Choosing Discretisation Techniques

Conservation equations, as explained in Section 8.2, are very complex and thus impractical or even impossible to be solved analytically. Therefore, partial differential equations need to be converted into discrete forms suitable for solving on computers. This conversion process is mathematically known as discretisation. Since the process of converting differential equations into algebraic equations is based on numerical approximations, CFD and multiphysics software tools can only approximately model physical systems. Commercially available software tools are based on different discretisation methods, such as finite difference method, finite volume method and finite element method. Finite difference method is the simplest and the oldest discretisation technique to approximate partial derivatives by using backward, forward and central differences. It is nowadays mostly used for modelling one-dimensional physical systems and a good starting point for CFD engineers who want to get an advanced knowledge of computational modelling. Finite volume method is the most popular discretisation technique in the CFD community with majority of CFD software tools based on this method. It solves partial differential equations as algebraic equations where variables, such as velocity, temperature or concentration are solved in each mesh element that represents a physical system. Finite element method uses approximate functions to solve partial differential equations. It is one of the well-established and powerful techniques for solving physical systems with combination of different conservation partial differential equations and thus appropriate for modelling multi physical systems. There exist some other discretisation techniques as well, such as spectral element method, boundary element method and meshless or meshfree methods, but they are still part of intense research activities and are not implemented in commercial CFD and multiphysics software tools yet.

TABLE 8.1
Mathematical Description of the Most Widely Used Turbulence Models

Models	Mathematical Description	Applications
$k-\varepsilon$	The $k-\varepsilon$ model solves two additional transport equations for the turbulent kinetic energy k $\rho\frac{\partial k}{\partial t}+\underbrace{\rho\vec{\bar{v}}'\cdot\vec{\nabla}k}_{convection}=\underbrace{\vec{\nabla}\cdot\left[\left(\mu+\frac{\mu_T}{\sigma_k}\right)\vec{\nabla}k\right]}_{diffusion}+\underbrace{P_k}_{generation}-\underbrace{\rho\varepsilon}_{dissipation},$ and the turbulent dissipation rate ε $\rho\frac{\partial \varepsilon}{\partial t}+\underbrace{\rho\vec{\bar{v}}'\cdot\vec{\nabla}\varepsilon}_{convection}=\underbrace{\vec{\nabla}\cdot\left[\left(\mu+\frac{\mu_T}{\sigma_\varepsilon}\right)\vec{\nabla}\varepsilon\right]}_{diffusion}+\underbrace{C_{\varepsilon 1}\frac{\varepsilon}{k}P_k}_{generation}-\underbrace{C_{\varepsilon 2}\rho\frac{\varepsilon^2}{k}}_{destruction}.$ The $k-\varepsilon$ model requires solving additional terms for the production of eddies $P_k=\mu_T\left[\vec{\nabla}\bar{v}':\left(\vec{\nabla}\bar{v}'+\left(\vec{\nabla}\bar{v}'\right)^T\right)-\frac{2}{3}\left(\vec{\nabla}\bar{v}'\right)^2\right]-\frac{2}{3}\rho k\vec{\nabla}\bar{v}',$ The turbulent kinetic energy $k=\frac{1}{2}\left[\overline{v_x'v_x'}+\overline{v_y'v_y'}+\overline{v_z'v_z'}\right],$ The turbulent dissipation rate $\varepsilon=\nu\overline{\frac{\partial v_t}{\partial x_j}\left(\frac{\partial v_t}{\partial x_j}+\frac{\partial v_j}{\partial x_i}\right)},$ and the turbulent viscosity $\mu_T=\rho C_\mu\frac{k^2}{\varepsilon}.$ $C_{\varepsilon 1}$, $C_{\varepsilon 2}$, C_μ, σ_k and σ_k in the above written equations are the empirical constants found in [10].	The $k-\varepsilon$ model is used in real-world industrial applications for a wide range of flows such as isotropic flows, flows with recirculation and separation zones and flows with strong gradients of pressure.

TABLE 8.1 (*Continued*)
Mathematical Description of the Most Widely Used Turbulence Models

Models	Mathematical Description	Applications
$k-\omega$	The $k-\omega$ model solves two additional transport equations for the turbulent kinetic energy k $\rho\frac{\partial k}{\partial t}+\underbrace{\rho\vec{v}\cdot\vec{\nabla}k}_{convection}=\underbrace{\vec{\nabla}\cdot\left[\left(\mu+\sigma^{*}\mu_T\right)\vec{\nabla}k\right]}_{diffusion}+\underbrace{P_k}_{generation}-\underbrace{\rho\beta^{*}k\varepsilon}_{dissipation}.$ and the turbulent specific rate of dissipation ω $\rho\frac{\partial \omega}{\partial t}+\underbrace{\rho\vec{v}\cdot\vec{\nabla}\omega}_{convection}=\underbrace{\vec{\nabla}\cdot\left[\left(\mu+\sigma\mu_T\right)\vec{\nabla}\omega\right]}_{diffusion}+\underbrace{\alpha\frac{\omega}{k}P_k}_{generation}-\underbrace{\rho\beta\omega^2}_{destruction}.$ The $k-\omega$ model also requires solving additional terms for the production of eddies P_k and the turbulent kinetic energy k, as similarly defined for the $k-\varepsilon$ model. It solves the turbulent dissipation rate with the auxiliary expression $\varepsilon=\beta^{*}\omega k$, and the turbulent viscosity $\mu_T=\rho\frac{k}{\omega}.$ $\alpha, \beta, \beta^{*}, \beta_0, \beta_0^{*}, f_B, f_{\beta^*}, \chi_\omega, \chi_k, \sigma$ and σ^{*} in the above written equations are the empirical constants and equations found in [10].	The $k-\omega$ model is used for modelling flows around airfoils, in turbine blades and for adverse pressure gradients due to good resolution of boundary layers in wall bounded flows.

Commercially available discretisation methods for solving complex real-world physical systems are predominantly based on finite volume method and finite element method. The most widely used commercial software tools, such as ANSYS® Fluent®, STAR CD® and STAR-CCM+® are based on finite volume method, whereas ANSYS® CFX® uses finite element-based control volume method. On the other hand, COMSOL Multiphysics® is based on finite element method. Both methods have advantages and disadvantages but nevertheless they give enough comparable computational results with real-world physical systems.

Further information of discretisation techniques can be found in literature [3], [4], [11] and [12].

8.3.1.2 Modelling Geometry

The first step in computational modelling and simulation incorporates generation of computational geometry that represents a real-world physical system. A

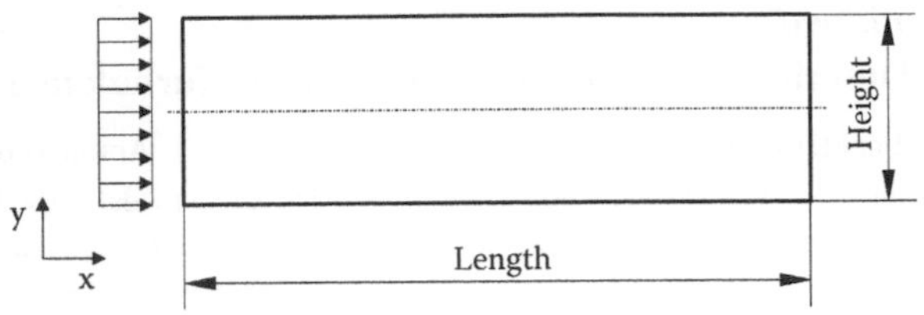

FIGURE 8.3 Two-dimensional computational geometry of a simple physical system.

two-dimensional or three-dimensional geometry is usually used, depending on the nature of the physical system.

Geometry can be generated in an appropriate computational fluid dynamics (CFD) or computer-aided design (CAD) software tools. Some of CFD based pre-processor units have limited functionality in geometry creation. If complex geometry is involved, the most common practise is to create the geometry in any CAD software tool and export it to a CFD based pre-processor unit for mesh generation. With the improvements of CAD software tools and easy-to-use graphical interfaces, it is possible to export complex geometries, made in commercially available CAD software tools, such as Autodesk® AutoCAD®, Autodesk® Inventor®, CATIA®, SolidWorks®, PTC® Creo® or ANSYS Discovery SpaceClaim., into CFD software tools to generate computational meshes.

Figure 8.3 represents a two-dimensional computational geometry of a simple physical system.

8.3.1.3 Modelling Physical System

8.3.1.3.1 Conservation Equations

This step consists of mathematically translating a real-world physical system into a set of conservation equations described in Section 8.2. For example, a hot flow in a pipe is numerically described by continuity equation, momentum equations and energy equation. A CFD engineer decides which conservation equations and terms are important to describe the computational model, representing the real-world physical system. One of the first decisions he needs to encounter is whether the flow is laminar or turbulent. Is it enough to design a steady-state or time-dependent model as well? Is the flow compressible or incompressible? Are viscous effects important to be considered or not? Is it enough to consider only thermal convection or thermal radiation effects as well? Is it enough to consider only one chemical reaction or mechanism of chemical reactions to accurately model a real-world physical system? These are some of the questions every CFD engineer needs to answer before applying the conservation equations to the computational model. Only a reasonable decision, based on good understanding of a real-world physical system, will enable the CFD engineer to design enough simplified computational model comparable with the real one. Therefore, a thorough description and understanding of conservation equations, as described in Section 8.2, is essential to design any computational model.

8.3.1.3.2 Physical Properties

Conservation equations include physical quantities, such as density, viscosity, thermal conductivity or specific heat. These quantities do not depend on conservation equations, but are unique to a specific fluid or material that needs to be used to represent a real-world physical system. Therefore, a CFD engineer has to choose appropriate physical properties from experimental data, literature, databases or build-in libraries embedded in CFD and multiphysics software tools. Only properly assigned physical properties will adequately describe conservation equations.

8.3.1.4 Selecting Initial and Boundary Conditions

Once the appropriate conservation equations are chosen, initial and boundary conditions must be applied to the computational model to mathematically describe a real-world physical system.

Every partial differential equation needs an initial value or guess for numerical solver to start computing the equations. On the other hand, boundary conditions are specific for each conservation equation, described in Section 8.2. The variable in the continuity equation and momentum equations is the velocity vector, the variable in the energy equation is the temperature vector and the variable in the species equation is the concentration vector. Therefore, appropriate velocity, temperature and concentration values, that represent real-world values, need to be prescribed on each computational boundary, such as inlet, outlet or wall at time zero. The prescribed values on boundaries are called boundary conditions. Each boundary condition needs to be prescribed on a node or line for two-dimensional system or on a plane for three-dimensional system. In general, there are several types of boundary conditions where the Dirichlet and Neumann are the most widely used in CFD and multiphysics applications. The Dirichlet boundary condition specifies the value on a specific boundary, such as velocity, temperature or concentration. On the contrary, the Neumann boundary condition specifies the derivative on a specific boundary, such as heat flux or diffusion flux. Once the appropriate boundary conditions are prescribed to all boundaries on the two-dimensional or three-dimensional model, the set of the conservation equations is closed and the computational model can be executed.

Figure 8.4 shows inlet, outlet and wall boundaries and example of associated boundary conditions applied to the flow through the pipe.

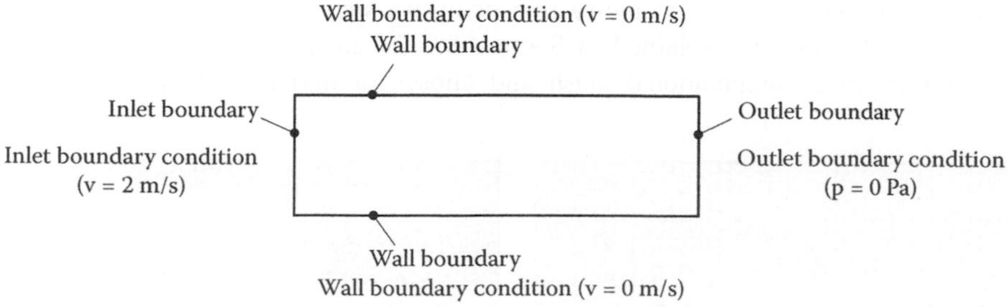

FIGURE 8.4 Inlet, outlet, and wall boundaries and associated boundary conditions applied to a physical system.

FIGURE 8.5 Boundaries on triangular and hexahedral mesh elements.

8.3.1.5 Generating Computational Mesh

Generating computational mesh represents partitioning the domain geometry into a geometric model. Primitive shapes consist of triangular or quadrilateral (square, rectangle) mesh elements for creating two-dimensional meshes and tetrahedral, hexahedral (cube, cuboid), prism or pyramid mesh elements for creating three-dimensional computational meshes. Each mesh element comprises of different boundaries shown on Figure 8.5.

In general, computational mesh can be structured or unstructured. A structured mesh usually consists of quadrilateral and hexahedral elements. It is used for generating simple geometries, such as straight pipes or rectangular rooms. On the other hand, an unstructured mesh usually consists of triangular and tetrahedral elements. This type of computational mesh is more appropriate for meshing complex geometries due to flexibility of elements to form different shapes. Structured mesh is usually preferred due to the simplicity of computation and less susceptible to numerical diffusion. Numerical models based on structured meshes are usually faster to execute and require less computer memory compared to numerical models based on unstructured meshes.

Figure 8.6 shows a difference between the structured and unstructured mesh of simple two-dimensional computational domain.

A computational mesh can be stationary or moving. In most CFD applications, meshes are stationary, such as flow in pipes or heat transfer through walls. Examples of moving meshes are pistons in internal combustion engines, blades in turbines or stirrers in batch reactors. Designing a computational model with a moving mesh is a challenging task compared to a stationary mesh and thus requires an advanced technique of mesh creation. A mesh ratio between element dimensions plays an important role when the mesh moves and thus squeezes individual element. If the element ratio is not appropriate, a computational model leads to convergence failure.

Dividing computational geometry into primitive shapes is needed to solve the conservation equations explained in Section 8.2 for each mesh element. Number of elements in a computational mesh and chosen numerical technique to solve

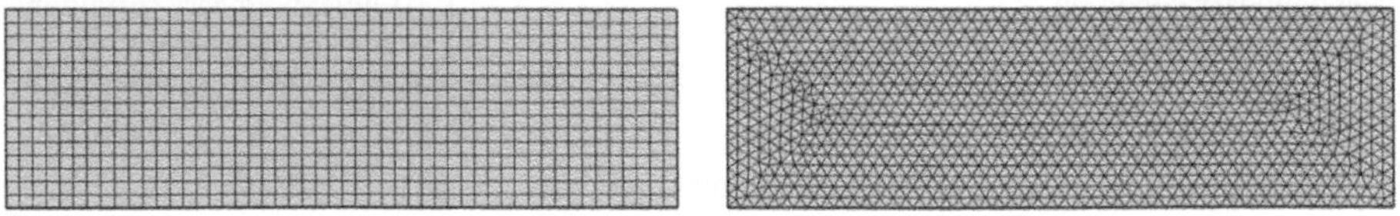

FIGURE 8.6 Difference between the two-dimensional structured and unstructured mesh (image made using COMSOL Multiphysics® and is provided courtesy of COMSOL®).

conservation equations, determines the accuracy of a computational model. If the computational mesh consists of many mesh elements, the computational solution will be accurate, but the computational time will be longer to solve conservation equations in each mesh element.

A CFD engineer spent a certain amount of time in the past to manually modify a computational mesh for specific applications. Since a badly generated computational mesh can lead to a non-converged solution, a lot of effort has been invested to improve mesh generation algorithms. Generating computational mesh is nowadays much easier process due to advanced mesh generation algorithms and self-adapting meshes. Many mesh generation software tools are today available on the market, such as ANSYS® ICEM CFD™ or Pointwise®.

8.3.2 Processing Unit

8.3.2.1 Choosing Solvers

An appropriate solver needs to be selected to numerically solve discretised conservation equations explained in Section 8.2. There are two main types of solvers, a stationary or steady-state solver for solving steady-state linear or nonlinear computational models and time-dependent or transient solver for solving transient linear or nonlinear computational models.

Computational fluid dynamics and multiphysics software tools are nowadays sophisticated enough that the software automatically suggests an appropriate solver based on conservation equations used. More experienced CFD engineers can still manually select a specific solver with additional settings applied to the particular solver. Since conservation equations used in computational models are complex in nature, solvers break down the equations into linear system of equations. Direct solvers solve the linear system of equations by using Gaussian elimination, while iterative solvers need an initial guess to find a solution by using successive approximation technique. Good initial conditions, usually based on experience, need to be selected to quickly achieve the converged solution. Iterative solvers are usually slower for solving one-dimensional and two-dimensional computational models and therefore more stable and reliable direct solvers are preferred. On the other hand, iterative solvers are more appropriate for solving complex three-dimensional computational models because they are faster and use less computer memory.

It is important to emphasise that choosing an appropriate solver requires some theoretical and practical knowledge of numerical mathematics prior working with CFD and multiphysics applications. Detailed explanation of solvers is beyond the scope of this chapter but further information can be found in literature [13].

8.3.2.2 Monitoring Convergence

Monitoring convergence is the most important test showing a computational progress of solving conservation equations. If a numerical solution called iteration step does not change from the previous step during the solving process, the solution converges and the numerical simulation terminates giving the final result. Another convergence criterion that determines whether a numerical solution converges or diverges is called tolerance. The tolerance value is integrated part of solvers and represents a dimensionless number with the typical value 10^{-4}. If the observed variables from the conservation equations, sometimes called residuals, such as velocity or temperature, fall below the tolerance limits, the numerical solution converges. A converged numerical solution does not necessarily mean an accurate solution due to several numerical errors, such as discretisation error, round-off error, convergence error, modelling error and human error. Thus, there are possible several errors if a solution diverges. The most common errors are an inappropriately chosen type of the mesh or an inappropriately chosen solver. The convergence error due to partially achieved convergence and modelling error due to simplification of conservation equations or an inappropriately chosen initial and boundary conditions are common as well.

8.3.2.3 Grid Independence Test

The grid independence test is another important indicator that determines a successful numerical solution. Any computationally modelled real-world physical system needs to give comparable results with different mesh density or mesh size. It means that the discretisation error, which is a difference between the exact numerical solution and the numerical solution due to different mesh resolution, needs to be within acceptable limits. A CFD engineer always needs to perform the grid independence test to get the correct computational results of any real-world physical system.

8.3.3 Post-Processing and Visualisation Unit

The post-processing and visualisation unit visualises computational results of a simulated real-world physical system. Visualisation results can include velocity field or temperature distribution images in two-dimensional or three-dimensional space as a result of numerically solved matrices of momentum or energy equations. Each colour in the result image represents different velocity or temperature values based on physical behaviour of the real-world physical system. These images are then thoroughly analysed by CFD engineers, who suggest improvements of an existing real-world physical system to product development engineers or collaborate with them to develop an entirely new product.

With the advance of software engineering, commercially available CFD and multiphysics based software modelling tools, such as ANSYS® CFX®, ANSYS® Fluent®, ANSYS® Multiphysics™, COMSOL Multiphysics®, STAR CD® or STAR-CCM+®, have sophisticated post-processing and visualisation units incorporated in their products. There are also available stand-alone visualisation units, such as ANSYS® CFD-Post™, AVS/Express™, EnSight® or Tecplot 360™, developed by software companies with a specialism in visualisation software tools.

8.4 CASE STUDIES WITH COMSOL MULTIPHYSICS

This section applies conservation equations, explained in Section 8.2, to three real-world physical systems. It shows a step-by-step computational modelling approach of three conventional chemical multiphysics systems. The first case study is a three-dimensional computational model of a laminar static mixer with twisted

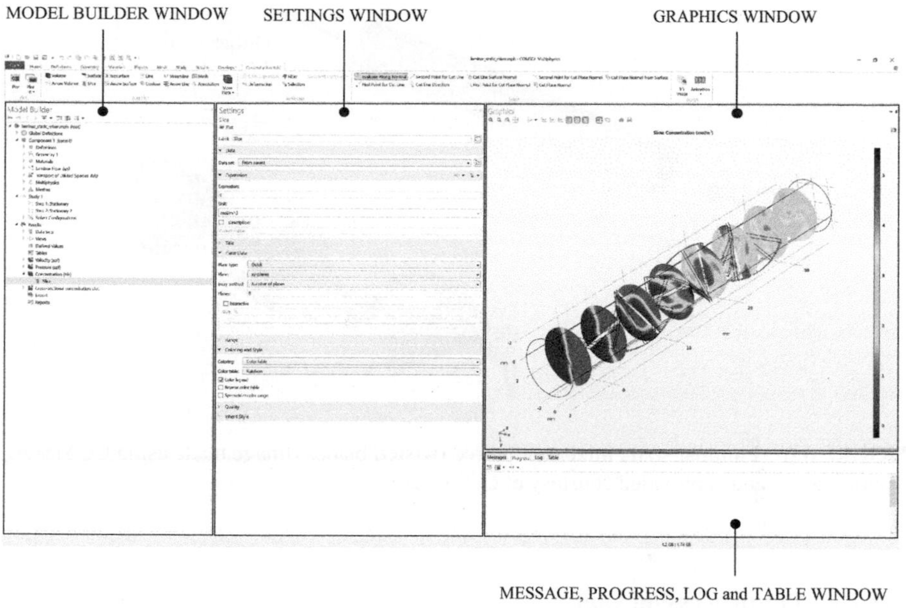

FIGURE 8.7 COMSOL Multiphysics® graphical user interface (image made using COMSOL Multiphysics® and is provided courtesy of COMSOL®) (See color insert.)

blades, the second case study is a three-dimensional computational model of a porous reactor with injection needle and the last case study is a three-dimensional model of an isothermal heat exchanger. All three case studies are computationally modelled and simulated with one of the commercial software tools COMSOL Multiphysics®, the CFD and multiphysics modelling software tool developed by COMSOL AB from Sweden.

Figure 8.7 shows the COMSOL Multiphysics® graphical user interface divided into three main sections. The first section on the left-hand side is called the "Model Builder Window", where pre-processing, processing and post-processing unit steps are defined. The second section in the middle is called the "Settings Window" and describes specific parameters for each of three units. The third section on the right-hand side is called the "Graphics Window", where the individual unit steps from modelling geometry or generating mesh to visualisation of results are shown.

8.4.1 Laminar Static Mixer

The first case study is an example of a laminar static mixer with twisted blades modelled by COMSOL Multiphysics® 5.3a, which is available in COMSOL® "Application Gallery" under the name "Laminar Static Mixer" with the Application ID number 245. This example studies the stationary mixing process of one species with water at room temperature. A flow through the blades is considered to be laminar and suitable for small pressure drops or losses. The laminar static mixer is shown on Figure 8.8.

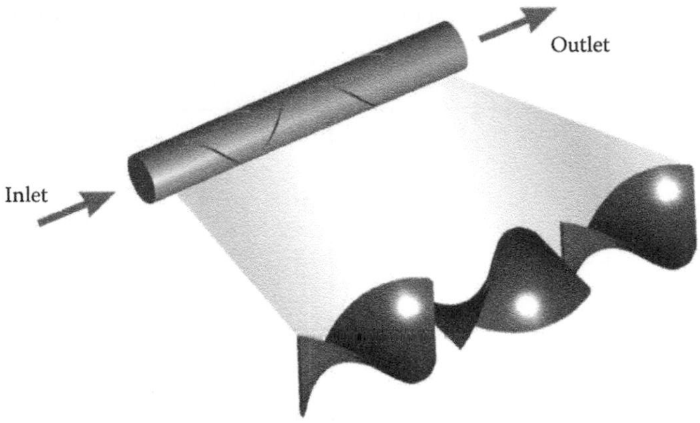

FIGURE 8.8 Laminar static mixer with three twisted blades (image made using COMSOL Multiphysics® and is provided courtesy of COMSOL®).

8.4.1.1 Pre-processing Unit

8.4.1.1.1 Choosing Discretisation Techniques

COMSOL Multiphysics® is based on finite element method that converts partial differential conservation equations described in Section 8.2 into discrete numerical form suitable for solving on computers.

8.4.1.1.2 Modelling Geometry and Generating Computational Mesh

The left-hand side of Figure 8.9 shows the "Model Builder Window" and process of modelling cylindrical geometry, representing static mixer, with appropriate dimensions shown in the "Settings Window". The right-hand side of Figure 8.9 shows the

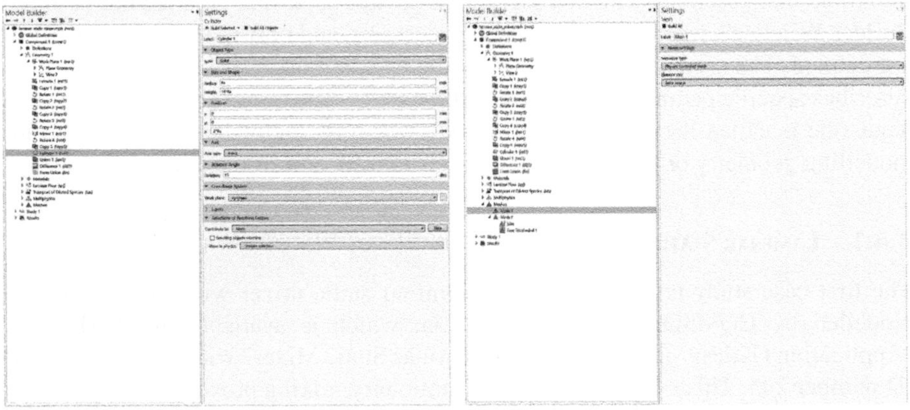

FIGURE 8.9 Geometry and computational mesh setup for the laminar static mixer (image made using COMSOL Multiphysics® and is provided courtesy of COMSOL®).

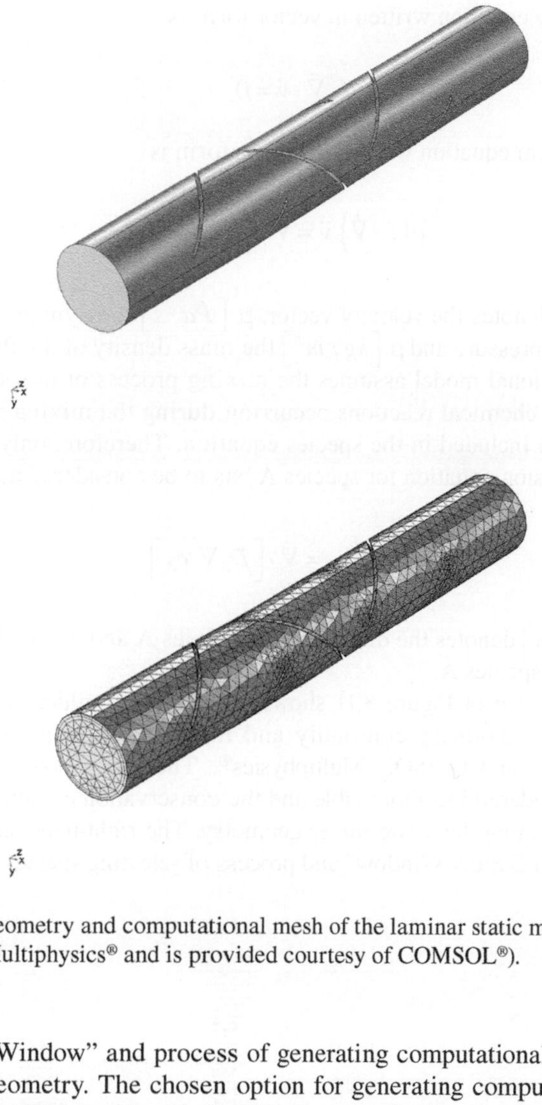

FIGURE 8.10 Geometry and computational mesh of the laminar static mixer (image made using COMSOL Multiphysics® and is provided courtesy of COMSOL®).

"Model Builder Window" and process of generating computational mesh based on the cylindrical geometry. The chosen option for generating computational mesh is the "Physics-controlled mesh" with the "Extra coarse" selection of mesh elements as shown in the "Settings Window".

The left-hand side of Figure 8.10 shows the laminar static mixer model consisting of three-dimensional geometry with three twisted blades of alternating rotations. The right-hand side of Figure 8.10 shows the computational mesh of the same laminar static mixer model.

8.4.1.1.3 Modelling Physical System

8.4.1.1.3.1 Conservation Equations The computational model consists of steady-state continuity equation and momentum equations describing the flow inside the laminar static mixer.

The continuity equation written in vector form is

$$\vec{\nabla} \cdot \vec{v} = 0 \tag{8.108}$$

and the momentum equation written in vector form is

$$\rho\left(\vec{v} \cdot \vec{\nabla}\right) \vec{v} = \vec{\nabla} \cdot \left[\mu \vec{\nabla}\, \vec{v} - p\right] \tag{8.109}$$

where $\vec{v}\ [m/s]$ denotes the velocity vector, $\mu\ \left[Pa \cdot s\right]$ the dynamic viscosity of the fluid, $p\ [Pa]$ the pressure and $\rho\ \left[kg/m^3\right]$ the mass density of the fluid.

The computational model assumes the mixing process of one species dissolved in water without chemical reactions occurring during the mixing and thus without the reaction term included in the species equation. Therefore, only the steady-state convection-diffusion equation for species A has to be considered as

$$\vec{v} \cdot \vec{\nabla} c_A = \vec{\nabla} \cdot \left[\mathcal{D}_A \vec{\nabla}\, c_A\right] \tag{8.110}$$

where $\mathcal{D}_A\ \left[m^2/s\right]$ denotes the diffusivity of species A and $c_A\ \left[mol/m^3\right]$ the molar concentration of species A.

The left-hand side of Figure 8.11 shows the "Model Builder Window" and process of selecting stationary continuity and momentum equations named as the "Laminar Flow" in COMSOL Multiphysics®. The flow shown in the "Settings Window" is considered incompressible and the conservation equations are applied to domain 1 representing the entire mixer geometry. The right-hand side of Figure 8.11 shows the "Model Builder Window" and process of selecting species equations named

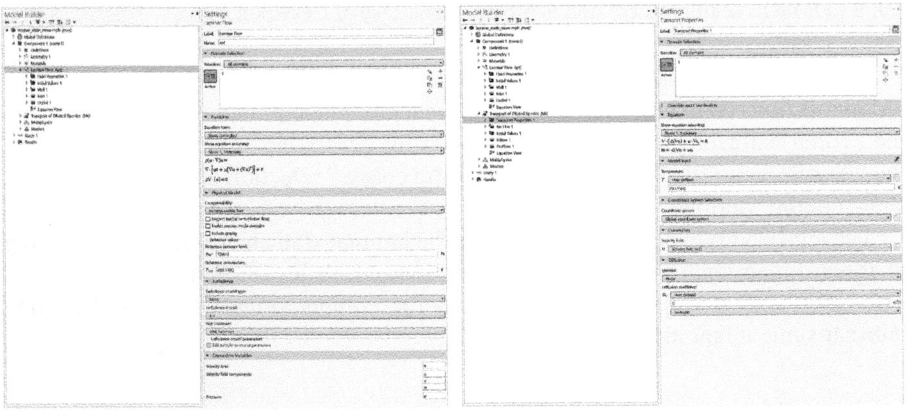

FIGURE 8.11 Momentum and species equations prescribed for the laminar static mixer (image made using COMSOL Multiphysics® and is provided courtesy of COMSOL®).

as the "Transport of Diluted Species" in COMSOL Multiphysics®. The velocity components needed in the species equations are applied to domain 1 and solved by the momentum equations in each mesh element. A CFD engineer defines the diffusivity coefficient in the "Diffusion" section. These steps are shown in the "Settings Window" of the "Transport of Diluted Species" physics.

8.4.1.1.3.2 Physical Properties The left-hand side of Figure 8.12 shows the "Model Builder Window" with material properties selected for water. Water with its properties is prescribed in domain 1 where it flows through the static mixer geometry. The water properties are shown in the "Settings Window". The right-hand side of Figure 8.12 shows the same water properties with some additional values.

8.4.1.1.3.3 Selecting Initial and Boundary Conditions Once the conservation equations are prescribed to the laminar static mixer computational model, the initial and boundary conditions must be applied to each geometry surface. Only appropriate boundary conditions completely describe the real-world physical system.

8.4.1.1.3.4 Initial Conditions Initial conditions need to be applied for all conservation equations used in the computational model. Thus, the initial velocity for the continuity equation and momentum equations in the axial direction is assumed as

$$v_{z,\,initial} = v_{mean} \tag{8.111}$$

and the initial pressure as

$$p_{initial} = 0 \tag{8.112}$$

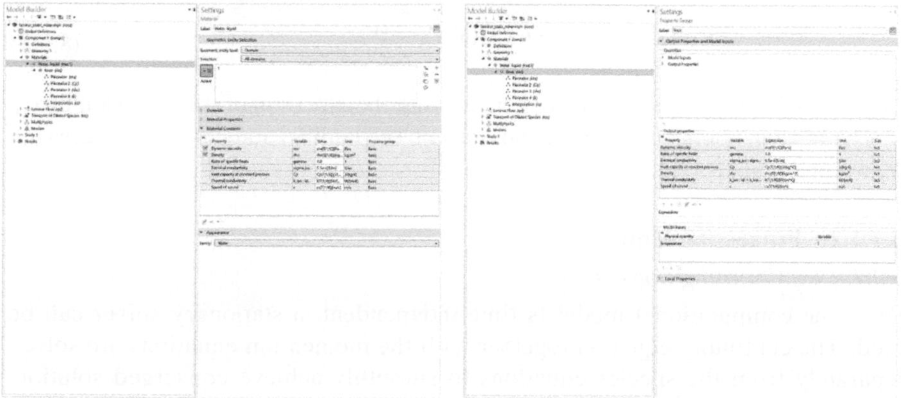

FIGURE 8.12 Material properties of water prescribed for the laminar static mixer (image made using COMSOL Multiphysics® and is provided courtesy of COMSOL®).

The species equations require an initial value for the concentration of species A assumed as

$$c_{A,initial} = 0 \quad (8.113)$$

8.4.1.1.3.5 Inlet Boundary Conditions We assume a parabolic inlet velocity profile for fully developed laminar flow in the axial direction of the inlet side as

$$\vec{v} = v_z = 2 \cdot v_{mean} \cdot \left(1 - \frac{x^2 + y^2}{r^2}\right) \quad (8.114)$$

where $v_{mean}\ [m/s]$ denotes the mean velocity scalar, $r\ [m]$ the tube diameter and $x\ [m]$ and $y\ [m]$ the coordinates of the static mixer.

We assume a constant and discontinuous concentration profile of species A as

$$c_{A,inlet} = c_{A0,\ inlet}\ for\ x < 0 \quad (8.115)$$

$$c_{A,inlet} = 0\ for\ x \geq 0 \quad (8.116)$$

where $c_{A,inlet}\ \left[mol/m^3\right]$ denotes the inlet concentration of species A.

8.4.1.1.3.6 Outlet Boundary Conditions We assume no pressure difference at the outlet side with no viscous stresses as

$$p = 0 \quad (8.117)$$

and no diffusion flux as

$$-\vec{n} \cdot \mathcal{D}_A \vec{\nabla}\ c_A = 0 \quad (8.118)$$

where $\vec{n}\ [-]$ denotes the normal vector showing the direction of the diffusion flux.

The inlet and outlet boundary conditions on the laminar static mixer surfaces are shown on Figure 8.13.

8.4.1.2 Processing Unit

8.4.1.2.1 Choosing Solvers

Since the computational model is time-independent, a stationary solver can be used. The continuity equation together with the momentum equations are solved separately from the species equations to smoothly achieve converged solution. Returned values from the continuity and Navier-Stokes equations are then automatically called by the chosen solver and used in the convection-diffusion equation.

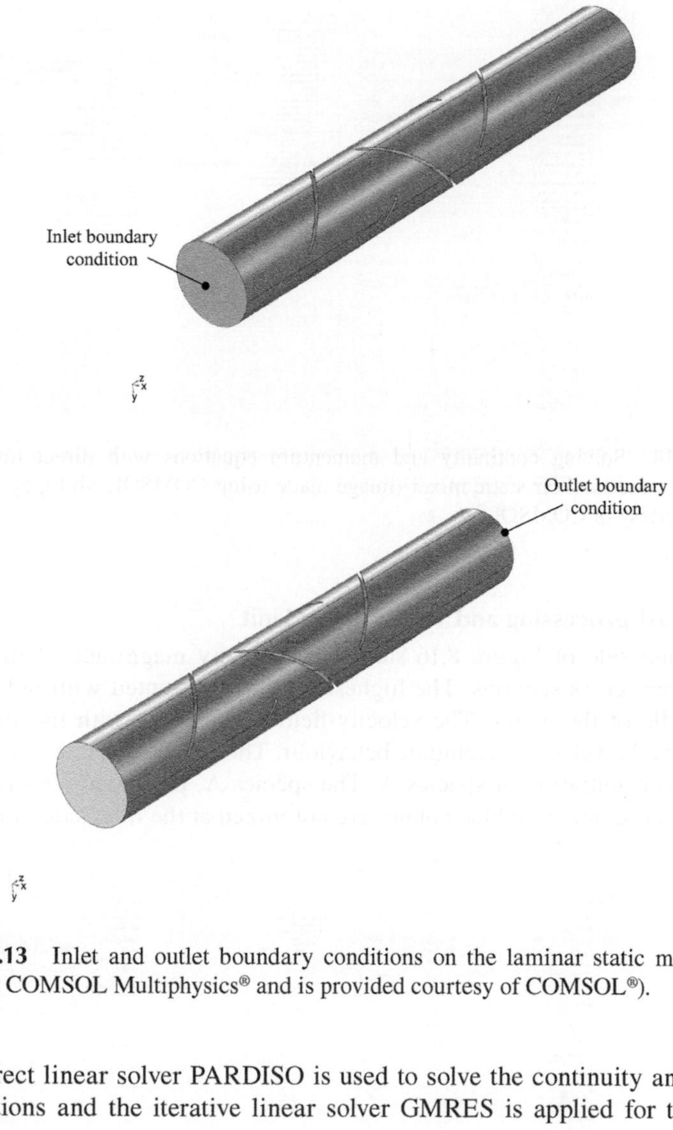

FIGURE 8.13 Inlet and outlet boundary conditions on the laminar static mixer (image made using COMSOL Multiphysics® and is provided courtesy of COMSOL®).

The direct linear solver PARDISO is used to solve the continuity and momentum equations and the iterative linear solver GMRES is applied for the species equations.

The left-hand side of Figure 8.14 shows the process of selecting stationary solver for the "Laminar Flow" physics, which includes continuity and momentum equations only. The right-hand side of the same figure shows the selected "Direct" linear solver PARDISO with the specific settings that need to be applied to solve the set of continuity and momentum equations.

The left-hand side of Figure 8.15 shows a similar process as described for Figure 8.14, but for the "Transport of Diluted Species" physics, which includes species equations. The chosen solver is the iterative linear solver GMRES with the specific parameters shown on the right-hand side of Figure 8.15.

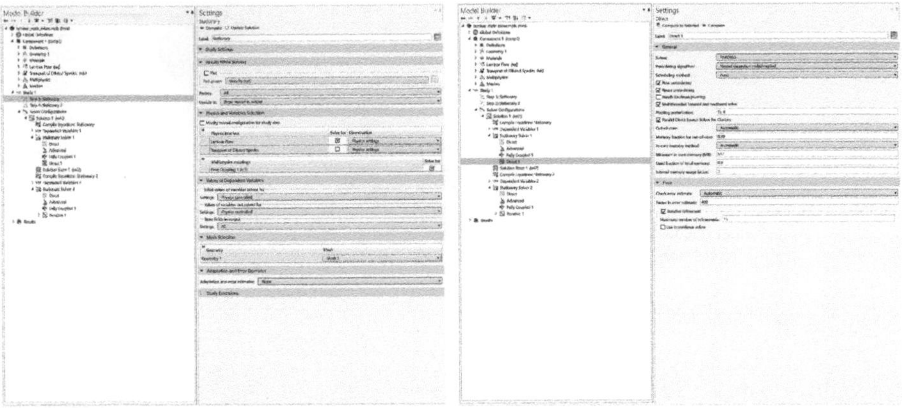

FIGURE 8.14 Solving continuity and momentum equations with direct linear solver PARDISO for the laminar static mixer (image made using COMSOL Multiphysics® and is provided courtesy of COMSOL®).

8.4.1.3 Post-processing and Visualisation Unit

The left-hand side of Figure 8.16 shows the velocity magnitude of the flow on eight different cross sections. The highest velocity, presented with red colour, is in the middle of the mixer. The velocity field is presented with the streamlines and supports the velocity magnitude behaviour. The right-hand side of Figure 8.16 shows the concentration of species A. The species A, presented with red colour, and water, presented with blue colour, are not mixed at the inlet side of the mixer.

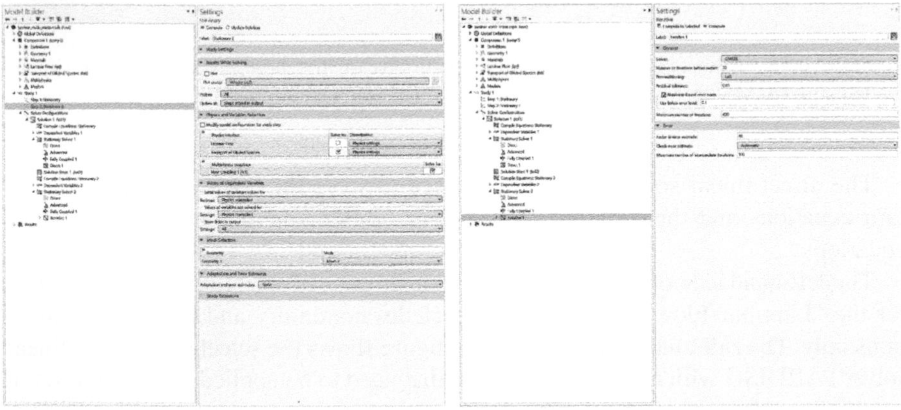

FIGURE 8.15 Solving species equations with iterative linear solver GMRES for the laminar static mixer (image made using COMSOL Multiphysics® and is provided courtesy of COMSOL®).

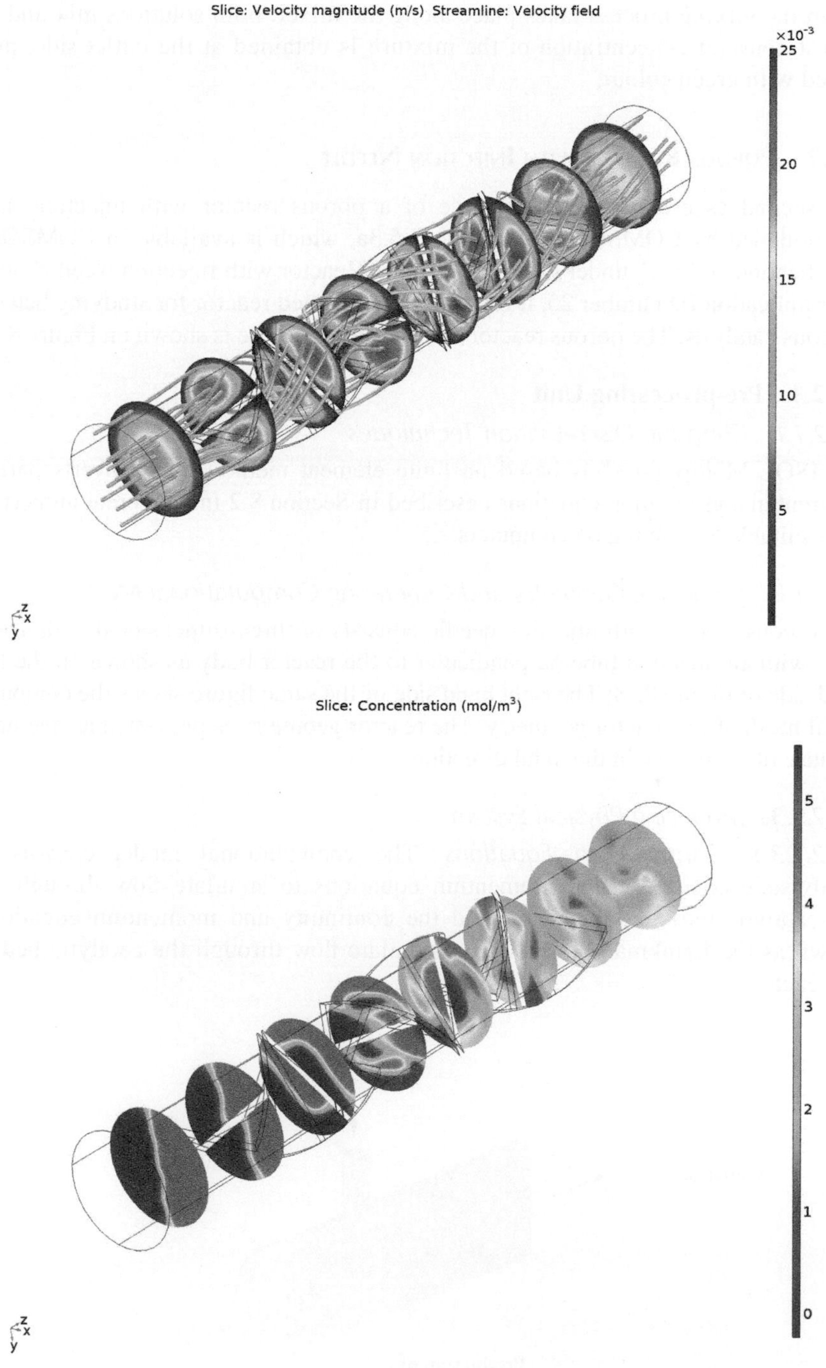

FIGURE 8.16 Velocity magnitude, velocity field and concentration of species A inside the laminar static mixer (image made using COMSOL Multiphysics® and is provided courtesy of COMSOL®). (See color insert.)

When the mixing process takes place along the mixer, both solutions mix and an almost constant concentration of the mixture is obtained at the outlet side, presented with green colour.

8.4.2 Porous Reactor with Injection Needle

The second case study is an example of a porous reactor with injection needle modelled by COMSOL Multiphysics® 5.3a, which is available in COMSOL® "Application Gallery" under the name "Porous Reactor with Injection Needle" with the Application ID number 25. It represents a fixed bed reactor for studying heterogeneous catalysis. The porous reactor with injection needle is shown on Figure 8.17.

8.4.2.1 Pre-processing Unit

8.4.2.1.1 Choosing Discretisation Techniques

COMSOL Multiphysics® is based on finite element method that converts partial differential conservation equations described in Section 8.2 into discrete numerical form suitable for solving on computers.

8.4.2.1.2 Modelling Geometry and Generating Computational Mesh

The porous reactor with injection needle consists of three-dimensional cylindrical body with an injection tube perpendicular to the reactor body as shown on the left hand side of Figure 8.18. The right hand side of the same figure shows the computational mesh of the reactor geometry. The reactor geometry is presented as one half, because of symmetry in the axial direction.

8.4.2.1.3 Modelling Physical System

8.4.2.1.3.1 Conservation Equations The computational model consists of steady-state continuity and momentum equations to simulate flow through the non-catalytic part of the reactor and the continuity and momentum equations, known as the Brinkman equations, to simulate flow through the catalytic bed of the reactor.

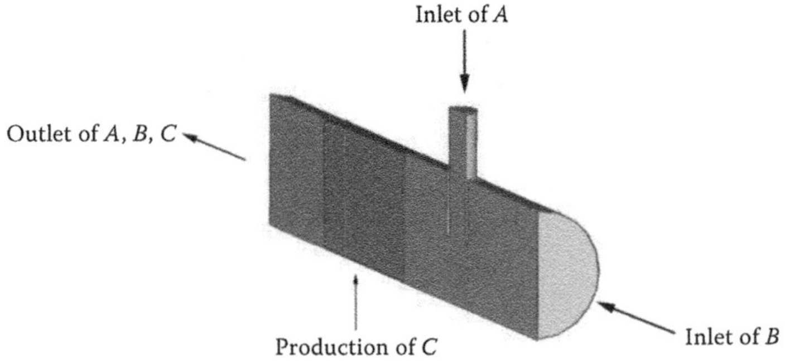

FIGURE 8.17 Distribution of chemical species in the porous reactor with injection needle (image made using COMSOL Multiphysics® and is provided courtesy of COMSOL®).

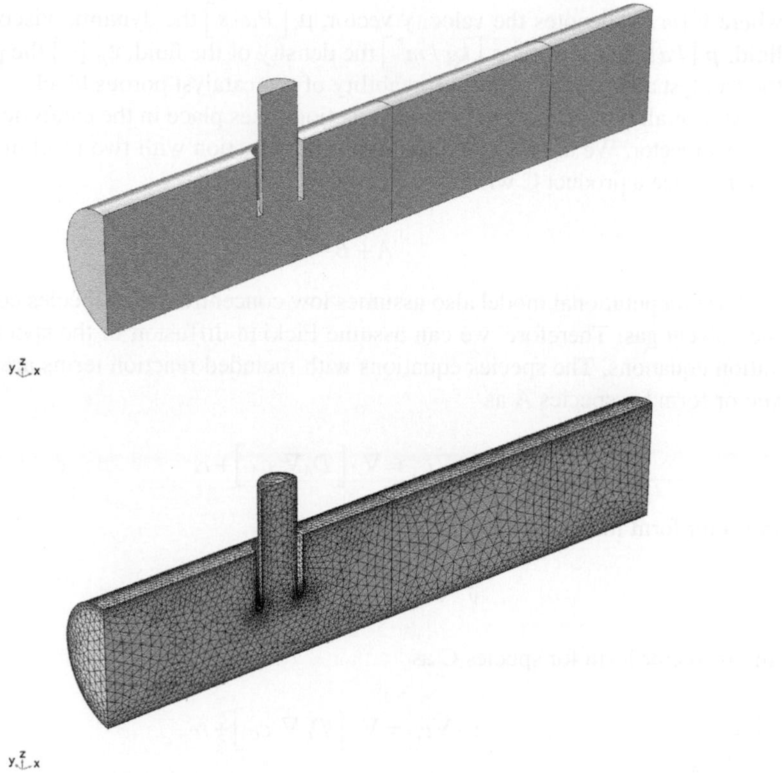

FIGURE 8.18 Geometry and computational mesh of the porous reactor with injection needle (image made using COMSOL Multiphysics® and is provided courtesy of COMSOL®).

The continuity equation for the non-catalytic part is written in vector form as

$$\rho\vec{\nabla}\cdot\vec{v}=0 \tag{8.119}$$

and the momentum equation for the non-catalytic part is written in vector form as

$$\rho\left(\vec{v}\cdot\vec{\nabla}\right)\vec{v}=\vec{\nabla}\cdot\left[\mu\vec{\nabla}\,\vec{v}-p\right] \tag{8.120}$$

The continuity equation for the catalytic bed is written in vector form as

$$\rho\vec{\nabla}\cdot\vec{v}=0 \tag{8.121}$$

and the momentum equation for the catalytic bed with added porosity and permeability terms is written in vector form as

$$\frac{\rho}{\varepsilon_p}\left(\vec{v}\cdot\vec{\nabla}\frac{\vec{v}}{\varepsilon_p}\right)=\vec{\nabla}\cdot\left[\frac{\mu}{\varepsilon_p}\vec{\nabla}\,\vec{v}-p\right]-\frac{\mu}{\kappa_{br}}\vec{v} \tag{8.122}$$

where $\vec{v}$ $[m/s]$ denotes the velocity vector, μ $[Pa \cdot s]$ the dynamic viscosity of the fluid, p $[Pa]$ the pressure, ρ $[kg/m^3]$ the density of the fluid, ε_p $[-]$ the porosity of the catalyst and κ_{br} $[m^2]$ the permeability of the catalyst porous block.

An overall homogeneous chemical reaction takes place in the catalytic bed of the porous reactor. We assume a simple first order reaction with two reactants A and B that produce a product C written as

$$A + B \xrightarrow{k} C \tag{8.123}$$

The computational model also assumes low concentration of species compared to the solvent gas. Therefore, we can assume Fickian diffusion in the species conservation equations. The species equations with included reaction terms are written in vector form for species A as

$$\vec{v} \cdot \vec{\nabla} c_A = \vec{\nabla} \cdot \left[\mathcal{D}_A \vec{\nabla}\, c_A \right] + r_A \tag{8.124}$$

in vector form for species B as

$$\vec{v} \cdot \vec{\nabla} c_B = \vec{\nabla} \cdot \left[\mathcal{D}_B \vec{\nabla}\, c_B \right] + r_B \tag{8.125}$$

and in vector form for species C as

$$\vec{v} \cdot \vec{\nabla} c_C = \vec{\nabla} \cdot \left[\mathcal{D}_C \vec{\nabla}\, c_C \right] + r_C \tag{8.126}$$

where D_A $[m^2/s]$ denotes the diffusivity of species A, D_B $[m^2/s]$ the diffusivity of species B, D_C $[m^2/s]$ the diffusivity of species C, c_A $[mol/m^3]$ the molar concentration of species A, c_B $[mol/m^3]$ the molar concentration of species B, c_C $[mol/m^3]$ the molar concentration of species C, r_A $[mol/(m^3 \cdot s)]$ the reaction rate of species A, r_B $[mol/(m^3 \cdot s)]$ the reaction rate of species B and r_C $[mol/(m^3 \cdot s)]$ the reaction rate of species C.

The reaction rates for species A is given as

$$r_A = -k\ c_A\ c_B \tag{8.127}$$

for species B as

$$r_B = -k\ c_A\ c_B \tag{8.128}$$

and for species C as

$$r_C = +k\ c_A\ c_B \tag{8.129}$$

where k $[m^3/(s \cdot mol)]$ denotes the reaction rate constant for the particular reaction defined with the Arrhenius law as

$$k = A\ exp\left(\frac{-E}{R\ T}\right) \tag{8.130}$$

where $E\ [J/mol]$ denotes the activation energy, $R\ \left[J/(mol \cdot K)\right]$ the universal gas constant and $T\ [K]$ the temperature of the flow.

8.4.2.1.3.2 Selecting Initial and Boundary Conditions Once the conservation equations are prescribed to the porous reactor with injection needle computational model, the initial and boundary conditions must be applied to each geometry surface. Only appropriate boundary conditions completely describe the real-world physical system.

8.4.2.1.3.3 Initial Conditions The initial conditions for the continuity and momentum equations are initial velocities in x, y and z directions prescribed as

$$v_{x,\ initial} = v_{y,\ initial} = v_{z,\ initial} = 0 \tag{8.131}$$

and the initial pressure as

$$p_{initial} = 0 \tag{8.132}$$

The species equations require initial values for concentration of species A, species B and species C prescribed as

$$c_{A,initial} = c_{B,initial} = c_{C,initial} = 0 \tag{8.133}$$

8.4.2.1.3.4 Inlet Boundary Conditions We assume a constant velocity profile for laminar flow at the inlet side of species A and species B as

$$\vec{v} = v_{inlet} \tag{8.134}$$

where $v_{inlet}\ [m/s]$ denotes the inlet velocity scalar.

We also assume a constant concentration of species A as

$$c_A = c_{A0,\ inlet} \tag{8.135}$$

of species B as

$$c_B = c_{B0,\ inlet} \tag{8.136}$$

and of species C as

$$c_C = c_{C0,\ inlet} \tag{8.137}$$

where $c_{A0,inlet}\ \left[mol/m^3\right]$ denotes the inlet concentration of species A, $c_{B0,inlet}\ \left[mol/m^3\right]$ the inlet concentration of species B and $c_{C0,inlet}\ \left[mol/m^3\right]$ the inlet concentration of species C.

8.4.2.1.3.5 Outlet Boundary Conditions We assume no pressure difference at the outlet side with no viscous stresses as

$$p = 0 \tag{8.138}$$

We also assume no diffusion flux of species A as

$$-\vec{n} \cdot \mathcal{D}_A \vec{\nabla}\ c_A = 0 \tag{8.139}$$

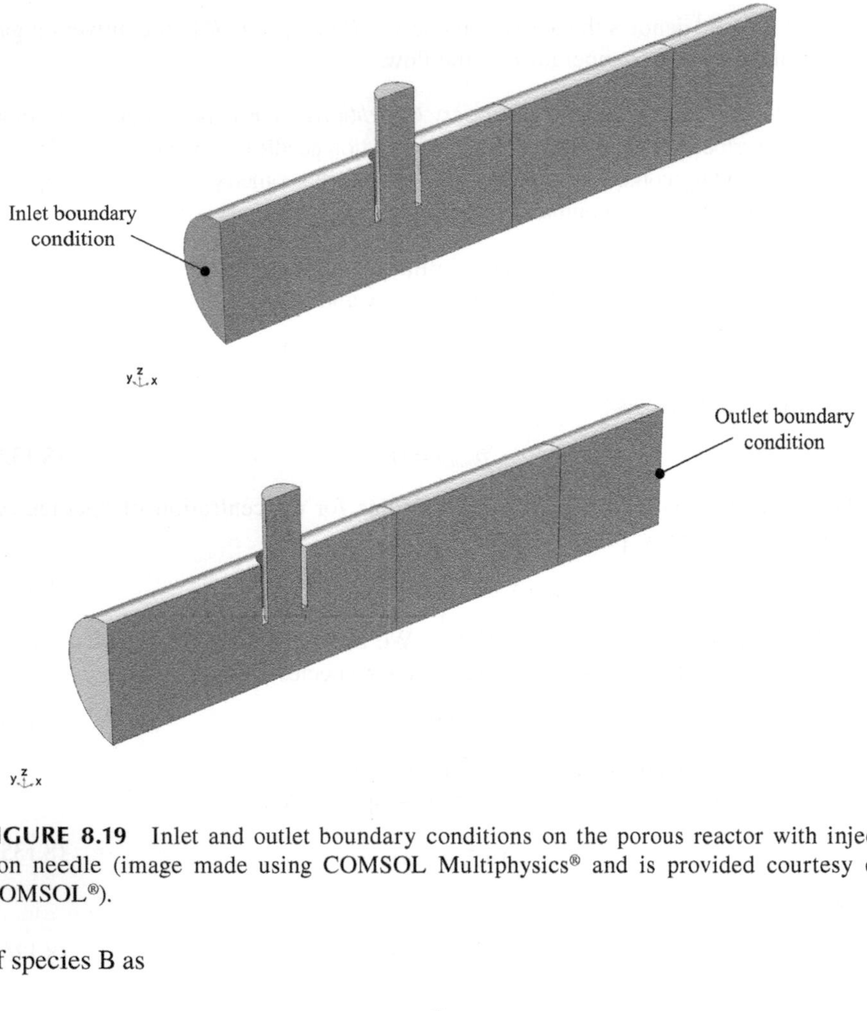

FIGURE 8.19 Inlet and outlet boundary conditions on the porous reactor with injection needle (image made using COMSOL Multiphysics® and is provided courtesy of COMSOL®).

of species B as

$$-\vec{n} \cdot \mathcal{D}_B \vec{\nabla}\ c_B = 0 \tag{8.140}$$

and of species C as

$$-\vec{n} \cdot \mathcal{D}_C \vec{\nabla}\ c_C = 0 \tag{8.141}$$

where $\vec{n}$ $[-]$ denotes the normal vector showing the direction of the diffusion flux.

The inlet and outlet boundary conditions on the porous reactor with injection needle are shown on Figure 8.19.

8.4.2.2 Processing Unit

8.4.2.2.1 Choosing Solvers

The computational model is time-independent and thus a stationary solver can be used. The continuity and momentum equations for the non-catalytic part and catalytic bed of the porous reactor with injection needle are solved simultaneously with the iterative linear solver GMRES.

8.4.2.3 Post-processing and Visualisation Unit

The left hand side of Figure 8.20 shows the velocity magnitude with the highest value, presented with red colour, where the inlet streams from species A and species B collide. The collision intensifies local mixing and thus increases local velocity gradients. Due to catalytic bed, flow slows down through the catalyst and increases again at the outlet side of the porous reactor. The right hand side of Figure 8.20 shows the concentration distribution of species reactant B. The lowest concentration of species B, presented with blue colour, is observed where the species reactant A enters the reactor. On the contrary, the highest concentration of species B, presented with red colour, is observed in the main stream of species B. Green colour presents the formation of species product C, mostly formed in the catalyst.

8.5 CASE STUDIES WITH ANSYS FLUENT

8.5.1 Introduction and ANSYS Workbench

ANSYS® was conceived as a product design tool that allows evaluating and predicting the performance of mechanical systems such as solids (ANSYS Mechanical®) and fluids (ANSYS CFX®, ANSYS Fluent®) or electromagnetics (ANSYS Antenna®). Furthermore, it also includes more specific packages for complex and common geometries such as engines and turbines. The prediction of different phenomena in the units enhances its capability as modelling and simulation tool, being applied to a wide number of units in chemical engineering processes such as stirred tanks, cyclones, towers (for distillation, absorption, etc.), combustors or heat exchangers.

ANSYS has a project structure where the management application is known as ANSYS Workbench®. It is structured as seen in Figure 8.21. It allows the selection and connection of the different tools that compose ANSYS®. In Figure 8.21 it is possible to appreciate the main tabs in the top related to:

- File, where one can open, save, import or export the project.
- The View option allows the selection of new tabs to be shown in the in the window such as toolbox extensions, project outputs like messages or progress updates.
- Tools bar, where one can refresh and update the project, check the license and select main options like the default path where the project is saved.
- Units bar to select the system of units desired.
- Extensions to be installed from the app store of ANSYS®.
- Jobs helps monitor any of the jobs that being oriented out.
- Help and documentation tab to access the user guide.

To the left, the toolbox bar contains all the possible applications that conform ANSYS® and it is structured in different sections such as Analysis systems, Component systems, Custom systems, Design exploration and External connections. The applications can

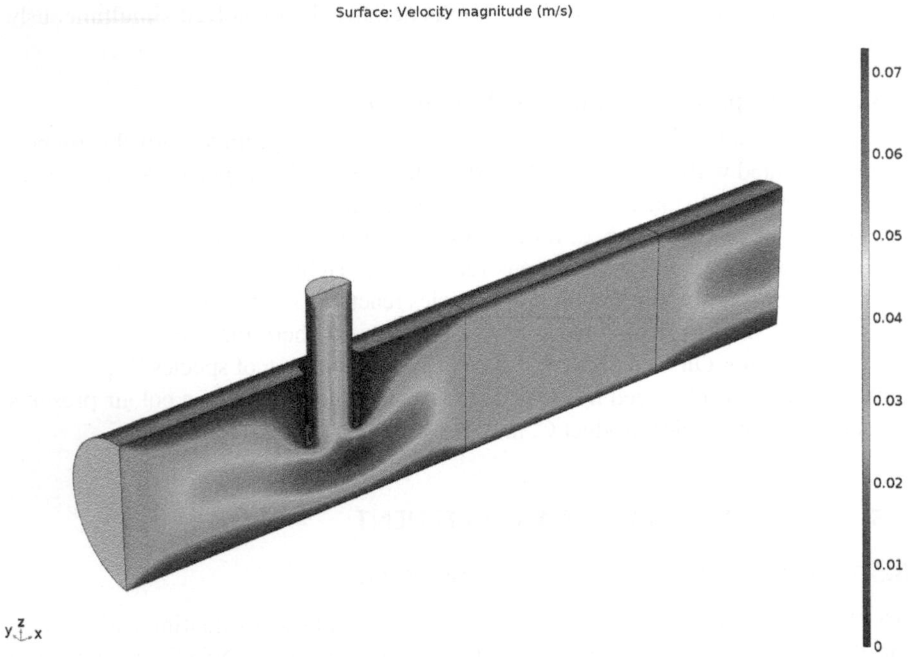

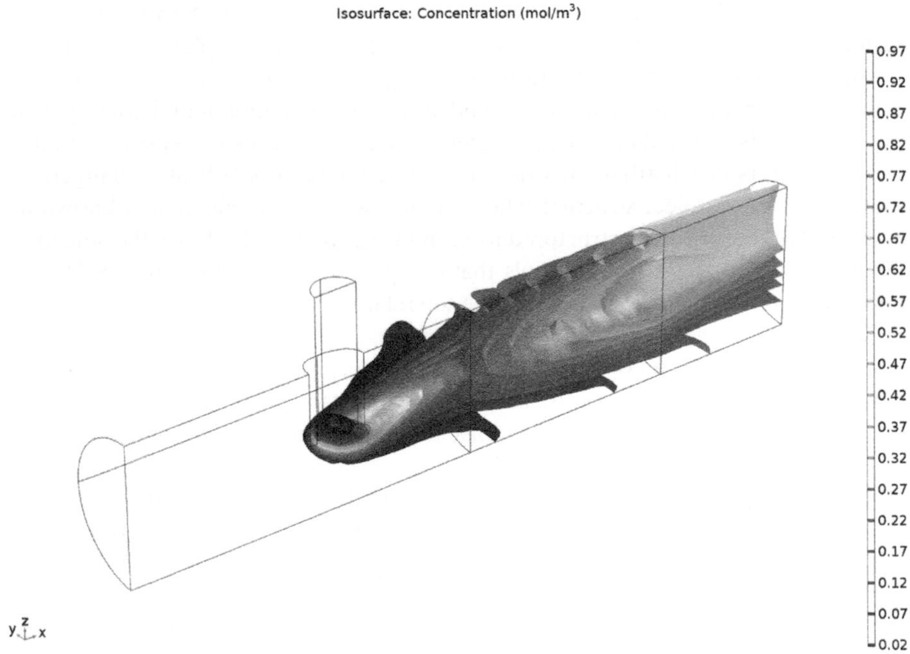

FIGURE 8.20 Velocity magnitude and concentration of species B inside the porous reactor with injection needle (image made using COMSOL Multiphysics® and is provided courtesy of COMSOL®). (See color insert.)

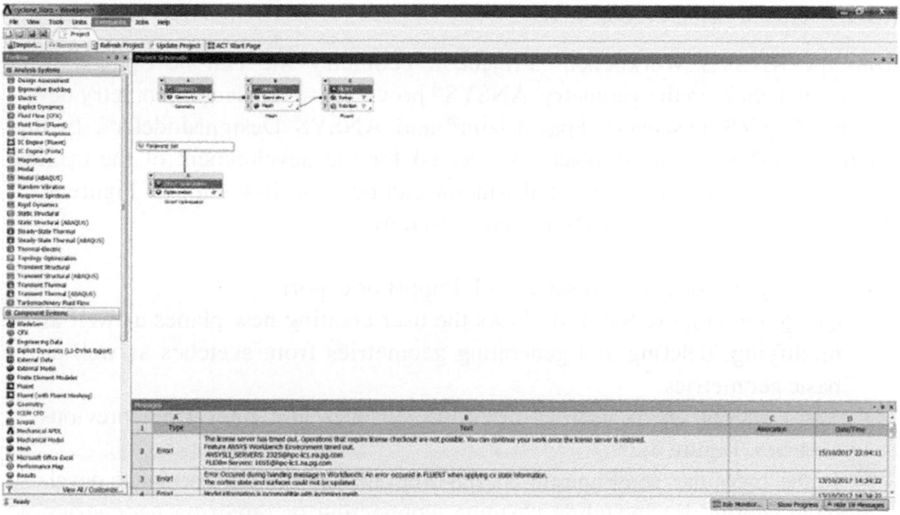

FIGURE 8.21 ANSYS Workbench® window.

be drag and dropped at the Project Schematic panel where they can be later opened and modified. They can be connected to each other since the applications are dependent on previous ones like the mesh dependency of the geometry as showed in Figure 8.21. It is also important to mention a helpful tool such as the Parameters that allows the modification of the project without the need to open each of the applications. The parameters are modified from an additional sheet, "Parameter Set". Furthermore, in versions later than 16, ANSYS® includes optimization and quality design tools which helps improving initial designs.

To present the use of this package we present a case of study, a cyclone, starting with of the geometry, continuing with the generation of the mesh, developing the ANSYS Fluent® simulation and analyzing the results obtained. The cyclone is used to recover bagasse and its specifications are described in [14]. It has the following dimensions from Figure 8.22.

De (m)	0.16
S (m)	0.5
Li (m)	0.15
a (m)	0.056
b (m)	0.25
h (m)	0.73
B (m)	0.1
H (m)	1.2
α	41°

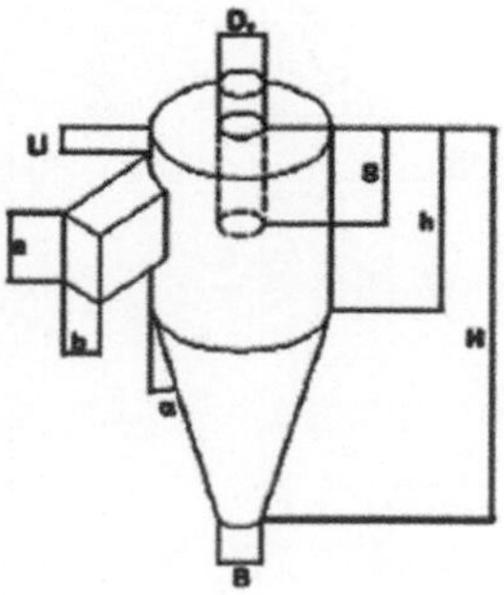

FIGURE 8.22 Cyclone dimensions. Adapted from Fernandes de Oliveira et al [14].

8.5.2 Geometry Development

From the ANSYS Workbench® Project, a geometry file can be opened by a right-button click on the geometry. ANSYS® provides two tools for geometry development: ANSYS Discovery SpaceClaim® and ANSYS DesignModeler®. In this chapter, ANSYS DesignModeler® is selected for the development of the cyclone geometry. In Figure 8.23, the initial window can be seen. In the top of Figure 8.23, the reader can appreciate the toolbar composed by:

- The File options such as save, load, import or export.
- Create tab (Figure 8.24). It allows the user creating new planes as well as modifying, deleting and generating geometries from sketches as well as basic geometries.
- The Concept tab is useful to develop surfaces and lines from previous sketches, Figure 8.25.
- In the Tools tab, see Figure 8.26, the user can operate with previous developed geometries including cleaning, connecting or repairing geometries, development of symmetries or fillings that simplifies building the geometry.
- The other three additional tabs are the ones related to the View, Units and Help which are quite similar to previous cases.

In the development of a cyclone geometry, different strategies can be developed. In this case, we start from a 2D sketch for half of the cyclone main geometry on the default XY plane. The lines are drawn connected by each point. Initially, only the shape is approached; then, the dimensions are given from the dimension tab where the dimensions are specified for each line (with horizontal, vertical and general tools from the Sketching toolbox panel of Figure 8.27) and also the angle between two lines.

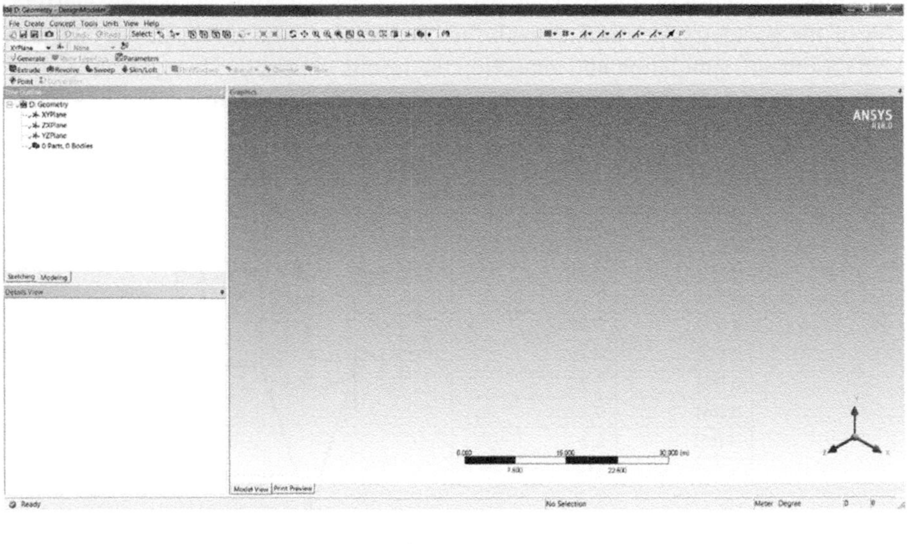

FIGURE 8.23 ANSYS DesignModeler® initial window.

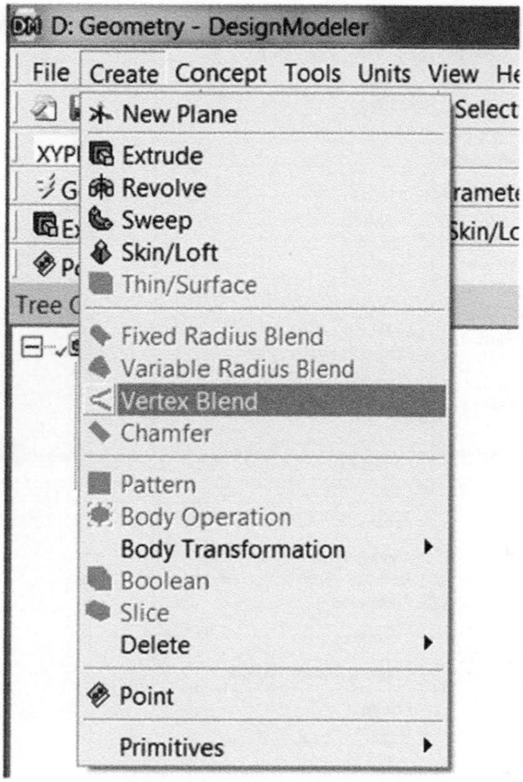

FIGURE 8.24 Create options in ANSYS DesignModeler®.

Once the main geometry is sketched, the 3D geometry is generated using "Revolve" from the Create tab. Once Revolve appears in the Tree, it is necessary to click on it and go to the "Details View" panel where the geometry selected is the Sketch previously developed. Finally, it is also important to specify in the

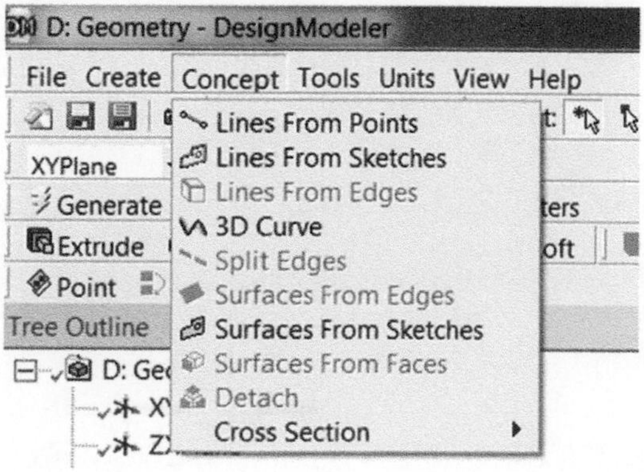

FIGURE 8.25 Concept options in ANSYS DesignModeler®.

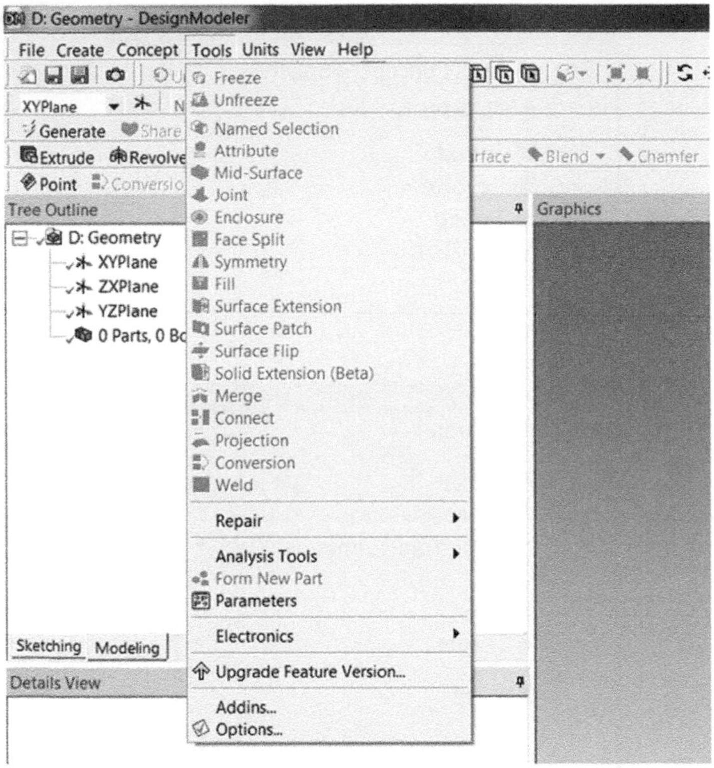

FIGURE 8.26 Tool options in ANSYS DesignModeler®.

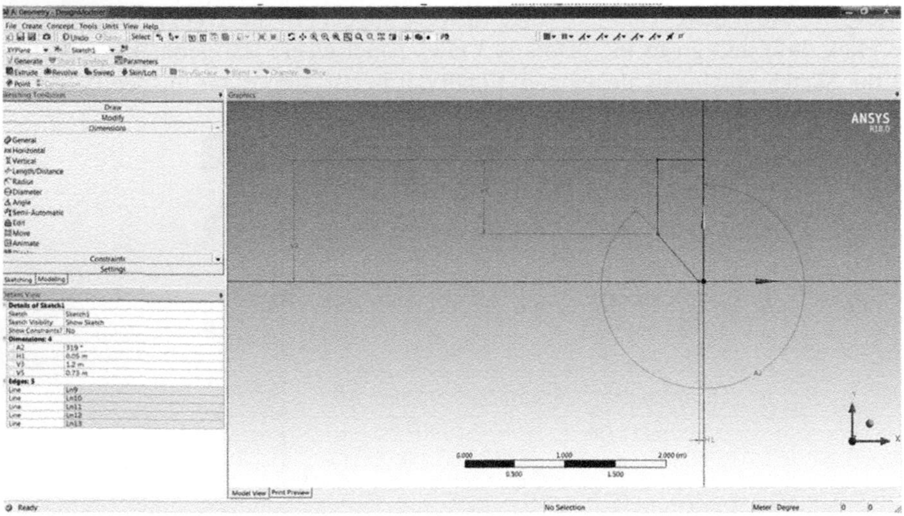

FIGURE 8.27 Sketch of the cyclone main geometry.

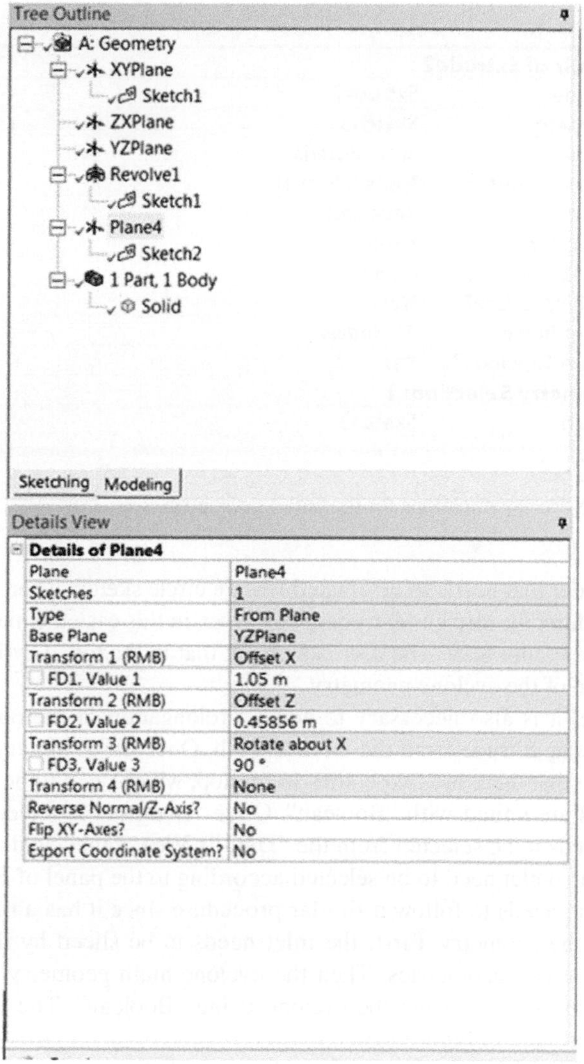

FIGURE 8.28 Specifications for the new plane created.

"Details View" panel, from the bottom-left hand side of Figure 8.28, the operation "Frozen" is added since then the region created is a fluid part instead of a solid material.

In order to continue with the development of the geometry, the inlet is designed. The inlet is created from a plane that is generated from the YZ Plane adding the following offsets of Figure 8.28.

Once the plane has been generated, a rectangle is sketched specifying the dimensions and then using "Extrude" a 3D tangential object, corresponding to the inlet, is generated. The outlet follows a similar procedure. A new plane is generated from plane ZX using an offset of 1.25 m in the Y axis. The geometry generated for the

Details View

Details of Extrude2	
Extrude	Extrude2
Geometry	Sketch3
Operation	Slice Material
Direction Vector	None (Normal)
Direction	Reversed
Extent Type	Fixed
FD1, Depth (>0)	0.55 m
As Thin/Surface?	No
Target Bodies	All Bodies
Merge Topology?	Yes
Geometry Selection: 1	
Sketch	Sketch3

FIGURE 8.29 Slice specifications for the outlet geometry.

outlet is a cylinder that needs to be created from a circle sketch. Then, following the previous procedure, the circle is extruded. However, in this case the material is sliced as it is shown in Figure 8.29. The user can check that in the top, a cylinder has been generated inside of the cyclone geometry.

Furthermore, it is also necessary to add a prolongation of the outlet. It is prolonged 5 cm using extrude from the circle sketch. Once the geometry is generated, the user can see that there are two bodies in the top, which need to be united as one. The geometries are united with "Boolean". Once "Boolean" is selected, the operation "Unite" needs to be selected from the "Details View Panel" and the two bodies that conform the outlet need to be selected according to the panel of Figure 8.30.

The inlet also needs to follow a similar procedure since it has a tube introduced inside the cyclone geometry. First, the inlet needs to be sliced by the wall of the cyclone, creating two geometries. Then the cyclone main geometry is united with the inlet part that is still inside the cyclone using "Boolean". The final geometry generated is presented in Figure 8.31.

Finally, it is also important to conform the parts that compose our model. Three bodies can be now appreciated in the Tree Outlet. The name of the bodies usually

Details View

Details of Boolean1	
Boolean	Boolean1
Operation	Unite
Tool Bodies	2 Bodies

FIGURE 8.30 Boolean options for unite two bodies.

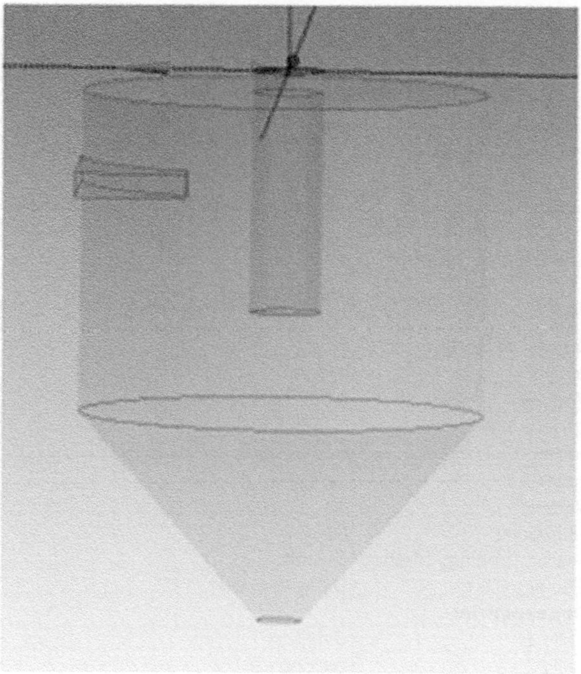

FIGURE 8.31 Final figure in ANSYS DesignModeler®.

appears as "Body". However, we can rename each of them as inlet, outlet and cyclone with: right click → Rename. Furthermore, the type of geometry also needs to be modified from Solid to Fluid for all of them since we will simulate later with ANSYS Fluent®. Another important definition that must be generated is the "Part" that will conform the future connections between two bodies, allowing that not all the external surfaces of the cyclone body will be taken as wall and, consequently, the connection between the inlet body and the cyclone is differentiated from the rest of the external wall of the cyclone. In Figure 8.32, it is showed how the tree outline and the "Details View" of one of the bodies finally looks.

8.5.3 Mesh Development

Once the geometry is generated, the following stage is to generate the finite volumes in which the discretization is performed with the mesh implementation. The meshing tool is opened from ANSYS Workbench®, obtaining the following window showed in Figure 8.33.

The menu available in ANSYS Meshing® are structured in:

- File. It has similar options than ANSYS DesignModeler®.
- Edit allows operating with the parts of the geometry cutting, copying or selecting.

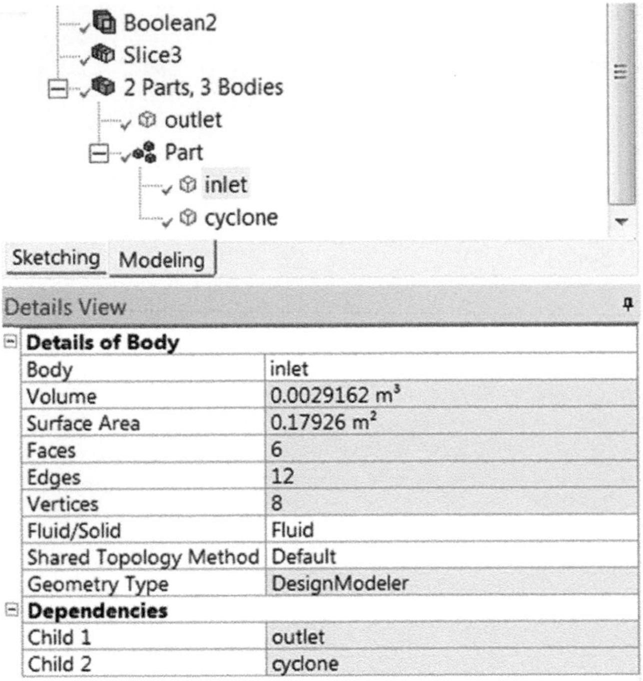

FIGURE 8.32 Creation of parts in ANSYS DesignModeler®.

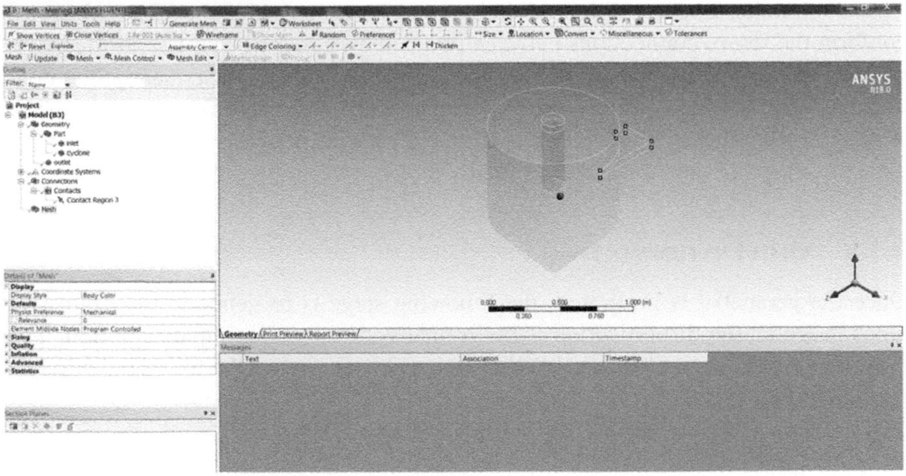

FIGURE 8.33 Initial screen in ANSYS Meshing® with cyclone reading.

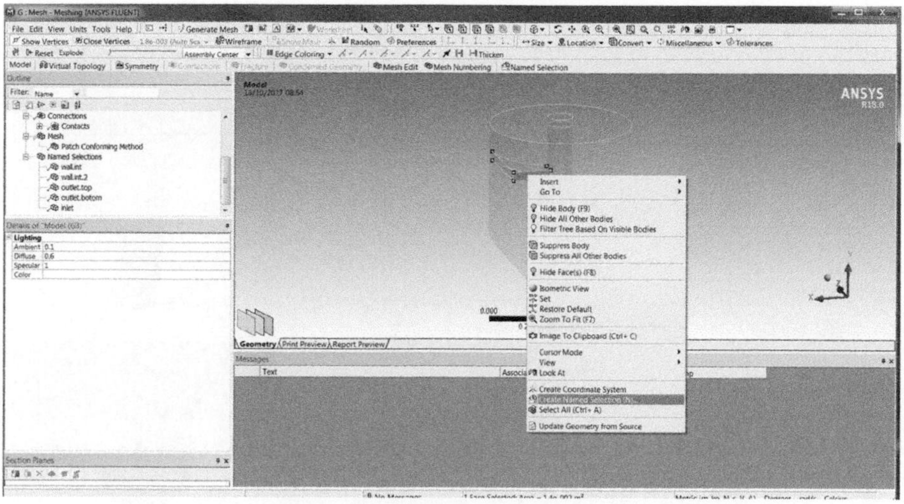

FIGURE 8.34 Generation of Named Selection in ANSYS Meshing®.

- View: From the View panel one can select the display options as well as the extra toolbars that can be used for fast procedures, see Figure 8.34 for the options.
- The Units tab is similar to the one in ANSYS DesignModeler®.
- Tools tab contains extra Add-ins that can be loaded for the mesh generation and Options that contain some of the options available in the Tree when Mesh is selected.
- Help.

In the mesh generation procedure, the first step required is the definition of the names of the surfaces that will be recognized later in ANSYS Fluent®. ANSYS Fluent® automatically defines the boundary condition from the names given in the mesh. The names required to be specified in this stage are the ones that cannot be easily identified by ANSYS Fluent®, such as inlets, outlets and inner walls. In Figure 8.34, a screenshot showing the Name generation is presented. It is created from right-click over the surface selected. Once the new name is defined, a new "Named Selection" appears on the Outline left panel. The same procedure is followed for the specification of the top and bottom outlets as well as for the internal wall of the outlet. It is important to note that the internal wall of the outlet needs to be specified in two boundary conditions so it needs to be defined in the body of the outlet and in the body of the cyclone.

Before continuing with the following steps, it is also important to check the contact regions between the geometries. By default, one contact region appears between the outlet and the cyclone. However, the contact region is referred to two faces for each of the two bodies, meanwhile, due to the existence of the wall, the contact region only exists for one face of each body. The implementation of the contact region is summarized on Figure 8.35. First, using the right-click one can select: Insert → Contact Region. Then in the details panel, the user needs to select the two faces that will

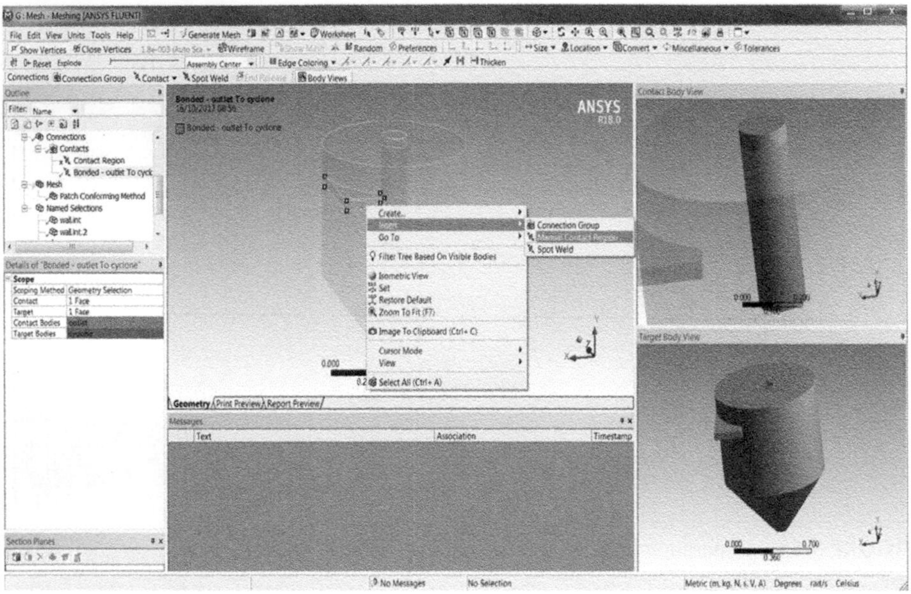

FIGURE 8.35 Generation of contact regions.

contact, one as contact and the other one as Target. The comparison between the actual zone of contact and the default one can be also appreciated in Figure 8.36.

Once the boundaries have been specified, we continue with the general specifications of the mesh. Clicking on Mesh from the Outline toolbar, the "Details panel" appear in Figure 8.37. In the "Details panel" is important to specify at least the following characteristics:

- In the Defaults section the Physics Preference must be changed from solid to CFD where it is necessary to check that the Solver preference is "Fluent". This will help to generate a mesh easier to compute with Finite Volume instead of Finite elements.
- From the Sizing, the cell size is defined. The default function is "Adaptative", which automatically generates the sizes for the cells. Alternatively, the user can determine the size, in our case "Uniform". The relevant centre can be kept as "Fine" since smaller elements are required in the centre than in the walls due to it is a transitional region and skewness can be generated between the cells. Transition can be also set as "Slow". The following parts to complete are the sizes where if the sizes are between 2 cm and 4 cm a final mesh with 80,000 elements is obtained. Note that it is a small number of elements but in this chapter an example that takes low computational cost wants to be developed since the user is starting to familiarize with the software and can check the results fast.
- The following section is Quality where a target to be achieved in the mesh generation can be provided and it is kept with the default values. A quite

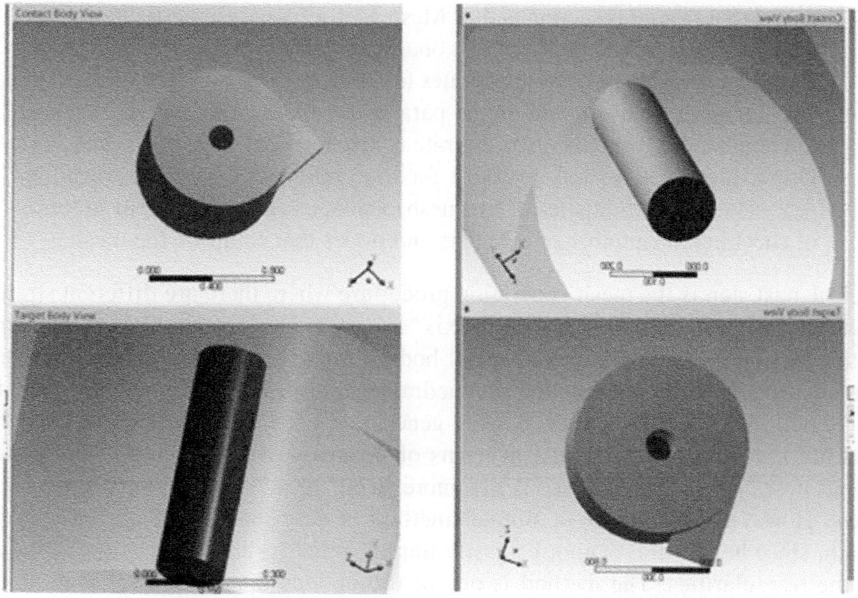

FIGURE 8.36 Comparison between the previous contact regions (left) and the new contact regions (right).

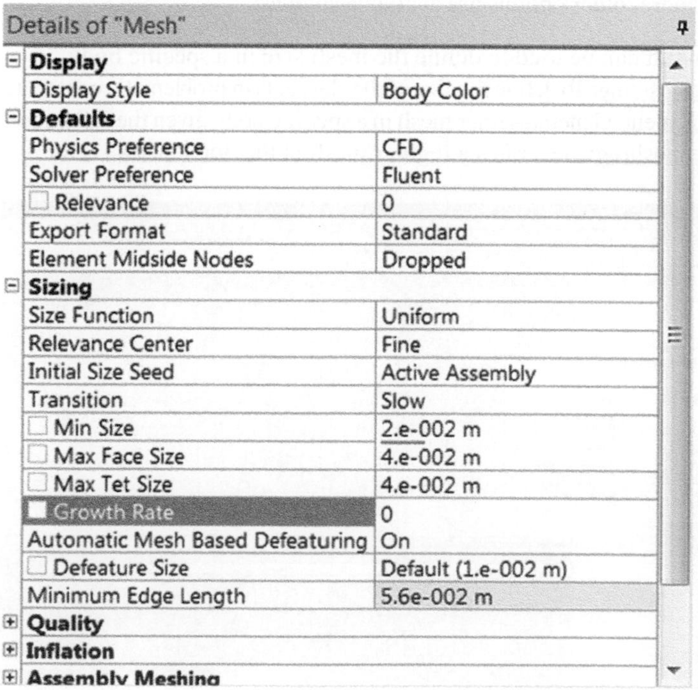

Details of "Mesh"		
Display		
Display Style	Body Color	
Defaults		
Physics Preference	CFD	
Solver Preference	Fluent	
☐ Relevance	0	
Export Format	Standard	
Element Midside Nodes	Dropped	
Sizing		
Size Function	Uniform	
Relevance Center	Fine	
Initial Size Seed	Active Assembly	
Transition	Slow	
☐ Min Size	2.e-002 m	
☐ Max Face Size	4.e-002 m	
☐ Max Tet Size	4.e-002 m	
☐ Growth Rate	0	
Automatic Mesh Based Defeaturing	On	
☐ Defeature Size	Default (1.e-002 m)	
Minimum Edge Length	5.6e-002 m	
Quality		
Inflation		
Assembly Meshing		

FIGURE 8.37 General specifications panel

useful tool from this section is the "Mesh Metric" where a mesh quality variable can be selected and a report is obtained with the minimum, maximum, average and standard deviation values for cells that compose the geometry.
- The Inflation section contains the parameters that can be defined such as the number of layers, the growing rate or the size.
- Other three sections are available for the options of Assembly Meshing, Advance and the Statistics of the mesh. Statistics section is useful in terms of checking the number of elements and nodes that compose the mesh.

The next step is the mesh generation procedure where there are different strategies for mesh generation like "median axis" or "multi-block meshes". In the current case, the cyclone can be sliced in small bodies where each of them is discretized with a different method, being the hexahedral preferred. The different blocks needs to be defined in the previous geometry generation with ANSYS DesignModeler® and are the most recommended in terms of accuracy and reduction of computational cost since hexahedral cells are more accurate and fast to solve than tetra ones. However, in this case a simpler method is described based on tetrahedral mesh, since hexahedral cannot be easily implemented for the whole geometry due to the irregularities. The method is one of the options that can be selected since the user makes right-click on Mesh (see Figure 8.38). The tetrahedral method is implemented to the main body meanwhile the outlet and inlets are given free, being automatically selected by the tool that in this case are set as hexahedral.

Furthermore, other options are available such as:

- Sizing: It can be used to define the mesh size in a specific body.
- Contact sizing: To define the size of the elements in problematic contact regions.
- Refinement: Generates finer mesh in a specific body given the refinement ratio.
- Face meshing: Generates a regular mesh at the boundaries.

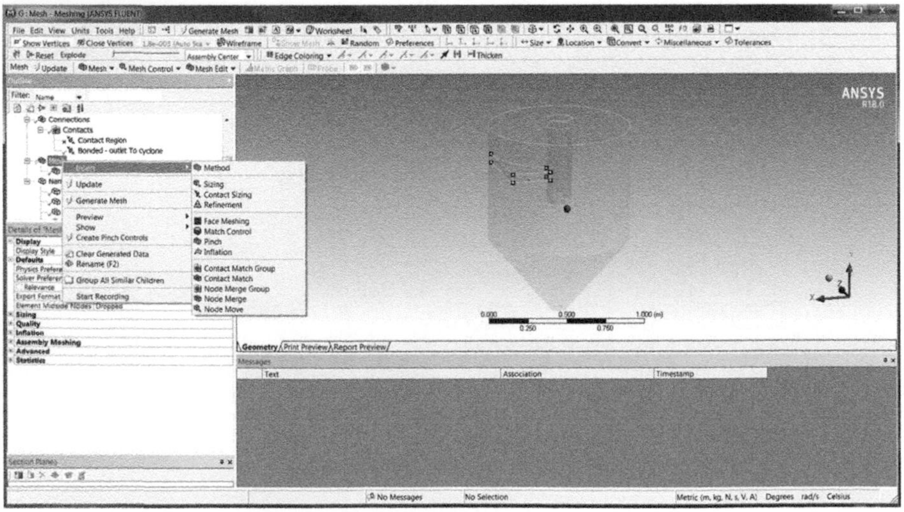

FIGURE 8.38 Application of mesh methods.

- Match control: Matches the mesh on two or more faces or edges in a model. There are two types of match control depending on the meshing applied. For 3D, volume meshing; meanwhile for 2D surfaces, surface meshing.
- Pinch is used to provide a virtual topology in problematic regions like edges, where the mesh needs to be readapted.
- Inflation is used to define the first cell size from the boundary. It is quite useful when further resolution is required near the walls and its dimension can be straightforward defined being parallel to the boundary.
- Contact and node operators.

Once the mesh is generated is necessary to evaluate its quality. Andersson et al. [15] recommended that the maximum skewness, which refers to the asymmetry degree between the cell faces, acceptable for the mesh is 0.95 and its average must be below 0.33. Meanwhile, the minimum orthogonal quality between the centroid of a cell and the faces recommended is 0.05 and an average should not be longer than 0.68. Finally, the last parameter to check is the aspect ratio, which is recommended to be lower than 30.

In our case the quality cannot achieve these specifications since the mesh used is so simple and also the number of cells that compose the system is very low in order to ensure that the system can be solve in minutes without the need of High Performance Computers or Workstations with multiple cores.

8.5.4 Simulation in ANSYS Fluent

From ANSYS Workbench®, with drag and drop the user can connect an ANSYS Fluent® app with the ANSYS Meshing® tool. Once ANSYS Fluent® is launched, the first window that appears is Figure 8.39. The dimension of the geometry is set by

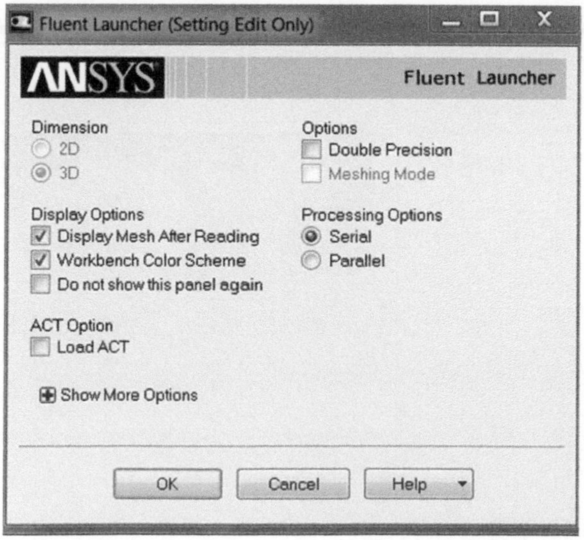

FIGURE 8.39 ANSYS Fluent® Launcher.

default and cannot be changed since the geometry to be solved comes from previous definitions. Other two options are important in terms of speed and resources available. First, the Precision desired, which can be single or double, more accurate. Second, the processing options that can be Serial, 1 process, or parallel where is necessary to specify the number of processes applied. For our case we will select the double precision, meanwhile the processing options depend on the computer for each user.

Once ANSYS Fluent® is started, the screen presented in Figure 8.40 is opened. ANSYS Fluent® is structured in the following areas:

- Toolbars. To the top of the screen different toolbars are showed such as:
    - File. It allows operating with the files as before.
    - Setting up Domain is composed of different tools that will modify our initial mesh and geometry. It contains a mesh section that allows to display the mesh and its quality, improve it or transform its coordinates. In addition, the tool Make Polyhedral is quite useful when one has tetrahedral meshes. The tool improves the accuracy of the simulation since it merges the tetrahedral cells in the boundaries creating regular polyhedrals.
    - Another section is the zones operation that allows to operate with the initial zones defined, combining, separating or deleting them.
    - Finally 4 additional tools are available for the definition of Interfaces, the generation of Dynamic meshes, the definition of cells region or the modification of the surfaces that compose the boundary conditions.
    - The Setting up physics panel contains direct connection to the definitions that the user can provide in the tree panel.
    - User Defined toolbar is key for the definition of User Defined Functions (UDFs) that allows the user to implement its own physics in a wide number of areas such as the turbulence modelling, the energy transfer, the

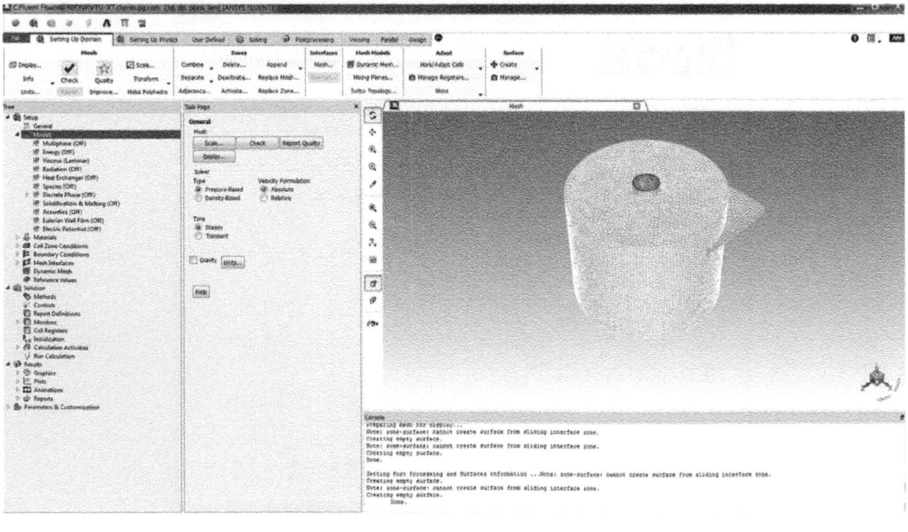

FIGURE 8.40 Window in *ANSYS Fluent®*.

models for discrete phase (momentum modelling or heat & mass transfer) or the properties of the materials. The User Defined Functions need to be coded externally and then from Functions → Compile are implemented in the ANSYS Fluent® code. To use them, in the functions there are hooks that need to be defined in the code text file, and allows to connect with the property/model specified. Then, in the adequate panel that is connected with the hook the user can select the function. For example, in viscous models, if the user wants to specify the turbulent Prandtl according to the properties, the model can be coded in a text file defining also the adequate hook and then in the viscous panel will be selected. The use of "User Defined Functions" is an advance part in the area of CFD modelling so it is not further described in this chapter. There are courses provided by ANSYS® with a focus on the development of "User Defined Functions" as well as more information and manuals can be found online.

- The solving toolbar provides direct access to the solving tools also available in the solution tree.
- The post-processing toolbar is available in versions 14 and later and it helps to visualize the results obtained from the simulation directly on ANSYS®. If more precessive analysis desire to be performed, ANSYS® has its own post-processing tool known as ANSYS CFD-Post®.
- The Viewing toolbar contains all the display options for the visualization window.
- The parallel panel helps to modify the resources applied to the simulation as well as how the partitions are performed in the computation.
- The Design toolbar can be used if design and simulation is performed simultaneously, being the results obtained from the simulation applied to the design in order to find an optimal.

- Visualization window: In this window the user can check the display of the geometry as well as the outputs from the simulation as the residuals or the results.
- Selection tree. The selection tree contains the key tools in order to perform a simulation. In order to avoid mistakes, it is recommended to go from the top to the bottom in the selection of the different options explained in the following paragraphs.
  - General setup contains the general options of the system in terms of display, mesh and solver specification for the system.
  - Models section includes models classified in different areas according to the physics. The areas available in ANSYS 18.0® are:
    - Multiphase that allows the simulation with volume of fluid methods, populations balances or dense discrete phase models.
    - Energy, activates the energy balance in the model.
    - Viscous model, that allows selecting/implementing the turbulent model required for the system that is being solved.
    - Radiation activates the energy transfer by radiation.
    - Heat exchanger is useful to define the bodies between the exchanges of energy happen and according to a defined model.

  - The species modelling bar is applied for the definition of mass transfer between phases or to define chemical reactions.
  - Discrete phase allows the introduction of discrete phases when they are in low concentrations, volume fraction lower than 10^{-3}.
  - Solidification & Melting
  - Acoustics
  - Eulerian Wall Film
  - Electric Potential
- Materials tab is composed of the phases that are included in our model such as solid, fluid, discrete phases or mixtures so as to include its properties.
- Cell zone is useful to operate with each of the bodies previously defined in the geometry, having the chance to modify the physics in each body once the adequate models have been previously selected.
- Boundary conditions tab is used to define the boundary conditions applied to solve the problem and are usually connected with the surfaces that compose the bodies of the system. For the simulation of fluid systems, the common boundaries are the inlets, outlets, walls and internal zones in case that the whole geometry is composed of different bodies.
- Mesh interfaces shows the connection between two surfaces and are referred to the connections previously defined in ANSYS Meshing®.
- Dynamic mesh is useful for the modification of mesh since the results generated are used to improve the mesh, helping to define and optimize the mesh in terms of accuracy and computational time with the use of the results provided.
- Reference values shows the reference values used for the computation as well as the surfaces used to reference our system from which the computation starts. If it is not defined, it will be automatically selected when the computation starts.
- Solution controls contains the solution methods and controls used to specify as well as the reports and activities desired to be performed through the simulation.
- Results contains some of the post-processing tools such as plots, reports or animations.

- Task page shows the possible options and settings to be activated or modified when each of the tabs from the Tree are selected.
- The Console shows all the activities ran in ANSYS Fluent®. Moreover, it also contains a Text User Interface (TUI) which can be used to code the settings available in the Graphic User Interface (GUI) such as mesh options, models defined, display or boundary conditions.

Once the possible options have been briefly defined, in the following paragraphs we continue with our example. The simulation of the cyclone is performed in two steps: first, the air is solved until the residuals are stable, ensuring that the solution obtained is stable; second, the particles are introduced over the previous air solution. The first action required to start the simulation is to check that the mesh has been read correctly. It can be done from the check and report quality options in the general tab.

Furthermore, in this general tab is necessary to specify the solver options that for our case are Pressure-Based, since we are solving an internal flow in a geometry, and absolute velocity formulation. The system is solved as steady considering that the operation simulated is the steady-state operation of the cyclone and no further resolution is required for the prediction of the transitional behaviors in the swirl generated in the cyclone. If higher accuracy is desired, the user can perform the simulation in transient; however, the computational cost required is larger and here only a brief example is described.

The next step for the development of the simulation is the specification of the model. In this first step we only simulate the flow. Thus, the turbulent model is the first to be selected. In the Tree, opening Models → Viscous. The user must select the Reynolds Stress Model (RSM) as in Figure 8.41. This model is selected for swirling flows since they are usually anisotropic and the RSM model is preferred versus the

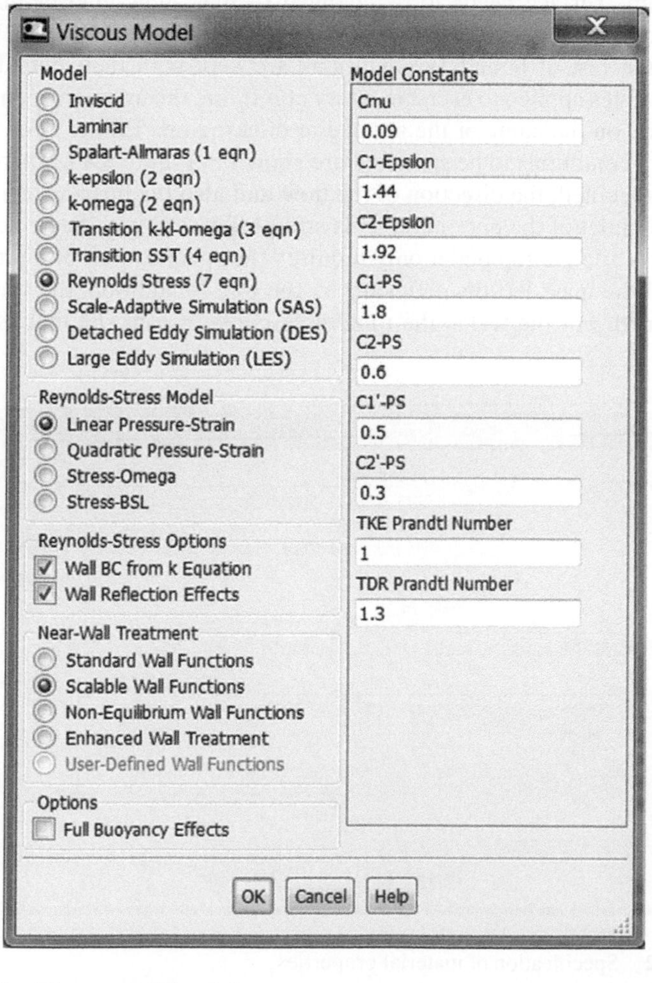

FIGURE 8.41 Viscous model panel.

k-ε ones [16]. Furthermore, it is also important to select the scalable wall function due to it ensures stable results because it avoids the influence of the diffusion in the turbulence balance, assuming a minimum y+ of 11 [17]. The model constants and the other options are kept as the default ones.

For the simulation of the continuous phase there are not more models required to be implemented.

In the materials panel, the properties of the air, such as viscosity and density, need to be in agreement for an exact temperature. For a temperature of 25°C, the density of the air is 1.25 kg/m^3 and its viscosity is $1.78{\cdot}10^{-5}$ Pa.s (see Figure 8.42).

The following step is the definition of the boundary conditions. Once the boundary conditions have been selected from the Tree, in the task page the user can modify the type of boundary conditions applied to each surface as it is showed in Figure 8.43. The inlet is changed from velocity-inlet to mass-flow-inlet. In the paper [14] the mass flow of operation is specified as 0.075 kg/s. Furthermore, the "outlet.bottom" surface needs to be changed to mass-flow-inlet defining a zero flow since zero cannot be set as outlet. The "outlet.top" is also changed to outflow, meanwhile the rest of boundary conditions are kept with their default type. To specify the values applied to each boundary condition, the user can open them with a double-click on the name of the surface or clicking on "Edit…". For the "inlet" the boundary conditions to be specified are shown in Figure 8.43, where the mass flow rate is specified, the direction of the flow and also the turbulence intensity in the inlet. The inlet of the bottom outlet is straightforward specified since its value is zero meanwhile the top outlet only requires the specification of a rate which is one since all the mass left the geometry by the top. In addition, the walls need to be also defined. For the walls, the roughness contains a model that assumes the

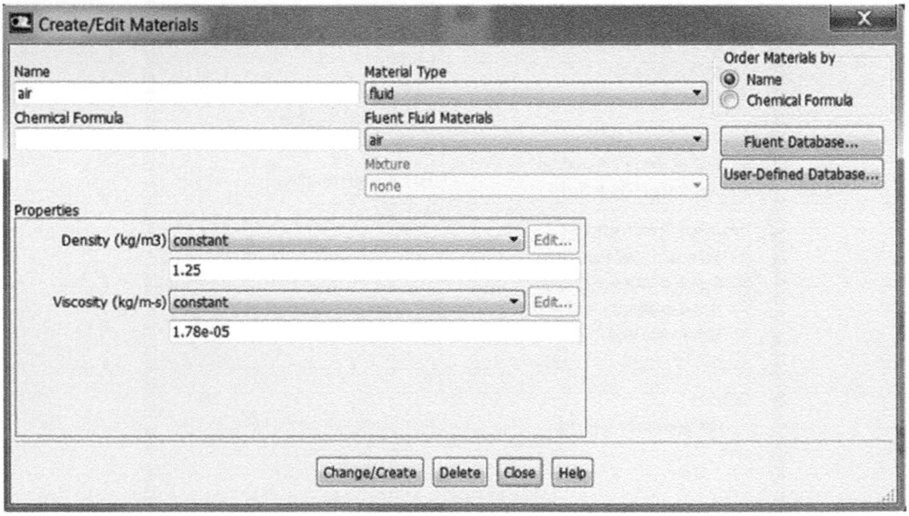

FIGURE 8.42 Specification of material properties.

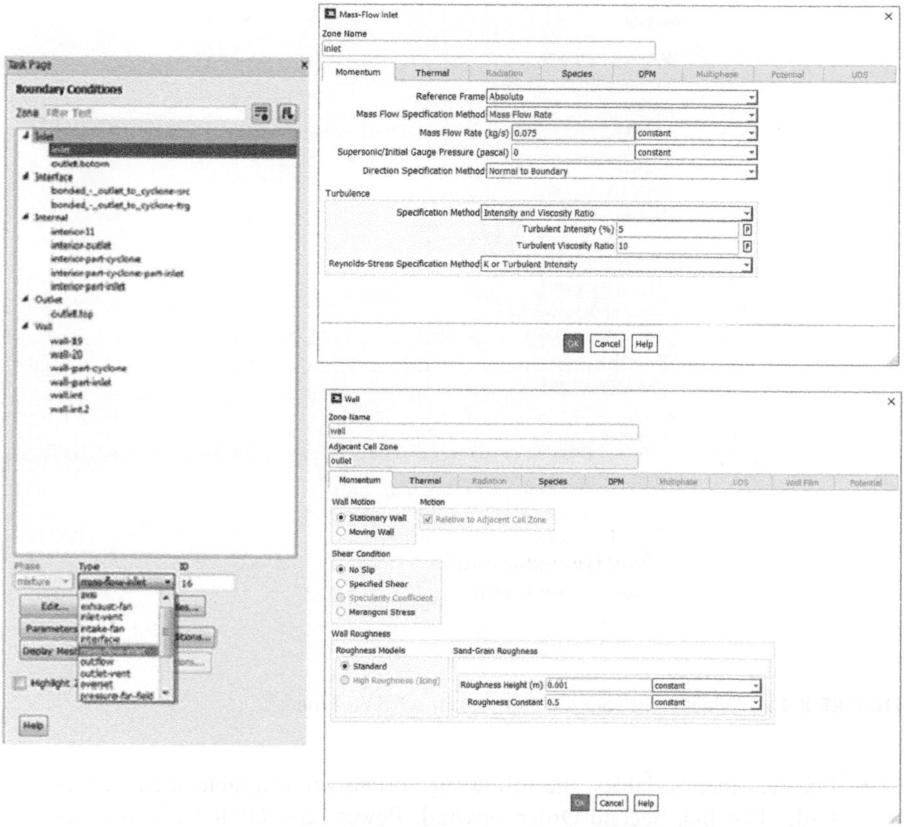

FIGURE 8.43 Boundary conditions specification (left), mass flow rate definition (top-right) and wall boundary conditions definition for momentum (bottom-right).

surface as a composition of grains. We can assume a value for the roughness of 1mm as it is showed in Figure 8.43.

In the last activity of this first step, the solution panel needs to be described. First, going to the Methods section, Figure 8.44. The method applied to the simulation is SIMPLE since is faster than PISO due to it has lower number of correction factors for the discretization in the equations that need to be re-computed. Furthermore, it is required to define the spatial discretization methods:

- The gradient defines how the value is estimated in each cell. It can be estimated according to the geometrical centre of the cell (Least Squares Cell Based), the average between two nodes (Green-Gauss Node Based) or the average between two cells (Green-Gauss Cell Based).
- The Pressure discretization is set to PRESTO! since the flow simulated is a swirl and contains strong body forces. For other types of flows, the other discretization methods can be applied.

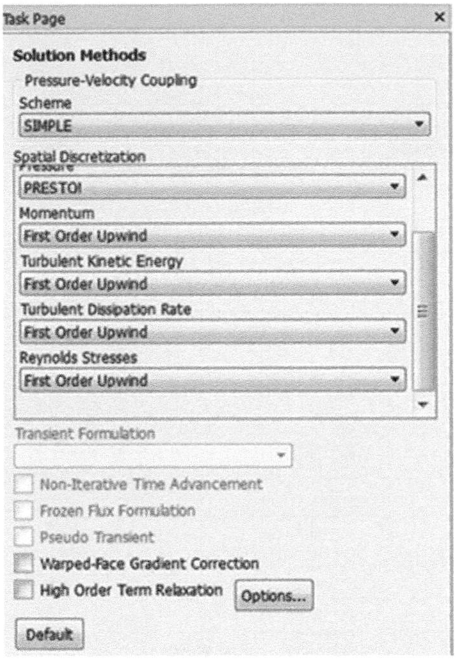

FIGURE 8.44 Selection of solution methods in ANSYS Fluent®.

- For the other variables, the following options are available such as First-Order Upwind, Second Order Upwind, Power Law, QUICK, Third order MUSCL

  Since we want to improve the speed of the computation, the simulation can be performed with First-Upwind solver because it only uses one order of discretization. A very useful solver when tetra and hexahedral meshes are combined is QUICK because it uses first discretization order for hexahedral cells, which are more regular than tetrahedral ones, and second order for tetrahedral. QUICK solver is a good approach between speed and accuracy.

The next part to specify in a simulation are the Solution Controls. Here the relaxation factors are specified which will be taken for the relaxation between iterations in the solution computation. These relaxation factors are recommended to be the default ones. Only if the simulation has problems of convergence can be reduced. The relaxation factors for the actual simulation are showed in Figure 8.45.

In order to check the convergence and stability in the solution, it is recommended to use some physical variable that needs to be computed and report its stability. From the Tree, the user can create the plot going to Monitors → Report Plots and a new panel is opened as in Figure 8.46. The user can create a new report by clicking on "New..." A new window appears as "New Report Plot" where the reporting interval can be selected from the left and the type of report is selected from the left. The

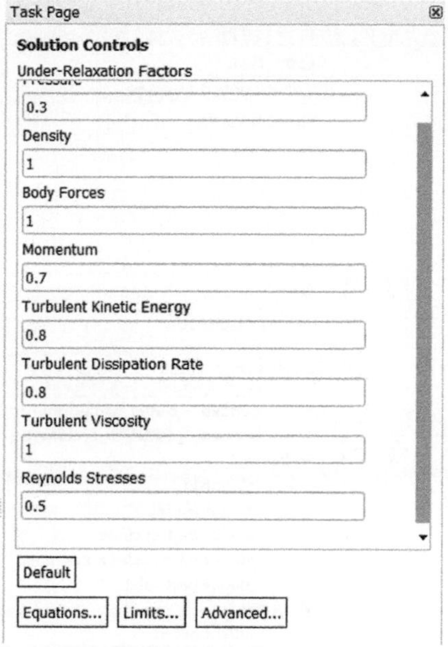

FIGURE 8.45 Selection of relaxation factors in ANSYS Fluent®.

report selected is Area-Weight average since we are going to check the convergence of our solution analyzing the area weight average velocity in the outlet. Then, the velocity and the outlet top surfaces are selected as in Figure 8.47.

At this stage, we have specified all the requirements and tools to run our simulation and the final steps are the initialization and computation. Initialization can

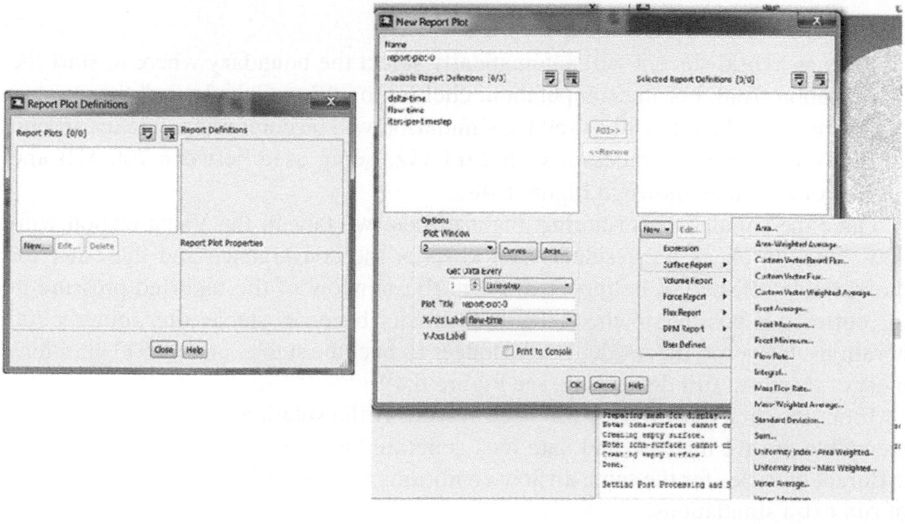

FIGURE 8.46 Report plot generation.

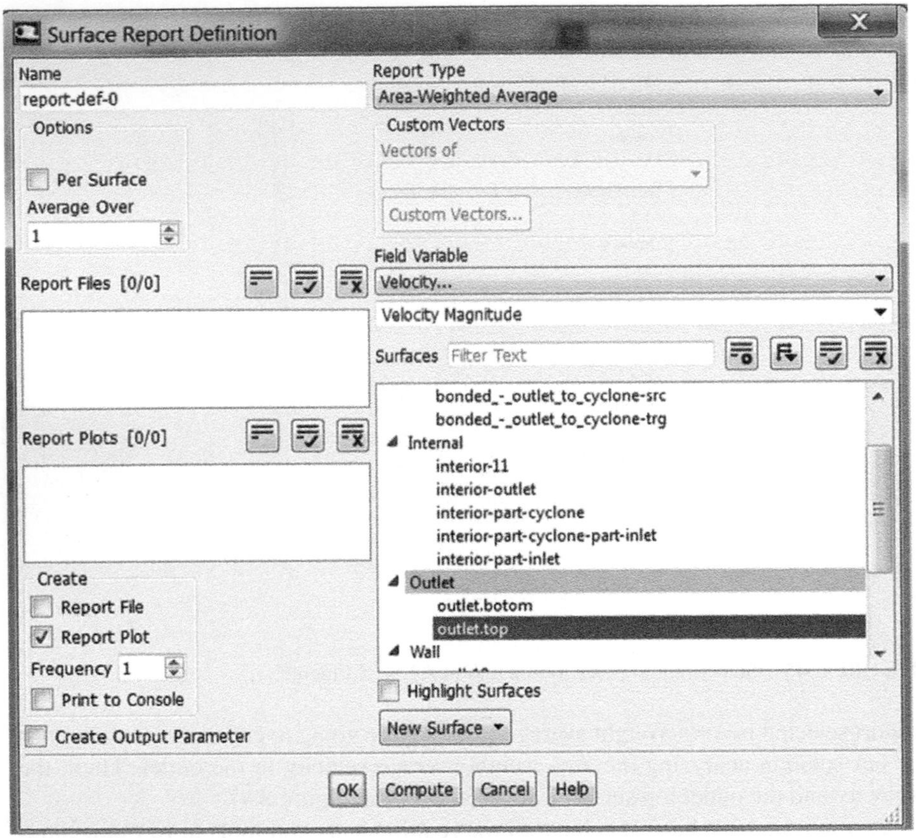

FIGURE 8.47 Selection of surfaces of monitors to track the variables.

be done as hybrid since it will automatically select the boundary where to start the computation from. For the computation, clicking on "Run Calculation" the number of Iterations can be set to 4000 and the simulation will be computing. It takes around 20 min on an Intel i7 processor with 2.6 GHz, being used between 700 MB and 1.2 GB of RAM, as shown in Figure 8.48.

Once the simulation is running, the user has two tabs in the Visualization window. Once to check the residuals that ensures the convergence and the other for the variable selected to be reported. From the window of the reported pressure at the outlet, it is possible to check that the velocity becomes stable after roughly 700 iterations. However, the residuals take longer to become stable, until 2,500 iterations most of them are still decreasing, see Figure 8.49.

From this first part, the air has been solved. If the solution is stable, it is recommendable to save the case and data files generated because it is possible to explore different particles for the same airflow conditions in a second stage without the need of run extra simulations.

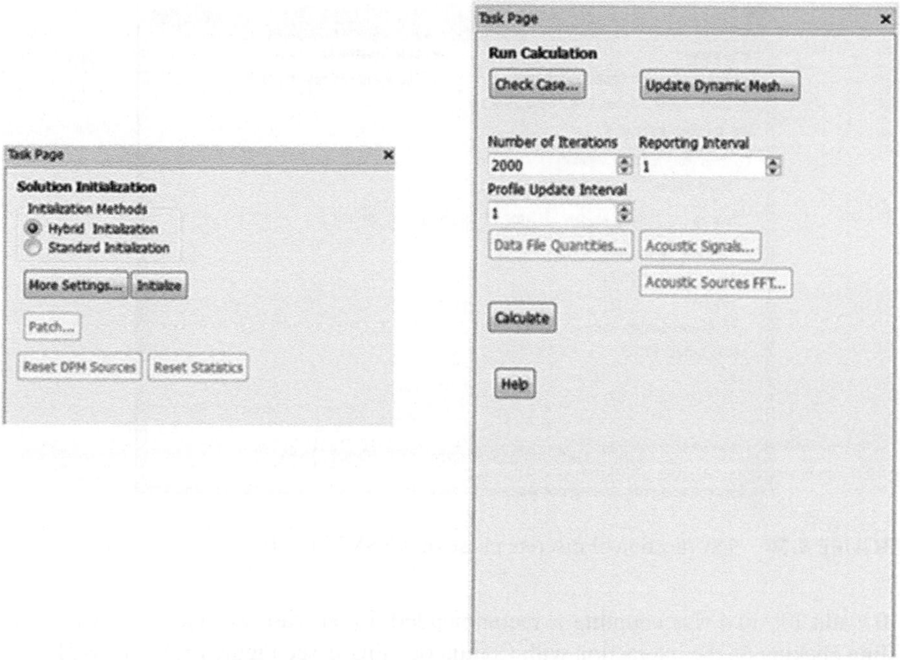

FIGURE 8.48 Solution initialization (left) and running (right).

For the particle simulation, the first thing required is the modification of the models in order to include the particles. The model used for cyclones depend on the volume fraction of the particles. For very disperse systems with a volume fraction lower than 10^{-6} only one-way coupling is recommendable, meanwhile for volume fractions between

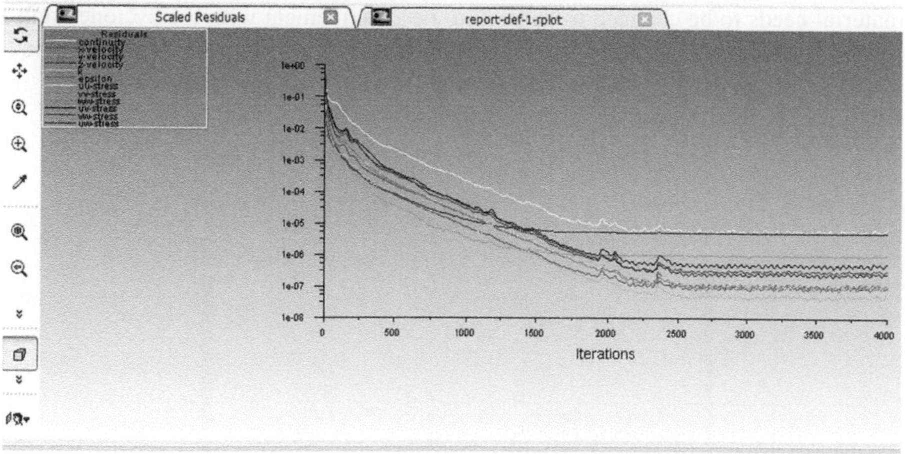

FIGURE 8.49 Residuals for the cyclone simulation when the continuous phase is run. (See color insert.)

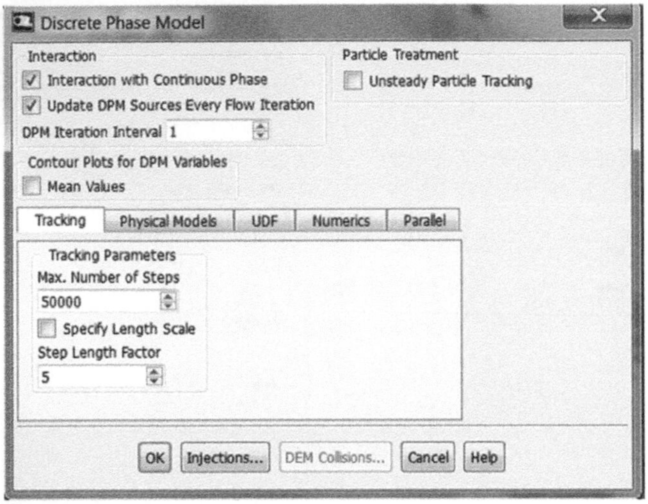

FIGURE 8.50 Specification of discrete phase in ANSYS Fluent®.

10^{-6} and 10^{-3} two way coupling is recommended. In our case we select two way coupling clicking on the interaction with Continuous Phase, see Figures 8.50 and 8.51.

The injection of particles consists of 4 injections since one injection only cannot represent the particle size distribution (PSD) from the paper [14]. One injection follows a Rosin-Ramler distribution and the other three ones are defined as uniform. The injections are created following Figure 8.52. Once the injection is created, it can be appreciated on the Tree and a new panel appear as Figure 8.53. The particles are injected from the inlet so "Surface" is selected as Injection Type and "inlet" selected. The particle type is set as inert and the material is kept as anthracite. However, the material needs to be changed to bagasse to be in agreement with the cyclone application. To change it, the user can go to: Material → Inert Particle → Anthracite, and the user can change the name to bagasse and the density to 120 kg/m^3 [18].

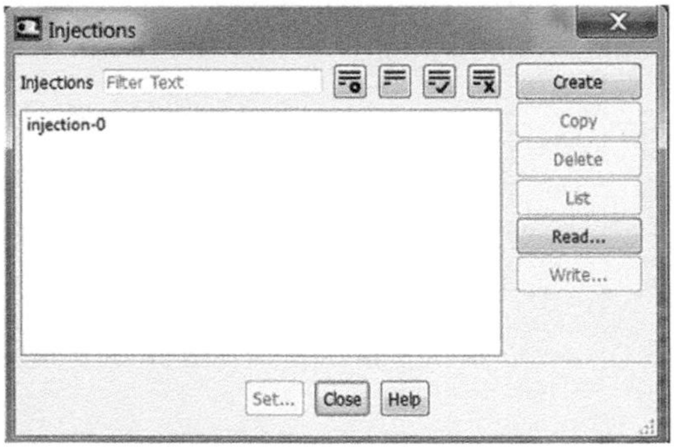

FIGURE 8.51 Generation of injections in ANSYS Fluent®.

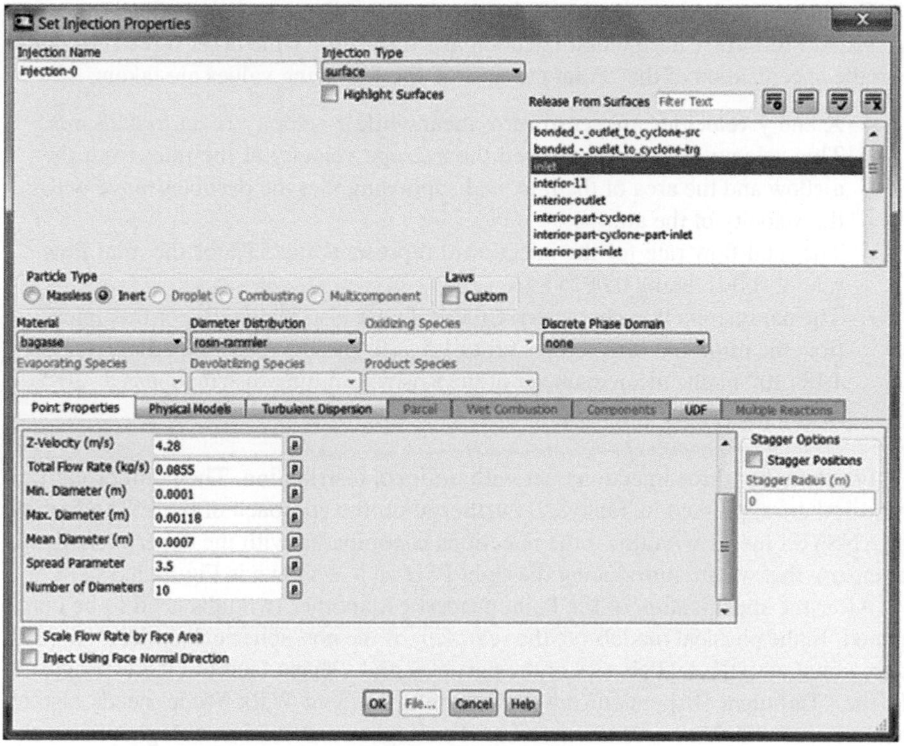

FIGURE 8.52 Specification of the injection in ANSYS Fluent®.

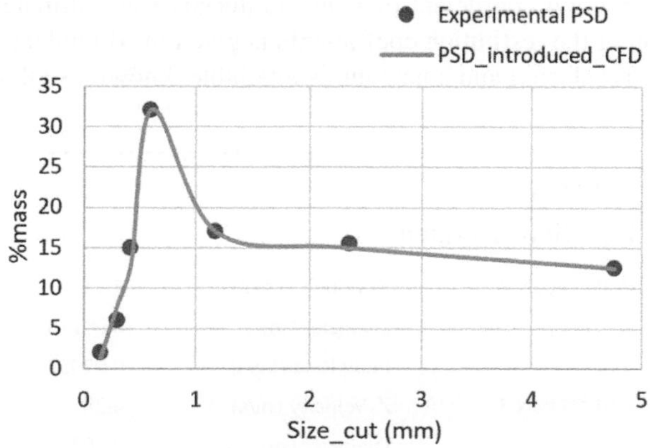

FIGURE 8.53 Comparison between the experimental PSD [1] and the PSD introduced with the four injections in ANSYS Fluent®.

Going back to the specification of the injection, the diameter distribution is changed to a Rosin-Rammler for the first injection and uniform for the other three injections. For the specification of the "Point properties" the following values are taken:

- X and y velocities are set to zero; meanwhile z-velocity is set to 4.28 m/s. This velocity is computed taken the average velocity at the inlet from the airflow and the area of the inlet and supposing that the particles move with the velocity of the air.
- The total flow rate for the injection-0 represents the 57% of the total flow rate of solids, being 0.0825 kg/s.
- The parameters that characterize the PSD are also included. For this injection, the minimum diameter is set to $1.5 \cdot 10^{-4}$ m, the maximum diameter is $1.18 \cdot 10^{-3}$ m, the mean diameter of the Rosin-Rammler distribution is $7 \cdot 10^{-4}$ m and the spread parameter is 3.5.

For the other three injections, set with uniform distribution. The point properties required are expressed in Table 8.2. Furthermore, the approach of the PSD adaption to ANSYS Fluent® with this three injections is compared with the experimental one to ensure that we are introducing the right PSD as it is shown is Figure 8.33.

After the specification of the Point properties, another two tabs need to be completed. In the physical models tab the selection of the non-spherical model is more in agreement with the real physics of the particles and a shape factor of 0.8 is provided. In the "Turbulent Dispersion" tab the Discrete Random Walk Model needs also to be activated and three tries are specified. For larger number of tries, higher accuracy is expected due to it is a stochastic model. However large computational time is required. For this example, since the speed is preferred, the number of tries is low.

In the next step, the restitution coefficients need to be specified for the particles in order that the particles will rebound when they hit the wall and will not be trapped. The restitution coefficient of wood materials was estimated as 0.6 [19]. The specification of the restitution coefficients is given in: Boundary conditions → Walls. One wall is selected and a new tab is available, known as DPM and showed

TABLE 8.2
Injections Defined

INJECTION 2	Z-Velocity (m/s)	4.28
	Diameter (m)	0.00118
	Flow Rate (kg/s)	0.020229
INJECTION 3	Z-Velocity (m/s)	4.28
	Diameter (m)	0.00236
	Flow rate (kg/s)	0.0225
INJECTION 4	Z-Velocity (m/s)	4.28
	Diameter (m)	0.00475
	Flow rate (kg/s)	0.01875

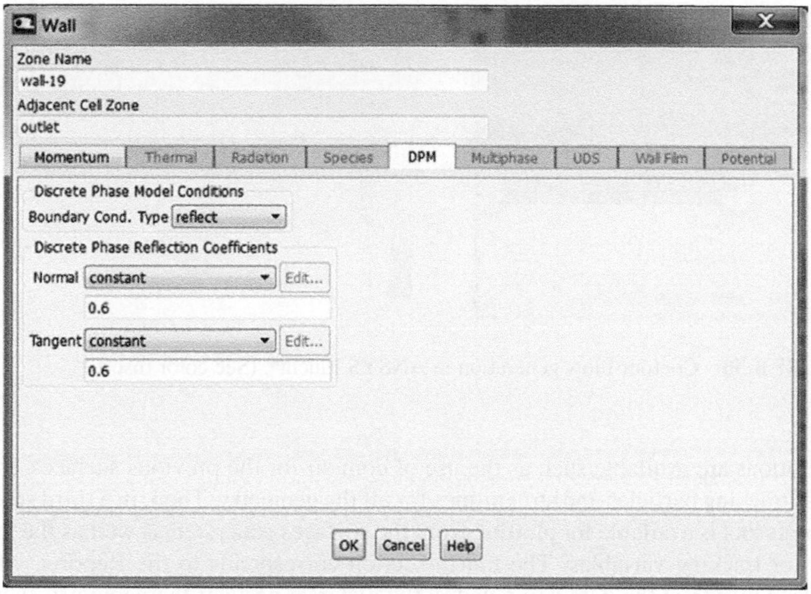

FIGURE 8.54 DPM boundary conditions in the wall.

in Figure 8.54. The walls are set with a boundary condition of reflect and the normal and tangential coefficients are set to constant. It is assumed that the restitution coefficient is not affected by the impact angle or particle properties, which is not true but simplify the definition for this problem. Once the restitution coefficients have been defined in one wall, it has to be copied to the other walls from the Task Page panel.

Now the user is ready to run the simulation. The number of iterations is set to 500, taking roughly 15 minutes to be completed. The residuals increase up to 1e-1 and then decrease until reach an equilibrium around 3e-2.

8.5.5 Postprocessing

Once we have completed the simulation, the last thing is to read and extract the results. There are two options for post-processing: do it from ANSYS Fluent® or open ANSYS CFD-Post®. In this chapter we will provide a brief description of both. First, starting with the post-processing tools available in ANSYS Fluent® (Figure 8.55), the first option to the left is the generation of lines, planes or other geometries for the variables that want to be analysed. In a second section, the options for the graphics

FIGURE 8.55 Postprocessing tools available in ANSYS Fluent®.

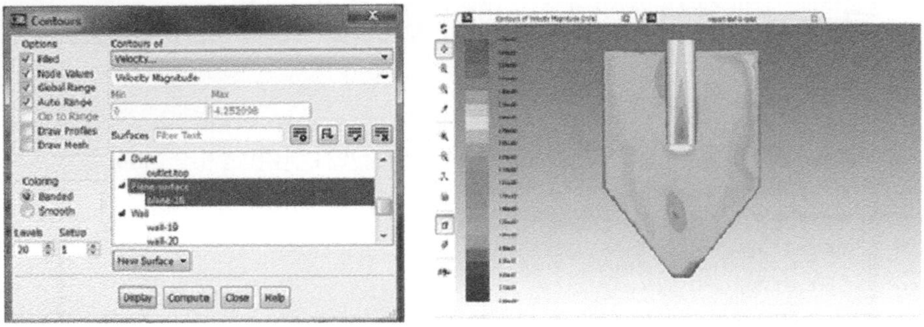

FIGURE 8.56 Contour Plots generation in ANSYS Fluent®. (See color insert.)

generations are available such as the use of contour for the previous surfaces or the use of tracking particles and streamlines for all the geometry. Then, in a third section the plots tool is available for plotting from the surfaces generated as well as the monitors for tracking variables. The fourth section corresponds to the Reports, which are used to extract the data computed in the different parts of the geometry such as the boundary conditions or interfaces. Finally, the "Animation" tool and the "Model Specific", which is used for the extraction of the particle variables, are available.

In the following paragraphs four examples are illustrated. The first one is the illustration of the velocities in a plane which can be performed generating a plane with: Surface → Create → Plane; and specifying the default locations the geometry is cut in a half. Once the plane has been created, it is available clicking on Graphics → Contours. In the new window, if the velocity is selected as in Figure 8.56 (left), the figure of the velocity profiles are obtained for that surface as in Figure 8.56 (right).

The second example is the illustration of the path followed by the particles. From Graphics → Particle Tracks, a new window is opened as in Figure 8.57. In the window the color of the particles plotted is set by Particle Diameter allowing us to check that the small particles are not recovered and the large ones are recovered in the

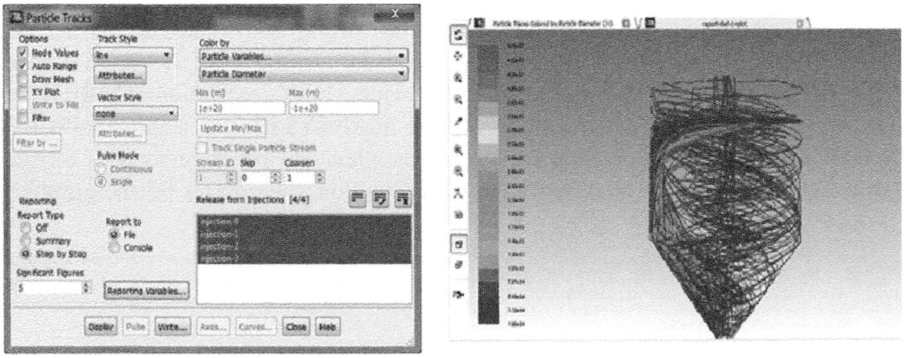

FIGURE 8.57 Particle tracks by diameter in ANSYS Fluent®. (See color insert.)

bottom. In addition, all the Injections have been selected to be reported. Finally, the Reporting options have been also modified selecting "Step by Step" in order to provide the particles tracks at different times (it provides the particles variables selected at different times) and "Report to a File" that will be created. Clicking on "Display", a .his file is saved in the folder desired and in the Display Window the particles tracks are showed.

The third example to be described here is the generation of the plots. In particular, the plot of velocity profiles. First of all, the generation of a line where velocity wants to be measured is necessary. The line is generated with Surface → Create → Line/Rake and specifying the points desired as it is showed in Figure 8.58. Then, going to Plots → XY Plot → Edit, the user can select Y Velocity to be plotted, the Curve Length for the X Axis and the line-17 needs to be selected. If "Write to File" is selected, the author can also export the data. Clicking on "Plot" in the Display Window, the velocity profile is plotted.

The last analysis is the extraction of the data with the use of Report. In this case the report is applied to compute the mass recovery in the bottom. The report can be selected in the Tree as: Report → Discrete Phase → Sample. Alternatively, it can be selected from the Post-processing bar as Model Specific → Discrete Phase → Sample. The report has the option to Append Files, which generates a text file whit the particle variables and can be used later for further computation of other data.

If we move now to ANSYS CFDPost®, the tool can be open from ANSYS Workbench® as ANSYS Fluent® or ANSYS DesignModeler®, selecting Results from the Toolbox tree. Once the tool is opened, the following window (Figure 8.58) appears.

The available tools in the top of the window are structured similarly than the previous tools explained. The main applications for the post-processing are located in "Insert", where the user can generate similar applications as in ANSYS Fluent® except the generation of Reports. However, there are other tools available as Iso-surfaces and Vortex core surfaces generation, the table reporting or the implementation of new variables to be reported by the use of formulas. These tools are presented in Figure 8.59.

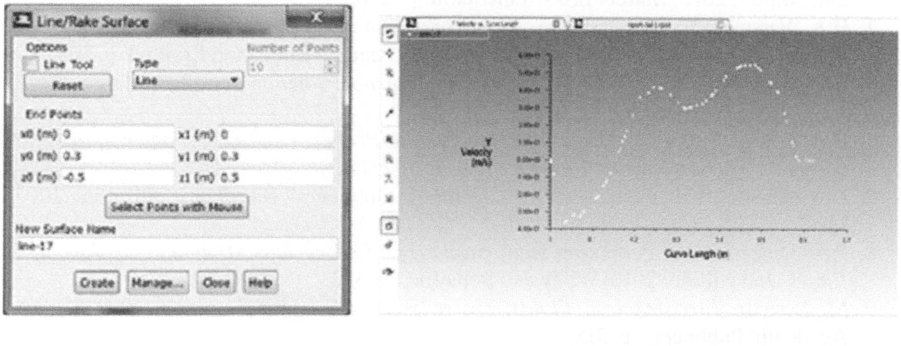

FIGURE 8.58 Generation of line in surfaces and plot of the Y-velocity.

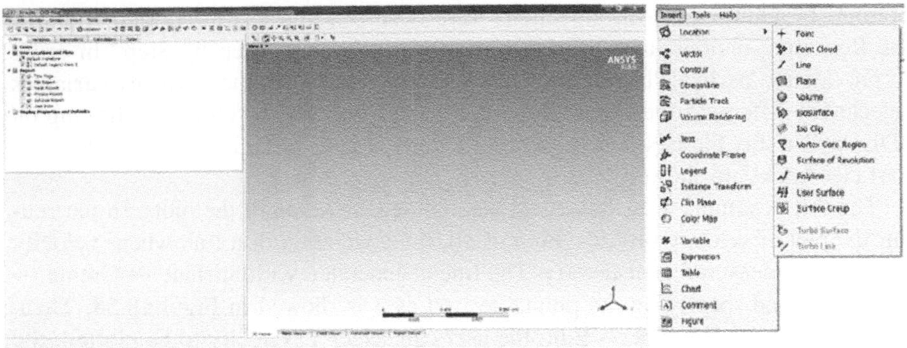

FIGURE 8.59 ANSYS CFD-Post® window (a) and post-processing tools available (b).

8.6 APPLICATIONS

Computational fluid dynamics and multiphysics modelling and simulation can be applied to many science and engineering disciplines. The main areas in chemical engineering are combustion processes, food process engineering, batteries, fuel cells and supercapacitors, microfluidic flows and devices, pipe flows and mixing, and reaction engineering.

Other engineering fields where computational fluid dynamics and multiphysics software tools are widely utilised are aerospace and defence engineering, automotive engineering, biological engineering, civil and environmental engineering, consumer goods engineering, electrical engineering and electronics, energy and power engineering, health care engineering, materials, and process engineering.

REFERENCES

1. J.D. Anderson, *Computational Fluid Dynamics: The Basics with Applications*, 1995, McGraw-Hill, p. 547.
2. *Hypersonic and High-Temperature Gas Dynamics*, 2nd edition, 2006, American Institute of Aeronautics and Astronautics, p. 811.
3. J. Tu, G.H. Yeoh, C. Liu, *Computational Fluid Dynamics: A Practical Approach*, 2nd edition, 2012, Butterworth-Heinemann, p. 250.
4. H.K. Versteeg, W. Malalasekera, *An Introduction to Computational Fluid Dynamics: The Finite Volume Method*, 2nd edition, 2007, Pearson Education Limited, p. 503.
5. R.B. Bird, W.E. Stewart, E.N. Lightfoot, *Transport Phenomena*, 2nd edition, 2002, John Wiley and Sons Inc., p. 895.
6. B.E. Poling, J.M. Prausnitz, J. O'Connell, *The Properties of Gases and Liquids*, 5th edition, 2001, McGraw-Hill Professional, p. 752.
7. D.W. Green, R.H. Perry, *Perry's Chemical Engineers' Handbook*, 8th edn, 2007, McGraw-Hill Professional, p. 2400.
8. P.A. Durbin, B.A. Pettersson Reif, *Statistical Theory and Modeling for Turbulent Flows*, 2nd edition, 2010, Wiley and Sons Inc., p. 372.
9. M. Lesieur, *Turbulence in Fluids*, 3rd revised and enlarged edition, 1997, Kluwer Academic Publishers, p. 515.
10. D.C. Wilcox, *Turbulence Modeling for CFD*, 2nd edition, 1998 DCW Industries.

11. R.J. LeVeque, *Finite Difference Methods for Ordinary and Partial Differential Equations: Steady-State and Time-Dependent Problems*, 2007, SIAM, p. 341.
12. O.C. Zienkiewicz, R.L. Taylor, J.Z. Zhu, *The Finite Element Method: Its Basis and Fundamentals*, 6th edition, 2005, Elsevier Butterworth-Heinemann, p. 734.
13. W. Frei, *Solutions to Linear Systems of Equations: Direct and Iterative Solvers*. Available at: https://www.comsol.com/blogs/solutions-linear-systems-equations-direct-iterative-solvers/
14. L. Fernandes de Oliveira, J.L. Gomes Correa, P. Gaspar Tosato, S. Vilela Borges, J. Guilherme, L.F. Alves, B. Elyezer Fonseca, Sugarcane Bagasse Drying in a Cyclone: Influence of Device Geometry and Operational Parameters. *Drying Technology*, 2011, 29: 946-952.
15. B. Andersson, R. Andersson, L. Hakansson, R. Sudiyo, B. van Wachem, *Computational Fluid Dynamics for Engineers*, 2012, Cambridge, United Kingdom.
16. K. Elsayed, K. Lacor, Numerical Modeling of the Flow Field and Performance in Cyclones of Different Cone-tip Diameters, *Computers and Fluids*, 2011, 51: 48-59.
17. ANSYS. 2018. ANSYS Help. Available under customer registration at: https://ansyshelp.ansys.com/
18. Z. Bubnik, P. Kadlec, D. Urban, M. Bruhns, *Sugar Technologists Manual*, 8th edition, 1995 Germany, Bartens.
19. J. Bennett, R. Meepagala, *Coefficients of Restitution*, 2005. Available at: https://hypertextbook.com/facts/2006/restitution.shtml

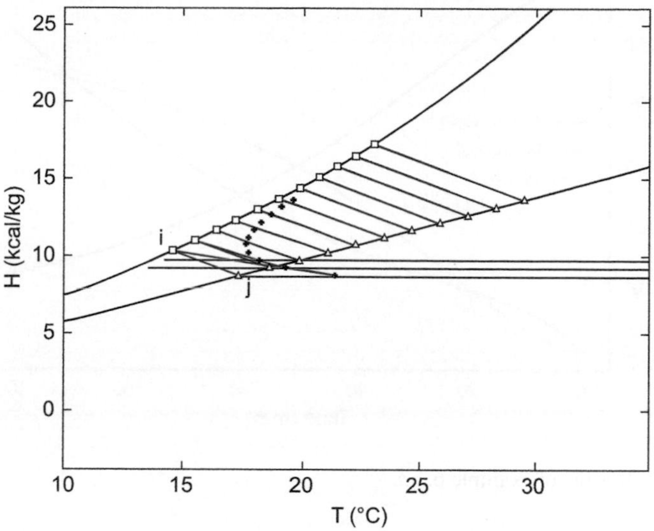

FIGURE 3.23 Enthalpy - temperature diagram.

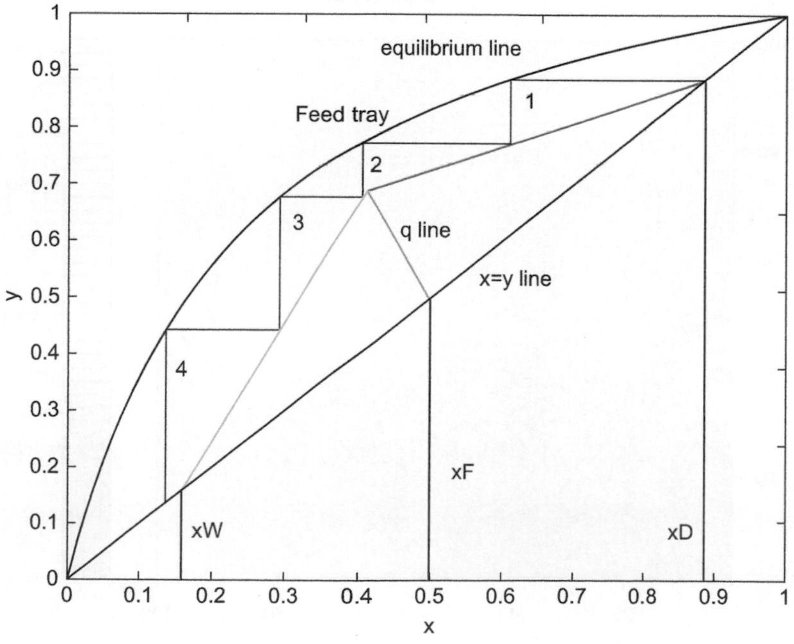

FIGURE 4.11 McCabe Thiele Method for Binary Distillations.

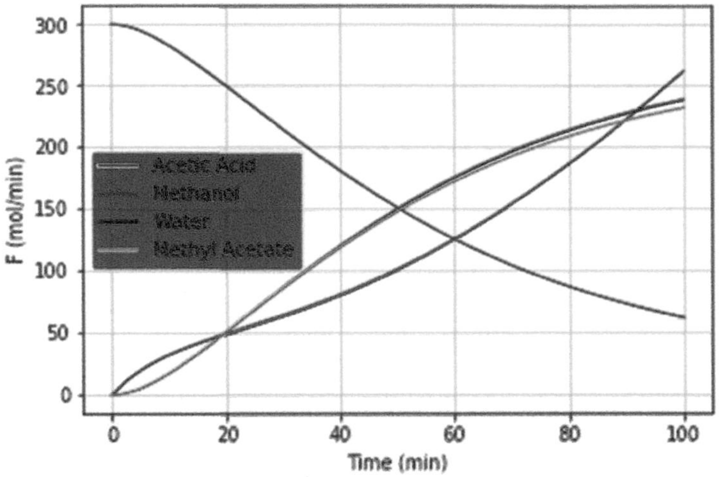

FIGURE 6.4 Results of example 6.3.3.

```
> windows(8,7)
> filled.contour(x,y,z,nlevels=30,xlab = 'x', ylab = 'y',color
= terrain.colors)
> title('Z contour')
```

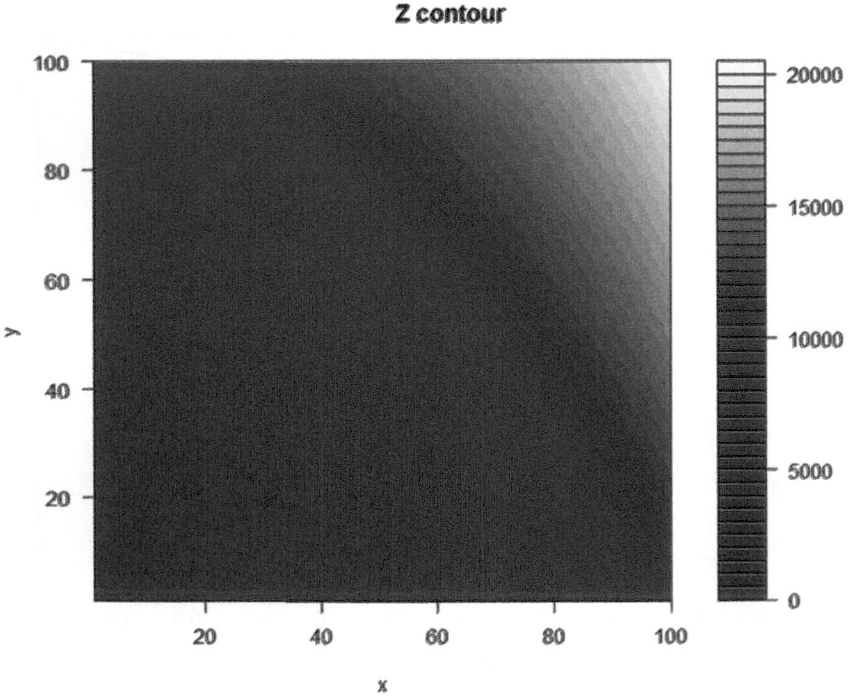

FIGURE 7.8 Filled contour plot.

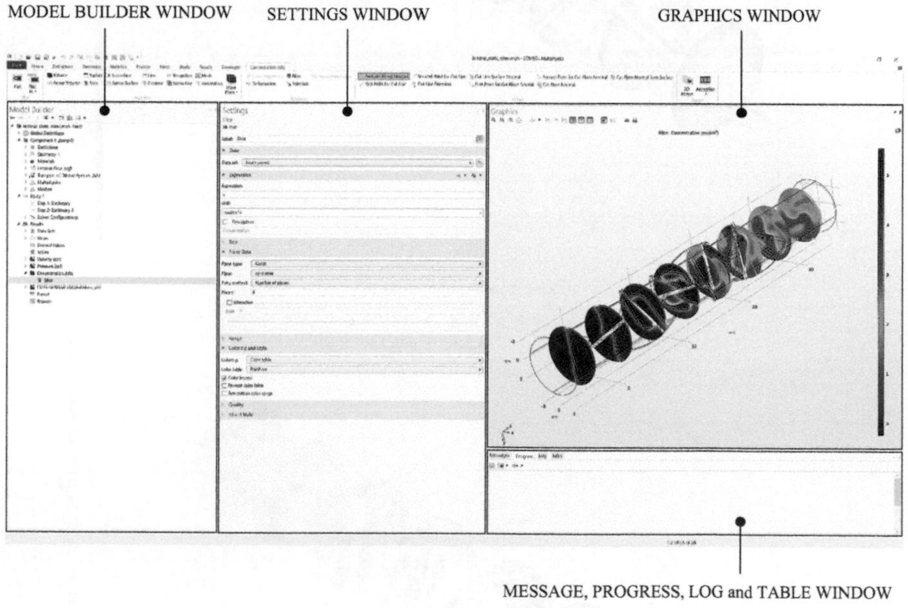

FIGURE 8.7 COMSOL Multiphysics® graphical user interface (image made using COMSOL Multiphysics® and is provided courtesy of COMSOL®)

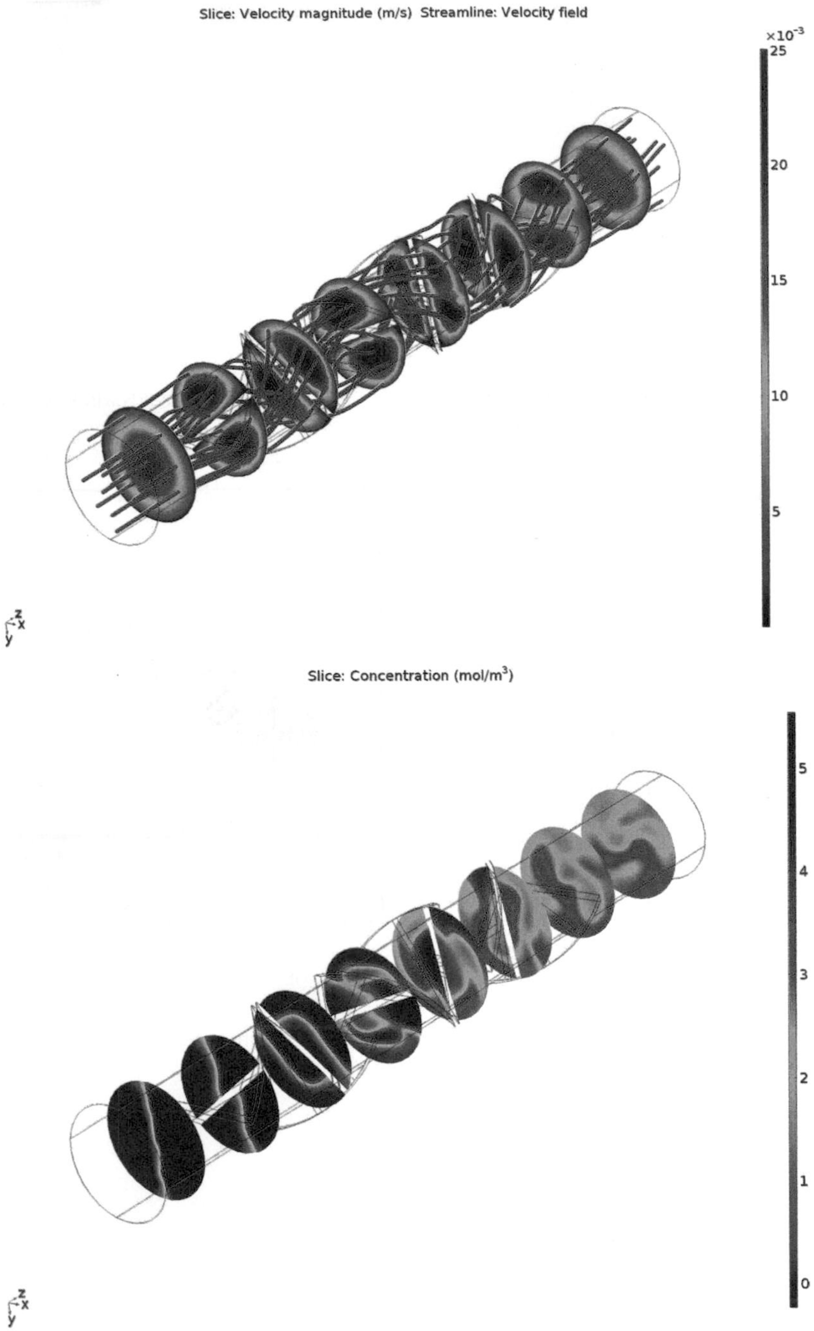

FIGURE 8.16 Velocity magnitude, velocity field and concentration of species A inside the laminar static mixer (image made using COMSOL Multiphysics® and is provided courtesy of COMSOL®).

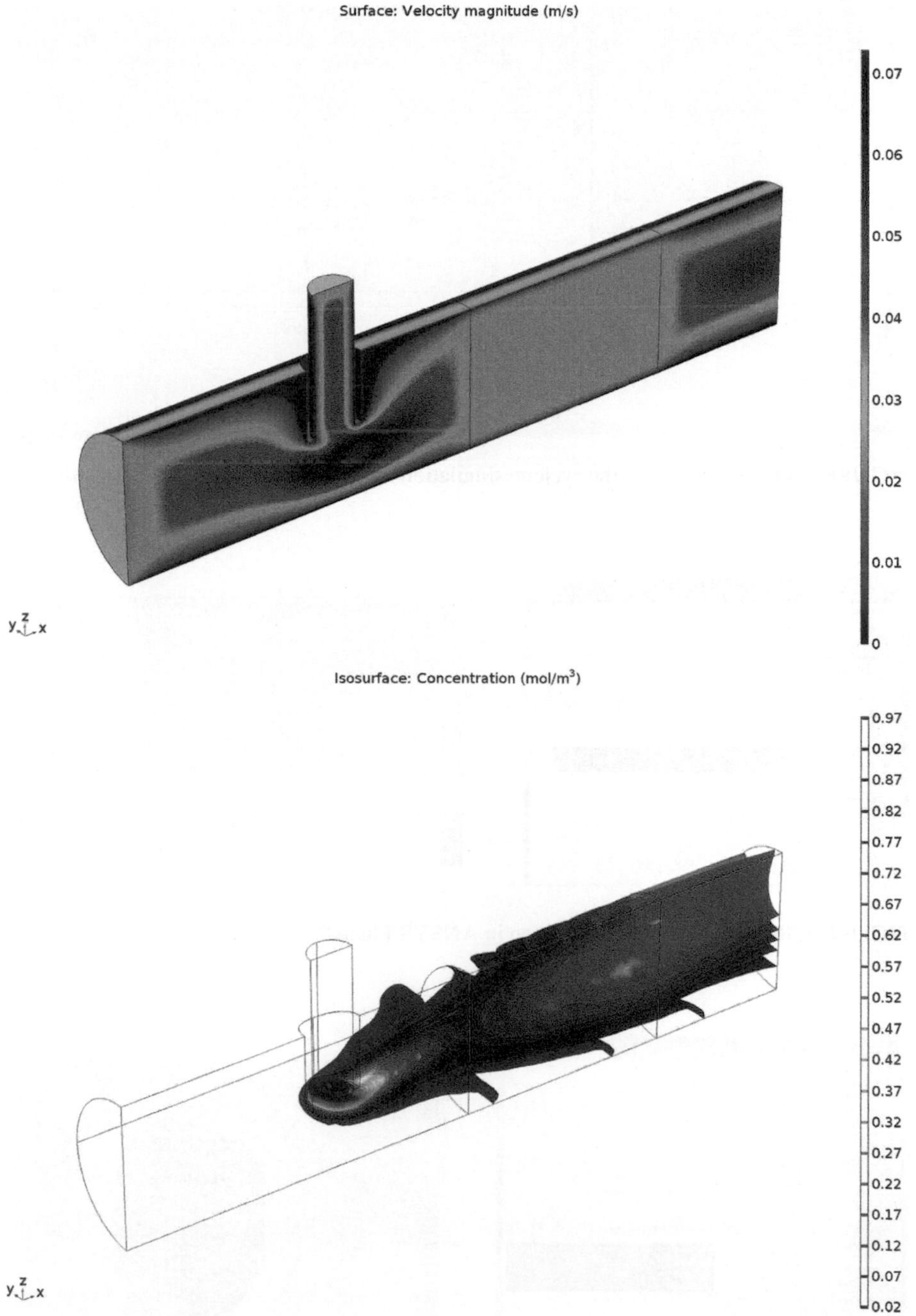

FIGURE 8.20 Velocity magnitude and concentration of species B inside the porous reactor with injection needle (image made using COMSOL Multiphysics® and is provided courtesy of COMSOL®).

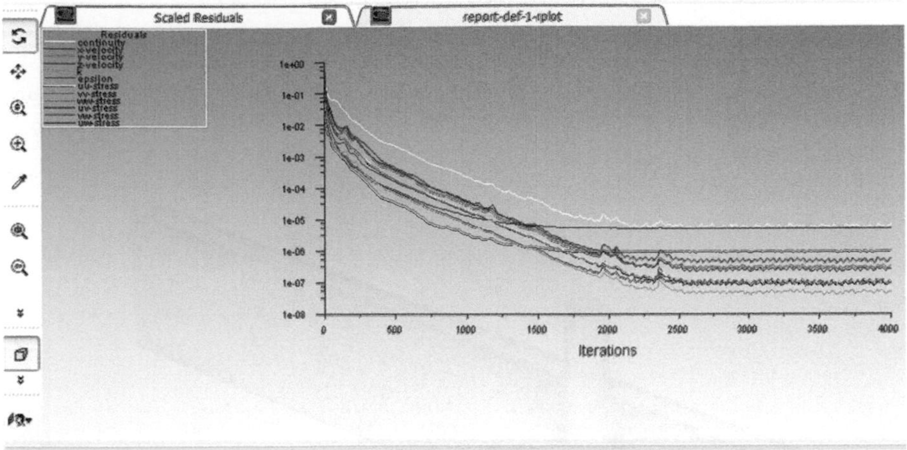

FIGURE 8.49 Residuals for the cyclone simulation when the continuous phase is run.

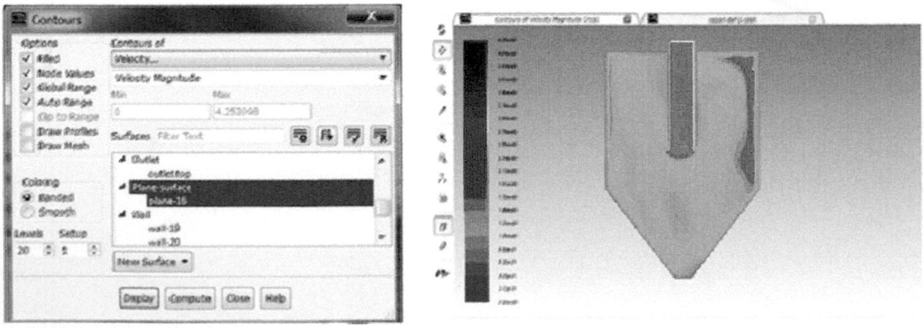

FIGURE 8.56 Contour Plots generation in ANSYS Fluent®.

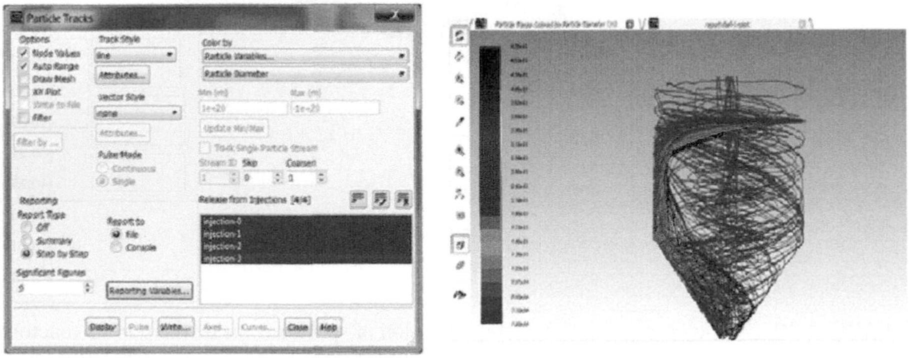

FIGURE 8.57 Particle tracks by diameter in ANSYS Fluent®.

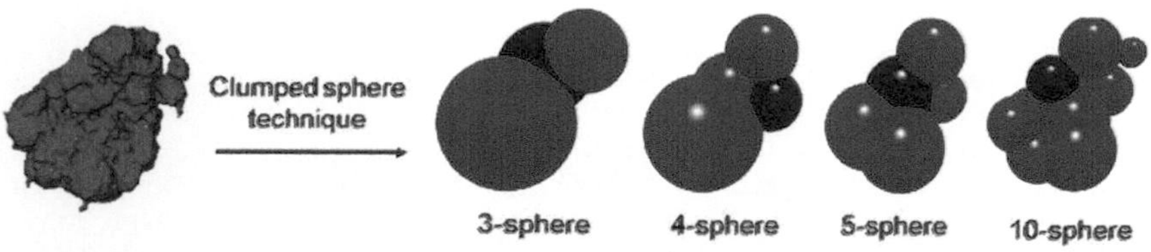

FIGURE 9.10 Particle shape representation using clumped-sphere approach; increasing the number of spheres in the clumps results in more accurate representation of the real particle [38].

FIGURE 9.11 Polyhedral-shaped particles [41].

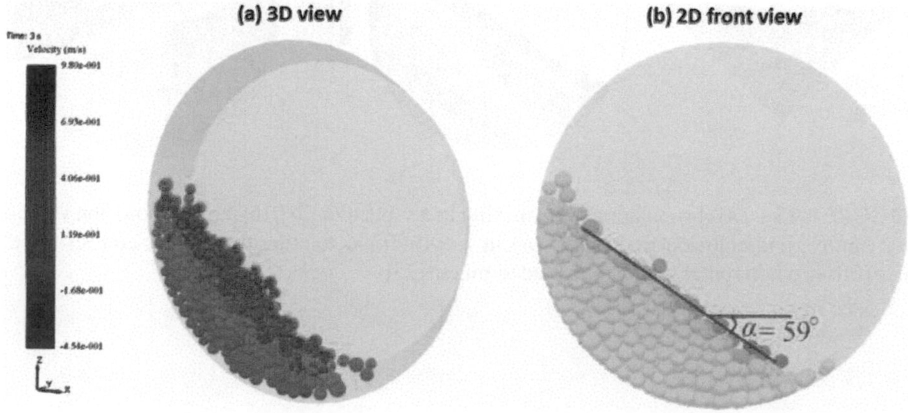

FIGURE 9.21 (a) 3D representation of the simulated rotating drum system with the particles coloured based on their x-component of velocity. (b) Avalanching angle measurement from 3D representation.

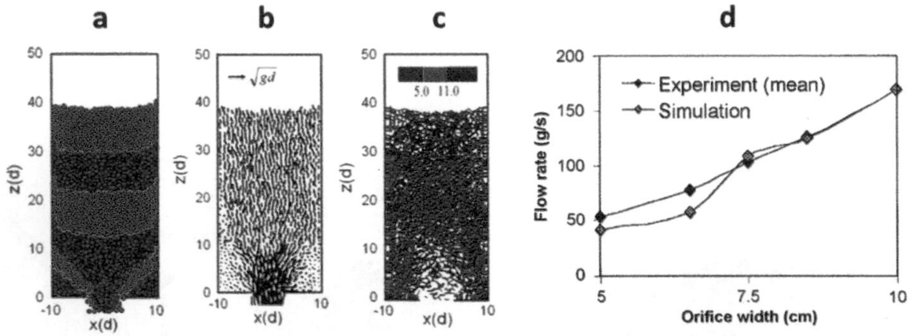

FIGURE 9.24 (a) flow pattern; (b) velocity field; and (c) force structure of the granular flow in a cylindrical hopper [66], and (d) discharge flow rate out of a hopper as a function of orifice width [63].

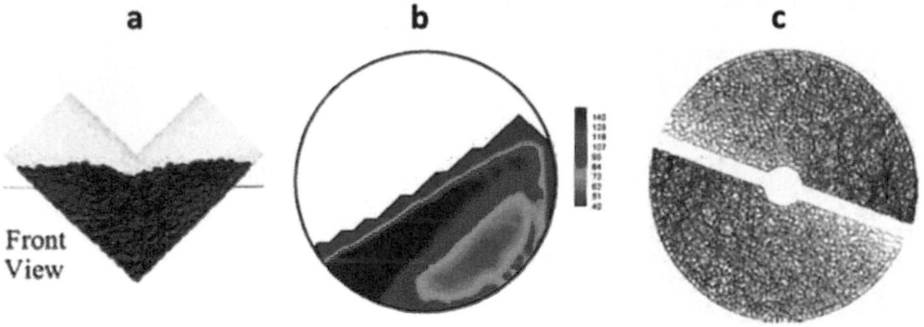

FIGURE 9.25 DEM simulation of (a) mixing in a v-blender [71], (b) particle collision velocity (mm/s), and collision frequency (1/s) in a cylindrical rotating drum [72], and (c) force network in a horizontal section of a bladed mixer [73].

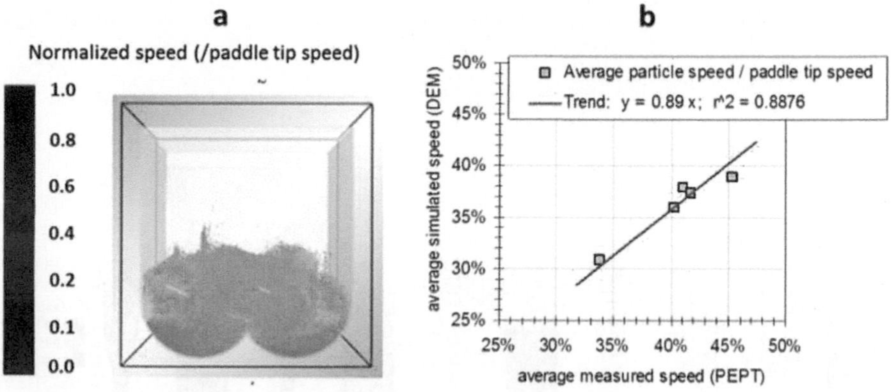

FIGURE 9.26 (a) The schematic flow fields inside the paddle mixer simulated by DEM. (b) Comparison of DEM simulation vs. PEPT measurement of time-averaged particle speed, normalized to paddle tip speed, for a range of conditions with different mixer speed, particle density and mixer fill level [69].

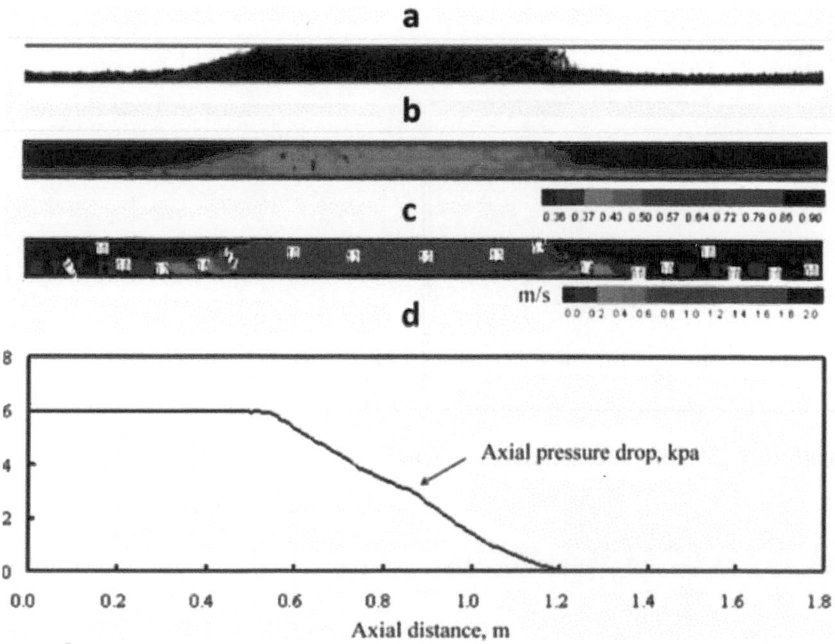

FIGURE 9.28 Simulation of distributions of flow properties: (a) particle configurations at 14.498 s, (b) porosity, (c) axial particle velocity, and (d) axial pressure drop [87].

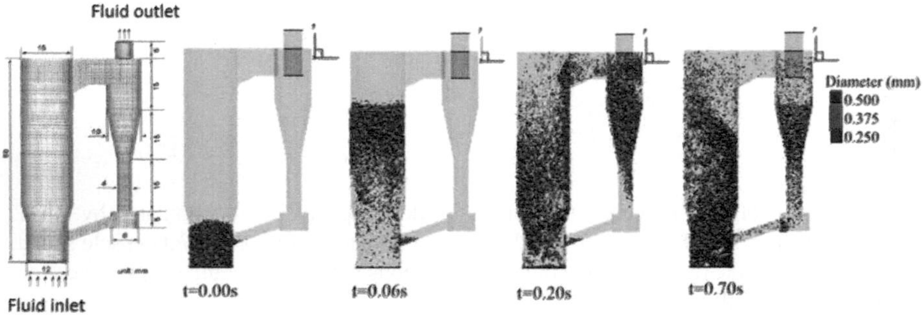

FIGURE 9.29 DEM-CFD simulation of particle size distribution (particles are coloured by size) in circulating fluidised bed [80].

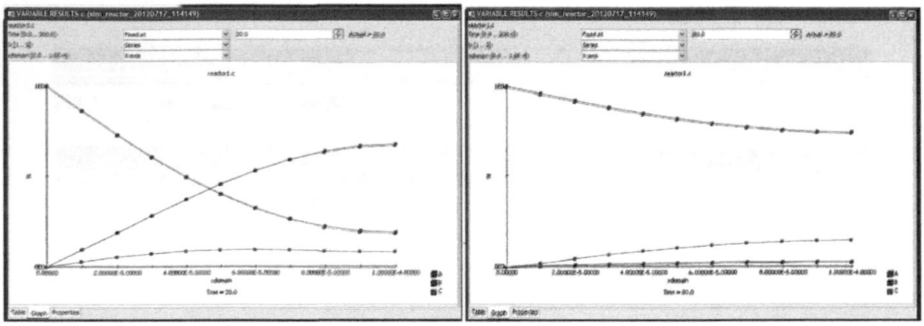

FIGURE 10.7 Plotting of concentration results.

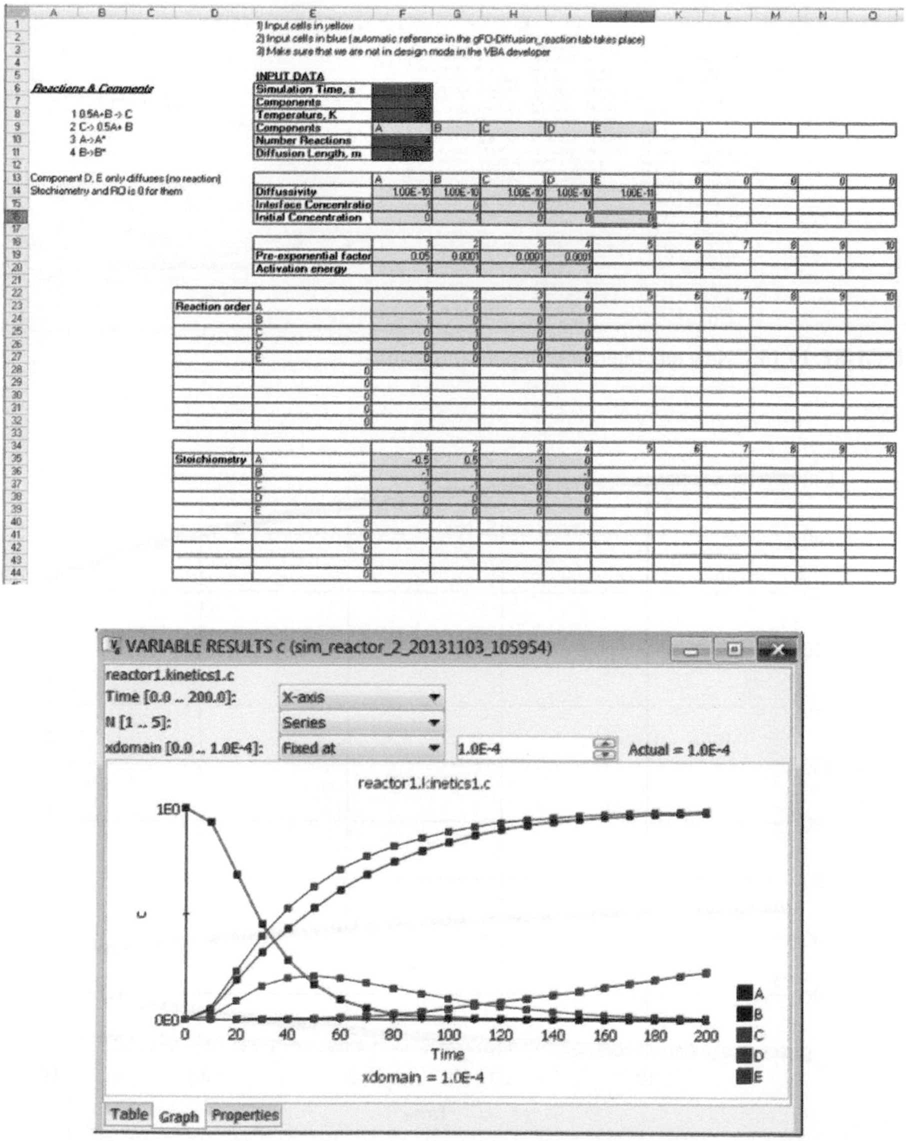

FIGURE 10.13 Input table ad results.

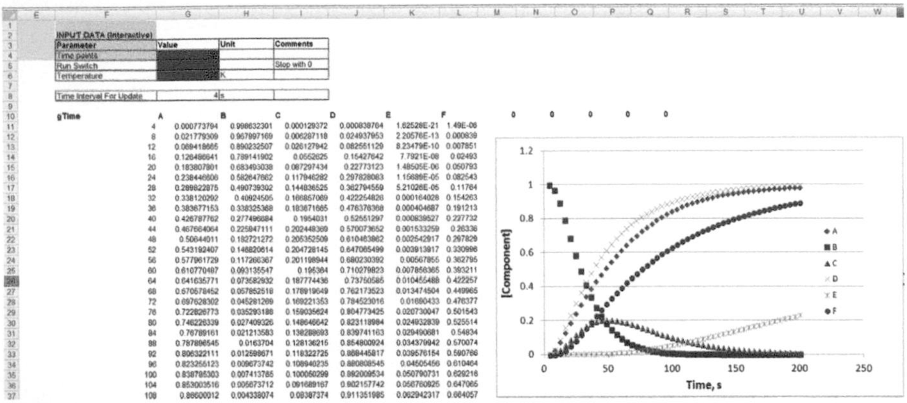

FIGURE 10.15 New tab: interactive inputs and results.

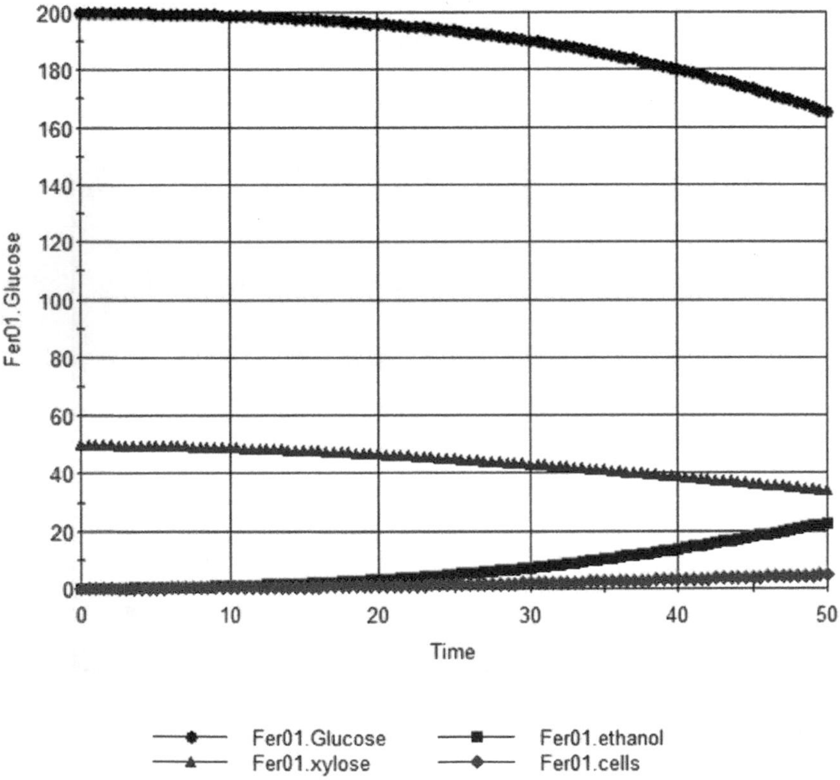

FIGURE 10.27 Results for the ethanol production example.

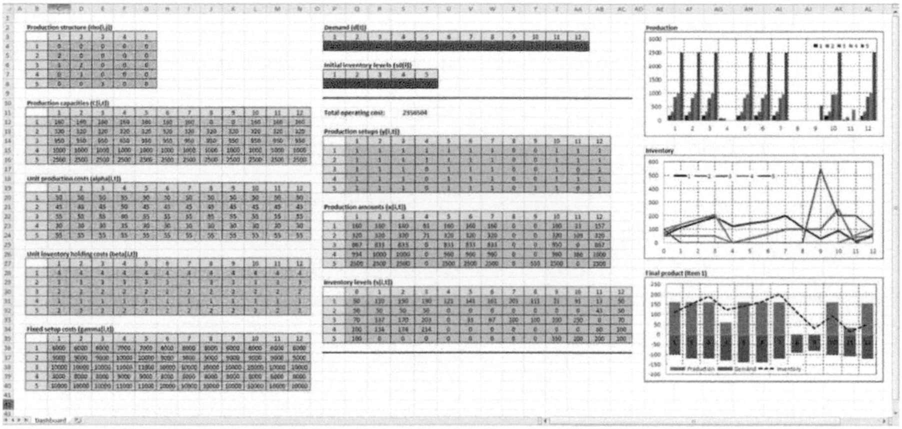

FIGURE 19.9 Dashboard for the end-user (not showing the model).

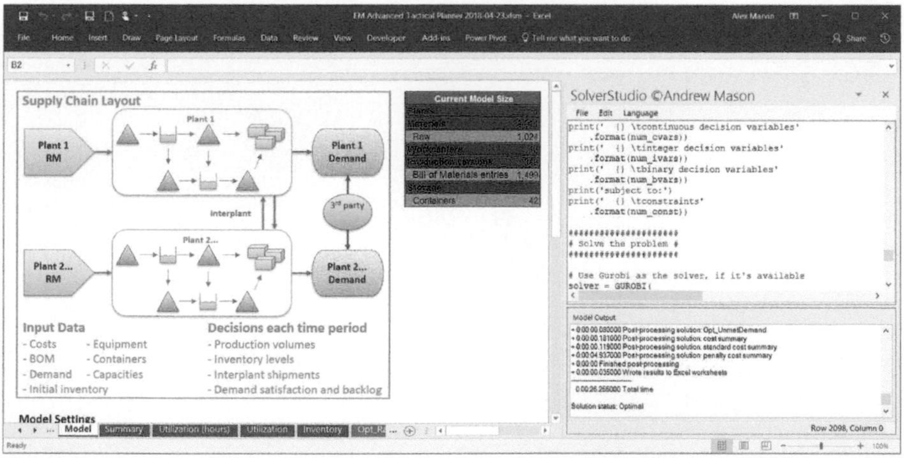

FIGURE 19.12 Advanced Tactical Planner (ATP).

9 Discrete Element Method Applications in Process Engineering

Ali Hassanpour, Massih Pasha, and Mohammadreza Alizadeh Behjani

CONTENTS

9.1 DISTINCT ELEMENT METHOD (DEM)

Understanding the bulk behaviour of particles is of strategic importance in the processing of particulate solids including, but not limited to, mixing, granulation, compaction, size reduction, flow out of storage vessels, fluidisation, dosing of small quantities of powder in capsules, and dispersion and sedimentation of particles in fluids. Common problems encountered in understanding these processes include: the parameters (e.g., internal flow and stresses) which cannot be readily measured or quantified experimentally; scale-up of particulate processes from laboratory to pilot plant and industrial scales requiring extensive trial and error; insufficient availability of material at early stages of development (e.g., pharmaceuticals); and difficulty in carrying out sensitivity analysis of process conditions as well as material properties. For all the above examples, modelling can provide an enhanced understanding of the process. However, rigorous analysis of these processes is scientifically very challenging, due to the heterogeneous nature of particles.

The macroscopic bulk behaviour of particles is governed by the microscopic activity of the individual particles in an assembly. This implies that in order to gain a better understanding of particulate systems and their functioning, the particle interactions must be analysed at the microscopic level. It is currently very difficult to investigate the behaviour of individual particles within a bulk assembly experimentally. Therefore, it is helpful to model the behaviour of particles by the use of numerical simulations. Furthermore, computer simulation provides a cost-effective method as an alternative to experiments since no material or process equipment is required, provided the simulation results are validated. Simulations are invaluable for cases where actual experiments are hazardous such as process handling of radioactive powders. For particulate solids, the most appropriate approach for this purpose is the use of computer simulation by the Distinct Element Method (DEM). The principal of DEM was introduced by Cundall and Strack [1]. In this technique, Newton's laws of motion are applied to describe particle motion and to describe the particle interactions with its neighbours contact mechanics is applied (Figure 9.1). Under the assumption that within a time-interval the velocity of the elements is unchanged and particle interactions do not go beyond its neighbour, the position of the elements is updated. This method is often referred as the soft-sphere approach since the spherical particles are allowed to have "deformation". Another method for simulating particulate systems is the hard-sphere approach,

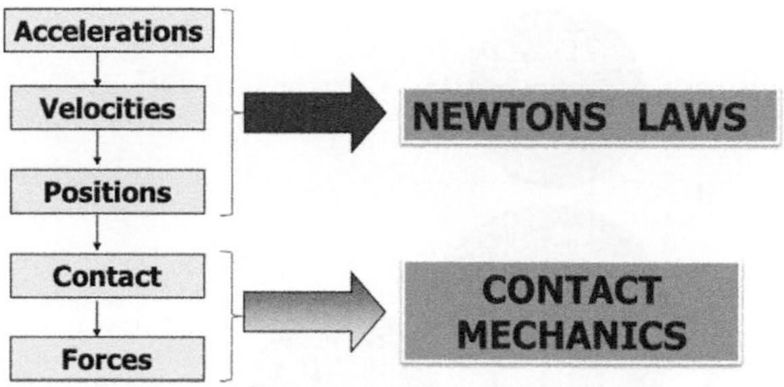

FIGURE 9.1 Flowchart of simulation for particle behaviour using DEM.

where the interaction forces are assumed to be impulsive and hence the particles only exchange momentum through collisions. In hard-sphere simulations often the forces between particles are not explicitly considered and it is used to simulate dynamic systems where the solid fraction is small or particle contact force details are not important. In contrast, the soft-sphere method evaluates the forces accurately, hence it can be used for quasi-static (slow) particulate systems where the solid fraction is relatively high, (i.e., dense systems).

A complete review of the methodology of the DEM and its applications are presented elsewhere [2, 3]. In this chapter a brief summary of the soft-sphere methodology is provided since it is more common.

9.1.1 Motion Calculations

Each particle within a granular flow can have two types of motion: translational and rotational. Newton's second law of motion is used to calculate the translational and rotational accelerations. By integrating the accelerations over a time-step, particle velocities and positions are updated. The rotational motion is calculated based on Equation (9.1),

$$I\frac{d\omega}{dt} = M \tag{9.1}$$

where I is the moment of inertia, ω is the angular velocity, M is the resultant contact torque acting on the particle and t is time. The translational motion is calculated based on Equation (9.2).

$$m\frac{dV}{dt} = F_g + F_c + F_{nc} \tag{9.2}$$

where V is the translational velocity of the particle, m is the mass of the particle, F_g is the resultant gravitational force acting on the particle and F_c and F_{nc} are the resultant contact and non-contact forces between the particle and surrounding particles or walls. A schematic representation of these forces can be seen in Figure 9.2.

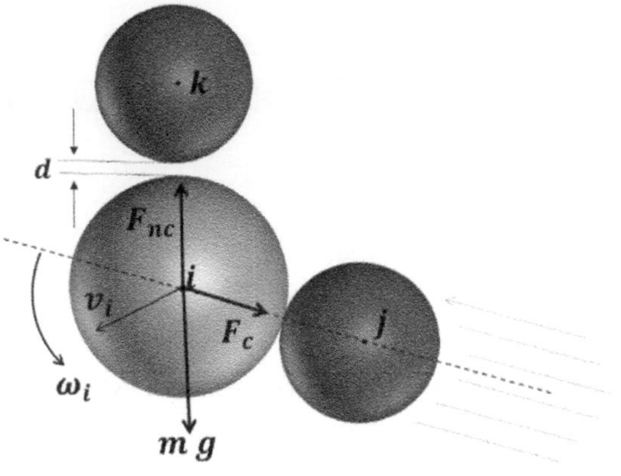

FIGURE 9.2 Schematic representation of the forces acting on a particle.

To solve the equations of motion, various integration methods could be used, as an example here the *forward Euler method* is presented for quantification of velocity and positions of the particles at different times:

$$x(t+\Delta t) = x(t) + v(t)\Delta t \tag{9.3}$$

$$v(t+\Delta t) = v(t) + a(t)\Delta t \tag{9.4}$$

where *v(t)* is velocity, *x(t)* is the position, and *a(t)* is the acceleration of a particle at a given time *t*, and *Δt* is the time interval or time steps (see Figure 9.3 for further illustrations).

9.1.2 Time-Step

Particulate systems are composed of distinct particles which displace independently from one another and interact only at contact points [1]. Therefore, the time step for the calculation of particle motion determines the resolution of the calculations. It should be sufficiently small to prevent excessive overlap with the neighbouring particles (Figure 9.4). Furthermore, movement of a particle within a granular flow is

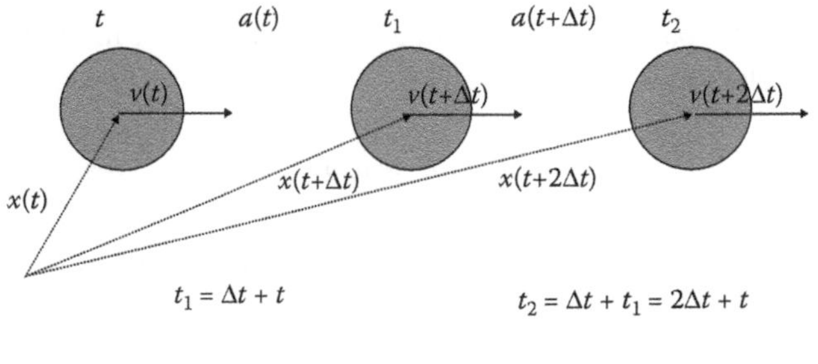

FIGURE 9.3 Schematic diagram of element motion calculation in terms of acceleration, velocity and position in DEM.

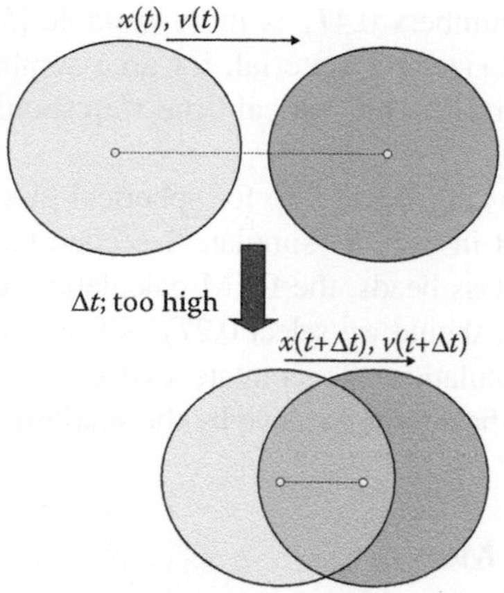

FIGURE 9.4 Illustration of time-step in motion calculation, in DEM.

affected not only by the forces and torques originated from contacts with its immediate neighbouring particles, but also by disturbance propagations from particles far away.

To avoid evaluation of effects of disturbance waves in DEM, the particle displacement calculations should be performed after a time-step, within which the disturbance cannot propagate from each particle further than its immediate neighbouring particles [1]. The speed of disturbance waves is approximated by Rayleigh surface wave propagation based on physical properties of the discrete medium. The time must be sufficiently smaller than Rayleigh time-step in order to ensure realistic force transmission rates in the assembly and to prevent numerical instability (Figure 9.5) [4]. The Rayleigh time-step for a spherical element is given by Equation (9.5),

$$T_R = \frac{\pi R \left(\frac{\rho}{G} \right)^{1/2}}{0.1631\upsilon + 0.8766} \tag{9.5}$$

where R is the particle radius, ρ is the density, G is the shear modulus and υ is the Poisson's ratio of the particle. In practice, some fraction of this maximum value is used for the integration time-step. For dense systems with high coordination numbers (average number of immediate neighbours around a particle in the system), i.e. 4 and above, a typical time-step of $0.2T_R$ has been shown to be appropriate.

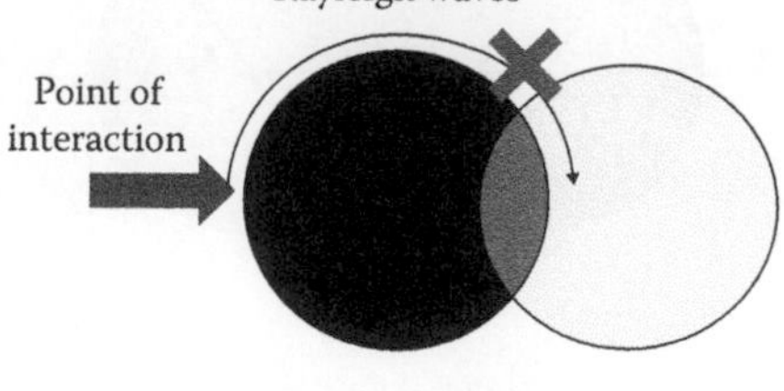

FIGURE 9.5 Schematic representation of Rayleigh wave propagation in particles.

For lower coordination numbers $0.4T_R$ is more suitable [5]. Since the time-step is dependent on the particle size and material, for an assembly consisting of different size and material type particles, the critical time-step should be the smallest among those determined [4].

As an example, the Rayleigh time-step for spherical glass beads of 1 μm is about 1 ns (T_R). This means that in order to simulate 1 second real time of a dense system which consists of 1 μm glass beads, the DEM calculation loops (Figure 9.1) must be performed 5 billion times, should we select $0.2T_R$ as time step. This highlights one of the limitations of DEM simulations. To simulate a system for a give real time, the speed of calculation will be significantly influenced by the smallest particle size in the system.

9.1.3 Contact Force Models

In order to consider particle interactions in the calculations, the resulting forces from particle collisions and contacts are modelled. There exists a number of force models which mostly allow particles to have deformation. The deformation is modelled as a virtual overlap between particles (Figure 9.6). In the general case of an assembly of many particles, the contact force model is applied at each contact and the vectorial sum of these contact forces is determined to yield the resultant contact force acting on that particle [1]. There has been extensive work in the literature on the development of accurate and/or computationally efficient contact models, based on theories of contact mechanics, to be employed in DEM simulations. Most of these models are developed for spherical contacts based on Hertz theory. Following is a brief review of the most commonly used contact models in DEM.

9.1.3.1 Elastic Contacts

9.1.3.1.1 Linear Spring Contact Model

In this simple model [1] the contact force between two perfectly elastic spheres is resolved into normal and shear components with respect to the contact plane,

$$F = \vec{F}_n + \vec{F}_s \tag{9.6}$$

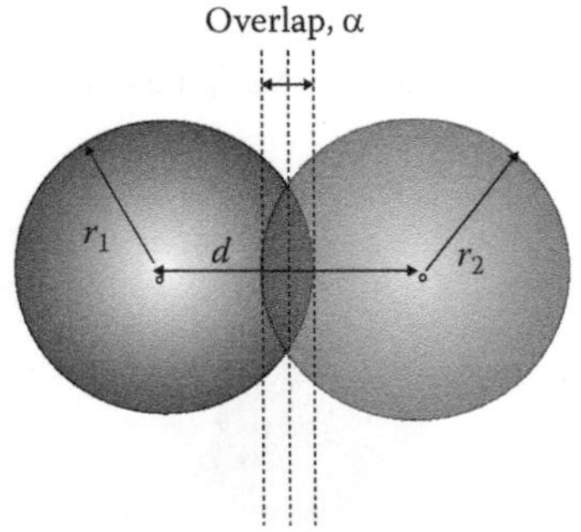

FIGURE 9.6 Illustration of virtual overlap between the elements in DEM.

where F is the contact force, and F_n and F_s are the normal and shear components of the contact force. The normal contact force is calculated from the overlap of the particles in contact,

$$F_n = k_n \alpha_n \tag{9.7}$$

where α_n is the normal overlap and k_n is the normal stiffness at the contact which can be calculated as follow,

$$k_n = \frac{k_n^1 k_n^2}{k_n^1 + k_n^2} \tag{9.8}$$

where k_n^1 and k_n^2 are the normal stiffness for the particles in contact. Figure 9.7 shows a schematic force-overlap response of linear-spring model.

The increment of shear force in this model is calculated from,

$$\Delta F_s = -k_s \Delta \alpha_s \tag{9.9}$$

where $\Delta \alpha_s$ is the increment of shear displacement of the contact and k_s is the shear contact stiffness which can be calculated as follow,

$$k_s = \frac{k_s^1 k_s^2}{k_s^1 + k_s^2} \tag{9.10}$$

where k_s^1 and k_s^2 are the shear stiffness for the particles in contact. The total shear force is found by summing up the previous shear force with the increment of shear force,

$$F_s = F_s' + \Delta F_s \tag{9.11}$$

where F_s' is the previous shear force [1].

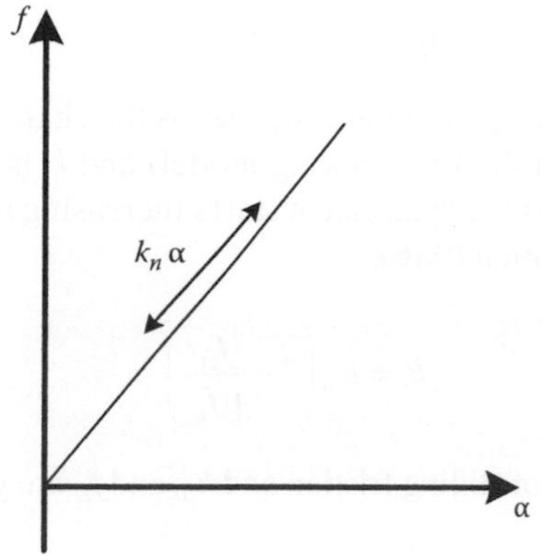

FIGURE 9.7 Schematic of force-overlap response of linear-spring model.

9.1.3.1.2 Hertz Normal Contact Model

Based on this model [6], the normal contact force between two perfectly-elastic spheres in contact is given by [7],

$$F_n = \frac{4}{3} E^* R^{*1/2} \alpha^{3/2} \tag{9.12}$$

where E^* and R^* are the equivalent Young's modulus and radius which are given by Equations (9.13) and (9.14), respectively, and α is the normal overlap.

$$E^* = \left(\frac{1-\upsilon_1^2}{E_1} + \frac{1-\upsilon_2^2}{E_2} \right)^{-1} \tag{9.13}$$

E_1 and E_2 are Young's moduli of the particles in contacts and υ_1 and υ_2 are Poisson's ratios of the particles in contact.

$$R^* = \left(\frac{1}{R_1} + \frac{1}{R_2} \right)^{-1} \tag{9.14}$$

R_1 and R_2 are the radii of the particles in contact.

For elasto-plastic contacts there are a number of contact models, notably models of Walton and Braun [8], Thornton [9], and Pasha et al. [10]. However, these models will not be discussed in this chapter.

9.1.3.1.3 Mindlin and Deresiewicz's Tangential Contact Model

Mindlin and Deresiewicz [11] developed a tangential model for perfectly elastic contacts. Due to the tangential slip at the contact, the tangential force-displacement relationship depends on the loading history and rate of change of the normal and tangential force. Following is a description of the model for cases where the normal displacement does not change (i.e., the radius of the contact area is constant) through the change of tangential force:

The tangential force is calculated using the following general equation,

$$f_t = f_{t0} + k_t(\alpha_t - \alpha_{t0}) \tag{9.15}$$

where f_{t0} is the previous tangential force, $\alpha_t - \alpha_{t0}$ is the change in tangential displacement (the same as $\Delta\alpha_s$ in the linear-spring model) and k_t is the tangential stiffness. Initially when the tangential displacement starts increasing the tangential stiffness is calculated based on Equation (9.16),

$$k_t = k_{t0} \left(1 - \frac{f_{t1}}{\mu f_n} \right)^{1/3} \tag{9.16}$$

where μ is the coefficient of sliding friction and k_{t0} and f_{tl} are given by Equations (9.17) and (9.19), respectively,

$$k_{t0} = 8G^* R^{*1/2} \alpha_n^{1/2} \tag{9.17}$$

where G^* is the equivalent contact shear modulus given by Equation (9.18),

$$G^* = \left(\frac{1-\upsilon_1^2}{G_1} + \frac{1-\upsilon_2^2}{G_2} \right)^{-1} \tag{9.18}$$

where G_1 and G_2 are the shear moduli of the particles in contact.

$$f_{t1} = \frac{2}{3} k_{t0} \alpha_t \tag{9.19}$$

The contact starts sliding (macro-slip) once the tangential force reaches Coulomb's limit of friction given by Equation (9.20).

$$|f_t| \leq \mu |f_n| \tag{9.20}$$

Before the macro-slip occurs, there are regions in the contact area where the shear stress exceeds the Coulomb's limit [12]. Therefore, small local tangential displacement occurs while the remainder of the contact area is not relatively displaced. This phenomenon is known as micro-slip. If the contact is unloaded in the tangential direction, the unloading path on the force-displacement curve would be different to that of the loading curve due to the micro-slip. Mindlin and Deresiewicz [11] considered a hysteretic behaviour to account for the micro-slip. Involving equations for the hysteretic behaviour are not mentioned here due to their complexity but can be found in [11].

9.1.3.1.4 Mindlin's No-slip Tangential Contact Model

Mindlin [13] proposed a tangential model for perfectly elastic contacts by neglecting the effects of micro-slip. The loading and unloading path of the tangential force is the same and the tangential stiffness is given by,

$$k_t = 8 G^* R^{*1/2} \alpha_n^{1/2} \tag{9.21}$$

Di Renzo and Di Maio [14] simulated the oblique impact of a particle to a flat wall at different impact angles using the linear spring, Mindlin's no-slip and Mindlin and Deresiewicz's models. The comparison of their results to the experimental findings of Kharaz et al. [15] is shown in Figure 9.8.

Figure 9.8 shows a very good agreement of the three models with the experimental results, except for shallow impact angles. The linear model surprisingly produced better results compared to Mindlin's no-slip model.

9.1.3.2 Elastic-Adhesive Contacts

9.1.3.2.1 JKR Elastic-Adhesive Normal Contact Model

The theory of Johnson et al. [16], referred to as the JKR model, assumes that the attractive forces are confined within the area of contact and are zero outside. In other words, the attractive inter-particle forces are within infinitely short range. The JKR theory extends the Hertz model to two elastic-adhesive spheres by using an energy

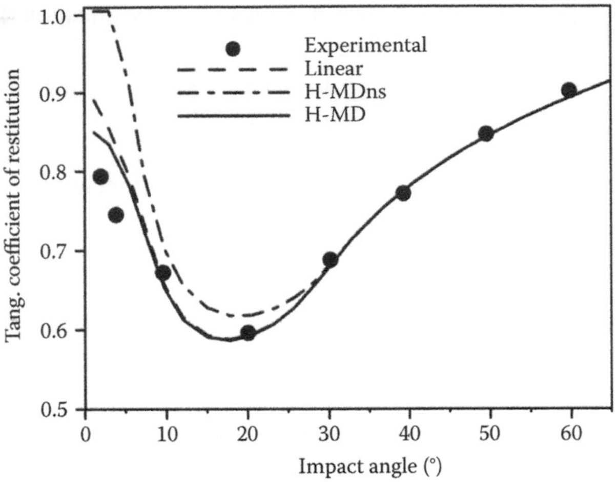

FIGURE 9.8 Comparison of linear-spring, Mindlin's no-slip and Mindlin and Deresiewicz's models with the experimental data for oblique impact of a particle at different impact angles [14].

balance approach. The contact area predicted by the JKR model is larger than that by Hertz. Consequently, there is an outer annulus in the contact area which experiences tensile stresses. This annulus surrounds an inner circular region over which a Hertzian compressive distribution acts [17]. Figure 9.9 schematically shows the force-overlap response of the JKR model.

When two spheres come into contact, the normal force between them will immediately drop to a certain value ($8/9 f_c$, where f_c is the pull-off force [7]) due to van der Waals attractive forces. The velocities of the spheres are then reduced gradually and part of the initial kinetic energy is radiated into the substrates as elastic waves. The particle velocity reduces to zero at a point where the contact force reaches a maximum value and the loading stage is complete. In the recovery stage, the stored elastic energy is released and is converted into kinetic energy which causes the spheres to move in the opposite direction. All the work done during the loading stage is recovered when the contact overlap becomes zero. At this point, the spheres remain

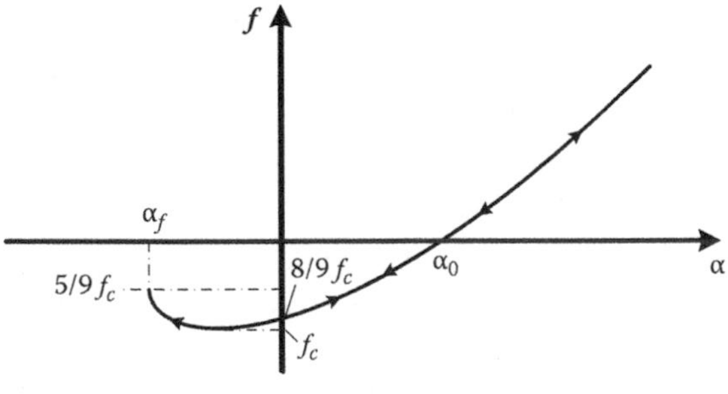

FIGURE 9.9 Schematic force-overlap response of the JKR model.

adhered to each other and further work (known as the work of adhesion) is required to separate the surfaces. The contact breaks at a negative overlap, α_f, with the contact force being 5/9 f_c [18]. The pull-off force, f_c, is the maximum tensile force that the contact experiences and is given by [16],

$$f_c = \frac{3}{2}\pi R^* \Gamma \tag{9.22}$$

where Γ is the interfacial energy. The governing equation for the force-overlap relation is given by [19],

$$F_n = \frac{4E^* a^3}{3R^*} - \left(8\pi\Gamma E^* a^3\right)^{1/2} \tag{9.23}$$

where a is the contact area. The overlap can be evaluated by Equation (9.24) [19],

$$\alpha = \frac{a^2}{R^*} - \left(\frac{2\pi\Gamma a}{E^*}\right)^{1/2} \tag{9.24}$$

For the tangential behaviour in presence of adhesion, the models of Savkoor and Briggs [20] and Thornton and Yin [17] could be used which will not be discussed here.

9.1.3.2.2 DMT Elastic-Adhesive Normal Contact Model

The theory of Derjaguin et al. [21], referred to as DMT model, assumes that the attractive forces all lie outside the contact zone. The attraction forces are also assumed not to change the deformed profile from that of the Hertz theory. However, outside the contact area, a tensile stress is assumed to be present which decreases with surface separation. At the point of contact prior to any deformation, the resultant attractive force is suddenly increased to a value known as the pull-off force which is given by,

$$F_c = 2\pi R^* \Gamma \tag{9.25}$$

As the spheres start deforming, the attractive force is rapidly reduced to half of its initial value. The normal contact force in this model is given by,

$$F_n = \begin{cases} F_c & \rightarrow \alpha_n = 0 \\ F_{Hertz} + \dfrac{F_c}{2} & \rightarrow \alpha_n > 0 \end{cases} \tag{9.26}$$

where F_{Hertz} is the Hertz elastic force given by Equation (9.12). Pashley [22] showed that the DMT model is valid for small, rigid spheres with a low surface energy, whilst JKR model is more applicable to larger, more elastic spheres with a high surface energy.

For elasto-plastic and adhesive contacts, the models proposed by Thornton and Ning [7] and Pasha *et al.* [10] could be used which are not discussed here due to their complexity.

9.1.4 Non-Contact Forces in DEM

Particle interactions are not just due to particle collisions and inter-particle contacts. Particles may interact with each other while they are apart from each other, e.g., in the case of charged particles, attractive or repulsive forces may influence the particle's motion and this can be considered in DEM (F_{nc} in Equation 9.2). The following section is only a brief review of common forces which may be present in particulate systems. For further readings please refer to the relevant references brought in each section.

9.1.4.1 Van der Waals Forces

Van der Waals force is a result of dipole interactions between molecules. Different electronic configurations of molecules give them a dipolar character which may result in attractions. Van der Waals force operates both in gaseous and in a liquid environment, although it is substantially reduced in liquid environments [23]. Hamaker [24] calculated the interaction force between a sphere and a semi-infinite body, by summarising all the possible individual molecular interactions, as follows [25],

$$F_{van} = \frac{A_H R}{12d^2} \tag{9.27}$$

where A_H is the Hamaker coefficient which is a material property related to the molecular properties of the particle with the radius of R, and d is the separation distance between the two bodies. The van der Waals forces become noticeable when particles can come sufficiently close together, that is at separation distances of the order of a molecule size (i.e., 0.2 to 1 nm). Moreover, the magnitude of van der Waals forces becomes negligible compared with that of the gravitational force when the particle size exceeds a few microns. This is due to the fact that the gravitational force is proportional to the cube of the particle diameter, but the van der Waals force is proportional to the diameter. Once the particles are in contact, the overall van der Waals attraction is increased significantly due to the increase in the contact area. This situation is enhanced when plastic deformation takes place [23].

9.1.4.2 Liquid Bridges

In humid systems (with a relative humidity of > 60%) capillary condensation of the fluid in the gap between the particles in close contact may take place resulting in an attraction. This can also occur when a liquid binder is deliberately added to a mixture of dry particles. The maximum attraction is achieved for two smooth spherical particles if the liquid covers the particle surfaces completely. For this case, the liquid bridge attraction force, F_{LB}, can be calculated as follow [23],

$$F_{LB} = 2\pi\gamma R \tag{9.28}$$

where γ is the surface tension of the liquid and R is the radius of the two particles. For further reading please refer to the works of Washino *et al.* [26] and Lian *et al.* [27].

9.1.4.3 Electrostatics

Electrostatic forces can form as a result of triboelectric charging or formation of a potential difference between particles. In the latter case, charged particles attract adjacent uncharged particles due to their own image charge. The attraction force can be evaluated by the classical Coulomb equation [23],

$$F_{t\_elec} = \frac{Q^2\left(1 - \frac{d}{\sqrt{R^2 + d^2}}\right)}{16\pi\varepsilon_0 d^2} \tag{9.29}$$

where Q and R are the charge and radius of the charged particle, d is the separation distance between the two bodies, and ε_0 is the permittivity of the vacuum. In the former case, (i.e., potential difference), particles with different work-functions can form a potential difference when they are brought together. This results in an attraction force which can be calculated as follows [23],

$$F_{V\_elec} = \frac{\pi\varepsilon_0 R}{d}(\Delta V)^2 \tag{9.30}$$

where ΔV is the potential difference formed by the contact. It should be noted that in humid environments the Coulomb attraction is reduced to zero due to the fact that particle surfaces are covered by the liquid, which mostly is a conductor, resulting in charge leakage and discharging. For further reading please refer to the works of Supuk et al. [28] and Pei *et al.* [29].

9.1.5 Contact Detection

For spherical particles, a contact is detected if the distance between two spheres in the system becomes less than the summation of their radii. Therefore, contact detection in DEM simulations involves checking the distance between all particles in the system. This process can be computationally extensive when a large number of particles are simulated. In DEM simulations, the calculation domain is usually discretised into 3-dimensional cells, which are sometimes referred to as 'grids'. The grids help the contact detection algorithms to be applied at a smaller scale, hence reducing the computational time. It also enhances the efficiency of parallel computations where a number of these grids can be passed to each of the processors. The grid size does not affect the accuracy of the calculation, however, determines the simulation speed. Very small grid sizes result in a large number of grids which may reduce the efficiency of contact detection, whereas very large grid sizes result in a very large number of particles inside each grid which again reduces the simulation speed. The grid size must be chosen based on the size distribution of the generated particles, dynamics of the process which is being simulated and the distribution of solid fraction or voidage across the simulation domain. A grid size of 3 to 5 times the smallest particle radius is found to be the optimum range for most of the simulations.

One of the limitations of DEM is in the number of particles that can be simulated. By increasing the number of particles, the contact detection requires to loop through a larger set of particles leading to extremely slow simulations. The relationship between the simulation time and the number of particles is not usually linear. Additionally, larger number of particles requires more hardware needs in terms of memory (for storing individual particle information) and in terms of hard disk storage (for storing the simulation files). With the recent advances in computer technology, specifically parallel-computations using multi-core processors, this limitation is being alleviated gradually. Back in 1979, when DEM concept was proposed by Cundall and Strack [1], the simulations were limited to a few hundreds of 2-dimensional particles (i.e., using circles instead of spheres). Nowadays 3-dimensional simulations of millions of non-spherical particles are achievable using workstation computers with a couple of multi-core processors (up to 24 cores per processor) with shared memory. Moreover, using high-performance computers, with interlinked processors and distributed memory, the performance can be significantly improved. Most recently, GPU-based (graphics processing unit) DEM codes have attracted many researchers, due to their promising capability to speed up simulations. GPU performance can lead to orders of magnitude faster simulations compared to that of a CPU (central processing unit). This also enables researchers to simulate the fluid-particle media more realistically [30]. For example, He et al. [30] have developed a GPU-based coupled SPH-DEM code whereby simulating 10 million fluid particles on a single GPU card is easily achievable.

9.1.6 Particle Shape

The rotation of spheres is restricted only by frictional forces between the particles and also with other surfaces, whereas for irregular particles, the rotation is affected by mechanical interlocking in addition to the frictional contacts. For spherical particles only tangential forces result in rotation of particles and normal contact forces do not contribute to the moment and rotation, since they always act through the centre of the spheres. This is not the case for irregular particles for which the rotation can be a result of both normal and tangential forces [31]. Therefore, spherical particles may not be a good representative for irregular particles in some cases. It has been shown that spherical particles tend to have a smaller angle of repose [32] and a reduced strength [33] as compared to non-spherical particles.

9.1.6.1 Reduction of Rotational Freedom of Spheres

Amongst all 3D shapes, spheres require the simplest method of contact detection in which a contact can be detected if the distance between two adjacent spheres becomes less than the sum of their radii. Another advantage of spheres is the efficient and accurate evaluation of the contact overlap, which provides a fast and reliable calculation of contact forces. Owing to these attributes of spheres, it is always of interest to simulate the particulate systems using spheres. In order to represent irregularity of particles while using spheres, Morgan [34] proposed a damped sphere rotation as a proxy for mechanical interlocking of irregular particles. It was shown that restricting the rotation of particles by this method can yield more realistic values of assemblage friction compared to free rolling spheres. Another approach is based on considering a rolling friction [35,36] that provides a threshold torque beyond which angular motion is allowed. In order to reduce the rotational freedom of spheres, the centre of mass of the spheres can also be moved away from the geometric centre. Although it has been reported that introducing rolling resistance and replacing the geometrical centre of mass lead to higher shear strength of the bulk, many believe that higher strength of bulk solids with irregular particles principally arises from the mechanical interlocking of particles which may not be well represented by rolling friction or spheres with a modified centre of mass [37,38].

9.1.6.2 Clumped Spheres

In this method, particle shape is approximated by a number of overlapping or touching spheres with different sizes whose centres are fixed in position relative to each other (Figure 9.10) [35]. This method provides an approximation of the actual irregularity while maintaining computational efficiency and accuracy of spheres. It must be noted that this advantage comes at the expense of increased total number of spheres, which increases the computational cost of the simulations [38]. Theoretically, any particle shape can be modelled; however, highly angular particles require a large number of small spheres to approximate their sharp edges. Approximation of shapes with this technique mostly produces inadvertent surface roughness for the modelled particles. The induced roughness can be controlled by increasing the number of spheres while the computations are within a reasonable range of complexity [38]. However, this may change the effective stiffness of the particles. Kodam et al. [39] showed that when a particle, made from clumped spheres, contacts a flat surface, the contact stiffness of the master particle is effectively higher than the stiffness of each spherical element. Therefore, the values of the spheres' stiffness should be tuned based on the number of spheres simultaneously contacting the wall and the degree of overlap for each spherical element, so that the clumped-sphere has the same effective contact stiffness as of the real particle [40].

9.1.6.3 Polyhedral Shapes

The geometry of polyhedrons is defined in terms of corners, edges, and faces; the location of corners is given by a series of vectors from the centre of mass and a unit outward normal vector is associated with each face. The location and orientation in space of each polyhedron are defined by the components of a vector to the centre of

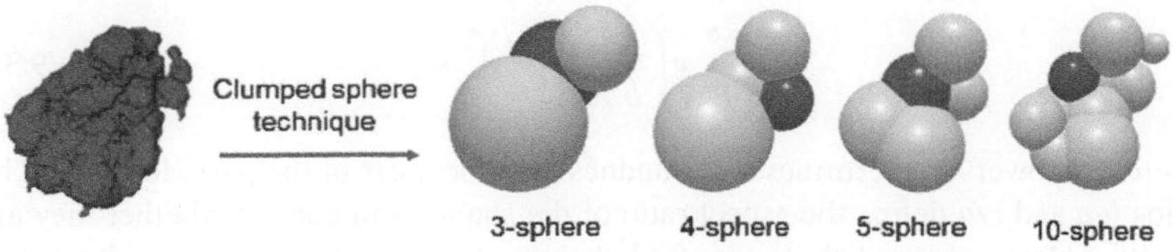

FIGURE 9.10 Particle shape representation using clumped-sphere approach; increasing the number of spheres in the clumps results in more accurate representation of the real particle [38]. (See color insert.)

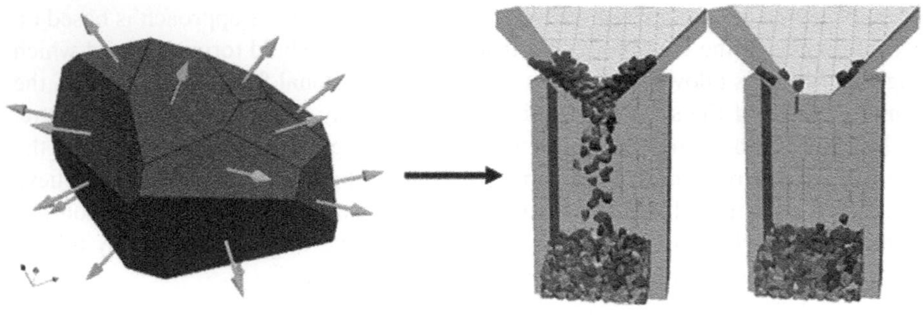

FIGURE 9.11 Polyhedral-shaped particles [41]. (See color insert.)

gravity (with respect to a fixed reference frame) and by the principal axes of inertia of the body. The advantage of this type of shape is that complex flat-faced particles can be accurately represented (Figure 9.11).

During a simulation, the coordinates of each corner and face need to be recomputed each time the collision detection algorithm is used. This is the main limitation of this approach which requires massive computational power to calculate these coordinates for each individual particle. For calculations of contact force, contacts are divided into two main types; corner-to-face and edge-to-edge contacts. There exists a number of force evaluation methods for these two types of contact. Mostly a linear contact model based on stiffness is used [42]. The stiffness varies for edge-to-edge and corner-to-face contacts. Contact detection involves calculation of the distance between all edges of a particle with all the faces of the surrounding particles which again increases the computational complexity of the simulations. This approach is not applicable for smooth irregular shapes since a large number of corners and edges are required to approximate curvatures [43].

Recently, a combination of polyhedra and spheres have been introduced to model complex curved shapes. These shapes also allow for simulations of spheres and irregular particles together. Spheropolyhedra (Figure 9.12) are obtained as Minkowski sums of polyhedra and spheres [44].

9.1.6.4 Continuous Super-quadric Function

Smooth irregular shaped particles can also be represented by using one or more continuous functions to describe the surface of a particle. Super-quadrics are mathematical shapes that can be generated using Equation (9.31) [37],

$$\left(\frac{x}{a}\right)^m + \left(\frac{y}{b}\right)^m + \left(\frac{z}{c}\right)^m = 1 \tag{9.31}$$

where the power m determines the roundness or blockness of the particle shape. The ratios b/a and c/a define the aspect ratio of the shapes and control whether they are prolate, oblate, or roundish. Figure 9.13 shows a range of shapes that can be generated by this method.

Contact detection between two super-quadric shapes can be determined from the intersection of the two functions. Due to the nonlinearity of the equations, this

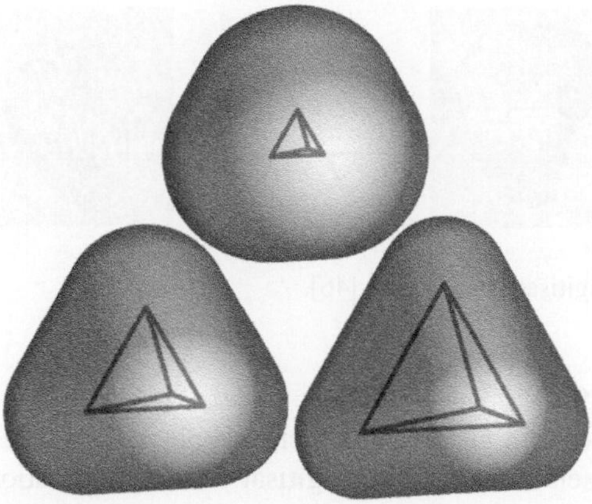

FIGURE 9.12 Spheropolyhedra [44].

process is computationally expensive, though more efficient than polyhedra. Similar to polyhedral, the lack of a well-defined contact model for this shape type is another disadvantage.

9.1.6.5 Discrete Function Representation (DFP)

Solving the nonlinear equations in a continuous function shape can be avoided by discretising the boundary of the mathematical shape. This will allow for the contact detection to calculate distances between nodes of the discretised elements. It must be noted that discretisation affects the smoothness [43]. By enhancing the discretisation the smoothness can be approximated; however, this will increase the computational complexity of contact detection. With DFP, super-quadrics perform better compared to the polyhedral shapes [37].

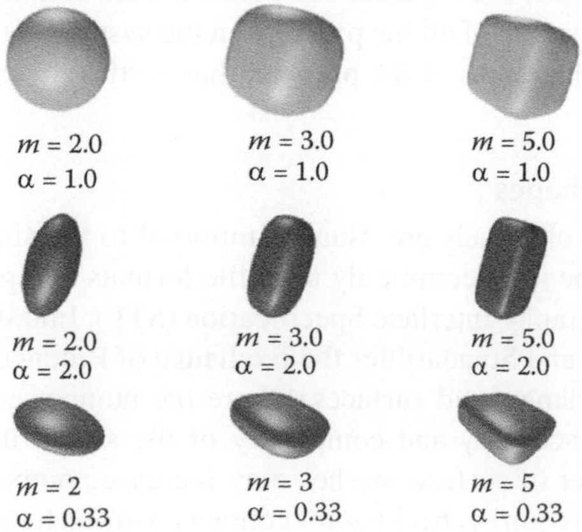

FIGURE 9.13 Super-quadric shapes [45].

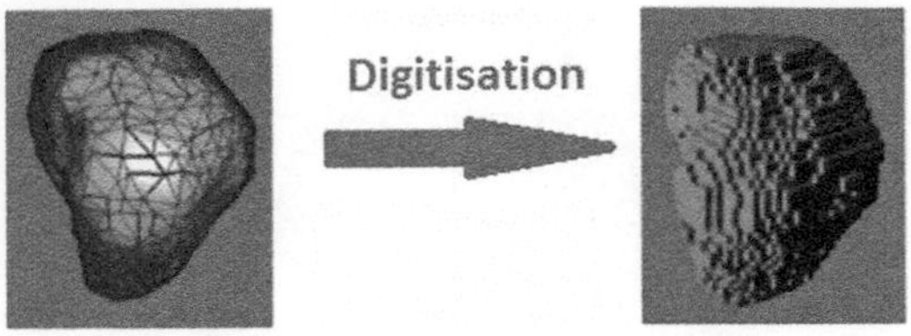

FIGURE 9.14 3D digitisation of shapes [46].

9.1.6.6 Digitisation

Jia et al. [46] developed a method in which complex shapes are digitised and represented by 3D pixels (voxels). The digitisation is computationally efficient and complex shapes can be approximated quite accurately by increasing pixel density. Digitisation translates each of the particles into a coherent collection of pixels as shown in Figure 9.14.

A contact is detected if one pixel is occupied by two or more particles. The overlaps can be accurately and efficiently calculated using the digitised pixels. The time taken to check overlaps is a linear function of the particle number and not of the complexity of particle shapes, making this technique preferable to other shape representations in terms of computational demand. A linear contact force as a function of overlapping volume is proposed for this technique [47].

9.1.7 Geometries in DEM Simulations

9.1.7.1 Simple Shapes

In DEM simulations, to model the vessels and containers in which particles exist simple geometries, such as cuboid boxes, 2D rectangular polygons, and cylindrical shapes are usually defined by their equations, where the contact detection (particle-geometry) is very efficient. For example, for a 2D polygonal plane parallel to z-plane, separated by 1 cm from the z-plane, the particle-wall contact detection involves subtraction of z-component of all the particles in the system. If the result of this subtraction is less than the radius of the particle, that particle is regarded as in contact to the polygon.

9.1.7.2 Complex Shapes

Complex geometries of vessels are usually imported to the simulations as 3D surface-meshed files. The most commonly used file formats accepted in DEM simulations are Stereolithography Interface Specification (STL), Initial Graphics Exchange Specification (IGES) and Standard for the Exchange of Product (STEP). The shapes are represented by triangulated surfaces, where the number and size of these surfaces determine the accuracy and complexity of the shape. It must be noted that increasing the number of surface meshes may increase computational needs since the contact detection requires checking for contacts with each mesh. Moreover, when the geometries have motion, updating their position and orientation would need more computational power.

9.1.7.3 Movement of the Geometries

The geometries are allowed to have dynamic motions. In DEM simulations it is common to have translational, rotational and sinusoidal motions with/without acceleration. For complex geometries with a large number of surface meshes, calculation and updating the position of complex geometries with large number of surface meshes can be computationally expensive leading to slow simulations.

9.2 A TYPICAL DEM SIMULATION

There are a number of DEM packages available both in public domain as open-source codes (e.g., LIGGGHTS [48], MercuryDPM, GranOO, MFIX, Yade) and commercial software (e.g., EDEM® [49], PFC3D™ [50], STAR-CCM+, Rocky, DEMpack). Additionally, many researchers have developed their own in-house codes. The advantage of open-source codes is that they are available at no cost; also there is typically a large community of developers and users who share their experience and help each other. On the other hand, the commercially available programmes provide full technical support, although the users have to pay for a licence. In this section, EDEM® Software Version 2.7 (provided by DEMSolutions, Edinburgh, UK) is used due to its user-friendly graphical interface. In this section, a simple simulation of a rotating drum is presented with an aim to demonstrate how a typical DEM simulation is carried out. Rotating drum is commonly used in powder processing such as coating, drying, mixing, granulation, as well as characterising powder flowability.

9.2.1 Simulation Setup

Figure 9.15 shows the EDEM® user interface once the programme is loaded. EDEM® has three general tabs (see A in Figure 9.15) known as Creator, Simulator and Analyst tabs. In the Creator tab, which by default is loaded, the user can define the physical and material properties of the particles and geometries as well as the contact properties. The geometries are also created in this tab. The user can also define the particle generation in Creator tab. Creator tab itself has four sub-sections (see B in Figure 9.15); Globals, Particles, Geometry, and Factories.

In the Globals sub-section, the contact models for particle-to-particle and particle-to-geometry interactions are defined (see C in Figure 9.15). EDEM® has a number of built-in contact models such as Hertz-Mindlin no-slip model (i.e. the Hertz model is used for normal contact force calculations see Section 9.1.3.1.2 and Mindlin no-slip model is used for tangential contact force calculations 9.1.3.1.4), linear-spring model (see Section 9.1.3.1.1) and JKR adhesive model (see Section 9.1.3.2.1). In the Globals sub-section, the user is able to add any number of materials, for each of which, the following properties must be defined; name, Poisson's ratio, shear modulus and density (see D in Figure 9.15). The interaction properties (coefficients of restitution, sliding friction, and rolling friction) between all the materials must also be set, as shown by E in Figure 9.15. Physical and mechanical properties of the particles, such as the shape, density, stiffness, and surface energy, must be calibrated before running a simulation. Alizadeh et al. [51] have proposed a rigorous methodology for calibration and scaling these parameters based on the real experimental values.

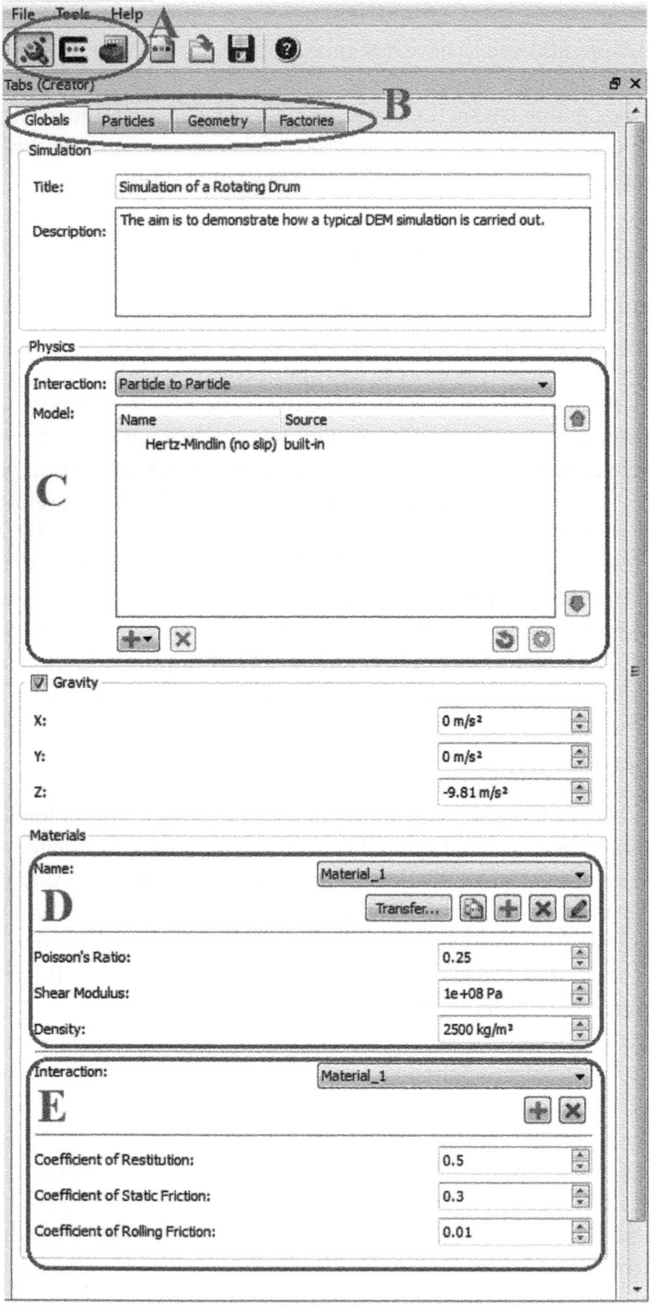

FIGURE 9.15 EDEM® user interface (Globals sub-section of Creator tab).

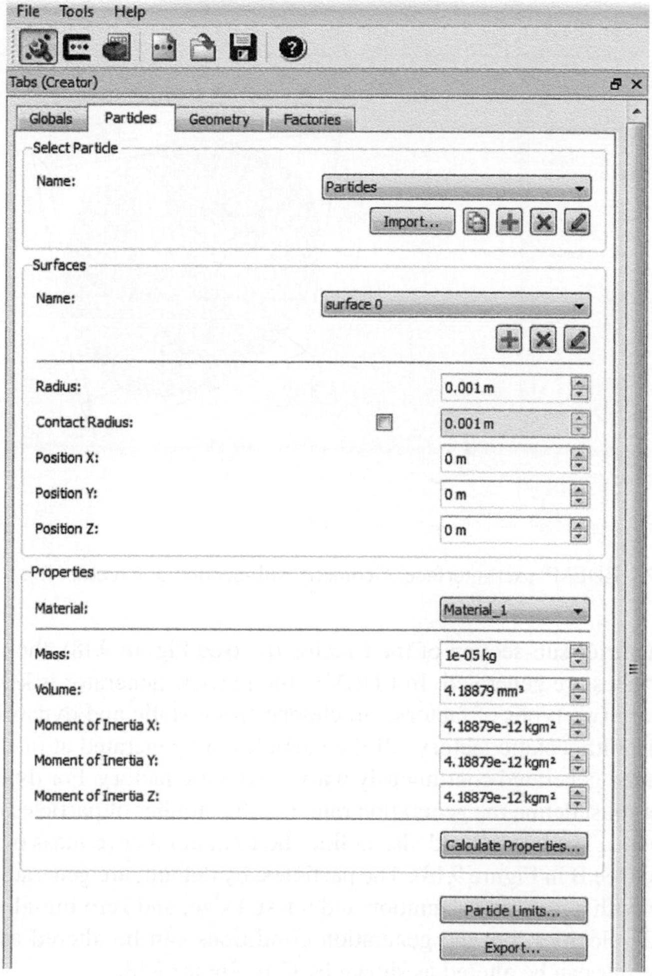

FIGURE 9.16 EDEM® user interface (Globals sub-section of Creator tab).

In the Particles sub-section of Creator tab, the user defines particle radius, material, mass, volume and moment of inertia (see Figure 9.16). It must be noted that particle size which is defined here is the average particle size. Later on in the Factories sub-section, the user is able to provide a distribution for particle size.

The Geometry sub-section in the Creator tab enables the user to define the geometries of the simulation. Geometries can be made by either using the built-in simple shapes (box, cylinder, and polygon) or for complex geometries by importing 3D CAD files. Figure 9.17 shows the Geometry sub-section in which a cylindrical shape is made to be used for our rotating drum simulation. In this sub-section, the dynamics of the geometries can also be defined. Geometries can be given translational, rotational and sinusoidal motions.

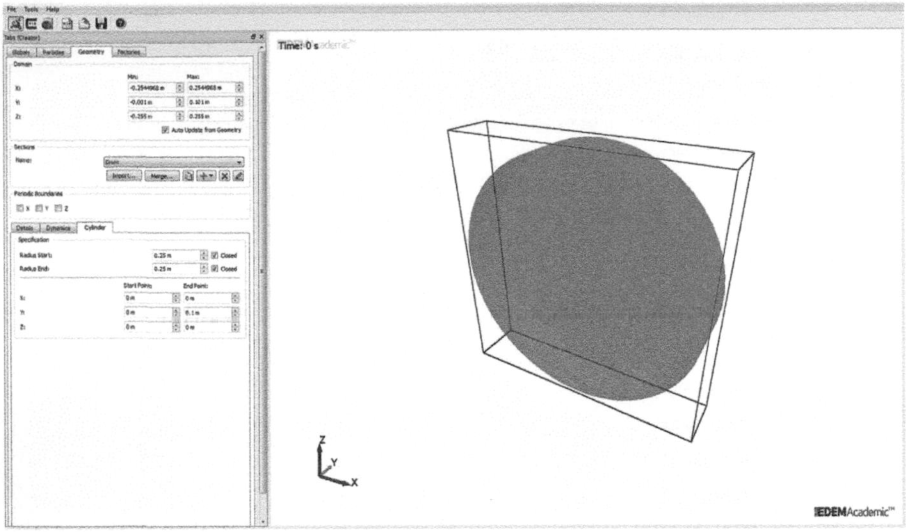

FIGURE 9.17 EDEM® user interface (Geometry sub-section of Creator tab).

In the Factories sub-section of the Creator tab (see Figure 9.18), the user defines how the particles are generated. In EDEM®, the particle generator is known as factory. There are two types of factories to choose from; static and dynamic (see A in Figure 9.18). With a static factory, all the particles are generated at once, while the particles can be generated continuously with a dynamic factory. For dynamic factories, the user must define the generation rate, i.e. the number of particles to be generated per second. The user should also define the total number or mass of particles to be generated (see B in Figure 9.18). The particles, by default, are generated randomly in the space with a random orientation and a fixed size, and zero initial angular and translational velocities, but the generation conditions can be altered as necessary. These properties can be altered as shown by C in Figure 9.18.

In Simulator tab (see Figure 9.19), the user defines the simulation properties such as time-step and total simulation time. The time-step can be set based on a fraction of the Rayleigh time-step (see Section 9.1.2). 'Target Save Interval' is the time-step for saving the information of the particles, contacts, and geometries. A small value for 'Target Save Interval' provides the information more frequently, but that will increase the size of data saved on a computer hard disk. This parameter must be set according to the dynamics of the system that is simulated. EDEM® and most of the DEM codes make use of parallel-computing in order to increase the computational power; hence providing faster simulations.

To this end, before running a simulation, the number of processors can be set. In Figure 9.19 the number of processors (cores) is set to 2. As presented in this figure, the grid size for contact detection can be set based on fractions of the smallest particle in the simulation. As discussed in Section 9.1.5, the grid size does not affect the accuracy of the calculation, however, determines the simulation speed. A grid size of 3 to 5 times the smallest particle radius is found to be the optimum range for most of the simulations.

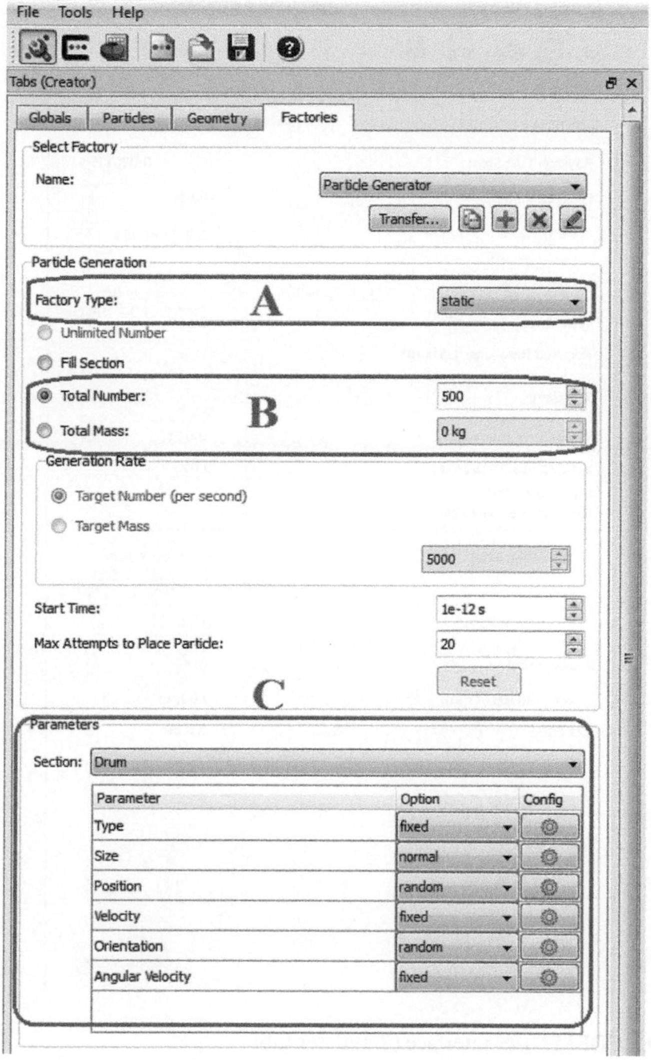

FIGURE 9.18 EDEM® user interface (Factories sub-section of Creator tab).

The Analyst tab is used for analysing the simulation results (see Figure 9.20). The particles and geometries can be visualised as a 3D image or movie. Particles and geometries can be coloured based on various attributes such as velocity, size, and force. The data is available for individual particles and individual contacts, which can be used for quantitative analysis. The particle data includes the number of particles, their mass, position, moment of inertia, force, torque, kinetic energy, potential energy, rotational energy, and translational and rotational velocities. The contact data includes the number of contacts, normal and tangential contact forces, normal and tangential overlaps, contact vectors, and position.

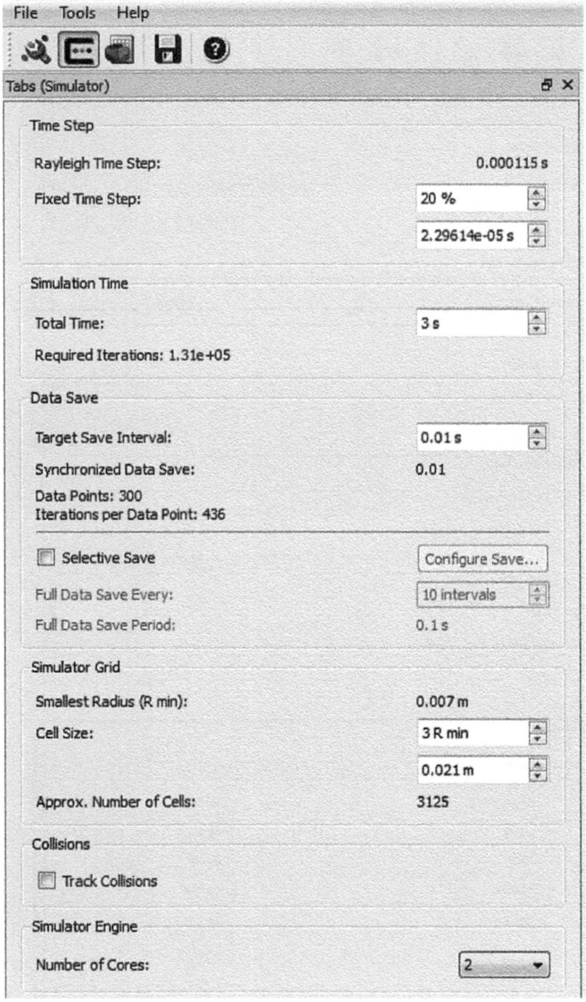

FIGURE 9.19 EDEM® user interface (Simulator tab).

9.2.2 Simulation of a Rotating Drum

9.2.2.1 Simulation Setup

For this simulation the same material properties for particles and geometry of the drum were used. These properties are summarised in Table 9.1.

The interaction properties were set as given in Table 9.2. It should be noted that these properties (both material and interaction) do not refer to a specific material and are only used as an example for the demonstration. The coefficient of sliding friction between the particles and the geometry (drum) is deliberately given higher than that of particle-particle so that the inner wall of the rotating drum could grip the particles.

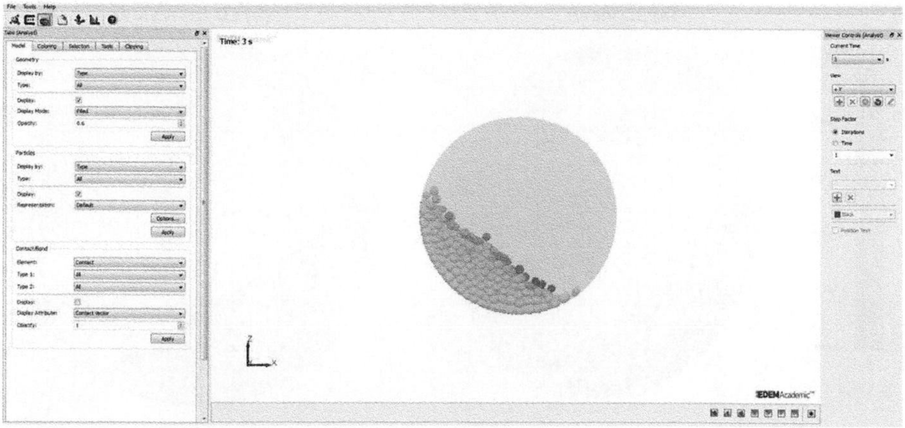

FIGURE 9.20 EDEM® user interface (Analyst tab).

500 particles with a mean size of 2 cm (in diameter) and with a normal size distribution of 0.2 cm standard deviation were generated inside a cylindrical drum with a radius and a width of 25 and 10 cm, respectively. The drum (with 0.25 m radius) was given a rotational motion with a constant rotational speed of 20 rpm around the y-axis. The simulation was performed with a time-step equal to 20% of Rayleigh

TABLE 9.1
Material Properties Used in Simulation of the Rotating Drum

Property	Value
Poisson's ratio (-)	0.25
Shear modulus (MPa)	100
Density (kg/m^3)	2500

TABLE 9.2
Interaction Properties Used in Simulation of the Rotating Drum

Interaction	Particle–Particle	Particle–Geometry
Coeff. of restitution	0.5	0.5
Coeff. of sliding friction	0.3	0.8
Coeff. of rolling friction	0.01	0.01

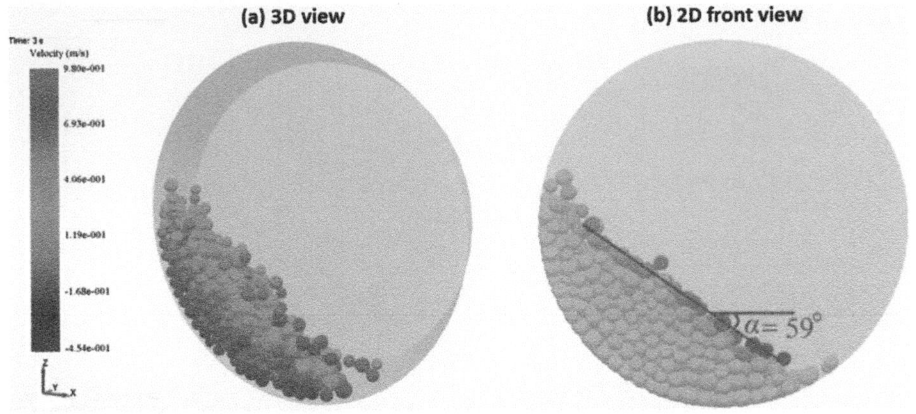

FIGURE 9.21 (a) 3D representation of the simulated rotating drum system with the particles coloured based on their x-component of velocity. (b) Avalanching angle measurement from 3D representation. (See color insert.)

time-step for a total time of 3 s. The 'Target Save Interval' was set to 0.01 s. This simulation took about 2 minutes to complete using 2 cores of a workstation having two Intel Xeon quad-core (3.0 GHz) processors.

9.2.2.2 Analysis

In order to analyse the simulation results, first the simulated system is visualised. Figure 9.21 (a) shows a 3D representation of the simulated rotating drum system with the particles coloured based on their x-component of velocity.

The angle that the surface of avalanching particle bed makes with the horizon inside a rotating drum could be regarded as the internal angle of friction [52], which could be an implication of powder flow behaviour. For cohesive and less free flowing powders this angle is larger than that for free-flowing powder. Using the 2D representation of the heap, one is able to measure this angle in DEM, as shown in Figure 9.22 (b).

For further investigation, the sensitivity of the avalanching angle and average particle velocity to the material properties such as size, cohesion, and Young's modulus, as well as interaction properties such as sliding friction between the particles and particles/geometry can be analysed.

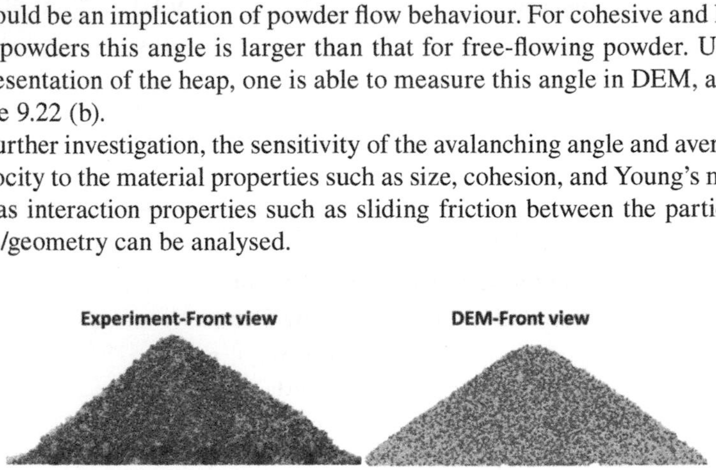

FIGURE 9.22 Heap formation of a binary mixture of a home washing powder, simulated experimentally and numerically [38].

9.3 APPLICATION OF DEM IN PROCESS ENGINEERING

Zhu et al. [3] have categorised the applications of DEM into particle packing; particle flow and particle fluid flow. In this section, case studies from all three major categories will be reviewed.

9.3.1 Particle Packing

Understanding the packing structure of particles and attaining certain structure for particulate assemblies is desirable for various applications such as thermal conductivity, permeability, and dispersion. Microscopically, at single particle level, packing is highly heterogeneous which greatly affects the properties of the assemblies. Particle packing can be dynamically achieved by gravitational pouring and followed by compression and compaction. Inter-particle forces, such as friction, elastic forces, and cohesion, can greatly affect the particle packing structure and its behaviour during compression and compaction. Numerous researchers have looked at various aspects of packing: e.g., packing under gravity [53-55]; compression [56-59] and vibration after deposition [55,60].

Packing under gravity could be classified as confined packing and unconfined packing. Heap formation is an example of unconfined packing as presented in Figure 9.22.

Analysis of packing usually involves macroscopic properties such as the angle of repose, packing fraction, and density or microscopic properties, such as the particle coordination number, packing density distribution, and stress distribution. Yang et al. [61] studied unconfined particle packing under gravity and investigated the effect of particle size on the packing structure. They showed that the bed porosity decreases, hence assembly becomes denser, as the particle size is increased. Liffman et al. [62] investigated the effect of particle size on the force network in sand piles. They predicted a stress dip in the sand pile, which could be observed when different particle sizes are simulated.

Particle confined packing have been studied by various researchers. Figure 9.23 shows an example of particles confined in a box where compression is imposed by an upper moving platen [59]. In this particular study, in addition to force network and porosity, deformation and damage to individual particles were also studied.

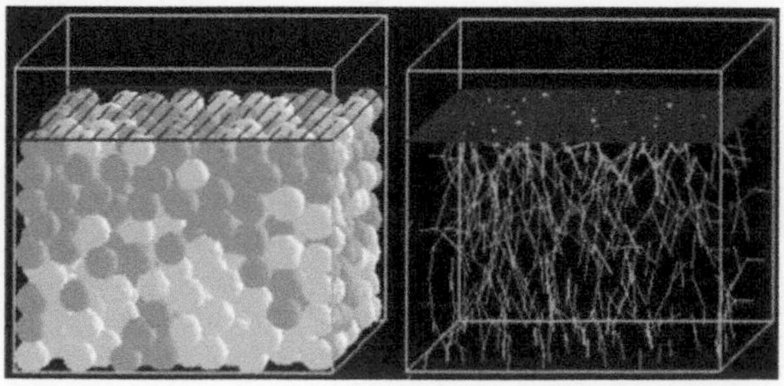

FIGURE 9.23 Simulated assemblies of particles (left image) where damaged particles are highlighted with grey colour and distribution of forces in beds (right image) [59].

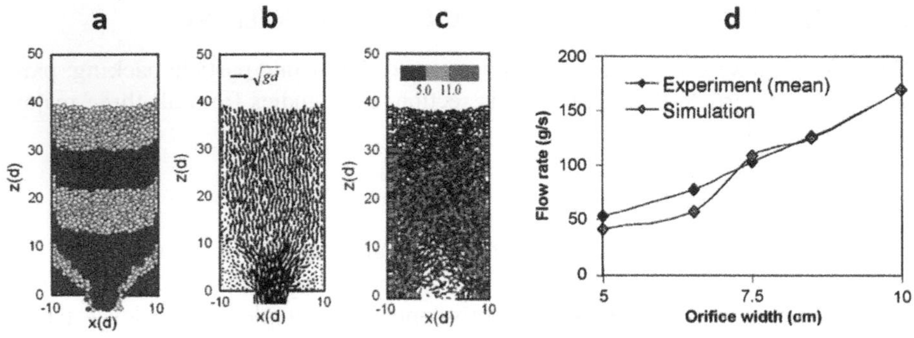

FIGURE 9.24 (a) flow pattern; (b) velocity field; and (c) force structure of the granular flow in a cylindrical hopper [66], and (d) discharge flow rate out of a hopper as a function of orifice width [63]. (See color insert.)

9.3.2 Particle Flow

Particle flow could be driven by gravity or external forces. Particle discharge from silos or hoppers is an example of particle flow under gravity. Common problems in such plants are non-uniform and/or the interruption of flow from the discharge orifice in the hopper, or converging section beneath a storage vessel for powders. DEM simulations have been extensively used to study such systems [63–68].

Parameters such as discharge rate, particle velocity patterns, and stress distribution (Figure 9.24 (a–c)) were studied as a function of orifice size and geometry of the vessel (Figure 9.24 (d)).

Particle motion in rotating drums and mixers are examples of particle flowing under external forces. The external force could be due to the frictional forces induced from the vessel walls such as v-blenders (Figure 9.25 (a)) and rotating drums (Figure 9.25 (b)) or using a mechanical stirrer or agitators such as blades (Figure 9.25 (c)) and paddles (Figure 9.26).

In such systems, common parameters to be analysed are particle mixing and segregation indices (Figure 9.25 (a)), stress distribution (Figure 9.25 (c)), residence

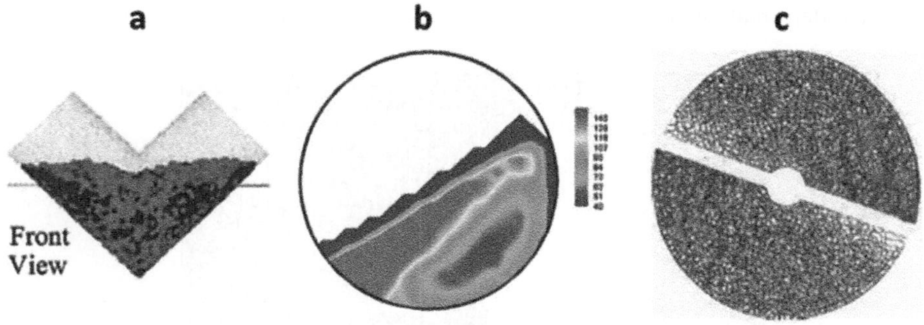

FIGURE 9.25 DEM simulation of (a) mixing in a v-blender [71], (b) particle collision velocity (mm/s), and collision frequency (1/s) in a cylindrical rotating drum [72], and (c) force network in a horizontal section of a bladed mixer [73]. (See color insert.)

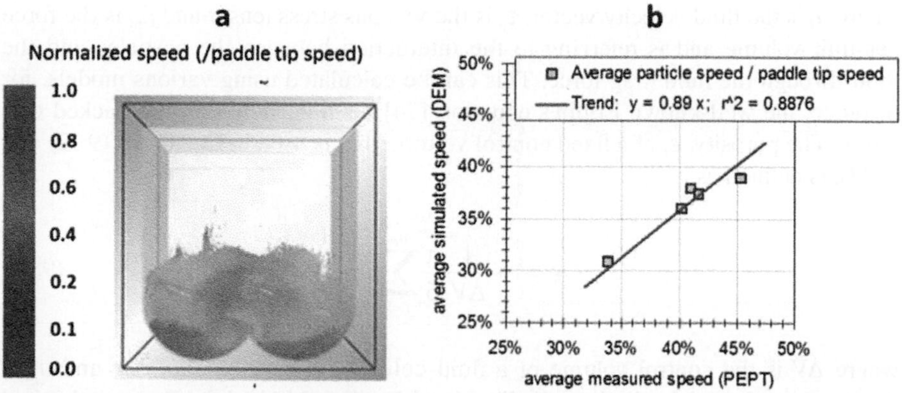

FIGURE 9.26 (a) The schematic flow fields inside the paddle mixer simulated by DEM. (b) Comparison of DEM simulation vs. PEPT measurement of time-averaged particle speed, normalized to paddle tip speed, for a range of conditions with different mixer speed, particle density and mixer fill level [69]. (See color insert.)

time distribution, collisions frequency and motion analysis (Figure 9.25 (b)). Hassanpour *et al.* [69] analysed the motion of particles in a commercial paddle mixer (Figure 9.26 (a)). The particle behaviour such as velocity pattern as well as quantitative powder dynamics in terms of particle mixing and velocity distribution were analysed under various process conditions (rotational speed, fill level and particle density). They compared the DEM prediction with the experimental measurements based on Positron Emission Particle Tracking (PEPT) [70], where an acceptable agreement was obtained (Figure 9.26 (b)).

9.3.3 Particle Fluid Flow

In recent years, DEM has been used in combination with computational fluid dynamics (CFD) aiming at investigating particulate behaviour in fluid phase. For a two-phase particle-fluid system, the solids motion and fluid mechanics are solved through the application of Newton's equations of motion for the discrete particles and Navier-Stokes equations for the continuum fluid [2]. The governing equations for the discrete phase (particles) have been already discussed in Section 9.1 and the description of the CFD method is given in Chapter 8. In brief, in this method the continuity and Navier-Stokes equations for the fluid phase in the fluid-solid two-phase model, for an incompressible fluid with constant density, are given by Equations (9.32) and (9.33), respectively:

$$\frac{\partial \varepsilon}{\partial t} + \nabla \cdot (\varepsilon u_f) = 0 \tag{9.32}$$

$$\frac{\partial (\varepsilon u_f)}{\partial t} = -\nabla \cdot (\varepsilon u_f u_f) - \frac{\varepsilon}{\rho_f} \nabla P - \frac{\varepsilon}{\rho_f} \nabla \cdot \tau_v + \varepsilon g + \frac{f_{\text{int}}}{\rho_f} \tag{9.33}$$

where u_f is the fluid velocity vector, τ_v is the viscous stress tensor and f_{int} is the force per unit volume and is referring to the interaction between the particles and the fluid through the fluid drag force. This can be calculated using various models, for instance, the well-known Ergun's equation [74] for fluid flow through packed columns. The porosity, ε, of a fixed control volume that is used in Equations (9.32) and (9.33), is defined as,

$$\varepsilon = 1 - \frac{1}{\Delta V}\frac{\pi}{6}\sum_{i=1}^{n_p} D_{pi}^3 \tag{9.34}$$

where ΔV is the control volume of a fluid cell, D_{pi} is particle diameter and n_p is the number of particles in the cell. The Semi-Implicit Method for Pressure Linked Equations (SIMPLE) algorithm [75] is used for the computation of pressure and velocity vectors of the fluid in each cell. Further details of the coupled DEM-CFD method can be found elsewhere [76].

As stated by Zhu *et al.* [3], the major applications of this method are in particle fluidisation, pneumatic conveying and process studies for which examples are provided below. DEM coupled with CFD has been used widely for analysing particulate behaviour in fluid systems such as gas fluidisation with non-cohesive [77, 78] and cohesive particles [79], pneumatic conveying [80], and particle dispersion in liquids [81–83] and gases [84, 85].

9.3.3.1 Fluidisation

One of the widely analysed particulate processes involving fluid in process engineering is fluidisation, where a bed of particle is mobilised using a fluid flow. An important feature of the fluidised bed is that several processing steps (e.g. mixing, reaction, coating and drying) may be carried out in the same vessel, making it very attractive unit operation in process engineering. Analysis and understanding of particle motion, mixing pattern, and inter-particle forces are important to design a better fluidisation unit. Combined DEM-CFD has been extensively used to model particle behaviour during fluidisation. Figure 9.27 shows the results of a study undertaken by Feng et al. [86]. In this work particle motion and particle segregation due to the size as well as regimes of fluidisation have been studied. Figure 9.27 clearly illustrates bubble formation during the fluidisation and complete segregation due to the difference in particle size which may not be desirable in these systems.

9.3.3.2 Pneumatic Conveying

Pneumatic conveying is a common method for transportation of particulate solids within or between processing plants. Particles are mobilised commonly using air and transported inside pipes or ducts. To attain a consistent flow of particles, particle mobilisation and fluid pressure drop should be understood in detail. Stationary particles and excessive pressure drop could halt the flow. Figure 9.28 shows a study by

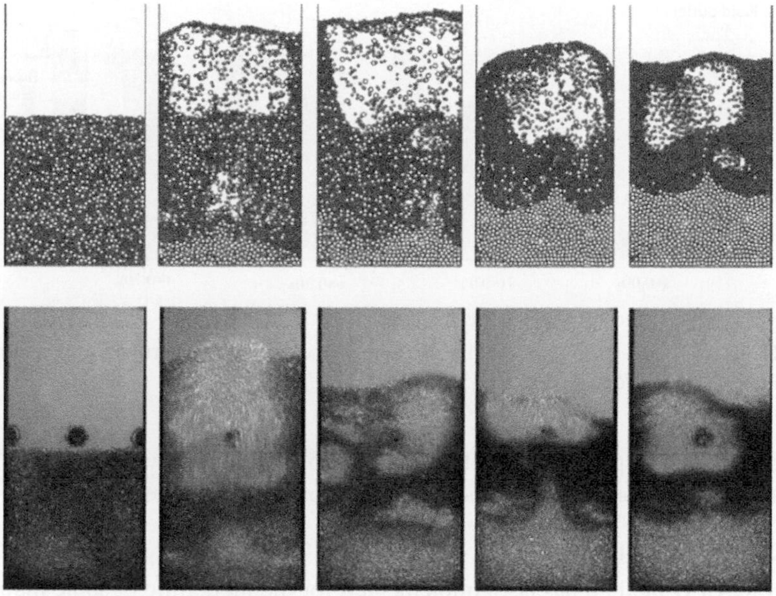

FIGURE 9.27 DEM-CFD simulation of segregation in gas fluidization of binary mixtures of particles (top) and comparison with experiments (bottom) [86].

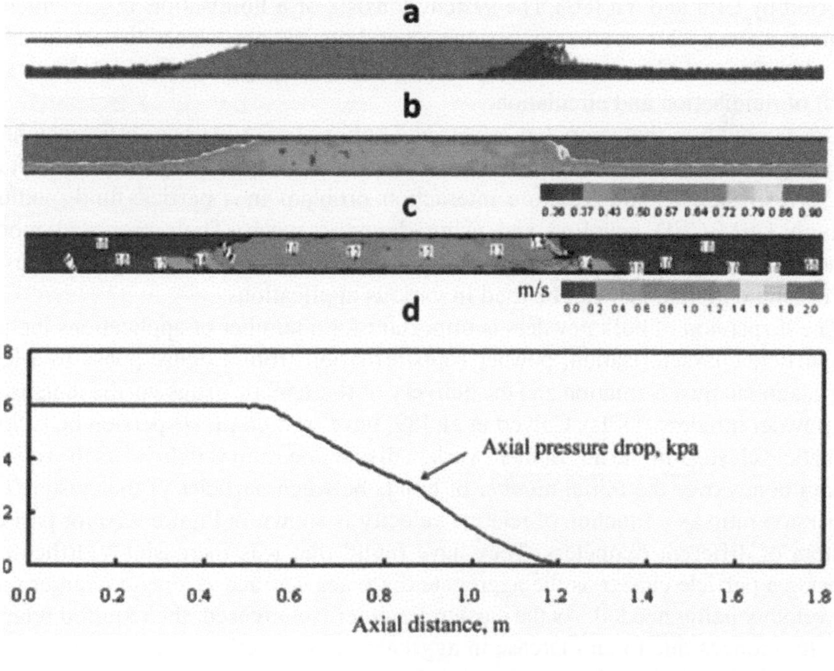

FIGURE 9.28 Simulation of distributions of flow properties: (a) particle configurations at 14.498 s, (b) porosity, (c) axial particle velocity, and (d) axial pressure drop [87]. (See color insert.)

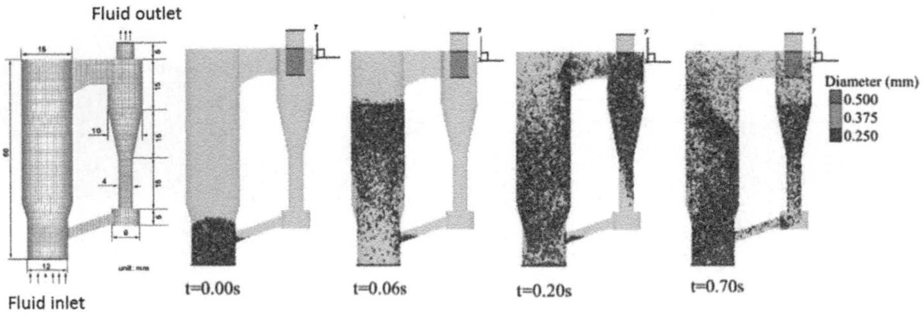

FIGURE 9.29 DEM-CFD simulation of particle size distribution (particles are coloured by size) in circulating fluidised bed [80]. (See color insert.)

Kuang et al. [87] on particle-gas behaviour in a horizontal pipe with a view to investigate the particle porosity and velocity distribution as well as the gas pressure drop.

9.3.3.3 Process Studies

The DEM-CFD method has been implemented for the modelling of some industrial processes such as complex processes, gas cyclones (for separating particles from fluids) and powder flow and dispersion. Figure 9.29 shows an example of a coupled DEM-CFD simulation of a complex process flow in a circulating fluidised bed reported by Chu and Yu [80]. The system consists of a fluidisation vessel which is connected to a cyclone at the top to separate and recycle particles with certain sizes. The aim of this study was to investigate the particle segregation due to the size as a result of fluidisation and circulation.

Development in the computers power and enhancement in codes efficiency have encouraged researchers to simulate more complex problems. For example, Ye et al. [88] investigated a fluid-structure interaction problem in a particle-fluid medium through DEM-CFD coupling and using dynamic mesh. Their proposed model demonstrates a great potential to simulate complex and free-moving geometries in particle-fluid media, and can be used in various applications.

The dispersion of bulk powders is important for a number of applications including particle characterisation, powder transformation from a dense phase to a lean phase, aeolian dust formation and the delivery of therapeutic drugs via the lung using dry powder inhalers (DPIs). Calvert et al. [89] have studied the dispersion behaviour of particle clusters in the air. In their work, a dispersion ratio is defined as the ratio of broken bonds over the initial number of bonds between particles of the cluster. The dispersion ratio as a function of relative velocity is shown in Figure 9.30 for particle clusters of different diameters. They have found that it is increasingly difficult to disperse a particle cluster as the aggregate diameter is reduced, due to a larger relative velocity being needed. As the cluster diameter is increased, the required relative velocity reduces due to an increase in aggregate cross-sectional area. Calvert *et al.* [89] report that for these large aggregates (the size ratio between the aggregates and primary particles being greater than 20), dispersion appears as though the particles are peeling away from the aggregate surface (Figure 9.30 (right image)).

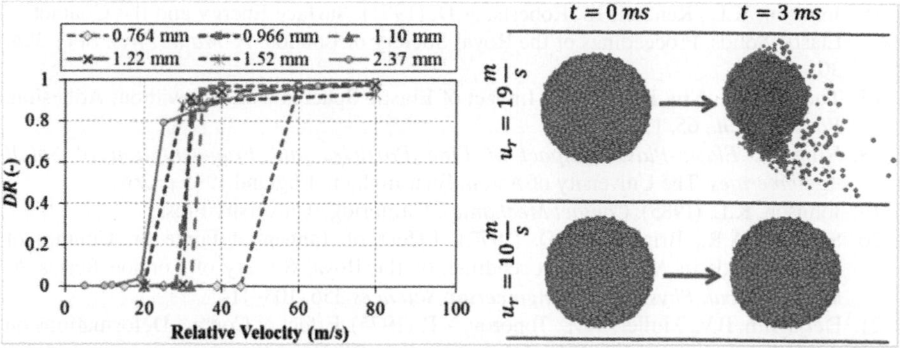

FIGURE 9.30 Simulation of the relationship between dispersion ratio (DR) and relative velocity between the fluid and particles for clusters with different diameters (graph on the left) and dispersion behaviour of a large aggregate in different air velocities (right image) [89].

REFERENCES

1. Cundall, P.A., Strack, O.D.L. (1979). A Discrete Numerical Model for Granular Assemblies. *Geotechnique*, 29, 47–65.
2. Zhu, H.P., Zhou, Z.Y., Yang, R.Y., Yu, A.B. (2007). Discrete Particle Simulation of Particulate Systems: Theoretical Developments. *Chem. Eng. Sci.,* 62, 3378–3396.
3. Zhu, H.P., Zhou, Z.Y., Yang, R.Y., Yu, A.B. (2008). Discrete Particle Simulation of Particulate Systems: A Review of Major Applications and Findings. *Chem. Eng. Sci.,* 63, 5728–5770.
4. Ning, Z., Ghadiri, M. (2006). Distinct Element Analysis of Attrition of Granular Solids under Shear Deformation. *Chem. Eng. Sci.,* 61, 5991–6001.
5. DEM Solutions (2009). *EDEM User Manual.* In: DEMSolutions (Ed.), DEMSolutions.
6. Hertz, H. (1882) Crellés J. 92, 156.
7. Thornton, C., Ning, Z. (1998). A Theoretical Model for the Stick/Bounce Behaviour of Adhesive, Elastic-Plastic Spheres. *Powder Tech.* 99, 154–162.
8. Walton, O.R., Braun, R.L. (1986). Viscosity, Granular-Temperature, and Stress Calculations for Shearing Assemblies of Inelastic, Frictional Disks. *J. Rheo* 30, 949–980.
9. Thornton, C. (1997). Coefficient of Restitution for Collinear Collisions of Elastic-Perfectly Plastic Spheres. *J. Ap. Mech. Transactions ASME* 64, 383–386.
10. Pasha, M., Dogbe, S., Hare, C., Hassanpour, A., Ghadiri, M. (2013). A New Linear Contact Model for Elasto-Plastic and Adhesive Contacts in Distinct Element Method. *Granular Matter.* 16 (1), 151–162.
11. Mindlin, R.D., Deresiewicz, H. (1953). Elastic Spheres in Contact under Varying Oblique Forces. *J. Ap. Mech.,* 20, 327–344.
12. Courtney-Pratt, J.S., Eisner, E. (1957). The Effect of a Tangential Force on the Contact of Metallic Bodies. Proceedings of the Royal Society of London. Series A. *Math. Phys. Sci.,* 238, 529–550.
13. Mindlin, R.D. (1949). Compliance of elastic bodies in contact. *J. Ap. Mech.,* 16 (1949), 259–268.
14. Di Renzo, A., Di Maio, F.P. (2004). Comparison of Contact-Force Models for the Simulation of Collisions in Dem-Based Granular Flow Codes. *Chem. Eng. Sci.,* 59, 525–541.
15. Kharaz, A.H., Gorham, D.A., Salman, A.D. (2001). An Experimental Study of the Elastic Rebound of Spheres. *Powder Tech.,* 120, 281–291.

16. Johnson, K.L., Kendall, K., Roberts, A.D. (1971). Surface Energy and the Contact of Elastic Solids. Proceedings of the Royal Society of London. A. *Math. Phys. Sci.,* 324, 301–313.
17. Thornton, C., Yin, K.K. (1991). Impact of Elastic Spheres with and without Adhesion. *Powder Tech.,* 65, 153–166.
18. Ning, Z. *Elasto-Plastic Impact of Fine Particles and Fragmentation of Small Agglomerates.* The University of Aston, Birmingham, England 1995 p. 269.
19. Johnson, K.L. (1985). *Contact Mechanics.* Cambridge University Press.
20. Savkoor, A.R., Briggs, G.A.D. (1977). Effect of Tangential Force on Contact of Elastic Solids in Adhesion. Proceedings of the Royal Society of London Series A: *Mathematical, Physical and Engineering Sciences* 356, 103–114.
21. Derjaguin, B.V., Muller, V.M., Toporov, Y.P. (1975). Effect of Contact Deformations on the Adhesion of Particles. *J. Coll. Interface Sci.,* 53, 314–326.
22. Pashley, M.D. (1984). Further Consideration of the DMT Model for Elastic Contact. *Colloids and Surfaces* 12, 69–77.
23. Visser, J. (1989). Van Der Waals and Other Cohesive Forces Affecting Powder Fluidization. *Powder Tech.,* 58, 1–10.
24. Hamaker, H.C. (1937). The London—Van Der Waals Attraction between Spherical Particles. *Physica* 4, 1058–1072.
25. Seville, J.P.K., Willett, C.D., Knight, P.C. (2000). Interparticle Forces in Fluidisation: A Review. *Powder Tech.,* 113, 261–268.
26. Washino, K., Tan, H.S., Hounslow, M.J., Salman, A.D. (2013). A New Capillary Force Model Implemented in Micro-Scale CFD–DEM Coupling for Wet Granulation. *Chem. Eng. Sci.,* 93, 197–205.
27. Lian, G., Thornton, C., Adams, M.J. (1993). A Theoretical Study of the Liquid Bridge Forces between Two Rigid Spherical Bodies. *J. Coll. Interface Sci.* 161, 138–147.
28. Šupuk, E., Hassanpour, A., Ahmadian, H., Ghadiri, M., Matsuyama, T. (2011). Tribo-Electrification and Associated Segregation of Pharmaceutical Bulk Powders. *Kona Powder Particle J.* 29, 208–223.
29. Pei, C., Wu, C.-Y., England, D., Byard, S., Berchtold, H., Adams, M. (2013). Numerical Analysis of Contact Electrification Using DEM–CFD. *Powder Tech.,* 248, 34–43.
30. He, Y., Bayly, A.E., Hassanpour, A., Muller, F., Wu, K., Yang, D. (2018). A GPU-Based Coupled SPH-DEM Method for Particle-Fluid Flow with Free Surfaces. *Powder Tech.,* 338, 548–562.
31. Favier, J.F., Abbaspour-Fard, M.H., Kremmer, M., Raji, A.O. (1999). Shape Representation of Axisymmetrical, Non-Spherical Particles in Discrete Element Simulation Using Multi-Element Model Particles. *Eng Comput.* 16, 467–480.
32. Cordelair, J., Greil, P. (2004). Discrete Element Modeling of Solid Formation During Electrophoretic Deposition. *J. Mat. Sci.,* 39, 1017–1021.
33. Ting, J.M., Meachum, L., Rowell, J.D. (1995). Effect of Particle Shape on the Strength and Deformation Mechanisms of Ellipse-Shaped Granular Assemblages. *Eng Comput.,* 12, 99–108.
34. Morgan, J.K. (2004). Particle Dynamics Simulations of Rate- and State-Dependent Frictional Sliding of Granular Fault Gouge. *Pure Ap. Geophys.,* 161, 1877–1891.
35. Zhou, Y.C., Wright, B.D., Yang, R.Y., Xu, B.H., Yu, A.B. (1999). Rolling Friction in the Dynamic Simulation of Sandpile Formation. *Physica A: Statistical Mechanics and Its Applications,* 269, 536–553.
36. Ai, J., Chen, J.-F., Rotter, J.M., Ooi, J.Y. (2011). Assessment of Rolling Resistance Models in Discrete Element Simulations. *Powder Tech.,* 206, 269–282.
37. Cleary, P.W. (2010). Dem Prediction of Industrial and Geophysical Particle Flows. *Particuology,* 8, 106–118.

38. Alizadeh, M., Hassanpour, A., Pasha, M., Ghadiri, M., Bayly, A. (2017). The Effect of Particle Shape on Predicted Segregation in Binary Powder Mixtures. *Powder Tech.,* 319, 313–322.
39. Kodam, M., Bharadwaj, R., Curtis, J., Hancock, B., Wassgren, C. (2009). Force Model Considerations for Glued-Sphere Discrete Element Method Simulations. *Chem. Eng. Sci.,* 64, 3466–3475.
40. Höhner, D., Wirtz, S., Kruggel-Emden, H., Scherer, V. (2011). Comparison of the Multi-Sphere and Polyhedral Approach to Simulate Non-Spherical Particles within the Discrete Element Method: Influence on Temporal Force Evolution for Multiple Contacts. *Powder Tech.,* 208, 643–656.
41. Nassauer, B., Liedke, T., Kuna, M. (2013). Polyhedral Particles for the Discrete Element Method. *Granular Matter,* 15, 85–93.
42. Boon, C.W., Houlsby, G.T., Utili, S. (2012). A New Algorithm for Contact Detection between Convex Polygonal and Polyhedral Particles in the Discrete Element Method. *Comput. Geotech.,* 44, 73–82.
43. Hogue, C. (1998). Shape Representation and Contact Detection for Discrete Element Simulations of Arbitrary Geometries. *Eng. Comput.,* 15, 374-+.
44. Pournin, L., Tsukahara, M., Liebling, T.M. (2009). Particle Shape Versus Friction in Granular Jamming, In: Nakagawa, M., Luding, S. (Eds.) *Powders and Grains*, pp. 499–502.
45. Delaney, G.W., Cleary, P.W. (2009). Fundamental Relations between Particle Shape and the Properties of Granular Packings, In: Nakagawa, M., Luding, S. (Eds.) *Powders and Grains*, pp. 837–840.
46. Jia, X., Gan, M., Williams, R.A., Rhodes, D. (2007). Validation of a Digital Packing Algorithm in Predicting Powder Packing Densities. *Powder Tech.,* 174, 10–13.
47. Caulkin, R., Jia, X., Xu, C., Fairweather, M., Williams, R., Stitt, H., Nijemeisland, M., Aferka, S., Crine, M., Léonard, A. (2008). Simulations of Structures in Packed Columns and Validation by X-Ray Tomography. *Ind. Eng. Chem. Res.,* 48, 202–213.
48. Kloss, C., Goniva, C., Hager, A., Amberger, S., Pirker, S. (2012). Models, Algorithms and Validation for Opensource DEM and CFD-DEM. *Progress in Computational Fluid Dynamics* 12, 140–152.
49. EDEM. DEMSolutions. Edimburgh, UK https://www.edemsimulation.com/software/
50. https://www.itascacg.com/software/pfc (2011). *Particle Flow Code in 3 Dimensions (Pfc3d).* Itasca Consulting Group Inc.
51. Alizadeh, M., Asachi, M., Ghadiri, M., Bayly, A., Hassanpour, A. (2018). A Methodology for Calibration of DEM Input Parameters in Simulation of Segregation of Powder Mixtures, a Special Focus on Adhesion. *Powder Tech,* 339, 789–800.
52. Pohlman, N.A., Roberts, J.A., Gonser, M.J. (2012). Characterization of Titanium Powder: Microscopic Views and Macroscopic Flow. *Powder Tech.,* 228, 141–148.
53. Latham, J.P., Munjiza, A. (2004). The Modelling of Particle Systems with Real Shapes. *Philosophical Transactions of the Royal Society of London Series A: Mathematical, Physical and Engineering Sciences,* 362, 1953–1972.
54. Munjiza, A., Latham, J.P. (2004). Comparison of Experimental and Fem/Dem Results for Gravitational Deposition of Identical Cubes. *Eng. Comp.,* 21, 249–264.
55. An, X.Z., Yang, R.Y., Dong, K.J., Zou, R.P., Yu, A.B. (2005). Micromechanical Simulation and Analysis of One-Dimensional Vibratory Sphere Packing. *Phys. Rev. Lett.,* 95(20): 205502.
56. Luding, S. (2004). Micro-Macro Transition for Anisotropic, Frictional Granular Packings. *Int. J. Solid. Struct.,* 41, 5821–5836.
57. Zhang, H.P., Makse, H.A. (2005). Jamming Transition in Emulsions and Granular Materials. *Physical Review E* 72, 011301, 1–12.
58. Hassanpour, A., Ghadiri, M. (2004). Distinct Element Analysis and Experimental Evaluation of the Heckel Analysis of Bulk Powder Compression. *Powder Tech.* 141, 251–261.

59. Samimi, A., Hassanpour, A., Ghadiri, A. (2005). Single and Bulk Compressions of Soft Granules: Experimental Study and DEM Evaluation. *Chem. Eng. Sci.,* 60, 3993–4004.
60. Yu, A.B., An, X.Z., Zou, R.P., Yang, R.Y., Kendall, K. (2006). Self-Assembly of Particles for Densest Packing by Mechanical Vibration. *Physical Review Letters* 97, 011301, 1–12.
61. Yang, R.Y., Zou, R.P., Yu, A.B. (2000). Computer Simulation of the Packing of Fine Particles. *Physical Review E* 62, 3900–3908.
62. Liffman, K., Nguyen, M., Metcalfe, G., Cleary, P. (2001). Forces in Piles of Granular Material: An Analytic and 3d Dem Study. *Granular Matter* 3, 165–176.
63. Favier, J.F., Abbaspour-Fard, M.H., Kremmer, M. (2001). Modeling Nonspherical Particles Using, Multisphere Discrete Elements. *J. Eng. Mech.—ASCE* 127, 971–977.
64. Wassgren, C.R., Hunt, M.L., Freese, P.J., Palamara, J., Brennen, C.E. (2002). Effects of Vertical Vibration on Hopper Flows of Granular Material. *Physics of Fluids* 14, 3439–3448.
65. Parisi, D.R., Masson, S., Martinez, J. (2004). Partitioned Distinct Element Method Simulation of Granular Flow within Industrial Silos. *J. Eng. Mech.—ASCE* 130, 771–779.
66. Zhu, H.P., Yu, A.B. (2005). Steady-State Granular Flow in a 3d Cylindrical Hopper with Flat Bottom: Macroscopic Analysis. *Granular Matter* 7, 97–107.
67. Zhu, H.P., Yu, A.B., Wu, Y.H. (2006). Numerical Investigation of Steady and Unsteady State Hopper Flows. *Powder Tech.,* 170, 125–134.
68. Datta, A., Mishra, B.K., Das, S.P., Sahu, A. (2008). A Dem Analysis of Flow Characteristics of Noncohesive Particles in Hopper. *Materials and Manufacturing Processes,* 23, 196–203.
69. Hassanpour, A., Tan, H.S., Bayly, A., Gopalkrishnan, P., Ng, B., Ghadiri, M. (2011). Analysis of Particle Motion in a Paddle Mixer Using Discrete Element Method (DEM). *Powder Tech.,* 206, 189–194.
70. Parker, D.J., Forster, R.N., Fowles, P., Takhar, P.S. (2002). Positron Emission Particle Tracking Using the New Birmingham Positron Camera. *Nuclear Instruments & Methods in Physics Research Section A: Accelerators Spectrometers Detectors and Associated Equipment,* 477, 540–545.
71. Moakher, M., Shinbrot, T., Muzzio, F.J. (2000). Experimentally Validated Computations of Flow, Mixing and Segregation of Non-Cohesive Grains in 3d Tumbling Blenders. *Powder Tech.,* 109, 58–71.
72. Yang, R.Y., Zou, R.P., Yu, A.B. (2003). Microdynamic Analysis of Particle Flow in a Horizontal Rotating Drum. *Powder Tech.,* 130, 138–146.
73. Zhou, Y.C., Yu, A.B., Stewart, R.L., Bridgwater, J. (2004). Microdynamic Analysis of the Particle Flow in a Cylindrical Bladed Mixer. *Chem. Eng. Sci.,* 59, 1343–1364.
74. Ergun, S. (1952). Fluid Flow through Packed Columns. *Chemical Engineering Progress* 48, 89–94.
75. Patankar, S.V. (1980). Numerical Heat Transfer and Fluid Flow / Suhas V. Patankar. Hemisphere Pub. Corp., McGraw-Hill, Washington, New York.
76. Afkhami, M., Hassanpour, A., Fairweather, M., Njobuenwu, D.O. (2015). Fully Coupled LES-DEM of Particle Interaction and Agglomeration in a Turbulent Channel Flow. *Comput. Chem. Eng.,* 78, 24–38.
77. Tsuji, Y., Kawaguchi, T., Tanaka, T. (1993). Discrete Particle Simulation of Two-Dimensional Fluidized Bed. *Powder Tech.,* 77, 79–87.
78. Di Renzo, A., Di Maio, F.P., Girimonte, R., Formisani, B. (2008). Dem Simulation of the Mixing Equilibrium in Fluidized Beds of Two Solids Differing in Density. *Powder Tech.,* 184, 214–223.
79. Moreno-Atanasio, R., Xu, B.H., Ghadiri, M. (2007). Computer Simulation of the Effect of Contact Stiffness and Adhesion on the Fluidization Behaviour of Powders. *Chem. Eng. Sci.,* 62, 184–194.

80. Chu, K.W., Yu, A.B. (2008). Numerical Simulation of Complex Particle-Fluid Flows. *Powder Tech.,* 179, 104–114.
81. Higashitani, K., Iimura, K., Sanda, H. (2001). Simulation of Deformation and Breakup of Large Aggregates in Flows of Viscous Fluids. *Chem. Eng. Sci.,* 56, 2927–2938.
82. Fanelli, M., Feke, D.L., Manas-Zloczower, I. (2006). Prediction of the Dispersion of Particle Clusters in the Nano-Scale—Part I: Steady Shearing Responses. *Chem. Eng. Sci.,* 61, 473–488.
83. Fanelli, M., Feke, D.L., Manas-Zloczower, I. (2006). Prediction of the Dispersion of Particle Clusters in the Nano-Scale—Part II, Unsteady Shearing Responses. *Chemical Engineering Science* 61, 4944–4956.
84. Iimura, K., Suzuki, M., Hirota, M., Higashitani, K. (2009). Simulation of Dispersion of Agglomerates in Gas Phase – Acceleration Field and Impact on Cylindrical Obstacle. *Adv. Powder Tech.,* 20, 210–215.
85. Iimura, K., Yanagiuchi, M., Suzuki, M., Hirota, M., Higashitani, K. (2009). Simulation of Dispersion and Collection Process of Agglomerated Particles in Collision with Fibers Using Discrete Element Method. *Adv. Powder Tech.,* 20, 582–587.
86. Feng, Y.Q., Xu, B.H., Zhang, S.J., Yu, A.B., Zulli, P. (2004). Discrete Particle Simulation of Gas Fluidization of Particle Mixtures. *AICHE J.,* 50, 1713–1728.
87. Kuang, S.B., Chu, K.W., Yu, A.B., Zou, Z.S., Feng, Y.Q. (2008). Computational Investigation of Horizontal Slug Flow in Pneumatic Conveying. *Ind. Eng. Chem. Res.,* 47, 470–480.
88. He, Y., Bayly, A.E., Hassanpour, A. (2018). Coupling CFD-DEM with Dynamic Meshing: A New Approach for Fluid-Structure Interaction in Particle-Fluid Flows. *Powder Tech.,* 325, 620–631.
89. Calvert, G., Hassanpour, A., Ghadiri, M. (2011). Mechanistic Analysis and Computer Simulation of the Aerodynamic Dispersion of Loose Aggregates. *Chem. Eng. Res. Des.,* 89, 519–525.

Section IV

Process Simulation

10 Introduction to gPROMS® for Chemical Engineering

Carlos Amador, Mariano Martín, Laura Bueno

CONTENTS

10.1 INTRODUCTION

gPROMS® stands for general PROcess Modeling System [1]. It is a custom modeling and flowsheeting environment whose main application lies in the modeling for process and equipment design and optimization. It is developed by Process Systems Enterprise, which has its roots in the Department of Chemical Engineering at Imperial College in London.

10.2 gPROMS® BASICS

For a detailed description of how to develop your own models in gPROMS, please see the "Model Developer Guide" included in your gPROMS installation. This document, as well as many others, can be accessed within gPROMS ModelBuilder from the menu Help > Documentation.

10.2.1 Getting Started

When we open gPROMS®, we find three sections on the screen: the main window; to the left side the project tree; and to the right side the Palette or Attributes pane. In the project tree we are to create the different entities of our model. The most basic model must contain:

- Definition of the "variable types," such as temperature, pressure, flow, etc., where a range for the variables is to be defined. This is important, since the solution must lie within these limits.
- Definition of our models. We create a new model by clicking with the left button of the mouse on the folder "Models." We can create as many as we need, and
- Definition of the process, which is the code where an instance of the model is defined, with specific inputs and an operating schedule for a simulation.

There are other more advanced elements such as "tasks", which are a way to generalize operating procedures, or "connection types" which are defined to connect models. We use the examples along this chapter to introduce the use of them.

To the right of the main window we open the different entities created—from the variables dialogue box to the models, tasks, and processes. We see in Figure 10.1 the dialogue box for a model. We can create an icon for the model so that we can use it in flowsheets (similar to the method for typical process simulators; see Chapter 12). Alternatively, we can click on the gPROMS® language to see the equations governing the model or to write our own model (i.e. mass and energy balances, etc.)

When running a calculation in gPROMS® and the tick box "Send results trajectory to gRMS" is ticked, a smaller window also opens; see Figure 10.2. It belongs to gRMS (gPROMS Results Management System). This is a tool that is very useful for plotting results, which will be introduced in the step-by-step example.

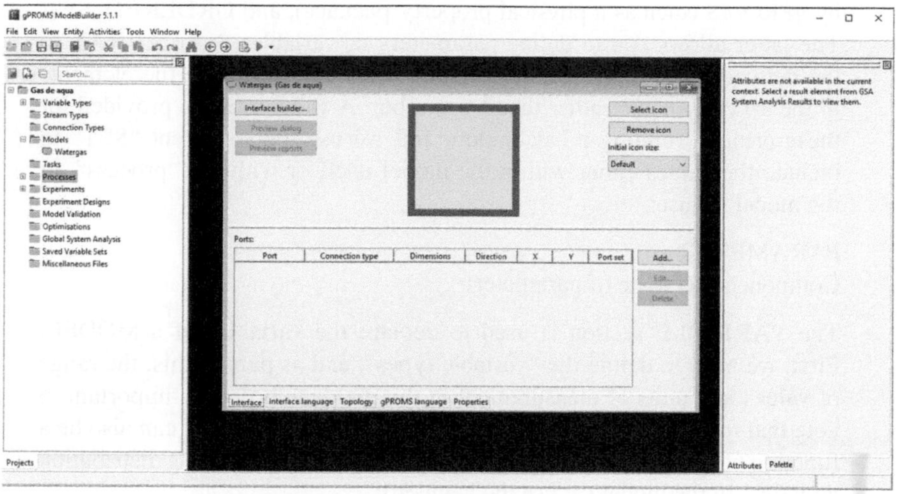

FIGURE 10.1 Initial screen of gPROMS®.

10.2.2 Writing the Models

When we open a new model in the gPROMS language tab, we get a new window that includes a template with the potential different sections depending on our needs. These are commented out by default.

- The PARAMETER section is used to declare the constants (parameters) of a MODEL. The parameters can be real, integer, boolean, each of the type scalar or array (tensor), a FOREIGN_OBJECT, referring to any program used

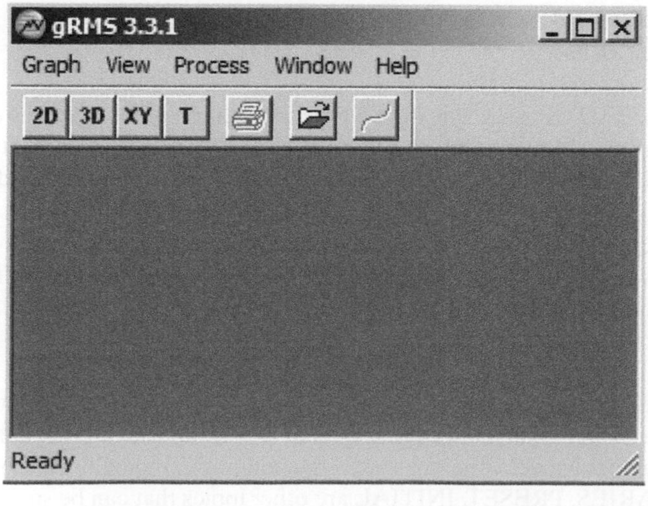

FIGURE 10.2 gRMS main screen.

by gPROMS (such as a physical property package), and ORDERED_SET. The latter allows you to define parameters or variables as an ARRAY of the defined ORDERED_SET, which means we can refer to the elements in the array by name rather than by number. A value must be provided for these prior to running a calculation, and we use the statement "SET" to include the values either within the model itself or within a "process". In the model we use:

PARAMETER
Components as (type of parameter)

- The VARIABLE section is used to declare the variables of a MODEL. First, we need to define the "variable types", and as part of this, the range of values and units of measurement in the dialogue box. It is important to note that in gPROMS® all variables are functions of time and can also be a function of space dimensions (distributed variables) by defining distribution domains. In the model we use the keyword:

VARIABLE
Flow as "variable type"
Flow as array (components) of variable type
Length as distribution (axial) of variable type

- This last one, as a distribution, requires the definition of a "DISTRIBUTION_DOMAIN" and it is particularly useful for distributed variables such as the ones used in tubular reactors. The syntax is as follows:

DISTRIBUTION_DOMAIN
Axial as (0:L)

We can use notype as a general variable too, although it is better to use specific variable types.

In order to provide a value to a variable we can use the statement "ASSIGN" either in the model section or in the process

- The SELECTOR section is used for the declaration of the system states that arise from a discontinuity such as open/close, turbulent/laminar, etc.
- BOUNDARY: This section is used for boundary conditions (additional equations for models containing partial differential equations).
- The EQUATION section is used to declare the equations. There is no need to provide an order of the equations, since all algebraic and differential equations are solved simultaneously. We use the symbol "$" before a variable to define a differential equation in time.

We can relate the variables of the different models within our project so that we can calculate them in one model and use them in a composite model. For doing that we simply can use Modelname.variable in the composite model to call that variable.

BOUNDARIES, PRESET, INITIAL are other topics that can be specified within the model and we refer to the examples for its use.

- UNIT: It is used to refer to define an instance of a component model.
- PORT: It is used to define a port to connect models in a flowsheet.

10.2.3 Tasks

Task: We need to provide a name of the task.
PARAMETER: Parameters in a task are the elements we can send to the task. Furthermore, each task must have a parameter of the type MODEL, to indicate on what model the task is operating on.
VARIABLE: Variables in a task only exist inside the task and cannot be used outside the task. They are used for intermediate calculations or for keeping track of an iteration number for example.
SCHEDULE: It is aimed to determine the order in which the models/tasks are evaluated or executed.

10.2.4 Process Section

UNIT: Name as (Modelname)
SET: As before to set parameters
ASSIGN: To provide values to degrees of freedom

We can use the statement WITHIN (Model) DO — END to provide values for the different variables in the model.
INITIAL: Include here the initial conditions for dynamic calculations.
SOLUTIONPARAMETERS: we can specify the solvers for initialization and integration (simulation activities), optimization (optimization activities) and parameter estimation (parameter estimation activities). A PROCESS entity may contain specifications for all three types. Furthermore, we can specify characteristic parameters of the solvers. Note that all solution parameters can also be accessed through the solution parameters tab in the PROCESS.

Linear equations	MA28 or MA48: based on LU decomposition
Non Linear equations	BDNLSOL: Block Decomposition NonLinear SOLver
	SPARSE: Newton-type method without block decomposition
Differential algebraic equations	DAEBDF. Based on variable time step/variable order Backward Differentiation Formulae
	SRADAU: Implements a variable time step, fully-implicit Runge-Kutta method
Optimization:	CVP SS can solve optimisation problems with both discrete and continuous decision variables (\mixed integer optimisation"). Both steady-state and dynamic problems are supported.
	"CVP_SS" ["DASolver" := "----";
	"MINLPSolver" := "--------"];
	The MINLP solver can be OAERAP: Based on outer approximation algorithms or NLPSQP: employs a sequential quadratic programming (SQP) method for the solution of the nonlinear programming (NLP) problem
Parameter estimation:	MAXLKHD. This is based on a general maximum likelihood approach.

We can also specify the REPORTINGINTERVAL: = (NUMBER): non-zero values will provide increasingly detailed solver diagnostic information.
SCHEDULE: can be used to define dynamic operating procedures. We can use as statements:

SEQUENCE (Task list) END
PARALLEL (Task list) END
WHILE (Expression) DO … END
IF (Expression) THEN … ELSE(optional) … END
CONTINUE FOR (Expression)
CONTINUE UNTIL (Expression)
CONTINUE FOR () and|| or ()
RESET Variable END
REPLACE Variable WITH ….. END
REINITIAL differential equation WITH … END
SWITCH Selector END

10.2.5 Arrays and Built-in Functions

Operators: +, –, *, ^, /, DIV, MOD

Built-in vector functions: ABS(x), ACOS (x), ASIN(x), ATAN(x), COS(x), COSH(x), INT(x), LOG(x), LOG10(x), SGN(x), SIN(x), SINH(x), SQRT(x), TAN(x), TANH(x).

OLD(x)

Built-in scalar functions: MAX(x,< >), Min (x, < >), Product (x, < >), SIGMA (x, < >)

Partial derivatives: PARTIAL (x, domain)

Integration: INTEGRAL(z:=Lowlimit:Upperlimit; Expression)

Arrays: They are tensors (vectors, matrices or higher order) and can be used as such to store data as parameters or variables in an arbitrary number of dimensions. They are interesting to write the same equation applied to all the units/components with little effort. A(i;j)

We can use a:b to define a number of elements from the row i or column j. We leave A(1,) the full row is selected.

10.2.6 Logic Statements

In this section we present the common syntax for the use of logic statement within a model in gPROMS®. We have as relational operators

=,<>,<,>,>=,<=

As logical values: TRUE, FALSE,
As logical operations: AND, OR, NOT
The logical expressions typically used have the following structure:

Mathematical notation	Interpretation	gProms notation
[a,b]	$z \in [a,b] \Leftrightarrow a \le z \le b$	a:b
(a,b]	$z \in (a,b] \Leftrightarrow a < z \le b$	a|+:b
[a,b)	$z \in [a,b) \Leftrightarrow a \le z < b$	a : b|-
(a,b)	$z \in (a,b) \Leftrightarrow a < z < b$	a|+:b|-

FIGURE 10.3 Interval schemes.

FOR i:= 1 TO () STEP(optional) DO
END

In order to define the application of the statement we can use also Figure 10.3:

```
IF  (Expression) THEN
      ...
ELSE
      ...
END
WHILE () DO
      ...
END
CONTINUE FOR ()
CONTINUE UNTIL ()
CONTINUE FOR () AND|| OR ()
CASE () OF
      WHEN ...
      WHEN...
END
```

For defining the state for case we use:

```
SELECTOR AS (a,b) DEFAULT (a)
```

10.3 A FIRST STEP-BY-STEP EXAMPLE–DIFFUSION AND CHEMICAL REACTION

The following example is used to explore all the basic functionality within gPROMS®: setting up of a time-dependent model with distributed variables, breaking up the model into submodels that are then connected, initialising the simulation with previous solutions, creating more complex tasks, connecting excel with gPROMS® for use as an input interface and output of results, creating a stand-alone application for deployment to users and learning to create a model flowsheet with the submodels developed. The example is typical in the field of reaction engineering / transport phenomena [2].

10.3.1 Basic Model Creation and Analysis of Results

Here we describe how to create a model for diffusion and reaction, implemented in gPROMS and stored in the file Diffusion_Reaction.gPJ. A mesh microreactor with a stagnant water liquid phase containing species B and of thickness 100 μm is put in contact with a gas phase, which contains species A, which absorbs into the liquid phase and reacts to produce main product C following the reactions below. A and B also degrade with time:

[1]	$0.5A + B \rightarrow C$	
[2]	$C \rightarrow 0.5A + B$	(reversible of 1)
[3]	$A \rightarrow A^*$	(decomposition of A)
[4]	$B \rightarrow B^*$	(decomposition of B)

A generic mass balance for the stagnant liquid phase is carried over which accounts for diffusion and reaction of N species subject to M different chemical reactions yielding:

$$\frac{\partial c_i(t,x)}{\partial t} = D_i \cdot \frac{\partial^2 c_i(t,x)}{\partial x^2} + \sum_{j=1}^{M} v_{i,j} \cdot R_j(t,x) \tag{10.1}$$

for all i = 1…N (species), $x \in (0, L)$

Where c_i (kmol/m^3) is the concentration of species i, D_i (m^2/s) is the diffusion coefficient of species i, $v_{i,j}$ is the stochiometric coefficient of species i in reaction j and R_j (kmol/m^{3}s) is the reaction rate for reaction j and L is the length of the x domain (thickness). The reaction rate can be written as:

$$R_j = k_{0,j} e^{-E_{aj}/(R \cdot T)} \cdot \prod_{i=1}^{N} c_i^{RO_{i,j}} \tag{10.2}$$

j = 1… M (Reactions)

The reaction stochiometry $v_{i,j}$ and reaction order $RO_{i,j}$ are written as a matrix:

$$v_{i,j} = \begin{pmatrix} v_{1,1} & v_{1,2} & v_{1,3} & \cdots & v_{1,M} \\ v_{2,1} & v_{2,2} & v_{2,3} & \cdots & v_{2,M} \\ v_{3,1} & v_{3,2} & v_{3,3} & \cdots & v_{3,M} \\ \cdots & \cdots & \cdots & \cdots & \cdots \\ v_{N,1} & v_{N,2} & v_{N,3} & \cdots & v_{N,M} \end{pmatrix} \tag{10.3}$$

$$RO_{i,j} = \begin{pmatrix} RO_{1,1} & RO_{1,2} & RO_{1,3} & \cdots & RO_{1,M} \\ RO_{2,1} & RO_{2,2} & RO_{2,3} & \cdots & RO_{2,M} \\ RO_{3,1} & RO_{3,2} & RO_{3,3} & \cdots & RO_{3,M} \\ \cdots & \cdots & \cdots & \cdots & \cdots \\ RO_{N,1} & RO_{N,2} & RO_{N,3} & \cdots & RO_{N,M} \end{pmatrix} \tag{10.4}$$

For this specific system these two matrices are:

$$v_{i,j} = \begin{pmatrix} -0.5 & 0.5 & -1 & 0 \\ -1 & 1 & 0 & -1 \\ 1 & -1 & 0 & 0 \end{pmatrix} \quad RO_{i,j} = \begin{pmatrix} 1 & 0 & 1 & 0 \\ 1 & 0 & 0 & 1 \\ 0 & 1 & 0 & 0 \end{pmatrix} \tag{10.5}$$

Note: there is no need to track species A* and B* since they do not react any further.

First, the specific case will be solved while later on excel will be used to allow to solve any generic case using an interface. The rest of model constants needed arc given below:

Diffusion coefficients: $D_A = 1e\text{-}10 m^2/s$; $D_B = 1e\text{-}10 m^2/s$; $D_C = 1e\text{-}10 m^2/s$;
Interfacial Concentrations: $c_{int,A} = 1$ kmol/m^3; $c_{int,B} = 0$ kmol/m^3; $c_{int,C} = 0$ kmol/m^3
Initial concentration in liquid: $c_{i,A} = 0$ kmol/m^3; $c_{i,B} = 1$ kmol/m^3; $c_{i,C} = 0$ kmol/m^3;
Pre-exp factors: $k_{0,1} = 5e\text{-}2$ m^3/kmols; $k_{0,2} = 1e\text{-}4$ s^{-1}; $k_{0,3} = 1e\text{-}4$ s^{-1}, $k_{0,4} = 1e\text{-}4$ s^{-1}.
Activation Energies: $E_{a,1} = 1$ J/kmol; $E_{a,2} = 1$ J/kmol; $E_{a,3} = 1$ J/kmol; $E_{a,4} = 1$ J/kmol

Note on variable types: There are two types of variables, namely, *differential variables* in both time and spatial domains, such as concentration, and *algebraic variables,* which are not described by means of a time derivative. Initial and boundary conditions are only needed for differential variables which are solved from differential equations.

Note on boundary conditions: The two *boundary conditions* needed to solve a second order differential equation per i species are shown below:

At $x = 0$; $c_i(0) = cint_i$ (interfacial concentration due to solubility from gas phase)

and

$$\text{At } x = L \frac{\partial c_i}{\partial x} = 0 \text{ (This implies there is no chemical flux out)}$$

An initial condition is also needed for each species:

$$\text{At } t = 0;\ c_A = 0, c_B = 0 \text{ and } c_C = 1 \text{ kmol/m}^3$$

Note on degrees of freedom: There are 3 differential equations that contain 6 variables (3 concentrations and 3 reaction rates). The reaction rates can be written as a function of the concentration using other 3 equations resulting in 6 equations and 6 variables (closed system). If there are other variables, they need to be assigned or calculated in new equations. For example, the reaction rate equation also contains temperature T. We can either assume an isothermal reactor, in which case we assign a constant value to T, or define an energy balance to compute T.

Note on units: SI units of measurement will be used throughout.

Variable Types (Diffusion_Reaction)

Name	Lower bound	Default value	Upper bound	Units
concentration	0.0	0.5	1.0	kmol/m3
reaction_rate	0.0	0.5	1.0	kmol/m3s
temperature	0.0	300.0	1000.0	K
<new>				

FIGURE 10.4 Table for input of variable types.

Below the step by step process:

1. *Create a new file:* Menu File→New→ Save as Diffusion_Reaction.gPJ. There is a new project created in the project tree area.
2. *Define variable types for concentrations, reaction rates and temperature:* Variable Types Folder right click → New Entity → Name: concentration. Repeat for Name: reaction_rate and Name: temperature. Choose lower and upper bounds for each variable, which should contain all possible values for the problem solution. The default value is used as an initial guess for those types of variables during model initialisation, see Figure 10.4.

   Note: Although T is initially constant, a variable as opposed to a parameter is used so that it is possible to later on input time profiles for temperature or changes in the SCHEDULE section.

   Note: It is good practice to use descriptive names for the variables vs. single letters above when using the search tool if find/replace is later on needed.
3. *Create "Reactor" model:* Model folder right click → New Entity → Name: Reactor.
4. *Define parameters within the PARAMETER section of the model:* N as number of species, M as number of reactions, gas constant, diffusion_length, diffusion coefficient D_i, concentrations at boundary $c_{int,i}$. pre-exponential factor k_{0j}, Activation Energy Ea_j, reaction order $RO_{i,j}$ and stochiometry $v_{i,j}$.

```
PARAMETER
 N                    as       Ordered_set # List of species
 M                    as       Integer # Number of reactions
 gas_constant as               REAL default 8314 # J/kmolK
 diffusion_length  as                 Real
 D    as Array (N) of    Real # Diffusion coefficient, m2/s
 cint as Array (N) of    Real # Interface concentration
   of species i, kmol/m3
 ci    as Array (N) of    Real # initial concentrations
   for all species.
 k0    as Array (M) of    Real # depends on the reaction
   order.
 Ea    as Array (M) of    Real # Joules/kmol
 RO    as Array (N,M) of   Real # kmol/(m3*s)
 v     as Array (N,M) of   Real # Stoichiometry coefficient
```

Note: An ordered_set is a parameter type allowing the user to list a series of strings and use these in the definition of variables and units.

Note: Sections UNIT and PORT can be removed or left commented out for this example.

5. *Define distribution domain:* Since concentrations are dependent on the position x, an x-domain needs to be defined:

```
Distribution_Domain
xdomain    as [0:diffusion_length] # discretization array.
```

6. *Define variables for reactor model:* concentrations of species in x domain, reaction rates in x domain and temperature (independent of x-domain).

```
VARIABLE
c as DISTRIBUTION (N, xdomain)  of concentration #kmol/m3
R as DISTRIBUTION (M, xdomain)  of reaction_rate # kmol/m3s
T as                temperature # K
```

7. *Boundary conditions:* for x = 0, $c_i = c_{int,i}$ and for x = diffusion_length $dc_i/dx = 0$. These equations only apply to two points in the x domain.

Note: that there are 2 equations per species. c(,0) is an array with the concentration of each species i at x = 0.

Note: The xdomain is discretized and these values are called by directly putting the x value vs. the index position in discretized domain. For example, if x = [0,3] and the domain is discretized in 10 elements (11 points), in order to call the last element c(,3) is used (and not c(,11)) (actual x value is written). If the x value is between two discretized points, the closest one is selected. For example c(,1) ~ c(,0.9) since the domain has been discretized as 0, 0.3, 0.6, 0.9, 1.2, 1.5 …

```
BOUNDARY
#Inlet Boundary
    c(,0) = cint;

#Outlet Boundary
    PARTIAL(c(,diffusion_length),xdomain) = 0;
```

8. *Input equations:* SIGMA and PRODUCT are the gPROMS® built-in functions for summation and product of variables, used in the mass balance eq. (10.1) and reaction rate eq. (10.2) equations respectively.

```
EQUATION
#Mass Balance with partial derivatives of second order.
  This is only valid for all x domain apart from
# x = 0 and x = diffusion_length where boundary condition
  equations are used
    For i In N DO # For ordered_sets we use different
      nomenclature for FOR
```

```
            For x:= 0|+ TO diffusion_length|- DO
                $c(i,x)= D(i)*Partial(Partial(c(i,x),xdomain),
                  xdomain)+SIGMA(v(i,)*R(,x));
            end
        end
    #reaction Rate with no partial derivatives (no boundary
      eq required so we define equation in full x domain)
        For j:=1 to M DO # For ordered_sets we use different
          nomenclature for FOR
            For x:= 0 TO diffusion_length DO
                R (j,x) = k0(j)*exp(-Ea(j)/(gas_constant*T))
                  *Product(c(,x)^RO(,j));
                # Equivalent Product(c('A':'C',x)
                  ^RO('A':'C',j));
            end
        end
```

Note: "For" is used differently for ordered_sets than for integers or distributed domains.

Note: Indexes i, j and x used in the "For" do not need to be declared.

Note: The mass balance differential equations used to calculate the values of the differential variables c(i,x) apply to all x domain apart from x = 0 and x = diffusion_length where boundary condition equations have already being defined. Thus, the expression "For x: = 0|+ TO diffusion_length|- DO" is used.

Note: The reaction rates are algebraic variables and thus need to be solved in the full x domain since boundary equations are not defined for them. Thus, the expression "For x: = 0 TO diffusion_length DO".

Note: For calling ranges of an ordered_set, their actual names are used vs. positions. In the code above "Product(c(1:N,x)^RO(1:N,j))" would be wrong and "Product(c('A':'C',x)^RO('A':'C',j))" should be used if calling the species A, B, C. The actual code uses full range by leaving space in blank "Product(c(,x)^RO(,j))"

With these sections, a basic definition of the model is implemented and now a PROCESS entity is created to create an instance of this model and run a simulation for specific values of the *parameters, initial conditions and discretization of the xdomain.*

9. *Create a new Process entity called "sim_reactor":* Process folder right click → new entity → Name: sim_reactor.
10. *Call unit reactor:* Basically, the process entity is the highest level that is run and the one that creates an instance of the model Reactor, called here reactor1.

```
UNIT
    reactor1 AS Reactor below UNIT section
```

The Parameter and Monitor sections can be removed for the moment.

11. *Set parameters and domains, assign variables and define initial conditions for differential variables for the created model instance reactor1:* The variables that need to be assigned are the degrees of freedom which are not being solved from previous equations (degrees of freedom = variables-equations) in the ASSIGN section. In this case, it is only T since c and R are being solved from the equations. For this, we need to access the parameters of a unit model, by writing "reactor1.PARAMETER_NAME" or using the construct "within … DO".

    The distribution domain xdomain is discretised within the SET section by "reactor1.xdomain: = [CFDM,2,100]", which means discretization of the xdomain into 100 elements (101 points) using a second order Centred Finite Difference Method.

    Initial conditions for differential variable c are defined in the section INITIAL section. This definition is done excluding the boundary points (internal domain (0,diffusion_length)) where a solution is already available for all times. Also note that initial conditions are generally distributed equations because we need to give the initial value of c in the whole domain where they apply.

    Note: It is possible to discretize different domains using different methods.

    Note: These SET and assignments could have been done within the model Reactor but in that case they will always be the same even if we have multiple instances of the same model combined together while by defining them from the Process entity, different model instances of the same model can have different parameters and assigned variables.

```
UNIT
 reactor1 AS Reactor
SET
    reactor1.N:=['A','B','C']; #3 components A, B and C
    # or we can access all parameters of a defined unit
      using "within"
    within reactor1 DO
        M:=4; # 4 reactions
        diffusion_length:=0.0001;
        D:=[1e-10, 1e-10,1e-10];
        cint:=[1,0,0];  #Interfacial concentrations for
          A, B and C
        ci:= [0,1,0];    # Initial concentrations for A, B
          and C
        k0:=[5e-2,1e-4,1e-4,1e-4];
        Ea:=[1, 1,1,1];
        RO(,1):=[1,1,0];     # For reaction 1
        RO(,2):=[0,0,1];     # for reaction 2
        RO(,3):=[1,0,0];     # for reaction 3
        RO(,4):=[0,1,0];     # for reaction 4
        v(,1):=[-0.5,-1,1]; # For reaction 1
```

```
            v(,2):=[0.5,1,-1];  # for reaction 2
            v(,3):=[-1,0,0];    # for reaction 3
            v(,4):=[0,-1,0];    # for reaction 4
            xdomain:=[CFDM,2,10];
        end

    ASSIGN
        reactor1.T:=300;
    INITIAL
        #Implicit definition and without FOR
        #reactor1.c('A',0|+:reactor1.diffusion_length|-)
          =reactor1.ci('A');
        #reactor1.c('B',0|+:reactor1.diffusion_length|-)
          =reactor1.ci('B');
        #reactor1.c('C',0|+:reactor1.diffusion_length|-)
          =reactor1.ci('C');

        #explicit definition
    Within reactor1 DO
        for i in N DO
            for x:= 0|+ TO diffusion_length|- DO # No initial
              conditions needed at the boundary
                c(i,x)=ci(i);
            end
        end
             end
```

Note: In case of using implicit definition, the boundary points need to be excluded since the variable is not differential at the boundaries, using

```
reactor1.c('A',0|+:reactor1.diffusion_length|-)=reactor1.
  ci('A');
```

12. *Run simulation for 100 s:* Run Simulation for 100 seconds with reporting interval every 10 seconds (use SCHEDULE section in the Process entity as shown below). Click the sim_reactor process and then the play button, green triangle on Figure 10.5. When running a simulation, a case file is created in the project tree, which also contains results for all variables, see Figure 10.6.

```
SCHEDULE
       Continue for 200
```

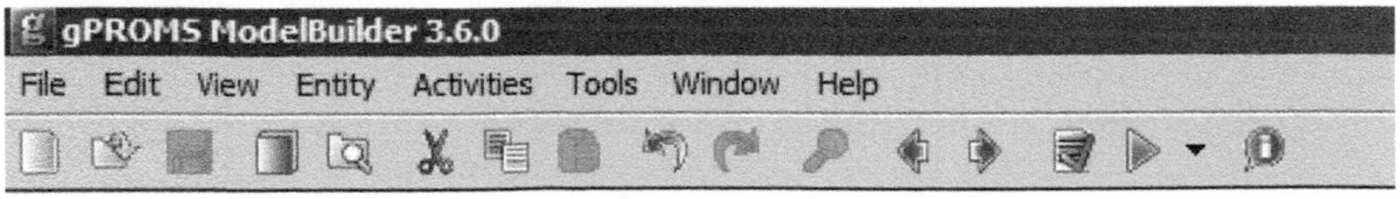

FIGURE 10.5 Executing the model.

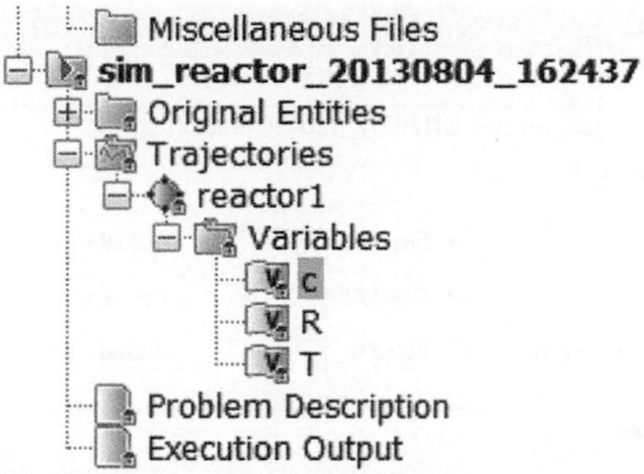

FIGURE 10.6 Access to variable results from the case file.

13. *Plotting results:* A simple way to plot standard results is to use the case folder shown below by just clicking the variables that needs to be plotted.

    By default, the case file contains the YearMonthDay_HoursMinutes Seconds in the name.

    Double click c. Under "time", select "fixed at" so results for a given time will be plotted.

    For N, select "series" so all three species concentration will be plotted.

    For xdomain, select x-axis so xdomain will be plotted in the x axis with c in the y axis. Graphs below show the results of concentration along the xdomain at 20 and 80 s. The steady state solution is zero C, zero B and 1 A in whole x domain towards the solution is converging to. See Figure 10.7.

    Results after 20 and 80 seconds.

    The gRMS software, which opens automatically when a gPROMS® calculation is run and the tick box "Send results trajectory to gRMS" is ticked, can also be used for plotting results, see Figure 10.8. Here, one line is added at a time: Click 2D button in Figure 10.2 → add line button → the screen below pops up (a fixed time of 20 s, N = 1 (A species) and leave x domain unchecked so that the concentration for component A is plotted along full x

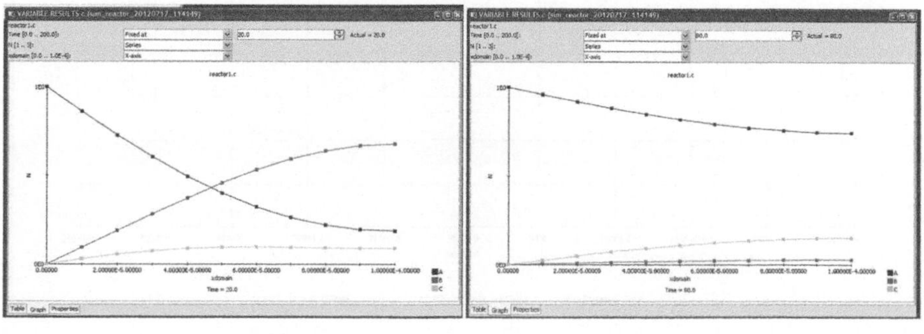

FIGURE 10.7 Plotting of concentration results. (See color insert.)

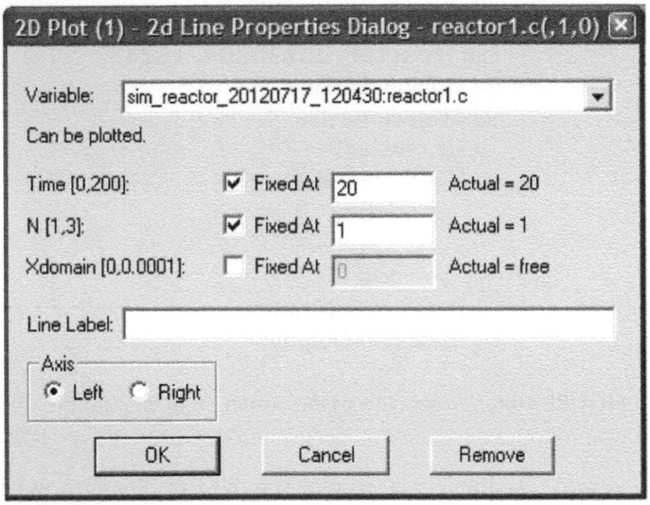

FIGURE 10.8 Use opf gRMS for plotting results.

domain). Repeat same steps to add lines for components B and C. We can see the results in Figure 10.9.

14. *Compare different cases:* It is possible to create multiple processes with different inputs, run them and plot results for both of them in the same graph to compare results. To compare the code in two processes or models, select them both and go to Tools → Compare.

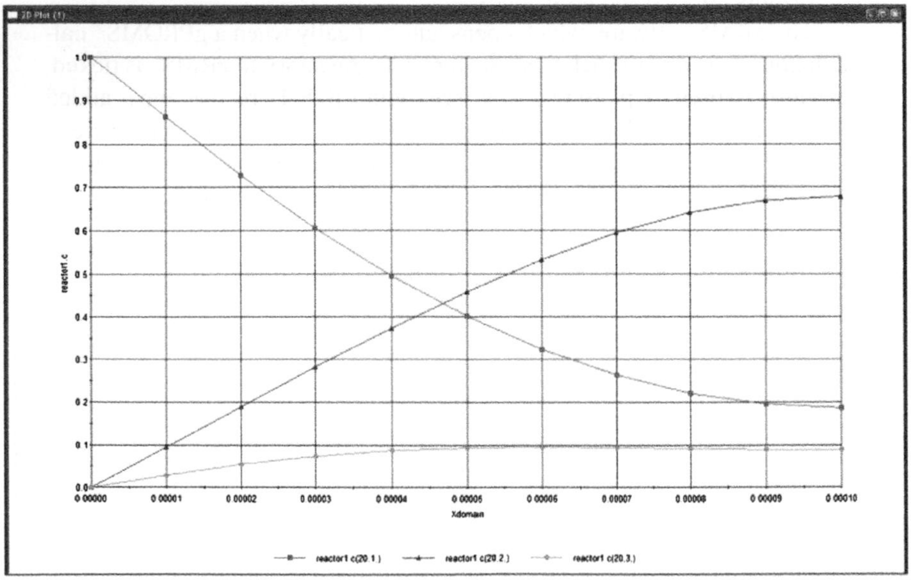

FIGURE 10.9 Graph generated in the gRMS software.

15. *Initialization with previous solutions:* gPROMS® first step is to solve the initialisation, which is solving a set of all discretised (and often nonlinear) equations at time 0. This process requires initial guesses for all variables to provide a starting point for the solution algorithm. gPROMS® uses the default values in the variable types for this purpose (the same for all variables of the same type). In some specific cases, it may be better to provide initial guesses for each variable, which override those ones. For this, the PRESET section can be used either within a model or in the process. A simple way to provide initial guesses for all variables is to save a variable set of a working simulation using the "save" command within the SCHEDULE section and then restore those values in the PRESET section:

```
...
ASSIGN
    reactor1.kinetics1.T:=300;
PRESET
    restore "NewSavedVariableSet";
INITIAL
    STEADY_STATE
SCHEDULE
    #Sequence
        #Continue for 10 # Now we don't need to run for
          long
        save "NewSavedVariableSet_nextproblem"
    #End Sequence
```

A saved variable set restored in the PRESET section will also override all the single variable PRESET definition within the models.

16. *Creating a gPROMS® project from a case file:* Often, when we have many processes, it can be useful to create a gPROMS® model with a specific process only. To do this; is Right click on the case file name → Create gPROMS® project.
17. *Diagnostic information:* Retain the license after execution by unchecking this box in the play button simulation screen. This allows to access additional diagnostic information in the execution output screen (in the case file).
    Problem Report: Right click in the Execution Output screen → Create problem report (this file can be accessed on the Results folder and is called ProblemReport.lst): This reports is a summary of the simulation and shows: variables types (3), all the parameters for all the arrays one by one (45 parameters), distribution domains, equations (77), variables (78 variables, out of which 27 are differential, 50 algebraic and 1 assigned (degree of freedom)). As previously mentioned, reaction rates and concentrations of A, B and C at the boundaries appear as algebraic. Also note that there are variables for each point of the x discretised domain since gPROMS® is solving a system of discretised equations with unknowns in each point of the xdomain. Figure 10.10 shows a snapshot for the report.

```
TEXT FILE ProblemReport.lst (sim_reactor_20130805_095445)
  33 : reactor1.RO("C",4) := 0
  34 : reactor1.v("A",1) := -0.5
  35 : reactor1.v("A",2) := 0.5
  36 : reactor1.v("A",3) := -1
  37 : reactor1.v("A",4) := 0
  38 : reactor1.v("B",1) := -1
  39 : reactor1.v("B",2) := 1
  40 : reactor1.v("B",3) := 0
  41 : reactor1.v("B",4) := -1
  42 : reactor1.v("C",1) := 1
  43 : reactor1.v("C",2) := -1
  44 : reactor1.v("C",3) := 0
  45 : reactor1.v("C",4) := 0

DISTRIBUTION DOMAINS
~~~~~~~~~~~~~~~~~~~~
  1 : reactor1.xdomain := [CFDM, 2, 10]

VARIABLES
~~~~~~~~~

ID  Name                       Value               TimeDeriv              Bounds   Form          Info
 1  reactor1.c("A",0.0)        1                   n/a                    [0:1]    Algebraic
 2  reactor1.c("A",1E-005)     0.974487472408      0.000581201231028      [0:1]    Differential
 3  reactor1.c("A",2E-005)     0.949385421794      0.0011549202172        [0:1]    Differential
 4  reactor1.c("A",3E-005)     0.925873264835      0.0016928748124        [0:1]    Differential
 5  reactor1.c("A",4E-005)     0.903977160764      0.00219468173486       [0:1]    Differential
 6  reactor1.c("A",5E-005)     0.884698212797      0.00263649035589       [0:1]    Differential
 7  reactor1.c("A",6E-005)     0.868087328384      0.00301747367069       [0:1]    Differential
 8  reactor1.c("A",7E-005)     0.854870221373      0.00332045586488       [0:1]    Differential
 9  reactor1.c("A",8E-005)     0.845119364105      0.0035440892922        [0:1]    Differential
10  reactor1.c("A",9E-005)     0.839215613083      0.00367940700755       [0:1]    Differential
11  reactor1.c("A",0.0001)     0.837247488161      n/a                    [0:1]    Algebraic
12  reactor1.c("B",0.0)        0                   n/a                    [0:1]    Algebraic
13  reactor1.c("B",1E-005)     0.00112771215644    -7.40944427485E-005    [0:1]    Differential
```

FIGURE 10.10 Snapshot of Problem Report file.

When getting error messages in relation to number of variables and equations, use a low number for the discretization of the domain so that it is easy to count all equations and variables easily and check what is missing. Also note that temperature appears as assigned in order to close the system since we have 78 variables and 77 equations (first 6 are boundary conditions, 27 discretised differential equations, 44 algebraic equations which are the equations for the reaction rate R). There is also scaling information for each algebraic and differential equation and all these equations apply to every single time. *Querying equations:* Right click on Execution Output screen → Query Equation → Equation 4 (boundary condition for A at x = diffusion_length) → OK. Figure 10.11 shows an example.

Once all diagnostics have been carried out, right click on the case file name and release licence. Refer to gPROMS® manual for greater detail on this topic.

10.3.2 Nesting of Submodels

Next, using the same previous model, the reaction rate equation is separated into a submodel that is then called from the main reactor model.

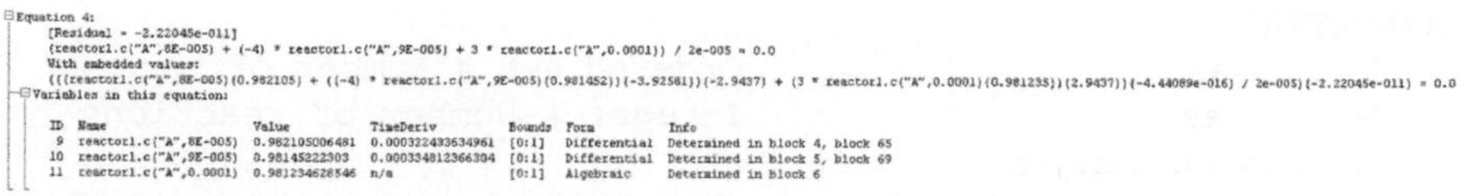
Equation 4:
[Residual = -2.22045e-011]
(reactor1.c("A",8E-005) + (-4) * reactor1.c("A",9E-005) + 3 * reactor1.c("A",0.0001)) / 2e-005 = 0.0
With embedded values:
(((reactor1.c("A",8E-005)(0.982105) + ((-4) * reactor1.c("A",9E-005)(0.981452))(-3.92581))(-2.9437) + (3 * reactor1.c("A",0.0001)(0.981235))(2.9437))(-4.44089e-016) / 2e-005)(-2.22045e-011) = 0.0
Variables in this equation:
ID Name Value TimeDeriv Bounds Form Info
9 reactor1.c("A",8E-005) 0.982105006481 0.000322433634961 [0:1] Differential Determined in block 4, block 65
10 reactor1.c("A",9E-005) 0.98145222303 0.000334812366304 [0:1] Differential Determined in block 5, block 69
11 reactor1.c("A",0.0001) 0.981234628546 n/a [0:1] Algebraic Determined in block 6

FIGURE 10.11 Query of diagnostic information.

1. Create a new model call MainReactor as a copy of Reactor model.
2. Copy the process sim_reactor to another process called sim_reactor_2.
3. The process sim_reactor_2 needs to call a unit of the model MainReactor instead of Reactor:

```
UNIT
reactor1 as MainReactor
```

4. *Create a new model called "Kinetics,"* which will contain the kinetics equations: Cut and paste the kinetics equations from MainReactor into Kinetics model.

   Note that now both the MainReactor and submodel Kinetics contain variables with distributed domains.

   Note that variables R(j,Axial), c(i,Axial) are used in both models as well as parameters diffusion_length, N, M, xdomain while parameters ko(j), Ea(j), gas_constant, T and RO are only used in the submodel Kinetics.

   Based on the rules of Parameter Propagation, we conduct the following changes:
5. *Changes in MainReactor Model*:
   - Delete ko(j), Ea(j), gas_constant, T and RO definitions in MainReactor and add them to the Kinetics model.
   - Create an instance of the Kinetics model called kinetics1 in the MainReactor model under the UNIT section.
   - Add the necessary equations to connect the Kinetics model variables to the MainReactor variables: e.g. R = kinetics1.R; kinetics1.c = c; (R and c are the two variable that are shared by both models). The order in the equations is irrelevant, but a simple way to remember whether a variable is calculated within a model is to put it on the right side of the equation if it is calculated in the model. For example, c is calculated within MainReactor so kinetics1.c = c while R is calculated in the Kinetics model so R = kinetics1.R. This is more useful when flowsheeting and connecting models via port connections.

Note: If we write R() = kinetics1.R() we get a mismatch error in dimensions because R is two-dimensional. R(,) = kinetics1.R(,) is the valid notation.

```
PARAMETER
    N       as                    Ordered_set # Number of species
    M       as                    Integer # Number of reactions
    diffusion_length        as                        Real
    D       as Array (N) of    Real # Diffusion coefficient, m2/s
    cint    as Array (N) of       Real # Interface concentration
      of species i, this could also be defined as a variable
      kmol/m3
    ci      as Array (N) of       Real # initial concentrations
      for all species.
    v       as Array (N,M) of     Real

Unit
    kinetics1 as Kinetics

Distribution_Domain
xdomain     as [0:diffusion_length] # discretization array.

VARIABLE
    c        as DISTRIBUTION (N, xdomain)  of  concentration
      #kmol/m3
    R        as DISTRIBUTION (M, xdomain)  of  reaction_rate #
      kmol/m3s
    #T       as                               temperature # K

BOUNDARY
#Inlet Boundary
    c(,0) = cint;

#Outlet Boundary
    PARTIAL(c(,diffusion_length),xdomain) = 0;

EQUATION
#Equation to Link Shared variables in MainReactor model with
  Kinetics model
    R = kinetics1.R ;
    kinetics1.c = c;

#Mass Balance with partial derivatives of second order. This
  is only valid for all x domain apart from x = 0 and x =
  #diffusion_length where boundary condition equations are
  used
    For i In N DO # For ordered_sets we use different
      nomenclature for FOR
        For x:= 0|+ TO diffusion_length|- DO
            $c(i,x)= D(i)*Partial(Partial(c(i,x),xdomain),
              xdomain)+SIGMA(v(i,)*R(,x));
        end
    end
```

Changes in Kinetics Model:

- Add ko(j), Ea(j), gas_constant, N, M, diffusion_length, RO parameter and xdomain distribution_domain.
- Add R(j,Axial), c(I,Axial) and T variable definitions.

```
PARAMETER
   N                          as                    Ordered_set #
     Number of species
   M                          as                    Integer # Number
     of reactions
   diffusion_length           as                    Real
   # Unique parameters for this model below
   gas_constant               as                    REAL default
     8314 # J/kmolK
   k0                         as Array (M) of       Real # depends
     on the reaction order.
   Ea                         as Array (M) of       Real #
     Joules/kmol
   RO                         as Array (N,M) of     Real #
     kmol/(m3*s)

Distribution_Domain
   xdomain                    as [0:diffusion_length] #
     discretization array.

VARIABLE
   c                          as DISTRIBUTION (N, xdomain)  of
     concentration #kmol/m3
   R                          as DISTRIBUTION (M, xdomain)  of
     reaction_rate # kmol/m3s
   T                          as                   temperature # K

EQUATION
#reaction Rate with no partial derivatives (no boundary eq
 required so we define equation in full x domain)
   For j:=1 to M DO # For ordered_sets we use different
     nomenclature for FOR
       For x:= 0 TO diffusion_length DO
           R (j,x) = k0(j)*exp(-Ea(j)/(gas_constant*T))
             *Product(c(,x)^RO(,j));
           # Equivalent Product(c('A':'C',x)^RO('A':'C',j));
       end
   end
```

6. *Changes in sim_reactor_2 process:*
   - SET and ASSIGN the only specific parameters-variables for Kinetics model (kinetics1) in the sim_reactor_2 process separately: SET ko(j), Ea(j), gas_constant, and RO and ASSIGN T. kinetics1 is called using “within reactor1.kinetics1 DO … End”.

- M, N, diffusion length does not need to be SET for Kinetics model again since they propagate from the parent model MainReactor.
- xdomain distribution domain also propagates from reactorl to kineticsl, because the PARAMETER has the same name, is of the same type and has the same size, so there is no need to set it in kineticsl as well.
- There is no need to create an instance of the Kinetics model in the sim_reactor_2 process since this already creates an instance of the MainReactor called reactorl model which in turn creates an instance of the Kinetics model called kineticsl.

```
UNIT
 reactor1 AS MainReactor # By calling the MainReactor model we
   are calling in
#all submodel instances defined in Main Reactor model as well.

SET
    reactor1.N:=['A','B','C']; #3 components A, B and C
    # or we can access all parameters of a defined unit using
      "within"
    within reactor1 DO
        M:=4; # 4 reactions
        diffusion_length:=0.0001;
        D:=[1e-10, 1e-10,1e-10];
        cint:=[1,0,0];  #Interfacial concentrations for A, B
          and C
        ci:= [0,1,0];   # Initial concentrations for A, B and C

        v(,1):=[-0.5,-1,1]; # For reaction 1
        v(,2):=[0.5,1,-1];  # for reaction 2
        v(,3):=[-1,0,0];    # for reaction 3
        v(,4):=[0,-1,0];    # for reaction 4

        xdomain:=[CFDM,2,10];
    end

    within reactor1.kinetics1 DO # Parameters that are only
      used in the kinetics model and do not need to be defined
      in the main model
        k0:=[5e-2,1e-4,1e-4,1e-4];
        Ea:=[1, 1,1,1];
        RO(,1):=[1,1,0];    # For reaction 1
        RO(,2):=[0,0,1];    # for reaction 2
        RO(,3):=[1,0,0];    # for reaction 3
        RO(,4):=[0,1,0];    # for reaction 4
    End

ASSIGN
    reactor1.kinetics1.T:=300;

INITIAL
#explicit definition
```

```
Within reactor1 DO
    for i in N DO
        for x:= 0|+ TO diffusion_length|- DO # No initial
          conditions needed at the boundary
            c(i,x)=ci(i);
        end
    end
end

SCHEDULE
    Continue for 200
```

10.3.3 Connecting Excel with gPROMS® for Inputs/Outputs (Files Diffusion_Reaction_Excel_FO.gPJ and Diffusion_Reaction_Excel_FP.gPJ)

When gPROMS® is connected to excel and Excel acts as a Foreign Object (FO), EXCEL® is used to provide a calculation service to gPROMS®. To interactively exchange information with EXCEL® at certain given times during the simulation (open/close valves, stop simulation, read results, etc), EXCEL® should be used as a Foreign Process rather than a Foreign Object.

In this example, EXCEL is used to read all parameters, initial conditions, boundary conditions and assigned variables (N, M, diffusion_length, D(N), cint(N), ci(N), k0(M), Ea(M), RO(N,M), v(N,M) and T). Then, the concentration variable is printed as a function of time.

1. *Changes in main reactor model:*
   In the PARAMETER section of MainReactor add a parameter called my_FO as Foreign_Object.
2. *Open the FO & FP macros excel template* from PSE folder and enable Macros when asked.
3. *Save Excel file:* Save the excel file as DiffusionReaction_EXCEL_FO.xls
4. *Create methods in the Excel file:*
   a. Hit ctrl+shft+O to create one method in the foreign object excel → click New and name: Diffusion_reaction. This creates a new tab for interfacing with gPROMS® called gFO-Diffusion_reaction
   b. Create following methods. These are the arrays in excel that will connect with gPROMS® to send information
      i. Name: Number_reactions; type: integer
      ii. Name: Diffusion_length; type: real
      iii. Name: Diffusivity; type: real
      iv. Name: Interfacial_concentration; type: real
      v. Name: Initial_concentration type: real
      vi. Name: Preexponential_factor type: real
      vii. Name: Activation_energy type: real
      viii. Name: Reaction_order type: real
      ix. Name: Stoichiometry type: real

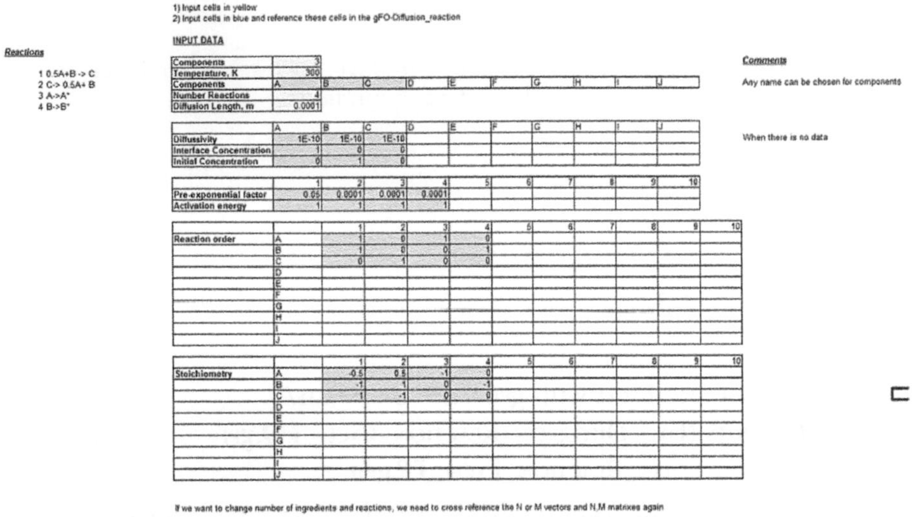

1) Input cells in yellow
2) Input cells in blue and reference these cells in the gFO-Diffusion_reaction

Reactions

1 0.5A+B -> C
2 C-> 0.5A+ B
3 A->A*
4 B->B*

INPUT DATA

Components	3									
Temperature, K	300									
Components	A	B	C	D	E	F	G	H	I	J
Number Reactions	4									
Diffusion Length, m	0.0001									

	A	B	C	D	E	F	G	H	I	J
Diffussivity	1E-10	1E-10	1E-10							
Interface Concentration	1	0	0							
Initial Concentration	0	1	0							

	1	2	3	4	5	6	7	8	9	10
Pre-exponential factor	0.05	0.0001	0.0001	0.0001						
Activation energy	1	1	1	1						

		1	2	3	4	5	6	7	8	9	10
Reaction order	A	1	0	1	0						
	B	1	0	0	1						
	C	0	1	0	0						
	D										
	E										
	F										
	G										
	H										
	I										
	J										

		1	2	3	4	5	6	7	8	9	10
Stoichiometry	A	-0.5	0.5	-1	0						
	B	-1	1	0	-1						
	C	1	-1	0	0						
	D										
	E										
	F										
	G										
	H										
	I										
	J										

If we want to change number of ingredients and reactions, we need to cross reference the N or M vectors and N,M matrixes again

Comments

Any name can be chosen for components

When there is no data

FIGURE 10.12 Inputs data tab and calculation tab.

5. *Add input interface:* Add a new tab to the excel file called "Input data" with the structure shown in Figure 10.12. This interface will be used by the user to input all parameters into gPROMS®. A generic format is chosen so that any number of reactions and chemical species can be used:
6. *Cross reference the user inputs with the methods linked to gPROMS®:* Go to the gFO_Diffusion_reaction tab and go to the Cell X-ref of a method (e.g. Reaction order) and input "=" then move to the inputs tab and select the 3X4 matrix that defines the Reaction order user data. Repeat procedure with all outputs (values being sent to gPROMS®). If the number of elements of the array changes, this procedure needs to be repeated. It is however possible to use VBA for excel to make that range selection automatic based on the number of reactions and components that the user has defined.
7. *Changes in process sim_reactor_2:*
   a. *SET Foreign Object (link to file):* In the process sim_reactor_2, SET my_FO:="ExcelFO::C:\Folder_path\DiffusionReaction_EXCEL_FO.xls;
   b. *SET parameters, ASSIGN variables, etc:* Change the value of all the parameters for their method in excel using my_FO foreign object (e.g. M:=my_FO.Number_reactions).

```
SET
    reactor1.N:=reactor1.my_FO.Components; #3 components
      A, B and C
    # or we can access all parameters of a defined unit
      using "within"
    within reactor1 DO

        my_FO:= "ExcelFO::C:\Folder_Path \DiffusionReaction_
          EXCEL_FO.xls";
```

```
        M:=my_FO.Number_reactions ; # 4 reactions
        diffusion_length:=my_FO.Diffusion_length;
        D:=my_FO.Diffusivity;
        cint:=my_FO.Interfacial_concentration;
          #Interfacial concentrations for A, B and C
        ci:= my_FO.Initial_concentration;   # Initial
          concentrations for A, B and C
        v:=my_FO.Stoichiometry;
        xdomain:=[CFDM,2,100];
    end
    within reactor1.kinetics1 DO
        k0:=reactor1.my_FO.Preexponential_Factor;
        Ea:=reactor1.my_FO.Activation_energy;
        RO:=reactor1.my_FO.Reaction_order;
    End

ASSIGN
    reactor1.kinetics1.T:=reactor1.my_FO.Temperature;
```

8. Now, it is simple to include new ingredients or change their properties. Below results when 5 ingredients A-E are considered where ingredients D, E only diffuse but do not react. See Figure 10.13.

10.3.4 Excel® Interactivity (Foreign Process and Standalone Applications with gorun) (Diffusion_Reaction_Excel_FP.gpj)

Next, EXCEL® as a Foreign Process is used in order to export results from gPROMS® to the Excel interface and enable the option to start/stop the simulation.

1. Open the Diffusion_Reaction_EXCEL_FO.gPJ file and save it as Diffusion_Reaction_EXCEL_FP.gPJ with models MainReactor and Kinetics and process sim_reactor_2 (rest of models and processes have been deleted).
2. Sampling interval and Run switch: Add a variable section (before the SET section) in the process sim_reactor_2 and define two new variables of type length: TimeInterval and RunSwitch. When RunSwitch is 0 the simulation is stopped and it runs while RunSwitch is 1. If TimeInterval is 100 s it means every 100 s we communicate with excel via the SCHEDULE section (send and receive values). For this, create a new variable type "length". These are local variables to the process because they are only used in the SCHEDULE section and for this reason they are defined here instead of inside a model, see Figure 10.14.

```
VARIABLE
    TimeInterval  as  length
    RunSwitch   as  length
```

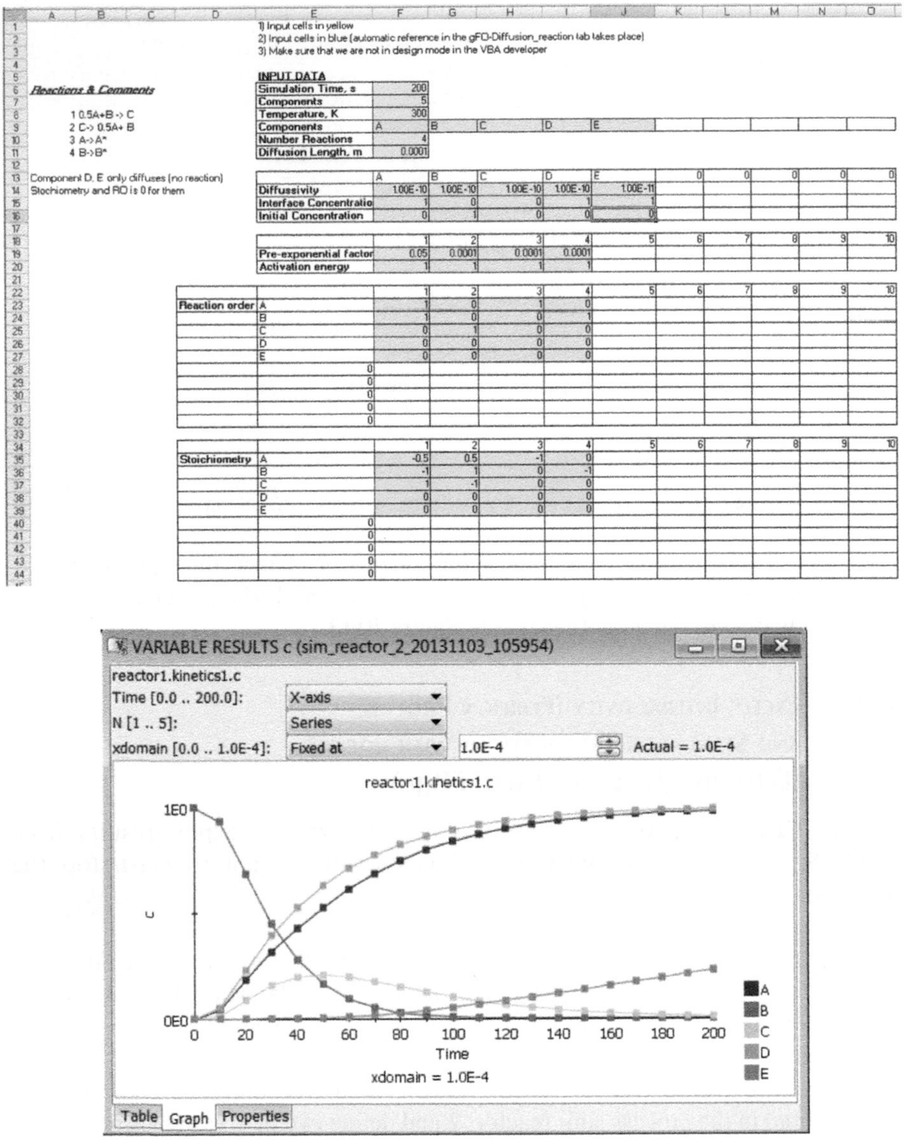

FIGURE 10.13 Input table ad results. (See color insert.)

3. In the Assign section, give them values of 10 s and 1 respectively (the run switch is active by default). Even if these values are read from the Foreign Process every TimeInterval, they still need to have an initial value. Similarly, the temperature assignment is also kept despite reading the temperature from the FO so that during initialization there are the same number of equations as unknown variables.

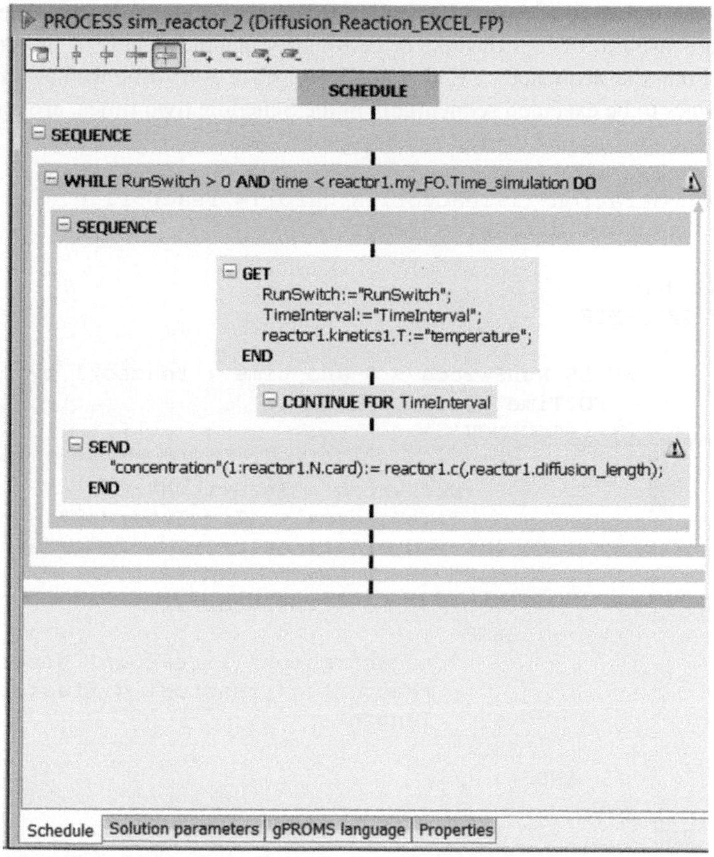

FIGURE 10.14 Schedule in schedule tab view.

```
ASSIGN
  reactor1.kinetics1.T       :=   reactor1.my_FO.Temperature;
  TimeInterval               :=   10;
  RunSwitch                  :=   1;
```

4. The Schedule operates while RunSwitch is larger than 0 and the time is smaller than the simulation time Time_simulation. Go to the schedule tab, see Figure 10.14, of the process sim_reactor_2 and change the projects view to Palette. Add a while construct with the condition RunSwitch>0 and time < reactor1.my_FO.Time_simulation.

   Select a Get … End construct from the Schedule palette, where the values of RunSwitch, TimeInterval and T are red from Excel every TimeInterval

   Add Continue for TimeInterval

   Select a "Send...End" construct to plot the concentrations of all species at x = diffusion_length (back of the reactor) everytime TimeInterval. Note that the range of all species is called by using 1:reactor1.N.card (method card is used to extract the size or cardinality (as an integer) of the array). Note that the Sequence … End construct appears everytime we have multiple lines to be executed sequentially in the Schedule (two times, around the While construct and the Get and Send constructs).

```
"concentration"(1:reactor1.N.card):= reactor1.c
  (,reactor1.diffusion_length);

SCHEDULE
    SEQUENCE

        WHILE RunSwitch > 0 and time < reactor1.my_
          FO.Time_simulation DO
            SEQUENCE
                GET
                    RunSwitch:="RunSwitch";
                    TimeInterval:="TimeInterval";
                    reactor1.kinetics1.T:="temperature";
                END
                CONTINUE FOR TimeInterval
                SEND
                    "concentration"(1:reactor1.N.card):=
                      reactor1.c(,reactor1.diffusion_
                      length);
                END
            END
        END
    END
```

5. Test Foreign Process is ok from the gPROMS® side using FPI: ="EventFPI" ; under the SOLUTIONPARAMETERS section just before the schedule section. Run the Process and click abort. IF SUCCESS is reported, next step can be followed.
6. *Foreign process Excel interface:* Open the previous excel interface DiffusionReaction_EXCEL_FO.xls and save it as Diffusion Reaction_EXCEL_FP.xls.

   In the process sim_reactor_2 in gPROMS®, change the path location of my_FO to the same DiffusionReaction_EXCEL_FP.xls file so this file will be used both as Foreign Object and as Foreign Process.

```
...
SET
    points:= reactor1.my_FO.points;# set from excel
      interface (Simulation_Control tab)
     reactor1.N:=reactor1.my_FO.Components; #3 components
       A, B and C
    # or we can access all parameters of a defined unit
      using "within"
```

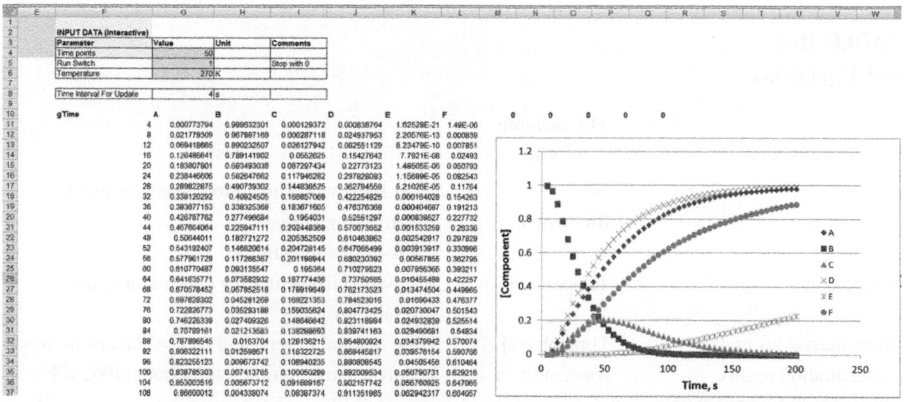

FIGURE 10.15 New tab:interactive inputs and results. (See color insert.)

```
    within reactor1 DO
  my_FO:= "ExcelFO::C:\Folder_Path \DiffusionReaction_
    EXCEL_FP.xls";
...
```

Create two new tabs with the inputs shown as presented in Figures 10.15 and 10.16.

Below are the new FPI variables that need to be defined in Excel by pressing Ctrl+Shift+P, see Table 10.1 and Figure 10.17.

Below there are two examples for definition. Note array selected for concentration output and table model r (different rows for different times). gTime is automatically created once the first variable is created. Define RunSwitch as Real since it is of type length in gPROMS®.

Now, cross reference the three inputs in tab InteractiveInput_and_Results (RunStatus, temperature and TimeInterval) and the 2 outputs (gTime in 1 cell and concentration in 3 columns or more depending on how many components have been included) in the gFPI tab as shown in Figure 10.18.

Replace FPI: = EventFPI by

```
FPI:="ExcelFP::C:\Folder_Path\ DiffusionReaction_EXCEL_
  FP.xls";
```

Note that "Time_simulation" and "points" are two new Foreign Object methods that need to be defined in DiffusionReaction_EXCEL_FP.xls as Real and Integer and cross-referenced to = Inputs!F6 and = Simulation_Control!H12 respectively.

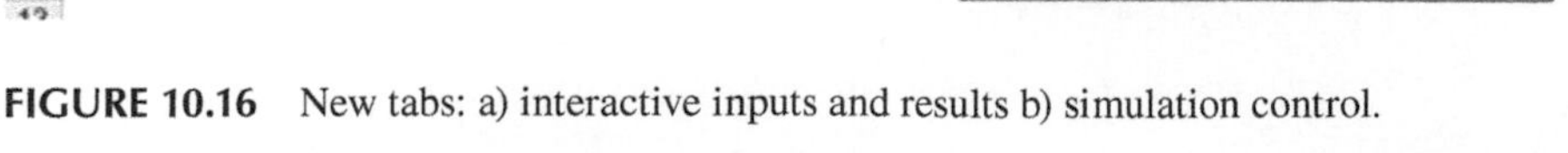

FIGURE 10.16 New tabs: a) interactive inputs and results b) simulation control.

TABLE 10.1
FPI Variables

Excel Caption	FPI variable Name	Use
Time points	NA	Time Interval = Simulation time/time points
Run Switch	RunSwitch	0 = stop simulation 1 = re-start simulation
Temperature	Temperature	Allows to change reactor temperature during simulation
Time interval for update	TimeInterval	Time interval to send and receive data from excel
Concentration results (columns G-L in example)	concentration	Receive concentration results from gPROMS®

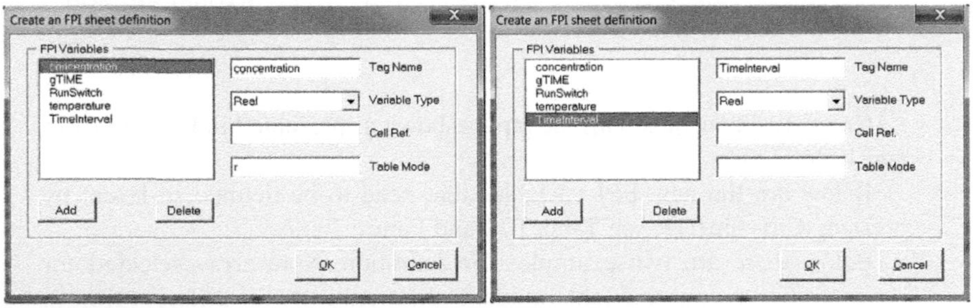

FIGURE 10.17 Creation of FPI variables in Excel.

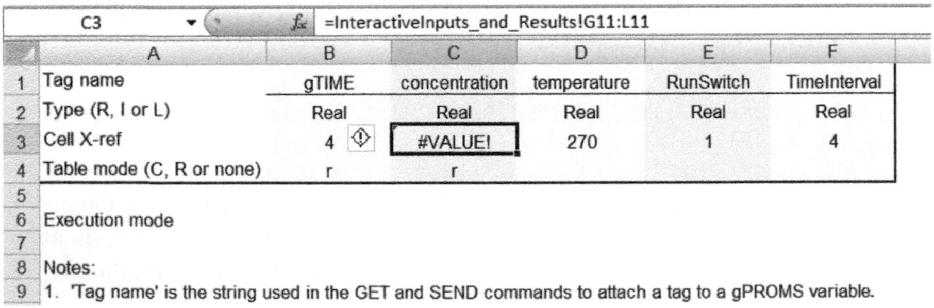
C3 =InteractiveInputs_and_Results!G11:L11

	A	B	C	D	E	F
1	Tag name	gTIME	concentration	temperature	RunSwitch	TimeInterval
2	Type (R, I or L)	Real	Real	Real	Real	Real
3	Cell X-ref	4	#VALUE!	270	1	4
4	Table mode (C, R or none)	r	r			
5						
6	Execution mode					
7						
8	Notes:					
9	1. 'Tag name' is the string used in the GET and SEND commands to attach a tag to a gPROMS variable.					

FIGURE 10.18 Cross-referencing the FPI variables in Excel with their input/ouput cells in the InteractiveInputs_and_Results tab.

A local parameter to the process is defined called "points" of type Integer and set to my_FO.points.

Refer to the gPROMS® manual for deploying an application with an excel interface using a goRUN license.

10.3.5 Basics of Model Flowsheeting

Here the basic concept of flowsheeting with user-developed model is presented using the same example already discussed. Please, refer to the gPROMS® manual to understand how to use already existing libraries of chemical engineering models.

Flowsheeting consists of developing a parent flowsheet model where unit operation models are dragged and dropped and connected to develop a composite (flowsheet) model. There are certain rules to follow to ensure that the modular models can be connected with each other. The concepts of Foreign Object and Process can still be applied when Flowsheeting so that data and results are shared with an EXCEL® interface. However, in the following example, the file Diffusion_Reaction.gPJ is used as a starting point. The basic process is highlighted below:

1. Open the Diffusion_Reaction.gPJ and save it as: Diffusion_Reaction_Flowsheeting.gPJ.
2. The idea is to create 2 models or topology entities, one is an instance of the MainReactor model where mass transfer and reaction takes places and the other one is an instance of the kinetics model where the reaction rates are calculated. These two instances will be called from a higher hierarchy flowsheet model.
3. Propagation of parameters across models: Parameters need to be defined in every single model that use them. However, they only need to be SET once in one model and get propagated from there to the other models that use them. Two main approaches are:
   a. *(recommended) Parameter propagation via connections or horizontal propagation:* The parameter can be set in a submodel of the flowsheet and be passed to another flowsheet submodel via a connection.
   b. *Hierarchal propagation from parent model to child model or vertical propagation* (e.g. as seen when nesting MainReactor model and Kinetics model on page 391). In a flowsheet model, all visible models are nested to the flowsheet model so if a parameter is SET in the flowsheet model it will propagate downwards to all models in the flowsheet where the parameter is defined if it has the same name, has the same size and is of the same type. It is possible to have submodels of the main models of the flowsheet and if the parameter is not defined in a middle level, the propagation will continue.
4. *Create "kinetics_connection" connection type to share variables, parameters and distribution domains across models with the same connection:* Connection Types Folder right click → New Entity → Name: kinetics_connection. This connection needs to make accessible variables concentration c and reaction rate R. Since c and R are dependent on N, M parameters

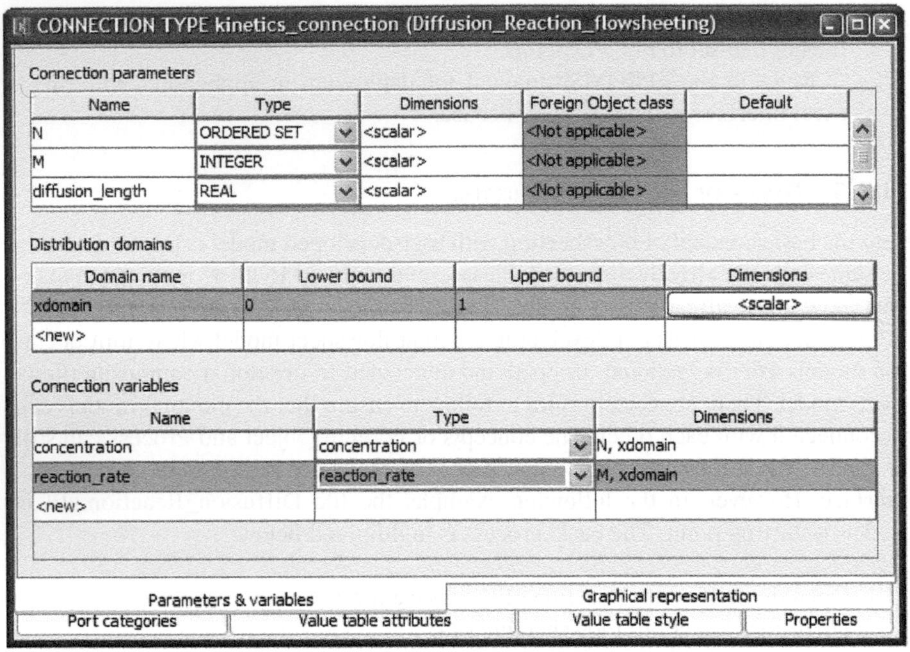

FIGURE 10.19 Definition of parameters, variables and distribution domains that belong to the connection type.

and xdomain distribution domain, they also need to be passed to the kinetics_connection. The parameter diffusion_length is also used in both submodels MainReactor and Kinetics. Here there are two options. Either SET this parameter in the flowsheet model so that it will propagate to both submodels or SET it in one of the submodels and pass via the connection (the recommended option is used here as shown in Figure 10.19).

It is important to note that the name of the variables and parameters within the connection do not need to be the same as those ones in the models. These names are the local definitions within the connection type which will be equated to the value of the variables/parameters in the models.

We change the process sim_reactor_2 to call an instance of the new flowsheet developed flowsheet1 and change the path of the variables being SET.

```
UNIT
 flowsheet1 AS flowsheet
 #kinetics1 as Kinetics

SET
# or we can access all parameters of a defined unit using
  "within"
# Parameters used in both MainReactor and kinetics model
  - Using top down propagation
```

```
#Parameteres used in MainReactor only
    within flowsheet1.reactor1 DO
        N:=['A','B','C']; #3 components A, B and C
        ...
    end
#Parameters used in kinetics model only
    within flowsheet1.kinetics1 DO # Parameters that are
      only used in the kinetics model and do not need to
      be defined in the main model
        k0:=[5e-2,1e-4,1e-4,1e-4];
```

5. *Creation of ports of the kinetics_connection type in both submodels:* Go to Kinetics model and create a port of the type kinetics_connection in both kinetics and MainReactor models using the PORT section after VARIABLE section:

```
...
VARIABLE
    c      as DISTRIBUTION (N, xdomain)  of concentration
      #kmol/m3
    R      as DISTRIBUTION (M, xdomain)  of reaction_rate
      # kmol/m3s
    #T     as                              temperature # K
PORT
    kinetics_connection1     as kinetics_connection
...
```

Note: The port creation can also be done in the interface tab in each of the models.

6. *Interface tab for each model (direction of connections and port location)*
   a. Go to the interface tab in each of the models and make sure that the direction type of kinetics_connection1 port is bi-directional (sends information in both directions, the kinetics model receives concentrations from MainReactor model and sends reaction rates R to it).
   b. Define the position of the ports in the interface with x and y relative to the left corner of the square.
7. *Edit specification input dialog box:* You can also select an icon or develop an input dialog box for the submodel using the edit specification button. Please refer to the gPROMS® manual for more information on this.
8. *Connections of variables and parameters of model to variables/parameters of connection:* Now, The equations to connect variables in the connection with variables in the models are written. The order of the equations is irrelevant. A similar task is due for the assignment of model parameters in models and connections but in this case the assignment direction is important and the parameter being SET must be on the left of the equation. Do similarly for kinetics1 model (**watch-out that the SET direction for the Parameter is reversed** since the parameters get the values from the connection parameters)

MainReactor Model

```
...
SET
# Since N is defined in reactor1, we assigned this N to
  the N in port
    kinetics_connection1.M := M;
    kinetics_connection1.N := N;
    kinetics_connection1.xdomain := xdomain;
    kinetics_connection1.diffusion_length := diffusion_
      length;

...
EQUATION
#  Connection to kinetics model
    kinetics_connection1.concentration = c;
    R = kinetics_connection1.reaction_rate;
...
```

And the Kinetic model

```
...
SET
# pass the value of the parameter M (which is SET by
  propagation to the parameter M in the connection
    M := kinetics_connection1.M;
    N:= kinetics_connection1.N;
    xdomain:= kinetics_connection1.xdomain;
    diffusion_length:= kinetics_connection1.diffusion_
      length ;

EQUATION

#connection of model variables to kinetics_connection
  port variables
    c = kinetics_connection1.concentration ;
    kinetics_connection1.reaction_rate = R;
...
```

9. *Remove calling of submodel kinetics 1 from MainReactor model:* remove UNIT section in MainReactor. Now, all submodels are called from a new Flowsheet model, which instance is created from the process sim_reactor_2.
10. *Create a new model called "flowsheet"* that will incorporate the other two models (flowsheet model). From its Topology tab, *drag and drop a MainReactor instance and a kinetics instance.* The name that appears in each of the models is the equivalent unit name in this flowsheet model (i.e. MainReactor001 is an instance of the MainReactor model). Click on the port of MainReactor001 and then on the port of Kinetics001 to create a connection. Because the connection is bidirectional there are no arrows.

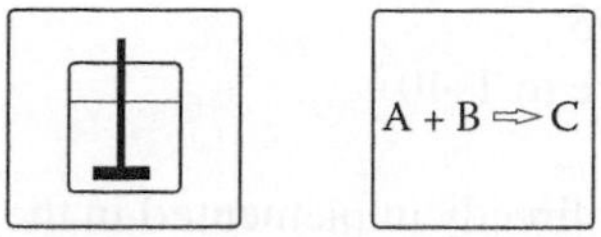

FIGURE 10.20 Topology diagram.

11. *Name topology instances as before flowsheeting*: To avoid having to redefine the calling path of all the parameters and DOFs for each of the models from the process, change the name of the MainReactor and kinetics instances dragged onto the flowsheet topology to the previous names reactor1 and kinetics1, see Figure 10.20. Instead of MainReactor001 we call it reactor1 as before and instead of Kinetics001 we call it kinetics1.
12. *Create flowsheet model instance in the process*: Go to the process sim_reactor_2 and in the UNIT section define an instance of the flowsheet model instead of an instance of the MainReactor model.

```
UNIT
flowsheet1 AS flowsheet
```

13. *Modify the path of parameters and variables depending on your propagation strategy:*

```
UNIT
 flowsheet1 AS flowsheet

SET

#Parameteres used in MainReactor only
    within flowsheet1.reactor1 DO
        N:=['A','B','C']; #3 components A, B and C
...
    within flowsheet1.kinetics1 DO # Parameters that are
      only used in the kinetics model and do not need to
      be defined in the main model
        k0:=[5e-2,1e-4,1e-4,1e-4];
```

14. Run Process sim_reactor_2

10.3.6 Creating Schedules and Tasks

Now, a procedure is followed to increase the temperature of the reactor by 5°C and 10°C after a certain concentration of species A at position x = L is achieved in order to avoid fast formation of side product B. Use the previous file and save it as Diffusion_Reaction_Flowsheeting_Operation.gPJ. The procedure is shown below:

[1] Change Temperature from Ambient to T (use ASSIGN)
[2] Wait until $[A]_L = 0.5$
[3] Change Temperature to T+5 (USE Re-ASSIGN)

[4] Wait until [A] = 0.75
[5] Change temperature to T+10

1. First, this procedure is directly implemented in the SCHEDULE Section of the process sim_reactor_2 and then a TASK (model of a specific procedure or task) is created to simplify the coding of the SCHEDULE:

```
SCHEDULE
    SEQUENCE
        continue until
flowsheet1.kinetics1.c('A',flowsheet1.reactor1.diffusion_
  length)>0.5
        REASSIGN
            #Note the use of old(x) function is needed to
              reassign a variable that uses its old value
            flowsheet1.kinetics1.T := old(flowsheet1.
              kinetics1.T)+5;
        END
        continue until
flowsheet1.kinetics1.c('A',flowsheet1.reactor1.diffusion_
  length)>0.75
        REASSIGN
            flowsheet1.kinetics1.T:=old(flowsheet1.
              kinetics1.T) + 10;
        END
        continue until flowsheet1.kinetics1.c('A',flowsheet1.
          reactor1.diffusion_ length) >0.98 or time> 1000
        # If conc is not reached before 1000 s it will stop
    END
```

Upon execution, every time a variable is reassigned, gPROMS® re-initializes the problem.

2. *Create a new process:* Copy the process sim_reactor_2 into a process called sim_reactor_2_tasks.
3. *Create a new task and define the PARAMETER and SCHEDULE sections:* Tasks folder right click → New Entity → Name: IncreaseTemperature. This task will increase the temperature by a user defined delta_T (parameter) and will act on a kinetics model (defined in this TASK as another parameter). In the current example, only the kinetics model uses the Temperature variable. See Figure 10.21.
4. In the SCHEDULE section, do the re-assign of the variable Temperature on any kinetic type model.

   The task code is:

```
PARAMETER
    modelkinetics as MODEL Kinetics
    delta_T AS REAL

SCHEDULE
```

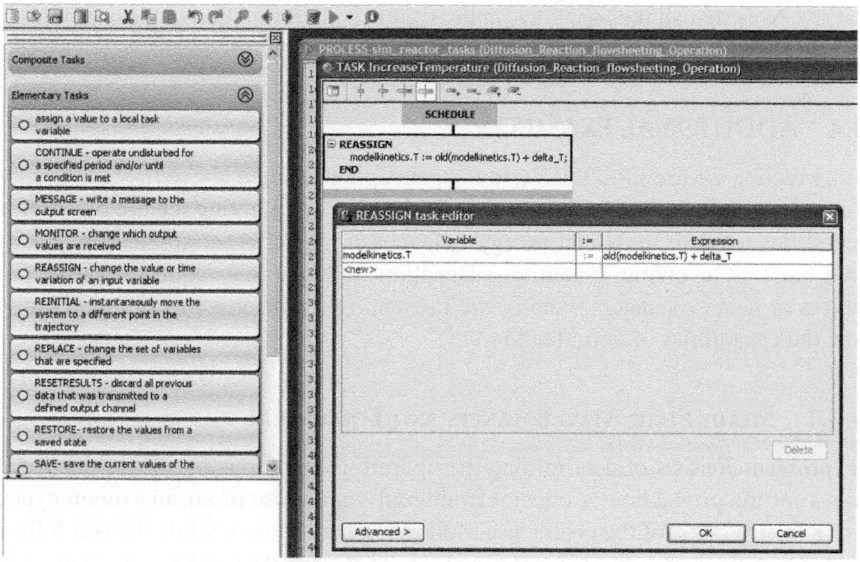

FIGURE 10.21 Defining the internal schedule of a task.

```
REASSIGN
    modelkinetics.T := old(modelkinetics.T) + delta_T;
END
```

5. Change the reassign blocks for the new task in process sim_reactor_2_tasks. The two input parameters to the task is the model where the Temperature is to be reassigned, which is flowsheet1.kinetics1 and the delta_T which is 5 C first, then 10C.

```
...
SCHEDULE
    SEQUENCE
        continue until
flowsheet1.kinetics1.c('A',flowsheet1.reactor1.diffusion_
  length)>0.5
        IncreaseTemperature(modelkinetics IS flowsheet1.
          kinetics1, delta_T IS 5)
        continue until
flowsheet1.kinetics1.c('A',flowsheet1.reactor1.diffusion_
  length)>0.75
        IncreaseTemperature(modelkinetics IS flowsheet1.
          kinetics1, delta_T IS 10)
        # If conc is not reached before 1000 s it will stop
        continue until flowsheet1.kinetics1.c('A',
          flowsheet1.reactor1.diffusion_length) >0.98 or
          time> 1000
    END
```

Note that all the basics learnt in section 3 can be applied to the examples presented in section 4 where often basic versions are presented.

10.4 ADDITIONAL EXAMPLES

In this section we use gPROMS® and the concepts presented above in order to solve problems in different areas of chemical engineering highlighting the use of different capabilities to solve mass and energy balances using different approaches, design batch reactors involving dynamics, adsorption columns that require iterative procedures or heat exchangers. Finally, we present a parameter estimation problem to show the capabilities of optimization.

10.4.1 Steady State: Mass Balances and Equilibrium

The problem consists of determining the operating conditions for the production of syngas for the production of ethanol from coal via the use of air and steam cycles. In this facility, 40% of the coal is used following the air cycle while the rest follows the steam cycle. Atmospheric air is used (T = 23 °C, $\varphi = 0.63$ and 1 atm). Determine the operating temperatures at each process in order to obtain a ratio of H_2:CO of 1, considering that the CO/CO_2 ratio in the air cycle is equal to 4. The pressure at the air cycle is 1 atm and we feed H_2O/C = 2 in a molar basis and the energy ratio available in the gases from each of the two processes before mixing compared to form the initial coal. The scheme is shown in Figure 10.22.

Data:

$$CO + \frac{1}{2}O_2 \rightarrow CO_2$$

$$\Delta h_{f,CO_2} = -94.052 kcal / mol$$

$$\Delta h_{f,CO} = -26.416 kcal / mol$$

$$H_2 + \frac{1}{2}O_2 \rightarrow H_2O$$

$$\Delta H_r = \Delta h_{f,H_2O} = -57.7979 \frac{kcal}{mol H_2}$$

Sinnot. R.K. (1999)

$$C + CO_2 \longleftrightarrow CO \Rightarrow \log(kp) = 9.1106 - \left(\frac{8841}{T}\right)[3] \tag{10.6}$$

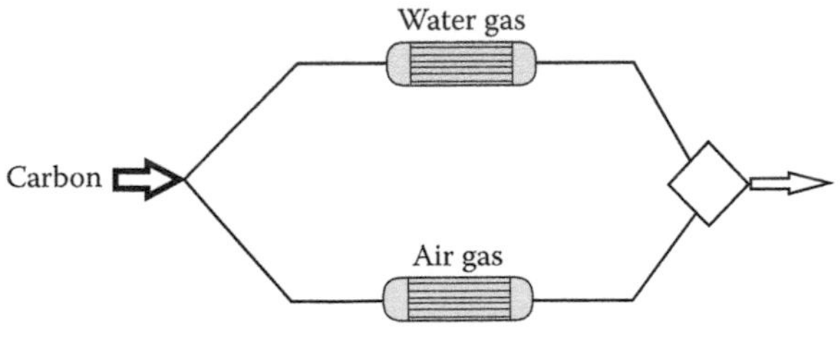

FIGURE 10.22 Scheme for the production of syngas.

$$\log_{10} Kp\left(H_2O+CO\longleftrightarrow CO_2+H_2\right)=\left(\frac{2073}{T}-2.029\right)[4] \tag{10.7}$$

The value of equilibrium constant is found out at constant temperature and pressure using the standard state Gibbs function of change:

$$Lnkp(C+H_2O\longleftrightarrow CO+H_2)=\left(\frac{-131000}{8.3143\mathrm{T(K)}}+\frac{134}{8.314}\right) \tag{10.8}$$

Solution:

a. Air cycle
   The average chemical reaction that occurs is as follows

$$C+a(O_2+3.76N_2+yH_2O)\rightarrow xCO+zCO_2+a3.76N_2+tH_2+kH_2O \tag{10.9}$$

Where we basically consider two main equilibria [3]

$$C_{(s)}+CO_{2(g)}\leftrightarrow 2CO_{(g)} \tag{10.10}$$

$$\begin{gathered} kp=\frac{P_{CO}^2}{P_{CO_2}}\Rightarrow \log(kp)=9.1106-\frac{8841}{T} \\ 10\wedge\left(9.1106-\frac{8841}{T[=]K}\right)=\frac{(n_{CO})^2(P_T)}{\left(n_{CO}+n_{CO_2}+n_{N_2}+n_{H_2}+n_{H_2O}\right)n_{CO_2}} \end{gathered} \tag{10.11}$$

$$CO+H_2O\longleftrightarrow CO_2+H_2\rightarrow 10\wedge\left(\frac{2073}{T}-2.029\right)=\frac{n_{CO_2}\cdot n_{H_2}}{n_{CO}\cdot n_{H_2O}} \tag{10.12}$$

The mass balances to the species involved are as follows, assuming that all the carbon is consumed

$$\begin{aligned} n_{C_{ini}} &= n_{CO_2}+n_{CO}\big|_{out} \\ n_{H_2O_{ini}} &= n_{H_2O}+n_{H_2}\big|_{out} \\ 2n_{O_2}+n_{H_2O}\big|_{in} &= n_{CO}+2\cdot n_{CO_2}+n_{H_2O}\big|_{out} \\ 4n_{CO_2}\big|_{out} &= n_{CO}\big|_{out} \end{aligned} \tag{10.13}$$

Together with them we know that per mole of oxygen, 0.79/0.21 moles of Nitrogen are fed to the system.

$$n_{N_2}\big|_{in}=3.76n_{O_2}\big|_{in} \tag{10.14}$$

$$n_{N_2}\big|_{in}=n_{N_2}\big|_{out} \tag{10.15}$$

And the water accompanying the air

$$y = \frac{0.63 \cdot Pv(23)}{760 - 0.63 \cdot Pv(23)} = 0.01846 \frac{mol_{H_2O}}{mol_{dryair}} \quad (10.16)$$

Per mol of oxygen we can have 1/0.21 moles of dry air, 4.76
Thus, the water with the air is 0.087moles per mol of oxygen

$$n_{H_2O}\big|_{in} = 0.087 n_{O_2}\big|_{in} \quad (10.17)$$

Pt = 1atm
9 eqs with variables, at the inlet $n_{N_2}, n_{O_2}, n_{H_2O}$ and at the outlet T, $n_{CO}, n_{CO_2}, n_{N_2}, n_{H_2}, n_{H_2O}$

b. Water cycle
In this case the main equilibria are

$$C + H_2O \longleftrightarrow CO + H_2 \quad (10.18)$$

Where

$$e^{\left(\frac{-131000}{(8.314 \cdot T(K))} + 134/8.314\right)} = \frac{(\text{Pressure} \cdot n_{CO} \cdot n_{H_2})}{(n_{H_2O} \cdot (n_{CO} + n_{CO_2} + n_{H_2O} + n_{H_2}))} \quad (10.19)$$

$$CO + H_2O \longleftrightarrow CO_2 + H_2 \quad (10.20)$$

And the kp is given as before
The atom mass balance for these equilibria are:

$$\begin{aligned} n_C &= n_{CO} + n_{CO_2}\big|_{out} \\ 2n_{H_2O} &= 2n_{H_2O} + 2 \cdot n_{H_2}\big|_{out} \\ n_{H_2O} &= n_{CO} + 2n_{CO_2} + 2n_{H_2O}\big|_{out} \end{aligned} \quad (10.21)$$

We also know that

$$\frac{n_{H_2}\big|_{water} + n_{H_2}\big|_{air}}{n_{CO}\big|_{water} + n_{CO}\big|_{air}} = 1 \quad (10.22)$$

All the carbon gets consumed in the reaction
6 eqs. with variables, at the outlet T, P, $n_{CO}, n_{CO_2}, n_{H_2}, n_{H_2O}$

a. As Equation solver

In this first example we use gPROMS® as an equation solver just by writing the equations that constitute the problem as modeled before.

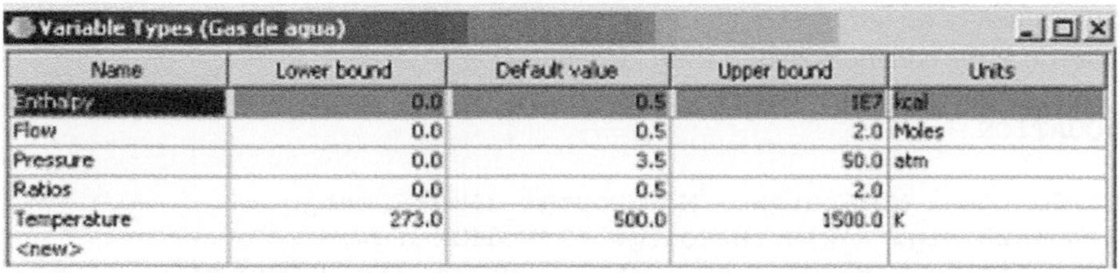
Variable Types (Gas de agua)

Name	Lower bound	Default value	Upper bound	Units
Enthalpy	0.0	0.5	1E7	kcal
Flow	0.0	0.5	2.0	Moles
Pressure	0.0	3.5	50.0	atm
Ratios	0.0	0.5	2.0	
Temperature	273.0	500.0	1500.0	K
<new>				

FIGURE 10.23 Range for the variables.

Thus we go to gPROMS® to write these equations. We create a new model in the project menu. We first define the parameters:

```
PARAMETER
MolCinigw                AS                         REAL
MolH2Oinigw              AS                         REAL
MolCoutgw                  AS                          REAL
MolCiniga    As                    Real
DeltaH2      As                    Real
DeltaCO      As                    Real
```

Next we declare the variables, as presented above. For each type we go to variable type in the left menu of gPROMS® and define them with the range of operation that will help the solver, see Figure 10.23. Be careful, the values of our variables must lie within these ranges otherwise an error message is reported that one variable is "stuck" on one of the bounds.

Now it is possible to write the list of variables that must be introduced by the word "Variable":

```
VARIABLE
MolH2Ooutgw                AS     Flow
MolH2outgw                 AS     Flow
MolCOoutgw                 AS     Flow
MolCO2outgw                AS     Flow
Pressuregw                 AS     Pressure
Temperaturegw              AS     Temperature
MolN2iniga                 As     flow
MolO2iniga                 AS     Flow
MolCOoutga                 AS     Flow
MolCO2outga                AS     Flow
MolH2Oiniga                AS     Flow
MolH2Ooutga                AS     Flow
MolH2outga                 AS     Flow
Pressurega                 AS     Pressure
Temperaturega              AS     Temperature
RatioH2CO                  AS     Ratios
EnergyGW                   AS     Enthalpy
EnergyGA                   AS     Enthalpy
```

Subsequently we write the equations of the model that has been presented above. We introduce the equations by the word "Equation":

```
EQUATION

MolCinigw = MolCoutgw + MolCOoutgw +MolCO2outgw ;
2*MolH2Oinigw = 2*MolH2Ooutgw +2*MolH2outgw;
MolH2Oinigw = MolCOoutgw +2*MolCO2outgw +MolH2Ooutgw ;
MolH2Ooutgw*(MolCOoutgw+MolCO2outgw+MolH2Ooutgw+MolH2outgw)
  *exp(-131000/((Temperaturegw+1)*8.314)+134/8.314) =
    Pressuregw*MolCOoutgw*MolH2outgw;
MolCOoutgw*MolH2Ooutgw*10^(2073/(Temperaturegw+1)-2.029)
  =MolCO2outgw*MolH2outgw;
MolCiniga = MolCOoutga +MolCO2outga ;
MolO2iniga*2+ MolH2Oiniga =MolCOoutga +2*MolCO2outga+
  MolH2Oiniga;
MolN2iniga=3.76*MolO2iniga;
MolH2Oiniga=0.087*MolO2iniga;
MolH2Oiniga= MolH2Ooutga + MolH2outga;
MolN2iniga+MolCOoutga+MolCO2outga+MolH2Ooutga+MolH2outga)
  *MolCO2outga*10^(10.1106-8841/Temperaturega) =
    MolCOoutga^2*Pressurega;
MolCOoutga*MolH2Ooutga*10^(2073/(Temperaturega+1)-2.029)
  =MolCO2outga*MolH2outga;
RatioH2CO = (MolH2outgw +MolH2outga )/(MolCOoutga+MolCOoutgw) ;
MolCOoutga =4*MolCO2outga ;
EnergyGW = DeltaH2*MolH2outgw+DeltaCO*MolCOoutgw;
EnergyGA = DeltaCO*MolCOoutga+DeltaH2*MolH2outga;
```

Once we have the model we go to processes where we define the initial conditions and assign parameters so that we can have a general model that can be used.

We create a new process and we define Unit, as the process. Set is used to declare the parameters and Assign to provide values for variables:

```
UNIT

Equilibrium AS Watergas
SET

    WITHIN Equilibrium DO

    MolCinigw    :=1;
    MolH2Oinigw  :=2;
    MolCoutgw    :=0;
    MolCiniga    :=0.67;
    DeltaH2      :=57.798;
    DeltaCO      := 67.633;
 END
Assign
    WITHIN Equilibrium DO
```

```
   Pressurega:=1;
   RatioH2CO =1;
END
```

Now we are ready to solve the problem by clicking on the green triangle in the menu bar. The results are reported in a file named after the model and with the date and minute that the model was run. In can be seen in the menu section of gPROMS® and can be also stored. The temperature and pressures at the water cycles are 970 K and 1 atm and for the air cycle it turns out to be 1,160 K and 30.3 atm.

b. As flowsheet

We write our model as if we are dealing with a flowsheet. This means that we model each unit independently so that the outlet of one unit is the feed for the next one. This formulation is more general than the previous one since the models are independent and can be used for any other application where we are dealing with the same units. On the other hand the formulation is more complex with larger number of variables and equations. We define one unit per equipment, splitter, reaction for air, reaction for water and mixer. We need to assign variables that link the units so that the outlet flow of one unit corresponds to the inlet flow of the next. We also have molar and mass basis since it is useful for flowsheeting analysis. The flow variables will have a different form, an array, instead of individual variable per species

To the left we create five models, see Figure 10.24, one per unit and a model process to run them all.

Next, we also create tasks so that we can evaluate the models in sequence, or in parallel, which is useful if we have to schedule the operation of a plant including batch units. The types of variables are the same as before adding the mass flow and "notype" for the split fraction.

Model splitter: The species we are dealing with are water, CO, CO_2, N_2, H_2, Coal, O_2. We define an ordered_set to include them. The properties of the species such as molecular weights are defined as arrays within parameters while the flows, molar and mass flows, are also arrays.

```
Parameter

Species                      As     ordered_set
Molecular                    AS     array(Species) of real
Feedstock                    AS     array(Species) of real
no_comp               As    integer

VARIABLE
Flow_in_split           As     array(Species) of Flow
Flow_out_spl_wat        As     array(Species) of Flow
Flow_out_spl_gas        As     array(Species) of Flow
Mass_Flow_in_split      As     array(Species) of Mass
Mass_Flow_out_spl_wat   As     array(Species) of Mass
Mass_Flow_out_spl_gas   As     array(Species) of Mass
split_frac    As       notype
```

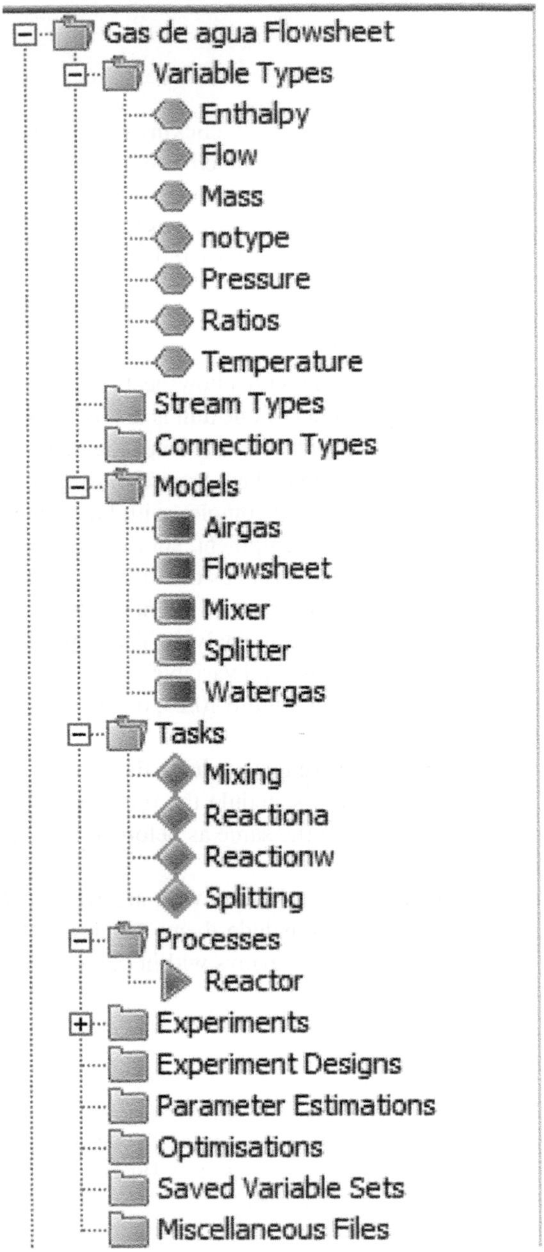

FIGURE 10.24 Browse.

```
SET
Feedstock:=[0, 0, 0, 0, 0, 100, 0];
Molecular :=[18, 28, 44, 28, 2, 12, 32];
species:=['Water','CO','CO2','N2','H2','Coal','O2'];
no_comp:=7;

Equation
split_frac=0.4;
for j in Species do
Feedstock(j)= Mass_Flow_in_split(j);
Mass_Flow_in_split(j)=Flow_in_split(j)*Molecular(j);
Flow_in_split(j)*split_frac =Flow_out_spl_gas(j);
Mass_Flow_out_spl_gas(j)=Flow_out_spl_gas(j)*Molecular(j);
Flow_in_split(j)*(1-split_frac) =Flow_out_spl_wat(j);
Mass_Flow_out_spl_wat(j)=Flow_out_spl_wat(j)*Molecular(j);
end
```

Model water gas: The particularity of this model is the use of logic functions so apply an equation to all the elements within a set such as the identification of the inlet flow as the outlet from the previous unit.

```
UNIT
Spl01                  As Splitter
AirReact               AS Airgas

Parameter
Species              As ordered_set
Molecular            AS array(Species) of real
no_comp              As integer

VARIABLE
Flow_in_Reac_wat            As array(Species) of Flow
Flow_out_Reac_wat           As array(Species) of Flow
Flow_in_wat                 As Flow
Mass_Flow_in_Reac_wat       As array(Species) of Mass
Mass_Flow_out_Reac_wat      As array(Species) of Mass
Pressuregw                  AS                   Pressure
Temperaturegw               AS                   Temperature
RatioH2CO                   AS                   Ratios

SET
Molecular :=[18, 28, 44, 28, 2, 12, 32];
species:=['Water','CO','CO2','N2','H2','Coal','O2'];
no_comp:=7;

EQUATION
for j in species do
Flow_in_Reac_wat (j)= Spl01.Flow_out_spl_wat(j) ;
Mass_Flow_in_Reac_wat(j) = Spl01.Mass_Flow_out_spl_wat(j) ;
end
```

```
Flow_in_wat = 2*Flow_in_Reac_wat('Coal');
Flow_out_Reac_wat('Coal')=0;
Flow_out_Reac_wat('O2')=0;
Flow_out_Reac_wat('N2')=0;
Flow_in_Reac_wat('Coal')=
Flow_out_Reac_wat('Coal')+Flow_out_Reac_wat('CO')+Flow_out_
  Reac_wat('CO2');

2*Flow_in_wat = 2*Flow_out_Reac_wat('Water')+2*Flow_out_
  Reac_wat('H2');

Flow_in_wat =Flow_out_Reac_wat('Water')+ Flow_out_Reac_
  wat('CO')+2*Flow_out_Reac_wat('CO2');

Flow_out_Reac_wat('Water')*(Flow_out_Reac_wat('CO')+Flow_out_
  Reac_wat('CO2')+Flow_out_Reac_wat('Water')+Flow_out_Reac_
  wat('H2'))*exp(-131000/((Temperaturegw+1)*8.314)+134/8.314) =
    Pressuregw*Flow_out_Reac_wat('CO')*Flow_out_Reac_wat('H2');
Flow_out_Reac_wat('CO')*Flow_out_Reac_wat('Water')*10^(2073/
  (Temperaturegw+1)-2.029) =Flow_out_Reac_wat('CO2')
  *Flow_out_Reac_wat('H2');

RatioH2CO =1;
RatioH2CO*(Flow_out_Reac_wat('CO')+AirReact.Flow_out_Reac_
  gas('CO')) = (Flow_out_Reac_wat('H2')+AirReact.Flow_out_
  Reac_gas('H2')  );

for j in species do
Flow_out_Reac_wat(j)*Molecular(j)=Mass_Flow_out_Reac_wat(j);
end
```

Model Mixer

You can see the code in 10.4.1. Steady state B (Flowsheet)

Model airgas

```
UNIT
WaterReact AS Watergas
AirReact AS Airgas

Parameter

Species                 As                      ordered_set
Molecular               AS array(Species) of real
no_comp                 As                      integer
DeltaH2                 As                      Real
DeltaCO       As                     Real

VARIABLE
Flow_in_mix_wat         As array(Species) of Flow
Flow_in_mix_gas         As array(Species) of Flow
Flow_out_mix            As array(Species) of Flow
Mass_Flow_in_mix_wat    As array(Species) of Mass
Mass_Flow_in_mix_gas    As array(Species) of Mass
```

```
Mass_Flow_out_mix        As array(Species) of Mass
EnergyGW                 As Enthalpy
EnergyGA                 AS Enthalpy
Set
    DeltaH2 :=57.798;
    DeltaCO := 67.633;

    Molecular :=[18, 28, 44, 28, 2, 12, 32];
    species:=['Water','CO','CO2','N2','H2','Coal','O2'];
    no_comp:=7;
Equation

for j in species do
Flow_in_mix_wat (j)= WaterReact.Flow_out_Reac_wat(j) ;
Mass_Flow_in_mix_wat(j) = WaterReact.Mass_Flow_out_Reac_wat(j) ;
Flow_in_mix_gas (j)= AirReact.Flow_out_Reac_gas(j) ;
Mass_Flow_in_mix_gas(j) = AirReact.Mass_Flow_out_Reac_gas(j) ;

Flow_out_mix(j) =Flow_in_mix_wat (j)+Flow_in_mix_gas (j);
Mass_Flow_out_mix(j) =Flow_out_mix(j)*Molecular(j);
end

EnergyGW = DeltaH2*Flow_in_mix_wat('H2')+DeltaCO*Flow_in_
  mix_wat('CO');
EnergyGA = DeltaCO*Flow_in_mix_gas('CO')+DeltaH2*Flow_in_
  mix_gas('H2');
```

Model process

```
UNIT

Spl01        As Splitter
WaterReact         AS Watergas
AirReact           AS Airgas
Mix01        AS Mixer
```

Tasks

All the units operate under steady state conditions. We use a parameter to call each of the models, the same as we have used for defining the units, but the assign a schedule with 0 time operation.

```
Parameter Mix01 as Model Mixer
Schedule
sequence
continue for 0
end
```

Process

Here we call all the different units. In this case, we operate in steady state conditions for all the units, specified in the tasks before. gPROMS® always solves the entire model at once. In case there is a unit or units that operate in unsteady state, we can define an operating procedure so that the products of the batch process are obtained and used in the next unit.

```
Unit
Spl01          As Splitter
WaterReact            AS Watergas
AirReact              AS Airgas
Mix01          AS Mixer
Schedule
    Parallel
    Splitting(Spl01 is Spl01)
    Reactiona(AirReact is AirReact)
    Reactionw(WaterReact is WaterReact)
    Mixing(Mix01 is Mix01)
    End
```

10.4.2 Batch Reactor: Ethanol Production from Glucose and Xylose

Nowadays the production of ethanol from lignocellulosic raw materials is one of the processes that are being developed to produce alternative fuels, bioethanol. Unlike traditional production of ethanol from one single substrate, typically glucose from corn, sugar, etc. we here have two substrates, glucose and xylose, from cellulose and hemicelluloses in the raw material. In this example we present the solution of differential equations, the use of logic statements (IF), as well as the use of the plotting tool. The kinetics is characterized by the inhibition due to the presence of ethanol in the liquid and the inhibition due to the substrate, typically modeled using Monod type equations. Based on the model and experimental work by Krishman et al and Bertielsson et al [5,6] we present here the use of gPROMS® to model batch processes. Thus the model basically consists of differential equations for the glucose, xylose, ethanol and cells concentrations that are also known as substrates, product and biomass. See Table 10.2 for the model.

The size of the inoculums is considered to have an effect on the maximum growing rates as long as X is lower than 5g/L.

The parameters of the model were obtained from the above mentioned paper. It is interesting to see that the parameters determining the inhibition due to the product presence in the glucose fermentation depend on the concentration of the product.

Thus we build our model as by defining “Parameters” and “Variables”. For the sake of the space we include them in the following Table 10.3 and the code in Table 10.4.

Those parameters in the models which actually depend on the concentration should be included using logic constraints thus we use the “**if then else end**” statement to define the value of these particular variables as function of the concentration.

To define the differential equations we use the $ symbol before the variable. We define two types of variables for this example, see Figure 10.25:

TABLE 10.2
Model for Ethanol Production from Lignocellulosic Raw Materials

Cells

$$\mu_g = \frac{\mu_{m,g} \cdot S}{K_{s,g} + S + S^2 / K_{i,g}} \left(1 - \left(\frac{P}{P_m}\right)^{\beta_g}\right)$$

$$\mu_x = \frac{\mu_{m,x} \cdot S}{K_{s,x} + S + S^2 / K_{i,x}} \left(1 - \left(\frac{P}{P_m}\right)^{\beta_g}\right)$$

$$\frac{1}{X}\frac{dX}{dt} = \frac{G}{G+X}\mu_g + \frac{X}{G+X}\mu_x$$

Sustrate

$$-\frac{dS}{dt} = \frac{1}{Y_{X/S}}\frac{dX}{dt} + mX = \frac{1}{Y_{P/S}}\frac{dP}{dt}$$

$$-\frac{dS}{dt} = \frac{1}{Y_{P/S}}\frac{dP}{dt}$$

$$-\frac{dxylo}{dt} = \frac{1}{Y_{P/S}}(\nu_{E,x}X)$$

$$-\frac{dglu}{dt} = \frac{1}{Y_{P/S}}(\nu_{E,g}X)$$

Product:

$$\nu_{E,g} = \frac{\nu_{m,g} \cdot S}{K_{s,g} + S + S^2 / K_{i,g}} \left(1 - \left(\frac{P}{P_m}\right)^{\gamma_g}\right)$$

$$\nu_{E,x} = \frac{\nu_{m,x} \cdot S}{K_{s,x} + S + S^2 / K_{i,x}} \left(1 - \left(\frac{P}{P_m}\right)^{\gamma_x}\right)$$

$$\frac{1}{X}\frac{dP}{dt} = (\nu_{E,x} + \nu_{E,g})$$

With

$$\mu_{m,g} = 0.152 \cdot X^{-0.461}$$

$$\mu_{m,x} = 0.075 \cdot X^{-0.438}$$

$$\nu_{m,g} = 1.887 \cdot X^{-0.434}$$

$$\nu_{m,x} = 0.075 \cdot X^{-0.233}$$

TABLE 10.3
Data for the Fermentation Model

Parameter	Glucose Fermentation	Xylose Fermentation
μ_m (h^{-1})	0.662	0.190
ν_m (h^{-1})	2.005	0.250
K_S (g/L)	0.565	3.400
K_S ' (g/L)	1.342	3.400
K_i (g/L)	283.700	18.100
K_i ' (g/L)	4890.000	81.300
P_m(g/L)	95.4 for $P \le 95.4$ g/L	
	1210.9 for $95.4 \le P \le 129$ g/L	510.040
P_m ' (g/L)	103 for $P \le 103$ g/L	
	136.4 for $103 \le P \le 136.4$ g/L	60.200
β	1.29 for $P \le 95.4$ g/L	
	0.25 for $95.4 \le P \le 129$ g/L	1.036
γ	1.42 for $P \le 95.4$ g/L	
m (h^{-1})	0.097	0.067
$Y_{P/S}$ (g/g)	0.470	0.400
$Y_{X/S}$ (g/g)	0.115	0.162

TABLE 10.4
Data in gPROMS® Syntax

Parameter	Set	Variable
K_s_x as real	K_s_x: = 3.4;	mu_g as notype
K_s_g as real	K_s_g: = 0.565;	mu_x as notype
K_sp_x as real	K_sp_g: = 1.342;	v_g as notype
K_sp_g as real	K_sp_x: = 3.4;	v_x as notype
K_i_x as real	K_i_g: = 283.7;	P_m_g as notype
K_i_g as real	K_i_x: = 18.1;	Beta_g as notype
K_ip_x as real	K_ip_g: = 4890;	gamma_g as notype
K_ip_g as real	K_ip_x: = 81.3;	P_mp_g as notype
P_m_x as real	P_m_x: = 510.04;	mu_m_g as notype
P_mp_x as real	Beta_x: = 1.036;	mu_m_x as notype
Beta_x as real	gamma_x: = 0.608;	v_m_g as notype
gamma_x as real	P_mp_x: = 60.2;	v_m_x as notype
m_g as real	m_g: = 0.097;	
m_x as real	m_x: = 0.067;	Glucose as concentration
Y_P_S_x as real	Y_P_S_x : = 0.4;	xylose as concentration
Y_P_S_g as real	Y_P_S_g: = 0.47;	ethanol as concentration
Y_X_S_x as real	Y_X_S_x: = 0.162;	cells as concentration
Y_X_S_g as real	Y_X_S_g: = 0.115;	

```
EQUATION

if ethanol <95.3 then
P_m_g=95.4;
Beta_g=1.29;
gamma_g=1.42;

else
P_m_g=129;
Beta_g=0.25;
gamma_g=0;
end

if ethanol <103 then
P_mp_g=103;
else
P_mp_g=136;
end
```

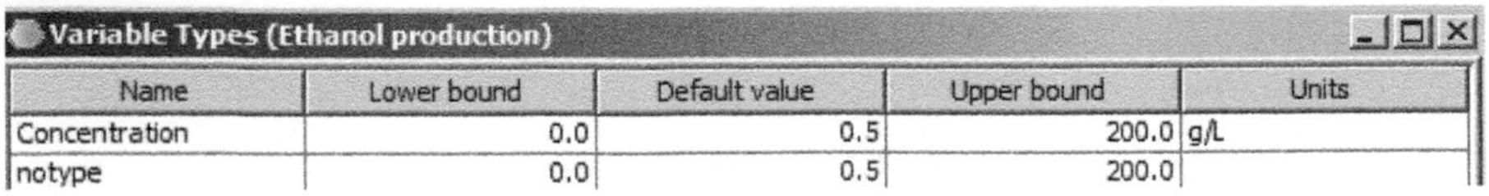
Variable Types (Ethanol production)

Name	Lower bound	Default value	Upper bound	Units
Concentration	0.0	0.5	200.0	g/L
notype	0.0	0.5	200.0	

FIGURE 10.25 Variable types for the model.

```
if cells < 5 then
mu_m_g=0.152*(cells+0.0001)^(-0.461);
v_m_g=1.887*(Glucose+0.0001)^(-0.434);
mu_m_x=0.075*(cells+0.0001)^(-0.438);
v_m_x=1.887*(cells+0.0001)^(-0.233);
else
mu_m_g =0.662;
v_m_g= 2.005;
mu_m_x=0.190;
v_m_x= 0.25;
end

mu_g=mu_m_g*Glucose*(1-(ethanol/P_m_g+0.0001)^Beta_g)/(K_s_g+
  Glucose+Glucose^2/K_i_g);
mu_x =mu_m_x*xylose*(1-(ethanol/P_m_x+0.0001)^Beta_x)/(K_s_x+
  xylose+xylose^2/K_i_x);
v_g=v_m_g*Glucose*(1-(ethanol/P_mp_g+0.0001)^gamma_g)/(K_sp_g+
  Glucose+Glucose^2/K_ip_g);
v_x=v_m_x*xylose*(1-(ethanol/P_mp_x+0.0001)^gamma_x)/(K_sp_x+
  Glucose+Glucose^2/K_ip_x);
$Glucose =-(1/Y_P_S_g)*v_g*cells;
$xylose =-(1/Y_P_S_x)*v_x*cells;
$ethanol =(v_g+v_x)*cells;
$cells = cells*((Glucose)*mu_g/(Glucose+xylose)+(xylose)*mu_x/
  (Glucose+xylose));
```

Now that we have the model we have to define the process. Here, we also provide the initial conditions for the variables. After defining the model that we are using by defining an instance Fer01 of the model "Reactor", we use INITIAL to provide the values for Glucose, xylose, ethanol and cells. Next, the values of time that are reported, using the statement "SolutionParameters". Finally, it is necessary to define the simulation time. We want to run the fermentation for 50 h.

```
Unit Fer01 as Reactor
INITIAL
  WITHIN Fer01 DO
    Glucose = 200;
    xylose  = 50;
    ethanol  = 0;
    cells  = 0;
  END # Within

SOLUTIONPARAMETERS
  ReportingInterval := 0.5 ;
SCHEDULE
  CONTINUE FOR 50
```

Now that we have the results we go to the left column where the solution is reported and we see the variables. Using the gRMS tool of gPROMS® we can easily

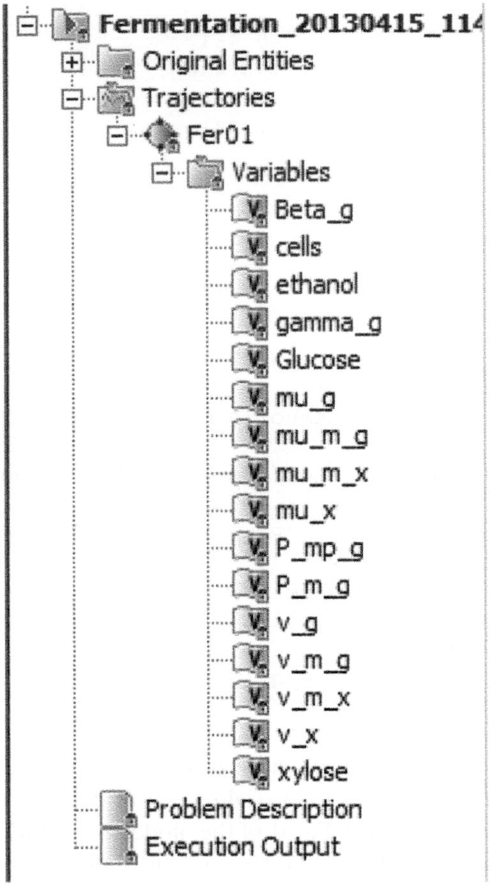

FIGURE 10.26 Variables and plotting ethanol example.

plot the results. Click on 2D and next on the add line button, see Figure 10.26, on the left and select the file where the results have been stored. We can browse there to find the variable and directly select the ones to be plotted. We add each one, Glucose, xylose, cells and ethanol one by one. We can use one or two different y axis. We plot the results in Figure 10.27.

QUESTIONS

- Evaluate the effect of the initial concentration of substrates. This is similar to the effect of different raw materials with different compositions in cellulose and hemicelluloses.
- Assume that there is no inhibition because of the product. What could be the results? Couple this result with the energy consumption in the ethanol dehydration in Chapter 8.

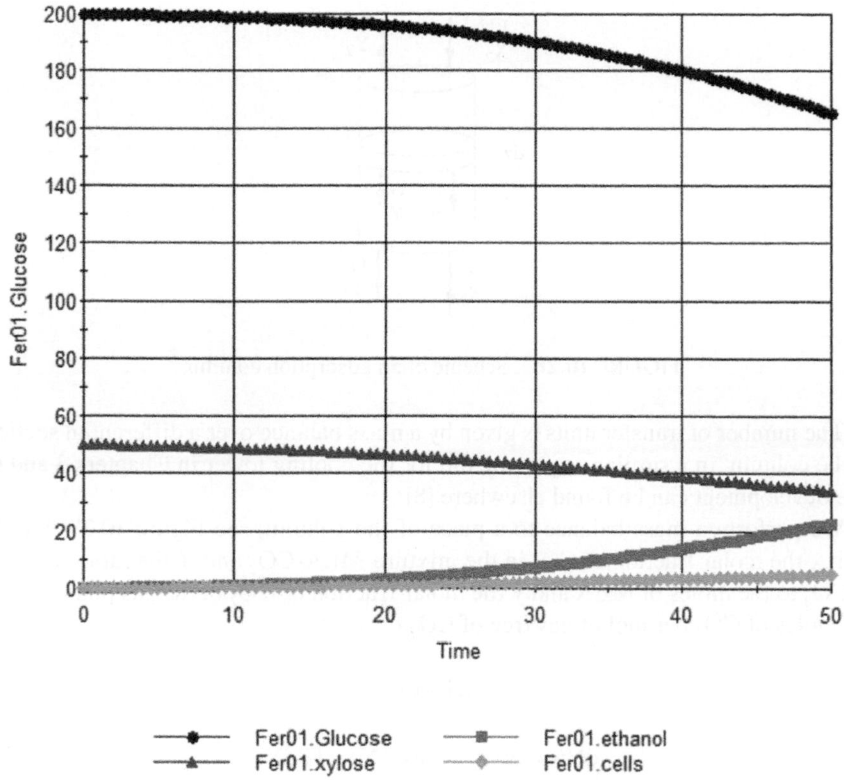

FIGURE 10.27 Results for the ethanol production example. (See color insert.)

10.4.3 Iterative Procedures; Packed Absorption Column

CO_2 emissions are of increasing concern nowadays. There are a number of technologies available for separating the CO_2 from syngas or from flue gases in power plants. Among them we can mention Pressure swing adsorption, membranes, use of CaO, physical adsorbents or chemical adsorbents. One of the most used belongs to the last of the groups. It consists of the use of a solution of ethanol amines (mono, di or tri), mixtures of some of them. In this example we present an iterative procedure that requires the solution of non-linear equations within the loops as well as the use of built in functions such as summation (SIGMA)

This technology is based on the chemical reaction and equilibrium between the four gases, CO_2 and H_2S, and the ethanol-amine. In this example we are going to determine the number of units of transmission for a column that treats a gas flow of 50000 kg/hm^2 with a mass composition of 0.045 H_2, 0.627 CO and 0.328 CO_2 in order to remove 910.9% of the CO_2 present for its use in the production of second generation bioethanol. The flow of solution is 1000 kmol/hm^2. The equilibrium data for the CO_2 absorption in MEA can be found in Joe et al [7].

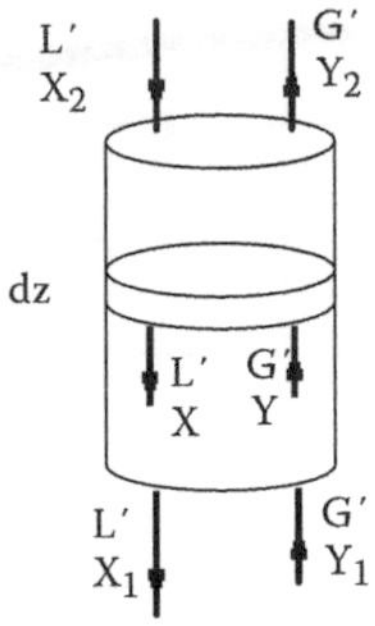

FIGURE 10.28 Scheme of an adsorption column.

The number of transfer units is given by a mass balance over a differential section of the column, in a similar way as we did for the cooling tower in Chapters 3 and 6. The development can be found elsewhere [8].

We perform a mass balance to a piece of the column, see Figure 10.28. We do with x the molar fraction of CO_2 in the mixture MEA-CO_2 and X the ratio of moles of CO_2 to the moles of MEA and y the molar fraction of CO_2 in the gas phase and Y the moles of CO_2 per mol of gas free of CO_2.

$$dG = dL \tag{10.23}$$

$$dN/s = d(Gy) = d(G'Y) = G'dY \tag{10.24}$$

where G' is the gas phase in absence of the species to be absorbed.

Since

$$Y = \frac{y}{1-y} \Rightarrow dY = \frac{dy}{(1-y)^2} \tag{10.25}$$

$$G' = G(1-y) \tag{10.26}$$

Thus

$$\frac{dN}{S} = G'dY = G(1-y)\frac{dy}{(1-y)^2} = \frac{G}{1-y}dy \tag{10.27}$$

The molar flow of gas absorbed is given by

$$N = k_y(y - y_i) \tag{10.28}$$

$$dN = k_y(y - y_i)aSdz \tag{10.29}$$

$$z = \int_{y_2}^{y_1} \frac{G}{k_y a} \frac{dy}{(1-y)(y-y_i)} = \frac{G}{k_y a(1-y)_{\log}} \int_{y_2}^{y_1} \frac{(1-y)_{\log} dy}{(1-y)(y-y_i)} \tag{10.30}$$

$$z = \int_{y_2}^{y_1} \frac{G}{k_y a} \frac{dy}{(1-y)(y-y_i)} = \frac{G}{k_y a(1-y)_{\log}} NTU \tag{10.31}$$

Where:

$$(1-y)_{\log} = \frac{(1-y_i)-(1-y)}{\ln\left(\frac{(1-y_i)}{(1-y)}\right)} \tag{10.32}$$

Operation line corresponds to a mass balance from the bottom of the column to any height:

$$\begin{gathered} G'Y_1 + L'X = G'Y + L'X_1 \\ G'\frac{y_1}{1-y_1} + L'\frac{x}{1-x} = G'\frac{y}{1-y} + L'\frac{x_1}{1-x_1} \end{gathered} \tag{10.33}$$

We assume negligible resistance in the gas phase.

$$\frac{(y-y_i)}{(x-x_i)} = \infty \tag{10.34}$$

From the experimental data we obtain y_1, x_1, L' and G' and we need to solve the equation for each point. We are going to use gPROMS® to model the column and determine the number of transfer units. We start with fitting the equilibrium line. We do in piece wise due to the complexity and the range of data. For the sake of the room we present the parameter and variable definition as well as the values in Table 10.5.

The model is as follows:

```
Equation

Gas_flow_p = Gas_flow*mass_comp('H2')/molecular('H2')+Gas_
  flow*mass_comp('CO')/molecular('CO');

ym_1=molar_comp('CO2');
ym_2=(1-rend)*ym_1;
Y_1=ym_1/(1-ym_1);
Y_2=ym_2/(1-ym_2);

X_2=0;

Gas_flow_p *(Y_1-Y_2) =Liquid_flow_p *(X_1-X_2);
```

TABLE 10.5
Variable and Parameter Definition for the Packed Column

```
PARAMETER
stages as integer
no_comp as integer
species as ordered_set
molar_comp as array(species) of
 real
mass_comp as array(species) of
 real
molecular as array(species) of
 real
Gas_flow as real
Liquid_flow_p as realrend as
 real

VARIABLE
Gas_flow_p as flow
Y_1 as concentration
Y_2 as concentration
X_1 as concentration
X_2 as concentration
ym_1 as concentration
ym_2 as concentration
xm_1 as concentration
xm_2 as concentration
x_co2 as array(stages) of  concentration
y_co2 as array(stages) of concentration
y_eq_co2  as array(stages) of  concentration
x_eq_co2  as array(stages) of  concentration
ylog as array(stages) of  notype
dy as array(stages-1) of  notype
dNog as array(stages) of  notype
meandNog as array(stages-1) of  notype
delta_x_co2 as notypeNTU as notype

SET
species:=['H2','CO','CO2'];
molar_comp:=[0.43,0.43,0.14];
mass_comp:=[0.045,0.627,0.328];
molecular:=[2,28,44];
rend:=0.999;
Liquid_flow_p:=1000;
stages:=11;
no_comp:=3;
Gas_flow:=50000;
```

```
xm_1=X_1/(1+X_1);
xm_2=X_2/(1+X_2);

x_co2(1)=xm_1;
delta_x_co2=(xm_1-xm_2)/(stages+1);

for st:=1 to stages-1 do
x_co2(st+1)=x_co2(1)-st*delta_x_co2;
end

for st:=1 to stages do

x_co2(st)/(1-x_co2(st)) =(Gas_flow_p*(y_co2(st)/(1-y_co2(st)))
  +Liquid_flow_p*(xm_1/(1-xm_1))-Gas_flow_p*(ym_1/(1-ym_1)))
  *(1/Liquid_flow_p);

x_eq_co2(st)=x_co2(st);

if x_co2(st) > 0.267 then
    If x_co2(st) > 0.357 then
        IF x_co2(st) >0.392 then
            y_eq_co2(st) =1.7908e5*x_eq_co2(st)^(1.6446e1);
        else
            y_eq_co2(st) =8.2808e-15*EXP
              (7.4047e1*x_eq_co2(st));
        end
    else

        y_eq_co2(st) =1.5245e-11*EXP(5.2659e1*x_eq_co2(st));
    end
else
    y_eq_co2(st) =10.0422e-8*EXP(2.060e1*x_eq_co2(st));
end

ylog(st) = ((1-y_eq_co2(st))-(1-y_co2(st)))/log((1-y_eq_
  co2(st))/(1-y_co2(st)));
dNog(st)=ylog(st)/((1-y_co2(st))*(y_co2(st)-y_eq_co2(st)));
end

for st:=1 to stages-1 do
meandNog(st)=0.5*(dNog(st+1)+dNog(st));
dy(st)=-y_co2(st+1)+y_co2(st);
end

NTU=sigma(dy()*meandNog());
```

In Figure 10.29 we present the computed operation line, in red diamonds, the approximation based on L/G given by the orange circles and the L/G_{min} in green diamonds and the equilibrium line in blue. We present the semi-logarithmic plot, where the equilibrium line it is better seen while the regular x-y plot can be misleading.

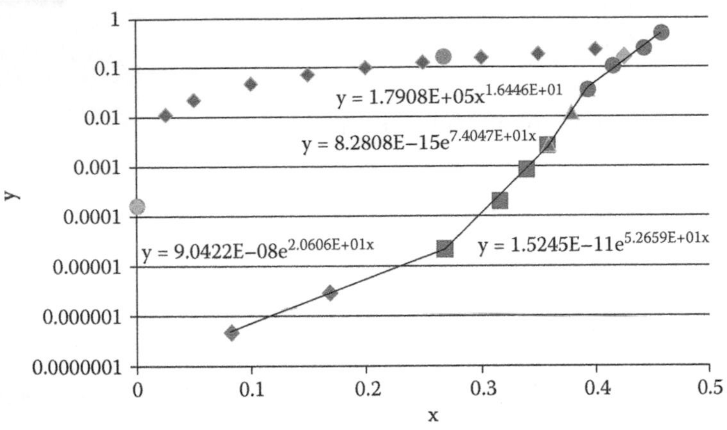

FIGURE 10.29 Results for the adsorption example.

The approximation of the equilibrium line is not that bad if we need to evaluate the number of transfer units by hand. The equilibrium line can also be simplified but it is more difficult to compute the results. By running the gPROMS® file created we obtain NTU = 2.

QUESTIONS

- Evaluate the effect of the mass transfer resistance on the number of transfer units.
- Evaluate the effect of the flow rates on the mass transfer units.
- Evaluate the effect of the concentration of CO_2 in the gas.
- Use the function Integral within gPROMS® to perform the integration

10.4.4 Heat Exchanger Design

Heat exchangers are one of the most commonly designed equipment in the chemical industry. Chemical facilities present hundreds of heat exchangers and the design procedure is typically iterative so that the area available for heat transfer is enough for the requirements. There are though mathematical based procedures, Sinnot 1999 [9] which reduce the dependency on reading tables and figures. In this section we use gPROMS® in order to help design heat exchangers by combining methods from the literature and by developing simple correlations out of the figures in order to systematize the design procedure. We use as an example a heat exchanger from an HDA plant.

10.4.4.1 Mass and Energy Balances and Thermodynamics

The different flow values as well as some physical properties of the streams are shown below, Table 10.6. It is considered a generic refrigerator RG for the cooling of the principal stream fed to the heat exchanger.

TABLE 10.6
Basic Stream Data for the Heat Exchanger

	Process		Utilities	
Flow (kmol/h)	**Inlet**	**Outlet**	**Inlet**	**Outlet**
Hydrogen	2429	2429		
Methane	1037.8	1037.8		
Benzene	300	300		
Toluene	267	267		
Difenile	1.55	1.55		
Total flow (kmol/h)	4035.35	4035.35		
Total flow (kg/h)	69670.18	69670.18	2037.84	2037.84
Temperature (K)	841	602		
Pressure (kPa)	3500	3400		
Vapor fraction	1	1		
Liquid fraction	0	0		
Enthalpy (MJ/h)	138345	79402		
Density(kg/m^3)	8.5	11.6	900	900
Heat capacity (Kcal/kg.K)	0.93	0.93	3.8	3.8
Viscosity (cp)	0.355	0.01	0.0585	0.12

10.4.4.2 Design Procedure

The prime objective in the design of an exchanger is to determine the surface area (A) required for the specified duty, Q, using the temperature difference available [9]. This could be achieved by dividing the design of the whole heat exchanger in the following steps:

- Step 1. Mean temperature difference.
- Step 2. Number of tubes and tube bundle diameter.
- Step 3. Overall heat transfer coefficient (Tube and Shell side)
- Step 4. Surface area required and available.
- Step 5. Pressure drop (Tube and Shell side)

10.4.4.3 gPROMS® Modeling

First of all, the design of the heat exchanger will be carried out by defining the parameters and variables as shown in previous examples. They include the specification of components, physical properties, etc. In the present case, all the flows and temperatures are also considered as parameters, as we already know them from the mass and energy balances.

We create a model and a process within gPROMS®. The PROCESS section is used to include the basic data of the streams and the main configuration feature of the heat exchanger.

```
UNIT

Heat_Exchanger                 AS  Heat_Exchanger

SET

WITHIN Heat_Exchanger DO

Components                       := ["Hydrogen", "Methane",
                               "Bencene","Toluene","Difenile"];

Refrigeration                    := ["Utilities"];

Mw("Hydrogen")                   := 2.016;
Mw("Methane")                    := 16;
Mw("Bencene")                    := 78;
Mw("Toluene")                    := 92;
Mw("Difenile")                   := 154.2;
Mw_aux("Utilities"):=200;

Flow_in(1,"Hydrogen")            := 2429; #kmol/h
Flow_in(1,"Methane")             := 1037.80;
Flow_in(1,"Bencene")             := 300.00;
Flow_in(1,"Toluene")             := 267.00;
Flow_in(1,"Difenile")            := 1.55;

Flow_in(2,"Hydrogen")            := 0; #kmol/h
Flow_in(2,"Methane")             := 0;
Flow_in(2,"Bencene")             := 0;
Flow_in(2,"Toluene")             := 0;
Flow_in(2,"Difenile")            := 0;

Flow_out(1,"Hydrogen")          := Flow_in(1,"Hydrogen"); #kmol/h.
Flow_out(1,"Methane")            := Flow_in(1,"Methane");
Flow_out(1,"Bencene")            := Flow_in(1,"Bencene");
Flow_out(1,"Toluene")            := Flow_in(1,"Toluene");
Flow_out(1,"Difenile")           := Flow_in(1,"Difenile");

Flow_out(2,"Hydrogen")           := 0; #kmol/h.
Flow_out(2,"Methane")            := 0;
Flow_out(2,"Bencene")            := 0;
Flow_out(2,"Toluene")            := 0;
Flow_out(2,"Difenile")           := 0;

Flow_in_aux                      := 2037.84; #kmol/h
Flow_out_aux                     := Flow_in_aux;

Density_in(1)                    := 8.5; #kg/m3
Density_out(1)                   := 11.6;
Density_in(2)                    := 900;#kg/m3
Density_out(2)                   := Density_in(2);
```

```
Heat_capacity_in(1)              := 0.93; #kcal/kg.k
Heat_capacity_out(1)             := 0.93;
Heat_capacity_in(2)              := 3.8;
Heat_capacity_out(2)             := 3.8;

Kf                               := 0.069; # W/m.k
Kf_s                             := 0.543;

T1                               := 841; #K
T2                               := 602;
t_1                              := 500;
t_2                              := 510;

END

ASSIGN

WITHIN Heat_Exchanger DO

De                              := 0.01755;
di                              := 0.01366;
L                               := 4.88;
C                               := 15/16;
tp                              := 2;

Cooling                         := 1;
{ Cooling = 1  for cooling,
  Cooling = 0  for heating}
tri                             := 1;
{  tri = 1 for triangular pitch
   tri = 0 for square pitch}
Head                            := 2;
{  Head = 1 = Pull through floating head
   Head = 2 = Split ring floating head
   Head = 3 = Outside packed head
   Head = 4 = Fixed and U-Tube}
b                               := 2;
{  b = 0 Non viscous liquids
   b = 1 Viscous liquids
   b = 2 Gases}

viscosity_in(1)                 := 0.355; #cp
viscosity_out(1)                := 0.01;
viscosity_in(2)                 := 0.0585;
viscosity_out(2)                := 0.12;

END
```

In the MODEL section, we define all the parameters and variables needed for the design of the heat exchanger as shown below.

```
PARAMETER
Components              AS ORDERED_SET
Refrigeration           AS ORDERED_SET
Mw                      AS ARRAY(Components)            OF   REAL
Mw_aux                  AS ARRAY(Refrigeration)         OF   REAL
no_in                   AS INTEGER
no_outs                 AS INTEGER
Flow_in                 AS ARRAY(no_in,Components)      OF   REAL
                           #Kmol/h
Flow_out                AS ARRAY(no_outs,Components) OF   REAL
                           #kmol/h
Flow_in_aux             AS REAL #kmol/h
Flow_out_aux            AS REAL
Density_in              AS ARRAY(no_in)                 OF   REAL
Density_out             AS ARRAY(no_outs)               OF   REAL
Heat_capacity_in        AS ARRAY(no_in)                 OF   REAL
Heat_capacity_out       As ARRAY(no_outs)               OF   REAL
Kf                      AS REAL #W/m.k
Kf_s                    AS REAL

T1                      AS REAL #Hot fluid temperature
T2                      AS REAL
t_1                     AS REAL #Cold fluid temperature
t_2                     AS REAL
pi                      AS REAL
       #NOTE. Input 2 and output 2 for utilities.


VARIABLE

Flow_in_total    AS ARRAY(no_in)  OF  molar_flowrate
Flow_out_total   AS ARRAY(no_outs)OF  molar_flowrate

# Mean temperature difference
Delta_T          AS temperature
LMTD             As temperature
F                   AS Dimensionless #correction factor
R                AS Dimensionless
S                AS Dimensionless
P                   AS Dimensionless
tp               AS Dimensionless # tube passes
tri              AS Dimensionless

# Tubes properties
De               AS length
di               AS length
Db                AS length #Bundle diameter
L                AS length
Nt               AS no_type
K1               AS no_type
```

```
n1                  AS no_type
Tpitch              AS no_type # Tube pitch
C                AS no_type # Clearance

#Reynolds
ut               AS velocity #Tube velocity
ut1              AS velocity
ut2              AS velocity
viscosity_in      AS ARRAY (no_in)  OF Viscosity #cp
viscosity_out     AS ARRAY (no_outs)OF Viscosity
Re_t             AS Dimensionless

#Overall heat transfer coefficient
#1  Tube side heat transfer coefficient
Nu               AS Dimensionless
Pr               AS Dimensionless
n                AS Dimensionless # coefficient
Cooling          AS Dimensionless
hi               AS Heat_Coefficient
b                AS no_type

#2  Shell side heat transfer coefficient
Re_s             AS Dimensionless
u_s               AS velocity
Pr_s             AS Dimensionless
Nu_s             AS Dimensionless
ho               AS Heat_Coefficient

#3  Fouling factors
ri      AS Resistence #m2.K/W
ro       AS Resistence
kw       AS Conductivity
  # wall thermal conductivity

#4  Overall heat transfer coefficient
U       AS Heat_Coefficient

#5  Surface area
A        AS area
Q        AS energy_rate
Ad       AS area

# Pressure drop
#1  Pressure drop in the tube side

dPt      AS pressure
f_t      AS no_type  # fanning friction factor
fd       AS no_type
  # Darcy friction factor
V        AS velocity
```

```
  # Average flow velocity through a single     tube.
#2  Pressure drop in the shell side.
dPs      AS pressure
f_s      AS no_type
Gs       AS velocity
   # mass velocity in the shell side
Ds_i     AS length # ID shell
Ds_e     AS length
Nb       AS no_type# no.baffles
Deq      AS length
   # Equivalent diameter
Re_s_eq AS Dimensionless
   # Reynolds number based on equivalent  diameter
fd_s     AS no_type
Sm       AS area
 # cross flow area measured close to the central symmetry
   plane of the shell containing its axis.
Lb       AS length
  # baffle spacing
Cp       AS no_type
Head     AS no_type
```

Once all parameters and variables have been defined, we write the equations in the model section. Equations are introduced by the word EQUATION as shown in previous examples. The development of the model will be carried out by following each of the steps specified previously in section 10.3.3.2.

Step 1. Mean temperature difference: The mean temperature difference, ΔT_m, can be calculated from the fluid temperatures at the inlet and outlet of the exchanger with a correction factor, F, as shown below [9].

$$\Delta T_m = F \Delta T_{lm}$$
$$\Delta T_{lm} = \frac{(T_1 - t_2) - (T_2 - t_1)}{ln \frac{(T_1 - t_2)}{(T_2 - t_1)}} \quad (10.35)$$

T_1, T_2 = hot fluid temperature, inlet and outlet respectively. t_1,t_2 = cold fluid temperature, inlet and outlet respectively.

Where ΔT_{lm} is the log mean temperature difference, and the temperature corrector, F, is a function of the shell and the tube fluid temperatures, and the number of tube and shell passes. It is normally correlated as a function of two dimensionless temperature ratios [9,10].

$$R = \frac{(T_1 - T_2)}{(t_2 - t_1)}; \quad S = \frac{(t_2 - t_1)}{(T_1 - t_1)}; \quad P = \frac{(T_2 - T_1)}{(t_2 - T_1)} \quad (10.36)$$

For 1 shell:2 tube pass exchanger:

$$F = \frac{\sqrt{(R^2+1)}\ln\left[\frac{(1-S)}{(1-RS)}\right]}{(R-1)\ln\left[\frac{2-S[R+1-\sqrt{(R^2+1)}}{2-S[R+1+\sqrt{(R^2+1)}}\right]} \tag{10.37}$$

For two shell passes and any multiple of four tube passes

$$F = \frac{\left[\frac{\sqrt{R^2+1)}}{2(R-1)}ln\frac{1-P}{1-PR}\right]}{\ln\left[\frac{\frac{2}{P}-1-R+\left(\frac{2}{P}\right)\sqrt{(1-P)(1-PR)}+\sqrt{R^2+1}}{\frac{2}{P}-1-R+\left(\frac{2}{P}\right)\sqrt{(1-P)(1-PR)}-\sqrt{R^2+1}}\right]} \tag{10.38}$$

```
#1  Mean temperature difference

Delta_T = LMTD*F;
    LMTD = ((T1 - t_2)- (T2 - t_1))/(LOG((T1-t_2)/(T2 -t_1)));
    R = (T1 - T2)/(t_2 - t_1);
    S = (t_2 - t_1)/(T1 - t_1);
    P = (T2 - T1)/(t_2 - T1);

IF tp = 2 THEN

  F = ((((R^2)+ 1)^0.5)*LOG((1 - S)/(1 - (R*S))))/
    ((R-1)*(LOG((2-(S*(R + 1 - ((R^2 + 1)^0.5))))/(2-(S*(R + 1
    + ((R^2 + 1)^0.5))))))));
ELSE # for tp = 4 and multiples of 4
  F = ((((R^2 + 1)^0.5)/(2*(R -1)))*(LOG((1-P)/(1-P*R))))/
    (LOG((2/P) - 1 - R + (2/P)*(((1-P)*(1-P*R))^0.5) +
    (((R^2) +1)^0.5))/((2/P) - 1 - R + (2/P)*(((1-P)*(1-
    P*R))^0.5) - (((R^2) +1)^0.5))));
END
```

Step 2. Number of tubes and tube bundle diameter: The characteristics of the tubes selected for the design of the heat exchanger have already been specified in the process section, as mentioned before. They include tube length, diameter, and thickness as well as tube layout. In this case will be used one of the most common tube sizes which is ¾ in outside diameter, 14 BWG, 4.88 m length, triangular pitch 15/16. It should be noticed that the user can specify any size of commercial tubes, and we propose as an exercise to the reader the development of the design of the heat exchanger with different tube configurations and sizes.

TABLE 10.7
Constants for use in Equations 10.39 and 10.40 [9]

	Triangular pitch, pt = 1.25 d_o				
No passes	1	2	4	6	8
K_1	0.319	0.249	0.175	0.0743	0.0365
n_1	2.142	2.207	2.285	2.499	2.675
	Square pitch, pt = 1.25d_o				
No passes	1	2	4	6	8
K_1	0.215	0.156	0.158	0.0402	0.0331
n_1	2.207	2.291	2.263	2.617	2.643

The number of tubes and tube bundle diameter could be obtained by applying equations 10.39 and 10.40, which are empirical based on standard tube layouts. The constants for use in these equations, for triangular and square patterns, are given in Table 10.6 [9].

$$N_t = K_1 \left(\frac{D_b}{d_o} \right)^{n1} \tag{10.39}$$

$$D_b = d_o \left(\frac{N_t}{K_1} \right)^{1/n1} \tag{10.40}$$

Where Nt is the number of tubes, D_b the bundle diameter (m) and d_o the tube outside diameter (m). The values of the parameters K_1 and n_1 depend on the tube passes (tp), and can be obtained from Table 10.7. We have developed a correlation from Table 10.6 for triangular pitch as it follows a polynomial tendency with a square factor of R^2 = 0.992. However, we cannot correlate the values for square configuration as they do not keep any relationship. The code will be expressed in gPROMS® as follows:

```
#2  Number of tubes and tube bundle diameter
Nt = K1*((Db/di)^n1);
Db = di*((Nt/K1)^(1/n1));

IF tri = 1 THEN
K1 = 0.0028*(tp^2) - 0.0658*tp + 0.3787;
n1 = 0.0053*(tp^2) + 0.0285*tp + 2.1128;
ELSE
IF tp = 1 THEN
K1 = 0.215;
n1 = 2.207;
ELSE
IF tp = 2 THEN
K1 = 0.156;
n1 = 2.291;
ELSE
IF tp = 4 THEN
```

```
K1 = 0.158;
n1 = 2.263;
ELSE
IF tp = 6 THEN
K1 = 0.0402;
n1 = 2.617;
ELSE
K1 = 0.0331;
n1 = 2.643;
END
END
END
END
END

Tpitch = De + C;
```

Step 3. Overall heat transfer coefficient: The overall heat transfer coefficient, U, can be expressed as [9]:

$$\frac{1}{U} = \frac{1}{h_o} + \frac{1}{h_{od}} + \frac{d_o \ln\left(\frac{d_o}{d_i}\right)}{2k_w} + \frac{d_o}{d_i}\cdot\frac{1}{h_{id}} + \frac{d_o}{d_i}\cdot\frac{1}{h_i} \qquad (10.41)$$

Where,

U_o = the overall coefficient based on the outside are of the tube, W/m^2.°C
h_o = outside fluid film coefficient, W/m^2.°C
h_i = inside fluid film coefficient, W/m^2.K
h_{od} = outside dirt coefficient, W/m^2.°C
h_{id} = inside dirt coefficient, W/m^2.°C
k_w = thermal conductivity of the tube wall material, W/m.°C
d_i = tube inside diameter, m.
d_o = tube outside diameter, m.

The tube side coefficient when there is no condensation can be obtained by using the Sieder-Tate equation. It must be noticed that different coefficients must be used depending on the type of fluid, being divided in non-viscous liquids, viscous liquids, and gases [9,10]. Table 10.8 presents the equations.

```
#3  Tube side coefficient method for sensible heat transfer.
Re_t=(((Density_in(1)+Density_out(1))*0.5)*ut*di)/
  (((viscosity_in(1)+viscosity_out(1))*0.5)/1000);

    ut1 = ((SIGMA(Flow_in(1, )*Mw()))/(Density_in(1)*(pi*(1/4)
      *(di^2))))/3600;
    ut2=((SIGMA(Flow_out(1, )*Mw()))/(Density_out(1)*(pi*(1/4)
      *(di^2))))/3600;
    ut  = (ut1 + ut2)/2;
```

TABLE 10.8
Nusselt Number for the Tube Side for Different Regimes

Laminar flow (100 < Re < 2100):

$$Nu = 1.86\left[(Re)(\mathrm{Pr})\left(\frac{d_e}{L}\right)\right]^{0.33}\left(\frac{\mu_B}{\mu_W}\right)^{0.14}$$

Turbulent flow (Re> 10000):

Non viscous liquids:

$$Nu = 0.023(Re)^{0.8}(Pr)^{0.33}\left(\frac{\mu_B}{\mu_w}\right)^{0.14}$$

Viscous liquids:

$$Nu = 0.027(Re)^{0.8}(Pr)^{0.33}\left(\frac{\mu_B}{\mu_w}\right)^{0.14}$$

Gases

$$Nu = 0.0231(Pr)^{0.33}\left(\frac{\mu_B}{\mu_w}\right)^{0.14}$$

Where, $Re = \frac{\rho u_t d_e}{\mu} = \frac{G_t d_e}{\mu}$ $\quad Nu = \frac{h_i d_e}{k_f}$ $\quad \mathrm{Pr} = \frac{C_p \mu}{k_f}$

Cp = fluid heat capacity, J/kg·K
de = external tube diameter, m
h_i = tube side heat coefficient, W/m^2·K
K_f = fluid conductivity, W/m·K.
L = tube length, m.
Nu = Nusselt number, dimensionless.
Re = Reynolds number, dimensionless.
Pr = Prandtl number, dimensionless
μ_B = viscosity of fluid in the bulk, Pa·s
μ_w = viscosity of the fluid in the tube wall, Pa·s

```
Pr=((Heat_capacity_in(1)*4.184)*((viscosity_in(1)+viscosity_
  out(1))*0.5))/Kf;

{ Cooling = 1  for cooling,
  Cooling = 0  for heating}

IF Cooling = 1 THEN
     n = 0.3;
ELSE n = 0.4;
END
  {b = 0 Non viscous liquids
   b = 1 Viscous liquids
   b = 2 Gases}
  {Considering the omission of the viscosity corrector
    factor}

IF Re_t > 10000 AND b = 0 THEN
    Nu = 0.023*(Re_t^0.8)*(Pr^n);
ELSE
IF Re_t > 10000 AND b = 1 THEN
```

Table 10.9
Nusselt Number for the Shell Side for Different Flow Regimes

Laminar flow (100 < Re < 2000)	Turbulent flow (2000 < Re < 10^6):
$Nu = 0.664(Re)^{0.5}(Pr)^{0.33}\left(\frac{\mu_B}{\mu_W}\right)^{0.14}$	$Nu = 0.36(Re)^{0.55}(Pr)^{0.33}\left(\frac{\mu_B}{\mu_W}\right)^{0.14}$

```
    Nu = 0.027*(Re_t^0.8)*(Pr^n);
ELSE
IF Re_t > 10000 AND b = 2 THEN
    Nu = 0.0231*(Pr^n);
ELSE
    Nu = 1.86*(((Re_t)*(Pr)*(di/L))^0.33);
END
END
END
Nu = (hi*di)/Kf;
```

The complex flow pattern on the shell-side, and the great number of variables involved, make it difficult to predict the shell-side coefficient and pressure drop with complete assurance. Typical methods for that prediction are those of Kern and Donohue among others. In the present case Kern method is selected for the shell-side heat coefficient prediction [9,10]. Table 10.9 shows the Nusselt number of the shell side for different flow regimes.

Reynolds, Prandtl and Nusselt dimensionless numbers can be obtained as defined previously for the calculation of the tube-side heat coefficient, applying the corresponding shell parameters.

```
#4  Shellside coefficient method for sensible heat transfer.

Re_s=(De*((Density_in(2)+ Density_out(2))*0.5)*u_s)/
   (((viscosity_in(2)+viscosity_out(2))*0.5)/1000);

u_s=((Flow_in_aux*Mw_aux("Utilities"))/(((Density_in(2)+
  Density_out(2))*0.5)*(pi*(1/4)*(De^2))))/3600;

Pr_s=(((Heat_capacity_in(2)+ Heat_capacity_out(2))*0.5*4.184)*
   ((viscosity_in(2)+ viscosity_out(2))*0.5))/Kf_s;

IF 2000 < Re_s AND Re_s < 10^8 THEN
    Nu_s = 0.36*(Re_s^0.55)*(Pr_s^0.33);
ELSE
    Nu_s = 0.664*(Re_s^0.5)*(Pr_s^(1/3));
END

Nu_s = (ho*De)/Kf_s;
```

We can observe in the gPROMS® code that the viscosity correction term has been considered insignificant in this example.

It should be noticed that fluids will foul the heat exchangers surfaces in an exchanger to a greater or lesser extent. The deposited material will normally have a relatively low thermal conductivity and will reduce the overall coefficient. The effect of fouling is allowed for in design by including the inside and outside fouling coefficients in equation 10.41. Fouling factors are usually quoted as heat-transfer resistances, rather than coefficients, they are difficult to predict and they are normally based on past experience. Typical values for the fouling coefficients and factors for common processes and service fluids are given elsewhere [9]. Once we select the fouling factors, we could obtain the overall heat transfer coefficient by applying equation 10.41.

```
#5  Fouling coefficients
ri = 1.7*(1/10^4);
ro = 1.3*(1/10^4);
kw = 16.3; # W/m.k
#5 Overall heat transfer coefficient

(1/U) = (1/ho) + (ro) + ((De*LOG(De/di))/(2*kw)) + (De/di)*ri
  + (De/di)*(1/hi);
```

Step 4. Surface area required and available: The surface area required by the heat exchanger could be obtained by applying the equation shown below. The available area of the heat exchanger should be at least as the required in order to validate the design of the exchanger.

$$A = QU\Delta T_m \tag{10.42}$$

```
#6  Surface area
Q = U*A*LMTD;
    # Duty
    Q=((SIGMA(Flow_in(1, )*Mw()))*(Heat_capacity_
      in(1)*4.184*1000)*(T1 - T2))/3600; #J/s
    #Available area
    Ad = L*(pi*De)*Nt;
```

Step 5. Pressure drop: The pressure drop in the tube side can be calculated using the following equation [11].

$$\Delta P_t = f\frac{L}{D}\left(\frac{1}{2}\rho V^2\right)Np \tag{10.43}$$

Where L is the length of the tubes, D is the inner diameter of the tubes, ρ is the density of the tube side fluid, and V is the average flow velocity through a single tube, Np is the number of tube passes. The fanning friction factor, f, can be calculated from Darcy friction factor, f_D.

$$f_D = \left(1.82 log_{10} N_{Re} - 1.64\right)^{-2} \tag{10.44}$$

The Darcy friction factor is related to the fanning friction factor by $f_D = 4f$.

```
# Pressure drop
#1  Pressure drop in the tube side.
dPt= f_t*(L/di)*(0.5*((Density_in(1) + Density_
  out(1))*0.5)*(V^2))*tp; # Pa
V = ut/Nt;
fd = (1.82*(LOG10(Re_t) - 1.64))^(-2);
fd = 4*f_t;
```

The pressure drop in the shell side can be calculated using the following equation,

$$\Delta P_s = \frac{2fG_s^2 D_s (N_B + 1)}{\rho D_e \left(\dfrac{\mu}{\mu_s}\right)} \tag{10.45}$$

Where f is the fanning friction factor for flow on the shell side, G_s the mass velocity on the shell side, D_s the inner diameter of the shell, N_B the number of baffles, ρ the density of the shell-side fluid, and D_e an equivalent diameter. The mass velocity $G_s = m/S_m$, where m is the mass flow rate of the fluid, and S_m is the cross flow area measured close to the central symmetry plane of the shell containing its axis. The cross flow area, S_m can be obtained from equation 10.46:

$$S_m = D_s L_B \frac{clearance}{pitch} \tag{10.46}$$

Where L_B is the baffle spacing, the equivalent diameter, D_e, is defined as follows:

$$D_e = \frac{4\left(C_p S_n^2 - \dfrac{\pi D_0^2}{4}\right)}{\pi D_o} \tag{10.47}$$

Where D_o is the outside diameter of the tubes and S_n is the pitch, center to center distance of the tube assembly. The constant Cp = 1 for square pitch, and Cp = 0.86 for a triangular pitch. The fanning friction factor is calculated using Reynolds number based on equivalent diameter as

$$Re = \frac{D_e G_s}{\mu_o} \tag{10.48}$$

```
#2  Pressure drop in the shell side.

dPs=(2*f_s*(Gs^2)*Ds_i*(Nb+1))/(((Density_in(2)+
  Density_out(2))*0.5)*Deq);

    Re_s_eq = (Deq*Gs)/((viscosity_in(2) +
      viscosity_out(2))*0.5);
    fd_s = (1.82*(LOG10(Re_s_eq) - 1.64))^(-2);
```

```
    fd_s = 4*f_s;
    Deq = (4*(Cp*(Tpitch^2) - (pi*(De^2)/4)))/(pi*De);
{ tri = 1 for triangular pitch
  tri = 0 for square pitch}
    IF tri = 1 THEN
        Cp = 0.86;
    ELSE Cp = 1;
    END
    Gs = (Flow_in_aux*Mw_aux("Utilities"))/Sm;
    Sm = Ds_i*Lb*(C/Tpitch);
```

The inner shell diameter used in previous equations for the calculation of the shell side pressure drop, must be selected to give as close as fit to the tube bundle as is practical; to reduce bypassing round the outside of the bundle. The clearance required between the outermost tubes in the bundles and the shell inside diameter will depend on the type of exchanger and the manufacturing tolerances. Typical values are given in Figure 12.10 in Sinnott [9]. In this Figure the bundle diameter and shell inside diameter are related to the type of head selected for the heat exchanger. We have developed correlations from the Figure in order to include this information in gPROMS®. The code is shown below:

```
{ Head = 1 = Pull trhough floating head
  Head = 2 = Split ring floating head
  Head = 3 = Outside packed head
  Head = 4 = Fixed and U-Tube}

IF Head = 1 THEN
    Ds_i - Db = 0.0091*Db + 0.0861;
ELSE
IF Head = 2 THEN
   Ds_i - Db = 0.0275*Db + 0.0447; #correlation
ELSE
IF Head = 3 THEN
   Ds_i - Db = 0.0377;
ELSE
IF Head = 4 THEN
   Ds_i - Db = 0.0098*Db + 0.0082;
ELSE
   Ds_i = 1;
END
END
END
END

Ds_e = Ds_i + 0.0095; # minimum thickness
IF 0.152 < Ds_i  AND Ds_i < 0.635 THEN
Lb = Ds_i - 0.0016;
ELSE
Lb = Ds_i - 0.0048;
END
```

10.4.4.4 Results

In this section the main parameters obtained for the design of the heat exchanger are detailed.

- Mean temperature difference: $\Delta T_m = 194.54$ K.
- Number of tubes and tube bundle diameter: Nt = 254, Db = 0.317 m.
- Overall heat transfer coefficient: U = 1963.10 W/m^2·K.
- Required surface area: A = 47.15 m^2
- Available surface area: $A_v = 68.15$ m^2

  Pressure drop:
  - Tube side pressure drop: $\Delta P_t = 3.86$ psi.
  - Shell side pressure drop: $\Delta P_s = 50.1$ psi.

As the available surface area is higher than the required one, we can conclude that the design of the heat exchanger is correct.

10.4.5 Optimization: Parameter Estimation

In this section, experimental data is used to estimate the coefficients of a simple kinetics model for two active species. The same approach can be used to search for specific reaction mechanism or decouple the effect of mass transfer and chemical reaction.

10.4.5.1 Batch Reaction Process

A standard industrial batch process contains chemical A and chemical B. Chemical A decomposes with time and also catalyses the decomposition of chemical B.

$$[1] \quad A \rightarrow A^* \quad [A^* \text{ is not active with B}]$$

$$[2] \quad B \xrightarrow{\textit{catalised by A}} C^*$$

Initially a pseudo first order deactivation of A is assumed while a pseudo second order reaction is assumed for A + B reaction

$$\frac{dc_A}{dt} = -k_1 \cdot c_A \tag{10.49}$$

$$\frac{dc_B}{dt} = -k_2 \cdot c_A \cdot c_B \tag{10.50}$$

A specific experiment has been designed with different initial levels of chemical A added to the reactor keeping chemical B outside the batch system. Samples at time 5 min and time = 10 min are taken from the batch (with different levels of decomposed chemical A for the 5 and 10 min samples) and mixed with a known concentration of chemical B which can be measured. Data is presented in Table 10.10.

TABLE 10.10
Data for the Parameter Estimation Problem

Sampling time, s	Times for titrations, s	Overall Time, s	Reactant A, kmol/m³	Reactant B, kmol/m³
300		0	0.042	
300	0	300		0.0099
300	600	900		0.0084
300	1800	2100		0.0068
300		0	0.083	
300	0	300		0.0099
300	600	900		0.0069
300	1800	2100		0.0045
300		0	0.167	
300	0	300		0.0099
300	600	900		0.0049
300	1800	2100		0.0022
300		0	0.333	
300	0	300		0.0099
300	600	900		0.0021
300	1800	2100		0.0004
600		0	0.042	
600	0	600		0.0099
600	600	1200		0.0094
600	1800	2400		0.0085
600		0	0.083	
600	0	600		0.0099
600	600	1200		0.0084
600	1800	2400		0.0067
600		0	0.167	
600	0	600		0.0099
600	600	1200		0.0067
600	1800	2400		0.0040
600		0	0.333	
600	0	600		0.0099
600	600	1200		0.0042
600	1800	2400		0.0016

The objective is to estimate k_1 and k_2 coefficients from the set of experimental measurements of chemical B and initial values of chemical A (no tracking of it) since this also depends on the decomposition reaction of A.

1. Before sampling, the reaction rate k2 is 0 since chemical B is not in contact with chemical A although this is already decomposing.
2. Define the 3 variables types below in Figure 10.30.
3. The parameters that need to be estimated have to be defined as a variable since their values change during the optimization algorithm. Similarly, any parameter that changes across experiments (e.g. temperature, pH, etc) needs

Variable Types (Kinetics_Catalase_fitting_alldata)

Name	Lower bound	Default value	Upper bound	Units
concentration	-0.05	-1.5E-6	1.0	
no type	0.0	1.0	1.0	
reaction_rate	0.0	0.5	1E10	
<new>				

FIGURE 10.30 Variable type in parameter estimation example.

to be defined as a variable so that it can be specified for every single experiment (no need in this example) in the controls tab of the experiment.

4. Create a model called kinetics of the equation (10.49) and (10.50).

```
VARIABLE
    rate_B_decomp                AS reaction_rate
    rate_A_decomp                as reaction_rate
    c_B                          as concentration
    c_A                          as concentration
    B_decomp_filter              as notype

EQUATION
    $c_B=-rate_B_decomp*c_B*c_A*B_decomp_filter;
    $c_A = -rate_A_decomp*c_A;
```

5. The process kinetics code is below. The concentrations of chemicals A and B are differential variables that require initial values. The initial concentrations can be different for each experiment but it is still necessary to have a generic value here (e.g. exp A_1 contains 0.083 kmol/m^3 of chemical A while experiment A_05 contains 0.0417 kmol/m^3).

   There are 5 variables (2 concentrations, 2 reaction rates and a filter that makes the rate of decomposition of chemical B equal to zero at times < sample time. This is needed because in a parameter estimation it is not possible to use the process schedule to perform reinitialisation. Since we only have 2 equations, the other 3 variables need to be assigned even if their values will be estimated by gPROMS® and these values will be used as initial guesses. In order to increase the likelihood of having a successful parameter estimation, the model can be run first mapping different potential values of both parameters to be estimated until a relatively close solution is found (e.g. rate_B_decomp : = 0.01 and rate_A_decomp : = 0.002)

```
UNIT
    kinetics AS kinetics

Assign
    Within kinetics DO
        rate_B_decomp    :=0.01;
        rate_A_decomp     :=0.002;
        B_decomp_filter := 1;
    End
```

```
INITIAL
    kinetics.c_A = 0.04167;
    kinetics.c_B =  0.009880698;

SOLUTIONPARAMETERS
    PESolver := "MAXLKHD" [
        "MINLPSolver" := "SRQPD" [
            "Scaling" := 1
        ]
    ]

SCHEDULE

  SEQUENCE
    CONTINUE FOR 2400
  END  # SEQUENCE
```

6. *Input experiments:* Performed experiments folder right click → New entity → Name: A_0.
7. *Fill general, control, and measured data tabs*

   Each experiment will contain the initial concentration of chemical A. No need to do this for chemical B since chemical B has always the same initial concentration already defined in the process kinetics and its time change will be included in the measured data tab, see Figure 10.31.

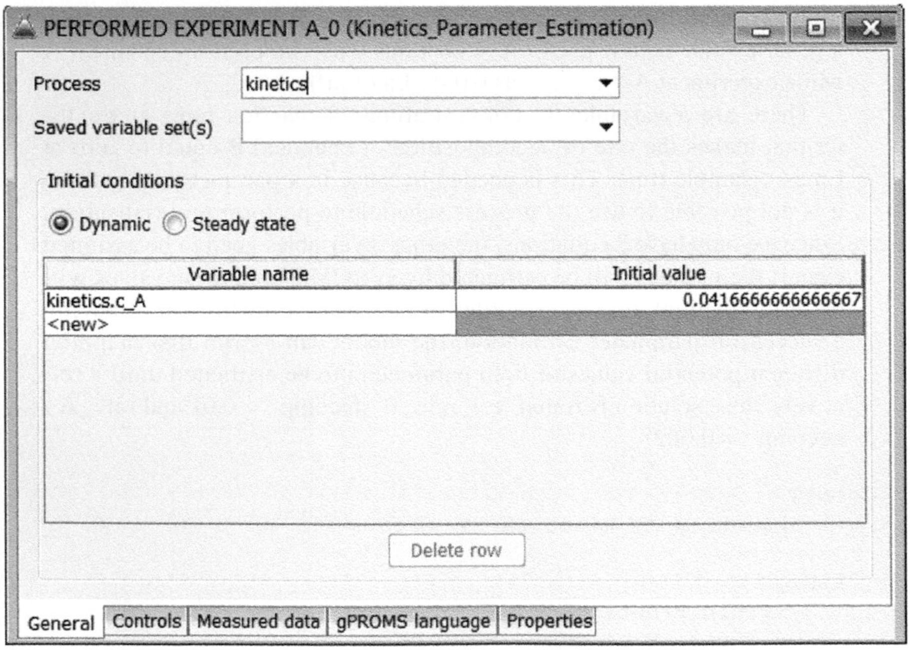

FIGURE 10.31 Input of initial conditions for experiments.

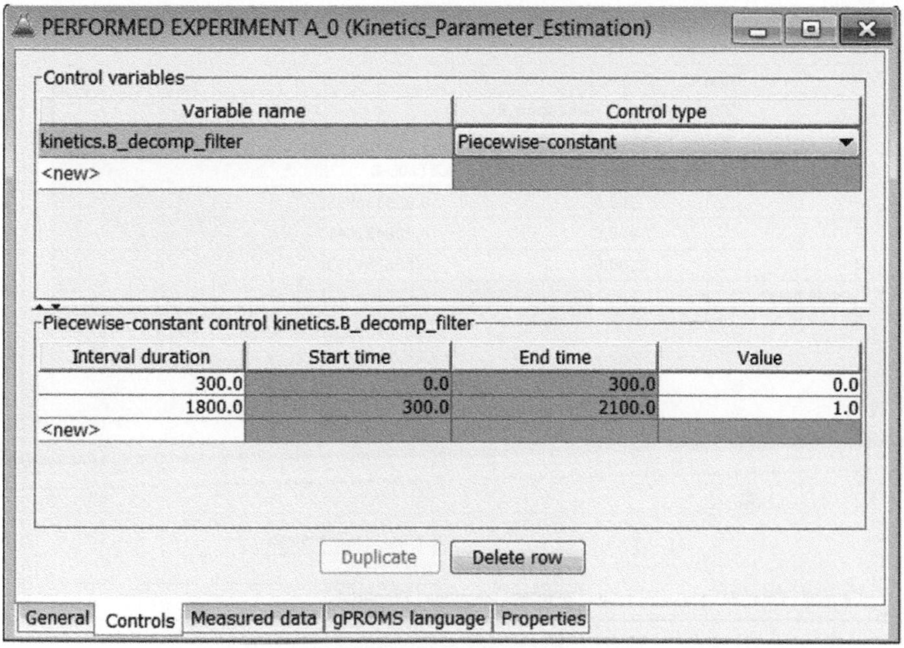

FIGURE 10.32 Input of control variable kinetics.B_decomp_filter.

In the control tab, the variable B_decomp_filter is defined as a control variable of piecewise-constant type which is 0 at time < sampling time (300 seconds for experiment A_0) and 1 at time > sampling time, see Figure 10.32:

Now, the concentrations of chemical B in time are added. Note that the first point is for time = 300 s which is the time at which a sample is taken and mixed with chemical B, see Figure 10.33:

The rest of experiments are added in a similar way.

8. *Check experiment specific variables have been input correctly:* Right click on one of the experiments (e.g. A_0) → simulate → this will create a new process kinetics_A_0 (see below) in which the values of the experiment specific variables are assigned from the experiment entity. Note that if some of the variables defined in each experiment are assigned in the model, their assignment will be repeated and gPROMS® will give an error (e.g. B_decomp_filter). In the example, B_decomp_filter was defined within the process to avoid this error.

```
UNIT
    kinetics AS kinetics

Assign
    Within kinetics DO
        rate_B_decomp   :=0.01;
        rate_A_decomp    :=0.002;
        B_decomp_filter := 0.0; { Original:  1; }
    End
```

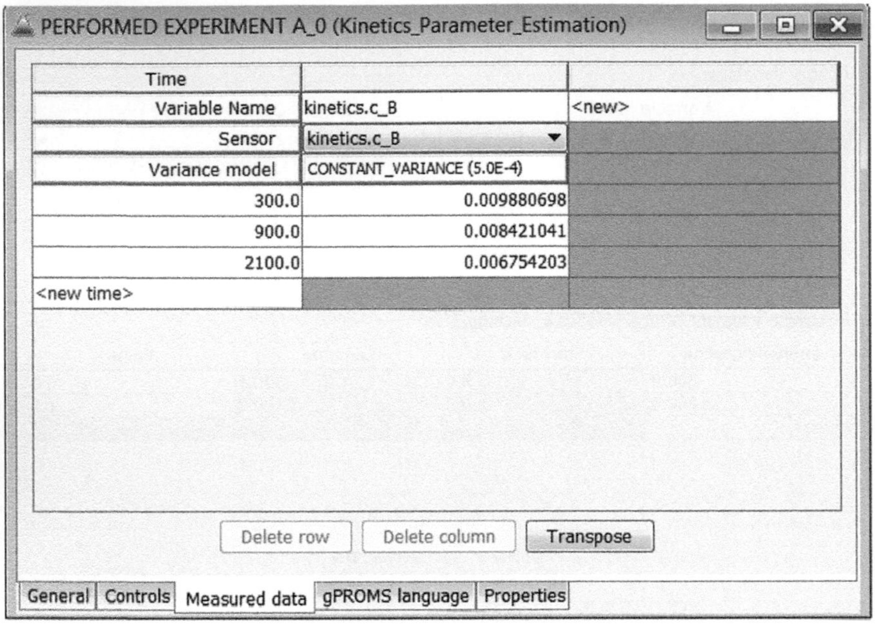

FIGURE 10.33 Input of measured data for concentration of chemical B.

```
INITIAL
    kinetics.c_A = 0.0416666666666667; { Original:
      0.04167; }
    kinetics.c_B =  0.009880698;

SOLUTIONPARAMETERS
    PESolver := "MAXLKHD" [
        "MINLPSolver" := "SRQPD" [
            "Scaling" := 1
        ]
    ]
  IntrinsicTasks := OFF

SCHEDULE
  SEQUENCE
    REASSIGN
      kinetics.B_decomp_filter := 0.0;
    END
    CONTINUE FOR 300.0
    REASSIGN
      kinetics.B_decomp_filter := 1.0;
    END
    CONTINUE FOR 1800.0
  END  # SEQUENCE
```

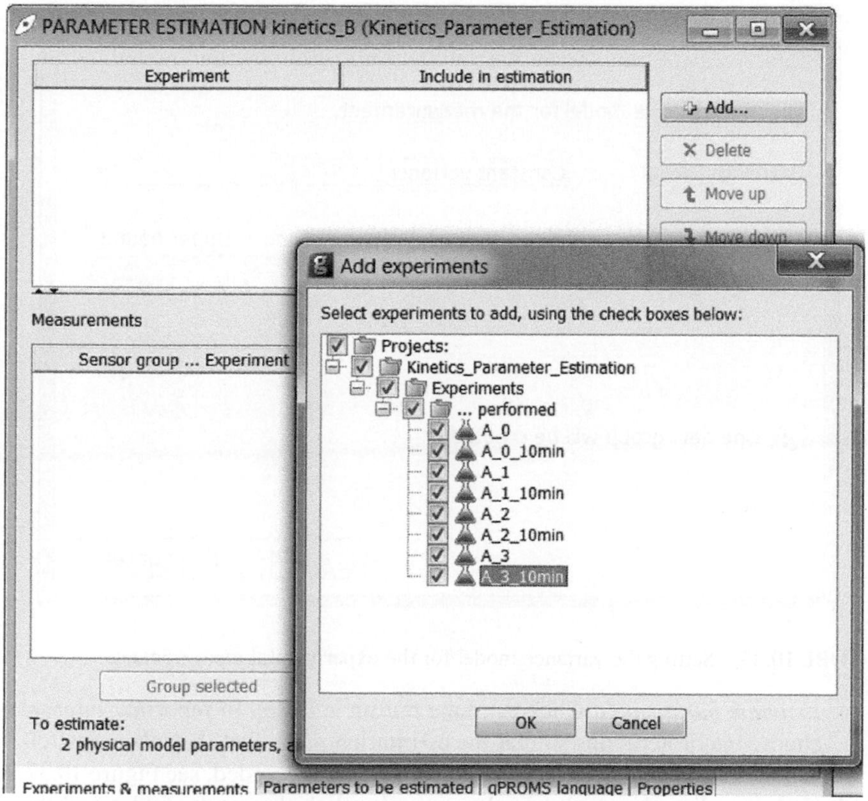

FIGURE 10.34 Select all the experiments to be included.

9. *Scale estimation parameters:* Important when the estimating parameters are of different order of magnitude. Go to the original process kinetics → Solution parameters → numerical solver → pESOlver → MINLPSolver → Scaling (change from 0 to 1)
10. Parameter estimations folder right click → New Entity → kinetics_B.
11. Click Add and select all the experiments to be included in the parameter estimation, see Figure 10.34.
12. Set the variance model to a constant value of 0.0005 kmol/m^3. This value is obtained by running multiple repeats of one condition and calculating what the variance of the data is. The smaller the variance, more of the total fitting error is associated to the model lack of fit. See Figure 10.35.
13. Input the parameters to be estimated, their upper and lower bounds, see Figure 10.36. When many parameters need to be estimated, it is good practice to fix some of them first to get good initial guesses for some of the parameters. Note that if the upper bounds are not large enough the coefficient may sometimes hit the bounds and stuck to that high value (check that the estimated parameters have not hit the bounds in the model parameters results table in Figure 10.36).

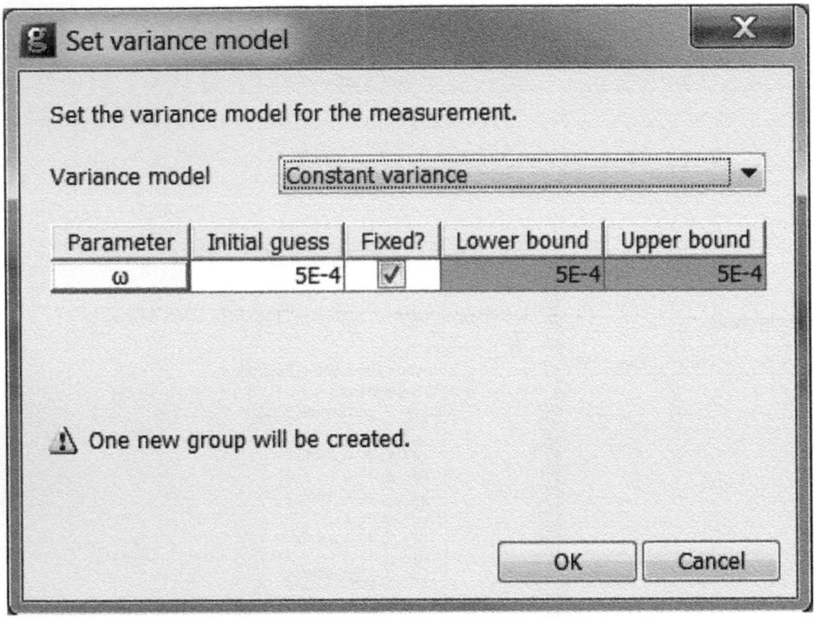

FIGURE 10.35 Setting the variance model for the experimental measurements.

14. *Estimate parameters:* Click estimate button (ellipse). In reporting interval choose major iterations so that the estimation paths that the solver has followed from the initial iteration to the final one is recorded, see Figure 10.37.
15. Understanding the results. In the case file, click the results folder and the file kinetics_B_params constraints the estimated parameters.

```
ESTIMATE
  KINETICS.RATE_A_DECOMP
  0.00131367 : 0 : 1

ESTIMATE
  KINETICS.RATE_B_DECOMP
  0.0160582 : 0 : 1
```

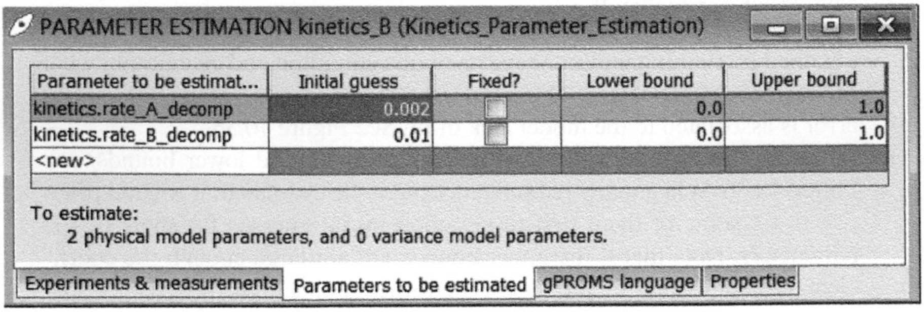

FIGURE 10.36 Input of parameters to be estimated.

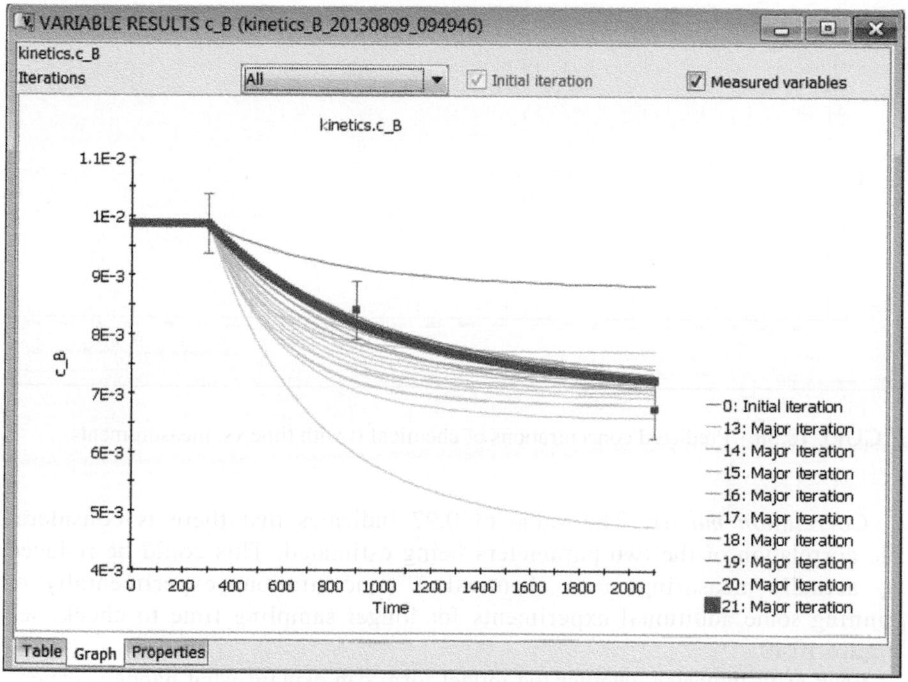

FIGURE 10.37 Iteration path for concentration of chemical B in experiment A_0.

16. Open the file kinetics_B (the file kinetics_B.stat-mr contains a more detailed description of the statistics for the fitting in excel format).

In the tab measurements, all the predictions for the different experiments are plotted vs. experimental data. Figure 10.37 shows the comparison for experiment A_0.

In the tab report, there is a summary of the statistics of fitting. Refer to gPROMS® manual for a more detailed description.

Objective function: Function of residuals between predicted and measured values of concentration of chemical B. Generally, the lower it is the better, but its absolute value is not a good indicator of how good the estimation is, see Figure 10.38.

Objective function contribution table: The residual terms give an idea of what experiments contribute most to the overall variance and can be used to quickly check irregularities or large lack of fit of the model for specific experimental conditions.

Model parameters: The table below is the most important contains a summary of the estimated parameters and their confidence intervals as well as their accuracy. Since the t-value (95%) for both parameters is larger than the reference t-value (95%), both parameters have been accurately estimated and thus the standard deviation is relatively small when compared to the estimated parameter. The t-value is highly dependent on the variance model used for the experimental measurements. The smaller the experimental variance, the larger the t-value but the larger the lack of fit of the model as well if the model does not capture all the physics. See Figure 10.39.

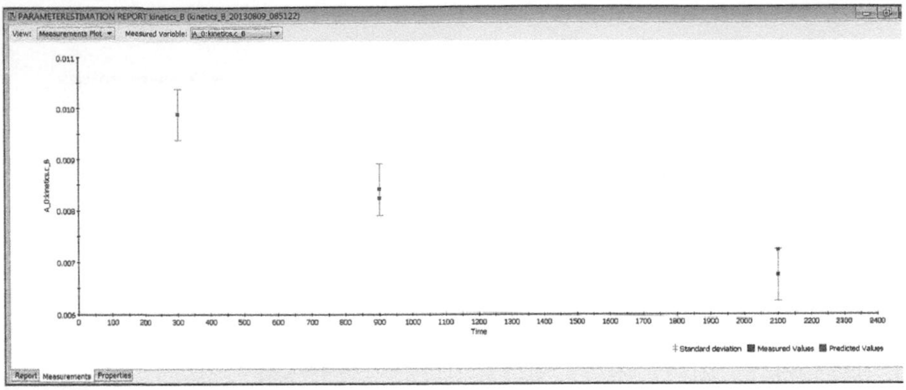

FIGURE 10.38 Predicted concentrations of chemical B with time vs. measurements.

Correlation matrix: The value of 0.97 indicates that there is considerable correlation of the two parameters being estimated. This could be reduced by actually measuring some chemical A concentrations experimentally or running some additional experiments for longer sampling time to check, see Figure 10.40.

Lack of fit indicates whether the model form can explain what happens experimentally. If the difference between predicted and experimental values is large but the experimental variance (error associated to the method/operator) is small, the lack of fit will be high because the model assumptions are not adequate (e.g. reaction order or mechanism chosen). In this case, the residual due to the model fitting are acceptable, see Figure 10.41.

10.5 BASICS ON SOLVING PROCESS AND PREVENTING ERROR MESSAGES

This is an extensive topic and the following are a few notes on minimising the amount of errors and/or how to solve them when they occur (refer to gPROMS Model Developer Guide for greater detail).

Model Parameters

- Probability of parameter lying between (Final Value -α% Confidence Interval) and (Final Value +α% Confidence Interval) = α%
- The t-value shows the percentage accuracy of the estimated parameters, with respect to the 95% confidence intervals.

Model Parameter	Final Value	Initial Guess	Lower Bound	Upper Bound	Confidence Interval 90%	95%	99%	95% t-value	Standard Deviation
kinetics. rate_A_decomp	0.0013137	0.002	0	1	0.00035	0.0004226	0.0005746	3.108	0.0002038
kinetics. rate_B_decomp	0.0160582	0.01	0	1	0.005049	0.006096	0.008288	2.634	0.00294
							Reference t-value (95%):	1.71729	

FIGURE 10.39 Model parameters results table.

Correlation Matrix			
Parameter	**No.**	**1**	**2**
kinetics.rate_A_decomp	**1**	1	
kinetics.rate_B_decomp	**2**	0.97*	1

FIGURE 10.40 Correlation matrix for estimated parameters.

Before the solver attempts to solve the system of equations, gPROMS® does the following checks:

- *System is square (same number of variables as equations):* You can use the Problem Report as presented in section 10.3.1 to quickly check if all degrees of freedom (variables-equations) have been assigned and system is closed.
- *System of differential and algebraic equations (DAE) is of index 1:* DAE index is defined as the number of time differentiations needed on the equations to obtain the time derivatives of all variables so that a system of Ordinary Differential Equations (ODEs) is obtained. Systems of index 1 are very similar to ODEs where the number of differential variables defines the number of initial conditions that can be set arbitrarily (in higher index systems this number is smaller) and are easily solved with standard methods. gPROMS can automatically reduce the index of a DAE system if the solution parameter IndexReduction is set to ON in the process section (solution parameters – Mathematical formulation & validation). However, it is preferable to define the model equations so that DAE index is 1 since the index reduction techniques are computationally intensive and bring more challenges on controlling error during integration. Often, small changes such as changing what variables are assigned versus calculated can yield index-1 systems.

```
SOLUTION PARAMETERS
IndexReduction := ON;
```

- *Initial conditions are consistent:* It is important to note that in DAE systems, initial conditions can be specified on the algebraic variables although they must be consistent (e.g. defining initial conditions for two variables that are related algebraically will give rise to inconsistent conditions). When possible, define initial conditions on the differential variables.

TOP

Lack of Fit Test

Weighted Residual	χ^2-Value (95%)	Comment
74.682	106.39	Good fit: weighted residual less than χ^2-Value

FIGURE 10.41 Lack of fit table for model with estimated parameters.

Every time the solver initializes (solution at time 0 or steady state problem) or integrates (solutions at different values of time), it solves a set of equations $f_j(x) = 0$ around an initial guess via an iterative process to a ***pre-specified accuracy*** (maximum error) for each variable x_i given by

$$\varepsilon_{tol,i} < \varepsilon_{abs} + \varepsilon_{rel} \left| x_i \right| \tag{10.51}$$

where x is the vector of solution variables. Default values of the ***relative*** (ε_{rel}) ***and absolute*** (ε_{abs}) ***tolerances*** are 1e-5.

For a given system of *m* non-linear equations $g_j(x) = 0$ with *n* variables x_i, the linearized equations $f_j(x)$ around the vector of approximate solutions x in each iteration *k* are obtained by

$$f_j\left(x^{(k+1)}\right) = f_j\left(x^k\right) + \sum_{i=1}^{n} \left(\frac{\partial f_j}{x_i}\right)_{xik} \left(x_i^{(k+1)} - x_i^{(k)}\right) = 0 \quad j = 1 - m \quad i = 1 - n \tag{10.52}$$

Therefore, the ***vector of change*** for each iteration in the direction of each variable $\left(x^{(k+1)} - x^{(k)}\right)$ can be calculated by solving the system of equations below,

$$\left(x^{(k+1)} - x^{(k)}\right) = J^{-1} \cdot \left(-f(x^{(k+1)}\right) \tag{10.53}$$

where J is the ***Jacobian matrix***, which has the derivatives of each function (equation) with respect to all variables being solved,

$$J = \begin{pmatrix} \frac{\partial f_1}{\partial x_1} & \frac{\partial f_1}{\partial x_2} & \cdots & \frac{\partial f_1}{\partial x_n} \\ \frac{\partial f_2}{\partial x_1} & \frac{\partial f_2}{\partial x_2} & \cdots & \frac{\partial f_2}{\partial x_n} \\ \cdots & & & \\ \frac{\partial f_m}{\partial x_1} & \frac{\partial f_m}{\partial x_2} & \cdots & \frac{\partial f_m}{\partial x_n} \end{pmatrix}^{(k)} \tag{10.54}$$

The ***residual of each equation*** *j* at iteration k is given by

$$\text{Residual}_j^{(k)} = g_j(x) - g_j\left(x^{(k)}\right) \tag{10.55}$$

The iteration process of the non-linear solver which uses the Jacobian is stopped when the largest equation residual (norm) is less than the ***convergence tolerance*** ε, that is $\max|\text{Residual}_j^{(k)}| < \varepsilon$.

Based on the basics aforementioned, the following considerations are important to reduce the number of error messages:

- *Select values of AbsoluteTolerance and RelativeTolerance:* In the Differential-Algebraic Solver DASOlver (either DASolv or SRADAU) you can change the default values of ε_{abs} and ε_{rel} from 10^{-5} to a smaller value if smaller variable maximum errors are required. This is needed when the value of the variables is very small. As the value of a variable gets closer to the default value of the absolute tolerance, its maximum error becomes of the same order as the variable which is unacceptable; e.g. for variable $x = 10^{-5}$ $\varepsilon_{tol,i} = 10^{-5} + 10^{-5} \cdot 10^{-5} \sim 10^{-5}$. In this case, absolute tolerance could be changed to 10^{-10}. The (recommended!) alternative is to change the units of the variables.
- *Select values of ConvergenceTolerance:* Similarly, you may need to decrease the ConvergenceTolerance below the default value of 10^{-5} or select a different convergence tolerance for each of the DASolver sub-solvers, smaller for the initialization sub-solver vs. The re-initialization sub-solver.
- *Avoid initial values of 0 for differential variables to avoid initialization errors:* Even if there is no term in any equation with a division by that variable, that variable may appear in the denominator of one of the Jacobian terms leading to same issue. You can use a small initial value above the absolute tolerance. For example if $\varepsilon_{abs} = 10^{-5}$ and the variable values in the relevant period are of the order of 1 you could use an initial value of 10^{-4}.
- *Control the vector of change for next iteration:* Instead of applying the full vector of change obtained from Equation (10.50) to calculate the new variable guess in next iteration, convergence can be improved by making a smaller change multiplying the vector of change by a number between 0 and 1 called SLRfactor within gPROMS®.

Below is an example with the implementation of some of these changes within two out of the three sub-solvers that are accessible from DASolv, the non-linear Initialisation and re-initialization solvers.

```
SOLUTIONPARAMETERS
DASolver := "DASOLV" [
        "InitialisationNLSolver" := "BDNLSOL" [
             "BlockSolver" := "SPARSE" [
                  "BoundsTightening" := 0.5,
                  "ConvergenceTolerance" := 1E-012,
                  "MaxFuncs" := 1000,
                  "MaxIterations" := 8000,
                  "MaxIterNoImprove" := 1000,
                  "NStepReductions" := 1000,
                  "OutputLevel" := 3,
                  "SLRFactor" := 0.8
             ]
        ],
```

```
        "ReinitialisationNLSolver" := "BDNLSOL" [
            "BlockSolver" := "SPARSE" [
                "BoundsTightening" := 0.5,
                "ConvergenceTolerance" := 1E-010,
                "MaxFuncs" := 1000,
                "MaxIterations" := 8000,
                "MaxIterNoImprove" := 1000,
                "NStepReductions" := 1000,
                "OutputLevel" := 3,
                "SLRFactor" := 0.8
            ]
        ],
        "AbsoluteTolerance" := 1E-010,
        "OutputLevel" := 2,
        "RelativeTolerance" := 1E-010
    ]
IndexReduction := ON
```

- *Variable bounds:* The definition of upper and lower bounds in variables types (see example in Figure 10.4 needs to be checked when an error "out of bounds" occurs. Often, same kind of variables can have very different values so that it is better to actually define two variable types of the same type (e.g. concentration_high for high concentrations and concentration_low for low concentrations) with different default values (used as initial guesses in iterative calculations) and bounds. Another consideration is that some algebraic variables are defined by the ratio of two variables with the one in the denominator having very low values at times close to 0 resulting in initial values of the variable out of bounds. For example, the mass concentration (mass of substance divided by volume of water) of an added substance to a tank can be very high when the tank starts filling if detail dissolution kinetics are not considered. In this case, we can increase the bounds or increase the initial value of the variable water volume so that mass concentrations will be inside the bounds while initial volumes are still negligible in relation to operation values.
- *Customize initial guesses of individual variables using PRESET:* In highly non-linear problems, using the default values found in the variable types may not be sufficient for convergence during initialization or when solving steady state problems. In these cases, we can use PRESET as previously shown on page XX, either restoring a set of saved variables or defining initial guesses for certain variables.

REFERENCES

1. PSE (2012) *gPROMS Introductory User Guide Release* 3.6.
2. Bird, R.B.; Stewart, W.E.; Lightfoot, E.N., (2002) *Transport Phenomena.* 2nd edn. John Wiley and Sons, New York.
3. Vian Ortuño, A. (1999) *Introducción a la Química Industrial.* Ed. Reverté, Barcelona.

4. Graaf, G.H.; Sijtsema, P.J.J.M.; Stamhuis, E.J.; Joosten, G.E.H. (1986) Chemical equilibria in methanol synthesis, *Chem. Eng. Sci.*, 41, 2883–2890.
5. Krishnan, M.S.; Ho, N.W.Y.; Tsa, G.T. (1999) Fermentation kinetics of ethanol production from glucose and xylose by recombinant Saccharomyces 1400 (pLNH33) *Appl. Biochem. Biotech.* 78 (1-3), 373–388.
6. Bertilsson, M.; Andersson, J.; Lidesonn, G. (2008). Modeling simultaneous glucose and xylose uptake in Saccharomyces cerevisiae from kinetics and gene expression of sugar transporters. *Bioprocess Biosyst. Eng.*, 31, 369–377.
7. Jou, F.Y.; Mather, A.E.; Otto. F.D. (1995) The solubility of CO_2 in a 30 mass percent monoethanolamine solution. *Can. J. Chem. Eng.*, 73, 140–147.
8. McCabe, W.L.; Smith, J.C.; Harriot, P. (1999) *Unit Operations of Chemical Engineering.* 7th edn. McGraw-Hill, New York.
9. Sinnott, R.K. (1999) Coulson and Richardson, *Chemical Engineering* 3rd edn. Butterworth Heinemann, Singapore.
10. Edwards, J. *Design and Rating Shell and Tube Heat Exchangers.* P & I Design Ltd, Teesside, UK. http://www.chemstations.com/content/documents/Technical_Articles/shell.pdf
11. Ghasem, N. (2011) *Computer Methods in Chemical Engineering.* CRC Press. Taylor & Francis.

11 EES® for Chemical Engineering

Ángel L. Villanueva Perales

CONTENTS

11.1 INTRODUCTION

EES ('Engineering Equation Solver') is a software tool for numerically solving algebraic and differential equations that are typically found in engineering problems. EES is not a programming language, and equations are naturally entered in EES as they would be entered in a word processor. In addition to this ease of use, EES provides built-in functions for the calculation of thermodynamic and transport properties of many chemical compounds and engineering fluids. This feature makes EES a powerful software tool for engineering applications, especially in energy systems. This chapter demonstrates some of EES's capabilities applied to solve problems in chemical engineering. For more examples and advanced features of EES the reader is referred to the books by the creator of the program, Prof. S Klein [1-3] and videos by F-Chart Software, the developer of EES [4]. An overview of the EES software can be found in the following video [5].

The reader is encouraged to download a demonstration version of EES from the F-Chart software website to do the problems of this chapter [4]. In addition, EES versions from F-Chart software are available to: (i) education institutions for classroom instruction and research (academic version); and (ii) single or multiple users for professional activity [4].

11.2 EES FUNDAMENTALS

11.2.1 Entering and Solving Equations

The graphical interface of EES is shown in Figure 11.1. The main window is the Equation window, where the user introduces the equations of the problem to be solved. Equations may be entered in any order and typically one equation per line. Some simple rules must be followed when writing equations. Variables must start with a letter and upper and lower case letters are not distinguished from one another by EES, that is, the variable X and the variable x are identical as far as EES is concerned [1]. EES will expect the use of a comma (European) or decimal point (U.S.) for the decimal separator character depending on the regional setting choice made in the operating system.

As an example, the following system of equations will be solved with EES: $y = x^2$; $x + y = 4$. Graphically, the solution of this system of equations is the interception of the parabola $y = x^2$ and the straight line $x + y = 4$, as shown in Figure 11.2. We want to compute numerically those solutions with EES, that for the sake of illustration, we will say in advance that they are (x,y) = (1.56, 2.44) and (x,y) = (−2.56, 6.56). First, enter the equations in the Equation window as shown in Figure 11.3. On the right side of the Equations window an information palette will display the variables

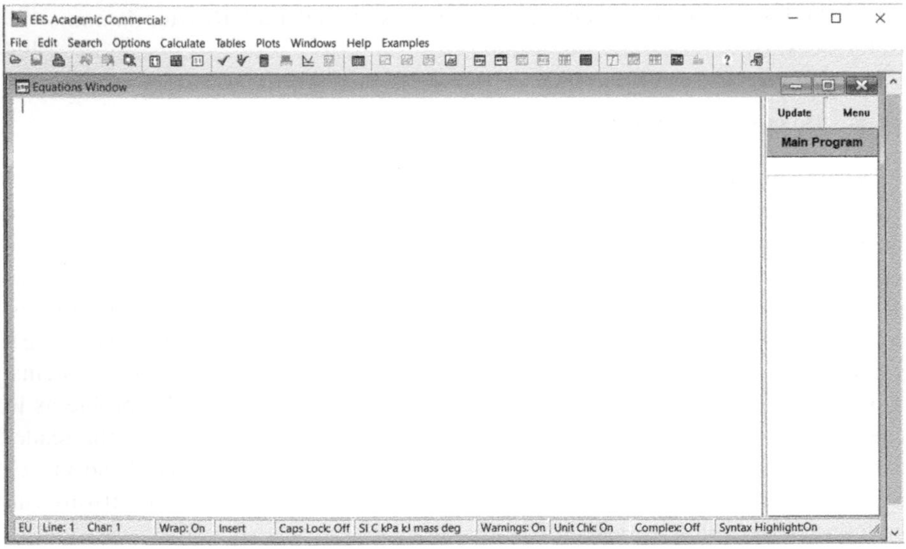

FIGURE 11.1 Graphical interface of EES.

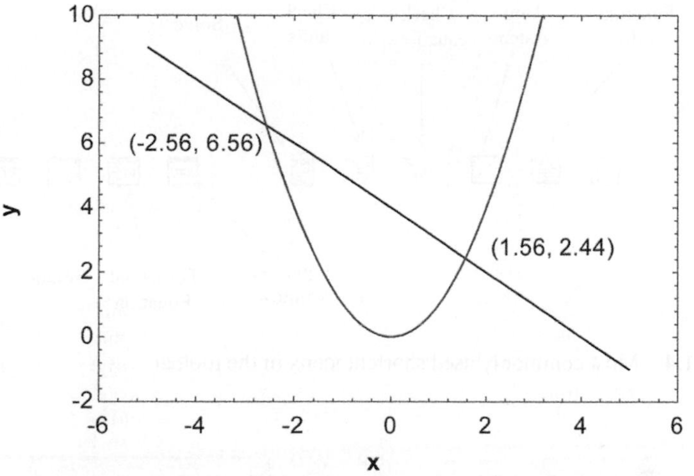

FIGURE 11.2 Interception of $y = x^2$ and $x + y = 4$.

of the equations in alphabetical order. Once equations have been entered and before solving them it is a good practice to check if the system is fully specified, i.e., the number of variable equals the number of equations. This is especially important when dealing with large system of equations. This can be done in EES by clicking on the check-mark icon (✓) of the toolbar (Figure 11.4) or alternatively selecting Check/Format from the Calculate menu. For our example EES will inform us that our system is fully specified so we can proceed to solve the system by clicking on the solve icon (Figure 11.4) or selecting Solve in the Calculate menu. EES will re-order and block the equations to initiate an iterative search for an actual solution to the equations, starting from a guessed solution. When the solution is found EES shows a dialog indicating that the calculations are completed and a Solution window will appear (Figure 11.5). In this case EES found the solution (x,y) = (1.56, 2.44).

The solution to which the search algorithm converges is the closest to the guessed solution from where the iterative search starts. By default, the guessed solution for each variable is 1—that means that in our case the iterative search started at (x,y) = (1,1). This explains that EES found the solution (x,y) = (1.56, 2.44), which is

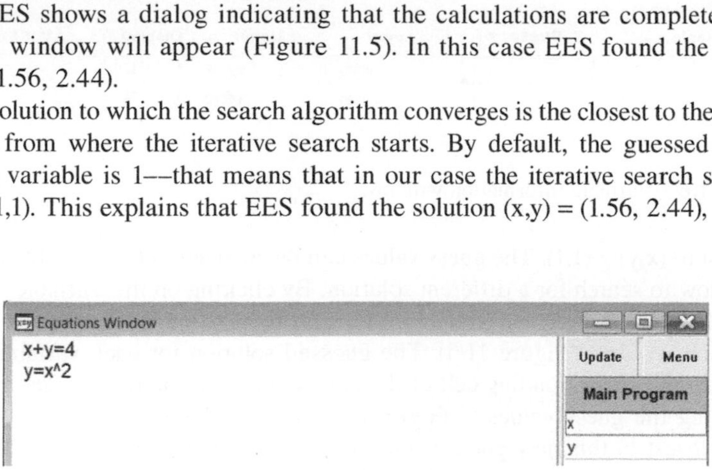

FIGURE 11.3 Entering $y = x^2$ and $x + y = 4$ in EES.

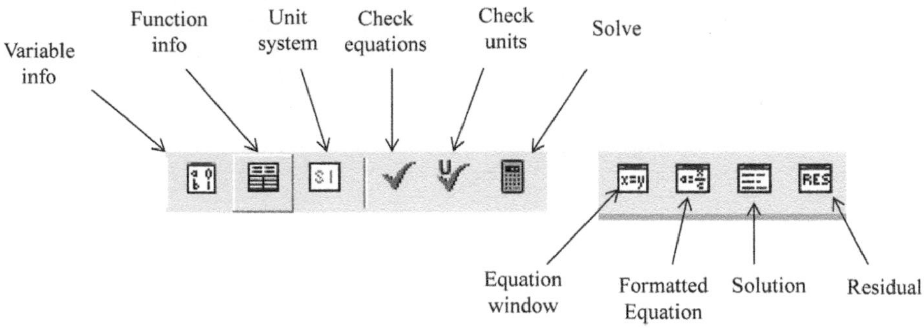

FIGURE 11.4 Most commonly used shortcut icons in the toolbar.

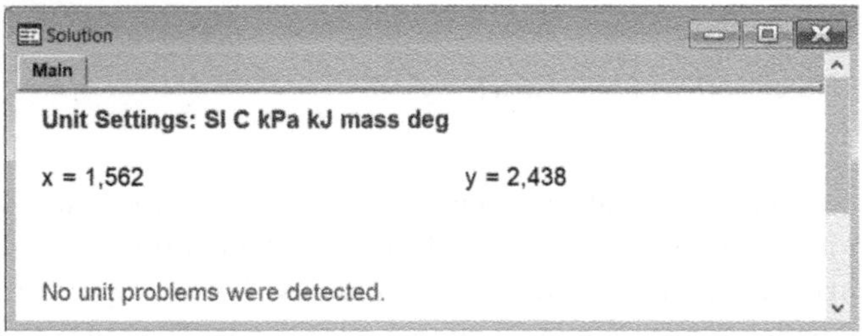

FIGURE 11.5 Solution window.

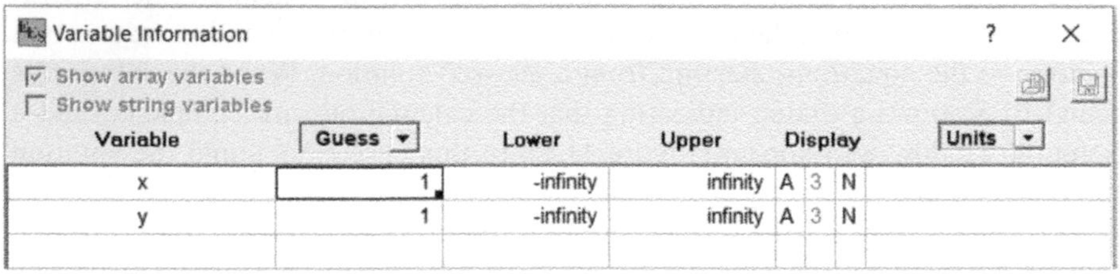

FIGURE 11.6 Variable Information window.

the closest to (x,y) = (1,1). The guess values can be changed in the Variable information window to search for a different solution. By clicking on the Variable info icon (Figure 11.4) or selecting Variable info in the Options menu we access the Variable Information window (Figure 11.6). The guessed solution for each variable can be modified in the corresponding cell of the Guess column. The reader can check that by changing the guess values to (x,y) = (–2, 1) and solving again the system, the solution closest to this new guess values, (x,y) = (–2.56, 6.56), is found. The main conclusion is that the solution found by EES largely depends on the guess values provided, which is also true for many search algorithms. In order to find a solution of interest, appropriate guess values should be always provided and the search space

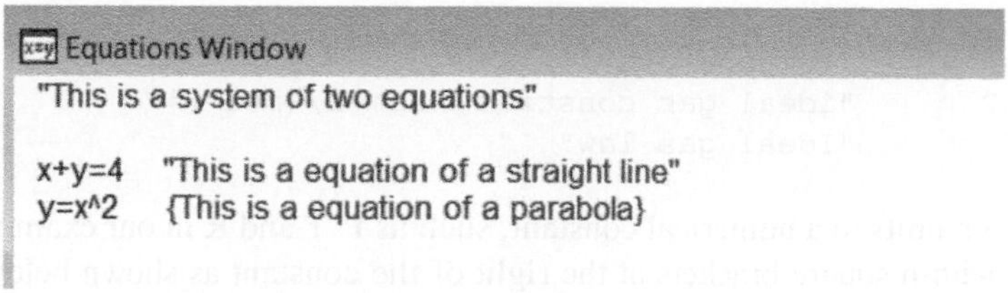

FIGURE 11.7 Annotating the code.

should also be bounded. The latter can be done by providing upper and lower limits for each variable in the Variable Information window (Lower and Upper columns).

Finally, a good practice is to include comments in the code that explain the meaning of the equations and how the problem is formulated. This will help both the user and others to understand the code and facilitate further modifications to it if necessary. Comments must be included in the equation window, at the right-hand of an equation, enclosed either in curly braces or quotes (Figure 11.7). Comments are ignored by EES and are only visible to the user.

A video explaining how to enter and solve equations with EES can be found here [6]. In addition, tips on how to solve equations effectively in EES are available here [7].

11.2.2 Unit Selection and Checking

Units can be assigned to constants and variables of equations. This is highly recommended because EES can check equations for unit consistency, preventing unit errors that can lead to wrong solutions. In addition, EES has built-in conversion factors that facilitate conversion between units.

As an example, let's calculate the mole volume of a gas at 2 atm and 323 K assuming it obeys the ideal gas law $P \cdot V = R \cdot T$. As we know the units of the ideal gas constant must be consistent with the units of pressure, temperature and volume. The following code correctly calculates the volume without assigning units to the variables and the ideal gas constant. In this case, the user has mentally checked the unit consistency of equations but with this practice we are prone to make mistakes, especially when dealing with many equations. Furthermore, without assigning units it is difficult for the user to interpret the solution in terms of physical variables (Figure 11.8-left).

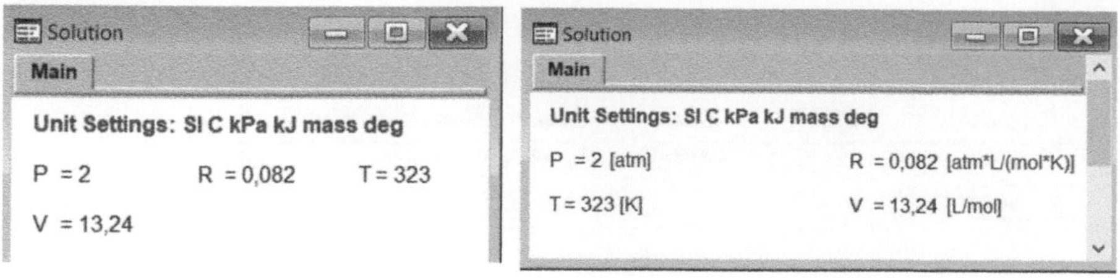

FIGURE 11.8 Solution window without (left) and with unit assignment (right).

```
P=2             "Pressure, in atm"
T=323           "Temperature, in K"
R= 0,082        "ideal gas constant, atm*L/(mol*K)"
P*V=R*T         "Ideal gas law"
```

To assign units to a numerical constant, such as P, T and R in our example, write the units within square brackets at the right of the constant as shown below. In the case of dimensionless constants write a hyphen within brackets, i.e., [-].

```
P=2          [atm]                "Pressure"
T=323       [K]                   "Temperature"
R= 0,082  [atm*L/(mol*K)]         "ideal gas constant"
P*V=R*T                           "Ideal gas law"
```

To assign units to a variable, such as V in our example, access the Variable Information window and enter the units in the Unit column. A shortcut to access the information of a variable is by clicking on that variable in the information palette at the right of the Equation windows, as shown in Figure 11.9. If we save the file after these unit assignments we will see the units of each constant and variable in the information palette. Before solving the equations we should check the unit consistency of equations by clicking on the Check unit icon (Figure 11.4). In our example EES informs that unit inconsistency was not detected so we can proceed to solve the equations. Now, the Solution window shows the units of the variables (Figure 11.8-right).

A question remains on what units are recognized by EES. This can be examined by selecting Unit Conversion Info in the Options menu. In the Unit Conversion Information dialog that shows up (Figure 11.10), the left window lists each dimension (e.g., pressure) while the right window lists the units that are recognized for the selected dimension. In addition, if you select two units in the right window (e.g., bar and kPa), then EES will display the unit conversion in red at the bottom of the dialog [1].

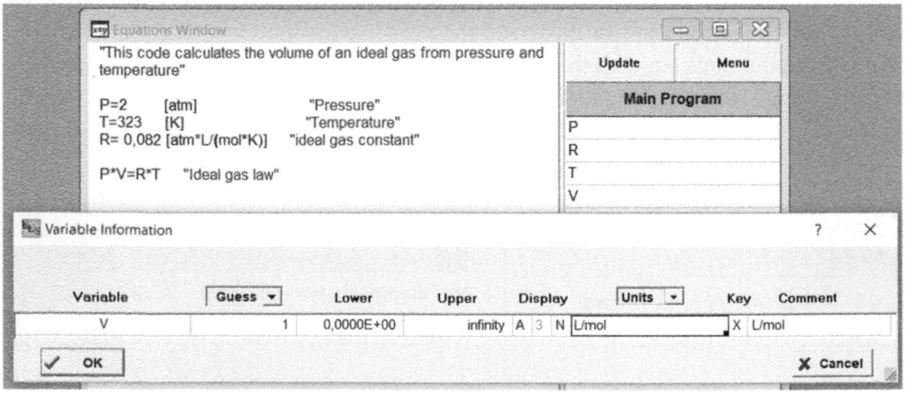

FIGURE 11.9 Assigning units to numerical constants in the code and to variables in the Variable Information window.

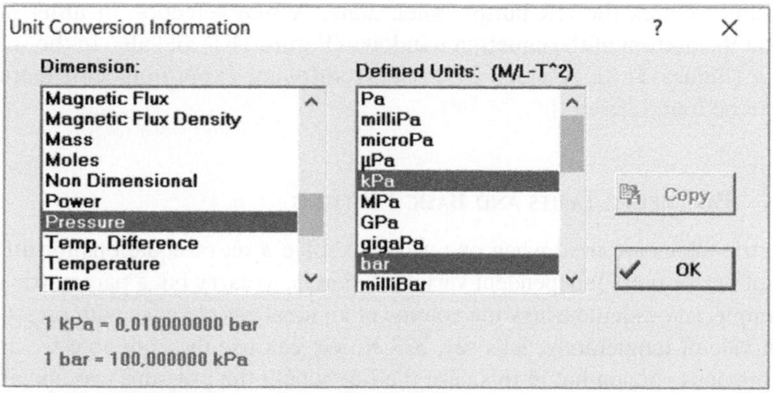

FIGURE 11.10 Unit Conversion Information window.

Finally, selection of a unit system, SI or English, is possible in EES but only important when calling trigonometric functions or functions for calculating property data (see Section 11.2.4) since EES will interpret that variables passed to those functions will be in the units selected in Unit System. Analogously, the results returned by those functions will always be in the units selected in Unit System. To select a unit system, click on the "Unit system" icon (Figure 11.4) or select Unit system from the Options menu and a window like Figure 11.11 will show up. For the unit system selected (SI or English) different unit choices are available for temperature, pressure, energy, angle and specific properties.

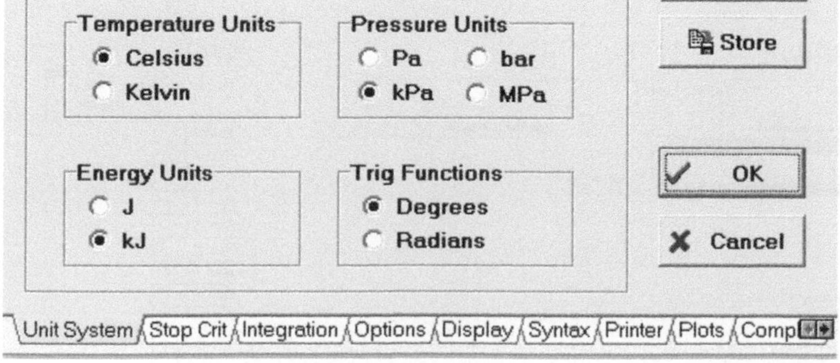

FIGURE 11.11 Selection of unit system.

Remember to click the OK button when done. Actual selection of units will be shown at the bottom of the equation windows (Figure 11.1) and also in the solution window (Figure 11.8). A video by F-Chart software explaining unit feature of EES can be found here [8]

11.2.3 Parametric Tables and Basic Plotting

Parametric tables are used when one wants to solve a set of equations for different values of one or more independent variables, that is, to carry out a parametric study. As example, let's calculate how the volume of an ideal gas changes with pressure for a given vale of temperature, let's say, 323 K. We can use the code that we created in the previous section but in this case the line where the pressure was specified at 2 atm should be deleted because the problem is going to be solved for many values of pressure that will be specified in a separate parametric table. Instead of deleting that line it is more convenient to temporarily remove it by commenting it out using square brackets, as shown below:

```
{P=2          [atm]                "Pressure"}
T=323       [K]                    "Temperature"
R= 0,082  [atm*L/(mol*K)]          "ideal gas constant"
P*V=R*T                            "Ideal gas law"
```

Now we have a system of four variables and three equations, that is, there is one degree of freedom (the pressure) that we need to specify to solve the system. To create a parametric table, select New Parametric Table in the Tables menu, which will bring up a dialog like Figure 11.12-left. The window on the left side of the dialog lists all the variables included in the Equations window. Highlight the independent (P) and dependent variables (V) of interest by clicking on them and then select Add to add the variables to the window on the right side, which lists variables to be included in the table. There will be a column for each of the variables included in the table. By default, the number of rows (runs) of the table is 10 but this can be specified in the field No. of Runs. Then, click OK and a new parametric table, named Table 1, will be created and shown in a new window (Figure 11.12-right).

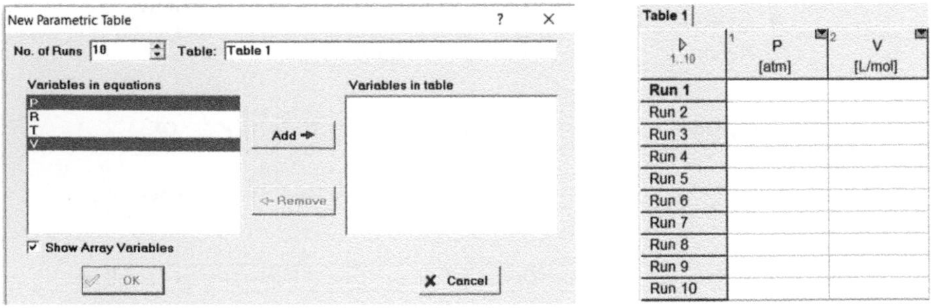

FIGURE 11.12 Left: Creation of a new parametric table. Right: Parametric table created.

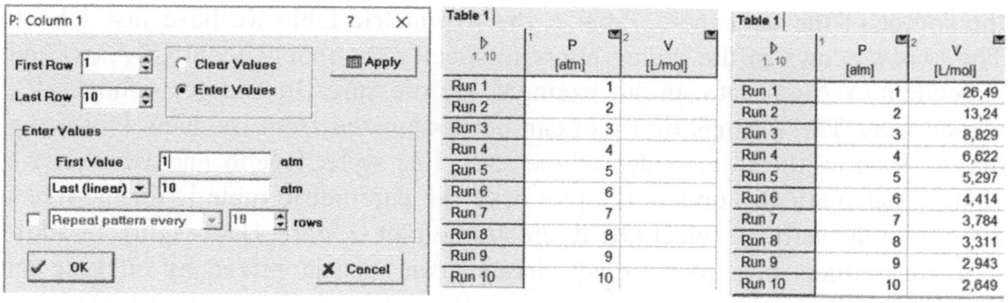

FIGURE 11.13 Specifying a parametric study.

In the pressure column we need to specify the values of pressure that will be used to solve the set of equations in sequence from Run 1 to Run 10, let's say we choose from 1 to 10 atm. We can fill in the table manually or with the help of EES. If we click on the triangular icon in the pressure column header a dialog like Figure 11.13-left will show up, where we can specify how the rows of that column are filled in. In Figure 11.13-left, the fields of the dialog are specified so that the pressure varies from 1 atm (row 1) to 10 atm (row 10) in equally spaced intervals, as shown in Figure 11.13-middle. Other options can also be selected by clicking on the drop-down menu under First Value. To quickly solve the parametric table we click on the green triangular icon in the upper-left of the table. EES will solve the set of equations from the first to the last run using the values of the variables specified in the columns of the parametric table. After each run, EES will fill in the values of the empty columns based on the solution (Figure 11.13-right). It is a good practice to inspect the results in a parametric study to see if the trends observed are in agreement with what we expected. In this case, for a given temperature, the volume of an ideal gas was expected to decrease with pressure, as observed in the solved parametric table.

The results of a parametric study are best analyzed when plotted in a graph. To create a plot in the Plot menu select New Plot window→ X-Y plot, and a dialog like Figure 11.14-left will show up. In the upper right side of the dialog we select

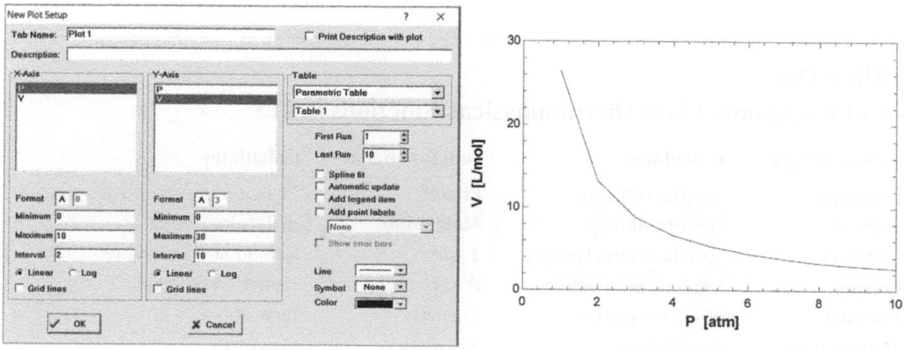

FIGURE 11.14 Left: Setup of a new plot. Right: Volume of and ideal gas as a function of pressure at a constant temperature.

the source of the data; here Table 1, the parametric table we have just solved. The two windows in the dialog allow to specify the independent (X-Axis) and dependent (Y-Axis) data, in our example, the pressure (P) and gas volume (V), respectively. The settings of a plot can be customized from the New Plot setup dialog, such as the title, scales of the axis, line style, legend and whether we want to automatically update the plot once the parametric table is rerun. When done with the settings select OK to create the plot (Figure 11.14-right). In addition, the settings of a plot already created can be customized by clicking on the plot.

In addition to X-Y plots, other types of two-dimensional plots (polar plots, bar plots) and three-dimensional plots (contour plots, gradient plots, surface plots) are available in EES [1]. You can create these plots by selecting the New Plot Setup window in the Plot menu. The data to be plotted must be available in a table previously generated by the user, typically from a parametric analysis, as we did before. The reader is encouraged to watch the videos by F-Chart software on parametric tables [9] and basic plotting [10] in EES.

11.2.4 Property Data

One of the most powerful features of EES is the library of built-in thermodynamic and transport properties functions, which can be applied to a built-in database of substances. The substances in EES are grouped in real fluids, ideal gases and brines. Real fluids and ideal gases comprise *pure substances* such as common hydrocarbons, aromatics, alcohols, inorganic compounds and refrigerants. Brines are refrigerants that consist of *mixtures* of water and another substance. In addition, the mixture of air and water vapour is available for psychrometric calculations.

In EES there are 62 thermophysical functions. Not all of them are available to the three groups of substances, e.g., some functions available for real fluids are not available for brines. The thermophysical functions most commonly used in chemical engineering calculations are shown in Table 11.1.

TABLE 11.1
Most Commonly Used Thermophysical Functions in EES

Function name	Calculates	Function name	Calculates
Enthalpy	specific enthalpy	**Phase$**	Phase (e.g., subcooled, saturated)
Entropy	specific entropy	**MolarMass**	molecular weight
IntEnerg	specific internal energy	**T_sat**	saturation temperature
Quality	vapour mass fraction	**P_sat**	saturation pressure
Pressure	absolute pressure	**Fugacity**	fugacity
Temperature	temperature	**Viscosity**	viscosity
Volume	specific volume	**CP**	constant pressure specific heat
Density	density		

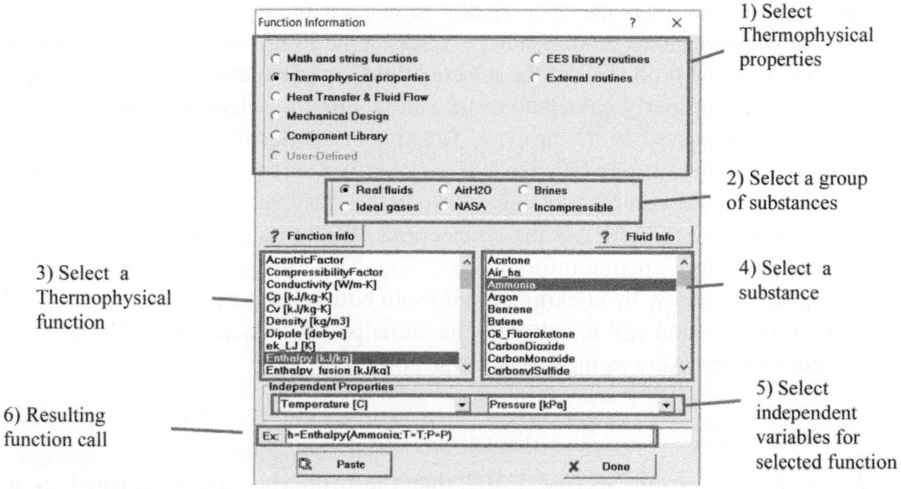

FIGURE 11.15 Function Information dialog to determine property function call.

The general form to call a property function in EES is:

```
Value = Function Name (Fluid Name; Property 1 = Value 1;
Property 2 = Value 2)
```

Input arguments are separated with a comma for the U.S. system and a semicolon for the European system. The first input argument required is the fluid name. As explained later, fluid names recognized by EES are shown in the Function information dialog (Figure 11.15). A general rule is that when a chemical formula is used to specify the substance, e.g., "N2" or "CO2", the fluid is modelled as ideal gas whereas spelled out names are used for real fluids, e.g., "Nitrogen" and "Carbon Dioxide." After the fluid name one or more arguments, which are intensive properties, are required to specify the state of the substance. Each property specification consists of a single letter (case-insensitive) that designates the property (parameters Property 1 and Property 2) followed by an equal sign and the value of that property (Value 1 and Value 2). The recognized single letter specifications are H or h for specific enthalpy, U o u for specific internal energy, S or s for specific entropy, V or v for specific volume, P or p for pressure, T or t for temperature and X or x for quality [2].

The call to a given property function, substance and property specifications can be found out through the Function information dialog (Figure 11.15), which can be accessed by selecting Function information in the Options menu or clicking on the Function information icon (Figure 11.4). First, select the Thermophysical properties radio button in the Function information dialog, as shown in Figure 11.15. Second, select the group of substances of interest: real fluids, ideal gases, brines or air-water mixture (AirH_2O). Two small windows will show up in the Function Information dialog. For the substance group selected, the left window lists the available thermophysical functions while the right window lists the substances whose properties can be calculated with the thermophysical functions listed in the left window. Click on a thermophysical function in the right window to select it and next click on a substance in the right window. (It is recommended to click on the Fluid info button to obtained information how the properties of the selected fluid are calculated and in what range of conditions the property calculations are valid). Finally, select what independent variables will be passed to the property functions in order to specify the state of the fluid. This is done in the Independent variable box of the Function information dialog. The possible combination of independent variables can be chosen from the drop-down menus. After all these selections, the function call will be shown at the bottom of the Function information dialog. This function call can be pasted in the Equation window by clicking on the Paste button. In particular, Figure 11.15 shows that the function call to calculate the enthalpy of ammonia as real fluid with temperature and pressure as independent variables is:

```
h=Enthalpy(Ammonia;T=T;P=P)
```

We must recall from Section 11.2.2 that the property values returned by a thermophysical function will be in the units defined in the Unit System dialog (Figure 11.11). Further, any values provided to the built-in property functions will be interpreted as if they have these units. It is a very common source of error to pass variables to a property function in units different to those defined in Unit System dialog. EES will identify this error only if Check Unit option has been selected in the Calculate menu.

As an example, let's calculate the enthalpy of pure ammonia in kJ/kg at 50 bar and 300°C. First, select Unit system in the Option menu and select specific properties in mass (kg), temperature in Celsius, pressure in bar and energy in kJ. Next in the Equation window write the function call to calculate the enthalpy of ammonia as real fluid as a function of temperature and pressure. In addition, to determine in what state pure ammonia is at the specified conditions we will use the Phase$ function, which returns a string ('superheated,' 'subcooled,' or 'saturated') that must be stored in a string variable (whose name must end with $).

```
T1=300 [C]                       "specified temperature assigned
                                 to auxiliary variable T1"
P1= 50 [bar]                     "specified pressure assigned to
                                 auxiliary variable P1"
h=Enthalpy(Ammonia;T=T1;P=P1)    "Enthalpy value returned in
                                 kJ/kg stored in variable h"
P$=Phase$(Ammonia;T=T1;P=P1)     "return a string informing on
                                 phase which is stored in P$"
```

Remember to assign units to the variable h, where we have stored the value of enthalpy returned by the property function, so EES can check unit consistency of the equation when clicking on Unit check. Finally, select Solve from Calculate menu. The calculated enthalpy is 2148 kJ/kg and the state is 'superheated.'

It is important to know that EES has different references for specific enthalpy and entropy. For real fluids, reference values and states for enthalpy and entropy usually differ between substances. For each substance, the references can be consulted by searching the substance name in the Help of EES (select Help index in the Help menu). For example, in the case of ammonia, the values of specific enthalpy and entropy are referred to 200 kJ/kg and 1 kJ/kg-K, respectively, for the state of saturated liquid at 273.15 K. For ideal gases, the reference for enthalpy is 0 at 298 K while reference for entropy is 0 at 0 K (Third Law of Thermodynamics).

The implication of different references is examined next for energy balances. Let's suppose an open system where there is heat and/or work exchange with the surroundings. *If there are not chemical reactions*, the mole flow of each substance (F_i) entering and leaving the system is the same and the energy balance can be rearranged as follows:

$$\sum_i F_i \cdot h_i(T_{in}, P_{in}) = \sum_i F_i \cdot h_i(T_{out}, P_{out}) + \dot{Q} + \dot{W} \rightarrow$$
$$\sum_i F_i \cdot \left(h_i(T_{in}, P_{in}) - h_i(T_{out}, P_{out})\right) = \dot{Q} + \dot{W} \tag{11.1}$$

As we can see in equation 11.1, for each substance the enthalpy terms can be rearranged as an enthalpy difference between the inlet and outlet conditions. Because the enthalpy difference between two states is independent of the reference there is not any problem if each substance has a different reference of enthalpy in the energy balance. Therefore, we can calculate the enthalpy terms directly with the Enthalpy function in EES. *If there are chemical reactions*, a common enthalpy reference is necessary for all substances. Examples of energy balances with and without chemical reactions are presented in Sections 11.3.2 and 11.3.1, respectively.

Finally, EES has the capability to automatically generate property plots, such as T-s, T-v, T-h, P-h, P-v, P-h and h-s. Select Property plot from the Plot menu to show the Property plot information dialog (Figure 11.16-left). At the top of the dialog select a group of substances and then click on a substance in the left window. Select the type of plot and then click OK. For each type of plot there is the option to

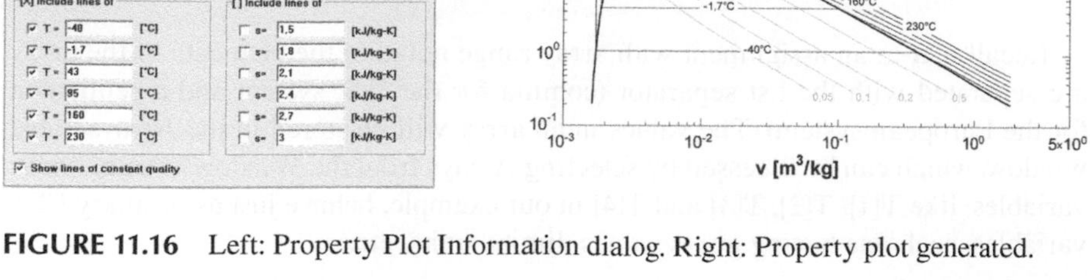

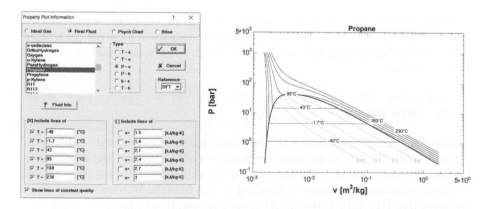

FIGURE 11.16 Left: Property Plot Information dialog. Right: Property plot generated.

add contour lines of predefined intensive properties, such as temperature, entropy, quality, etc. Figure 11.16-left shows the set up to plot a P-v diagram of propane as real fluid with lines of constant quality and isotherms. Figure 11.16-right shows the corresponding plot generated. Psychrometric plots can also be generated by selecting Psych Chart at the top of the Property plot information dialog. The reader is encouraged to watch the videos by F-Chart software on property data [11] and property plots [12] in EES.

11.2.5 Arrays

An array stores under the same name a group of variables with the same units. In EES only one dimensional (vector) and two-dimensional arrays (matrix) can be used. The notation to refer to an element of an array is the name of the array followed by the number of the element within square brackets, e.g., v[2] refers to the second element of array v.

An array is a convenient way to store information in an ordered manner. Let's suppose that we are designing a simple process that has four streams, named from 1 to 4. We can store information on those streams, such as temperature, pressure, mass flow, etc., in different arrays. For instance, we can store the temperature of the streams in an array named T where each element of the array corresponds to the temperature of each stream. If the temperature of streams 1 to 4 is 100, 200, 50, 150°C, respectively, we can create the array T by assigning the temperature of stream i to the i-th element of the array:

```
T[1]=100 [C]  "temperature of stream 1 assigned to first element
               of array"
T[2]=200 [C]  "temperature of stream 2 assigned to second element
               of array"
T[3]=50 [C]    "temperature of stream 3 assigned to third element
                of array"
T[4]=150 [C]    "temperature of stream 4 assigned to fourth
                 element of array"
```

A more compact way to do this assignment is with an array range notation whereby we can refer to the continuous range of indices in an array by writing within square brackets the first and last index of that range separated by two dots as shown below:

```
T[1..4]=[100 [C]; 200 [C]; 50 [C]; 150 [C] ]  "assignment with
array range notation"
```

Recall that in an assignment with array range notation the elements of the array are separated with the list separator (comma for the U.S. system and a semicolon for the European system). The values in an array will be stored in the Arrays Table window, which can be accessed by selecting Arrays from the Windows menu. Array variables, like T[1], T[2], T[3] and T[4] in our example, behave just as ordinary EES variables, each one having guess values, limits, units, etc.

To create a two-dimensional array (matrix) use two indices for array assignments. For example, the statement below creates a matrix T with two rows and three columns by assigning values to each row at a time:

```
T[1;1..3]=[100; 200; 300]  "assigning values to the first
                             row"
T[2;1..3]=[50; 150; 250]  "assigning values to the second
                            row"
```

A video by F-Chart software explaining arrays in EES can be found here [13]

11.2.6 Integration

EES can numerically solve systems of ordinary differential equations (ODE) by using the Integral function. Let's suppose that we want to solve the following initial value problem of a first order ODE:

$$\frac{dy}{dx} = 5x - y^2 \quad y(0) = 0 \quad x[0,20] \rightarrow y = y(0) + \int_0^{20} \frac{dy}{dx} dx \tag{11.2}$$

The purpose of the integral function in EES is to numerically solve the integral term of equation 11.2. The code to solve this initial value problem is shown below:

```
dydx=5*x-y^2          "Ordinary differential equation"
yini=0                " Initial condition"
xini=0                "start of integration"
xend=20               "end of integration"
y=yini+integral(dydx; x; xini; xend) "Call to integral
function"
$IntegralTable x;y
```

The first argument to the integral function is a variable that represents the rate of change of the dependent variable (y) with the independent variable (x), i.e., dy/dx. The second argument is the independent variable (x). The third and fourth arguments specify the range of integration in the independent variable. For each integration step EES solve the set of equations and the solution variables of interest along all integration steps can be stored in an integral table. To do so, at the end of the code the statement $IntegralTable must be added followed in first place by the independent variable and then the variables of interest to be stored. To access the integral table, shown in Figure 11.17-left, select Integral Tables from the Windows menu. To generate a plot from the integral table (Figure 11.17-right) just follow the steps to plot data from a table explained in 11.2.4. Make sure that the integral table is selected at the right side of the New Plot Setup dialog.

Main

	x	y
Row 1	0	1,000E-994
Row 2	0,005	0,0000625
Row 3	0,01	0,00025
Row 4	0,0225	0,001266
Row 5	0,035	0,003062
Row 6	0,0475	0,00564
Row 7	0,06	0,008999
Row 8	0,0725	0,01314
Row 9	0,085	0,01806
Row 10	0,0975	0,02375
Row 11	0,11	0,03023

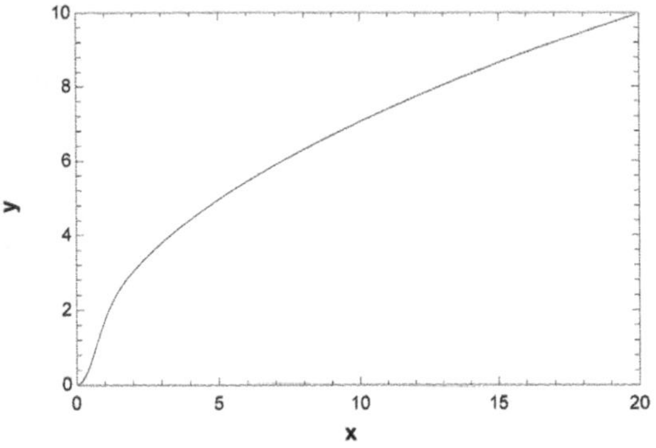

FIGURE 11.17 Left: Integral Table. Right: Plot generated from integral table.

To solve a set of coupled differential equations the integral function is called once for each dependent variable. As an example, the code to solve the initial value problem of an ODE system (equation 11.3) is shown below.

$$\begin{cases} \dfrac{dy}{dx} = -5y \\ \dfrac{dz}{dx} = 5y - 2z \end{cases} \qquad y_{ini} = 10, \quad z_{ini} = 0, \quad x[0,5] \tag{11.3}$$

```
dydx=-5*y               "First ODE"
dzdx=5*y-2*z            "Second ODE"
yini=10                 "Initial condition for y at xini"
zini=0                  " Initial condition for z at xini"
xini=0                  "start of integration"
xend=5                  "end of integration"
y=yini+integral(dydx;x;xini;xend) "Call to integral function
for y"
z=zini+integral(dzdx;x;xini;xend) "Call to integral function
for z"
$IntegralTable x; y; z
```

Second or higher order ODE and partial differential equations (PDE) can also be solved in EES by transforming them into a set of first order differential equations [14,15]. Videos by F-Chart software on how to solve ODE in EES can be found here [16,17]

11.2.7 Programming

EES is an equation-solver but its capabilities can be expanded by programming your own routines, which can be called from the Equation window. Within EES, *functions, procedures, subprograms* and *modules* are differentiated. A *function* is a code which performs a task from input arguments and returns only one output argument while a *procedure* is similar to a function but returns two or more output arguments. The codes of functions and procedures are created with assignments, similar to any other programming language, rather than equality equations, like those used in the main EES program. Control flow statements, like well-known if-then-else, while loops, GoTo, etc can be used within functions and procedures. On the other hand, *subprograms* and *modules*, which can be also called from the Equation window, serve the same purpose than functions and procedures but are written in terms of equations. Therefore, control statements are not allowed within subprograms and modules. The difference between subprograms and modules is subtle, in terms of execution, and the reader is referred elsewhere for details [1]. Functions and procedures can be programmed in the Equation window (*internal function or procedure*) or in any compiled language (*external function or procedure*), such as C++, Pascal or Fortran, and then added to EES as library files. For detailed material on how to program in EES, the reader is referred to the book Mastering EES [1]. For the basics on programming functions in EES, the reader is referred to a video by F-Chart software [18].

11.2.8 Optimization

EES can solve one and multidimensional optimization problems, i.e., problems where one or more variables (degrees of freedom) are varied in order to maximize or minimize an objective function subjected to constraints. EES use different optimization algorithms depending on the number of degrees of freedom. According to EES Help: "If there is one degree of freedom, EES will use either a Golden Section search or a Quadratic Approximations method. The Quadratic Approximation method is usually faster, but the Golden Section method is more reliable. Multidimensional optimization can be done using the Conjugate directions method or the Variable metric method. The Variable Metric method, which uses numerical derivatives, seems to be the most efficient method. The Conjugate Directions is best for problems in which the optimum is constrained to be on a bound or when numerical derivatives are not reliable." In the Commercial version of EES optimization problems with up to 20 degrees of freedom can be solved and up to 80 in the Professional version. Also in the Professional version optional optimization algorithms are available: the Nelder-Mead Simplex method, the DIRECT algorithm and the Genetic method.

As an example, let's solve the following optimization problem: Min $3x^2 + 2xy + y^2$ subjected to $x - y \leq 5$ and $x \geq 1$. This is an optimization problem with two variables (x and y) and zero equality constraints so there are two degrees of freedom. Figure 11.18 shows that the optimization problem is concave since the objective function is concave as well as the region of feasible solutions defined by the constraints. Therefore, there is only one global minimum in the region of feasible solutions, which is the point $(x,y) = (1, -1)$ where the value of the objective function is 2 (Figure 11.18). Obviously one does not know the exact solution of an optimization beforehand but some inspection of where the solution might be can be helpful to select guess values as close as possible to the global optimum. This is usually carried out by performing parametric analysis in the independent variables. This feature is available in EES in the Min/Max table from the Calculate menu and explained elsewhere [1].

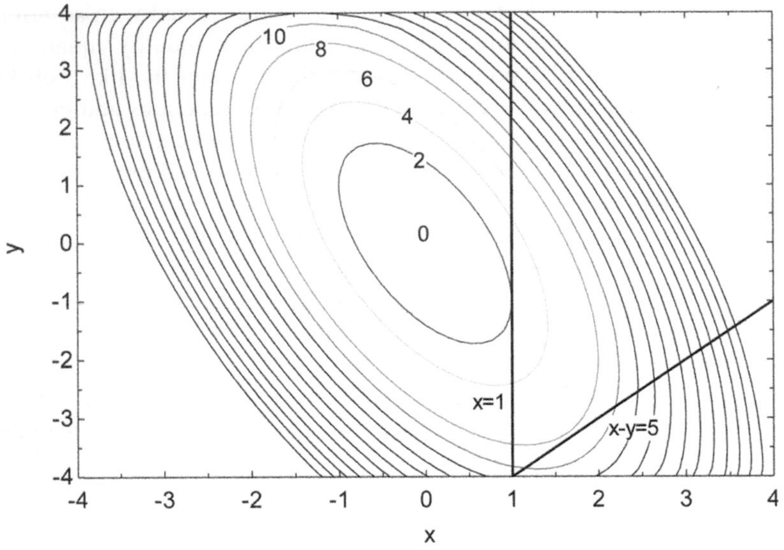

FIGURE 11.18 Contour lines of objective function $3x^2 + 2xy + y^2$ and boundaries formed by constraints $x - y \leq 5$ and $x \geq 1$.

The first step to solve the optimization problem in EES is to define the objective function and the constraints in the Equation window (see code below). Because only equalities can be defined in the Equation windows to reproduce the linear constraint $x - y \leq 5$ we will use an auxiliary variable $z = x - y$ whose upper bound will be later set 5. The addition of this auxiliary variable z does not change the degrees of freedom as we have also added a new equality constraint, the definition of z.

```
f=3*x^2+2*x*y+y^2 "Objective function"
z=x-y             "z as auxiliary variable to define the
                  linear constraint x-y<=5"
```

Next we select Min/Max from the Calculate menu and a dialog like Figure 11.19 will show up. At the top of the dialog we select Minimize and from the list below we select our objective function (f). From the list at the right side we select the independent variable or degrees of freedom, in this case, just two, x and z. Then we click on the Bounds button, which brings up the variable information dialog (Figure 11.20), to set the upper and lower bounds and guess values for the selected independent variables. From the constraints of the optimization problem we know that the lower bound of x is 1 and the upper bound of z is 5. The other bounds can be set to an arbitrary relative large number so the search region is large enough in order to include the optimum. The guess values should be set within the bounds specified previously and as close as possible to the expected location of the optimum, which typically is unknown. In this case we choose $x = 3$ (≥ 1) and $z = 3$ (≤ 5). Finally, in the dialog shown in Figure 11.19 we select the Conjugate direction method,

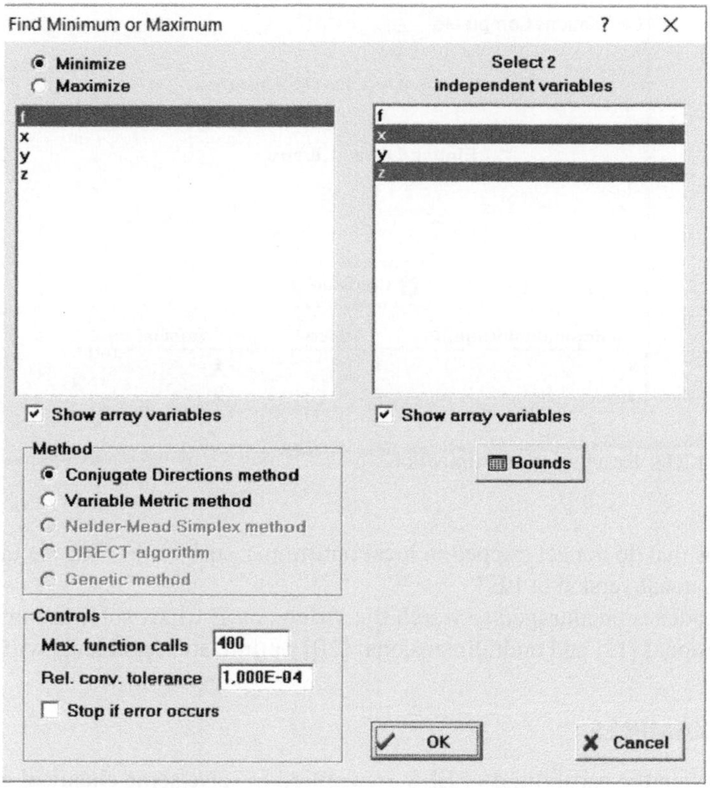

FIGURE 11.19 Setup dialog to solve an optimization problem.

which is appropriate for cases where the solution is on a boundary. By clicking OK the iterative search of the minimum starts which is found at x = 1 and z = 2 (or x = 1 and y = –1) where the objective function value is 2 (Figure 11.21).

It should be noted that, unlike the example we have solved, an objective function typically has multiple optimum points within a feasible region and the aim is to find the global optimum. Depending on the guess values different local optimums will be found. This usually forces to solve the optimization problem from different guess values in order to find the global optimum or to use global optimization

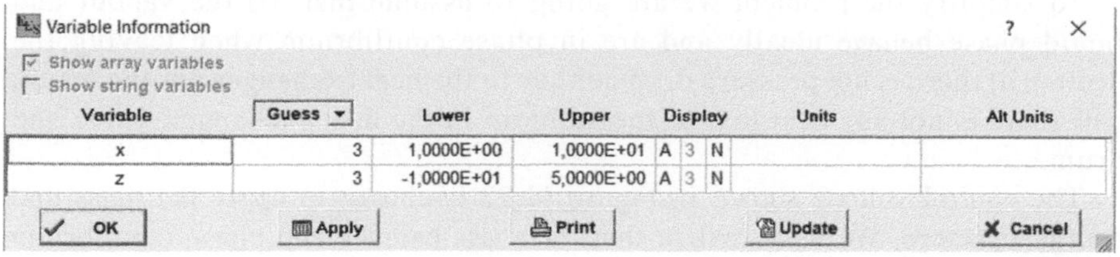

FIGURE 11.20 Dialog to set bounds and guess values of the selected independent variables.

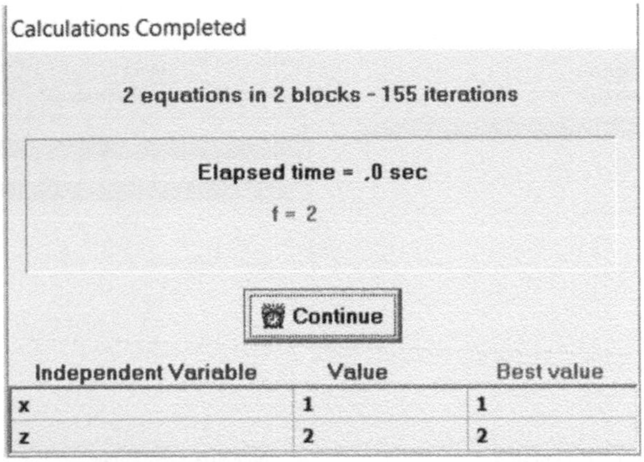

FIGURE 11.21 Result of the optimization.

algorithms that do not get trapped in local optimums, such as the Genetic method in the Professional version of EES.

The reader is encouraged to watch the videos by F-Chart software on solving unidimensional [19] and multidimensional [20] optimization problems with EES.

11.3 EXAMPLES

In this section the capabilities of EES are applied to solve some chemical engineering problems.

11.3.1 Phase Equilibrium and Energy Balance without Reaction: Flash Separation

Let's solve the following problem in EES. A 100 kmol/h stream (F) of 40% mole of methanol in water at 5 bar and 30°C is fed to a flash separation system as shown in Figure 11.22. The liquid stream is first heated, then partially flashed by expansion to 1 bar in a valve and finally the vapour (V) and liquid (L) formed are physically separated in a drum. We are asked to calculate how much heat (kW) must be supplied to the heat exchanger so the mole flow ratio between vapour and feed streams (V/F) is 0.4.

To simplify the problem we are going to assume that: (i) the vapour and liquid phase behave ideally and are in phase equilibrium when leaving the drum; (ii) there is not pressure drop neither in the heat exchanger nor the drum; (iii) there is not any heat loss to the ambient in the heat exchanger, valve and drum.

The control volume shown in Figure 11.22 is chosen to apply the mass and energy balances. We must realize that the mass balance and phase equilibrium calculations are decoupled from the energy balance. Thus, we can first calculate

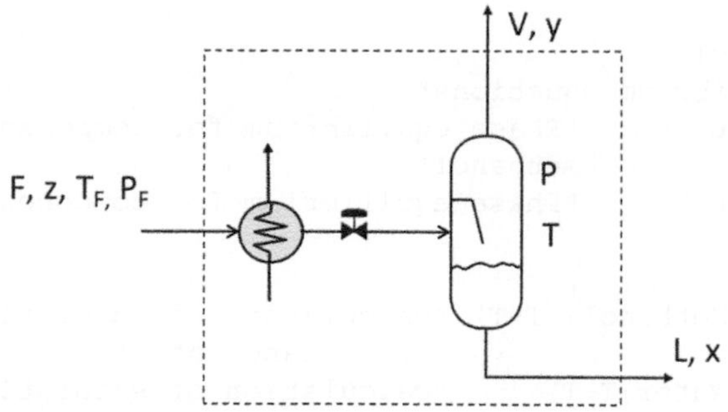

FIGURE 11.22 Flash separation system. Control volume is shown in dotted line.

the flash temperature T from the phase equilibrium equations and the mass balance (equations 11.4–9) and next solve the energy balance.

$$P \cdot y_1 = P_1^s(T) \cdot x_1 \tag{11.4}$$

$$P \cdot y_2 = P_2^s(T) \cdot x_2 \tag{11.5}$$

$$x_1 + x_2 = 1 \tag{11.6}$$

$$y_1 + y_2 = 1 \tag{11.7}$$

$$F = L + V \tag{11.8}$$

$$F \cdot z_1 = L \cdot x_1 + V \cdot y_1 \tag{11.9}$$

where components 1 and 2 are methanol and water, respectively, P is the flash pressure and P_i^s is the saturation pressure of component i at flash temperature T. In addition to these 6 equations, we have four specifications (V/F = 0.4, F = 100 kmol/h, P = 1 bar, z_1 = 0.4), so in total we have 10 equations with 12 variables (P, T, x_1, x_2, y_1, y_2, P_1^s, P_2^s, F, L, V,z_1). Therefore we need two more equations to solve the problem: those which provide the saturation pressure of each component as a function of temperature $P_i^s = f(T)$. We will use the P_sat function of EES to calculate saturation pressure for each component. The code to solve the flash temperature is:

```
"Specifications"
z1=0,4 [-] "Mole concentration of methanol in the feed,
dimensionless"
F=100/3600 "Feed mole flow, kmol/s"
P=1 [bar] "Presión en kPa"
V=0,4*F  "Vapour/feed ratio"
"Mass balance equations"
```

```
F=L+V
F*z1=L*x1+V*y1
"Phase equilibrium equations"
P*y1=x1*P1sat          "Phase equilibrium for component 1,
                       methanol"
P*y2=x2*P2sat          "Phase equilibrium for component 2, water"
x1+x2=1
y1+y2=1
P1sat=P_sat(Methanol; T=T) "calculation of saturation pressure
                            for methanol at T"
P2sat=P_sat(Water;T=T)      "calculation of saturation pressure
                            for water at T"
```

The more appropriate choice of units is bar, C, kmol, s and kJ given the units of the input data and required results. Remember that the units of the constants are set in the code while the units of variables are set in the Variable information dialog (see Section 11.2.3). Click on the Check units icon in the toolbar to detect any unit error. In addition, the variables should be bounded: mole concentrations between 0 and 1, and the lower bound of the rest of variables should be set to 0. Finally, we need to provide guessed values to help EES to converge to a valid solution. The most critical variable in the calculation is the flash temperature T, because it largely affects the equilibrium composition of the vapour and liquid phase, and consequently, the mass balance. A rough upper and lower bound for T is the saturation temperature at the flash pressure of water (99.6°C), the heaviest component of the mixture, and methanol (64.7°C), the lightest component, respectively. These temperatures can be calculated with the T_sat function of EES as shown below. For that purpose you can previously comment out the rest of the code and when done, uncomment it again.

```
T1sat=T_sat(Methanol;P=P) "saturation temperature of methanol
                          at flash pressure"
T2sat=T_sat(Water;P=P)      "saturation temperature of water
                            at flash pressure"
```

Set the calculated upper and lower bounds for T and choose arbitrarily a guess value for T within the bounds, let's say, 70°C. Solve the mass balance and phase equilibrium equations and EES will converge to the solution T = 84.85°C. Now we can calculate the heat to be supplied to the feed by solving the energy balance applied to the control volume, where ideal behavior is assumed when calculating the enthalpy of any mixture:

$$\sum_i F \cdot z_i \cdot h_i^l(T_F, P_F) + \dot{Q} = \sum_i V \cdot y_i \cdot h_i^v(T, P) + \sum_i L \cdot x_i \cdot h_i^l(T, P) \qquad (11.10)$$

To calculate the enthalpy of the components in the feed, overhead vapour and bottom liquid we will use the Enthalpy function of EES. Remember that because there is not any chemical reaction the reference state and value for enthalpy of each component can be different (see Section 11.2.4). But we must proceed with caution. The Enthalpy function provides enthalpy for a substance as a pure fluid, but in our case, we are dealing with mixtures. For instance, if we use the enthalpy function to calculate the enthalpy of water in the vapour phase at the flash temperature (84.85°C) and pressure (1 bar) we will make a mistake, as EES will return the enthalpy of water as pure fluid in those conditions (liquid). This same problem is associated to the calculation of the enthalpy of methanol in the liquid phase. As a workaround we are going to calculate the enthalpy of both components in the overhead vapour and bottom liquid as that corresponding to the saturated state (vapour or liquid) as pure components at the flash temperature. This decision is based on the fact that enthalpy is much more sensitive to temperature than pressure, and the difference between the flash pressure and the saturated pressure for each component at the flash temperature has a negligible impact of enthalpy. Consequently, the enthalpy of each component is calculated as a function of temperature T and quality x (0 for saturated liquid and 1 for saturated vapour). The code for the energy balance that must be added to the previous code (mass balance and phase equilibrium) in order to solve the whole problem is:

```
"Energy balance"
H_F+Q=H_V+H_L   "Q, heat to be supplied"
H_F=F*(z1*Enthalpy(Methanol;T=TF;
P=PF)+(1-z1)*Enthalpy(Water;T=TF;P=PF))
H_V=V*(y1*Enthalpy(Methanol;T=T;x=1)+y2*Enthalpy
(Water;T=T;x=1))
H_L=L*(x1*Enthalpy(Methanol;T=T;x=0)+x2*Enthalpy
(Water;T=T;x=0))
```

Setting the units of the new variables and then solving all the equations, the calculated heat to be supplied in the heat exchanger is 490.7 kW.

11.3.2 Chemical Equilibrium and Energy Balance: Autothermal Reforming

In this section we are going to learn how to solve in EES chemical equilibrium calculations and energy balances with reaction by modelling the catalytic autothermal reforming of natural gas for producing synthesis gas ($CO + H_2$) (Figure 11.23).

Natural gas (methane), steam and pure oxygen are fed to an authothermal reformer, which can be considered an adiabatic reactor. A fraction of methane is burned with oxygen (reaction 1: equation 11.11), supplying heat to the endothermic steam methane reforming (reaction 2: equation 11.12), which takes place over the reforming catalyst. Also, the water gas shift reaction (reaction 3: equation 11.13) takes place, increasing the hydrogen content of the syngas generated at the expense of carbon monoxide.

$$CH_4 + 2O_2 \leftrightarrow CO_2 + 2H_2O \tag{11.11}$$

$$CH_4 + H_2O \leftrightarrow CO + 3H_2 \tag{11.12}$$

$$CO + H_2O \leftrightarrow CO_2 + H_2 \tag{11.13}$$

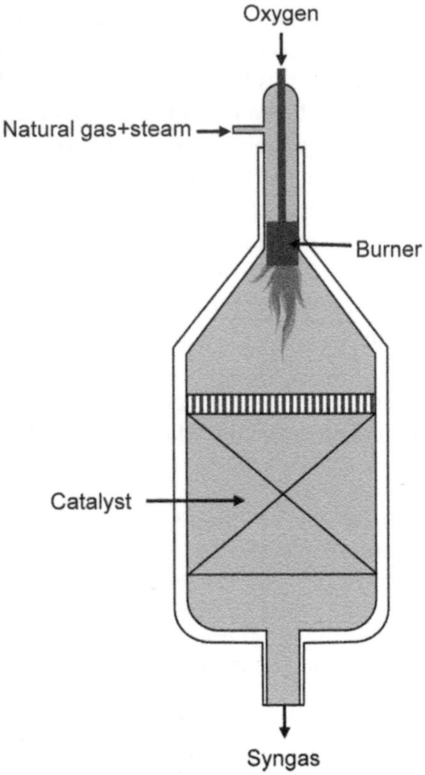

FIGURE 11.23 Autothermal reformer.

The authothermal reformer operates at 30 bar. Methane (36000 Nm^3/h) and steam (mole ratio $H_2O/CH_4 = 1.2$) are fed to the reformer at 650°C and 30 bar. The flow rate of oxygen, at 25°C and 30 bar, is regulated so the temperature of the syngas leaving the reactor is 1000°C. We are asked to calculate the flow rate of oxygen to be fed to the reformer and the flow rate of syngas (Nm^3/h) produced in terms of CO and H_2.

To simplify the problem we can assume that the syngas composition at the outlet conditions is that corresponding to chemical equilibrium because the operating temperature is so high that the rates of reactions are fast enough to reach equilibrium. To solve the chemical equilibrium we need to know the flow rate of methane, steam and oxygen fed to the reformer but the latter is unknown because it is coupled to the energy balance. Therefore, we must solve simultaneously the chemical equilibrium equations and the energy balance. For the sake of illustration we will group the equations and the related code in EES to solve them. The code to solve the whole problem can be obtained by merging the fractions of code presented along the text.

To solve the chemical equilibrium we will use the stoichiometric approach, that is, the chemical equilibrium constant for each reaction will be imposed when solving the mass balance. In this case, the mass balance will be applied to the conservation of the elements involved: C, H and O, equations 11.14, 11.15 and 11.16, respectively.

$$C\,atom\,balance: F_{CH4}^{out} + F_{CO}^{out} + F_{CO2}^{out} = F_{CH4}^{in} \tag{11.14}$$

$$H\,atom\,balance: 4\cdot F_{CH4}^{out} + 2\cdot F_{H2}^{out} + 2\cdot F_{H2O}^{out} = 4\cdot F_{CH4}^{in} + 2\cdot F_{H2O}^{in} \tag{11.15}$$

$$O\,atom\,balance: F_{CO}^{out} + 2\cdot F_{CO2}^{out} + F_{H2O}^{out} + 2\cdot F_{O2}^{out} = F_{H2O}^{in} + 2\cdot F_{O2}^{in} \tag{11.16}$$

where F_i is the mole flow of each substance and the superscripts in and out mean entering and leaving the reformer, respectively. The code in EES related to the mass balance equations, together with the input data for the whole problem, is shown below:

```
-----------------------INPUT DATA-----------------"
Q_CH4=36000 [Nm^3/h] "Flow rate of methane"
"Reference state for enthalpy of formation"
Tref= 25 [C]
Pref=1 [bar]
"Temperature and pressure of methane+steam feed"
T_feed=650 [C]
P_feed=30 [bar]
"Temperature and pressure of oxygen feed"
T_ox=25 [C]
P_ox=30 [bar]
"Operating temperature and pressure of ATR"
T=1000 [C]
P=30 [bar]
"Feed to ATR"
FCH4_in=Q_CH4/22,4  "methane, kmol/h"
FH2O_in=1,2*FCH4_in  "water, kmol/h"
"Constants"
R=8,314 [kJ/kmol K]
"-----------------------EQUATIONS--------------------------"
"MASS BALANCE (ELEMENTS)"
"Fi is mole flow (kmol/h) of substance i leaving
the reformer"
"Fi_in is mole flow (kmol/h) of substance i entering
the reformer"
FCH4_in=FCH4+FCO+FCO2 "Carbon balance"
4*FCH4_in+2*FH2O_in=4*FCH4+2*FH2O+2*FH2 "Hydrogen balance"
2*FO2_in+FH2O_in=FCO+2*FCO2+FH2O+FO2 "Oxygen balance"
```

The chemical equilibrium constant for each reaction *i* is calculated with equations 11.17–20:

$$K_i(T) = \exp\left(-\frac{\Delta G_i^o}{RT}\right) \tag{11.17}$$

$$\Delta G_i^o(T) = \Delta H_i^o(T) - T\Delta S_i^o(T) \quad (11.18)$$

$$\Delta H_i^o(T) = \sum_i \nu_i H_i^o(T) \quad (11.19)$$

$$\Delta S_i^o = \sum_i \nu_i S_i^o(T) \quad (11.20)$$

where the superscript *o* means evaluated at the standard state (1 bar, pure ideal gas) and equilibrium temperature and ν_i is the stoichiometric coefficient for substance *i* (positive for products and negative for reactants). We will calculate the enthalpy and entropy terms in EES with the functions Enthalpy and Entropy, respectively. The code to calculate the chemical equilibrium constants in EES is shown below:

```
"CHEMICAL CONSTANT CALCULATION"
"Ideal gas enthalpy at 1 bar and operating temperature T,
kJ/kmol"
h0_CH4=enthalpy(CH4;T=T)
h0_CO=enthalpy(CO;T=T)
h0_CO2=enthalpy(CO2;T=T)
h0_H2=enthalpy(H2;T=T)
h0_H2O=enthalpy(H2O;T=T)
h0_O2=enthalpy(O2;T=T)
"Ideal gas entropy at 1 bar y and operating temperature T,
kJ/kmol K."
s0_H2O=entropy(H2O;T=T;P=Pref)
s0_CH4=entropy(CH4;T=T;P=Pref)
s0_CO2=entropy(CO2;T=T;P=Pref)
s0_CO=entropy(CO;T=T;P=Pref)
s0_O2=entropy(O2;T=T;P=Pref)
s0_H2=entropy(H2;T=T;P=Pref)
"Combustion of methane CH4+2*O2<==>CO2+2*H2O"
DELTAH1=h0_CO2+2*h0_H2O-2*h0_O2-h0_CH4 "standard enthalpy of
reaction, kJ/kmol"
DELTAS1=s0_CO2+2*s0_H2O-2*s0_O2-s0_CH4 "standard entropy of
reacion, kJ/kmol K"
DELTAG1=DELTAH1-(T+273,15[C])*DELTAS1 "standard free energy
of reaction, kJ/kmol"
K1=exp(-DELTAG1/(R*(T+273[C]))) "equilibrium constant"
"Reforming reaction CH4+H2O<==>CO+3H2"
DELTAH2=h0_CO+3*h0_H2-h0_H2O-h0_CH4 "standard enthalpy of
reaction, kJ/kmol"
DELTAS2=s0_CO+3*s0_H2-s0_H2O-s0_CH4 "standard entropy of
reacion, kJ/kmol K"
DELTAG2=DELTAH2-(T+273,15[C])*DELTAS2 "standard free energy
of reaction, kJ/kmol"
K2=exp(-DELTAG2/(R*(T+273[C]))) "equilibrium constant"
```

```
"Water gas shift reactions CO+H2O<==>CO2+H2"
DELTAH3=h0_CO2+h0_H2-h0_H2O-h0_CO "standard enthalpy of
reaction, kJ/kmol"
DELTAS3=s0_CO2+s0_H2-s0_H2O-s0_CO "standard entropy of
reacion, kJ/kmol K"
DELTAG3=DELTAH3-(T+273,15[C])*DELTAS3 "standard free energy of
reaction, kJ/kmol"
K3=exp(-DELTAG3/(R*(T+273[C]))) "equilibrium constant"
```

The chemical equilibrium constants can be calculated by executing the above code without the need of the mass and energy balance equations. The calculated constants are $K_1 = 6.6 \cdot 10^{32}$, $K_2 = 8789$, $K_3 = 0.6$. We see that the equilibrium constant for the combustion reaction of methane (reaction 1) is very high which means that the oxygen will be fully consumed in that reaction ($F_{O2}^{out} = 0$).

We must relate the chemical equilibrium constants with the composition of the mixture in chemical equilibrium. To do this the chemical equilibrium constant is expressed in terms of the fugacity of each component in the mixture, $\hat{f}_i(T,P,y)$:

$$K_i(T) = \prod_i \hat{f}_i^{\nu_i}(T,P,y) \tag{11.21}$$

For the sake of simplicity we will calculate the fugacity of each component in the mixture with the Lewis rule for fugacity (equation 11.22) [21]:

$$\hat{f}_i(T,P,y) = y_i \cdot f_i(T,P) \tag{11.22}$$

where f_i is the fugacity of *i* as pure component, which can be calculated in EES with the function fugacity. This approximation is expected to be valid because at the high operating temperature intermolecular forces in the gas will probably be negligible compared to the kinetic energy of the molecules, and the gaseous mixture will behave as an ideal gas. The composition in equation 11.22 is calculated from the mole flow of each substance in the syngas leaving the reactor. Thus, the code to solve the chemical equilibrium by the stoichiometric approach is shown below:

```
"STOICHIOMETRIC APPROACH"
"Fugacity calculation of pure components, bar"
fgCH4=Fugacity(Methane;T=T;P=P)
fgH2=Fugacity(Hydrogen;T=T;P=P)
fgH2O=Fugacity(Steam;T=T;P=P)
fgCO2=Fugacity(CarbonDioxide;T=T;P=P)
fgCO=Fugacity(CarbonMonoxide;T=T;P=P)
"Fugacity in mixture for each substance calculated with Lewis
rule, bar"
fgCH4m=fgCH4*yCH4
fgH2m=fgH2*yH2
fgH2Om=fgH2O*yH2O
fgCO2m=fgCO2*yCO2
```

```
fgCOm=fgCO*yCO
"Equilibrium compositions"
yCH4=FCH4/Ftot
yCO=FCO/Ftot
yCO2=FCO2/Ftot
yH2=FH2/Ftot
yH2O=FH2O/Ftot
yO2=FO2/Ftot

Ftot=FCH4+FCO+FCO2+FH2+FH2O+FO2 "Total mole flow reactor
outlet, kmol/h"
"Chemical equilibrium constraints"
FO2=0 "equivalent equation to chemical equilibrium of
combustion of methane due to very high equilibrium constant"
K2*fgCH4m*fgH2Om=fgCOm*fgH2m^3 "reforming reaction"
K3*fgCOm*fgH2Om=fgCO2m*fgH2m     "water gas shift reaction"
```

Note that we have replaced the imposition of the chemical equilibrium constant of reaction 1 with the equation $F_{O2}^{out} = 0$, as explained before. This avoids numerical problems in the convergence because otherwise, if we had imposed equation 11.19 for that reaction, we would have had a division by zero as y_{O2} is in the denominator of the fraction of fugacities for reaction 1.

Finally, we need to apply the energy conservation law to the adiabatic reactor in terms of an enthalpy balance. Because there are chemical reactions, as explained in Section 11.2.4, a common enthalpy reference must be used for all substances. The most common enthalpy reference for reaction mixtures is to assign zero state of enthalpy for the elements at its simplest and most thermodynamically stable state at 25°C and 1 bar [22]. The enthalpy of molecular species relative to their constituent atoms at this reference is by definition the enthalpy of formation (h_f), which we use to express the energy balance (equation 11.23).

$$\sum_i F_i^{in} \cdot h_{f,i}^{in}(T,P) = \sum_i F_i^{out} \cdot h_{f,i}^{out}(T,P) \tag{11.23}$$

Usually, the enthalpies of formation of substances are only available at the reference state, h_f(25°C, 1 atm). To calculate the enthalpy of formation at any state (T,P) we add to h_f(25°C, 1 atm) the change of enthalpy from the reference state to the state of interest (equation 11.24). This change of enthalpy will be calculated in EES with the Enthalpy function.

$$h_{f,i}(T,P) = h_{f,i}(25^\circ C, 1\,\text{bar}) + (h_i(T,P) - h_i(25^\circ C, 1\,\text{bar})) \tag{11.24}$$

The code in EES related to the energy balance is shown below:

```
"ENERGY BALANCE"
"Energy balance in terms of enthalpy of formation"
"Enthalpy of formation at reference state, 25°C, 1 bar, kJ/
kmol"
```

```
h0f_O2=0
h0f_H2=0
h0f_CH4=Enthalpy_formation(CH4)
h0f_H2O=Enthalpy_formation(H2O)
h0f_CO=Enthalpy_formation(CO)
h0f_CO2=Enthalpy_formation(CO2)
"Enthalpy of formation of each substance at inlet conditions,
kJ/kmol"
hf_CH4in=h0f_CH4+(enthalpy(Methane;T=T_feed;P=P_feed)-
enthalpy(Methane;T=Tref;P=Pref))
hf_H2Oin=h0f_H2O+(enthalpy(Steam;T=T_feed;P=P_feed)-
enthalpy(Steam;T=Tref;P=Pref))
hf_O2in=h0f_O2+(enthalpy(Oxygen;T=T_ox;P=P_ox)-
enthalpy(Oxygen;T=Tref;P=Pref))
"Enthalpy of formation of each substance at outlet conditions,
kJ/kmol"
hf_CH4=h0f_CH4+(enthalpy(Methane;T=T;P=P)-
enthalpy(Methane;T=Tref;P=Pref))
hf_CO=h0f_CO+(enthalpy(CarbonMonoxide;T=T;P=P)-enthalpy(Carbon
Monoxide;T=Tref;P=Pref))
hf_CO2=h0f_CO2+(enthalpy(CarbonDioxide;T=T;P=P)-enthalpy(Carbo
nDioxide;T=Tref;P=Pref))
hf_H2=h0f_H2+(enthalpy(Hydrogen;T=T;P=P)-
enthalpy(Hydrogen;T=Tref;P=Pref))
hf_H2O=h0f_H2O+(enthalpy(Steam;T=T;P=P)-
enthalpy(Steam;T=Tref;P=Pref))
hf_O2=h0f_O2+(enthalpy(Oxygen;T=T;P=P)-
enthalpy(Oxygen;T=Tref;P=Pref))
"Enthalpy balance"
FCH4_in*hf_CH4in+FH2O_in*hf_H2Oin+FO2_in*hf_O2in=(FCH4*hf_
CH4+FCO*hf_CO+FH2*hf_H2+FH2O*hf_H2O+FCO2*hf_CO2+FO2*hf_O2)
"Methane conversion"
Conv_CH4=1-FCH4/FCH4_in
"Oxygen stoichiometric ratio"
O2rat=FO2_in/(FCH4_in*2)
"Syngas produced"
Syngas=FCO+FH2
H2COr=FH2/FCO "H2/CO ratio"
```

Before solving the whole problem remember to bound the variables (e.g., the lower bound for mole flows should be set to zero and mole fractions should be bounded within 0 and 1) and provide guessed values for the most critical variables. In this case, the most important variable is the mole flow of oxygen feed, which strongly affects the chemical equilibrium and the energy balance. In industrial autothermal reformers the ratio of oxygen feed to the stoichiometric oxygen needed to burn all methane (oxygen stoichiometric ratio) is around 0.3, which we can use to get a good guess value for the oxygen feed. For our example with a simple hand calculation we get $F_{O_2}{}^{in} \approx 964$ kmol/h O_2. By providing this guess value the problem converges in miliseconds. The calculated oxygen feed is 877.4 kmol/h, which corresponds to a stoichiometric ratio of 0.27. The conversion

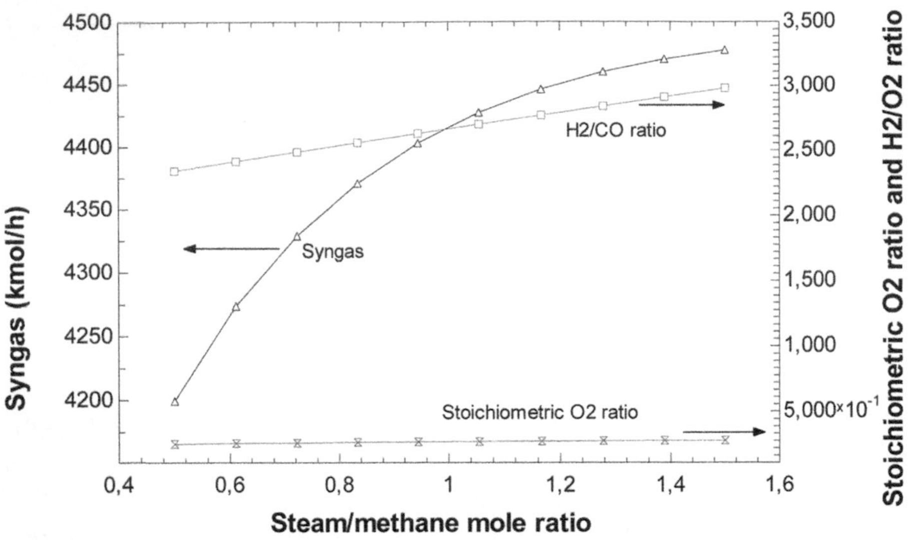

FIGURE 11.24 Syngas mole flow, H_2/CO ratio and oxygen stoichiometric ratio as a function of steam/methane mole ratio for an autothermal reformer operating at 1000°C and 30 bar.

of methane in the reactor is 96.5% and the mole flow and H_2/CO ratio of the syngas produced (CO + H_2) is 4451 kmol/h and 2.8, respectively.

Once the model of the reformer has been built and converged we can carry out sensitivity analysis in the operating variables to understand its performance. Remember that you can do this with a parametric table in EES (Section 11.2.4). For instance, Figure 11.24 shows oxygen stoichiometric ratio, mole flow (kmol/h) and H_2/CO ratio of syngas as a function of the steam/methane mole ratio in the feed when operating temperature and pressure of the reformer are fixed at 1000°C and 30 bar, respectively. Think if these results make sense.

11.3.3 Dynamic Modelling: Continuous Stirred Tank Reactor

Usually the final purpose of the dynamic modelling of a process is the design of a control strategy for that process. In this example we will model and solve the dynamics of a continuous stirred tank reactor (CSTR) where an exothermic reaction A→B takes place in liquid phase (Figure 11.25) [23]. The heat of reaction is removed with a cooling jacket where a coolant (water) circulates. In this process we want to control the concentration of A in the product stream, which is a measure of the conversion of A, by manipulating the flowrate of coolant that circulates through the cooling jacket. To this end we need to know the dynamics between the concentration of A in the product stream (C_A) and the coolant flowrate, around a steady-state (nominal) operating point. It is also of interest to study how some important disturbances, such as the feed flowrate and concentration, affect the conversion of A in the reactor.

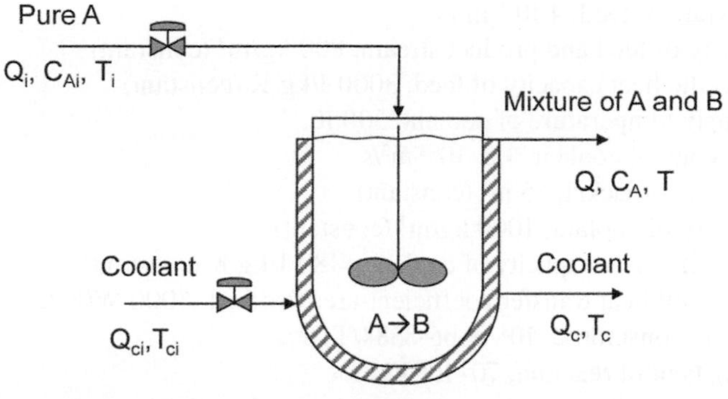

FIGURE 11.25 CSTR with cooling jacket and overflow line.

For modelling the CSTR we need to apply mass and energy balances. If we assume that the density of the feed (ρ_i) and product (ρ) streams are equal, the overall mass balance equation (equation 11.25) results in the equality of the feed (Q_i) and product (Q) flowrates at any time since the volume in the reactor (V) is kept constant by an overflow line.

$$\frac{d(V\rho)}{dt} = Q_i\rho_i - Q\rho \quad (11.25)$$

With this in mind, the dynamic model of the CSTR with cooling jacket can be formulated with equations 11.26–28 [23], where equation 11.26 is the mass balance of A in the reactor and equations 11.27 and 11.28 are the energy balances in the reactor and cooling jacket, respectively. In the formulation of the model it is considered that: (i) the rate of reaction is given by a first order law kC_A; (ii) the reactor and the cooling jacket are perfectly mixed; (iii) the volume of the cooling jacket is constant, so the inlet (Q_{ci}) and outlet (Q_c) flowrates of coolant are equal at any time; (iv) physical properties are constants.

$$V\frac{dC_A}{dt} = Q(C_{Ai} - C_A) - VkC_A \quad (11.26)$$

$$\rho C_p V\frac{dT}{dt} = Q\rho C_p(T_i - T) + (-\Delta H_R)VkC_A - UA(T - T_c) \quad (11.27)$$

$$\rho_c C_{pc} V_j\frac{dT_c}{dt} = Q_c\rho_c C_{pc}(T_{ci} - T_c) + UA(T - T_c) \quad (11.28)$$

where:

- V, the reactor volume, 20 m^3 (constant)
- C_{Ai}, concentration of reactant A in feed, 8 kmol/m^3
- T_i, Temperature of the feed, 350 K

- Q, flowrate of feed, $4 \cdot 10^{-3}$ m^3/s
- ρ, density of feed and product stream, 800 kg/m^3 (constant)
- C_p, specific heat capacity of feed, 3000 J/kg K (constant)
- T_{ci}, supply temperature of coolant, 300 K
- Q_c, flowrate of coolant, $1.2 \cdot 10^{-2}$ m^3/s
- V_j, volume of jacket, 15 m^3 (constant)
- ρ_c, density of coolant, 1000 kg/m^3 (constant)
- C_{pc}, specific heat capacity of coolant, 4180 J/kg K (constant)
- UA, Overall heat transfer coefficient-area product, 8000 W/K (constant)
- k, kinetic constant, $2 \cdot 10^5 \cdot \exp(-8384/T)$ s^{-1}
- $(-\Delta H_R)$, heat of reaction, $70 \cdot 10^6$ J/kmol

The nominal operating point can be calculated by solving the steady-state equations of the CSTR in EES with the code below. The calculated nominal values of C_A, T and T_c are 2.855 $kmol/m^3$, 416.4 K and 316 K, respectively. The nominal conversion of A is 66.3%.

```
"-------------------INPUT DATA---------------------------"
V= 20 [m3] "reactor volume"
CAi=8 [kmol/m3] "concentration of reactant A in feed"
Ti= 350 [K] "Temperature of the feed"
Q=4*10^(-3) "flowrate of feed"
rho=800 [kg/m3]  "density of feed and product stream"
Cp= 3000 [J/kg K] "specific heat capacity of feed"
Tci= 300 [K] "supply temperature of coolant"
Qc= 1,2*10^(-2) "flowrate of coolant"
Vj= 15 [m3] " volume of jacket"
rhoc= 1000 [kg/m3] "density of coolant"
Cpc=4180 [J/kg K] "specific heat capacity of coolant"
UA= 8000  [W/K] "heat transfer coefficient"
k=2 [s-1]*10^5 *exp(-8384 [K]/T) "kinetic constant,  s-1"
Hr=70*10^6 "heat of reaction, J/kmol"

"------------Steady state equations------------"
Q*(CAi-CA)-k*CA*V=0
Q*rho*Cp*(Ti-T)+Hr*k*CA*V-UA*(T-Tc)=0
Qc*rhoc*Cpc*(Tci-Tc)+UA*(T-Tc)=0
 "Conversion"
X=(1-CA/CAi)*100
```

From the nominal operating point we will simulate in EES a step change in the flowrate of coolant to study the dynamic response of C_A, T and T_c by solving the ODE system that model the dynamics of the CSTR (see code below). The initial conditions for C_A, T and T_c are their nominal values as we assume that the process is operating at steady state when the step change is made. To simulate a step change in the coolant flowrate, let's say from 0.012 to 0.02 m^3/s at 1000s, we use the IF function of EES which changes the value of the coolant flowrate once the time of integration is greater than 1000 s (see comment in the code). Because the value of the coolant flowrate along the integration is now set by the IF function we must comment out the equation in the input data section of the code that set the coolant flowrate to its nominal value.

```
"-------------------INPUT DATA----------------------------"
V= 20 [m3] "reactor volume"
CAi=8 [kmol/m3] "concentration of reactant A in feed"
Ti= 350 [K] "Temperature of the feed"
Q=4*10^(-3) "flowrate of feed"
rho=800 [kg/m3]  "density of feed and product stream"
Cp= 3000 [J/kg K] "specific heat capacity of feed"
Tci= 300 [K] "supply temperature of coolant"
{Qc= 1,2*10^(-2) "flowrate of coolant" }
Vj= 15 [m3] " volume of jacket"
rhoc= 1000 [kg/m3] "density of coolant"
Cpc=4180 [J/kg K] "specific heat capacity of coolant"
UA= 8000  [W/K] "heat transfer coefficient"
k=2 [s-1]*10^5 *exp(-8384 [K]/T) "kinetic constant,  s-1"
Hr=70*10^6 "heat of reaction, J/kmol"

"Conversion"
X=(1-CA/CAi)*100

"--------------Dynamic equations-----------------"
V*dCdt=Q*(CAi-CA)-k*CA*V "mass balance of A in reactor"
V*rho*Cp*dTdt=Q*rho*Cp*(Ti-T)+Hr*k*CA*V-UA*(T-Tc) "enthalpy
balance in reactor"
Vj*rhoc*Cpc*dTcdt=Qc*rhoc*Cpc*(Tci-Tc)+UA*(T-Tc) "enthalpy
balance in cooling jacket"

CA=CAini+integral(dCdt;time;tinit;tend)
T=Tini+integral(dTdt;time;tinit;tend)
Tc=Tcini+integral(dTcdt;time;tinit;tend)

 "Simulation time"
tinit=0
tend=40000 "s"

"Steady-state initial conditions"
CAini=2,855 "kmol/m3"
Tini=416,4 "K"
Tcini=316 "K"

"Step change in coolant flowrate from 0.012 to 0.02 m3/s at
1000 s using IF function
IF(A, B, X, Y, Z) allows conditional assignment statements in
the Main part of an EES program.
```

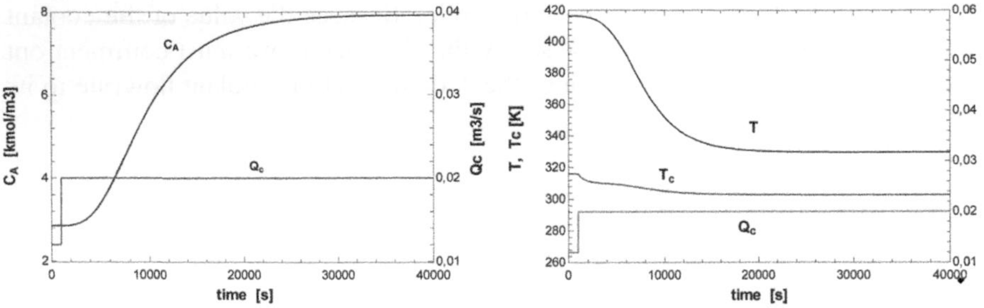

FIGURE 11.26 Dynamic response of the CSTR for a step change in the coolant flowrate from 0.012 to 0.02 m^3/s at 1000 s. Left: Concentration of A; Right: Reactor and cooling jacket temperatures.

```
        If A<B, the function will return a value equal to the
value supplied for X;
        if A=B, the function will return the value of Y;
        if A>B, the function will return the value of Z"
Qc= if(time; 1000; 1,2*10^(-2) ;2*10^(-2); 2*10^(-2))

 $IntegralTable time; CA; T; Tc; X; Qc
```

Figure 11.26 shows the dynamic response of the CSTR for the step change in the coolant flowrate. As expected, when the coolant flowrate increases more heat is removed from the reactor, which decreases the reactor temperature and thus the rate of reaction. The latter lowers the conversion of A, and thus the concentration of A increases with time until a new steady state is reached.

It is left as an exercise for the reader to simulate step changes in disturbance variables, such as the feed flowrate and temperature, the concentration of A in the feed and inlet temperature of the coolant.

REFERENCES

1. Klein S.A; Nellis G.F. (2017). *Mastering EES*. F-Chart Software, Madison, WI.
2. Klein S.A; Nellis G.F. (2011). *Thermodynamics*. Cambridge University Press. (1st ed.), NY, USA.
3. Klein S.A; Nellis G.F. (2012). *Heat Transfer*. Cambridge University Press. (1st ed.), NY, USA.
4. F-Chart website. EES. http://www.fchart.com/ees/
5. F-Chart Software. Overview of EES. https://www.youtube.com/watch?v=4Jta4wHCRXw
6. F-Chart Software. Entering and Solving Equations in EES. https://www.youtube.com/watch?v=T8GhPFKvnjg
7. F-Chart Software. Using EES effectively. https://www.youtube.com/watch?v=pmlv30t3cLk
8. F-Chart Software. Units in EES. https://www.youtube.com/watch?v=ON4jqyiJxns
9. F-Chart Software. Parametric tables. https://www.youtube.com/watch?v=5f6E945mP1g
10. F-Chart Software. Plotting in EES. https://www.youtube.com/watch?v=cemaNI9ts_Y
11. F-Chart Software. Properties in EES. https://www.youtube.com/watch?v=Spkc6CFtbQI

12. F-Chart Software. Property plots in EES. https://www.youtube.com/watch?v=ptkqjl8ZJ7o
13. F-Chart Software. Arrays. https://www.youtube.com/watch?v=T7xpFdq9WDc
14. Simmons G.F. (2016). *Differential Equations with Applications and Historical Notes* (3rd ed.). Chapman and Hall/CRC.
15. Strauss W.A. (2008). *Partial Differential Equations: An Introduction*. (2nd ed.). John Wiley & Sons.
16. F-Chart Software. ODE solver. https://www.youtube.com/watch?v=JOAmYonXKPw
17. F-Chart Software. Solving coupled ODEs using EES. https://www.youtube.com/watch?v=s6rtpH7zemM
18. F-Chart Software. Functions in EES. https://www.youtube.com/watch?v=bLICbSIf7vc
19. F-Chart Software. Optimization in onc dimension. https://www.youtube.com/watch?v=3xVJ782g42I
20. F-Chart Software. Optimization in multiple variables. https://www.youtube.com/watch?v=oNGxMNBgPMM
21. Prausnitz, J.M. (1998). *Molecular Thermodynamics of Fluid-Phase Equilibria* (3rd ed.). Prentice-Hall.
22. Sandler S.I. (2006). *Chemical, Biochemical, and Engineering Thermodynamics* (4th ed.). Wiley.
23. Luyben W.L. (2007). *Chemical Reactor Design and Control*. (1st ed.). Wiley-Interscience.

12 Modular Process Simulators

Rubén Ruiz-Femenía, César Ramírez-Márquez, Luis G. Hernández-Pérez, José A. Caballero, Mariano Martín, José M. Ponce Ortega, and Juan Gabriel Segovia

12.1 INTRODUCTION

The simulation, design and optimization of a chemical process plant, which comprises several processing units interconnected by process streams, are the core activities in Process Engineering. These tasks require performing material and energy balancing, equipment sizing and costing calculation. A computer package that can accomplish these duties is known as a computer aided process design package or simply a process simulator (also known as process flowsheeting package, flowsheet simulator or flowsheeting software). The capabilities of a process simulator include an accurate description of physical properties of pure components and complex mixtures, rigorous models for unit operations, as well as numerical techniques for solving large systems of algebraic and differential equations. By a process simulator, it is possible to obtain a comprehensive computer image of a running process, which is a valuable tool in understanding the operation of a complex chemical plant, and on this basis can serve for continuous improving the process, or for developing new processes.

The main activities process engineers have to face, result in three types of problems that can be solved with the aid of a process simulator (Figure 12.1). In the simulation problem, the variables associated with the input streams for each unit of the process flowsheet, and the parameters of all the units should be specified (the right statement of a simulation problem reduces the degrees of freedom for the system to zero). The unknowns are the variables associated with the output streams of the units. The design problem is similar to the simulation problem, except that some of the design variables are left unspecified and constraints are imposed on some of the stream variables (only equality constraints), and then the unknown operating conditions and equipment parameters are calculated. The statement of the design problem demands that the number of equality constraints is equal to the number of unspecified design variables. In the last type, the optimization problem, not only some of the design variables may be left unspecified, but also some variables associated with the streams can be undefined. In this case, an objective function is added to the model to assess the performance of the plant according to economic, environmental or social

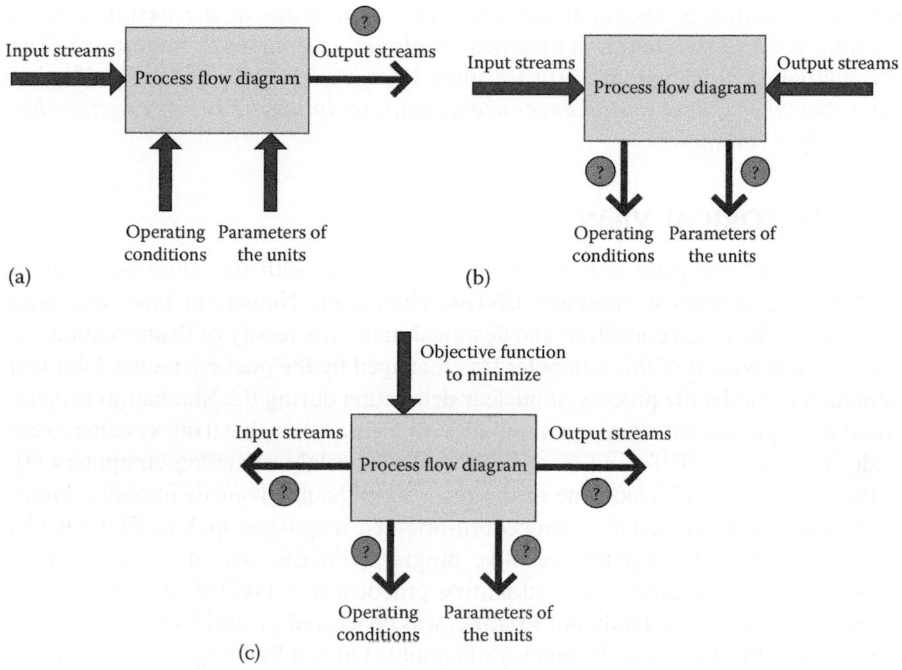

FIGURE 12.1 Differences among (a) simulation, (b) design and (c) optimization.

criteria. The unspecified variables are determined by optimizing the objective function. In this type of problem, both equality and inequality constraints may be present, and their number differ from the number of the unspecified variables.

The main advantage drawn from the usage of Process Simulators is to carry out the longer, repetitive and time-consuming everyday tasks of a process engineer. Specific advantages derived from a computer simulation of the plant are the following:

- Deliver a comprehensive report of material and energy streams.
- Investigate the formation and separation of by-products and impurities.
- Analyze how to reduce the environmental impact.
- Evaluate the behavior of the plant under changes in raw materials or products' demand.
- Enhance process safety and control.
- Optimize the economic performance of the plant.

Concerning Chemical Engineering education, the usefulness of computers and simulation programs is undoubtedly. In 1959, the College of Engineering of the University of Michigan recommended that all engineering students must be introduced to computing techniques and develop a proficiency in programming languages [1]. The use of flowsheet programs in the classroom was already introduced in 1974 as an innovative teaching technique [2]. Nowadays, as the process simulator is the key tool to develop the design of chemical processes, and the education of chemical engineers just culminates in the design course, teaching process simulators in chemical engineering courses is imperative. The introduction of this type of software must fit to the premise of Gaddy [2] that states that "*the student should never use a computer unless he knows exactly what the machine is doing.*"

12.2 HISTORICAL VIEW

The history of computer simulation started in 1946 with the announcement of the first general-purpose computer ENIAC (Electronic Numerical Integrator And Computer), which was conceived and designed in the University of Pennsylvania [3]. The first deployment of this computer was managed by the mathematician John von Neumann to model the process of nuclear detonation during the Manhattan Project. Initial attempts toward computer simulation of chemical engineering systems, were made in the early 1950's and were limited almost solely to analog computers [1]. In the late 1950's took place one of the most siginificant advances in digital simulation with the development of procedure-oriented languages such as FORTRAN. In 1964, the digital computer executive program PACER, one of the ancestor of today's process simulators, received its first practical test. PACER, an acronym for Process Assembly Case Evaluator Routine, was conceived in 1961 by Professor Paul T. Shannon while he was in the employ of Humble Oil and Refining Company. It was written in FORTRAN for an IBM 7090 computer and consists of 1000 FORTRAN statements [4]. The first case study was a 300 ton/day contact sulfuric acid plant. The sulfuric acid plant simulation involved the use of 42 equipment subroutines, which were interconnected by approximately 70 information streams. This initial undertaking involved the solution of 500 simultaneous equations with almost 1000 stream variables and 200 equipment parameters. A complete simulation run lasted 8 minutes of CPU time on the IBM 7040 computer [1]. In 1966, the next step was carried out by the company Simulation Science (located in Los Angeles, USA) that commercialized a computer program for simulating distillation columns. This program was the core of the flowsheeting package PROCESS. Three years later ChemShare (Houston, USA) released DESIGN, a flowsheeting program for gas and oil applications. In the following decade, the first process simulators based on the equation oriented approach appeared, as SpeedUp at Imperial College in London (UK) and TISFLO at DSM in The Netherlands. In 1976, US Department of Energy and MIT launched jointly the ASPEN project. In 1985 Aspen Tech released ASPEN PLUS. During late 80s PRO II was upgraded from PROCESS by Simulations Science, and major packages migrated to PCs. In the Early 90s all major packages underwent periodic upgrades (major revision per 3 years period). Furthermore, advanced applications as pinch analysis were released. In Mid 90s, major vendors convert the Graphical User Interface into a central part in the software development. During this period of time HYSIM became HYSYS. Process simulators became matured in late 90s (upgrades became minor revisions) and the development companies consolidated. In the early 2000s appeared the free-of-charge Cape-Open to Cape-Open (COCO) simulation environment. Table 12.1 summarizes the evolution of commercial process simulators.

TABLE 12.1
Process Simulation Packages Development

Name	Year	Developer	Type	Reference	Observations
PACER Process Assembly Case Evaluator Routine	1963	Paul T. Shannon, Dartmouth College	Sequential modular	[4] [1]	No commercial. Written in FORTRAN
PROCESS (precursor of PRO II*1)	1966	Simulation Science. Los Angeles USA PRO II: Invensys Inc.	Sequential modular	[5]	Steady-state simulator PRO II: nearly 200 pure component library
DESIGN (in 1975 was renamed DESIGN 2000 and in 1984 DESIGN II was created)	1969	University Computing Company and Chemshare, Houston, USA WinSim Inc. (since 1995), Houston, USA?	Sequential modular		For gas and oil applications
CHESS Chemical Engineering Simulation System (in 1983 becomes MicroCHESS and in 1985 is renamed CHEMCAD*2)	1969	University of Houston (Chemstations for CHEMCAD)	Sequential modular	[6]	Flowsheeting, sizing, costing
FLOWTRAN	1970	Monsanto, USA		[7]	
CHESSE Chemical Engineering Simulation System with Economics	1974	University of Missouri-Rolla		[2]	A modification of the CHESS system

(*Continued*)

TABLE 12.1 (*Continued*)
Process Simulation Packages Development

Name	Year	Developer	Type	Reference	Observations
FLOWPACK II		ICI Corporate laboratory and LINDE AG	Sequential modular (also simultaneous modular)	[8] [9]	Commercial flowsheeting program
DYSCO DYnanic Simulation and COntrol	1981	The Institute of Paper Chemistry	Dynamic modular simulation	[10]	Dynamic simulator
ASCEND	1970–1980	Carnegie Mellon University	Equation oriented	[11]	
ASPEN (ASPEN PLUS[*3] released in 1985)	1976	US Dept. of Energy MIT, USA Aspen Plus: Aspen Tech			All-purpose flowsheeting system, and specialized Packages
TISFLO	1970–1980	DSM. The Netherlands	Equation oriented		
FLOWSIM	Early 1980s	University of Connecticut	Equation oriented		
HYSIM HYSYS (Mids 90) (nowadays ASPEN-HYSYS[*4])		Hyprotech. Calgary-Canada In 2002 HyproTech merged with AspenTech.			Bi-directional transmission of information Steady state and dynamic simulator
ABACUS II			Equation oriented		Dynamic simulation of chemical processes
SpeedUp (Nowadays, Aspen Custom Modeler)	1986	Imperial College, London, UK	Equation oriented		Dynamic simulator Programmed in PASCAL language that produces a FORTRAN for execution
ProSimPlus[*5]	1989	ProSim, Labège, Toulouse, France	Sequential modular		Steady-state simulation CAPE-OPEN thermo and unit operation interfaces.

SuperPro Designer*6		Intelligen, Inc. Boston. USA.	Sequential modular		
Design II		WinSim, Houston, USA.			Flowsheeting, sizing
QUASILIN	Late 1980s	CAD Centre (UK)	Equation oriented	[12]	Allows both simulation and optimization
gPROMS*7		Imperial College, London, UK.	Equation oriented		**steady-state** and **dynamic** modelling within the same environment
BatchPro Designer		Intelligen, Scotch Plains. NJ, USA			Scheduling and design of batch processes
COCO*8 (Cape-Open to Cape-Open) simulator*		AmsterCHEM	Sequential modular		CAPE-OPEN flowsheeting environment

*1 http://iom.invensys.com/en/pages/simsci-esscor_processengsuite_proii.aspx
*2 http://www.chemstations.com/
*3 http://www.aspentech.com/products/aspenONE/
*4 http://www.aspentech.com/hysys/
*5 http://www.prosim.net/en/index.php
*6 http://intelligen.com/superpro_overview.html
*7 http://www.psenterprise.com/gproms.html
*8 http://www.cocosimulator.org/

12.3 CURRENT PROCESS SIMULATORS

The process simulation market has known severe transformations in the 1985–1995 decade. Relatively few systems have survived; they are CHEMCAD, Aspen Plus, Aspen-Hysys, PRO/II, ProSimPlus, SuperPro Designer and gPROMS. Nowadays, most of the current process simulators are developed following an object-oriented approach using languages like C++ or Java. This shift in paradigm, from procedural to object-oriented, has no doubt benefited and will continue to benefit the process engineering community immensely.

The chemical engineering community is progressing on interfacing process simulation software components through the CAPE-OPEN (CO) project [13]. The standard interfaces allow CAPE (Computer Aided Process Engineering) applications to interoperate. The CAPE-OPEN Laboratories Network (CO-LaN) [14] is the international organization for the management of CO standards, which define the rules to be implemented in flowsheeting software [15]. CO standard is open, multiplatform and available free of charge. It is described in a formal documentation covering areas such as unit operations, physical properties and numerical solvers, and enables components supplied by third parties to be used in "plug and play" mode in commercial process modelling software tools. The COCO (CAPE-OPEN to CAPE-OPEN) steady state flowsheeting suit is entirely based on this open standard. It is available free-of-charge for noncommercial use). As it is an open flowsheet ng environment, allows the user to add new unit operations or thermodynamic packages. COCO comprises four main components:

1. COFE (CAPE-OPEN Flowsheet Environment): the graphical user interface to process flowsheeting based on a sequential solution algorithm with automatic tear streams. COFE flowsheets can be used inside other simulators.
2. TEA (Thermodynamics for Engineering Applications): it is based on the code of the thermodynamic library of ChemSep LITE, which is a free equilibrium column simulator for distillation columns and liquid-liquid extractors [16], and includes a data bank of over 430 commonly chemicals. The package offers more than 100 property calculation methods with their analytical or numerical derivatives.
3. COUSCOUS: the unit-operations simple package is shipped with COCO. It consists of a set of unit-operations such as stream splitters/mixers, heat-exchangers, compressors, pumps and reactors.
4. CORN (CAPE-OPEN Reaction Numeric): a package enclosed in COCO to specify kinetic or equilibrium reactions.

12.4 ARCHITECTURE OF PROCESS SIMULATORS

12.4.1 Classification of Process Simulation Packages

Any process model belongs to one of the eight classes defined from three major groups: discrete or continuous, dynamic or steady state, and deterministic or stochastic [10]. Discrete processes are those that operate on individual or separate items. In continuous processes the unknowns can take a continuous range of values. For steady state processes the inputs, outputs and processing conditions do not change with time. Although real processes do not run under steady state conditions, most are designed under this assumption and then control systems are used to keep the process at steady state. On the contrary, dynamic processes are those that vary with time. All real processes are dynamic. Modeling of dynamic processes involves the solution of differential equations. In the last major group, deterministic models assume that for a fixed input the output is known without uncertainty. Stochastic, or random, processes are more like the real world. Given a fixed set of inputs and process variables, the output is a random variable. The output of a stochastic simulation must offer information on the mean and variance, and higher moments, of the process variables. The majority of Process simulators available are designed to work with deterministic, continuous and either steady state or dynamic processes.

Another way to classify process simulators is according how the process is described (modular vs nonmodular) and the method used to solve the resulting equations (sequential vs simultaneous). According to this, three types of process simulators arise: sequential modular, equation oriented (or simultaneous nonmodular) and simultaneous modular. The sequential modular approach is based on the concept of modularity, which extends the chemical engineering concept of unit operation to a "unit calculation" of the computer code (i.e. subroutine) responsible for the calculations of an equipment [4]. This method is similar in principle to the traditional method of hand calculation of unit operations. The equations for each equipment unit are grouped together in a subroutine or module. Thus, each module calculates the output streams for the given input streams and parameters for that equipment, irrespective of the source of input information or the sink of output information. In the equation-oriented type (Chapters 10 and 11), the complete model of the plant is expressed in the form of one large sparse system of nonlinear algebraic equations that is simultaneously solved for all the unknowns. The simultaneous modular approach combines the modularizing of the equations related to specific equipment with the efficient solution algorithms for the simultaneous equation solving technique. For each unit an additional module is written, which approximately relates each output value by a linear combination of all input values. Accordingly, rigorous models are used at units' level, which are solved sequentially, while linear models are used at flowsheet level, solved globally. The linear models are updated based on results obtained with rigorous models.

12.4.2 Sequential Modular Approach

The simulation of a steady state operation of a chemical plant can be represented by a system of nonlinear algebraic equations,

$$\mathbf{f}(\mathbf{x}) = \mathbf{0} \tag{12.1}$$

where $\mathbf{f}$ is a vector of functions and $\mathbf{x}$ is the vector of variables describing the input and output streams of a particular unit, and the design parameters of that unit (for example: reactor volume, reflux ratio for distillation column or area of heat exchanger).

This vector of functions **f** is derived from the mass, momentum and energy laws of conservation, kinetic laws for reactions systems, and phase and chemical equilibrium. The system of Eqs. (12.1) has a strong nonlinear character, particularly due to the relationship among physical properties and state variables. It is remarkable that the physical properties estimation may consume up to 90% from the computation time [17].

In the simulation mode, the system of Eqs. (12.1) may contain several thousand equations and variables. As it may be very difficult to solve a unique model that includes the entire plant, the solution procedure demands a systematic and modular approach. The strategy adopted by sequential modular process simulator is to write balance equations separately for each unit. This leads to rewrite the system of equations (12.1) in the following form:

$$\mathbf{f}_i(\mathbf{x}) = \mathbf{0}, \quad i = 1,2,\ldots,n_u \tag{12.2}$$

where $\mathbf{f}_i$ is a subset of vector **f** with the functions associated with unit i and n_u, is the number of units in the plant.

To solve Eq. (12.2) the sequential modular approach are by far the most widely used in process simulators. The modular strategy relies on the following assumptions:

1. The design variables for the units should always be defined, i.e. they are not unknowns. The same assumption is made for variables associated with streams entering the plant from the outside (feed streams).
2. The information flow in the mathematical model coincides with the material flow in the process (the process simulator AspenHysys is a remarkable exception). This assumption allows us to gather the variables associated with the streams entering the unit and define the functions that calculate the variables for the output streams.

These two assumptions lead to rewrite the equations (12.3) as follows:

$$\mathbf{x}^{out} = \mathbf{g}(\mathbf{x}^{in}), \quad i = 1,2,\ldots,n_u \tag{12.3}$$

where $\mathbf{x}_i^{out}$ and $\mathbf{x}_i^{in}$ are subsets of the variables associated with the output and the input streams of unit i, respectively. The output stream for a unit i becomes the input stream for a downstream unit and the calculation proceeds as before until all the units are visited. Thus, an entire flowsheet without recycle streams makes possible to compute it by using sequentially Eq. (12.3) (i.e., it is required only one flowsheet iteration to attain a converged solution). Nevertheless, the existence of recycle streams hinders a sequential calculation of the flowsheet, and then an iterative strategy should be adopted. The solution methodology for flowsheets containing recycle streams is more complicated and includes a topological analysis to determine the computational sequence and the corresponding tear streams. This strategy also requires iterative calculations for all units in the recycle loop, as the calculated value for the recycle may differ from the initial estimation. Moreover, the techniques to handle recycles in a sequential modular simulator need to set a convergence criterion and a method to update the estimated values of the tear streams for each iteration. The most common techniques for recycle loop convergence are direct substitution, Wegstein, Broyden and Newton-Raphson techniques. These are discussed latter. Commercial process simulators are able to identify the recycle loops and automatically select the streams to tear and the convergence method. Many simulation packages also allow the user to select the tearing of streams and method of convergence.

12.4.2.1 Advantages and Disadvantages of the Sequential Modular Approach

Sequential modular approach has some clear advantages for process flowsheeting that explain why it still dominates the technology of steady state simulation over the simultaneous or equation oriented approach. Table 12.2 shows a list of pros and cons about sequential modular process simulators. In order to cope with the disadvantages, a few process simulators have improved the flow of information and avoid redundant computations. As an example, Aspen-Hysys has implemented the bi-directional transmission of information technology.

12.4.3 The Structure of a Process Simulator

Figure 12.2 shows the interconnections among the basic components of a simulation package, which are concisely described below:

1. Component databank: a database with the required parameters to calculate the physical properties needed.
2. Thermodynamic property prediction methods: a set of thermodynamic methods to estimate the physical and thermodynamic property data.
3. Flowsheet Builder (Graphical User Interface): provides the user an interface to generate the flowsheet of the plant under a graphical environment.

TABLE 12.2
Advantages and Disadvantages of Sequential Modular Approach

Advantages	Disadvantages
1. The flowsheet architecture is most easily understood as it closely follows the process flow diagram.	1. They are particularly suited for process simulation not dynamic simulation, optimization or design since an iterative procedure is required to meet the constrains, increasing the CPU time
2. The unit blocks can be easily added to or removed from the flowsheet.	2. Present difficulties with flowsheets involving a number of recycles.
3. The models of the different units can be prepared and tested separately (as a computer program subroutine) leading to a library of unit modules.	
4. Specific solution methods, including the initialization, are developed for each process unit.	
5. Because of rigid requirements, user input data can be quite easily checked for completeness and consistency.	
6. Easy control of convergence, both at the units and flowsheet level.	

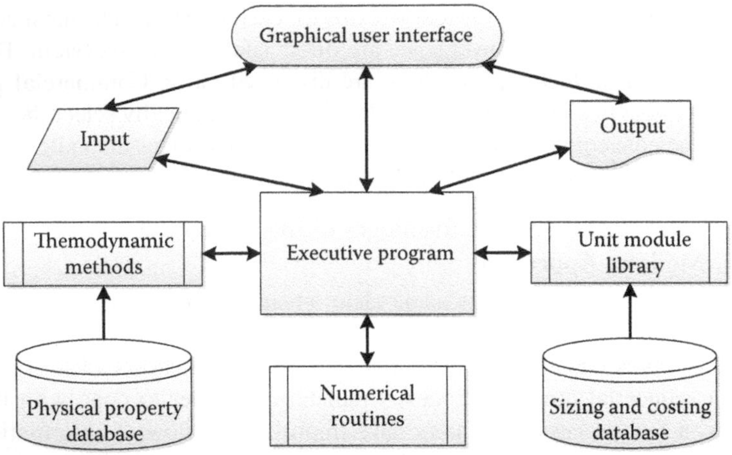

FIGURE 12.2 Components of process simulator software.

4. Unit module library: subroutines to perform energy and material balances, and design calculations for the typical process engineering units.
5. Numerical routines: a collection of mathematical methods for solving systems of linear, nonlinear and differential equations.
6. Data output generator: report the results of the simulation by tables and graphical displays.
7. Executive program (flowsheet solver): is the heart of any process simulator and controls the sequence of the calculations and the overall convergence of the simulation.

12.4.4 Component Data Bank

An essential part of a simulation package is the component data bank, which commonly contains more than a thousand chemical compounds. For chemicals not available in the databank, process simulators provide a user-added component facility to include the required compounds in the simulation. A short overview of the data needed in the simulation and design of processes is given in Table 12.3. The properties required for the design of a chemical process depends upon the temperature, pressure and concentration. The most crucial and least available physical property data are for phase equilibrium. The key parameters for the design of separation units that are based on equilibrium stages are called binary interaction parameters (BIPs).

12.4.5 Phase Equilibrium Calculaton

As many chemical process simulations include distillation, stripping or evaporation, solving the heat and material balances for these unit operations necessitates predicting the separation of chemical mixtures between the liquid and vapor phases, namely the vapor/liquid equilibrium (VLE). Liquid/liquid equilibrium (LLE) also

TABLE 12.3
Types of Property Data and Their Corresponding Specific Properties

Property type	Specific properties
Phase equilibrium	Boiling and melting points, vapor pressure, fugacity and activity coefficients, solubility (Henry's constants, Ostwald or Bunsen coefficients), binary interaction parameters
PVT behavior	Density, molar volume, compressibility, critical properties, acentric factor
Thermal properties	Heat capacity, latent heat, ionic conductivity, enthalpy, entropy
Transport properties	Viscosity, thermal conductivity, diffusion coefficients
Chemical reaction equilibrium	Equilibrium constants, association/dissociation constants, enthalpy of formation, enthalpy of combustion, heat of reaction, Gibbs free energy of formation, reaction rates
Boundary properties	Surface tension
Molecular properties	Virial coefficients, ion radius and volume, molecular weight and dipole moment
Safety characteristics	Flash point, explosion limits, toxicity, maximum working place concentration, lower and upper flammability limits

plays a key role in solvent extraction and extractive distillation [18]. To foretell phase equilibrium, a thermodynamic property method should be chosen, which often turns into a crucial decision in the simulation. Current process simulators offer a model selection wizard to guide the user to the proper method. The decision becomes complex since different thermodynamic packages can be used for different parts of the flowsheet.

In the calculation of phase equilibrium, the basic relationship for every component i in the vapor and liquid phases of a system at equilibrium is:

$$f_i^v = f_i^l \tag{12.4}$$

where f_i^v and f_i^l are the fugacity of component i in the vapor and liquid phase, respectively. There are two methods for calculating the fugacities from the phase equilibrium relationship in terms of measurable state variables, the equation-of-state (EOS) method and the activity coefficient method.

12.4.5.1 Equation of State Method

An equation of state predicts the behavior of a fluid through a mathematical relationship for pressure, volume and temperature:

$$f(P,V,T,a_k,k=1,\ldots,n_p)=0 \tag{12.5}$$

where a_k is a parameter of the EOS and n_p is the number of parameters, which determine the complexity of the model. Although the EOSs have some theoretical basis, many of them are semi-empirical because the parameters a_k must be adjusted. Most EOSs have different terms to represent attractive and repulsive forces between molecules. Any thermodynamic property, such as fugacity coefficients and enthalpies, can be calculated from the equation of state. Equation-of-state properties are calculated relative to the ideal gas properties of the same mixture at the same conditions.

In the EOS method, fugacities are calculated as follows:

$$f_i^v = \varphi_i^v y_i P \tag{12.6}$$

$$f_i^v = \varphi_i^v y_i P \tag{12.7}$$

where P is the total pressure of the system; y_i and x_i are the molar fraction of component i in the vapor and liquid phase, respectively; and φ_i^v and φ_i^l are the fugacity coefficients for the vapor and liquid phase, respectively, which are computed with the following expressions:

$$\ln \varphi_i^v = -\frac{1}{RT}\int_{\infty}^{V^v}\left[\left(\frac{\partial P}{\partial n_i}\right)_{T,V,n_{j\neq i}} - \frac{RT}{V}\right]dV - \ln Z^v \tag{12.8}$$

$$\ln \varphi_i^l = -\frac{1}{RT}\int_{\infty}^{V^l}\left[\left(\frac{\partial P}{\partial n_i}\right)_{T,V,n_{j\neq i}} - \frac{RT}{V}\right]dV - \ln Z^l \tag{12.9}$$

In Eqs. (12.8) and (12.9) Z^v and Z^l stand for the compressibility factors in the vapor and liquid phase, respectively, and are given by:

$$Z^v = \frac{PV^v}{RT} \tag{12.10}$$

$$Z^l = \frac{PV^l}{RT} \tag{12.11}$$

For a vapor at moderate pressure, fugacity coefficient φ_i^v is close to unity. By contrast, the behavior of a liquid differs from an ideal gas much more than a real gas differs from an ideal gas, and therefore fugacity coefficient for a liquid are very different from unity.

EOS models are derived from PVT relationships. Many are modifications of $PV = nRT$ based on pure component parameters, such as the normal boiling point (T_b), critical temperature (T_c), critical pressure (P_c), and acentric factor (ω). The default EOS is normally either the Soave-Redlich-Kwong (SRK) or the Peng-Robinson (PR). Both are cubic EOS and hence derivations of the van der Waals EOS, and, like most equations of state, they use three pure component parameters per substance and one binary-interaction parameter per binary pair. There are other more complex EOS (see Table 12.4). EOS models are appropriate for modeling ideal and real gases (even in the supercritical region), hydrocarbon mixtures and light-gas mixtures. However, they are less reliable when the sizes of the mixture components are significantly different, or when the mixture comprises nonideal liquids, especially polar mixtures.

TABLE 12.4
Thermodynamic Property Models Available in Commercial Process Simulators

Model Category		Model Name	Phase(s)	Guidelines
Equation of state (EOS)	**Ideal**	Ideal gas law	V	Petroleum pseudo-components Similar hydrocarbons Light gases.
		Nothnagel	V	For systems that exhibit strong vapor phase association.
	Virial equations of state	Hayden-O'Connell	V	
		Benedict-Webb-Rubin(BWR)-Lee-Starling	V,L	
		Lee-Kesler (LK)	V,L	
		Lee-Kesler-Plöcker	V,L	
	Cubic equations of state	Redlich-Kwong (RK)	V	
		Soave-Redlich-Kwong (SRK)	V,L	For polar systems, SRK makes a better prediction than PR.
		Peng-Robinson (PR)	V,L	PR obtains better liquid densities than SRK.
		Peng-Robinson-Wong-Sandler	V,L	
		Redlich-Kwong-Soave (RKS)	V,L	Gas processing, refinery, and petrochemical processes.
		Lee-Kesler	V,L	Good description of the thermodynamic properties of mixtures containing nonpolar and slightly polar components.
		Lee-Kesler-Plöcker	V,L	Hydrocarbon systems that include the common light gases, such as H_2S and CO_2. It can be used in gas processing, refinery, and petrochemical applications.
		Schwartzentruber-Renon	V,L	
	Mixing rules	PSRK mixing rules	V,L	
		Wong-Sandler (WS) mixing rules	V,L	
		Modified Huron-Vidal mixing rules (MHV2)	V,L	
	Steam	ASME Steam Tables	V,L	For water or steam (no parameter requirements).

(Continued)

TABLE 12.4 (*Continued*)
Thermodynamic Property Models Available in Commercial Process Simulators

Model Category		Model Name	Phase(s)	Guidelines
Activity coefficient Models	**Binary Interaction Parameter (BIP)**	NRTL (Non-Random-Two-Liquid)	L,L1-L2	Recommended for highly non-ideal chemical systems. It can be used for VLE and LLE applications
		Wilson	L	It is recommended for highly nonideal systems, especially alcohol-water systems. It cannot be used for liquid-liquid equilibrium calculations.
		van Laar	L	It describes non-ideal liquid solutions with positive deviations from Raoult's law.
		Scatchard-Hildebrand	L	
		UNIQUAC	L,L1-L2	For highly non-ideal chemical systems VLE and LLE applications.
	Group Contribution (predictive)	UNIFAC (UNIQUAC Functional-group Activity Coefficients)	L,L1-L2	
		Predictive SRK (PSRK)	V,L	For mixtures of nonpolar and polar compounds, in combination with light gases.
	Electrolyte Models	Electrolyte NRTL	L,L1-L2	Aqueous Electrolyte.
		Pitzer	L	Not suitable for a non-aqueous solvent.
		Bromley-Pitzer	L	
Special models	**Vapor pressure models**	API Sour	V	For correlating NH_3, CO_2, and H_2S volatilities from aqueous sour water system (from 20 to 140°C).
		Braun K10	V	For heavy hydrocarbons at low pressures (it can be used for pseudo components).
	Liquid fugacity model	Chao-Seader	L	Pure component fugacity coefficients for liquids.
		Grayson-Streed	L	
		Kent-Eisenberg	L	For liquid mixture component fugacity coefficients (amines).

12.4.5.2 Activity Coefficient Method

This method uses the same expression, Eq. (12.6), for the fugacity of component i in the gas phase, but the fugacity of component i in the liquid phase is related to the composition in that phase (x_i) by the activity coefficient γ_i:

$$f_i^l = \gamma_i x_i f_i^0 \tag{12.12}$$

where f_i^0 the standard-state fugacity of component i, which is commonly considered to be equal as the fugacity of pure component i at the system temperature and pressure. Here the activity is used to fit Gibbs excess energy deviation of mixtures. The calculation of the activity coefficient as function of temperature and composition of the mixture requires many binary interaction parameters (BIPs), which are highly temperature dependent.

The following points are recommendations if a liquid-state activity coefficient model is selected:

1. Check if BIPs for all the binary pairs in the system are available in the simulator databank (the most common source is DETHERM thermophysical database [14]). For the three most common models (Wilson, NRTL, UNIQUAC) the values of BIPs are different, and they are not correlated from one model to another.
2. For missing BIPs with available phase equilibrium data, a regression can be done with the simulator to find their values.
3. For binary pairs without measured phase equilibrium use the predictive UNIFAC [19] property method to estimate the BIPs. There are two UNIFAC methods: one for VLE and another for LLE. A predictive model requires that group interaction parameters be available for the various subgroups of a chemical's structure.
4. The Wilson model cannot describe liquid-liquid separation.
5. To estimate the uncertainty of the property method prediction, it is recommended to compare the same flowsheet while using the different property methods that fit better the experimental data.

The activity coefficient model is a reasonable choice if the conditions are far from the critical region of the mixture and if experimental data are available for the phase equilibrium (VLE or LLE).

Table 12.4 gathers the property models commonly available in commercial process simulators. These four factors should be considered for choosing the right property method:

1. Type of the thermodynamic property of interest
2. Composition of the mixture
3. Range of pressures and temperatures
4. Availability and reliability of the parameters.

Figure 12.3 shows a decision tree to help in the choice for the thermodynamic property model. Besides the four factors mentioned above, this decision tree takes

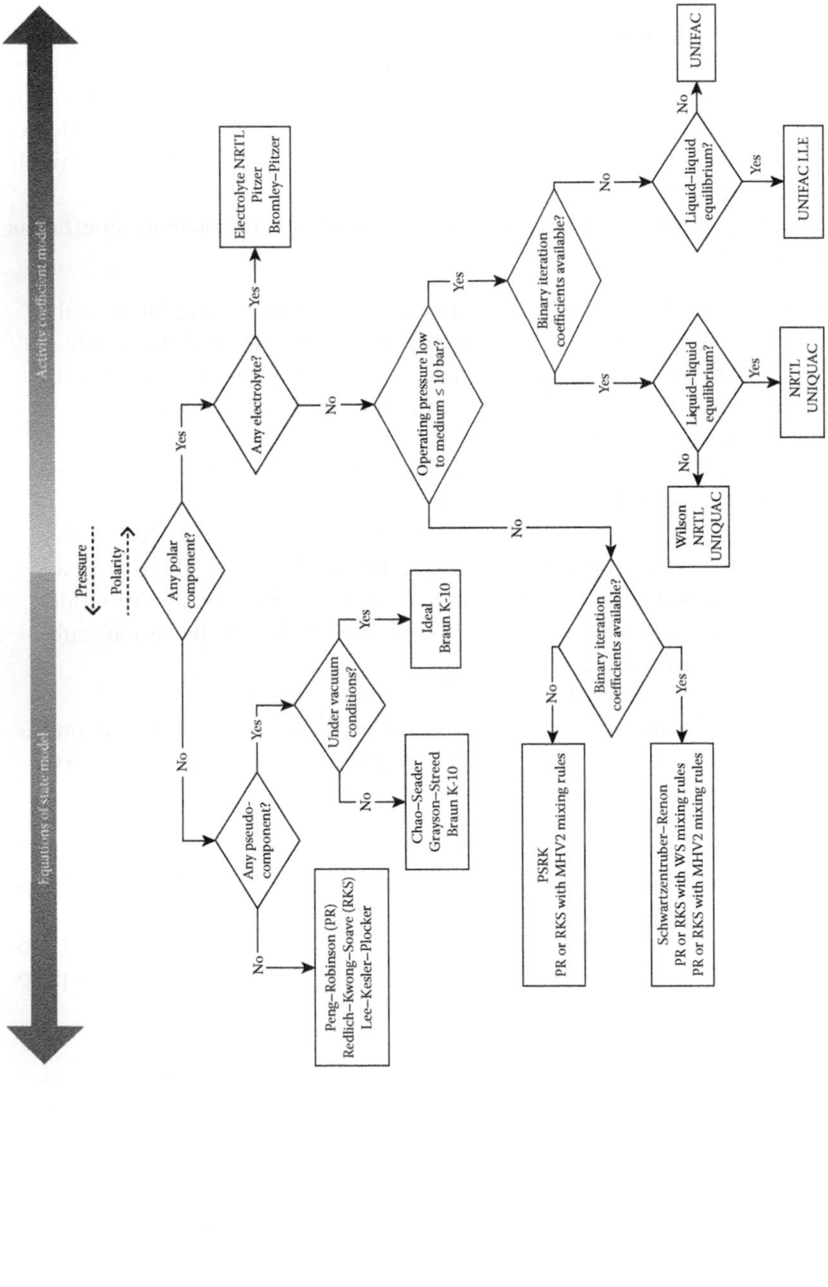

FIGURE 12.3 Decision tree for the selection of the thermodynamic property model.

into account the polarity of the mixture. Another feature of the mixture considered is the existence of pseudocomponents and the possibility of some of the components being electrolyte. The most common electrolyte methods are the Pitzer model, and the electrolyte NRTL.

Unfortunately, it does not exist a perfect thermodynamic property model, neither EOS nor activity coefficient. Despite both are developed from theoretical principles, they contain parameters that must be fitted, and hence they turn into semi-empiric models.

12.5 DEGREES OF FREEDOM IN A FLOWSHEET

A difficult problem in flowsheeting is to specify the problem so that the process simulator solves the desired problem. Tied up with this problem is one of correctly identifying the number of degrees of freedom and finding a legitimate means to satisfy them. We will first give the most basic view of the number of degrees of freedom, and then establish how many must exist for a process stream. Next we will consider the same problem for a given unit operation and show a simplified approach taking advantage of the previous result on process streams. Finally, we extend the approach to deal with complete process flowsheets.

12.5.1 Degrees of Freedom

In steady state simulation the degrees of freedom are the number of variables that must be assigned to solve the (non)linear algebraic system describing the operational unit. Even though in modular simulators these units appear as black boxes, the user should have at least a general knowledge of that unit to correctly set the design variables. If our problem is one with n independent equations in m ($m > n$) variables, then the number of degrees of freedom is given as:

$$d = m - n \tag{12.13}$$

This expression is based on the idea that n well-behaved independent equations can be used to solve for exactly n variables. Thus m-n of the variables must receive their values from another means before the problem can be solved. While the principle behind equation (12.13) is straightforward, the application of it is far from simple. In general it is not an easy task to write the set of equations which model a unit operation even if the theory underlying the equations is not a problem [20]. It is not uncommon to write too many equations not realizing that one or more of them can be derived algebraically from the others. For a flowsheet the problem can be very difficult and one's intuition or logic may easily fail.

12.5.2 Independent Stream Variables

What is the minimum number of variables to specify fully a stream? A stream can be defined as the flow of material between two units in a flowsheet. The variables normally associated with a stream are its temperature, pressure, total flow, overall mole fractions, phase fractions and phase mole fractions, total enthalpy, phase enthalpies, entropy, etc. Assuming phase and chemical equilibrium, how many of those variables must be specified to fix completely the stream? Without further considerations, for this case intuition gives us the correct answer. We know without writing equations that if we specify temperature, pressure and individual component flows the stream is fully specified. Of course, a priori we cannot know the final state of the stream (i.e. multiphase or single phase; liquid, vapor, solid or a mixture of them). If we are interested in a stream with some specific conditions like saturated liquid we cannot specify simultaneously pressure and temperature but pressure (or temperature) and phase fraction. A convention in process simulators is that when vapor (liquid) phase fraction is specified to zero or one, saturated conditions are assumed (bubble point or dew point). However, when vapor or liquid phase fractions are calculated a value of one (zero) does not mean saturated conditions but that the stream is in vapor (liquid) phase.

The question now arises as to whether we can prove our intuitive response. Duhem's theorem [21] provides us with a needed result. *Duhem's theorem* states:

> *Whatever the number of phases, of components or of chemical reactions, the equilibrium state of a closed system, for which we know the total initial masses of each component, is completely determined by two independent variables*

Again a question arises. Can we classify a stream as a closed system? According to Westerberg et al. [20] we can do it by taking a fixed amount of it and trapping it into a container. Because we know the composition and total flow of the stream, we know the total initial masses we have trapped. The theorem says that we only need fix two additional independent variables (i.e. pressure and temperature) to fix the equilibrium state of the system. A detailed proof of the theorem can be found in [21].

12.5.3 Degrees of Freedom in a Unit

As commented above, the degrees of freedom in a unit operation can be calculated by counting the independent equations and variables that fully define the unit. However, we can take advantage of the previous result on streams and simplify the procedure. Let us first use the classical approach for a rather simple unit: a mixer (Figure 12.4).

We have $c + 8$ equations and $3c + 12$ variables ($x_{i,1}, x_{i,2}, x_{i,3}\,[i = 1 \ldots c]$, F_1, F_2, F_3, P_1, P_2, P_3, T_1, T_2, T_3, H_1, H_2, H_3) the number of degrees of freedom is therefore:

$$d = 3c + 12 - (c + 8) = 2c + 4$$

This number corresponds with our intuition, which says if the feed streams are fully specified, the outlet is fully specified. However, in the previous approach we wrote equations that strictly speaking belong to the streams, and not to the unit operation (mole fraction sum and physical properties calculation; compositions and enthalpies) these equations and variables cancel each other. It is convenient to take into account only the relations that define the unit operation and not those that belong

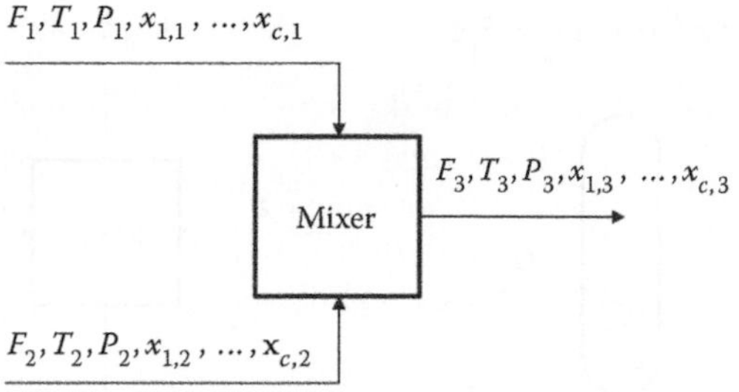

Material balance for each component:

$$x_{i,1}F_1 + x_{i,2}F_2 = x_{i,3}F_3 \quad i = 1,2, \ldots, c$$

Heat balance:

$$H_1F_1 + H_2F_2 = H_3F_3$$

Pressure balance:

$$P_3 = \min(P_1, P_2)$$

Mole fraction sum:

$$\text{å}_i \; x_{i,1} = 1; \quad \text{å}_i \; x_{i,2} = 1; \quad \text{å}_i \; x_{i,3} = 1;$$

Physical properties calculation:

$$H_1 = f_H(T_1, P_1, x_{1,1}, \ldots, x_{c,1})$$
$$H_2 = f_H(T_2, P_2, x_{1,2}, \ldots, x_{c,2})$$
$$H_3 = f_H(T_3, P_3, x_{1,3}, \ldots, x_{c,3})$$

FIGURE 12.4 A simple mixer unit.

to the stream. The following basic block (Figure 12.5) represents the mixer. The number inside the block corresponds to the net number of relationships among the fewest number of stream variables needed.

One additional observation is that we are not at liberty to choose any combination of $2c + 4$ stream variables to be the decision variables. It is obviously permissible to choose the $2c + 4$ variables for the input streams. We could also choose to specify as decision variables $c + 2$ variables for the input and stream and $c + 1$ variables, excluding the pressure of the outlet stream. The pressure balance equation as written above cannot be reversed in general, since if P_1 is specified, and $P_1 = P_3$, we can say nothing

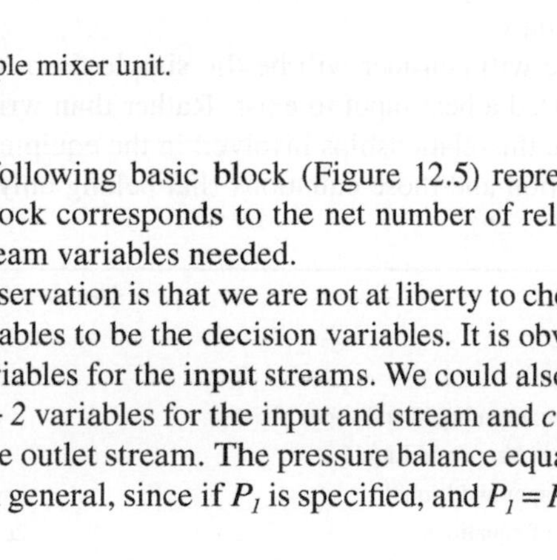

$$d.o.f = 3(c + 2) - (c + 2) = 2c + 4$$

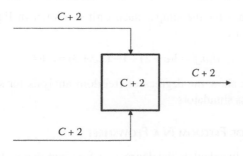

FIGURE 12.5 Basic block diagram for the mixer unit.

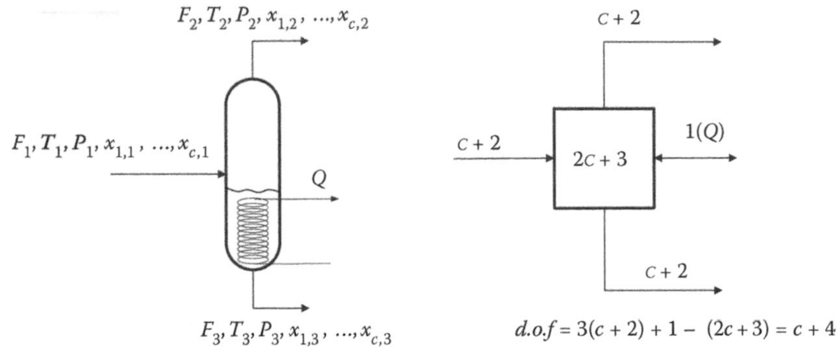

FIGURE 12.6 Simple liquid-vapor flash unit and its block diagram for degrees of freedom analysis.

about P_2 except that it is greater than or equal to P_1. We also cannot specify the total flows of all three streams. These flows are related by the overall mass balance. *We can choose a set of 2c + 4 stream variables to be decisions for our mixer only if we can then produce a solution procedure for the c + 2 model equations to solve for the remaining c + 2 variables.*

The second unit we will consider will be the simple flash unit in Figure 12.6 in which we have permitted a heat input to exist. Rather than write the flash equation we just will enumerate the relationships involved in the equipment but not the physical properties estimation and those equations that belong only to the streams. The relationships are:

Relation	Number of equations
Mass balance:	c
Energy Balance	1
Equilibrium between liquid and vapor phases	c
Temperatures in Vapor and Liquid	1
Pressures in Vapor and Liquid	1
Total number of equations	$2c + 3$

The block diagram for the simple flash unit is shown in Figure 12.6 then the degrees of freedom are:

$$d.o.f = 3(c+2) + 1 - (2c+3) = c + 4$$

In Table 12.5 we show the degrees of freedom analysis for some common unit operations in process simulators

12.5.4 Degrees of Freedom in a Flowsheet

Once we know how to calculate the degrees of freedom in a material stream and in a unit operation, to calculate the degrees of freedom in a flowsheet is straightforward, we just have to add the degrees of freedom of each unit and subtract those streams that have been counted twice by connecting two different units. If there

TABLE 12.5
Degrees of Freedom of Some Common Unit Operations

Unit	Unit Relationships	Block	Degrees of Freedom	Typical Valid Assignment
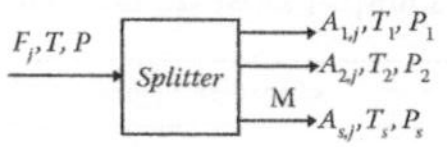	$A_{i,j} = \alpha_i F_j$ *sc* $\sum_i \alpha_i = 1$ 1 $T_i = T$ *s* $P_i = P$ *s* *Total* $sc + 2s + 1$	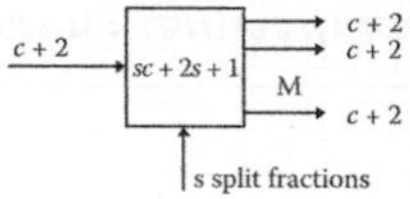	$(c + 2)(s + 1) + s -$ $[sc + 2s + 1] =$ $s + c + 1$	*Feed * * + (s − 1) Split fractions*
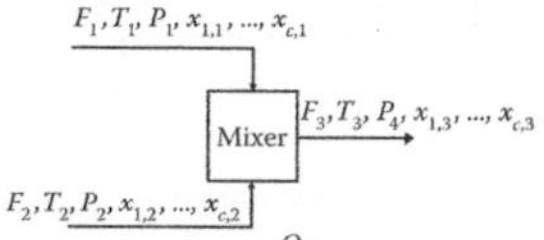	*Mass balance* c *Energy balance* 1 *Pressure relations* 1 *Total* $c + 2$	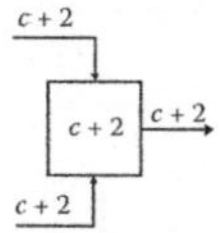	$3(c + 2) - [c + 2] =$ **2c + 4**	*Feeds*
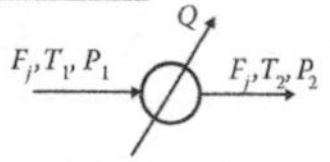 Heater/cooler	*Mass balance* c *Energy balance* 1 *Total* $c + 1$	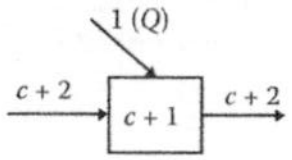	$2(c + 2) + 1 -$ $[c + 1] =$ **c + 4**	*Feed + ΔP + Q or Tout*
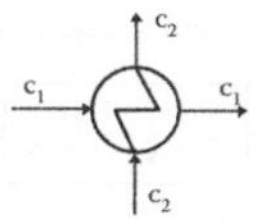 Adiabatic heat exchanger	*Mass balance hot fluid* c_1 *Mass balance cold fluid* c_2 *Energy balance hot fluid* 1 *Energy balance cold fluid* 1 *Design* ($Q = UA\Delta T_{LM}$) 1 *Total* $c_1 + c_2 + 3$	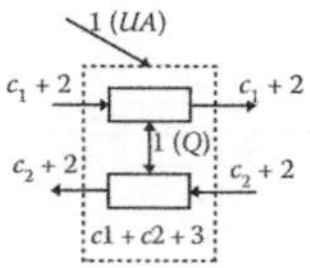	$2(c1 + 2) + 2(c2 + 2)$ $+ 2 - [c1 + c2 + c3] =$ **c1 + c2 + 7***	*Feeds +* $\Delta P_1 + \Delta P_2 +$ *Q or UA*
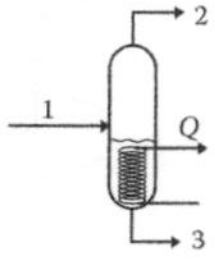 Flash	*Mass balance* c *Energy balance* 1 *Equilibrium* c $P_2 = P_3$ 1 $T_2 = T_3$ 1 *Total* $2c + 3$	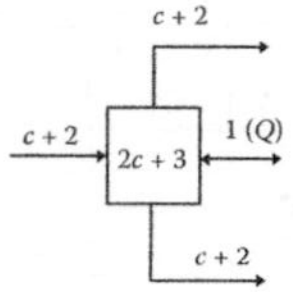	$3(c + 2) + 1 -$ $[2c + 3] =$ **c + 4**	*Feed + P + Q or Vapor fraction*
 Valve	*Mass balance* c *Isoenthalpic* ($H_1 = H_2$) 1 *Total* $c + 1$	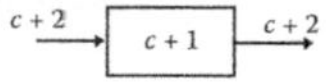	$2(c + 2) - [c + 1] =$ $c + 3$	*Feed +* ΔP *or* P_2

(Continued)

TABLE 12.5 (*Continued*)
Degrees of Freedom of Some Common Unit Operations

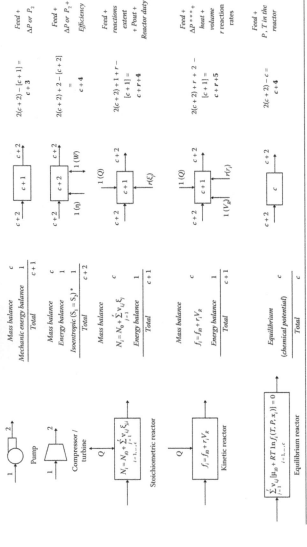

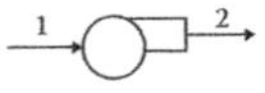 Pump	*Mass balance* c *Mechanic energy balance* 1 *Total* $c+1$	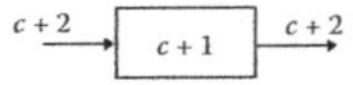	$2(c+2) - [c+1] =$ $\mathbf{c+3}$	*Feed* + ΔP *or* P_2
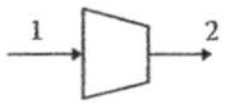 Compressor / turbine	*Mass balance* c *Energy balance* 1 *Isoentropic* $(S_1 = S_2)$* 1 *Total* $c+2$	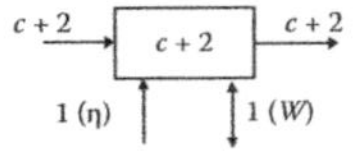	$2(c+2) + 2 - [c+2]$ = $\mathbf{c+4}$	*Feed* + ΔP *or* P_2 + *Efficiency*
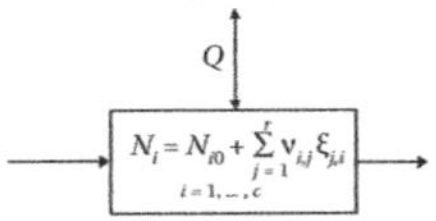 $N_i = N_{i0} + \sum_{j=1}^{r} \nu_{i,j}\xi_{j,i}$, $i = 1,\ldots,c$ Stoichiometric reactor	*Mass balance* $N_i = N_0 + \sum_{j=1}^{r} \nu_{i,j}\xi_j$ c *Energy balance* 1 *Total* $c+1$	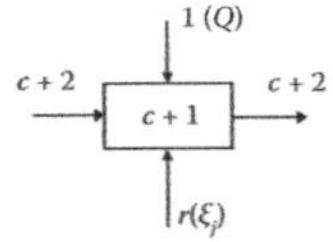	$2(c+2) + 1 + r -$ $[c+1] =$ $\mathbf{c+r+4}$	*Feed* + *reactions extent* + *Pout* + *Reactor duty*
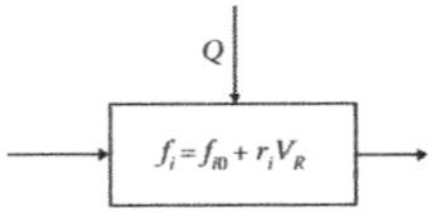 $f_i = f_{i0} + r_i V_R$ Kinetic reactor	*Mass balance* $f_i = f_{i0} + r_i V_R$ c *Energy balance* 1 *Total* $c+1$	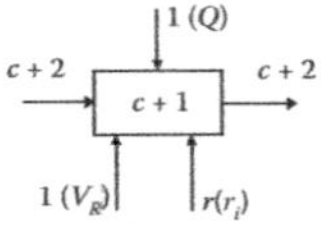	$2(c+2) + r + 2 -$ $[c+1] =$ $\mathbf{c+r+5}$	*Feed* + ΔP*** + *heat* + *volume* *r* reaction rates
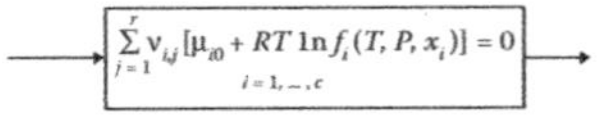 $\sum_{j=1}^{r} \nu_{i,j}[\mu_{i0} + RT \ln f_i(T, P, x_i)] = 0$, $i = 1,\ldots,c$ Equilibrium reactor	*Equilibrium (chemical potential)* c *Total* c	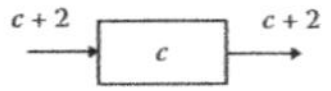	$2(c+2) - c =$ $\mathbf{c+4}$	*Feed* + *P, T in the reactor*

* Actual compressors turbines are not really isentropic. Efficiency $\eta = (\Delta H)s/(\Delta H)$ *actual* is usually assumed.
** Feed refers to the complete feed specification (component molar flows plus P and T = c + 2 specifications)
*** In a kinetic reactor is possible specify the P in the reactor (i.e. CSTR) or a relation for pressure drop along the reactor (typical in PFR).

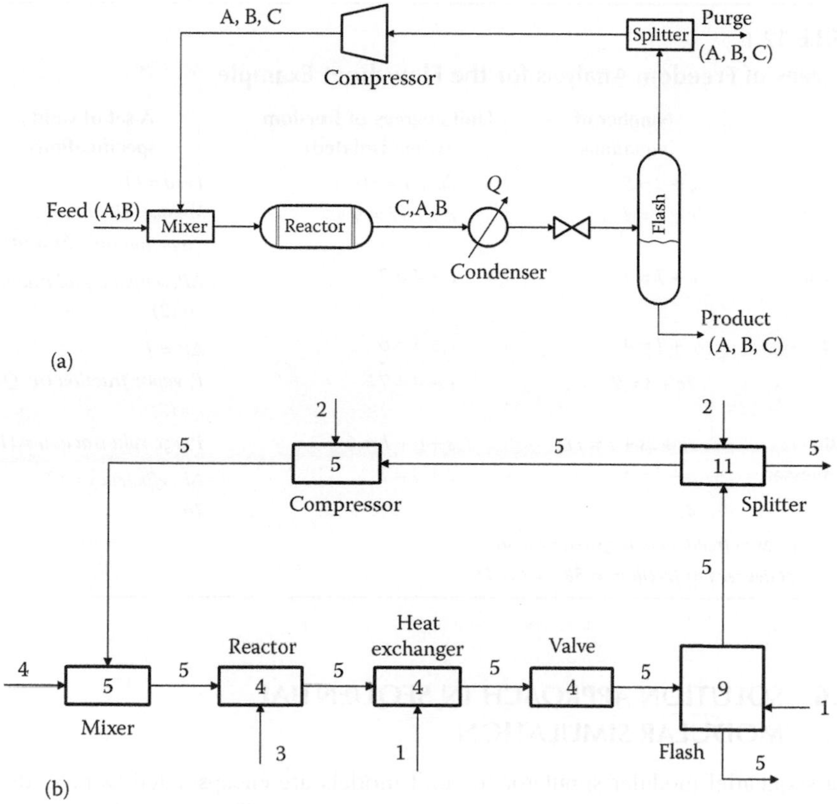

FIGURE 12.7 Flowsheet example (a) and basic block diagram for the flowsheet (b).

are any other relationship (i.e by a design specification in ASPEN a ADJUST in HYSYS or a CONTROLLER in CHEMCAD), we must subtract also these extra specifications. Alternatively, we can simply count the degrees of freedom of each stream plus the number of parameters in each unit and subtract the equations of the unit.

We will consider this problem by analyzing the simple flowsheet in Figure 12.7a [20]. The flowsheet has a high pressure feed stream of gaseous component A contaminated with a small amount of B. It mixes first with a recycle stream consisting mostly of A, and passes into a kinetic reactor where an exothermic reaction to form C from A takes place. The stream is cooled to condense component C and passes through a valve into a flash unit. Here most of the unreacted A and the contaminant B are flashed off, leaving a near pure C to be withdrawn as the liquid stream. The vapor from the flash is recycled previous compression and a purge is withdrawn to avoid accumulation of impurity B.

Figure 12.7b shows the blocks scheme of the flowsheet. Table 12.6 shows the degrees of freedom analysis for the flowsheet.

TABLE 12.6
Degrees of Freedom Analysis for the Flowsheet Example

Unit	Number of equations	Unit degrees of freedom (when isolated)	A set of valid specifications
Mixer	$c+2=5$	$2c+4=10$	*Feed* = *(4)*
Reactor	$c+1=4$	$c+r+5=9$	*Volume; Heat Flow, reaction rate,* ΔP= *(4)*
Condenser	$c+1=4$	$c+4=7$	ΔP, *output liquid fraction* = *(2)*
Valve	$c+1=4$	$c+3=6$	$\Delta P=1$
Flash	$2c+3=9$	$c+4=7$	*P, vapor fraction (or Q)* = *(2)*
Splitter	$sc+2s+1=11$	$s+c+1=6$	*Purge split fraction* = *(1)*
Compressor	$c+2=5$	$c+4=7$	ΔP, *efficiency* = *(2)*
Total	*42*		***16***

Total degrees of freedom in the streams = 58

Flowsheet degrees of freedom = 58 – 42 = 16

12.6 SOLUTION APPROACH IN SEQUENTIAL MODULAR SIMULATION

In a sequential modular simulator the unit models are encapsulated as procedures where the output streams are desired design parameters. These procedures are then solved in a sequence that generally parallels de flow of material of actual process. A remarkable exception is the process simulator HYSYS, for which most unit operations are solved as soon as all the degrees of freedom are satisfied independently of whether those specifications are introduced.

In an acyclic system the sequential modular approach is ideal from computational viewpoint. However, most of the process plants involve recycles, particularly those highly integrated, in that case the problem cannot be solved directly and an iterative process must be used. Consider, for example, the simple flowsheet in Figure 12.7. All the degrees of freedom are satisfied however, to calculate the mixer we need to completely know both, the feed stream and the recycle stream but we only know the feed. To calculate the recycle stream we should solve the compressor, to solve the compressor we need previously solve the splitter, for the splitter we need the flash, for the flash the valve, for the valve the condenser, for the condenser the reactor and for the reactor the mixer. But we cannot solve the mixer! To solve that problem we need to 'tear' a stream inside the loop (formed by mixer, reactor, condenser, valve, flash, splitter and compressor) and introduce a convergence block. Initial values must be provided for the output streams of the convergence block, then the complete flowsheet can be calculated and the inputs to the convergence block compared with its outputs, if the convergence criteria is met then the calculation stops in other case a suitable numerical method is used to provide new estimates for the tear stream and the iteration is repeated. Later we will discuss the numerical methods for convergence blocks in process simulators.

In a complex flowsheet with large number of units and recycles some questions arises. Which units operations must be solved simultaneously (we need the information provided by one of those units to calculate the other)? This is known as partitioning in the Chemical Engineering literature. In which order must we solve those groups? This is known as precedence ordering. Which streams must we tear in each group? This is known as tearing. In the following sections we discuss these subjects.

The literature contains a large number of articles related to partitioning and tearing flowsheets, most of these algorithms were developed in the last 1960s and 1970s. A sample includes [22–33].

12.6.1 Partitioning and Precedence Ordering

For the purposes of partitioning and tearing the information flowsheet may be represented by a directed graph (digraph). The chemical equipment represented by nodes (vertices) and the streams by directed edges. Here it is important to remark that the edges must represent the information flow between the different unit operations in a flowsheet. Usually the information flow and the mass flow coincide but this is not always the case.

Methods for partitioning have been developed based on path searching and Boolean operations performed on adjacency matrices. A detailed description of these methods may be found in [22, 26, 30, 34]. Path searching algorithms have proved to be more efficient from a computational point of view that algorithms based on Boolean operations over adjacency matrices. Among the path searching algorithms, the most efficient is the Tarjan's algorithm. A detailed description of the Tarjan's algorithm follows [35].

12.6.1.1 Tarjan's Algorithm

The basic idea of the algorithm consists of tracing paths and identifying information cycles. The algorithm uses a stack of visited nodes. The stack is built using a depth-first search and records both the current path and all the closed paths so far identified. Each cycle eventually appears as a group of nodes at the top of the stack and is then removed from it. We illustrate the algorithm following the approach presented by Duff et al. [35] through two examples.

Figure 12.8 shows the digraph and the stack at all the steps of the algorithm. Starting from node 1, in the first steps, the stack simply records the growing path 1-2-3-4-5-6. At step 6 we find and edge connecting the node at the top of the stack (node 6) to one lower down (node 4). This is recorded by adding a link, shown as a subscript. Since we know there is a path in the stack (4-5-6) this tells us that these three nodes lie on a closed path. Similarly at step 7, we record the link 7-3 and this indicates that nodes (3-4-5-6-7) lie on a closed path. There are no more edges from node 7 so it is removed from the path. However, it is not removed from the stack because it is part of the closed path (3-7). We indicate this in Figure 12.8 by showing node 7 in bold. Now node 6, which is immediately below node 7 on the stack, is at the path end so we return to it, but the only relabeling we do is to make that its link point to 3 and we discard the link from 7, thereby recording the closed path 3-7.

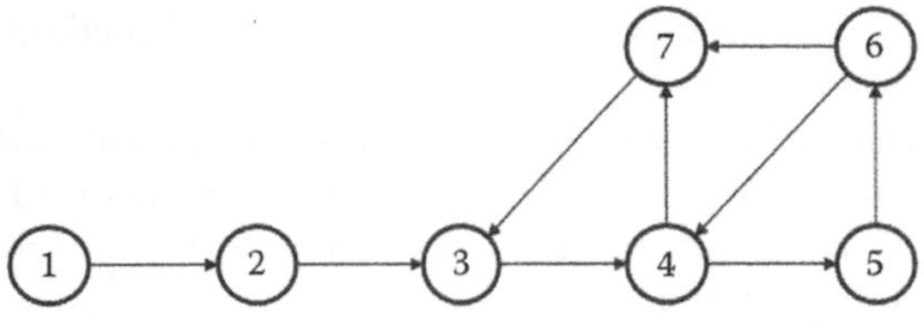

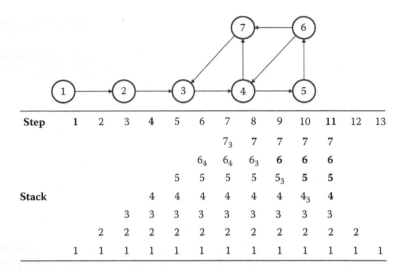

Step	**1**	2	3	**4**	5	6	7	8	9	10	**11**	12	13
							7_3	**7**	**7**	**7**	**7**		
						6_4	6_4	6_3	**6**	**6**	**6**		
					5	5	5	5	5_3	**5**	**5**		
Stack				4	4	4	4	4	4	4_3	**4**		
			3	3	3	3	3	3	3	3	3		
		2	2	2	2	2	2	2	2	2	2	2	
	1	1	1	1	1	1	1	1	1	1	1	1	1

FIGURE 12.8 Digraph and stack for the Tarjan's algorithm.

We now look for unsearched edges at node 6 and find that there are none, so we label 6 in bold, set the link from 5 to 3, and discard the link from 6 (step 9 in Figure 12.8). In the next step we move to node 5 that has no unsearched edges then we bold this node and pass the link to node 3 from node 6 to node 5. Then we move to node 4 and repeat the procedure (step 11) note that the link 3 is pointed to itself. Next we find that node 3 has not unsearched edges. Since it has not a link to a node below it, there cannot be any path from it or any of the nodes above it to a node below it. We have found a cycle form by units 3-4-5-6-7. And those units can be removed from the stack. Node 2 has neither unsearched edges nor links so it can be removed from the stack (we could say that is a cycle formed by a single unit) and finally node 1 must be also removed.

The previous example was chosen to be simple in the sense that the path always consists of adjacent nodes in the stack. This is not always the case. The following example (Figure 12.9) illustrates the more general situation.

Starting from node 1 we move to node 2. Here we have two alternatives: go to node 4 or to node 6. The final result will be the same independently of our election. We decide to go to node 4 (step 3). Node 4 has a link to node 2 that is already in the stack and therefore is shown as a subscript. From node 4 we move to node 5 that has a link to node 1, marked as a subscript. Node 5 has not unexplored edges and it is removed from the path, but not from the stack because of its link, so it is labeled bold in the stack. We backtrack in the path (move from node 5 to node 4) and transfer the link to the new node (step 5). Node 4 had a link to node 2 therefore we only retain the link to the node lower in the stack –in this case node 1-. Node 4 has not new unsearched edges so it is removed from the path but not from the stack due to its link to node 1, and labeled in bold in the stack. Again backtrack in the path, move to node 2 and transfer the link from node 4 to node 2 (step 6). Node 2 has the unexplored edge (2–6) so we move to node 6 (step 7). Node 6 has two links (to node 5 and to

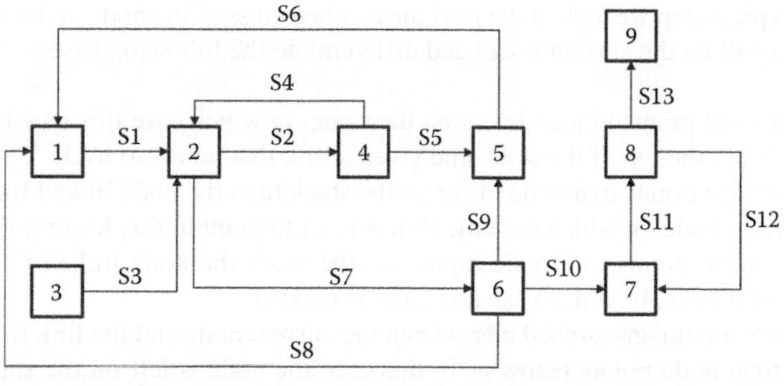

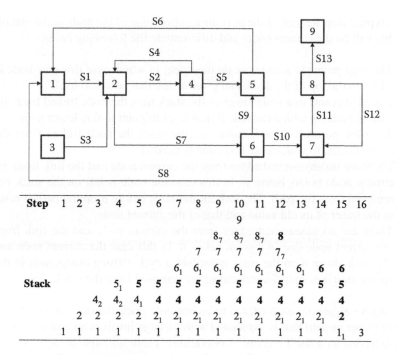

Step	1	2	3	4	5	6	7	8	9	10	11	12	13	14	15	16
										9						
									8_7	8_7	8_7	**8**				
								7	7	7	7	7_7				
							6_1	6_1	6_1	6_1	6_1	6_1	6_1	**6**	**6**	
Stack				5_1	**5**	**5**	**5**	**5**	**5**	**5**	**5**	**5**	**5**	**5**	**5**	
			4_2	4_2	4_1	**4**	**4**	**4**	**4**	**4**	**4**	**4**	**4**	**4**	**4**	
		2	2	2	2	2_1	2_1	2_1	2_1	2_1	2_1	2_1	2_1	2_1	**2**	
	1	1	1	1	1	1	1	1	1	1	1	1	1	1	1_1	3

FIGURE 12.9 Digraph and stack for the Tarjan's algorithm.

node 1). We only need to mark the link in the lower position on the stack (node 1 in this case) that is shown as a subscript. In the next step we add node 7 to the stack because the edge 6–7. And then we add the node 8 to the stack (edge 7–8). Node 8 has a link to node 7 that is in the stack and is marked as a subscript. From node 8 we move to node 9 (step 10). Node 9 has neither unsearched edges nor links to any node down itself in the stack and consequently node 9 is removed from both the path and the stack (step 11). Node 8 has not unsearched edges and it is removed from the path but not from the stack due to its link to node 7, so node 8 is labeled in bold. The link in node 8 is transferred to node immediately below (in this case node 7), that now is linked to itself (step 12).

Since node 7 has not unsearched edges and it has not a link to a node below it, there cannot be any path from it or any of the nodes above it to a node below it. We have found a cycle form by units 7–12. And those units can be removed from the stack. At this point (step 13) we verify that node 6 has not unsearched edges and can be removed from the path. Node 6 is label in bold and its link transferred to node 2. Node 2 had already a link to node 1 so no further action is needed. Node 2 has not unsearched edges, so it must be removed from the path but not from the stack due to its link. Node 2 is labeled in bold and its link transferred to node 1 (step 15). At this point we locate another cycle form by nodes 1-2-4-5-6.

After step 15, the stack is empty. However, the node 3 has not been visited and so it must be included in the stack. Node 3 has no unsearched edges and therefore the algorithm ends.

At a typical step we look at the next unsearched edge of the node at the end of the path (this will be the current node) and differentiate the following cases:

1. The edge points to a node not on the stack, in which case this new node is added to the top of the stack and given a link that points to itself.
2. The edge points to a node lower on the stack than the node linked from the current node, in which case the link is reset to point to this lower node.
3. The edge points to a node higher on the stack the node linked from the current node, in which case no action is needed.
4. There are no unsearched edges from the current node and the link from the current node points below it. In this case the node is left on the stack but removed from the path. The link for the node before it on the path is reset to the lesser of its old value and that of the current node.
5. There are no unsearched edges from the current node and the link from the current node does not point below it. In this case the current node and all those above it on the stack constitute a cycle (strong component in the sparse matrix nomenclature) and so are removed from the stack.

In Tarjan's algorithm each edge is reference only once and that each of steps (1-5 in previous paragraph) that is associated with an edge involves and amount of work that is bounded by a fixed number of operations. There will also be some $O(n)$ cost associated with initialization and accessing the rows, so the overall cost is $O(n) + O(p)$ for a matrix of order n with p entries. Details of the implementation can be found in Tarjan [34]; Duff and Reid [36,37].

Once the partition has been complete generate a precedence order for solving the problem is really straightforward. We know that all the units in a loop must be solved simultaneously (i.e. by tearing methods we will comment in next section). So we can condensate all the nodes in a new 'super-node'. The resulting flowsheet must be acyclic and therefore just starting from a node that does not receive information and following the information paths we must be able of solving the problem. For example, the calculation sequence for the digraph in Figure 12.9 is 3 – C1(1, 2, 4, 5, 6) – C2(7, 8) – 9.

12.6.2 Tearing a Loop of Units

The next issue is how we might solve each of the partitions containing more than a single unit. The first think we have to realize is that there are a large number of alternatives for tearing the recycle loop, i.e. an extreme situation consist of tearing all the streams, but of course this is clearly inefficient. In the literature different algorithms have been proposed for different tear criteria. The most important are:

1. Chose the order giving by the fewest number of tear streams [27]
2. Assign a weight to each stream. This weight can reflect the expected difficulty associated with tearing the stream and may, for example, be equal to the number of significant torn variables en each stream. Find the order which minimizes the sums of the weights associated with the torn streams [24,29].

3. Tear to give the best convergence properties for fixed point methods (non redundant set of tear streams). In other words, select a set of tear streams that avoid cutting a 'minor loop' twice. [28,38]

Instead of showing a detailed description of each of the previous algorithms we consider a single general tearing approach that can be adapted for covering the three possibilities. The approach treats tear selection as an optimization problem with binary (0-1) variables. This particular integer programming formulation devised by Pho and Lapidus [29] is known as a set covering problem and it allows for considerable flexibility in selecting tear sets.

We therefore treat the selection of tear streams as a minimization problem (e.g. minimize the number of tear streams or tear variables or time minor loops are cut) subject to the constraint that all recycle loops must be broken at least once. Before formulating the problem we first need to identify all the process loops in order to formulate the constraints. We present the loop finding through an example:

Consider the flowsheet partition in Figure 12.10 (Loop C2 in the previous example of Figure 12.9). Note that all the streams entering from an external node to the partition or exiting to any node outside the partition has been removed.

We start with any unit operation (node) in the partition and follow a path until a unit operation repeats. For example if we start from node 1 we can follow the following path:

$$1 \xrightarrow{S1} 2 \xrightarrow{S2} 4 \xrightarrow{S4} 2$$

We note that unit 2 repeats and the two streams (S2, S4) which connect the two appearances of the unit 2 are placed in a list of loops: $List:\{S2, S4\}$.

We then start with the unit just before the repeated one and trace any alternative path from it.

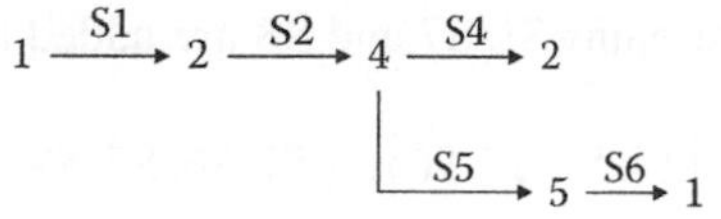

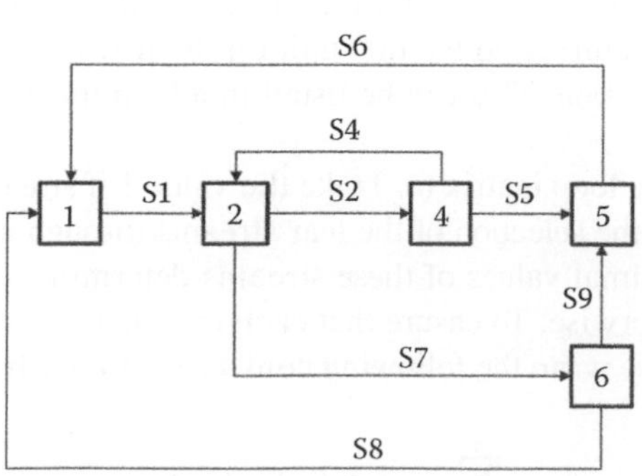

FIGURE 12.10 Strong component (maximum cycle) in a flowsheet.

Now the unit 1 repeats and we place streams S1, S2, S5 and S6 in the list of loops:

$$List:\{S2,S4\},\ \{S1,S2,S5,S6\}$$

If we back up to node 5 we realize that there is not any unexplored path. If we back up to unit 4, again there are no unexplored paths. On the other hand if we back up to node 2 we find another path that is given by:

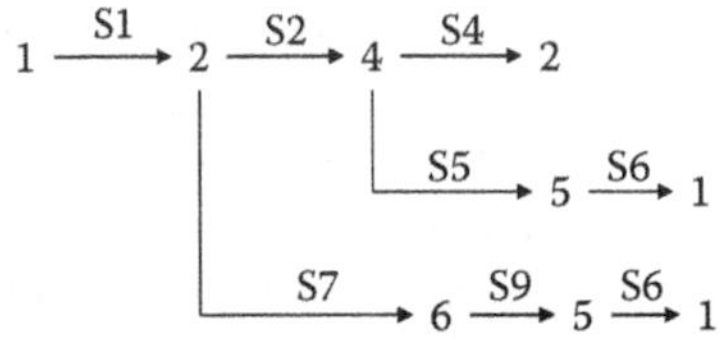

Unit 1 is repeated and the streams S1, S7, S9 and S6 are added to the list of loops:

$$List:\{S2,S4\},\ \{S1,S2,S5,S6\},\ \{S1,S6,S7,S9\}$$

If we back up to node 5 there is no any unexplored path, so we move to node 6 and we find another path that is given by:

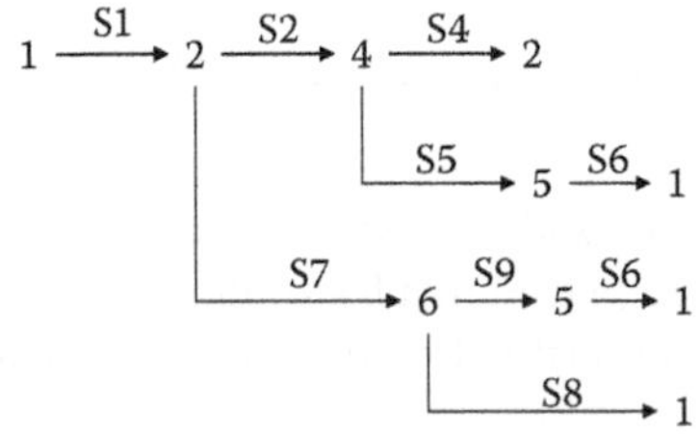

Unit 1 is repeated and streams S1, S7 and S8 are added to the loop list:

$$List:\{S2,S4\},\ \{S1,S2,S5,S6\},\ \{S1,S6,S7,S9\},\ \{S1,S7,S8\}$$

Backtracking to nodes 6, 2, and 1 we realize that there are no new unexplored paths. Since we have returned to the first unit on the list, we are done and there are four loops for this partition. This can be listed in a loop incidence matrix as shown in Table 12.7.

The elements of the loop matrix ($a_{i,j}$) take the value 1 if stream j is in loop i and 0 otherwise. We define the selection of the tear streams through an integer variable y_j, for each stream j: optimal values of these streams determine: $y_j = 1$ if stream j is a tear stream and 0 otherwise. To ensure that each recycle loop is broken at least once by the tear streams, we write the following constraints for each loop i:

$$\sum_{j=1}^{n} a_{i,j}\, y_j \geq 1 \quad i = 1 \ldots L \tag{12.14}$$

TABLE 12.7
Loop Incidence Matrix

Loop	S1	S2	S4	S5	S6	S7	S8	S9
1		1	1					
2	1	1		1	1			
3	1				1	1		1
4	1					1	1	

Where L is the number of loops and n the number of streams. We also formulate the following cost function for tear selection:

$$\sum_{j=1}^{L} w_j y_j \tag{12.15}$$

And we assign a weight w_j to the cost of tearing stream j. Three alternative choices for the weights are:

- Weight all the streams equally so that we minimize the number of tear streams $\left(w_j = 1\right)$. Note that the solution is not unique.
- Choose $w_j = \sum_{i=1}^{L} a_{i,j}$. If we sum over the loop constraints, we obtain coefficients that indicate the number of loops that are broken by a tear stream j. Breaking a loop more than once causes a delay in the tear variable iteration for the fixed point algorithms and much poorer performance. By minimizing the number of multiple broken loops we seek a nonredundant set of tear equations for better performance.
- Choose $w_i = n_j$ where n_j is the sum of variables in the jth stream.

The set covering problem is then given by:

$$\begin{aligned} \min: & \sum_j w_j y_j \\ s.t. & \sum_j a_{i,j} y_j \geq 1 \quad i = 1 \ldots L \\ & y_j = \{0,1\} \end{aligned} \tag{12.16}$$

If we solve the previous equation looking for a minimum set of tear streams we found three alternative sets formed by two streams: [S1, S2]; [S1, S4]; [S2, S7]. Although these three sets are valid sets, the set form by streams S1, S2 is redundant: the second loop is torn twice (See Figure 12.11).

If we solve for a set of nonredundant streams we found six alternative solution sets: [S1, S4]; [S4, S6, S8]; [S4, S5, S8, S9]; [S2, S8, S9]; [S4, S5, S7]; [S2, S7].

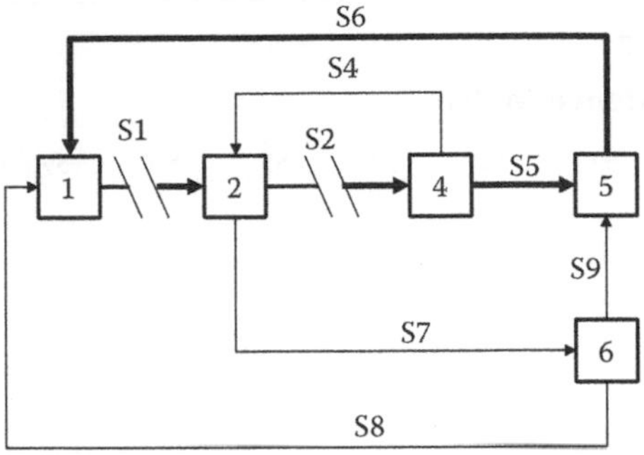

FIGURE 12.11 Double tearing a loop.

12.7 CONVERGENCE METHODS IN SEQUENTIAL MODULAR SIMULATORS

Modular process simulators are very robust solving each unit operation with numerical methods tailored to the specific characteristics to each one of these units. These include from specific inside-out algorithms to 'flash' a material stream going through detailed methods for reactors, heat exchangers until complex methods for distillation. However, one drawback of the approach is that some unit operations introduce numerical noise (this is also simulator and unit dependent). In other words, if we solve the same problem starting from different initial points we will get, for some variables, slightly different values. The difference can be in the second or third decimal point, what is not significant from a simulation point of view but it is a really large error if we try to estimate a derivative (derivative information is not provided by the simulator even though some unit operations internally use it to solve the module). This problem is magnified by information loops that could act as 'error accumulators'. In general, derivative information must be obtained perturbing some variables but if the numerical noise is relatively large, the derivative information could have 'poor quality' and the methods based on derivative information could show an unpredictable behavior. For these reasons most modular process simulators use derivative free methods for converging the tear streams.

12.7.1 First Order Methods

We develop these methods in a fixed form $x = F(x)$ where x and $F(x)$ are vectors of n stream variables. These methods are the most commonly used to converge recycle stream variables. Here x represents a guessed tear and $F(x)$ is the calculated value after executing the units around the flowsheet.

12.7.1.1 Direct Substitution Method

The simplest fixed point method is direct substitution. Here we define $x^{k+1} = F(x^k)$ with an initial guess x^0. For the fixed point function consider the Taylor series expansion:

$$F(x^k) = F\left(x^{k-1}\right) + \left(\frac{\partial F}{\partial x}\right)^T_{x=x^{k-1}} (x^k - x^{k-1}) + \ldots \qquad (12.17)$$

Assuming that $\left(\frac{\partial F}{\partial x}\right) \approx cte \neq 0$ is the dominant term near the solution, then:

$$x^{k+1} - x^k = F(x^k) - F(x^{k-1}) = \left(\frac{\partial F}{\partial x}\right)^T (x^k - x^{k-1}) \tag{12.18}$$

or

$$x^{k+1} - x^k = \Delta x^{k+1} = \left(\frac{\partial F}{\partial x}\right)^T \Delta x^k \tag{12.19}$$

Taking the norms in both sides of the previous equation and remembering the norms properties, we can write:

$$\left\|\Delta x^{k+1}\right\| \leq \left\|\left(\frac{\partial F}{\partial x}\right)^T\right\| \left\|\Delta x^k\right\| \tag{12.20}$$

If we use the Euclidean norm then $\|\partial F / \partial x\| = |\lambda^{\max}|$ which is the largest eigenvalue of $(\partial F / \partial x)$ in magnitude. From this expression we can show a linear convergence rate, but the speed of these iterations is related to $|\lambda^{\max}|$. Now by recurring the iterations for k we can develop the following relationship:

$$\left\|\Delta x^k\right\| = \left(\left|\lambda^{\max}\right|^k\right) \left\|\Delta x_0\right\| \tag{12.21}$$

A necessary and sufficient condition for convergence is that $|\lambda^{\max}| < 1$. This relation is known as a *contraction mapping*. However, the speed of convergence depends on how close $|\lambda^{\max}|$ is to zero. The number of iterations to reach a given tolerance $\|\Delta x^n\| \leq \delta$ can be estimated from the following relation:

$$n_{iter} \geq \frac{L_n\left(\delta / \left\|\Delta x^0\right\|\right)}{L_n\left(\left|\lambda^{\max}\right|\right)} \tag{12.22}$$

12.7.1.2 Relaxation – Acceleration Methods

For problems where $|\lambda^{\max}|$ is close to one, direct substitution is limited and converges slowly. In those cases, we can modify the fixed point function *F(x)* so that it reduces $|\lambda^{\max}|$. The general idea is to modify the fixed point function to:

$$x^{k+1} = wF(x^k) + (1-w)x^k = g(x) \tag{12.23}$$

Where w is chosen adaptively depending on the changes in x and *F(x)*. The two more common fixed point methods for recycle convergence, available in almost all chemical process simulators, are the dominant eigenvalue method –DEM- [39] and the Wegstein method [40].

In the DEM we obtain an estimate of $|\lambda^{max}|$ by monitoring the ratio:

$$|\lambda^{max}| = \frac{\|\Delta x^k\|}{\|\Delta x^{k-1}\|} \tag{12.24}$$

after a number of direct substitution iterations (typical values are between 3 and 8). Now from the transformation of the fixed point equation, we have:

$$\Delta x^{k+1} = x^{k+1} - x^k = g(x^k) - g(x^{k-1}) \approx \frac{\partial g}{\partial x}(x^k - x^{k-1}) = \Theta\left(x^k - x^{k-1}\right) \tag{12.25}$$

Where $\Theta = \partial g / \partial x = w\left\|\frac{\partial F}{\partial x}\right\| + (1-w)I$. We now chose the relaxation factor w to minimize $|\lambda^{max}|$ for Θ. If w is one, we have direct substitution, for $0 < w < 1$ we have an interpolation or damping factor, and for $w > 1$ we have extrapolation. To choose an optimum value for w we consider the largest eigenvalue for Θ, given by:

$$\det(\Theta - \theta I) = \det\left[w\left(\left\|\frac{\partial F}{\partial x}\right\| - (w-1+\theta)/wI\right)\right] = 0 \tag{12.26}$$

In the above expression, we note that $(w-1+\theta)/w$ corresponds to the eigenvalues and so we have: $\theta = 1 + w(\lambda - 1)$. To find $|\theta^{max}|$, we realize that this value is determined by the largest and smallest eigenvalues for $\left\|\frac{\partial F}{\partial x}\right\|$ as well as the relaxation factor. The optimum w^* occurs when:

$$\left(1 + w\left(\lambda^{min} - 1\right)\right)^2 = \left(1 + w\left(\lambda^{max} - 1\right)\right)^2 \rightarrow w^* = \frac{2}{2 - \lambda^{max} - \lambda^{min}} \tag{12.27}$$

While λ^{max} can be estimated from changes in x, λ^{min} is not easy to estimate. For DEM there is an important assumption: $\lambda^{max} \approx \lambda^{min} > 0$. So we have:

$$w^* = \frac{1}{1 - \lambda^{max}} \tag{12.28}$$

Note that if this assumption is violated and the minimum and maximum eigenvalues of Θ are very different, DEM may not converge. This approach has also been extended to the generalized dominant eigenvalue method (GDEM) [41] where several eigenvalues are estimated and used to determine the next step.

The Wegstein method obtains the relaxation factor by applying a secant method independently to each component of x. Then from each component we have:

$$x_i^{k+1} = x_i^k - f_i(x^k)\frac{x_i^k - x_i^{k-1}}{f_i(x^k) - f_i(x^{k-1})} \tag{12.29}$$

Defining $f_i(x^k) = x_i^k - g_i(x^k)$, we get:

$$\begin{aligned} x_i^{k+1} &= x_i^k - f_i(x^k)\frac{x_i^k - x_i^{k-1}}{f_i(x^k) - f_i(x^{k-1})} \\ &= x_i^k - \left[x_i^k - g_i(x^k)\right]\frac{\left[x_i^k - x_i^{k-1}\right]}{x_i^k - g_i(x^k) - x_i^{k-1} + g_i(x^{k-1})} \\ &= x_i^k - \frac{\left[x_i^k - g_i(x^k)\right]}{1 - s_i} = w_i g_i(x^k) + (1 - w_i)x_i^k \end{aligned} \tag{12.30}$$

Where $s_i = \dfrac{\left[g_i(x^k) - g_i(x^{k-1})\right]}{x_i^k - x_i^{k-1}}$ and $w_i = \dfrac{1}{1 - s_i}$

In practice DEM and Wegstein methods suffer from instability problems, since large acceleration factors are encountered in most problems. To stabilize the problem it is common to use only and acceleration each 2 to 8 direct substitution steps and to bound the acceleration factor as follows:

$$q_i = 1 - w_i = \frac{s_i}{s_{i-1}} \mid \ 0 \geq q_i \geq -n \tag{12.31}$$

typically the lower bound –n is between –2 and –5.

12.7.2 Broyden Method

As previously commented the standard method for solving equations is the Newton's method. But this requires the calculation of a Jacobian matrix at each iteration. Even assuming that accurate derivatives can be calculated, this is frequently the most time consuming activity for some problems, especially if nested nonlinear procedures are used. On the other hand, we can also consider the class of *Quasi-Newton methods* where the Jacobian is approximated based on differences in *x* and *f(x)*, obtained from previous iterations. Here the motivation is to avoid evaluation of the Jacobian matrix.

Here we follow the derivation presented by Biegler et al. [42] The basis for the derivation can be seen by considering a single equation with a single variable, as shown in Figure 12.12.

If we apply Newton's method to the equation starting from x_a, we obtain the new point x_c:

$$\text{Newton step}: x_c = x_a - \frac{f(x_a)}{f'(x_a)}$$

If the derivative is not readily available, we can approximate this term by a difference between two points, say x_a and x_b. The next point is x_d and this results from a secant that is drawn between x_a and x_b. The secant formula is given by:

$$\text{Secant step: } x_d = x_a - f(x_a)\left[\frac{x_b - x_a}{f(x_b) - f(x_a)}\right]$$

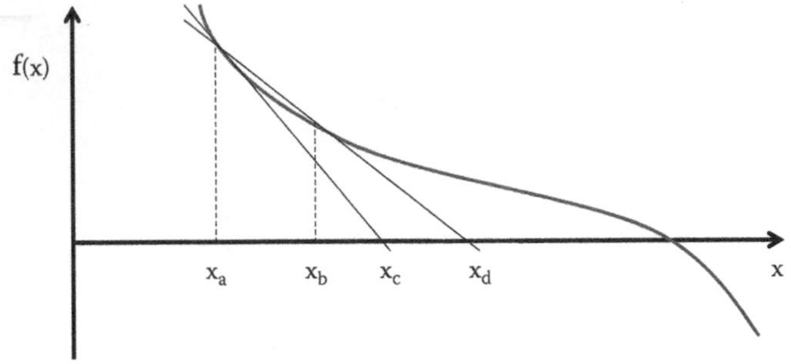

FIGURE 12.12 Comparison of Newton and Secant methods for a single equation.

We can define a secant relation so that for some scalar B we have:

$$B(x_b - x_a) = f(x_b) - f(x_a); \quad x_d = x_a - B^{-1} f(x_a) \tag{12.32}$$

For the multivariable case, we need to consider additional conditions to obtain a secant step. Here we define a matrix B that substitutes the Jacobian matrix and satisfies the secant relation:

$$B^{k+1}(x^{k+1} - x^k) = f(x^{k+1}) - f(x^k) \tag{12.33}$$

and therefore

$$x^{k+1} = x^k - \left(B^k\right)^{-1} f(x^k) \tag{12.34}$$

However, for the multivariable case, the secant relation alone is not enough to define B. A possibility consists of given a matrix B^k, calculate the least change for B^{k+1} that satisfies the secant formula [42]. This is a constrained minimization problem that can be written as follows:

$$\begin{aligned} &\min: \left\| B^{k+1} - B^k \right\|_F \\ &s.t. \quad B^{k+1} s = y \end{aligned} \tag{12.35}$$

Where $y = f(x^{k+1}) - f(x^k)$; $s = x^{k+1} - x^k$ and $\|B\|_F = \left(\sum_i \sum_j B_{ij} \right)^{1/2}$ is the Frobenius norm.

According to Biegler et al. [42] this problem can be stated and solved more easily with scalar variables. We use the following definitions: $b_{ij} = (B^k)_{ij}$; $\overline{b}_{ij} = (B^{k+1})_{ij}$ and y_i; s_i be the elements of vectors y and s respectively. Then we have:

$$\begin{aligned} &\min: \quad \sum_i \sum_j \left(\overline{b}_{ij} - b_{ij}\right)^2 \\ &s.t. \quad \sum_j \overline{b}_{ij} s_j = y_i \quad i = 1 \ldots n \end{aligned} \tag{12.36}$$

This problem is strictly convex and therefore has a unique minimum:

$$B^{k+1} = B^k + \frac{\left(y - B^k s\right)s^T}{s^T s} \tag{12.37}$$

With this relationship we can calculate the new search direction by solving the linear system:

$$B^{k+1} p^{k+1} = -f(x^{k+1}) \tag{12.38}$$

However, we can also calculate p^{k+1} explicitly by updating the inverse of (B^{k+1}) through a modification of Broyden's formula. Here we apply the Sherman Morrison Woodbury formula for a square matrix, and we get:

$$H^{k+1} = H^k + \frac{\left(s - H^k y\right)s^T H^k}{s^T H^k y} \tag{12.39}$$

A basic Broyden's algorithm can be stated as follows:

1. Guess x^0; B^0 $(e.g.\ J^0\ or\ I)$ and calculate H^0 $(e.g.(J^0)^{-1})$
2. If $k = 0$ go to step three, otherwise calculate $f(x^k)$; $y = f(x^k) - f(x^{k-1})$; $s = x^k - x^{k-1}$ and H^k or B^k
3. Calculate the search direction by $p^k = -H^k f(x^k)$ or by $B^k p^k = -f(x^k)$
4. If $\left\|p^k\right\| \le \varepsilon_1\ and \left\|f(x^k)\right\| \le \varepsilon_2$ stop. Else, find a step size α and update the variables so that $x^{k+1} = x^k + \alpha p^k$
5. Set $k = k + 1$ got to step 2.

The rank one update formulas for Broyden's method that approximate the Jacobian ensure superlinear convergence which is slower than Newton`s method but significantly faster than direct substitution.

12.7.3 The Cavett Problem: A Benchmark Problem for Testing the Convergence and Tear Set Selection

In order to illustrate the convergence of the numerical methods discussed above and the effect of the set of tear streams on the convergence of the system, we will solve the benchmark problem presented by Cavett [43] (see Figure 12.13). This problem represents a typical flowsheet problem for the petroleum industry. It consists of mixers TP-Flash units. The heat exchangers are 'fictitious' unit operations used to modify the P and T conditions to the Flash specifications. The problem is interesting because tear stream convergence is not easy and process could be sensitive to changes to the condition of operation. Data for this problem is shown in Table 12.8.

If we study the recycle structure of the problem we can see that all the units form a cycle, that is formed by three single loops [S1, S2, S6]; [S1, S3, S4, S5]; [S4, S7, S8]. It is possible select a minimum tear set or a nonredundant set. For the sake of comparison and to show the importance of the selection of a correct tear set we will

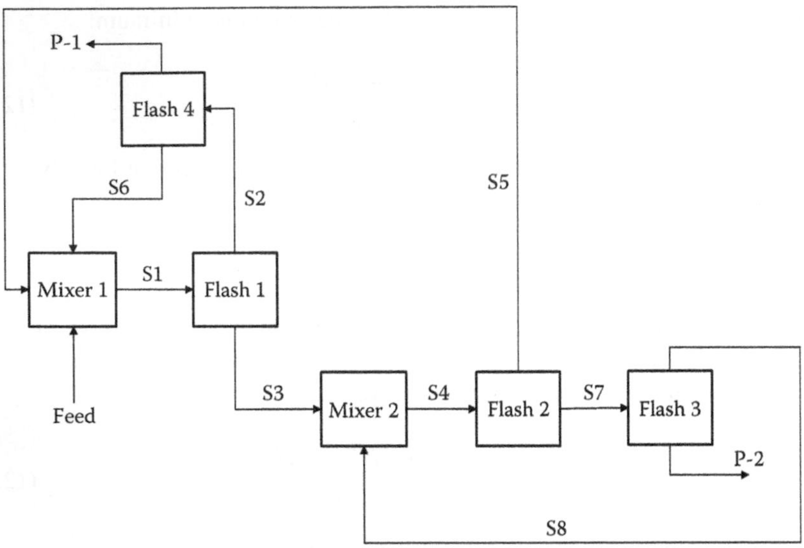

FIGURE 12.13 Block diagram for the Cavett problem.

TABLE 12.8
Data for the Cavett Problem and Results for Product Streams

	Streams				Flash Units	
Components	**Feed (kmol/h)**	**P-1 (kmol/h)**	**P-2 (kmol/h)**		**P (kPa)**	**T (°C)**
Nitrogen	45.23	45.23	0.0016	**Flash 1**	1960	412.85
CO_2	626.97	621.15	5.82	**Flash 2**	439	35.85
H_2S	42.85	312.58	4.27	**Flash 3**	191	29.85
Methane	3712.22	377.90	0.32	**Flash 4**	5620	37.85
Ethane	302.46	284.78	17.68	**Other Data**		
Propane	289.27	162.71	126.56	*Peng Robinson Equation of state.*		
i-Butane	76.28	112.83	57.45	*Default Aspen-Hysys parameters*		
n-Butane	194.43	33.98	160.45			
i-Pentane	99.80	6.56	93.24			
n-Pentane	142.66	6.91	135.75			
n-Hexane	222.82	2.88	219.94			
n-Heptane	329.13	1.10	3212.03			
n-Octane	232.89	0.19	232.70			
n-Nonane	210.73	0.043	210.69			
n-Decane	105.01	0.0055	105.00			
n-Undecane	153.35	0.0019	153.35			
P (kPa)	439	5620	191			
T (°C)	412.85	37.85	29.85			

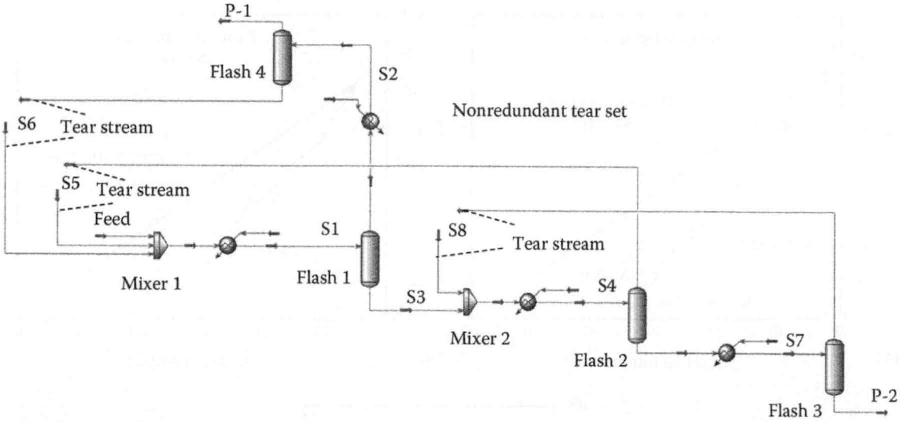

FIGURE 12.14 Aspen-Hysys implementation of the Cavett problem: example of a nonredundant and nonminimum tear set.

solve the problem with three different sets. The first one is a minimum set formed by streams S1 and S4. Note that this set is redundant (Scenario 1). Then we use the set [S5, S6, S8], that is nonredundant but is not minimum (Scenario 2). Finally we use the set formed by streams [S6, S4] that is both minimum and nonredundant.

As a process simulator we used Aspen-HYSYS (See Figure 12.14). In all the scenarios, as initial values, we use the streams values obtained when the simulation is done in open loop. Moreover, instead of using the simulator tools for converging the system (the recycle unit operation in HYSYS) we connect the simulator with external modules developed in Matlab. In that way we have a complete control over the numerical methods used for converging the system. In all the cases a termination tolerance equal to 10^{-4}, using a norm 1.

Using the direct substitution method, we can see that the convergence is faster for a nonredundant set of tear streams (Figure 12.15a) even though in scenario 2 we have a considerably larger number of variables (i.e. 48 vs. 32 component molar flows). It is worth remarking that the speed of convergence in scenarios 2 and 3 is the same—same slope in Figure 12.15a—independently of the number of tear streams or variables.

When we use a minimum number of tear streams and the set is redundant, we observe that the redundancy produce some instabilities in the convergence of both the Wegstein's and DEM methods (See Figure 12.15b). This instability is not observed in nonredundant sets (Figure 12.15c). Even though the number of tear variables (and total variables) is large in the scenario 2 the convergence is considerably faster when the redundancies are minimized.

The problem has also been solved for the three scenarios using the Broyden's method and the Newton's method (The nature of this problem, formed only by mixers and flash units, do not introduce numerical noise, so Newton's method can be applied). The scenario 1 required 17 iterations; scenario 2 required 11 iterations and scenario 3 required 10 iterations. The effect of redundancy is still apparent although

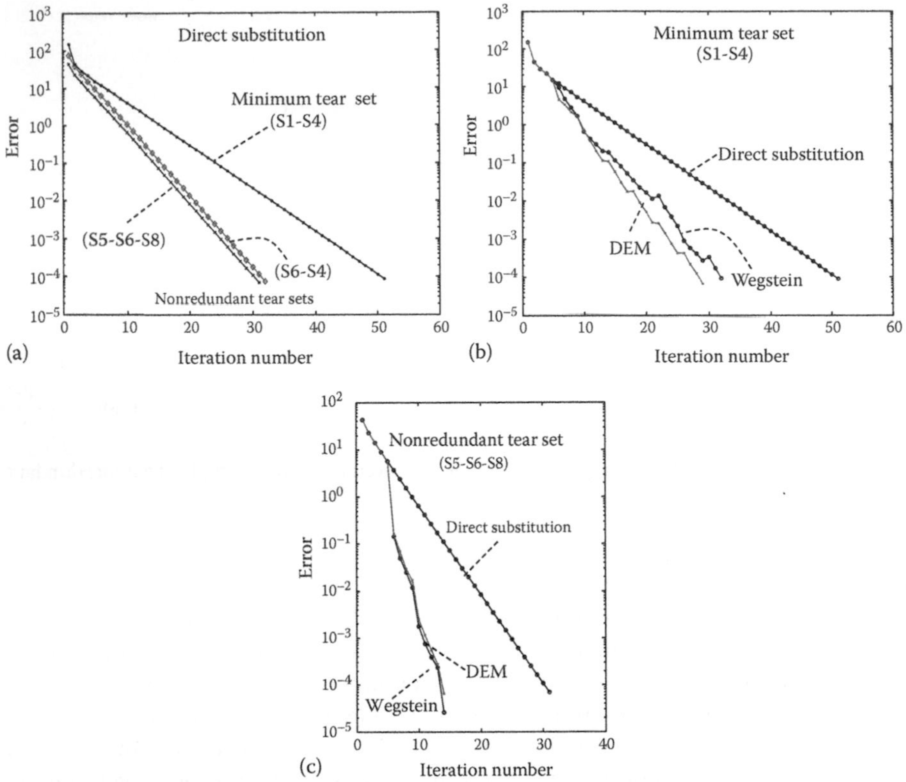

FIGURE 12.15 Comparison of the different strategies for tear selection and numerical methods performance.

it is less important than with fixed point methods. Finally with the Newton's method all three scenarios converged in 4 major iterations. However, most of the numerical effort is dedicated to evaluate derivatives. In the scenario 2 due to its larger number of variables the number of flowsheet evaluations was equal to 197 while in the other two was equal to 133 flowsheet evaluations. With Newton's method the effect of redundancy disappears an in this case it is better to select a minimum tear set.

12.8 EXAMPLES

We have included two example sections. The first one is dedicated to the modular process simulator CHEMCAD®. In the second one we will show an example using Aspen Hysys®. CHEMCAD follows a 'classical input-output' structure that is the most common approach in modular simulator. HYSYS, however, calculates a unit as soon as all its degrees of freedom are satisfied. In other words, it is able of calculating inputs in terms of outputs in most of the unit operations, which gives the user more flexibility specifying the problem, but at the same time the responsibility of correctly place the recycles to avoid inconsistencies in the information flow. Examples in ASPEN PLUS are also provided for comparison. We leave gPROMS® and ESS for Chapters 10 and 11 due to the differences presented in the Introduction.

12.8.1 CHEMCAD®

The screen, Figure 12.16, shows the five different sections when we open CHEMCAD®. The lop line is the menu, including graphic tools. To the left is the explorer where the different actions of the simulation are stored, as well as recent files that have been used and access to visual basic. In the middle is the main window. Just below it lies the message window and, to the right, the units' palette. In version 7 we can find three-color versions of the units, grayscale, color, and wireframe. To start a simulation, we follow a simple procedure:

1. **We start a new simulation using the tab New File of the menu:** Assign a name for your file and save it. Now the main active window is operative for you to start drawing your flowsheet.
2. **Select the units:** Before anything else, go to format, engineering units and select the system in which CHEMCAD® is going to perform the calculations including the cost estimations.
3. **Select the thermodynamic models:** This stage is one of the most important since it determines the reliability of the results. In this point we are not going to discuss again the selection procedure but only the stages.

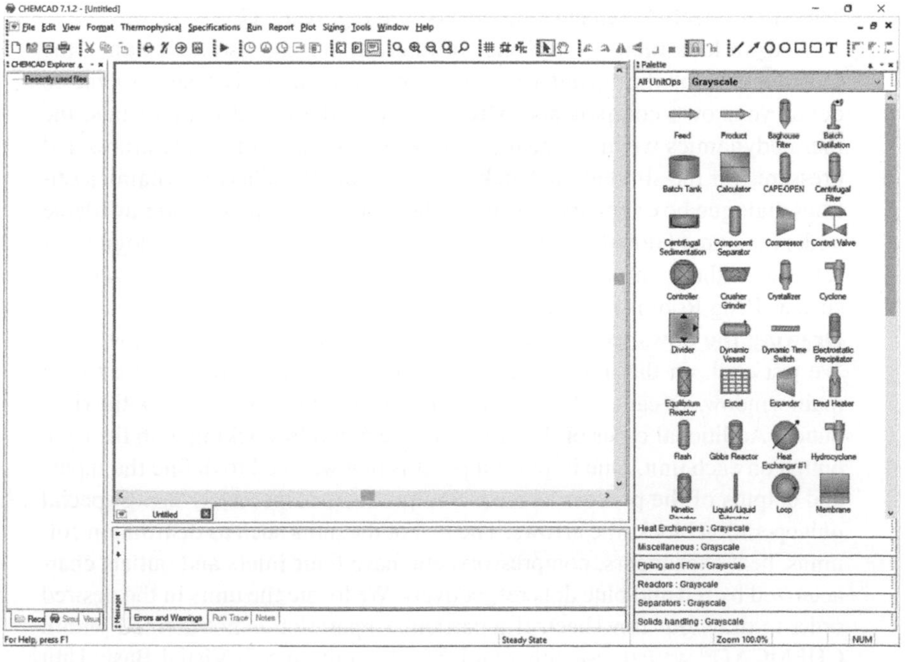

FIGURE 12.16 Initial screen CHEMCAD®.

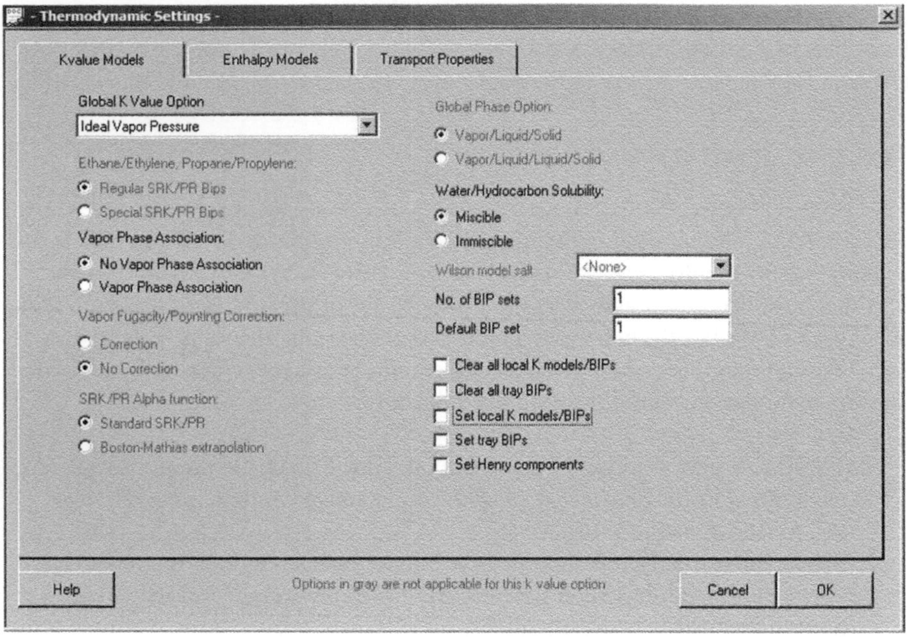

FIGURE 12.17 Thermodynamics wizard.

We first decide the components that are involved in the process. We go to Thermophysical → Select Components, see Figure 12.17.

And we look there for the ones of interest. We can use their name, the CAS number or the formula to select among them. It is also possible to define your own components. After accepting the list of components, the thermodynamics wizard appears, and we set the range of temperatures and pressures for the simulation. Click "Accept," and the Thermodynamics settings dialogue box appears. The main thermodynamic models are available under the name Global k Value Option. The second tab of the dialogue box refers to enthalpy models, where different equations of state are available such as Peng Robinson, Soave-Redlich-Kwong.

4. **Drawing the flowsheet:** Now we are ready to start drawing the flowsheet. We just click on the units we need and drag them from the palette to the main window. To cancel the selection of that particular unit we click the right button. Additional types of the same unit are found by clicking with the right button on each unit. One important point is that we need to define the inputs and outputs of the process as feeds and products respectively using special unit operation block, the arrows. The rest of the units such as distillation columns, heat exchangers, compressors, etc. have their inlets and outlets characterized by red and blue dots respectively. We locate the units in the desired order to draw our flowsheet. If a particular operation is not defined within CHEMCAD® we can use either an EXCEL® unit op or a Visual Basic Unit Op to build our own model. Both appear as units in the palette.

FIGURE 12.18 Top tab.

Now we link the different units. In version 7, to link two units it is enough to click on the outlet (red point) of the units and drag to the inlet (blue) of the next unit. This feature changes from previous versions where a unit "steam" was needed to link two units. In order to stop drawing we just click the right button.

5. **Units features:** The next step is problem dependent. For each of the units we click twice on it and the specifications dialogue box appears on the screen. It is important to understand what is its purpose; it is also useful to know about the operation it is going to perform, the components involved, and the operating conditions. In the examples we will pay more attention to this point. CHEMCAD® also performs economic evaluations. In the dialogue box for the unit specifications we see a particular tab, "Cost Estimations", where we have to provide further information on the material, type and size of the unit for CHEMCAD® to compute a cost. Most of the times equipment must be sized beforehand to compute the cost.
6. **Running the model:** We can run each of the unit operations by clicking with the left button on the unit. A list of actions appears. With this action we can also see the results of the streams once we have run the unit. It is convenient to run each of the units separately as initialization of the whole flowsheet before running it all. For running the entire flowsheet we go to the menu bar and look for the triangle, see Figure 12.18. Alternatively, we can go to the Run tab of the menu, see Figure 12.19. There we can modify the convergence procedure, run selected units of the entire flowsheet, optimization, sensitivity analysis and, finally, data reconciliation. New drawing tools have been added to the toolbar of version 7.

   Even before running CHEMCAD ® can give you different error and warning messages related to missing information, or, after trying to run, the convergence or not of the simulation
7. **Results:** We can see the results of particular units by clicking on it in the flowsheet window. Alternatively, we can use the Report of the main menu. We have some extra tools related to unit sizing for distillation columns, heat exchangers, pipes and tools for costing; we can update the cost index and the economics of the plant.

   Furthermore, we can run sensitivity analysis and optimization. The sensitivity study has two parts. We go to the Run ribbon in the general menu and select "Sensitivity Analysis" → "New analysis." After providing a name, we get a dialogue box, see Figure 12.20, there we define the variable under study "Adjusting" in the Figure below. The variables can be selected from the streams or from the equipment. We provide a range of values and CHEMCAD® performs the sensitivity study. We use the "Recording" tab to define the variables we store from the sensitivity study.

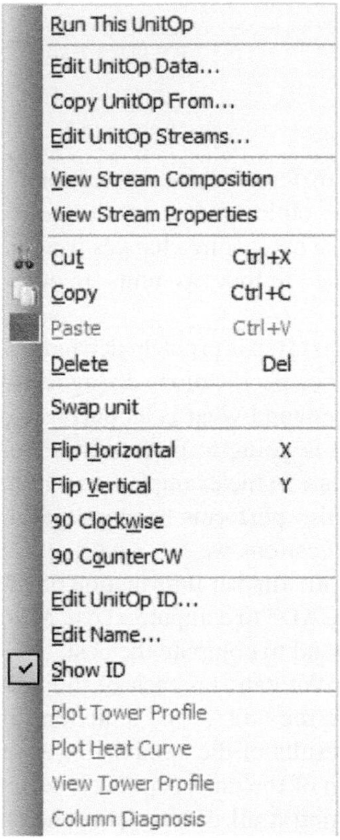

FIGURE 12.19 Options.

We can go a step further and optimize one or a few variables. In the Run ribbon we select "optimization." We get the dialogue box below, Figure 12.21a and b:

First, we define the objective function (see Figure 12.21b), such as energy consumption, the number of trays of a column, the conversion of a reactor.

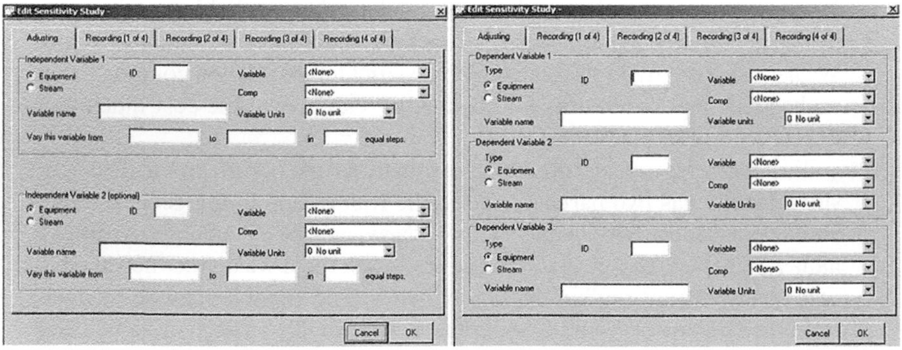

FIGURE 12.20 Sensitivity study screens.

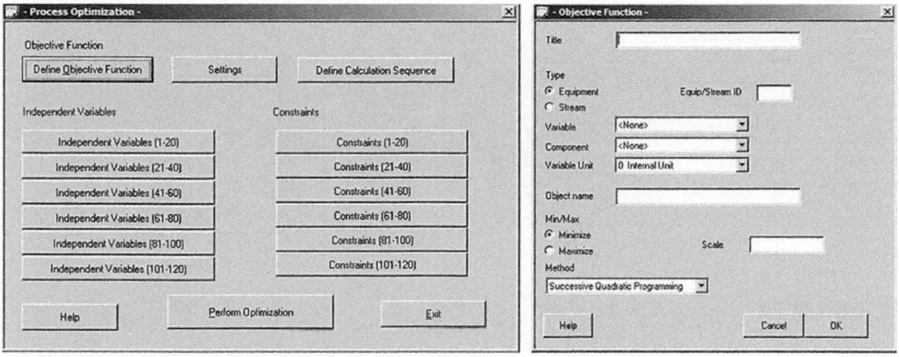

FIGURE 12.21 Dialogue box for optimization within CHEMCAD®.

We have up to three optimization algorithms for this case, SQR, General Reduced Gradient and Simultaneous Modular SQP. The particularities of the methods can be found elsewhere [44,45]. We define the independent variables and the constraints over variables from the streams or the equipment just by selecting the unit or the stream and the variable and the range of values of operation and an initial value.

12.8.1.1 Fluid Flow in a Pipe

As an example for fluid flow in pipes we propose a system similar to the one we solved in EXCEL® in Chapter 3. We like to feed two tanks with a total flow of 100000 kg/h. The information related to the system is given in Table 12.9 as shown the initial pressure is atmospheric and the final pressure at the tanks should also be atmospheric. The question is to determine the flow rate that reaches each tank, the pump outlet pressure and its cost. In other words, compute the pump energy and the split fraction.

We follow the procedure presented in the introduction. We select "Alt SI." Our only component will be water and for this particular operation, the thermodynamic model by default is good enough. We locate all the units, three pipes, a pump, one feed and two products and a stream splitter. Next, we link them using the "unit"

TABLE 12.9
Data for Piping Example in CHEMCAD®

Pipes	**1**	**2**	**3**
h (m)	0	27	42
d (m)	0.1524	0.0625	0.1016
Leq (m)	600	200	100
Fittings	5 elbows 90°	5 elbows 90°	1 elbows 90°
	10 elbows 45°	10 elbows 45°	10 elbows 45°
Material	Carbon steel	Carbon steel	Carbon steel

Stream No.	1
Stream Name	
Temp C	20
Pres bar	1
Vapor Fraction	0
Enthalpy MJ/h	-1589755
Total flow	100000
Total flow unit	kg/h
Comp unit	kg/h
Air	0
Water	100000

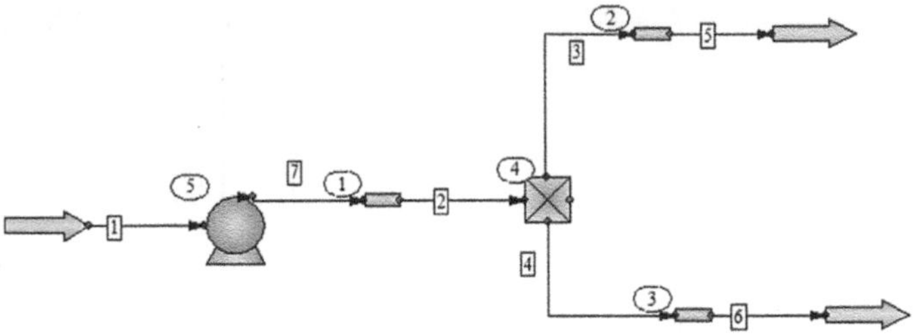

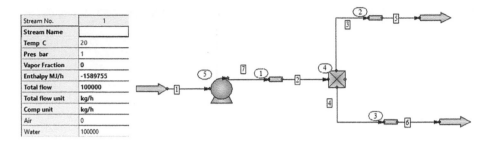

FIGURE 12.22 Piping flowsheet and input data.

stream. The flowsheet looks like the one in Figure 12.22. Stream numbers are represented in squares while unit numbers are given within circles. Next, we input the data. We start with stream [1]. We click twice and input the temperature, the pressure and the flowrate such as in Figure 12.22.

We now go the pump and select the pressure outlet and the efficiency; see Figure 12.23. The outlet pressure will be one of our variables. In the cost estimation tab we just click on the box so that CHEMCAD® returns the cost once it has obtained the results from the simulation. We use a centrifugal pump with the specification by default. The install factor is set to be 1.7.

For all the pipes we follow the same procedure; see Figure 12.24. We select the method. Single phase flow is selected since we only have water. The sizing option is left as default because we have diameter data. We input the diameter, the length of the pipe and the type of material. We alternatively can input a roughness factor if needed. The height of the discharge of the pipe is given by Elevation in the specifications dialogue box. Next, we click on the fittings tab (Figure 12.24) and input the number of elbows that our pipe has from the data of the problem. We can also add valves, which will also have an effect on the pressure drop. Finally, in the heat transfer tab, we assume adiabatic process.

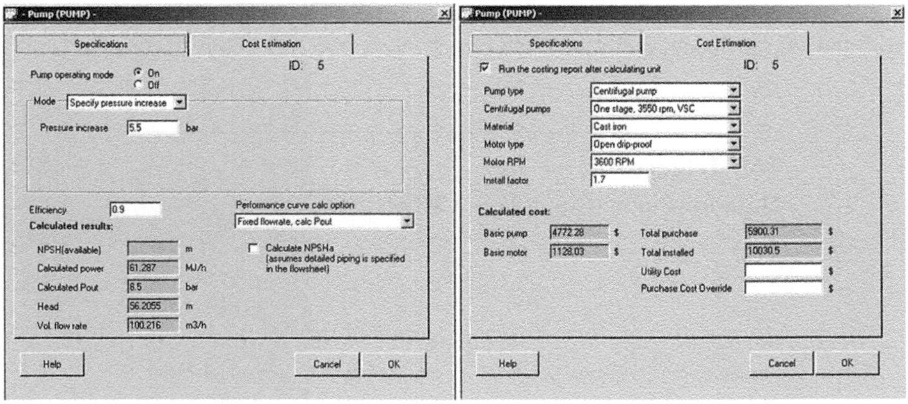

FIGURE 12.23 Pump input data.

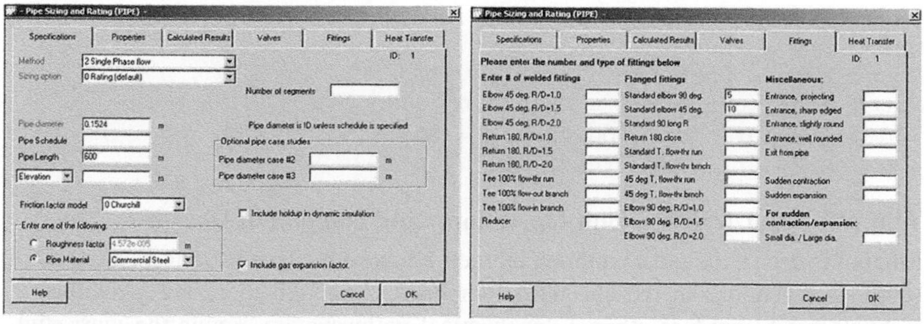

FIGURE 12.24 Detail of pipes input dialogue box.

The last unit is the splitter. The split fraction is another variable; see Figure 12.25. We now have to modify the split fraction and the input pressure so that the output is at 1 bar for both pipes. We try several values until we obtain the following results: The split fraction is 0.3 and 0.7 for streams 3 and 4 respectively. The full report of the streams can be obtained by selecting Report→Stream compositions → All streams.

To see the cost of the pump we can either click twice on the pump and go to costing tab or we can go to the menu on the top and click on the costing tab. Select Tools→ Costing→ Select equipment and we click on the pump and CHEMCAD® returns:

```
CHEMCAD ® Page 1
Job Name: Pipe Date: Time:
Preliminary Pump Cost Estimation
Pump Cost for Equip. 5
Base cost index = 421.1
Current cost index = 699.9
Centrifugal pump
Cb = $2871.28
Material = Case iron
Fm = 1
One stage, 3550 rpm, VSC
Ft = 1
```

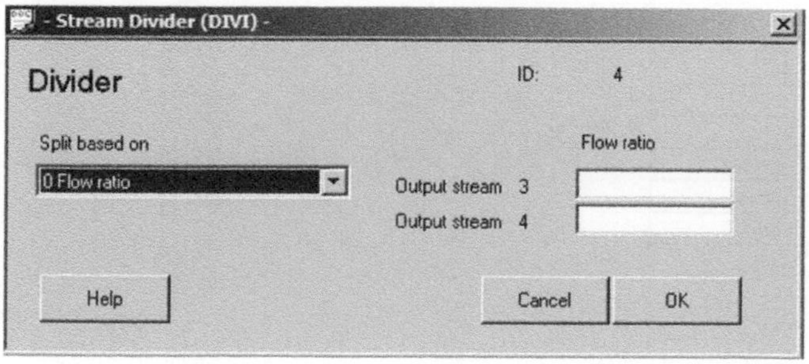

FIGURE 12.25 Detail of splitter input dialogue box.

```
Pump cost (purchase) = $4772.28
Motor rmp = 3600
Motor_type = Open, drip-proof
Motor cost (purchase) = $11212.03
(Pump + Motor) cost (purchase) = $5900.31
(Pump + Motor) cost (installed) = $10030.5
```

12.8.1.2 Heat Transfer, Unit Operations, and Equipment Design

Ethanol is one of the most common biofuels because it can substitute gasoline with small modifications in the current automobiles. Although there are a number of technologies to produce ethanol, biochemical pathways are among the most studied [46]. The main challenge in the production of bioethanol is the large amount of energy spent in dehydration. The mixture of ethanol and water exiting the fermentor typically contains ethanol from 6% to 15% depending on the raw material and the bacterium used [47,48]. The first stage for water removal after solid separation is the so-called beer column, a distillation column. The main feature that limits the dehydration of ethanol is the azeotrope of the water-ethanol system, at 96% w/w of ethanol, which prevents from simple purification of the bioethanol. Thus, we consider the study of such a column. The example we present here is the evaluation of the energy consumption in the beer column for the dehydration of ethanol. Apart from the simulation we include heat exchanger design topics using the tool CC-therm within CHEMCAD®.

We start our simulation following the steps presented in Section 12.4.1. The initial feed can be seen in Figure 12.26. Typically the fermentation takes place at 38°C and around 1bar and for the initial case we consider 8% ethanol in water [46]. Apart from these two main components, during the fermentation other species are generated in small amounts such as glycerol, acetic acid, succinic acid, etc. We do not include them in our analysis. Next, we draw the piece of flowsheet from the solid separations to the beer column. See Figure 12.26. We heat up the stream to become saturated liquid, using saturated steam that condenses and then we feed it to column (TOWER). In , we can use vapor fraction 10^{-7} to estimate the temperature but for design purposes this approximation results in the fact that phase change takes place.

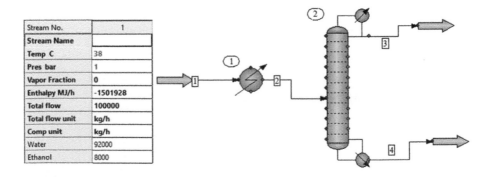

Stream No.	1
Stream Name	
Temp C	38
Pres bar	1
Vapor Fraction	0
Enthalpy MJ/h	-1501928
Total flow	100000
Total flow unit	kg/h
Comp unit	kg/h
Water	92000
Ethanol	8000

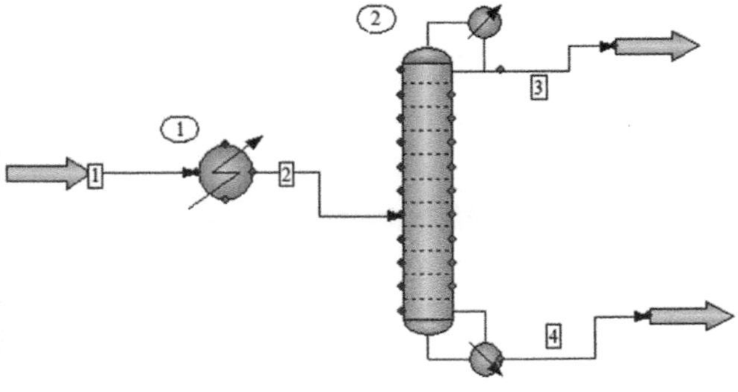

FIGURE 12.26 Flowsheet for the example and input data.

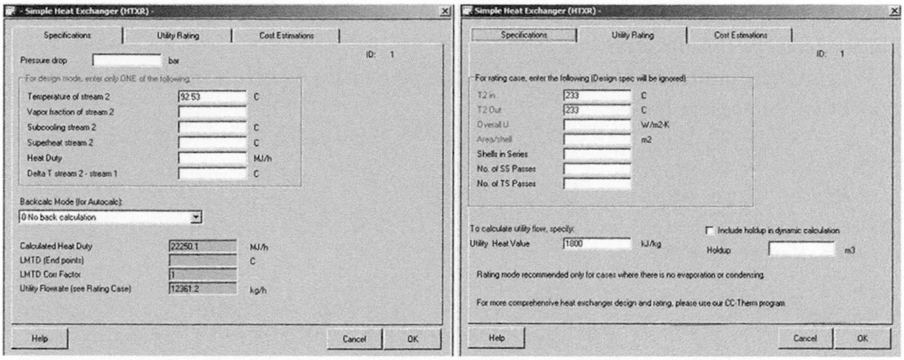

FIGURE 12.27 Data for the operation of the heat exchanger.

We start by inputting data to the heat exchanger, Figure 12.27. We use saturated steam at 233°C. For the moment we stop there since for cost calculations we need the design of the heat exchanger and it requires the solution of the mass and energy balances first.

Next, we focus on the Column; see Figure 12.28. We can select total or partial condenser depending on the next dehydration stage. For our case we select partial condenser since the next step involves molecular sieves. We estimate 11 stages based on rules of thumb and, since the feed is mostly water, we feed the column closer to the bottom. Alternatively, we can use a "Short-Cut" column in the design mode to obtain initial estimates. The feed plate is a source of lack of convergence. It is important to notice that CHEMCAD® considers the condenser and reboiler as stages. The next tab is specifications. Since the problem is asking for the evaluation of the reflux ratio, we select that as parameter while for the bottom we can chose between flow, temperature, and composition. In fact, temperature and composition are related to each other since the bottom temperature is the boiling point of the mixture. We select bottom temperature since we know that is should be close to water boiling point, but we can specify the bottom composition. Alternatively, we can define the recovery of one of the components, the flow rate or the stream composition. The next tab is convergence. We should provide data that will help solve the MESH equations of the column such as flow rates and temperature or else use

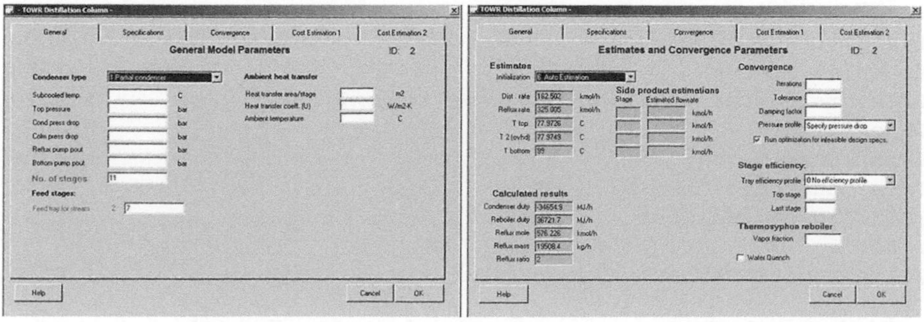

FIGURE 12.28 Dialogue box for input data for the operation of the distillation column.

"auto estimation" (Figure 12.28). For the moment we do not consider the costing so that we have a running solution.

We run the simulation, to report → Stream properties → all stream and we get the report that we present here

CHEMCAD Page 1

Job Name: Wate ethanol Date: Time:

Stream No.	1	2	3	4
Stream Name				
Temp C	312.0000*	92.5300	81.4418	99.0000
Pres bar	1.0000*	1.0000	1.0000	1.0000
Enth MJ/h	-1.5019E+006	-1.4797E+006	-66711.	-1.3995E+006
Vapor mole fraction	0.00000	0.00000	1.0000	0.00000
Total kmol/h	5280.5080	5280.5080	283.4980	4997.0105
Total kg/h	100000.0016	100000.0016	9670.7922	90329.2138
Total std L m3/h	102.0466	102.0466	11.5878	90.4587
Total std V m3/h	118355.44	118355.44	6354.22	112001.23
Flowrates in kg/h				
Water	92000.0000	92000.0000	2176.6954	89823.3095
Ethanol	8000.0000	8000.0000	7494.0972	505.9039

We click once to the heat exchanger, go to Report → Unit OPS → Select Unitops and we get the report. Note that only after the design we would be able to provide the transfer area to compute the cost and cost information will then be available.

CHEMCAD Page 1

Job Name: Wate ethanol Date: Time:

Heat Exchanger Summary

Equip. No.	1
Name	
1st Stream T Out C	92.5300
2nd Stream T Out C	233.0000
Calc Ht Duty MJ/h	22250.1484
LMTD Corr Factor	1.0000
Calc Area m2	33.9000
1st Stream Pout bar	1.0000
Cost estimation	1
Shell and tube	2
Install factor	1.7000
Material factor	1.7623
Pressure factor	1.0710
Type factor	0.5593
Basic cost $	11736
Total purchase cost $	12863
Total installed cost ($)	21867
Design pressure bar	1.0000

We do the same for the tower. Again, cost information will only be available after sizing.

CHEMCAD Page 1

Job Name: Wate ethanol Date: Time:

Towr Rigorous Distillation Summary

Equip. No. 2

Name

No. of stages	11
1st feed stage	7
Condenser type	1
Condenser mode	1
Condenser spec.	2.0000
Reboiler mode	3
Reboiler spec.	99.0000
Initial flag	6
Calc cond duty MJ/h	-23319.5391
Calc rebr duty MJ/h	36752.0000
Est. Dist. rate (kmol/h)	162.5024
Est. Reflux rate (kmol/h)	325.0049
Est. T top C	77.9738
Est. T bottom C	99.0000
Est. T 2 C	77.9749
Column diameter m	2.1336
Tray space m	0.6096
Thickness (top) m	0.0040
Thickness (bot) m	0.0040
Material density (kg/m3)	8030.0000
Actual no of trays	9.0000
Install factor	1.7000
Column purchase $	59036
Column installed $	100362
Cost estimation flag	1
Shell weight kg	1554
Cost of shell $	21465
Cost of trays $	17844
Platform & ladder $	7127
No of sections	1
Cond material	2
Condenser area m2	912.0000
Cond P design bar	1.0000
Cond inst. factr	1.7000
Rebl exchanger type	1
Rebl material	2
Reboiler area m2	280.0000
Rebl P design bar	30.0000
Rebl inst. factr	1.7000
Cond purchase $	29747
Cond installed $	50570
Rebl purchase $	213203
Rebl installed $	362445
Total purchase $	301986

```
Total installed $                                    513377
Calc Reflux ratio                                    2.0000
Calc Reflux mole (kmol/h)                          566.9963
 Calc Reflux mass kg/h                           14787.8291
Optimization flag                                         1
```

We can also plot the temperature profile as well as the flows across the column by using plot → UnitOp Plots → Column profiles.

Now we go to sizing the equipment. We start with the heat exchanger. Sizing → Heat exchangers →. We select the first one, equipment number 1 in the flowsheet. The first thing is to include the utilities specifications as another stream; see Figure 12.29. A high pressure saturated steam 233°C and 30 bar is used. After including the outlet temperature of the utility or its vapor fraction, we obtain the following dialogue box where the heat curve parameters are specified; Figure 12.30.

The heat exchanger is calculated by discretizing it. In the discretization points either the bubble—dew point method or de equal enthalpy methods can be used to compute the heat transfer. The first one reduces to the second if there is not condensation or evaporation within the heat exchanger. Thus, we use the default method and obtain Figure 12.31. Related to the wet or dry wall, it is related to the condensation or not of the streams when hitting the tubes.

Next, we get a dialogue box where you can see the heat diagram and edit it. There are also a number of options to compute the curve and here we refer to the user guide for further discussion.

Next, we go to the design phase. It is the next dialogue box that pops up. By clicking into "General specifications" it is possible to select the proper type of heat exchanger among the standard ones as well as the operation in the tube and the shell sides. Among the calculation modes, Design, Rating and Fooling, the first one,

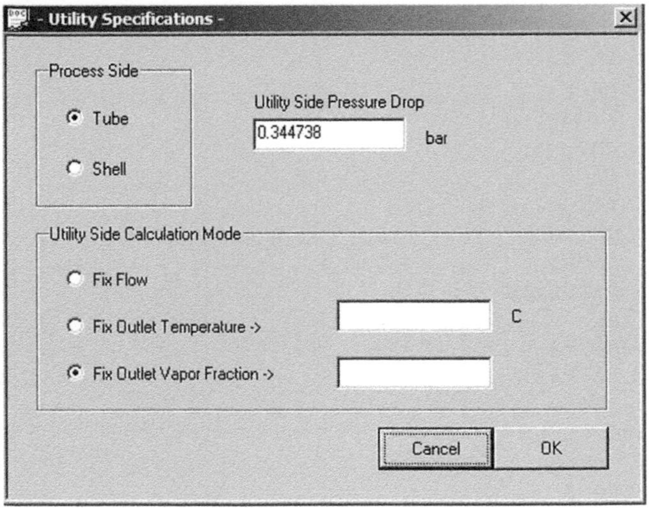

FIGURE 12.29 Utility specifications.

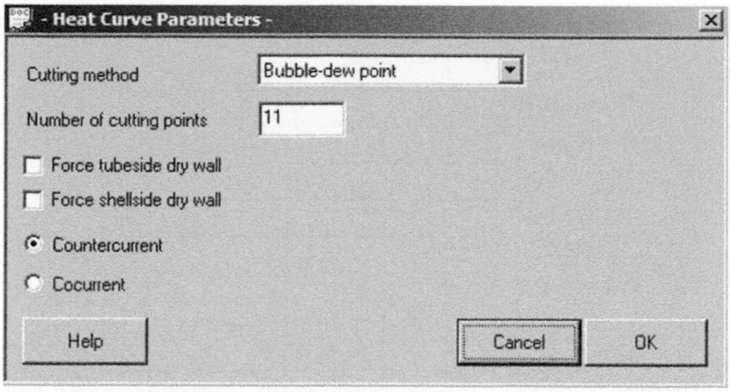

FIGURE 12.30 Heat curve parameters.

Design, optimizes the design, the second one, Rating, calculates whether we have excess of area or not. Finally, the third one, Fooling, adjusts the fouling factor to match the transfer area; see Figure 12.32. We select and AEP type of heat exchanger since it is a typical one for one pass on shell and outside packed fit head. The selection of the type determines the tubes configuration. For our case we have sensible flow in the tubes, to heat up the water ethanol mixture, and condensation in the shell for the steam. The software will calculate the heat transfer coefficient unless we input one in the last boxes of Figure 12.32.

The second tab of the dialogue box contains the models to be used for calculation. We are not going to discuss them here and refer to the User guide for further reference. We only check here whether the heat exchanger is vertical or horizontal and maintain the default methods. When the design mode is on, the next dialogue box is related to Design options (see Figure 12.33) to determine the pressure drop in tubes

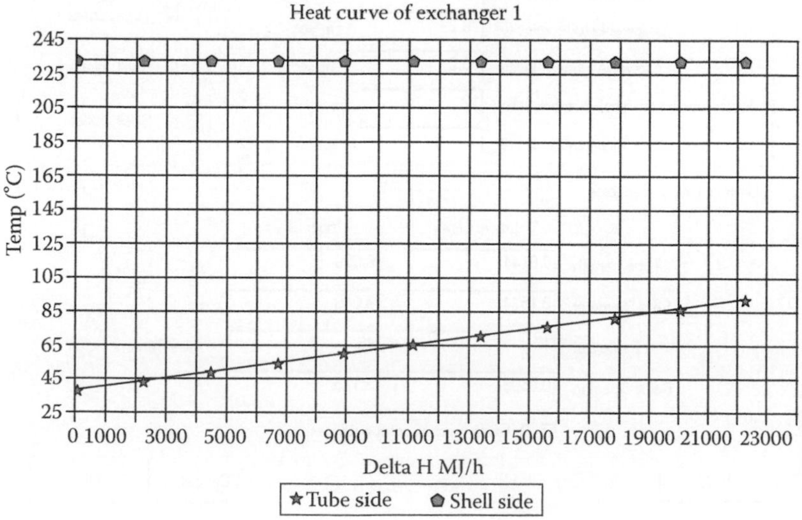

FIGURE 12.31 Heat curve.

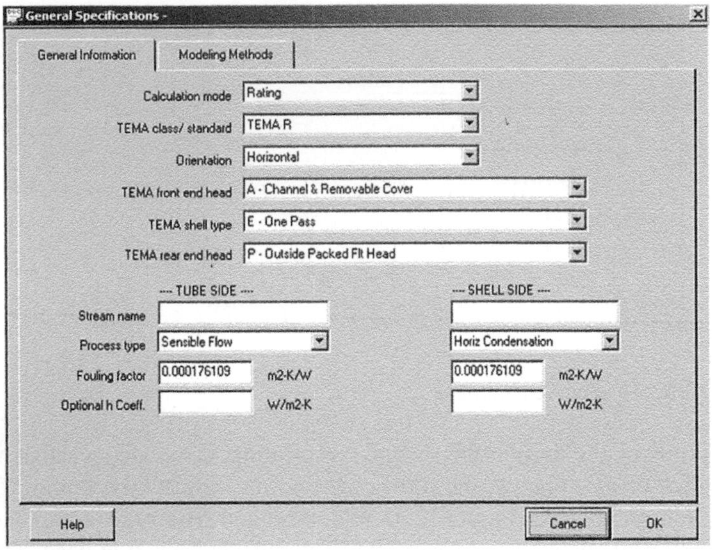

FIGURE 12.32 Heat exchanger design (I).

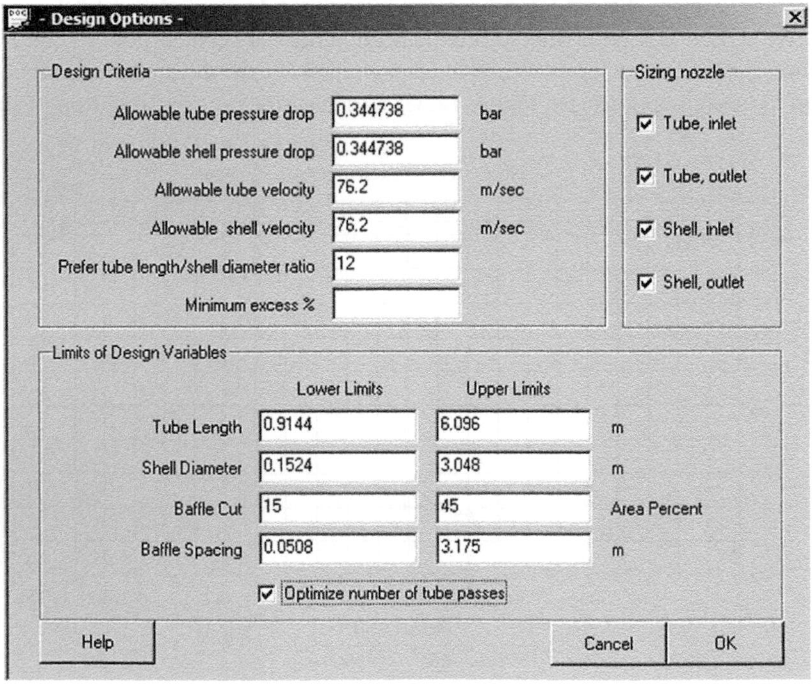

FIGURE 12.33 Heat exchanger design (II).

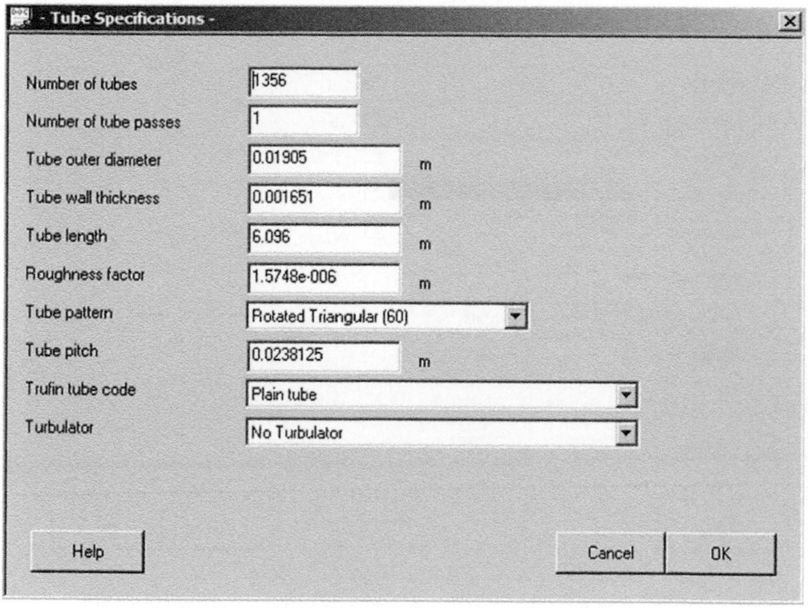

FIGURE 12.34 Heat exchanger design (III).

and shell that we allow, as well as the velocity of the flows on both sides. In terms of size, we provide the length of the tubes and shell. The software will try to meet the preferred ratio between the tube length and shell diameter. We chose to optimize the tube passes.

By accepting the data set go to "Exchanger Geometry," where the actual size of tubes, their diameter and thickness, and configuration are inputted. We also consider the number of tube passes and the roughness factor; see Figure 12.34.

Similarly, we can define the shell main features; see Figure 12.35. The shell diameter is related to the number of tubes so that we can actually fit them in the design. Thus, with the design mode both are calculated. We can input the number of shells

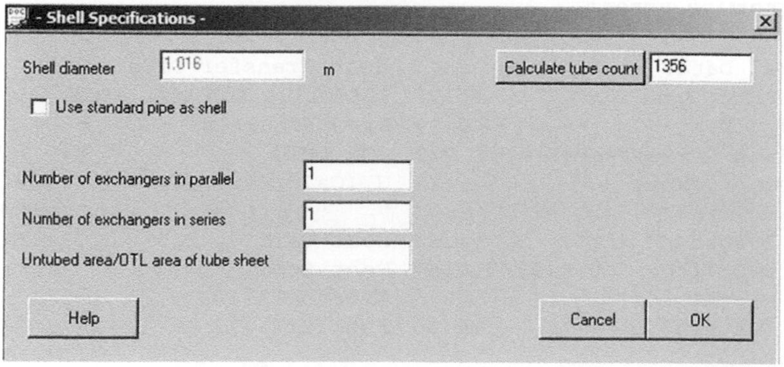

FIGURE 12.35 Heat exchanger design (IV).

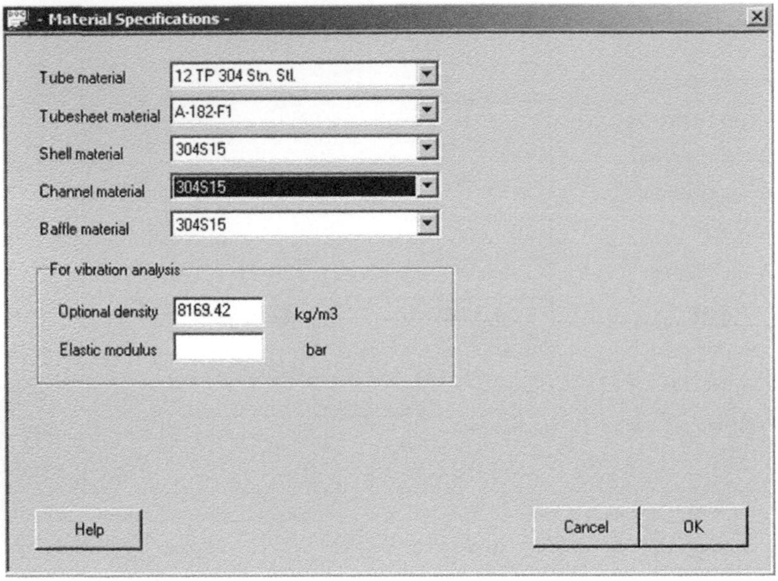

FIGURE 12.36 Heat exchanger design (V).

in parallel, if the size of the exchanger is too large, and the shells in series, to avoid temperature cross within the shell.

We can also specify baffles, nozzles and clearances. The baffles control the pressure drop across the shell and thus the heat transfer coefficient. Some data on the typical thickness are provided by default. Regarding the nozzles and clearances, we leave then by default.

Next, we decide on the material; see Figure 12.36. We select stainless steel for the tubes, due to the presence of water and ethanol, and for the shell we also select stainless steel because we use high temperature steam.

We compute and then go now to see Results and in particular click on "Summary Results." We include them into the "Cost Estimations" tab of the heat exchanger.

```
         SUMMARY REPORT
         --------------
General Data:                               Heat Transfer Data:
Exch Class/Type                      R/AEP  Effective Transfer Area      33.89
Shell I.D.                            0.34  Area Required                29.24
Shell in Series/Parallel               1/1  COR LMTD                    165.87
Number of Tubes                        119  U (Calc/Service)          1274.50/
                                                                       1099.42
Tube Length                           4.88  Heat Calc                 25793.37
Tube O.D./I.D.           0.0191/0.0157  Heat Spec                 22250.19
                                            Excess %                     15.92
Tube Pattern                         TRI60  Foul(S/T)              1.761E-004/
                                                                    1.761E-004
Tube Pitch                            0.02  Del P(S/T)               0.26/0.18
Number of Tube Passes                    1  SS Film Coeff             10589.85
```

```
Number of Baffles                 13  SS CS Vel              5.16
Baffle Spacing                  0.27  TW Resist          0.000110
Baffle Cut %                      23  TS Film Coeff       6322.03
Baffle Type                     SSEG  TS Vel                 1.25
Thermodynamics:
 K: NRTL
 H: Latent Heat
 D: Library
Number of Components: 2
 Calculation Mode: Design
Engineering Units:
 Temperature          C
 Flow/Hour            (kg/h)/h
 Pressure             bar
 Enthalpy             MJ
 Diameter/Area        m/m2
 Length/Velocity      m/(m/sec)
 Film                 W/m2-K
 Fouling              m2-K/W
```

But we can be more specific by viewing all the parameters. Now we do the same for the condenser and reboiler of the column. In this case before sizing we need to select the column and then go to sizing heat exchangers.

For the condenser we use cooling water from 20°C to 30°C. And with the energy to be removed, the utility is calculated. For the reboiler we use saturated steam, the same utility as for the first heat exchanger; see Figure 12.37.

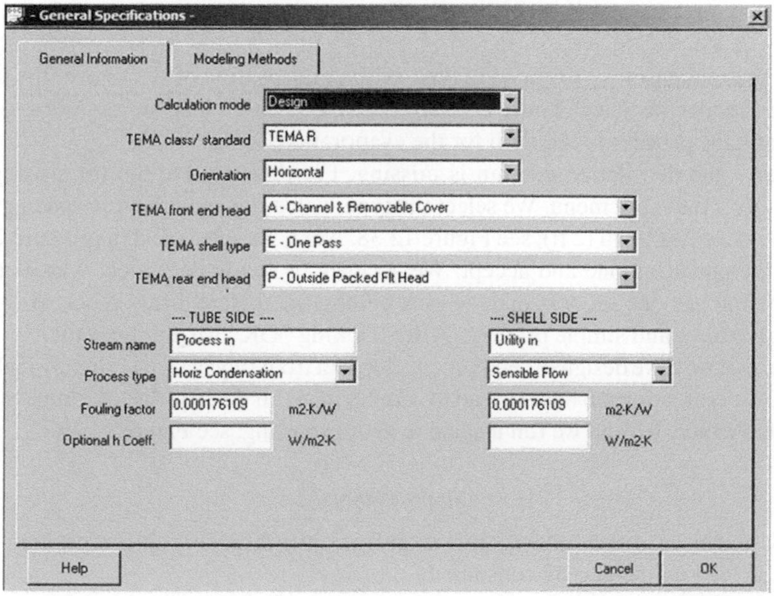

FIGURE 12.37 Heat exchanger design (VII).

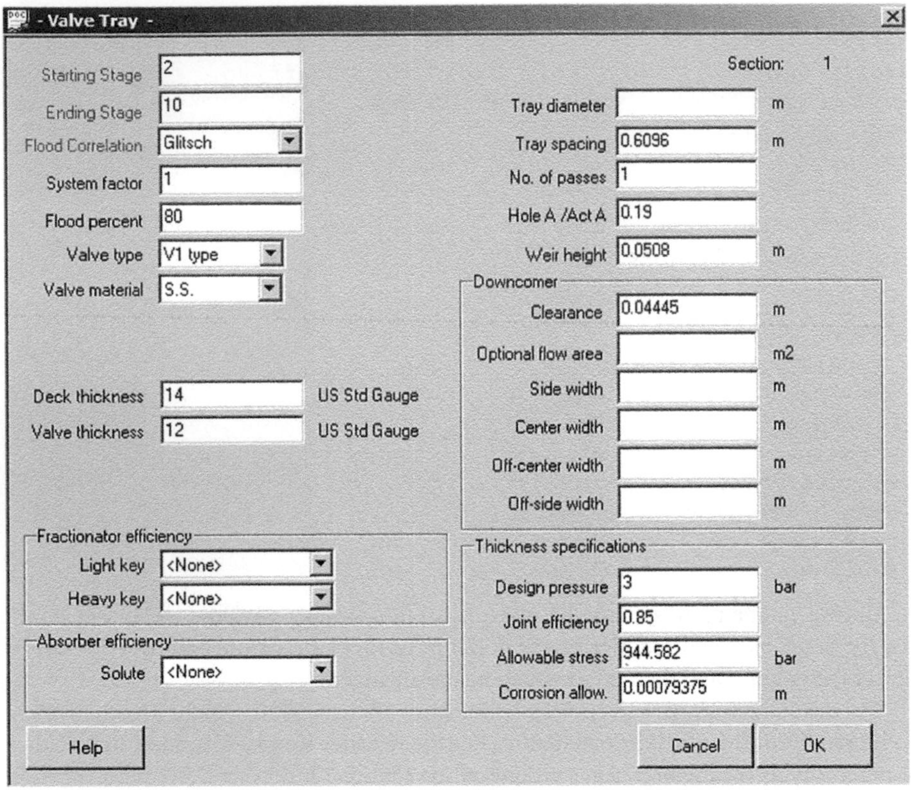

FIGURE 12.38 Tray sizing.

For the condenser we select AEP type, with condensation of steam over the tubes. For the reboiler we select Kettle reboiler and AKT, with the steam on tubes to condense and the process in the shell for the evaporation.

Finally, the distillation column is missing. For that we size the trays using the sizing tab of the main menu. We select the type, valve tray, and the tray spacing. The typical one is 0.6096 m (2 ft); see Figure 12.38. Next, we select design pressure 2 atm above the operating one and accept. We go back to sizing and select "distillation." Our column has one section, and we pick among the different tray types. We select here valve trays and simple printout. After clicking "OK" we get the results.

Now that we have design the equipment, the data from tray design is already included in the Cost estimation tab but we need to add the material density, that of stainless steel, the install factor, 1.7, and we run it again to get the costing; see Figure 12.39.

QUESTIONS

- Use the sensitivity analysis tool to evaluate the effect of the composition of the feed on the energy consumed.
- Use the optimization tool to optimize the reflux ratio of the column for minimum energy consumption.

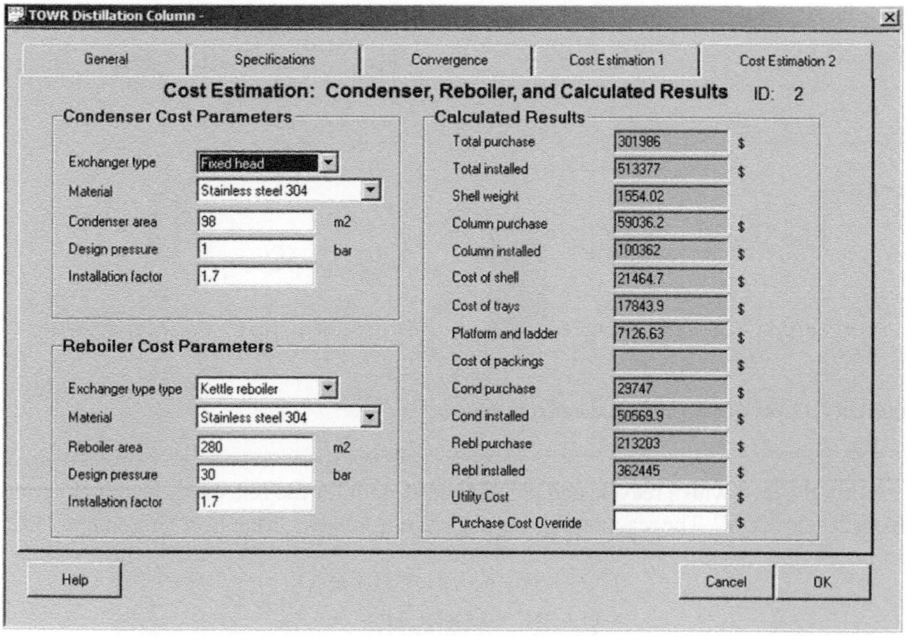

FIGURE 12.39 Costing distillation column.

CHEMCAD Page 1

Job Name: Wate ethanol Date: Time:

Vapor load is defined as the vapor from the tray below.

Liquid load is defined as the liquid on the tray.

Valve Tray Sizing for Equip. 2

Section: 1

Flood correlation: Glistch

Tray	Vapor kg/h	Liquid kg/h	Space m	NPass	Diameter m	%flood	PresDrop bar
2	20896.90	11226.11	0.61	1	1.98	72.04	0.0067
3	20312.64	10641.85	0.61	1	1.98	70.75	0.0066
4	20244.04	10573.25	0.61	1	1.98	70.60	0.0066
5	202312.41	10567.62	0.61	1	1.98	70.59	0.0066
6	20237.77	10566.99	0.61	1	1.98	70.59	0.0066
7	19162.65	109491.86	0.61	1	2.13	76.11	0.0093
8	18185.32	108514.54	0.61	1	2.13	75.75	0.0093
9	17271.64	107600.85	0.61	1	2.13	75.25	0.0093
10	16665.12	106994.33	0.61	1	2.13	75.14	0.0093

Total column pressure drop = 0.070 bar

12.8.1.3 Process Design: Methanol Process

The production of methanol from syngas is one of the best known processes and dates back to the beginning of 1900 when CO was hydrogenized to liquids. The process consists of a compression stage, since the methanol is produced in a reactor at 50 to 100 bar.

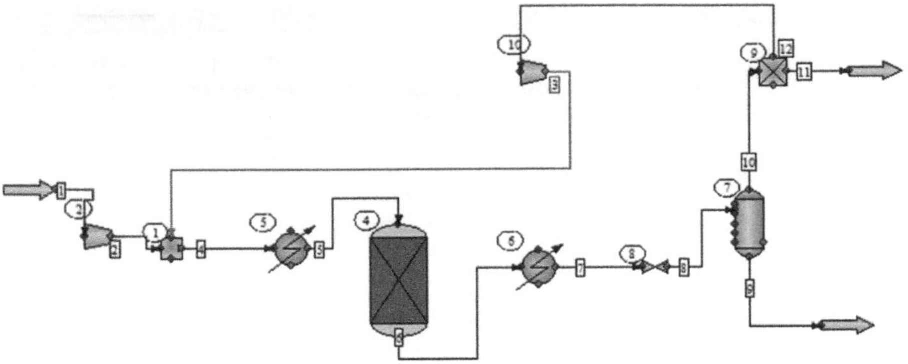

FIGURE 12.40 Methanol production flowsheet.

We mix the feed with a recycle and prepare the stream for the reactor by heating it up to from 200 to 300°C. The reactions taking place are as follows, Martín & Grossmann [49]:

$$CO_2 + 3H_2 \leftrightarrow CH_3OH + H_2O$$
$$CO + 2H_2 \leftrightarrow CH_3OH$$
$$CO_2 + H_2 \leftrightarrow CO + H_2O$$

Where only two out of the three reactions are independent. It is important to maintain from 2 to 8% of CO_2 in the feed to the reactor. The mixture is condensed and expanded in a valve and the phase, liquid (methanol and water) and gas (unreacted materials) are separated in a flash. We need to avoid accumulation of compounds in the systems, thus a purge is needed. Finally, we recycle the unreacted gases, compress them again and mix them with the feed; see Figure 12.40.

This process has the particularity of the recycle. To help converge the flowsheet it is convenient to tear the recycle stream and compute the exit of compressor 10 as a product and a new feed so that the systems is initialized; see Figure 12.41.

The thermodynamic model of choice is the NRTL since it works fine with polar compounds and can manage azeotropes. The feed stream is given as Figure 12.42. We compress it using a polytropic compressor until 50 bar assuming an efficiency of

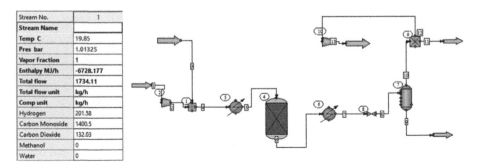

Stream No.	1
Stream Name	
Temp C	19.85
Pres bar	1.01325
Vapor Fraction	1
Enthalpy MJ/h	-6728.177
Total flow	1734.11
Total flow unit	kg/h
Comp unit	kg/h
Hydrogen	201.58
Carbon Monoxide	1400.5
Carbon Dioxide	132.03
Methanol	0
Water	0

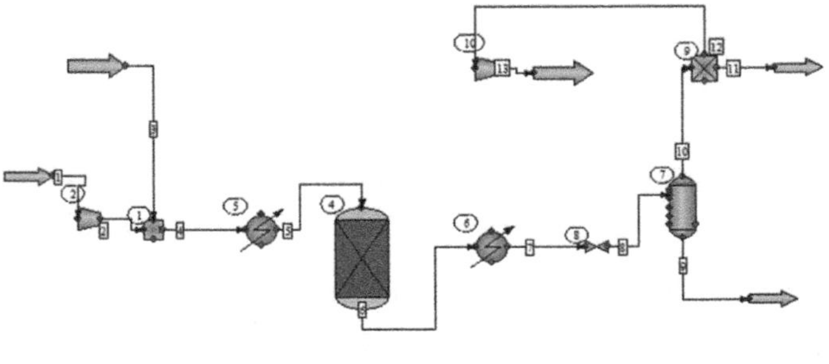

FIGURE 12.41 Tearing the recycle in methanol synthesis and input data.

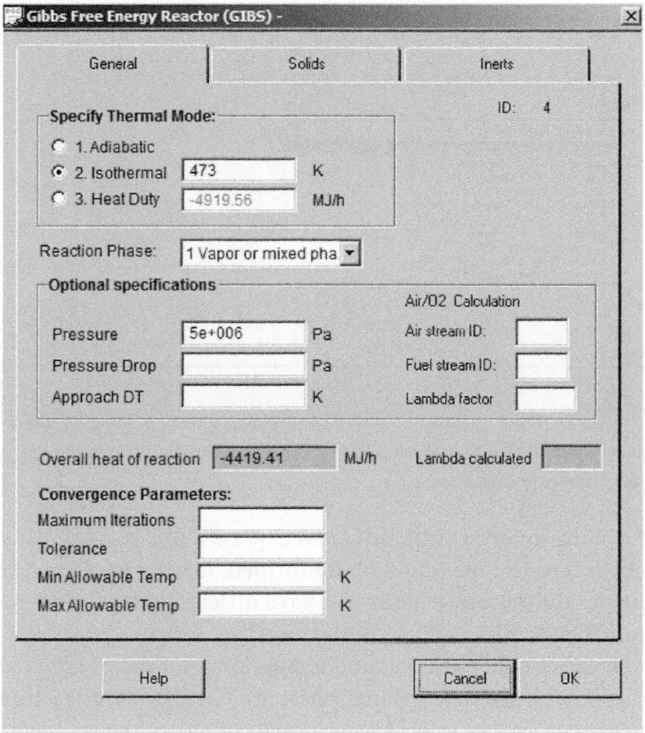

FIGURE 12.42 Equilibrium reactor specifications.

80%. We tick the box to perform the cost estimation assuming an installation factor of 1.7. We use a mixer and feed the recycle and the fresh feed to a heat exchanger to heat up the stream up to 200°C.

The selection of the reactor depends on the data available. We can use an equilibrium reactor and include the equilibrium constants or a Gibbs equilibrium reaction. For this particular case, we select a Gibbs reactor and we leave the use of the equilibrium reactor as exercise for the reader. Se select isothermal operation at the inlet temperature and the reaction will take place in vapor or mixed phase at 50 bar. We can select the presence of inerts in the third tab of the dialogue box in Figure 12.42.

When accepting, a component matrix appears representing the atomic balance (Figure 12.43).

	H	C	O
Hydrogen	2	0	0
Carbon-Monoxide	0	1	1
Carbon-Dioxide	0	1	2
Methanol	4	1	1
Water	2	0	1

FIGURE 12.43 Atomic balance coefficients in methanol synthesis.

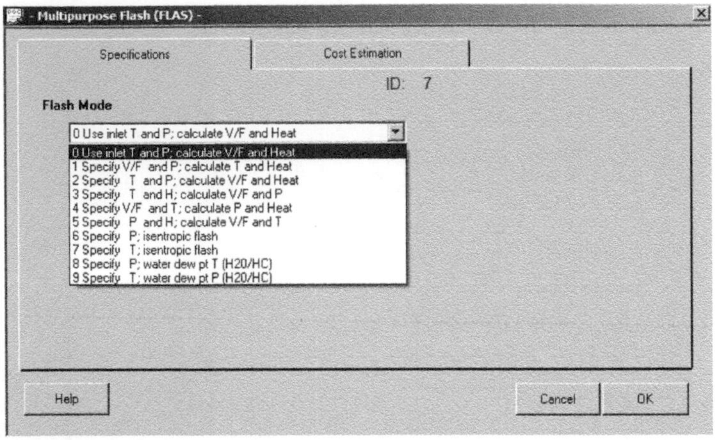

FIGURE 12.44 Flash specifications.

We cool down the mixture until 30°C and expand it to 2 bar. We use a flash after these two units to separate the liquid phase formed, mainly methanol, from the gas phase consisting of the untreated gases. We have different modes of operation for the flash; see Figure 12.44.

We select the use of inlet pressures and temperature for calculating the separation factor. The gas phase is sent to a splitter where we have the split fraction as a variable for convergence of the recycle. We use one to initialize the problem. In order to converge the recirculation, we follow the next procedure.

With the recycle open we run unit after unit to get a recycle. We input the results as a feed to the initial mixer and run it again. Once the streams are initialized, we erase the extra feed and product and close the loop of the recycle and run it again. We obtain the streams.

In order to price the equipment we need to size them. We go to the menu and size one equipment after the other. The compressors are done without further sizing due to the fact that the price depends on the energy for compression itself; see Figure 12.45.

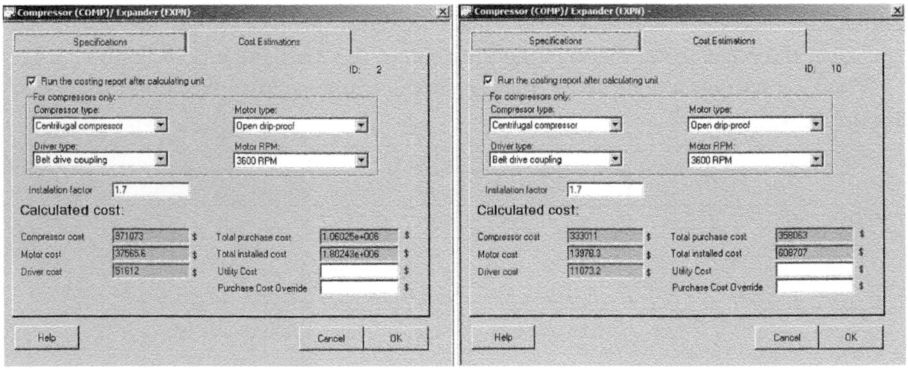

FIGURE 12.45 Compressor specifications.

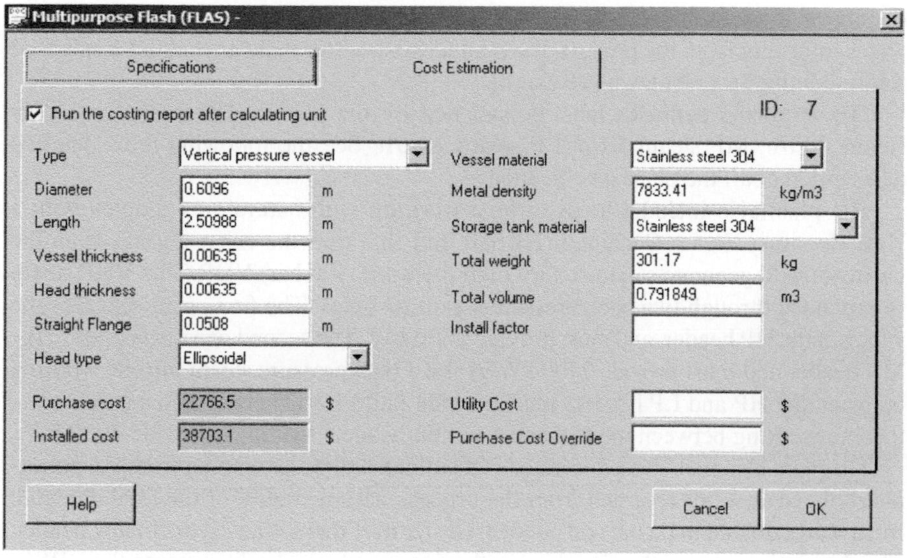

FIGURE 12.46 Flash costing.

Due to the compression rate, a series of compressors with intercooling should be considered. For the flash, we select sizing vessels LV (Liquid – Vapor). We use the data by default; see Figure 12.46. For the heat exchangers we follow the same steps as presented in the previous case.

QUESTIONS

- Use an equilibrium reactor to model the methanol synthesis. Hint: You need to transform the equilibrium constant equations from the literature [49] into the expression within CHEMCAD®.
- Use multistage compression with intercooling and compare the investment cost.

12.8.2 Implementation of a Utility System Power Plant in Aspen-Hysys™

We present the simulation of a combined heat and power plant. There are different commercial tools that can perform the simulation (or the synthesis) very efficiently (for example, Aspen Utility Planner™; or Ariane™ by ProSim™) using databases and cost correlations obtained directly from industrial applications. The objective here is to show that it is possible to simulate a complex system taking advantage of some of the particularities of Aspen-Hysys.

The plant we want to simulate uses fours steam headers: Very High Pressure (VHP) at 100 bar; High Pressure header (HP 40 bar); Medium Pressure header (MP 20 bar); Low Pressure header (LP 3 bar). Moreover, the plant includes a vacuum header for the condensing turbines working at 10 kPa that guarantees we can use cooling water as refrigerant for the condenser of those turbines, (the temperature of saturated water at 10 kPa is 46.1°C), and an atmospheric header that collects the water returning from the process, the treated fresh water makeup and the condensed steam coming from letdown valves.

The following demands must be satisfied by the plant: VHP steam 1 t/h; HP steam 20 t/h; MP steam 5 t/h; LP steam 10 t/h. See the following for a detailed description of all the plant specifications.

The plant also includes a gas turbine working with natural gas coupled with a Heat Recovery Steam Generator (HRSG) unit that uses the exhaust gases from the gas turbine to generate steam at very high pressure (100 bar) VHP. This VHP steam is expanded through two back pressure steam turbines. The first one discharges the vapor in the HP header, and must provide 2600 kW. The second one discharges in the MP header and must provide 2700 kW. A third backpressure steam turbine working between the HP and LP headers must provide 2800 kW. There is also a condensing turbine working between the MP and Vacuum headers that must provide 4000 kW.

The plant also includes a deaerator that collects the water coming from the atmospheric header: water returned from the process assumed 30% of the total demand; fresh water treated makeup; and condensed steam if there is an excess in any header. The deaerator removes the air solved in the water and returns the water to the HRSG. The deaerator is driven with LP steam from the own plant and it is assumed a vent rate of 5% of all the steam introduced. See Figure 12.47.

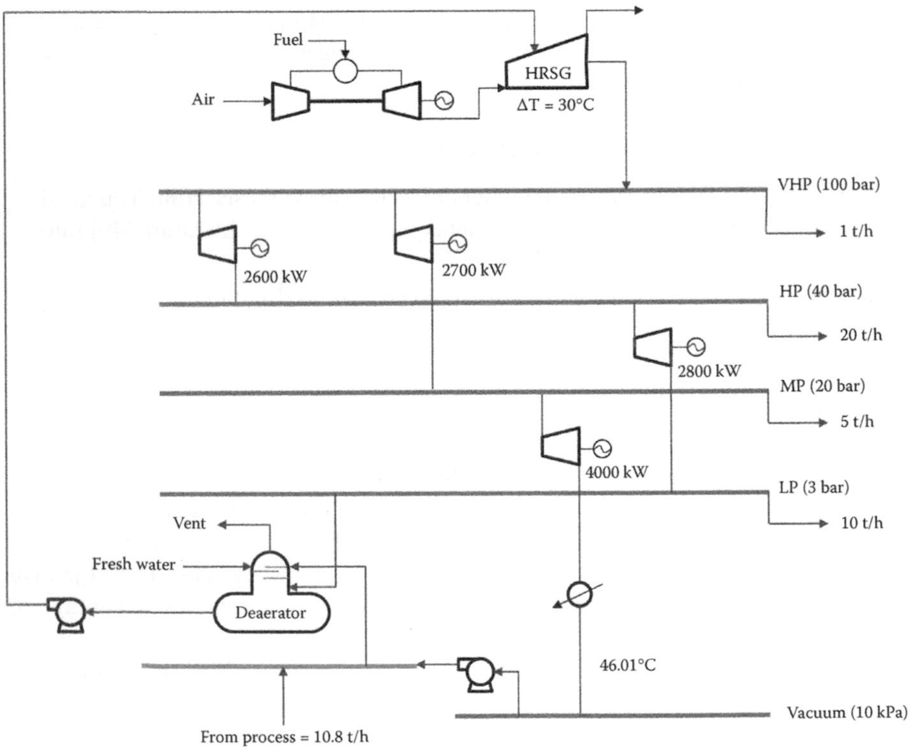

FIGURE 12.47 Utility power system scheme.

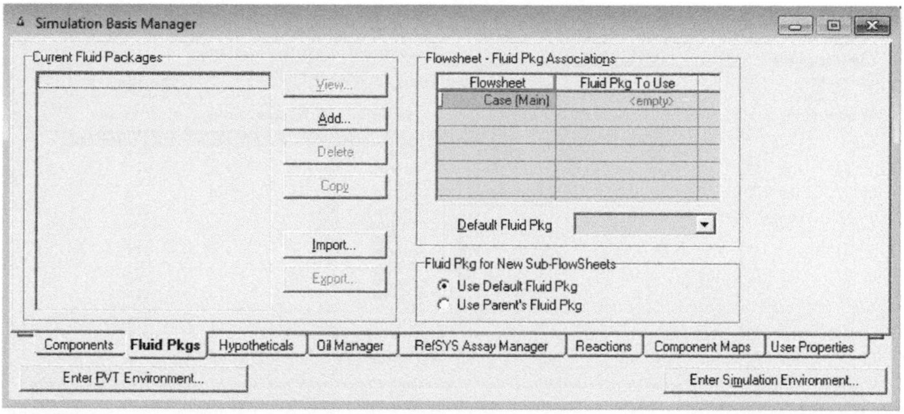

FIGURE 12.48 Initial window for selecting the fluid package in Hysys.

The major problem with this simulation is that we do not know the steam mass flow rate in the system. Moreover, the shaft power of a turbine depends on the pressure drop, the steam flowing through the turbine and on the inlet temperature of the steam. Besides, part of the LP steam is used in the deaerator but we do not known which is that flow.

Let us show step by step how to simulate the utility system. The first step is common to all the process simulators and consists of selecting the components and thermodynamics. When the user selects a "New" case in Hysys a window appears in which the user must select the "Fluid package"—Select the thermodynamics—and the components associated to that "Fluid Package." Figure 12.48 shows the initial window.

By clicking on the 'Add' button, a new window appears that allows selecting the thermodynamic package. In this case we use the Peng Robinson equation of state (Figure 12.49).

The components associated to a given fluid package can be selected by clicking on the "View" button under "Component List Selection." The components can be

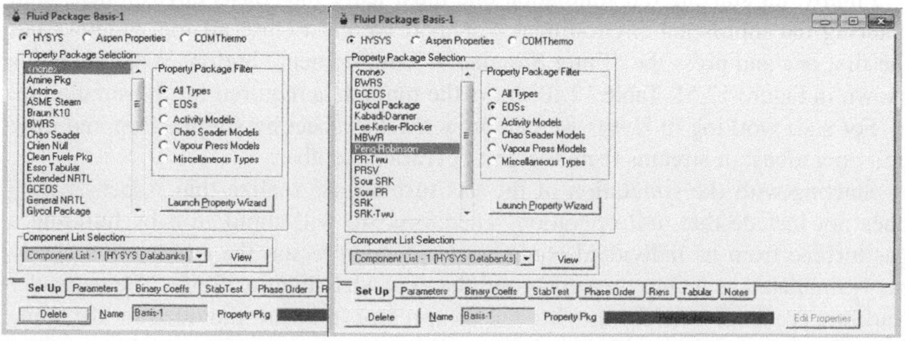

FIGURE 12.49 Selection of the Thermodynamic model.

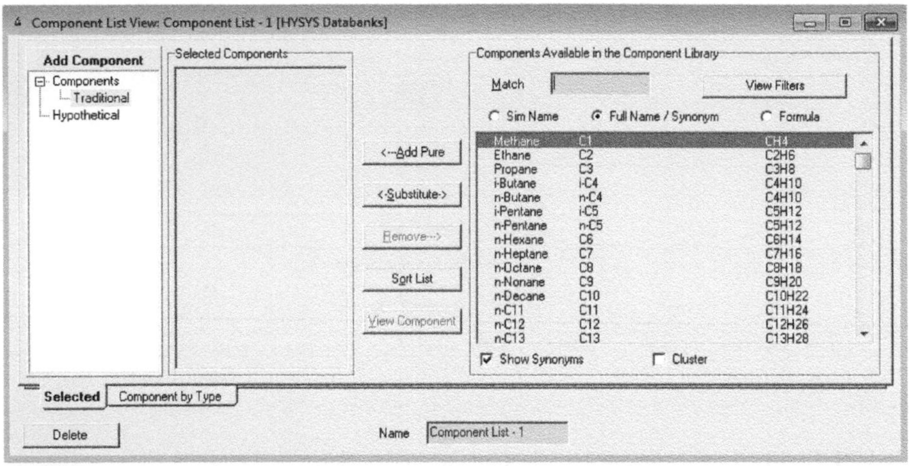

FIGURE 12.50 Window for component selection.

selected directly from the list (by clicking on the "Add Pure" button, removed from the list, sorted or substituted. It is possible to search components by name or formula and activate filters for searching by family (e.g. alcohols, amines, ketone, solids, etc). See Figure 12.50.

The HRSG can be simulated as a heat exchanger. Its function consists of recovering the energy from the exhaust gases and generating steam. HRSGs can be designed with or without extra fuel. In this example we assume that no extra fuel is used. The minimum approach temperature is fixed to 30°C. And the stack temperature for the exhaust gases fixed at 160°C (lower temperatures could produce corrosive condensations). It is also assumed that the water enters the HRSG at 80°C and 100 bar.

Here we must remark that with this information Aspen-Hysys is capable of calculating the steam flowrate (or water flowrate). Classical modular process simulators cannot perform this calculation because we are calculating an input in terms of outputs.

Finally, for starting the simulation one must leave the 'Basis environment' and entering the simulation environment. For that, we must close all the windows but the first one and press the "Enter Simulation Environment" button. What we get is shown in Figure 12.51. Table 12.10 shows the main data required for the simulation.

For start working in Hysys we can draw the flowsheet by simple drop and drag unit operations, or streams from the unit operations toolbar.

Starting with the simulation of the gas turbine, we realize that Aspen-Hysys® does not include that unit operation. Therefore, we will simulate it by building a gas turbine from its individual components: A compressor, the combustion chamber—simulated as a Gibss reactor—and the expander (Figure 12.52). The expander (turbine) provides work for moving the compressor (they are mounted on the same axis). The extra work provided by the gas turbine (net power) can be used to generate

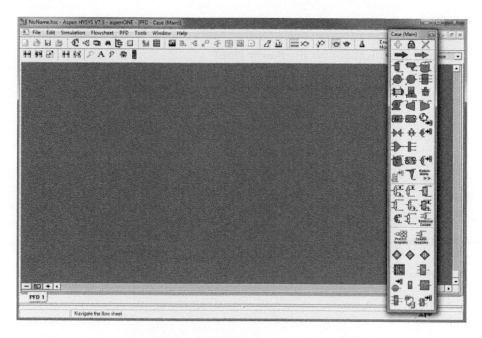

FIGURE 12.51 Simulation environment in Hysys.

electricity. This is simulated by an energy balance (the unit operation balance performs mass and/or energy balances).

The rest of specifications needed to simulate the gas turbine are the compression ratio (fixed at 20 kPa in this example) the ratio air to fuel (fixed at 53.888 in this example) and the fuel mass flow rate. At this moment we do not know how much fuel is needed. Therefore, provisionally we use any value (i.e. 1 t/h). We will go back to this point latter. A scheme of the simulation of the gas turbine is depicted in Figure 12.53.

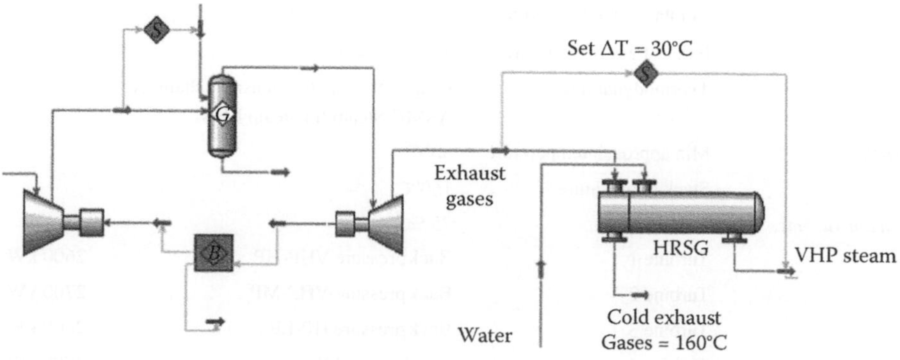

FIGURE 12.52 Gas turbine using HYSYS.

TABLE 12.10
Data and Specifications for the Power Plant

Utilities			
Fuel	Natural Gas		
	Composition (wt fraction)		
	Methane	0.8405	
	Ethane	0.1278	
	Propane	0.0203	
	n-Butane	0.0033	
	Ethylene	0.0081	
	Temperature	25°C	
	Pressure	101.325 kPa (1 atm)	
Cooling Water	Temperature	25–35°C	
	Pressure	101.325 kPa (1 atm)	
Demineralized Water	Temperature	25	
	Pressure	101.325 kPa (1 atm)	
Unit			**Demand**
Headers	VHP Pressure	100 bar	1 t/h
	HP Pressure	40 bar	20 t/h
	MP Pressure	20 bar	5 t/h
	LP Pressure	3 bar	10 t/h
	Condenser Pressure	1.01325 bar (1 atm)	
	Vacuum Pressure	0.1 bar (10 kPa)	
	Vacuum Temperature	46.01°C	
Gas Turbine	Compressor efficiency	83%	
	Turbine efficiency	85%	
	Mechanical efficiency	98 %	
	Compression ratio (*r*)	20	
	Combustion temperature	$T_{combustion}^{out} \leq 1200°C$	
	Exhaust gas temperature	$T_{Exhaust} \leq 600°C$	
	Thermodynamics	Peng Robinson (combustion chamber) ASME Steam for steam lines.	
HRSG	Min approach temperature	30°C	
	Stack temperature	160°C	
Steam Turbines	Efficiency	75 %	
	Turbine 1	Backpressure VHP-HP	2600 kW
	Turbine 2	Backpressure VHP-MP	2700 kW
	Turbine 3	Backpressure HP-LP	2800 kW
	Turbine 4	Condensing MP- Vac.	4000 kW
Deaerator	Pressure	1.01325 bar (1 atm)	
	Vent	5% LP introduced	

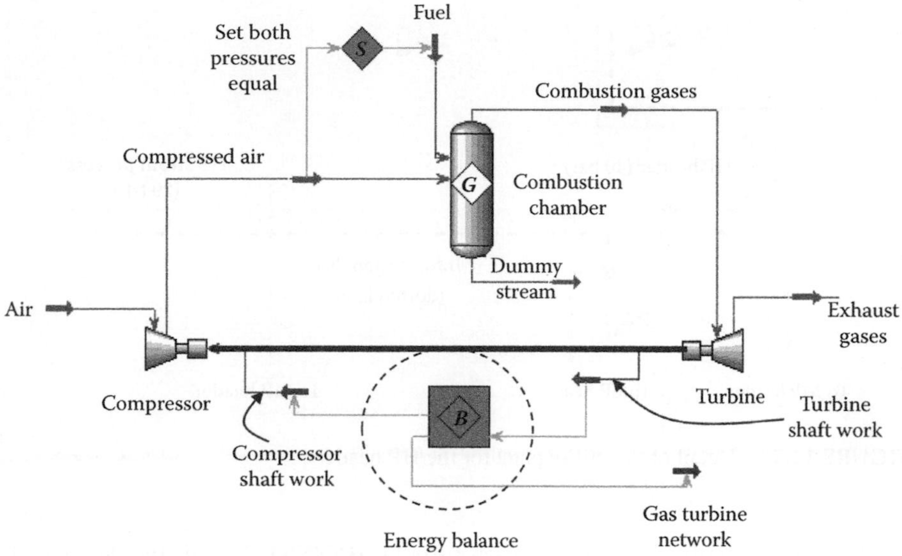

FIGURE 12.53 Detail of the gas turbine and the HRSG.

The VHP steam flow calculated depends on the fuel mass flow rate specified in the gas turbine. Note, however, that independently of the flow, the pressure, and temperature (and composition: pure water) are the VHP header conditions. So we will take advantage of this fact and create a new stream that is an exact copy of the "VHP Steam" stream except by the mass flow that will not be specified. We call it "VPH header."

The VHP header is simulated by a splitter with three streams (Figure 12.54). The first one is the VHP steam sent to the process to satisfy the demand. The other two are sent to the backpressure steam turbines.

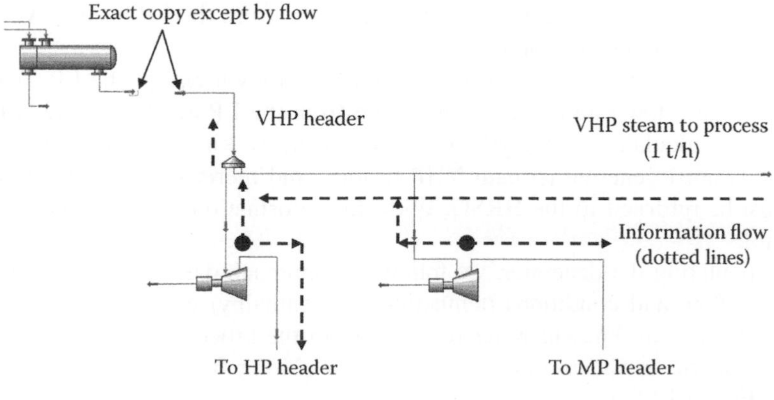

FIGURE 12.54 Scheme of the VHP header.

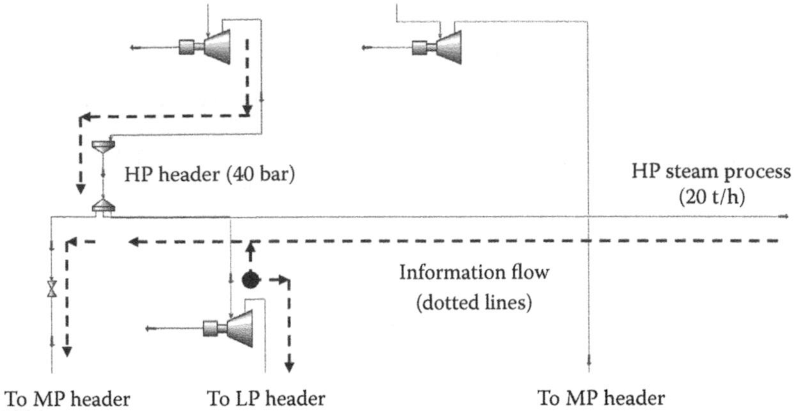

FIGURE 12.55 Detail of the utility plant for the HP headers.

If we now specify the power in each turbine, HYSYS® has all the degrees of freedom in each turbine satisfied and can calculate the steam flows. Note that again, the material flow and the information flow follow opposite directions.

To simulate the HP header we introduce a mixer that collects all the streams entering the HP header and a splitter that distributes the HP to turbines and process (Figure 12.55). The calculation is similar to the previous one: The flow is calculated, and the work is specified. Note that we would be very lucky if the HP steam flow requirements be exactly equal to the vapor entering the HP header. So, we introduce the possibility of sending part of the vapor by a letdown valve. (Note that if the flow through this valve is negative this means that we need more flow coming from the VHP header and we should introduce this extra stream between the VHP and HP headers).

The MP and LP headers are simulated equal than the HP header (Figure 12.56). The steam from the condensing turbine is pump to the atmospheric header that collects also the water returning from the process (fixed in a 30% of the total consumption) the treated fresh makeup water and the excess from the LP header. All these streams are sent to the deaerator.

As shown in Figure 12.56, all the streams are known except the LP steam to deaerator mass flow rate and the condensate from the LP header (if one is fixed, the other is calculated by Hysys). At this moment, we know the total steam that the HRSG must generate (stream VHP header) and therefore the water flowrate that must be returned to the HRSG. With this information, we can calculate the deaerator.

For simulating the deaerator, we follow the approach described by Smith [50]. The mass flow and conditions (temperature or enthalpy) of the following water (steam) streams are known: water returned from the process, water coming from the atmospheric or vacuum headers, and water exiting from the deaerator to the HRSG (Figure 12.57).

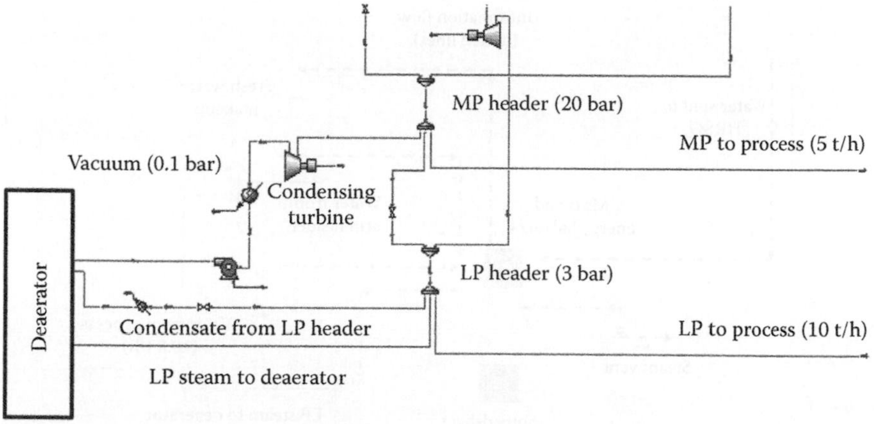

FIGURE 12.56 Detail of the utility plant for the MP and LP headers.

Assuming that the deaerator works with LP steam extracted from the LP main, and that a fixed percentage (5%) of the steam is vented, we can calculate the treated water makeup and the LP steam requirements just by mass and energy balances (Figure 12.58).

The final step consists of calculating the fuel needed to get the steam flow rate in the VHP header. This can be done by an "adjust" unit operation that simple varies the fuel flow until the mass flow rate of steam in the VHP Steam stream is equal to the VHP header steam mass flow rate.

Figure 12.59 shows the final complete flowsheet.

The following Table 12.11 shows the main results.

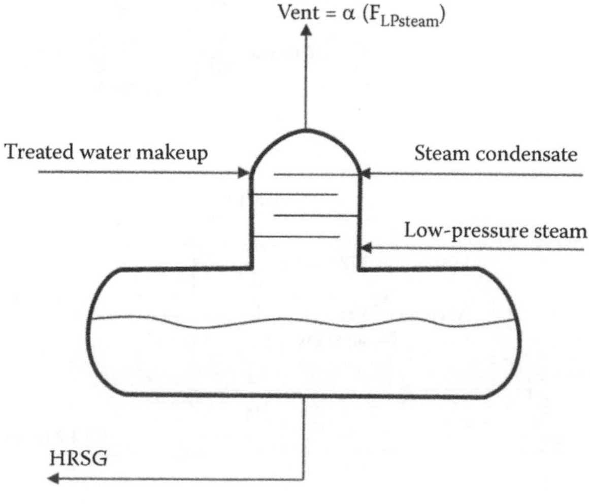

FIGURE 12.57 Deaerator detail.

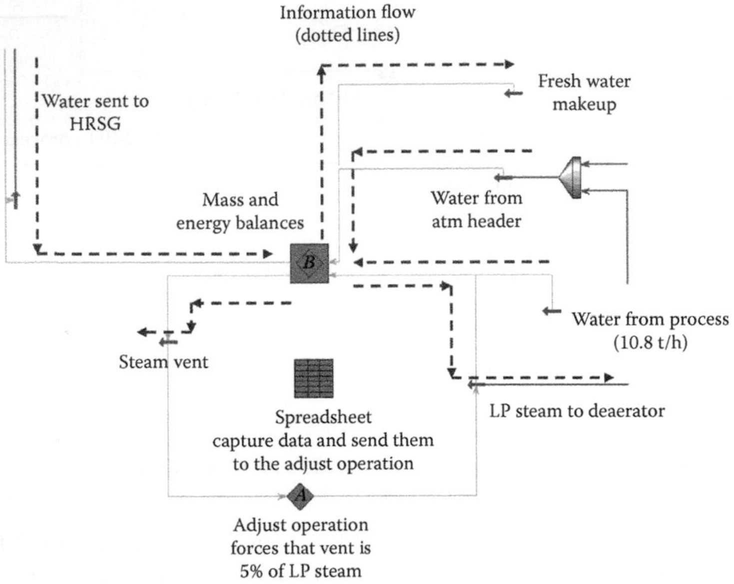

FIGURE 12.58 Simplified diagram.

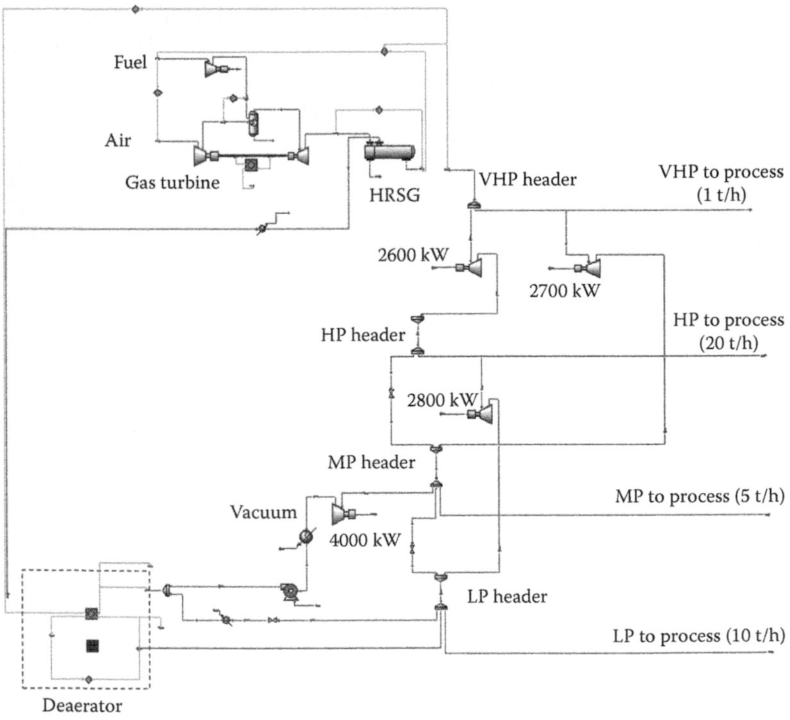

FIGURE 12.59 Final flowsheet implementation in Aspen Hysys.

TABLE 12.11
Results for Power Plant Example

Equipment	Variable	Value
Gas Turbine	Power (MW)	41.2 MW
	Fuel (t/h)	12.79
	Compression ratio	20
	Air to fuel ratio	53.89
	Exhaust gas temperature (°C)	575.1
HRSG	Water inlet temperature (°C)	80
	Steam mass flow rate (t/h)	71.57
Backpressure turbine 1	PIn – Pout (bar)	VHP(100) – HP (40)
	Work (kW)	2600
	Efficiency	0.75
	Steam Flow (t/h)	43.11
Backpressure turbine 2	PIn – Pout (bar)	VHP(100) – MP (20)
	Work (kW)	2700
	Efficiency	0.75
	Steam Flow (t/h)	27.45
	Installed cost (M$)	1.82
Backpressure turbine 3	PIn – Pout (bar)	HP (40) – LP(3)
	Work (kW)	2800
	Efficiency	0.75
	Steam Flow (t/h)	26.63
Condensing Turbine	PIn – Pout (bar)	MP (20) – Vacuum (0.1)
	Work (kW)	4000
	Efficiency	0.75
	Steam Flow (t/h)	20.72
	Condenser heat load (MW)	12.99
Deaerator	LP Steam flow rate (t/h)	12.14
	Fresh water makeup (t/h)	25.61
	Vent (t/h)	0.41
	Water from atm. header (t/h)	27.43

12.8.3 Process Simulation in Aspen Plus™

12.8.3.1 Start of Aspen Plus®

The following instructions will help to start the work:

1. **Start Aspen Plus V8.8:** Aspen's initial interface shows a window (Figure 12.60), where we have to choose "New" in "Installed Templates" then choose the option "Blank Simulation" to open a new "Template."
2. **Select the components:** Once the window is open, the first option that will appear will be loading the components that we will require in our process. To insert a component, just know its name and insert it in "Component Name" or there is also the option to enter the component by its chemical

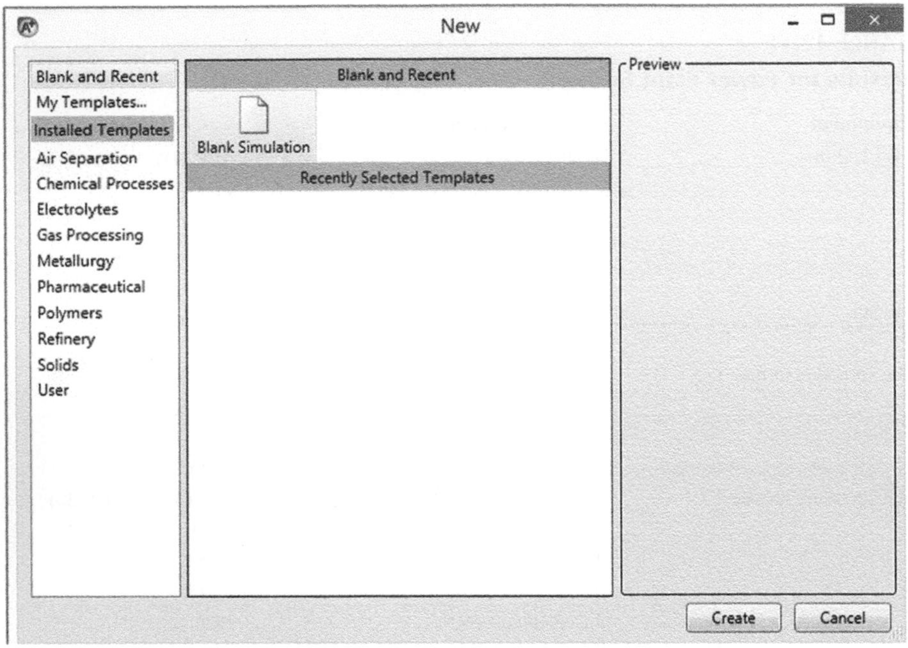

FIGURE 12.60 Start window of Aspen Plus V8.8.

formula in the "Alias" inside the dialogue box. If we enter it correctly, the "Conventional" label appears in the "Type" column. To proceed to the next stage of the thermodynamic method, we can do it with the "Next" icon or simply in the upper left corner look for the "Methods" icon.

3. **Start to assemble the process diagram:** Once the database of the components is generated, we will move on to the assembly of the process. To start, we have to search for the "Simulation" icon under "Properties" icon on the left side of the screen. Now we must have a look into the "Model Palette" to see the range of equipment that Aspen Plus V8.8 provides (Figure 12.61), as well as the streams (material, heat, and work) that will unite the processes.

12.8.3.2 Methanol Production

The conventional process for methanol production from gas synthesis, itself derived from oil, coal or, biomass. The process is adequately described in Chapter 12 (Section 12.8.1.3), where essentially the three reactions that are carried out are shown, the kinetics, in addition to the units and conditions necessary to carry out the reaction.

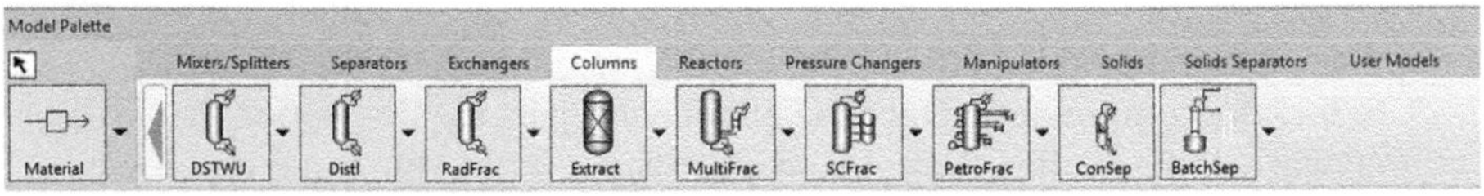

FIGURE 12.61 Model Palette window.

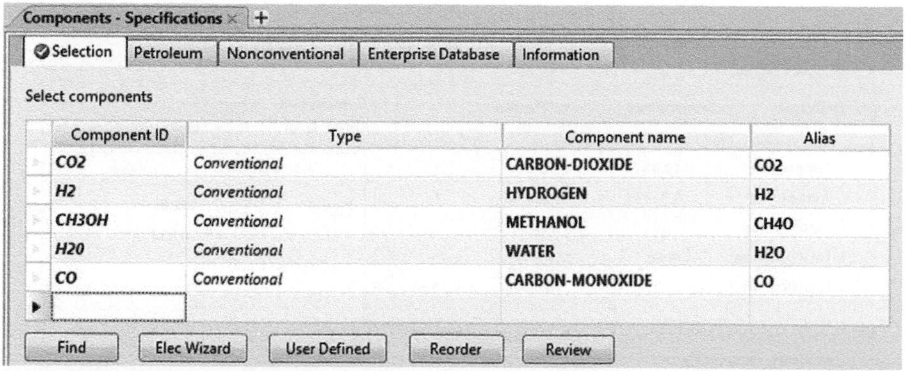

FIGURE 12.62 Components for the methanol production process.

12.8.3.2.1 Description of the Methanol Process using Aspen Plus

As mentioned in Section 12.8.1.3, the process consists basically in: 1) A compression stage, 2) A methanol production reactor, 3) A separation, 4) A reuse of the gases to close the process. Section 12.8.3.1 of this chapter provides the basic instructions for starting the use of Aspen Plus V8.8. With the above in mind, we will begin to load the components in the "Components" window, the necessary components are shown in Figure 12.62.

The next phase is the choice of the appropriate thermodynamic method, see Figure 12.63. For the components that are in the process, the NRTL method is

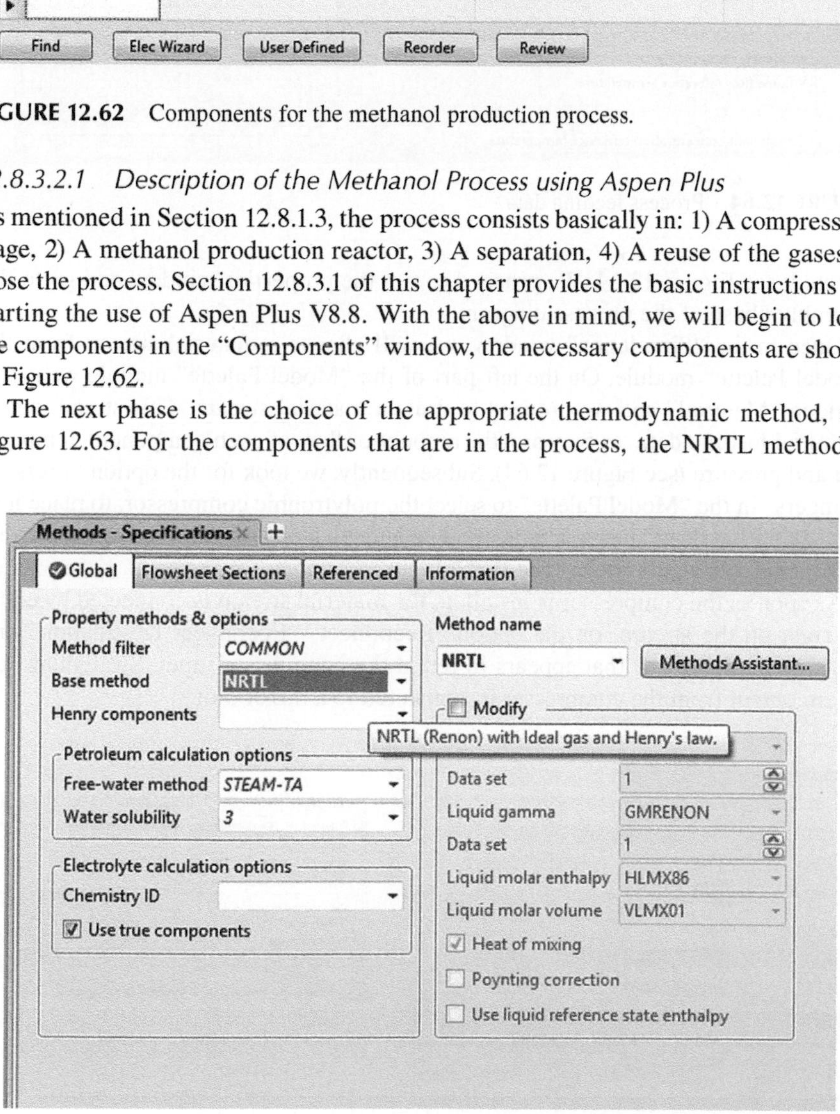

FIGURE 12.63 Selection of the thermodynamic method.

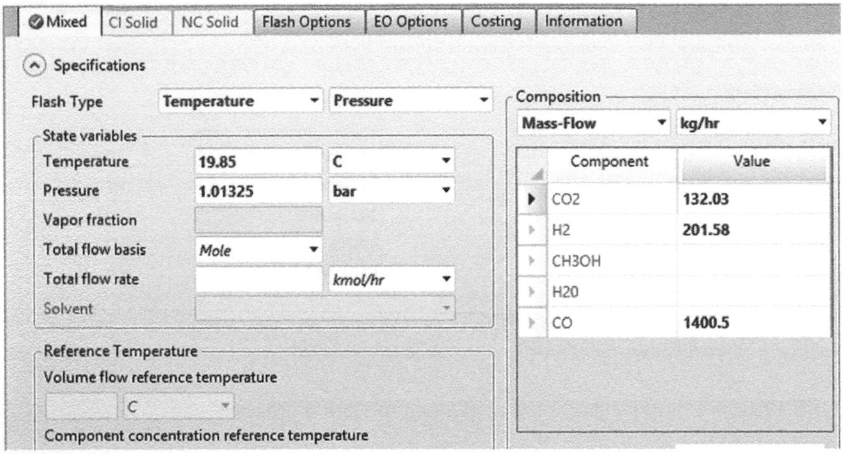

FIGURE 12.64 Process feeding data.

selected; see Figure 12.64. We proceed to run the properties window and go to the assembly phase of the process diagram.

Once in the "Simulation" window, we will choose a material stream from the "Model Palette" module. On the left part of the "Model Palette" module we select the icon "Material" and we proceed to draw a material stream. This stream needs the initial feeding data, as they are: the amount of flow of each component, temperature and pressure (see Figure 12.64). Subsequently, we look for the option "Pressure Changers" in the "Model Palette" to select the polytrophic compressor, to place it in the "Main Flowsheet" just select the module and click on the white screen, and select the specifications than we can see in the Figure 12.65.

As soon as the compressor is installed, the material stream is connected by clicking right on the stream, on the option "Reconnect"-"Reconnect Destination" and releasing on the arrow that appears in red at the compressor input. Straightaway, a stream output from the compressor is connected to a mixer unit.

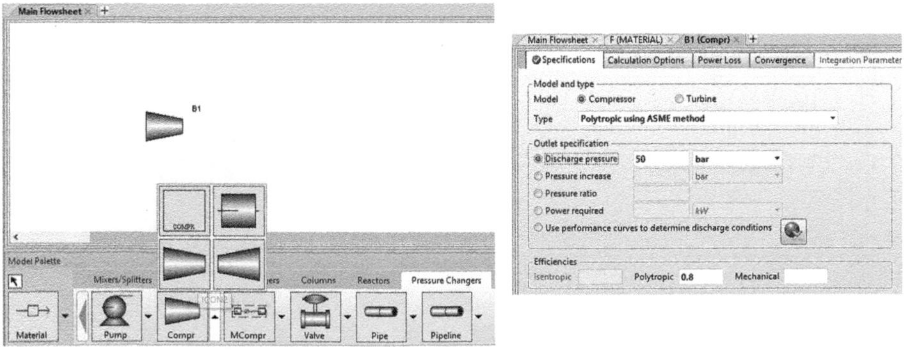

FIGURE 12.65 Compressor data.

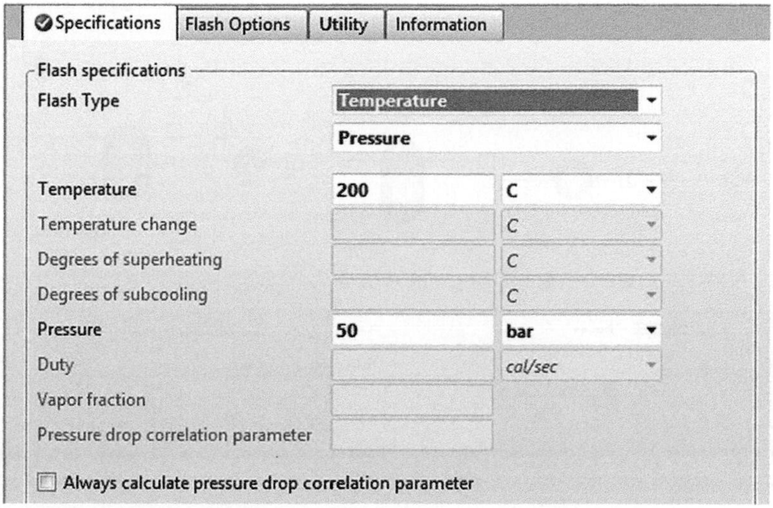

FIGURE 12.66 Operation conditions of the heater.

Then comes an exchanger to heat the mixture up to 200°C, to select the exchanger in the "Model Palette" we search for the option "Exchangers"-"Heater" and place it in the Main Flowsheet, double click on the "Heater" to place the operating conditions, and later we connect the output stream. See Figure 12.66.

The next phase is the choice of the reactor, we selected the Gibbs reactor. The Gibbs reactor is in "Reactors"-"RGibbs"-"ICON1", the reactor is connected with the material streams. In the window of the reactor we must have the option "Calculate phase equilibrium and chemical equilibrium"; the conditions of the reactor are $5{\cdot}10^6$ Pa of pressure and temperature of 473 K. It is important to ensure the results of the reactor. For this we run the simulation and we will see the results, as shown in Figure 12.67.

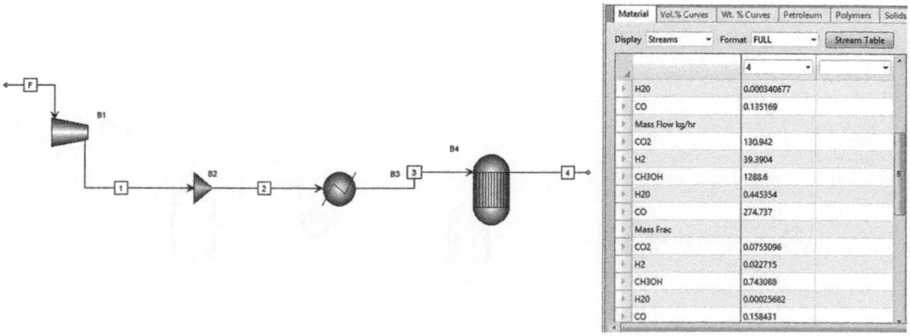

FIGURE 12.67 Gibbs reactor results.

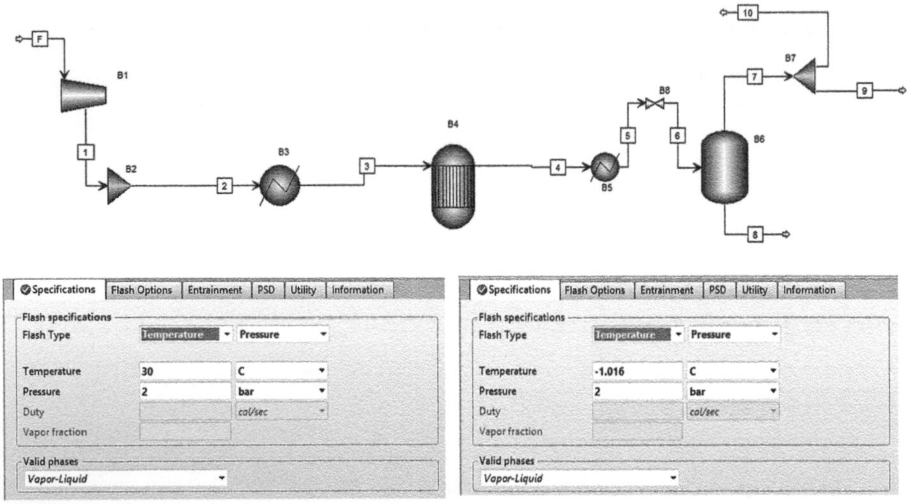

FIGURE 12.68 Flash separator and its operating conditions.

Then, an exchanger at 30°C and 2 bar, and an expansion valve with an outlet pressure of 2 bar are introduced. The process continues with a flash separation, which can be found in "Separators"-"Flash2"-"VDrum1." Figure 12.68 shows the operating conditions of the flash.

Immediately connect the output stream of the flash separator (the gas phase by the dome and the liquid by the bottom). The gas phase of the dome is divided into an outlet and a recycle stream, the second one requires a compressor with a discharge pressure of 50 bar, and at the end it is reconnected to the initial mixer-B2, the diagram of the process can be seen in Figure 12.69.

The results of the process are shown by clicking right on any stream we can see all the streams when you select them. In the table we can observe the mass flow, molar flow, temperatures, pressure, and other process conditions.

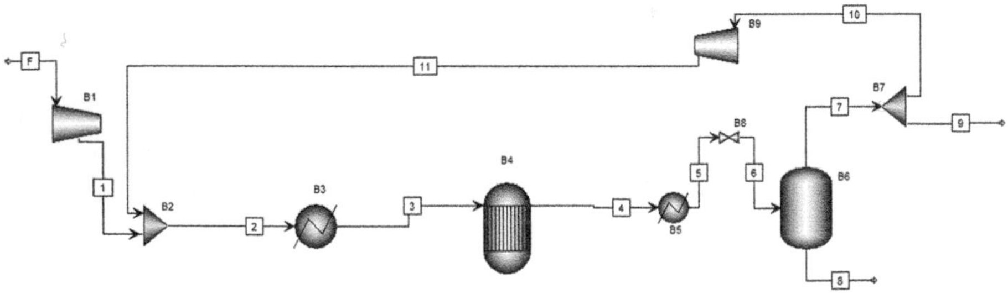

FIGURE 12.69 Complete flow chart.

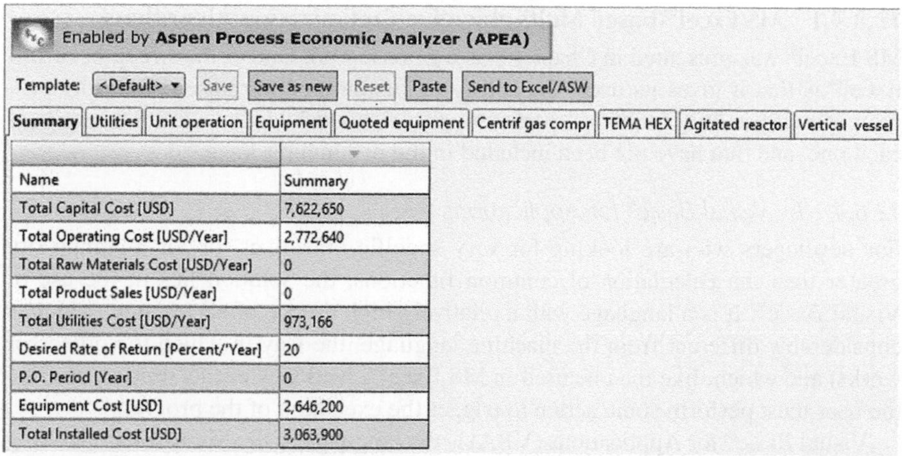

Name	Summary
Total Capital Cost [USD]	7,622,650
Total Operating Cost [USD/Year]	2,772,640
Total Raw Materials Cost [USD/Year]	0
Total Product Sales [USD/Year]	0
Total Utilities Cost [USD/Year]	973,166
Desired Rate of Return [Percent/'Year]	20
P.O. Period [Year]	0
Equipment Cost [USD]	2,646,200
Total Installed Cost [USD]	3,063,900

FIGURE 12.70 Process costs results.

Figure 12.70 shows the results of the process cost, to visualize them we must only activate the economic module that is above the "Main Flowsheet." The dimensions of the equipments are given automatically, just activate the option "Economics Summary."

QUESTIONS

- Use a Separator type "Sep"- "Icon1" and see the difference in process costs.
- Use a "RYield" reactor and verify the molar flows from the reactor outlet.

12.8.4 MS Excel® as Optimization Framework

Process optimization consists of finding the best operating conditions, the selection of the best technology or the best configuration of the process flowsheet. General search and optimization techniques are classified into three categories: enumerative, deterministic, and stochastic [51].

Computational software that allows to optimize is based on the use of mathematical tools to find the optimum of a specific problem. Two of the most commonly used commercial optimization programs are General Algebraic Modeling System (GAMS®), see Chapters 13, 14, and Matrix Laboratory (MATLAB®), see Chapter 4. MS Excel® can also be used to implement a stochastic optimization program so it can be used by itself for linking with the simulation software and optimizing the problem using parameters specified by the user, see Chapter 3 for generic use of Excel.

In this way, a stochastic optimization code can be developed in MS Excel® to solve the problem of the simulation program, and at the same time, interact repeatedly for the new values of each simulation. An interface between MS Excel® and the simulator software, based on COM technology, using MS Excel®-VBA scripts can be implemented [52].

12.8.4.1 MS Excel®-based Multi-objective Optimization Algorithms

MS Excel® was presented in Chapter 3 as a generic tool. One of the strengths of MS Excel® is that it gives its users the ability to customize their spreadsheets by programming their own functions, performing specific tasks, adjusted to the needs of each one, and that have not been included in the original package.

12.8.4.1.1 Visual Basic® for Applications

For developers who are looking for very specific objectives, or for a complexity greater than the calculation of common functions, the solution lies in the use of Visual Basic®. It is a language with a relatively high degree of abstraction (which is considerably different from the machine language, the way in which the processor works) and which, like the one used in MS Excel®, works by events (this means that the user must perform some action to trigger the execution of the program).

Visual Basic® for Applications (VBA) is the language of the Microsoft® operating system (Windows®) that is used for program applications. Many of the programs and add-ons that are used in Windows® are developed in this language so there are common elements that can be manipulated through this platform.

The combination of MS Excel® and Visual Basic® also has its limitations, and in no way can exceed the level of customization and precision possible through the creation of a program from scratch; but it is very effective and comfortable to the user. Due to the characteristics mentioned above, MS Excel® represents a viable alternative for implementing stochastic optimization algorithms.

12.8.4.1.2 Stochastic Optimization Algorithms

Deterministic methods are often ineffective when applied to nonconvex or other high-dimensional problems because they are limited by their requirements associated to the problem domain, knowledge (heuristics), and the search space, which can be exceptionally large [51].

There are optimization processes of industrial interest that involve functions that present a large number of local solutions and therefore it is very difficult to determine the optimal solution using deterministic optimization techniques. To solve these problems, there have been proposed stochastic search algorithms based on natural phenomena such as Simulated Annealing [53] and Genetic Algorithms [54]. These algorithms allow one to search for solution within a large number of local solutions.

12.8.4.2 MS Excel® as Linker Program

Microsoft® COM (Component Object Module) Technology in the Microsoft® Windows®-family of Operating Systems (OS) enables software components to communicate (microsoft.com), for this reason it is the technology used to achieve the link between the simulation software and the program in which the optimization algorithm is based. The details of the use of the COM Technology are described by [55] and [56].

A client-server interface based on COM Technology can be implemented. Using COM Technology, it is possible to add code so that the applications behave as an Object Linking and Embedding (OLE) Automation Server [57]. The use of the methods of this library to interoperate with other Windows® applications (such as MS Excel®) requires the use of a common scripting language, and VBA is the best option to this.

During the optimization process, a decision vector of design variables is sent from MS Excel® to the simulator software, in this process simulator, such as the ones describes along this chapter, rigorous calculations for the data that identify a particular design are obtained (e.g., temperature and pressure in the boiler, split fraction in the splitter, etc.) via solving the phase equilibria along with mass and energy balances and design relationships. These data are returned from the simulator software to MS Excel® to compute the objective functions, the obtained values for the objective functions are evaluated and new vectors of design variables are generated according to the stochastic procedure of this method.

12.8.4.3 Example

For this example, it was taken a simple simulation presented by Sandler [58], which was developed in the process simulator software Aspen Plus®. The considered process corresponds to the liquefaction of propane. This process starts with propane vapor at ambient conditions (298 K and 1 bar), which is compressed to 15 bar, cooled back down to 298 K, expanded through an adiabatic valve to 1 bar, and then the resulting gaseous and liquid streams are separated. Since propane is not an ideal gas throughout the process because of the conditions (high pressure and the presence of both a gas and a liquid), the Peng-Robinson equation of state will be used as the thermodynamic model, though other equations of state could have been chosen.

12.8.4.3.1 Constructing the Process Flowsheet

We start by opening the Aspen Plus User Interface. N-propane will be chosen as the single component, and the Peng-Robinson equation of state as the thermodynamic model. Next click on "Simulation," which brings up a blank Main Flowsheet window with the Model Palette.

Place the process units of this liquefaction example, starting with the Compressor, the Heater, the Valve and finally add the Flas2 (two-phase) separator to the flow sheet. Next, we have to connect the material flow streams to the process equipment. Rename the blocks (process units) and streams with names easier to recognize in the output. The revised flowsheet based on the instructions appears in Figure 12.71.

To proceed, the user must now provide the specifications for the process, which includes the inlet stream component(s) and conditions, pressures and temperatures where needed, and the type of each process unit or block according to Table 12.12.

After running the simulation, going to Results Summary and the Streams brings up a window containing the table of stream results.

12.8.4.3.2 Selecting the Stochastic Optimization Algorithm

After we have constructed the process flowsheet, we proceed to select the optimization algorithm. Due to the rigorous solution and the numerical methods used in the process simulator software, the nature of the problem is commonly highly nonconvex.

As mentioned before, deterministic methods are ineffective when applied to nonconvex or high-dimensional problems and stochastic searches have been developed as alternative approaches for solving this type of problems.

The multi-objective optimization hybrid method namely the Improved Multi-Objective Differential Evolution (I-MODE), developed by Sharma and Rangaiah [59],

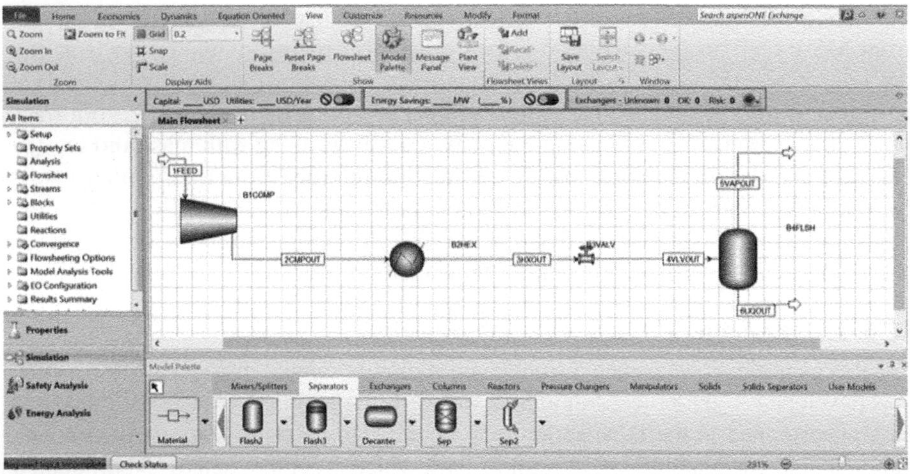

FIGURE 12.71 Complete flowsheet for the example.

was chosen as the stochastic optimization algorithm in this example, but another metaheuristic technique can be selected.

The first step to use the I-MODE is to open the User Interface sheet (Figure 12.72). The I-MODE algorithm has four fundamental parts in its Main Program Interface, which are: Objective Functions, Design Variables, Inequality Constraints and Algorithm Parameters. For a detailed description of all the sheets that compose the I-MODE algorithm, it is recommended to review the User Guide [59]. And for

TABLE 12.12
Specifications of the Process Units

Stream/Block	Variable	Value	Units
1FEED	Temperature	298	K
	Pressure	1	bar
	Total Flow Rate	1	kg/min
	Composition	1	kg/min
B1COMP	Type	Isentropic	-
	Discharge Pressure	15	bar
	Efficiency Isentropic	1	-
	Efficiency Mechanical	1	-
B2HEX	Temperature	298	K
	Pressure	15	bar
B3VALV	Calculation Type	Adiabatic Flash	-
	Outlet Pressure	1	bar
B4FLSH	Pressure	1	bar
	Duty	0	cal/sec

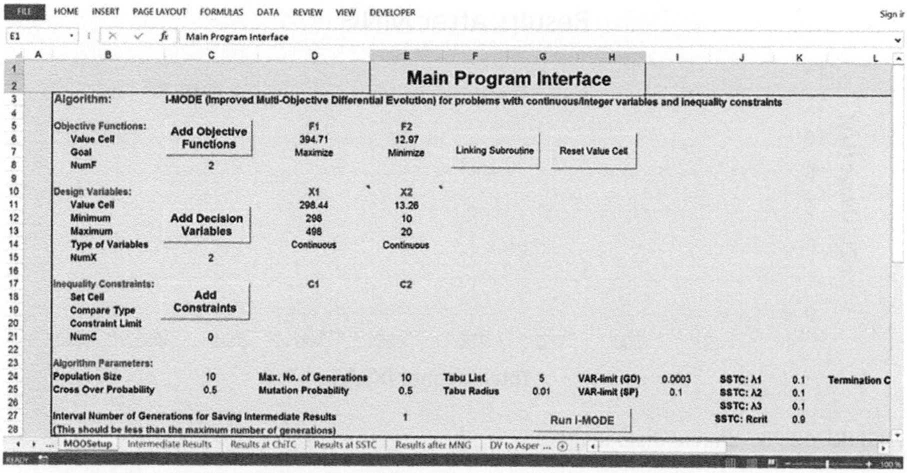

FIGURE 12.72 User interface for IMODE.

a detailed description of the procedure to link Aspen Plus® and MS Excel®, please consult Ponce-Ortega & Hernández-Pérez [56].

12.8.4.3.3 Model Formulation

This example presents the application of an optimization scheme using metaheuristic techniques in the case of the Liquefaction of Propane. It is considered the optimization of two decision variables, which correspond to the temperature in the heater and the pressure in the compressor (which must be the same as the pressure in the heater). These search variables directly and significantly impact on the objective functions.

There are considered two objective functions, one economic and the other environmental. The economic objective function consists in the maximization of the Total Annual Income. The economic objective function is calculated using a response variable calculated in the simulation software. Aspen Plus offers the total flow (kg/h) of the stream 6LIQOUT, which corresponds to the LP Gas, whit this value is possible calculate the Total Annual Income multiplying by 2.48 \$/gal (price of LP Gas), 1/3.7854 (gal/l), 1.92 l/kg, 24 h/d and 360 d/y. The economic objective function is expressed in thousands of dollars per year (k\$/y).

The environmental objective function consists in the minimization of the Total Annual H_2O Usage associated with the cooling utility needed in the heater. The environmental objective function is calculated using a response variable calculated in the simulation software. Aspen Plus offers the total utility usage (kg/h) of water for cooling utility in the heat exchanger. This value is obtained assigning a utility (cooling water) to the block B2HEX. With this value is possible to calculate the Total Annual H_2O Usage multiplying by 24 h/d and 360 d/y. The environmental objective function is expressed in tons per year (t/y).

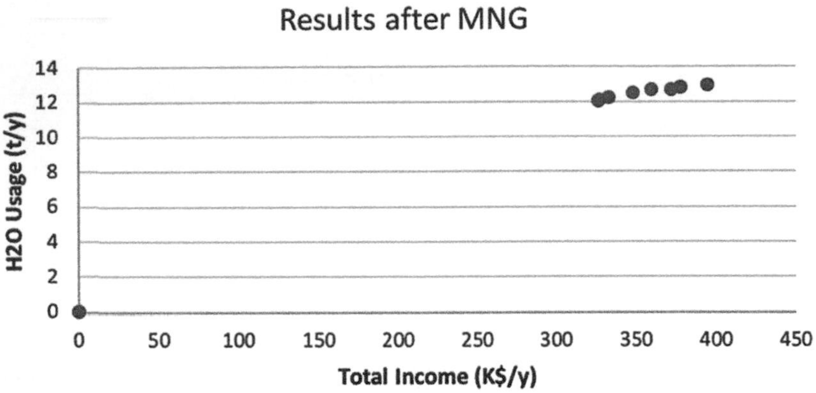

FIGURE 12.73 Pareto diagram.

TABLE 12.13
Results for the Simulation after Optimization

Heat and Material Balance Table							
Stream ID		1FEED	2CMPOUT	3HXOUT	4VLVOUT	5VAPOUT	6LIQOUT
From			B1COMP	B2HEX	B3VALV	B4FLSH	B4FLSH
To		B1COMP	B2HEX	B3VALV	B4FLSH		
Phase		VAPOR	VAPOR	LIQUID	MIXED	VAPOR	LIQUID
Substream: MIXED							
Mole Flow	kmol/hr						
PROPA-01		1.360652	1.360652	1.360652	1.360652	.5370488	.8236028
WATER		0.0	0.0	0.0	0.0	0.0	0.0
Total Flow	kmol/hr	1.360652	1.360652	1.360652	1.360652	.5370488	.8236028
Total Flow	kg/hr	60.00000	60.00000	60.00000	60.00000	23.68198	36.31802
Total Flow	l/min	552.5534	50.31697	2.033110	166.9861	165.9467	1.039394
Temperature	C	24.85000	121.0469	25.28980	-42.53807	-42.53807	-42.53807
Pressure	bar	1.000000	13.26451	13.26451	1.000000	1.000000	1.000000
Vapor Frac		1.000000	1.000000	0.0	.3946997	1.000000	0.0
Liquid Frac		0.0	0.0	1.000000	.6053003	0.0	1.000000
Solid Frac		0.0	0.0	0.0	0.0	0.0	0.0
Enthalpy	cal/mol	-25031.77	-23349.12	-28848.87	-28848.87	-26144.56	-30612.27
Enthalpy	cal/gm	-567.6587	-529.5003	-654.2209	-654.2209	-592.8940	-694.2105
Enthalpy	cal/sec	-9460.978	-8825.005	-10903.68	-10903.68	-3900.252	-7003.430
Entropy	cal/mol-K	-64.35947	-64.35946	-81.36750	-80.30731	-68.58067	-87.95393
Entropy	cal/gm-K	-1.459513	-1.459513	-1.845214	-1.821171	-1.555240	-1.994578
Density	mol/cc	4.10413E-5	4.50693E-4	.0111541	1.35805E-4	5.39379E-5	.0132064
Density	gm/cc	1.80978E-3	.0198740	.4918573	5.98852E-3	2.37847E-3	.5823587
Average MW		44.09652	44.09652	44.09652	44.09652	44.09652	44.09652
Liq Vol 60F	l/min	1.976215	1.976215	1.976215	1.976215	.7800115	1.196203

12.8.4.3.4 Results

After running the optimization algorithm, we can analyze the results. Figure 12.73 shows a pareto graphic with the possible solutions. The Graphic after Maximum Number of Generations (MNG) represents a Termination Criterion specified by the user in the Main Program Interface of the I-MODE algorithm.

The results of the simulation with the optimum value of the decision variables are shown in Table 12.13.

REFERENCES

1. White, J.A., (1968) Development of a modified pacer simulation program for using a chemical engineering curriculum. Department of Chemical and Petroleum Engineering. University of Kansas.
2. Gaddy, J.L., (1974) The use of flowsheet simulation programs in teaching chemical engineering design. *Chem. Eng. Educ.* 8(3), 124–129.
3. Goldstine, H.H., (1993) *The Computer from Pascal to von Neumann.* Princeton, NJ: Princeton University Press.
4. Shannon, P.T., (1963) The Integrated Use of the Digital Computer. *Chem. Eng. Educ.* March, 11–22.
5. Brannock, N.F.; Verneuil, V.S.; Wang, Y.L., (1979) PROCESSSM Simulation Program a comprehensive flowsheeting tool for chemical engineers. *Comput. Chem. Eng.* 3(1–4), 329–352.
6. Motard, R.L., Lee, H.M., Barkley, R.W., (1969) *Chess: chemical engineering simulation system. user's guide.* Houston, TX: University of Houston, Dept of Chemical-Petroleum, Engineering Defense Technical Information Center.
7. Seader, J.D., Seider, W.D., Pauls, A.C.; Hughes, R.R., (1974) *Flowtran Simulation—An introduction.* Ann Arbor, MI: National Academy of Engineering CACHE Committee, Monsanto Company.
8. Aylott, M.R., Ponton, J.W., Lott, D.H., (1985) *Development of a dynamic flowsheeting program.*
9. Berger, F., Perris, F.A., (1979) Flowpack II—a new generation of system for steady state process flowsheeting. *Comput. Chem. Eng.* 3(1–4), 309–317.
10. Parker, P.E., (1981) *DYSCO, a modular, dynamic simulation for the pulp and paper industry.* Vol. IPC technical paper series number 103. Wisconsin: The institute of paper chemistry, Appleton.
11. Locke, M.H., Kuru, S., Clark, P.A., Westerberg, A.W., (1980) ASCEND-II: An Advanced System for Chemical Engineering Design. *Modeling and Simulation, Proceedings of the Annual Pittsburgh Conference,* (pt 2), 409–413.
12. Gorczynski, E.W., Hutchison, H.P., Wajih, A.R.M., (1979) Development of a modulary organised equation-oriented process simulator. *J. Microwave Power.* 1, 568–575.
13. Braunschweig, B.L., Pantelides, C.C., Britt, H.I., Sama, S., (2000) Process modeling: The promise of open software architectures. *Chem. Eng. Prog.* 96(9), 65–76.
14. CO-LaN. *the CAPE-OPEN Laboratories Network.* 2003–2013; Available from: http://www.colan.org/.
15. Barrett Jr, W.M.; Yang, J., (2005) Development of a chemical process modeling environment based on CAPE-OPEN interface standards and the Microsoft .NET framework. *Comput. Chem. Eng.* 30(2), 191–201.
16. Kooijman, H.A., Taylor, R., (2006) *The ChemSep Book.* 2nd ed, http://chemsep.org/book/docs/book2.pdf.
17. Dimian, A.C., (2003) Integrated design and simulation of chemical processes. 1st ed. *Computer-aided chemical engineering.* Amsterdam, The Netherlands: Elsevier.

18. Carlson, E.C., (1996) Don't gamble with physical properties for simulations. *Chem. Eng. Prog. October,* 35–46.
19. Fredenslund, A., Jones, R.L., Prausnitz, J.M., (1975) Group-contribution estimation of activity coefficients in nonideal liquid mixtures. *AIChE J.* 21(6), 1086–1099.
20. Westerberg, A.W., Hutchison, H.P., Motard, R.L., Winter, P., (1979) *Process Flowsheeting*. Cambridge: Cambridge University Press.
21. Prigogine, I., Defay, R., (1954) *Chemical Thermodynamics* (transl. D. H. Everett). London: Logmans.
22. Sargent, R.W.H., Westerberg, A.W., (1964) 'SPEED UP' in chemical engineering deisng. *Trans. Inst. Chem. Eng.* 42, 190–197.
23. Lee, W., Rudd, D.F., (1966) On the ordering of recycle calculations. *AIChE J.* 12(6), 1184–1190.
24. Christensen, J.H., Rudd, D.F., (1969) Structuring design computations. *AIChE J.* 15(1), 94–100.
25. Forder, G.J., Hutchison, H.P., (1969) The analysis of chemical plant flowsheets. *Chem. Eng. Sci.* 24(4), 771–785.
26. Ledet, W.D., Himmelblau, D.M., (1970) Decomposition procedures for the solving of large scale systems. *Adv. Chem. Eng.* 8, 186.
27. Barkley, R.W., Motard, R.L., (1972) Decomposition of nets. *Chem. Eng. J.* 3(C), 265–275.
28. Upadhye, R.S., Grens, E.A., (1972) An efficient algorithm for optimum decomposition of recycle systems. *AIChE J.* 18(3), 533–539.
29. Pho, T.K., Lapidus, L., (1973) Topics in computer-aided design: Part I. An optimum tearing algorithm for recycle systems. *AIChE J.* 19(6), 1170–1181.
30. Kehat, E., Shacham, M., (1973) Chemical Process Simulation Programs. II. Partitioning and tearing system flowsheets. *Process Technol.* 18, 115–1112.
31. Lap-Kit, C., Kuh, E.S., (1974) The bordered triangular matrix and minimum essential sets of a digraph. *Circuits and Systems, IEEE Transactions on.* 21(5), 633–639.
32. Guardabassi, G., (1974) An indirect method for minimal essential sets. *Circuits and Systems, IEEE Transactions on.* 21(1), 14–17.
33. Smith, G., Walford, R., (1975) The identification of a minimal feedback vertex set of a directed graph. *Circuits and Systems, IEEE Transactions on.* 22(1), 9–15.
34. Tarjan, R., (1972) Depth-First Search and Linear Graph Algorithms. *SIAM J. Computing.* 1(2), 146–160.
35. Duff, I.S., Erisman, A.M., Reid, J.K., (1997) *Direct Methods for Sparse Matrices.* Oxford: Oxford Science Publications.
36. Duff, I.S., Reid, J.K., (1978) An Implementation of Tarjan's Algorithm for the Block Triangularization of a Matrix. *ACM Trans. Math. Softw.* 4(2), 137–147.
37. Duff, I.S., Reid, J.K., (1978) Algorithm 529: Permutations To Block Triangular Form [F1]. *ACM Trans. Math. Softw.* 4(2), 189–192.
38. Motard, R.L., Westerberg, A.W., (1981) Exclusive tear sets for flowsheets. *AIChE J.* 27(5), 725–732.
39. Orbach, O., Crowe, C.M., (1971) Convergence promotion in the simulation of chemical processes with recycle-the dominant eigenvalue method. *Canadian J. Chem. Eng.* 49(4), 509–513.
40. Wegstein, J.H., (1958) Accelerating convergence of iterative processes. *Commun. ACM.* 1(6), 9–13.
41. Crowe, C.M., Nishio, M., (1975) Convergence promotion in the simulation of chemical processes—the general dominant eigenvalue method. *AIChE J.* 21(3), 528–533.
42. Biegler, L.T., Grossmann, I.E., Westerberg, A.W., (1997) Systematic methods of chemical process design. *Prentice Hall international series in the physical and chemical engineering sciences.* Upper Saddle River, NJ: Prentice Hall PTR.

43. Cavett, R.H., (1963) Application of numerical methods to the convergence of simulated processes involving recycle loops. *Am. Pet. Inst., Repr. No 04-63.*
44. Nocedal, J., Wright, S.J., (2006) *Numerical Optimization.* 2nd ed. Springer series in operations research and financial engineering. New York: Springer.
45. Biegler, L.T., Grossmann, I.E., (2004) Retrospective on optimization. *Comput. Chem. Eng.* 28(8), 1169–1192.
46. Martín, M., Grossmann, I.E., (2012) Energy optimization of bioethanol production via hydrolysis of switchgrass. *AIChE J.* 58(5), 1538–1549.
47. Karuppiah, R., Peschel, A., Grossmann, I.E., Martín, M., Martinson, W., Zullo, L., (2008) Energy optimization for the design of corn-based ethanol plants. *AIChE J.* 54(6), 1499–1525.
48. Martín, M., Grossmann, I.E., (2011) Energy optimization of bioethanol production via gasification of switchgrass. *AIChE J.* 57(12), 3408–34212.
49. Martín, M., Grossmann, I.E. (2013) ASI: Towards the optimal integrated production of biodiesel with internal recycling of methanol produced from glycerol. *Env. Prog. Sus. Energ.*, 32(4), 791–801.
50. Smith, R., (2005) *Chemical Process Design and Integration.* Chichester, UK: Wiley.
51. Coello-Coello, C.A., Van-Veldhuizen, D.A. & Lamont, G. B., (2002) *Evolutionary Algorithmns for Solving Multi-Objective Problems.* New York: Kluwer Academic.
52. Woinaroschy, A., (2009) Simulation and Optimization of Citric Acid Production with SuperPro Designer Using a Client-Server Interface. *CHIM*, Volume 9, pp. 979–983.
53. Kirkpatrick, S., Gelatt, C.D., Vecchi, M.P., (1983) Optimization by Simulated Annealing. *Science.* 13 May, 220(4598), pp. 671–680.
54. Goldberg, D. E., (1989) Genetic Algorithms and Walsh Functions: Part I, A Gentle Introduction. *Complex Systems*, Volume 3, pp. 129–152.
55. Segovia-Hernández, J. G., Gómez-Castro, F. I., (2017) *Stochastic Process Optimization using Aspen Plus®.* s.l.:s.n.
56. Ponce-Ortega, J.M., Hernández-Pérez, L.G., (2018) *Optimization of Process Flowsheets through Metaheuristic Techniques.* 1st ed. Morelia, Mexico: Springer.
57. Birnbaum, D., (2005) Excel VBA Programming for the Absolute Beginner. *Thomson Course Technology PTR.*
58. Sandler, S. I., (2015) *Using Aspen Plus in Thermodynamics Instruction.* 1st ed. Hoboken, NJ: John Wiley & Sons.
59. Sharma, S., Rangaiah, G.P., (2013) An improved multi-objective differential evolution with a termination criterion for optimizing chemical processes. *Comp. Chem. Eng.* 13 September, Volume 56, pp. 155–173.

Section V

Process Design & Optimization

13 Algebraic Modeling and Optimization

Ricardo M. Lima, Ignacio E. Grossmann

CONTENTS

13.1 INTRODUCTION

The aim of this chapter is to provide an overview on algebraic modeling systems and their application in Chemical Engineering. Algebraic modeling systems are a class of computational modeling tools that are mainly oriented for the development of optimization models. One important characteristic of these systems is the representation of the model using a mathematical modeling notation using algebraic equations, which provides great flexibility in modeling. This flexibility has enabled the application of algebraic modeling systems in many different areas, namely agricultural economics, engineering, finance, management science and operations research, contract theory, economic development, power systems, energy economics, and energy markets, among others. In Chemical Engineering they have also been applied in many application areas, for example process optimization, process synthesis, planning and scheduling of plant operations, and supply chain design and operation. Chemical Engineers have a long tradition on building equation models based on first principles for the simulation and optimization of chemical processes, which has contributed to the development of computational tools for the systematic simulation and optimization of chemical processes design and operation. Examples of these computational tools include process simulators, either sequential based modular or equation-oriented, software that targets specific processing applications, modeling systems, and spreadsheets. Specific examples of these computational tools are described in other chapters of this book. However, modeling systems include gPROMS [1], ASCEND [3], and the object-oriented modeling language Modelica [4].

This chapter focuses on a specific class of modeling systems that are mainly applied for optimization applications. It therefore, gives an integrated overview of algebraic modeling systems, and basic notions about the mathematical formulation of optimization problems. In algebraic modeling systems, as with other computational tools for simulation or optimization, the Chemical Engineer modeler has a paramount role during the modeling and solution phase, and at the iterative process of analyzing the results and improving the model, or the solution strategy.

13.2 OPTIMIZATON MODELING

From a conceptual perspective, the optimization methodology is a tool to help in the decision making process of complex problems. Optimization as an improvement tool is based on the search for a decision or action that will lead to the best performance index, subject to the constraints of the problem. In this section, the focus is on the formulation of optimization problems using mathematical notation in the context of Chemical Engineering.

In algebraic modeling systems the optimization problem is formulated in a similar way to a mathematical program with the following compact representation:

$$\text{Maximize}\left\{f(x): x \in X, g(x) \leq 0, h(x) = 0\right\}, \tag{13.1}$$

where the goal is to maximize the objective function $f(x)$ subject to the inequality constraints $g(x) \leq 0$, the equality constraints $h(x) = 0$, and the variables represented by x that belong to the set X, which is a subset of $\mathbb{R}$, and defines the domain of the functions f and g. The above formulation can be extended for specific problems, by incorporating multiple objective functions, stochastic parameters, or a formulation with embedded optimization problems. However, the aforementioned formulation is general enough to cover a wide range of problems. In order to give an overview of the mapping between this formulation and a Chemical Engineering problem, let us consider a process design problem, where the objective function may be related with economic aspects such as profit, revenues, cost, or with the characteristics of a product in terms of quantity, or quality or selectivity. The constraints of the optimization problem establish the relation between the operating conditions and other variables of the process, where the equality constraints may represent material balances, energy balances, thermodynamic relations, stoichiometric equations, sizing of equipment, and performance equations of units, while the inequality constraints may involve bounds on the selectivity of one product or efficiency of one processing unit. The set of variables x includes the variables of the process, for example operating conditions such as temperature or pressure, flowrate of streams, compositions or concentrations of components, dimensions of equipment, or conversions. To further extend the scope of the formulation, the set of variables x can be partitioned in two sub-sets, one with variables $x \in \mathbb{R}$, and another sub-set $y \in \mathbb{Z}$, resulting in the formulation:

$$max\left\{f(x,y): x \in R, y \in Z, g(x,y) \leq 0, h(x,y) = 0\right\}. \tag{13.2}$$

The utilization of integer variables, or more specifically variables that can take the value of 0 or 1, called binary variables, opens the possibility to model a wide range of problems that involve for example the selection of units in a process flowsheet, the determination of the number of trays of a distillation column, the assignment of processing tasks or resources to units, or the logical relations between the existence of units in a flowsheet. For interested readers and in the context of Chemical Engineering, important aspects of modeling with binary variables and solution methods are covered in [5]. The two formulations presented are quite general, and do not provide much information about their structure, which turns out to be very important to classify them, and identification of possible convergence problems and decision over the solvers to apply. Therefore, based on an analysis of their mathematical structures these formulations can be broadly classified as a function of the domain of the variables and the relation between the variables in the constraints and objective function. Thus, the optimization problem can be classified as a Linear Programming (LP) problem if all the variables are in $\mathbb{R}$, and the objective function and equations are linear. It is classified as a Nonlinear Programming (NLP) problem, if all the variables of the problem are in $\mathbb{R}$, and there is at least one nonlinear equation; if in addition to the continuous variables, the problem exhibits integer variables then the problem is a Mixed Integer Linear Programming (MILP) problem, or a Mixed Integer NonLinear Programming (MINLP). There are some additional variants, but for the sake of simplicity they are not covered here.

Linear optimization problems are the easiest class of problems to solve nowadays. There are efficient LP solvers that with standard computers can handle models with thousands of equations and variables in a short time, which makes LP problems very reliable from the point of view of convergence and computational time. This reliability has been a good reason for some applications to have relied on and continued to use LP problems, even though these models do not capture some nonlinearities of the respective processes. However, developments on modeling and nonlinear solvers technology have been motivating the study of nonlinear models for cases where typically LP problems are used, one of these examples is the model developed in [6] for the planning of refinery production.

Nonlinear problems—NLP and MINLP—can be further classified as convex or nonconvex, depending on the convexity of the objective function and feasible region. The understanding of the type of problem in terms of classification and convexity is very important in the utilization of modeling systems, since there are specific solvers and solution techniques for each type of problem, and depending on the problem there may be local and global solutions. A more comprehensive study on mathematical programming topics and continuous nonlinear optimization is out of the scope of this chapter. The interested reader is directed to references [5], [7], and [8].

MILP problems are widely used in Chemical Engineering for two main purposes: 1) the optimization of supply chain, planning and scheduling of operations of both continuous and batch processes; and 2) the optimization of simplified models for preliminary process design. MILP models provide the flexibility of using binary or integer variables that are very powerful to model the following situations: a) in the scope of scheduling models the association of tasks to equipment, where a task can be a reaction or a separation; the assignment of tasks or products to slots of time, for example in a model where the time is discretized in time intervals, and the production of one product is assigned with a binary variable to a specific interval; or

b) in the scope of process synthesis using linear models, for example for the synthesis of distillation sequences, where a binary variable is associated with a distillation column, and the value of this binary variable defines the existence of the column in the flowsheet. Detailed examples of these types of applications can be found in [5].

MILP models are not as easy as LP models to solve and their complexity and computational time to solve may increase exponentially with the number of binary variables. The development of efficient MILP models is a very active area of research because of the capabilities of MILP models to represent reality, and also due to the challenges to build MILP models that are equivalent in terms of the solution, but much more efficient from the computational point of view. There is a wide body of knowledge on modeling and solving MILP problems, which requires concepts and integer programming theory that is out of the scope of this chapter. For additional information the reader is directed to the book [9]. Algebraic modeling systems, considerable advances in MILP solvers, and faster computers have been contributing for a wider application of MILP models [10], [11], [12]. Modeling systems have been helping by providing an easy access to different MILP solvers, and providing modeling features that allow to quickly test different alternative MILP formulations.

The specific characteristics of the problem to address, the level of detail desired, and the capability of available solvers to solve the problem, will determine the type of problem to use. For example, a well-known problem related with the optimization of blending of gasolines can be captured by linear equations, and therefore an LP model is good enough. But, the optimization of a process synthesis problem involving operation units with nonlinear relationships, and requiring the choice between different units to execute equivalent tasks suggests the definition of an MINLP problem. The MINLP approach would fully capture the essence and type of decisions of the problem. However, MINLP problems can be very difficult to solve, and therefore, the modeler may take the action to approximate the MINLP with an MILP model, with the disadvantage of missing some accuracy regarding the real system. Therefore, the trade-off between the level of detail desired or required and the capability of solving the problem will guide the choice about the type of problem to use. From the practical point of view, this trade-off is highly influenced by the size of the model, existence of nonlinearities, and number of integer variables.

In real problems the application of optimization methodologies is driven by the complexity of the system, i.e., when it is difficult to find optimal solutions by hand, or it is not obvious to capture the trade-offs involved in the process just by running simulations; and by the potential gain obtained by optimal solutions instead of a sub-optimal solution. Examples of applications of optimization in Chemical Engineering problems include process optimization in general, process design, process synthesis in general, reactor network synthesis, distillation network synthesis, design of reactive distillation columns, water network synthesis, heat exchanger synthesis, assessment of process flexibility, design and scheduling of batch processes, scheduling of continuous processes, energy systems, design of utility systems, supply chain design and optimization. An optimization problem may be represented by more than one type of model, depending on the level of detail used by the modeler. However, it is possible to make a typical correspondence between applications and the type of problem used, see Table 13.1.

TABLE 13.1
Summary of Applications and the Corresponding Type of Problem

Type of Problem	Application
LP	Blending†
	Economic planning
	Refinery planning
NLP	Process optimization
MILP	Scheduling
	Planning
	Supply chain
MINLP	Process synthesis
	Process network synthesis

† Some blending problems involve bilinear terms leading to nonconvex NLP problems.

13.3 ALGEBRAIC MODELING SYSTEMS

Algebraic modeling systems are a class of modeling languages that are oriented to the solution of large-scale optimization problems. Schichl (2004) divides the modeling languages in algebraic modeling systems, procedural languages, functional languages, and logic programming languages. From these modeling languages, the algebraic modeling systems and the procedural languages are the most widely used in Chemical Engineering for optimization. The class of procedural languages includes the programming languages C, C++, FORTRAN, and Java, and also the numerical computational environments MatLab, Mathematica, Mathcad, and Maple [14]. One of the first references to an algebraic modeling language is the report "Mathematical Programming Language" from 1968 [15], where a mathematical programming language is proposed, which is intended to be familiar with mathematical notation, and to reduce the time required to develop and debug large-scale optimization models. Some years later in 1970, GAMS, the first algebraic modeling system was developed at the World Bank [16] as a tool for optimization of planning and economic problems. Meanwhile, several algebraic modeling systems have been developed. The main motivation for the development of these systems is based on three main needs: 1) a modeling language with a syntax close to mathematical notation; 2) a system that avoids and overcomes some of the difficulties inherent to procedural languages; and 3) the automatic and standardization of connections to the solver libraries, including nonlinear problems for which derivatives are automatically generated by automatic differentiation [2]. Currently, most of the algebraic modeling systems available are commercial software, from which some have special versions for academic and research purposes, and demo versions for students. Below there is a list of algebraic modeling systems, and a list of integrated environments for optimization.

- ***AIMMS*** It stands for Advanced Interactive Multidimensional Modeling System. It is a commercial software for optimization with an algebraic modeling language to declare optimization problems, and graphical user interfaces to help on the definition and organization of the models. This system includes tools to build graphical user interfaces to display the results, and to help on creating decision support systems without resorting on third party software.
- ***AMPL*** It is an optimization software that includes an algebraic modeling language and modeling environment to formulate optimization problems with a structure close to mathematical programming problems.
- ***GAMS*** It stands for General Algebraic Modeling System and it was developed in 1970 at the World Bank. This is a commercial system widely used in academia and industry for research in Chemical Engineering and Power Systems. In order to build decision support systems GAMS can be linked to Microsoft Excel spreadsheets that work as front ends.
- *JuMP* It is a recent open-source modeling language [17], developed at the MIT Operations Research Center. JuMP is available under the programming language Julia [18]. It is not as mature as other modeling systems, but it is built over the efficient performance of Julia.
- ***LINGO*** It is a commercial optimization software that includes an algebraic modeling system and links to several optimization solvers.
- ***MPL*** It stands for Mathematical Programming Language and the oldest references to it point to [15].
- ***Pyomo*** It is an open-source software suitable for optimization problems [19]. Pyomo is based on the programming language Python and it was developed at the Sandia National Laboratories. As a result of being based on Python, it has access to a wide range of packages and parallel communication libraries.

In addition to the above modeling systems there are some implementations of modeling languages that have fewer features, but they also address the conversion of optimization problems to input files in a proper format for specific solvers:

- ***ZIMPL*** It was developed in academia by Koch, T. [20], and it is a command line program with a modeling language to translate mathematical programming models to input files for LP and MIP solvers.
- **LPL** It stands for Linear Programming Language. LPL was developed by Hurliman T. [23], and it converts LP models into input files for LP or MILP solvers.
- **FLOPC++** This is not a modeling system, but rather a group of libraries that allow the formulation of optimization models using a declarative approach similar to algebraic modeling systems within C++ programs [21].

As an alternative to the above modeling systems, there are also available integrated environments for optimization that use specific programming languages for modeling and provide tools to build graphical user interfaces for end-user decision support systems.

- ***IBM ILOG CPLEX Optimization Studio*** This software features a specific language denoted by Optimization Programming Language (OPL), and it has interfaces to several programming languages and applications to help on the deployment of business solutions.
- ***FICO Xpress Optimization Suite*** This is an integrated environment for optimization, which included the modeling and programming language Xpress-Mosel to build interactive graphical environments for decision support.

A comprehensive overview of some of the features of these systems is given in [14] and [22], where very interesting graphical and modeling features of systems like GAMS and AIMMS are discussed. The systems above are a sample of the many systems available for optimization, a more extended list can be found in the web server called NEOS guide [24], [25].

Algebraic modeling systems provide modeling features and an easy access to optimization solvers that have simplified the development and maintenance of large scale models when compared with procedural languages, and hence, allowing savings on the time necessary to build and maintain models. In general, algebraic modeling systems encompass two types of language: a) a modeling language that is used to describe the optimization problem; and b) a programming language that provides flow control structures such as loops or if-then, which are for example used in more advanced implementations to develop solution approaches based on decomposition methods. However, although these modeling systems are based on programming languages, these are conceptually different from procedural, functional, or logic programming languages. For example, in procedural languages the implementation of the model is not always independent of the solver, or solution algorithm, what means that there is not a clear separation between model and algorithm. Furthermore, the model or the interface of the model needs to be adapted to the requirements of the solver. However, tailored strategies using procedural languages can be very efficient solving specific problems, at the price of having to execute development tasks that can be very time consuming, and lack some flexibility for changes in the model. An example where concepts from algebraic modeling systems are integrated with procedural programming languages is the work developed in [26], where a methodology was developed to try to overcome the difficulties of modeling and solving optimization problems in FORTRAN.

In general, all algebraic modeling systems encompass the following components: a) a modeling language to declare the optimization problem; b) a programming language with some flow control features; c) a parser to convert the optimization model into an input format with the required information for the solvers; d) a library of solvers for different types of problems; and e) a system to report the results. In addition, some the algebraic modeling systems feature graphical visualization tools and libraries of interfaces to exchange information with other software. These additional features are constantly evolving, and provide a competitive edge between the different modeling systems vendors.

In algebraic modeling systems, the modeling is expressed using a declarative language, which means that the language defines the optimization problem but does not specify how to solve the problem. Therefore, the model is completely independent of the solver. In fact, the optimization problem is defined using mathematical semantics, and it is conceptually declared as an optimization problem similar to the formulations presented before. This independence of the problem from the solver, together with the existence of multiple solvers available in the library of optimization solvers for the same type of problem, allows the solution of a given model by different solvers without the need to change the model or generate new information for the solver. This is very important to tackle difficult problems, where if it is hard to find an initial or feasible solution, or finding the global solution in a reasonable computational time, different solvers can easily be tested.

An extremely important feature and a good programming practice in algebraic modeling systems is to build a model completely independent from the data of the problem. This means that the model can be solved for one set of data, and later the model can be solved with another set of data without the need to change the equations of the model. This is quite common during the development phase of a model. In this phase, the model is built and tested using small instances, for example with a production system with a small number of products, and then at a later development phase when the model is stable enough, the data set can be changed to solve real instances. This scale up feature of algebraic modeling systems is supported by the structure used to define the optimization problems, which is based on the following items:

a. Sets;
b. Parameters;
c. Variables;
d. Constraints;
e. Objective function.

The elements above resemble the components of the formulations presented in equations (13.1) and (13.2). The main function of the sets is to group identities in the problem that are similar and preferentially are involved in the same type of parameters, variables, and equations. Therefore, the first step to build an optimization model is to define the most convenient sets that will allow writing the most compact and elegant model. Basic sets are composed by only one dimension, typical examples are a) one set called GP that groups the types of gasolines in a blending problem, GP: = {alkylate, catalytic-cracked, straight-run, isopentane}; b) one set for the group of reactors in a flowsheet, RTR: = {Reactorl, Reactor2, Reactor3}; or c) one set for all the streams in the flowsheet, STR: = {strl, str2, str3}. More complex sets involve maps between elements of different sets, leading to multi-dimensional sets. Two-dimensional sets are very useful in building models for process synthesis in Chemical Engineering. For example a two-dimensional set may group for each reactor of the set RTR its output stream that is a member of the set STR, resulting in the set ROUT: = RTR × STR.

With the set structure defined, the parameters, variables, and equations of the problem can be declared for each member of the sets. This structure enables the definition of large-scale optimization problems using a compact as possible formulation. These concepts are better understood with examples, and therefore, two of them are given below in order to give an overview of the implementation of an optimization model in an algebraic modeling system. For each example, the problem statement is given, and their implementation in the modeling system GAMS is presented. Both examples are small, but they have all the components of an optimization problem. The first is a blending problem modeled as an LP problem, and the second is a process synthesis problem involving a reactor network design, modeled as an NLP problem.

13.4 EXAMPLE 1

This example is adapted from the book Linear Programming [27], which is a simplified blending problem based on a real case studied in [28]. The problem statement is as follows: given are four types of raw gasoline: alkylate, catalytic-cracked, straight-run, and isopentane produced by an oil refinery. Each gasoline is characterized by a quantity denoted by Performance Number (PN) that is related with octane ratings and its vapor pressure Reid Vapor Pressure (RVP). For each raw gasoline the values for PN and RVP, as well as the number of barrels available are presented in Table 13.2. In this example the original units of the problem are kept in order to keep some coherence. Each raw gasoline can be sold at \$4.83 per barrel, or it can be blended into one of two aviation gasolines. Each aviation gasoline has specifications in terms of PN and RVP that must be met, see Table 13.3. In this table it is also available the price per barrel of each aviation gasoline. The objective is to determine the number of barrels of each raw gasoline to sell, and the blending into each aviation gasoline, as well as its corresponding sales that maximize the revenues. Figure 13.1 illustrates the blending and selling operations considered in this problem.

This blending optimization problem can be solved using an LP model based on the following equations:

$$\text{Maximize} \quad \sum_i \left(VALUE\_g_i \times br\_sold\_raw_i\right) + \sum_j \left(VALUE\_a_j \times br\_aviation_j\right), \quad (13.3)$$

$$\text{Subject to:} \qquad br\_aviation_j = \sum_i br\_raw\_blended_{ij} \quad \forall j, \quad (13.4)$$

$$br\_raw\_sold_i + \sum_j br\_raw\_blended_{ij} = BR\_PRODUCED_i \quad \forall i, \quad (13.5)$$

TABLE 13.2
Data for Example 1

Raw Gasoline	PN	RVP (lbs/in^2)	Barrels Produced
Alkylate	107	5	3814
Catalytic-cracked	93	8	2666
Straight-run	87	4	4016
Isopentane	108	21	1300

TABLE 13.3
Data for Example 1

Aviation Gasoline	PN	RVP (lbs/in²)	Price Per Barrel
Avgas A	≥ 100	≤ 7	\$6.45
Avgas B	≥ 91	≤ 7	\$5.91

$$\sum_i \left(br\_raw\_blended_{ij} \times PN_i\right) \geq \sum_i \left(br_{rawblended_{ij}}\right) \times PNA_j \quad \forall j, \tag{13.6}$$

$$\sum_i \left(br\_raw\_blended_{ij}\right) \times RVPA_j \geq \sum_i \left(br\_raw\_blended_{ij} \times RVP_i\right) \quad \forall j, \tag{13.7}$$

$$br\_aviation_j,\ br\_sold\_raw_i,\ br\_raw\_blended_{ij} \geq 0, \tag{13.8}$$

where equation (13.3) is the objective function to maximize, which represents the revenues from selling raw gasolines plus aviation gasolines, equation (13.4) defines that the total barrels of aviation gasoline j is equal to the sum of barrels of each raw gasoline i blended in aviation gasoline j, equation (13.5) is the mass balance for each raw gasoline stating that the total number of barrels produced of each gasoline is equal to the number of barrels of raw gasoline sold plus the number of barrels blended, equation (13.6) and (13.7) are specifications for the properties of the blended aviation gasoline. The former enforces that for each aviation gasoline, its *PNA* must be greater or equal than the average weight based on the compositions of the PN_i of the blended raw gasolines. And equation (13.7) sets that the *RVPA* of each aviation gasoline j must be greater or equal than the composition average weight of the $RVPA_i$.

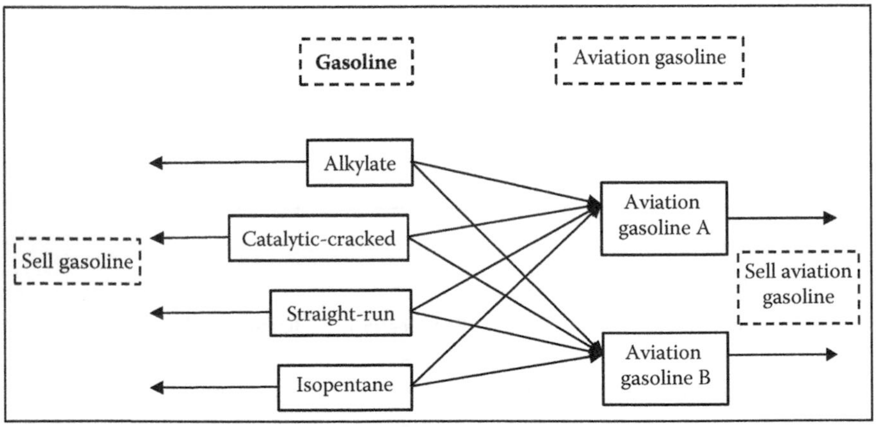

FIGURE 13.1 Diagram of the blending and selling operations.

This LP model can be easily implemented in all algebraic modeling systems, and can be solved with any available LP solver. Figure 13.2 displays the full model written in GAMS, which corresponds to the full extract of a GAMS input file. For illustration purposes the different sections of the modeling structure are limited by boxes with dash lines. These boxes limit the declaration of: a) the sets; b) the parameters; c) the list of variables and equations; d) the mathematical model; and e) the instruction to call the solver to solve the model declared in the

```
SETS
i               raw gasolines /alkylate, catalytic-cracked,
                              straight-run, isopentane/
j               aviation gasolines  /a, b/;

PARAMETERS
PN(i)           antiknock properties
                / alkylate          107
                  catalytic-cracked  93
                  straight-run       87
                  isopentane        108 /
RVP(i)          vapor pressure
               / alkylate             5
                 catalytic-cracked    8
                 straight-run         4
                 isopentane          21 /

BR_PRODUCED(i)  barrels produced
                / alkylate          3814
                  catalytic-cracked 2666
                  straight-run      4016
                  isopentane        1300 /

PNA(j)          minimum aviation pn
                / a 100
                  b  91 /

RVPA(j)         maximum aviation rvp
                / a    7
                  b    7 /

VALUE_G(i)      value of raw gasolines
                / alkylate           4.83
                 catalytic-cracked  4.83
                 straight-run       4.83
                 isopentane         4.83 /

VALUE_A(j)      value of aviation gasolines
                / a 6.45
                  b 5.91/;

POSITIVE VARIABLES
br_raw_sold(i)         barrels of raw gasoline i sold
br_raw_blended(i,j)    barrels of raw gasoline i blended in aviation gasoline j
br_aviation(j)         barrels of aviation gasoline j blended;

VARIABLES
profit;

EQUATIONS
    fobj
    ineq_pn
    ineq_rvp
    capacity_raw
    capacity_aviation;

fobj..                  profit =e= sum(i, VALUE_G(i) * br_raw_sold(i))
                               + sum(j, VALUE_A(j) * br_aviation(j));

ineq_pn(j)..            sum(i, br_raw_blended(i,j) * PN(i)) - sum(i, br_raw_blended(i,j)) * PNA(j) =g= 0;

ineq_rvp(j) ..          sum(i, br_raw_blended(i,j)) * RVPA(j) =g= sum(i, br_raw_blended(i,j) * RVP(i));

capacity_raw(i)..       br_raw_sold(i) + sum(j, br_raw_blended(i,j)) =e= BR_PRODUCED(i);

capacity_aviation(j).. br_aviation(j) =e= sum(i, br_raw_blended(i,j));

model lp1
        /fobj, ineq_pn, ineq_rvp, capacity_raw, capacity_aviation/;

solve lp1 maximizing profit using LP;
```

FIGURE 13.2 GAMS input file for Example 1.

TABLE 13.4
Equivalences between Two Different Notations

Matrix Notation		Index Notation	
Maximize	$c^T x$	Maximize	$\sum_{j=1}^{n} c_j x_j$
Subject to:	$Ax \le b$	Subject to:	$\sum_{j=1}^{n} a_{ij} x_j \le b_i \quad i = 1,2,\ldots,m$
	$x \ge 0$		$x_j \ge 0 \quad j = 1,2,\ldots,n$

previous box. The structure presented below follows the principle of separation of data from the model. The data is declared first together with the definition of the parameters, and there are no numeric values within the declaration of equations. However, the values of the parameters can still be re-defined in other parts of the input file. For example in this case, the prices of the aviation gasolines can be modified after the solve statement and the solver can be invoked again to solve the problem for the new prices.

We would like to highlight that in the mathematical programming approach with algebraic modeling systems the declaration of models is expressed with an index notation, and not matrix notation, as seen in Table 13.4, where both notations are shown for an LP problem.

In Table 13.4, the variables are denoted by the vector $\boldsymbol{x}$ on the left, and by the variables $x_j, j = 1,2,\ldots,n$ on the right. However, the input and definition of data in algebraic modeling systems can be done using matrices, for example, the definition of the parameters $a_{ij}, = 1,2,\ldots,m; j = 1,2,\ldots,n$ can be inserted using a matrix format. But the definition of the optimization problems uses a mathematical programming notation, as it can be observed by comparing the model presented in equations (13.1) to (13.6) with the model implemented in the GAMS input file.

13.5 EXAMPLE 2

This example involves the design of a reactor network structure involving two reactors, three components, and four reactions. The problem was proposed in [29], and it combines the reaction scheme proposed in [30] with the network structure defined in [31]. The scheme of the process is illustrated in Figure 13.3.

In reactor 1 only the chemical reactions referenced as 1 and 3 may occur, while in reactor 2 only the reactions 2 and 4, where the reactions schemes are as follows:

$$A \xrightarrow{k_1} B \xrightarrow{k_3} C$$

$$A \xrightarrow{k_2} B \xrightarrow{k_4} C$$

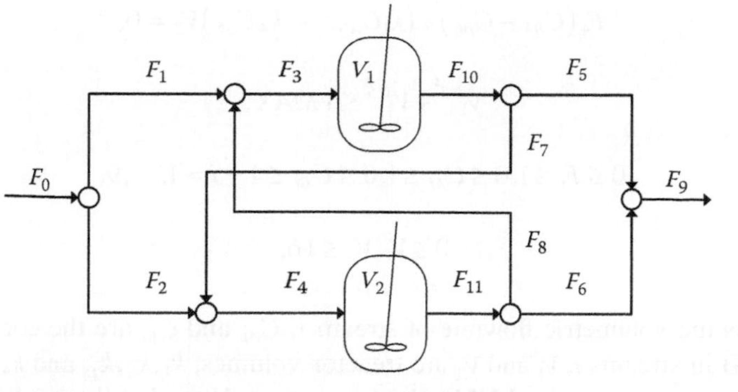

FIGURE 13.3 Diagram of the reactor network from Example 2.

The reaction rates involve first order rates, and the reactors are considered to be continuous stirred tank reactors, and the density of the mixture is assumed to be constant. The objective is to maximize the concentration of product B in the output stream. The model describing this optimization problem can be stated as

Maximize $$cbout = C_{B9}, \tag{13.9}$$

Subject to $$F_1 + F_2 - F_0 = 0, \tag{13.10}$$

$$F_1 + F_8 - F_3 = 0, \tag{13.11}$$

$$F_2 + F_7 - F_4 = 0, \tag{13.12}$$

$$F_5 + F_7 - F_{10} = 0, \tag{13.13}$$

$$F_6 + F_8 - F_{11} = 0, \tag{13.14}$$

$$F_1 C_{A0} + F_8 C_{A6} - F_3 C_{A3} = 0, \tag{13.15}$$

$$F_1 C_{B0} + F_8 C_{B6} - F_3 C_{B3} = 0, \tag{13.16}$$

$$F_2 C_{A0} + F_7 C_{A5} - F_4 C_{A4} = 0, \tag{13.17}$$

$$F_2 C_{B0} + F_7 C_{B5} - F_4 C_{B4} = 0, \tag{13.18}$$

$$F_3 (C_{A3} - C_{A5}) - k_1 C_{A5} V_1 = 0, \tag{13.19}$$

$$F_3 (C_{B3} - C_{B5}) + (k_1 C_{A5} - k_3 C_{B5}) V_1 = 0, \tag{13.20}$$

$$F_4 (C_{A4} - C_{A6}) - k_2 C_{A6} V_2 = 0, \tag{13.21}$$

$$F_4\left(C_{B4}-C_{B6}\right)+\left(k_1C_{A6}V_1-k_4C_{B6}\right)V_2=0, \tag{13.22}$$

$$V_1^{0.5}+V_2^{0.5}\le VMAX, \tag{13.23}$$

$$0\le F_i\le 1, 0\le C_{Ai}\le 1, 0\le C_{Bi}\le 1 \quad i=1,\ldots,9, \tag{13.24}$$

$$0\le V_1, V_2\le 16, \tag{13.25}$$

where F_i is the volumetric flowrate of stream i, C_{Ai} and C_{Bi} are the concentrations of A and B in streams i, V_1 and V_2 are treactor volumes, k_1, k_2, k_3, and k_4 denote the first order rate constants, and VMAX is a constant. Note that the model above has already been algebraic manipulated in order to reduce some variables and equations. For example, the variables F_{10}, F_{11} are not included in the model because $F_{10}=F_3$ and $F_{11}=F_4$. Thus, they have been replaced in the mass balances for the splitters. Similar substitutions were made for some variables representing concentrations. In the above formulation there is a clear relationship between the variables and equations and the network structure, which reduces the flexibility of the model to cope with different configurations of the network.

A more general approach to write down the model is based on the separation of the model from the data, and in this specific case the separation of the network structure from the model as much as possible. Based on this separation concept, a more general and compact formulation can be written as follows:

Maximize
$$cbout=C_{B9}, \tag{13.26}$$

Subject to:
$$F_s=\sum_{s'\in MI}F_{s'} \quad \forall s\in MO, \tag{13.27}$$

$$F_sC_{is}=\sum_{s'\in MI}F_{s'}C_{is'} \quad \forall s\in MO, \tag{13.28}$$

$$F_s=\sum_{s'\in SO}F_{s'} \quad \forall s\in SI, \tag{13.29}$$

$$C_{is}=C_{is'} \quad \forall s\in SO,\ \forall s'\in SI, \tag{13.30}$$

$$F_s=F_{s'} \quad \forall s\in RO,\ \forall s'\in RI, \tag{13.31}$$

$$F_{s'}C_{is'}-F_sC_{is}-V_n\sum_{c\in RR}\beta_{ic}\left(-r_{nc}\right)=0 \quad \forall s\in RO, \forall s'\in RI, \tag{13.32}$$

$$r_{nc}=k_cC_{is} \quad \forall s\in RO, \forall n\in NR, \forall c\in C, \tag{13.33}$$

$$V_1^{0.5} + V_2^{0.5} \leq VMAX, \tag{13.34}$$

$$0 \leq F_s \leq 0 \forall s;\ 0 \leq C_{is} \leq 1\ \ \forall i, s; 0 \leq V_n \leq 16, \forall n. \tag{13.35}$$

The model above represents a typical approach in Chemical Engineering modeling, where the model is built using specific blocks of equations for each unit. Therefore, the equations of the model for the reactors, mixers, and splitters are built independently of the network structure. The above formulation is based on the definition of sets that group similar identities, for example, in this case a set for each type of units, one for the products, one for the streams, and additional sets with two dimensions. These additional sets with two dimensions are the basis for defining the network structure of the problem, establishing the connections between the different units. Two-dimensional sets are defined in the model above by *MI, SI,* and *RI* that denote the mapping between the mixers, splitters, and reactors, respectively, and the corresponding input streams, and by the sets *MO, SO,* and *RO* that represent the mapping between the mixers, splitters, and reactors, respectively, and the output streams. In order to help to understand the main equations in the above formulation, Table 13.5 shows on the left the equations declared based on an index notation, and on the right the corresponding equations for each index combination. Within the equations on the right, it is possible to identify some of the equations from the formulation declared in equations (13.9) to (13.25). For example, equation (13.27) is a mass balance for the mixers representing the three linear equations in the right cell of the first row of the column. In fact, the list of equations on the right is automatically built by the modeling system based on the equations declared using the compact formulation above. The implementation and further details of this approach are presented in the GAMS input file in Figure 13.4 and Figure 13.5. For illustration purposes the different sections of the modeling structure are limited by boxes with dash lines. These boxes limit the declaration of: a) the sets; b) the parameters; c) the list of variables and equations; d) the mathematical model; e) the definition of upper bounds and initial values; and f) the instruction to call the solver to solve the model declared in the previous box.

This problem is implemented with six single dimension sets, from which there is one set for each type of unit and one set for the streams. The two-dimension sets are then used to define the conditional expressions in the equations of each block of units. For example, the set *MI(m,str)* defines for each mixer *m* the input streams *str*, while the set *MO(m,str)* defines for each mixer *m* the output stream. These sets are then used in the declaration of the equations corresponding to the mixers, as seen in Figure 13.6.

This implementation provides a great flexibility based on the fact that the equations are independent of the labels of the connections, and therefore, it is possible to cover different configurations of the network with the same model, without the need of changing the equations. In order to fully explore this approach, algebraic modeling systems provide dynamic sets, which allow removing members of the sets, and thus, changing the domain of the equations

TABLE 13.5
Comparison between Two Equivalent Modeling Approaches for the Same Problem

$F_s = \sum_{s' \in MI} F_{s'} \quad \forall s \in MO,$	$F_1 + F_8 - F_3 = 0,$ $F_2 + F_7 - F_4 = 0,$ $F_5 + F_6 - F_9 = 0,$
$F_s C_{is} = \sum_{s' \in MO} F_{s'} C_{is'} \quad \forall s \in MO, \forall i \in I,$	$F_1 C_{A0} + F_8 C_{A6} - F_3 C_{A3} = 0,$ $F_1 C_{B0} + F_8 C_{B6} - F_3 C_{B3} = 0,$ $F_2 C_{A0} + F_7 C_{A5} - F_4 C_{A4} = 0,$ $F_2 C_{B0} + F_7 C_{B5} - F_4 C_{B4} = 0,$
$F_s = \sum_{s' \in SO} F_{s'} \quad \forall s \in SI,$	$F_1 + F_2 - F_0 = 0$ $F_5 + F_7 - F_{10} = 0,$ $F_6 + F_8 - F_{11} = 0,$
$C_{is} = C_{is'} \quad \forall s \in SO, \forall s' \in SI$	$C_{A1} = C_{A0},$ $C_{A1} = C_{A0},$ $C_{B1} = C_{B0},$ $C_{B1} = C_{B0}.$ $C_{A2} = C_{A0},$ $C_{A2} = C_{A0},$ $C_{B2} = C_{B0},$ $C_{B2} = C_{B0}.$ $C_{A5} = C_{A10},$ $C_{A5} = C_{A10},$ $C_{B5} = C_{B10},$ $C_{B5} = C_{B10},$ $C_{A7} = C_{A10},$ $C_{A7} = C_{A10},$ $C_{B7} = C_{B10},$ $C_{B7} = C_{B10},$ $C_{A85} = C_{A10},$ $C_{A8} = C_{A11},$ $C_{B8} = C_{B11},$ $C_{B8} = C_{B11},$ $C_{A6} = C_{A11},$ $C_{A6} = C_{A11},$ $C_{B6} = C_{B11},$ $C_{B6} = C_{B11},$
$F_s = F_{s'} \; \forall s \in RO, \forall s' \in RI,$	$F_3 - F_{10} = 0,$ $F_4 - F_{11} = 0,$
$F_{s'} C_{is'} - F_s C_{is} - V_n \sum_{c \in RR} \beta_{ic}(-r_{nc}) = 0 \; \forall s \in RO, \forall s' \in RI,$	$F_3(C_{A3} - C_{A5}) - k_1 C_{A5} V_1 = 0,$ $F_3(C_{B3} - C_{B5}) + (k_1 C_{A5} - k_3 C_{B5}) V_1 = 0,$
$r_{nc} = k_c C_{is} \; \forall s \in RO, \forall n \in NR, \forall c \in C,$	$F_4(C_{A4} - C_{A6}) - k_2 C_{A6} V_2 = 0,$ $F_4(C_{B4} - C_{B6}) + (k_1 C_{A6} V_1 - k_4 C_{B6}) V_2 = 0,$

```
$TITLE Reactor Network Optimization
SETS
    i         'Components'                    /A,B/
    m         'Mixers'                        /m1,m2,mout/
    s         'Splitters'                     /sin,s1,s2/
    r         'Reactors'                      /r1,r2/
    rc        'Reactions'                     /1*4/
    str       'Streams in the process'        /str0*str11/;

ALIAS(str,str1);

* The following sets define the structure of the flowsheet
SETS
    MI(m,str)  'Definition of the inlet streams of each mixer'
    /m1.str1,   m1.str8
     m2.str2,   m2.str7
     mout.str5, mout.str6/

    MO(m,str) 'Definition of the outlet stream of each mixer'
    /m1.str3
     m2.str4
     mout.str9/

    SI(s,str)  'Definition of the inlet stream of each splitter'
    /sin.str0
     s1.str10
     s2.str11/

    SO(s,str) 'Definition of the outlet streams of each splitter'
    /sin.str1, sin.str2
     s1.str5,  s1.str7
     s2.str6,  s2.str8/

    RIN(r,str)  'Definition of the inlet stream of each reactor'
    /r1.str3
     r2.str4/

    ROUT(r,str) 'Definition of the outlet streams of each reactor'
    /r1.str10
     r2.str11/

* Reactions 1 and 3 occur in reactor 1
* Reactions 2 and 4 occur in reactor 2
    RR(r,rc) 'Assignment of reactions to reactors'
    /r1.1, r1.3
     r2.2, r2.4/

* Reactants of each reaction path
    REACT(rc,i)
    /1.A
     2.A
     3.B
     4.B/;

PARAMETERS
    k(rc)    'Kinetic constants [1/h]'
    /1 0.09755988
     2 0.096584202
     3 0.0391908
     4 0.03527172/;

SCALAR
    VMAX          'Value used in the constraint for the volume      [m3]' /16/;

* Stoichiometric coefficients of product i in reaction rc
TABLE beta(i,rc)
    1   2   3   4
A  -1  -1   0   0
B   1   1  -1  -1;
```

FIGURE 13.4 GAMS input file for Example 2 (to be continued in the next figure).

```
POSITIVE VARIABLES
    F(str)          'Volumetric flow rates                              [m3/h]'
    C(i,str)        'Concentrations                                   [mol/m3]'
    V(r)            'Reactor volume                                      [m3]'
    rct(rc)         'Reaction rate of reaction path rc                 [mol/h]'
VARIABLE
    CBout           'Output concentration of product B to maximize [mol/m3]';
EQUATIONS
    MXRBAL          'Total balance for the mixers'
    MXRCOMP         'Component balance for the mixers'
    SPLBAL          'Total balance for the splitters'
    SPLCOMP         'Component balance for the splitters'
    RCTBAL          'Total balance for the reactors'
    REACTOR         'Component balance for the reactors'
    REACTION        'Reaction rates'
    VOL             'Upper bound on the reactor volumes'
    FOBJ            'Objective function';
```

```
FOBJ.. CBout =e= C('B','str9');

MXRBAL(m,str) $(MO(m,str)).. F(str) =e= SUM(str1 $(MI(m,str1)), F(str1));

MXRCOMP(i,m,str) $(MO(m,str))..
        F(str) * C(i,str) =e= SUM(str1 $(MI(m,str1)), F(str1) * C(i,str1));

SPLBAL(s,str) $(SI(s,str))..  F(str) =e= SUM(str1 $(SO(s,str1)), F(str1));

SPLCOMP(i,s,str,str1) $(SI(s,str) AND SO(s,str1)).. C(i,str) =e= C(i,str1);

RCTBAL(r,str,str1) $(RIN(r,str) AND ROUT(r,str1)).. F(str) =e= F(str1);

REACTOR(i,r,str,str1) $(RIN(r,str) AND ROUT(r,str1))..
        F(str) * C(i,str) -  F(str1) * C(i,str1) - V(r) * SUM(rc $(RR(r,rc)),
        beta(i,rc)                        * (-rct(rc))) =e= 0;

REACTION(r,rc,i,str) $(REACT(rc,i) AND RR(r,rc) AND ROUT(r,str))..
        rct(rc) =e= k(rc) * C(i,str);

VOL.. SUM(r, SQRT(V(r))) =l= SQRT(VMAX);

MODEL RNET /MXRBAL, MXRCOMP, SPLBAL, SPLCOMP, RCTBAL, REACTOR, REACTION, VOL,
FOBJ/;
```

```
* Upper bounds on the variables
F.UP(str) = 1.0;
C.UP(i,str)  = 1.0;
V.UP(r) = 16.0;
* Initial values for the variables
V.L(r) = 4.0;
F.L(str) = 1.0;
C.L(i,str) = 0.5;
* Fix known values for the input and output stream
F.FX('str0') = 1.0;
F.FX('str9') = 1.0;
C.FX('A','str0') = 1.0;
C.FX('B','str0') = 0.0;
```

```
* Set Solver
$SET NSOLVER BARON

OPTIONS NLP = %NSOLVER%;

SOLVE RNET maximizing CBout using NLP;
```

```
* Write additional output file
file outres /results.out/;
outres.ap = 1;
put outres;
put 'Solver    ','Objective Function    ','CPU (s)'/;
put "%NSOLVER%":<10, RNET.objval:>18:6, RNET.resusd:>10:1 /;
putclose outres;
```

FIGURE 13.5 GAMS input file for Example 2 (continuation).

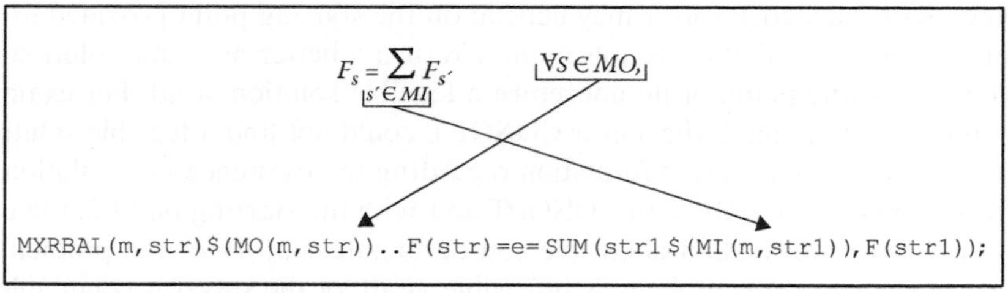

FIGURE 13.6 Equivalence between the equations and the GAMS input.

declared. This approach can be useful in process synthesis, whenever using a superstructure to integrate in one flowsheet the alternative configurations, and the goal is to optimize several process designs based on the model of the superstructure. However, changing the sets to modify the network must be done carefully in order to avoid undesirable configurations. An example of application of this modeling strategy was implemented in the solution of a large-scale optimization process design for the p-Xylene separation processes based on crystallization technology [32], [33].

Returning to Example 2, it involves continuous variables, linear equations, and bilinear terms in some equations, resulting in a nonconvex NLP problem, with potentially multiple local optima. For this type of problems, local NLP solvers (eg CONOPT, IPOPT, SNOPT, KNITRO) will return a solution that may not be the global optimum, and these solvers do not provide any quantification about how far the solution is from the global optimum. There are however global NLP solvers such as BARON that are guaranteed to obtain the global optimum, although at a greater computational expenses, especially for large problems. The values of the objective function and CPU time obtained with different solvers for this problem using two sets of starting points are presented in Table 13.6. From this table, as indicated before, the first four solvers are NLP solvers that do not guarantee the global optimum, while the last solver is a global optimization solver that for this specific problem can find the global optimum in a reasonable time. The solution

TABLE 13.6
Results Obtained for Example 2 with Different Solvers

	Starting Point 1		Starting Point 2	
Solver	Objective Function	CPU (s)	Objective Function	CPU (s)
IPOPT 3.11	0.3888	0.4	0.3746	1.3
CONOPT 3.15L	0.3881	0.1	INFES	0.0
SNOPT 7.2-12	0.3867	0.1	0.3746	0.0
KNITRO 8.1.1	0.3881	0.5	0.3867	0.2
BARON 12.3.3	0.3888	55.1	0.3888	58.7

Starting point 1: `V.L(r) = 4.0;F.L(str) = 1.0;C.L(i,str) = 0.5;`
Starting point 2: `V.L(r) = 1.0;F.L(str) = 0.0;C.L(i,str) = 0.0;`

obtained with each NLP solver may depend on the starting point provided to the solver, which means that these solvers may return a better or worse solution for a different starting point, or do not return a feasible solution at all. For example using the starting point 2, the solver CONOPT could not find a feasible solution, and it does not provide any information regarding the existence of a solution. If this problem is solved only with CONOPT and with the starting point 2, the user would not have information about the source of infeasibility of the problem, it would not be clear if the problem is infeasible at all, or the solver was not able to find a feasible solution because of the starting point. The definition of an infeasible problem may occur if unreachable specifications are set, for example the definition of a minimum selectivity constraint that the model cannot reach due to the specifications of the equipment. However, it is easy to apply other solver and/or starting point and obtain some additional information about the infeasibility of the problem. Note that some objective function values present small differences between them, which are not due to numerical errors, but they do represent different configurations of the reactor network.

Example 2 is a small process synthesis problem that allow us to focus on some important topics on the solution of NLP and MINLP problems, namely: a) the importance of a good initialization scheme and bounds on the variables; b) analysis of the results in terms of quality of the solution (feasible vs. infeasible, and local vs. global); and c) the importance of having the model completely independent of the solver, which easily allows the model to be run with multiple solvers and compare solutions. For some NLP or MINLP problems the corresponding solvers may have difficulties finding a feasible solution for the problem. This may be overcome by developing initialization schemes tailored for the specific problem to solve. For example distillation or reactive distillation columns may require specific initialization strategies to help the NLP solvers to find a solution. The development of efficient strategies is built over the combination of a good knowledge of the physical and chemical aspects of the system to study, and the characteristics of the solver to use. Regarding the analysis of the quality of the solution, the algebraic modeling systems discussed are oriented for optimization based on principles from optimization theory, and thus, they require some knowledge about optimization, formulation of problems into mathematical programs, classification and properties of different formulations, analysis of the results, and an understanding of the quality of the solution including infeasible or sub-optimal solutions.

13.6 SUMMARY

Algebraic modeling systems are one of the tools available for Chemical Engineers for the application of optimization methods in process design and operation. Independent of the specific algebraic modeling system chosen, they all share and provide tools on their own way based on the following ideas: a) the modeling language uses a mathematical programming notation and structure based on sets, parameters, variables, constraints, and objective function; b) the model structure should be separated from the data; c) the model is independent of the solvers, giving that only a specific class of problems can be solved by a subset of solvers. These characteristics clearly contribute to the main advantages of algebraic modeling systems, namely easy implementation of models and advanced solution algorithms (not discussed in this work), flexibility of scaling up the models to accommodate larger sets of data, and a trouble free procedure to call the optimization solvers.

The main contributions of this chapter are reflected in the process synthesis problem analyzed in the second example. The implementation of the optimization model takes the concept of separation to the limit, by separating the structure of the flowsheet from the equations of the model. This separation is extremely useful for more complex process synthesis problems using superstructures.

A very helpful method to learn and improve to build optimization models is to read the models developed by others. There are open libraries available with optimization models for the system GAMS, see for example a) the classic reference CACHE's 6th Process Design Case Study, Chemical Engineering Optimization Models with GAMS [34]; b) the model library provided by GAMS, where there are some examples related with Chemical Engineering; c) the web server http://newton.cheme.cmu.edu/interfaces/ available from the Department of Chemical Engineering from the Carnegie Mellon University with several examples of programs built with GAMS; d) the web site www.minlp.org that is dedicated to MINLP problems with several optimization problems available; and e) the book [35], which has some examples about the implementation of advanced decomposition strategies. Obviously each modeling system has its own languages and tools, and therefore, the manuals of each of them are the first source of information to look for.

REFERENCES

1. Process Systems Enterprise, 1997–2009. gPROMS. www.psenterprise.com/gproms.
2. Averick, B. M., Moré, J. J., Bischof, C. H., Carle, A., Griewank, A., 1994. Computing large sparse jacobian matrices using automatic differentiation. *SIAM J. Sci. Comput.* 15 (2), 285–294.
3. Piela, P. C., Epperly, T. G., Westerberg, K. M., Westerberg, A. W., 1991. ASCEND—an object-oriented computer environment for modeling and analysis—the modeling language. *Comput. Chem. Eng.* 15 (1), 53–72.
4. Fritzson, P., Bunus, P., 2002. Modelica–a general object-oriented language for continuous and discrete-event system modeling. In: *Proceedings of the 35th Annual Simulation Symposium.* pp. 14–18.
5. Biegler, L. T., Grossmann, I. E., Westerberg, A. W., 1999. *Systematic Methods of Chemical Process Design.* Prentice Hall PTR, New Jersey.
6. Alattas, A. M., Grossmann, I. E., Palou-Rivera, I., 2011. Integration of nonlinear crude distillation unit models in refinery planning optimization. *Ind. Eng. Chem. Res.* 50 (11), 6860–6870.
7. Bazaraa, M., Sherali, H., Shetty, C. M., 2006. *Nonlinear Programming: Theory and Algorithms*, 3rd ed. John Wiley & Sons, New Jersey.
8. Edgar, T., Himmelblau, D., Lasdon, L., 2001. *Optimization of chemical processes.* McGraw-Hill.

9. Wolsey, L., 1998. *Integer Programming*. Wiley.
10. Bixby, R., Rothberg, E., 2007. Progress in computational mixed integer programming—a look back from the other side of the tipping point. *Ann. Oper. Res.* 49 (1), 37–41.
11. Rothberg, E., 2007. An evolutionary algorithm for polishing mixed integer programming solutions. *INFORMS J. Comput.* 19 (4), 534–541.
12. Lima, R. M., Grossmann, I. E., 2011. Computational advances in solving mixed integer linear programming problems. In: *Chemical Engineering Greetings to Prof. Sauro Pierucci*. AIDIC, Milano, Italy, pp. 151–160.
13. Schichl, H., 2004. Theoretical concepts and design of modeling languages for mathematical optimization. In: *Modeling Languages in Mathematical Optimization*. Springer Massachusetts, pp. 45–62.
14. Kallrath, J. (Ed.), 2012. *Algebraic Modeling Systems: Modeling and Solving Real World Optimization Problems*. Applied optimization. Springer, Massachusetts.
15. Bayer, R., Bigelow, J., Dantzig, G., Gries, D., McGrath, M., Pinsky, P., Schuck, S., Witzgall, C., 1968. *MPL Mathematical Programming Language*. Technical report CS 119. Computer Science Department, Stanford University, California.
16. Bussieck, M., Meeraus, A., 2004. General Algebraic Modeling System (GAMS). In: *Modeling Languages in Mathematical Optimization*. Springer, Massachusetts pp. 137–158.
17. Dunning I., Huchette J., Lubin M., 2017. JuMP: a modeling language for mathematical optimization. *SIAM Rev.* 59 (2), 295–320.
18. Bezanson J., Edelman A., Karpinski S., Shah V.B., 2017. Julia: a fresh approach to numerical computing. *SIAM Rev.* 59 (1), 65–98.
19. Hart W. E., Laird C. D., Watson J. P., Woodruff D. L., Hackebeil G. A., Nicholson B. L., Siirola J. D., 2017. *Pyomo-optimization Modeling in Python*, 2nd ed., vol. 67. Springer Science & Business Media.
20. Koch, T., 2004. *Rapid mathematical programming*. Ph.D. thesis, Technische Universität Berlin, zIB-Report 04–58.
21. Hultberg, T., 2006. FLOPC++ *a*n algebraic modeling language embedded in C++. In: Waldmann, K.-H., Stocker, U. M. (Eds.), Operations Research Proceedings pp. 187–190.
22. Kallrath, J. (Ed.), 2004. *Modeling Languages in Mathematical Optimization* Applied Optimization. Springer, Massachusetts.
23. Hurlimann, T., 1993. LPL: a mathematical programming language. *OR Spektrum* 15, 43–55.
24. Czyzyk, J., Mesnier, M., More, J., 1998. The neos server. *IEEE Comput. Sci. Eng.* 5 (3), 68–75.
25. Buhmann, M., Iserles, A. (Eds.), 1997. Optimization environments and the NEOS server. In: *Approximation Theory and Optimization*. Cambridge University Press.
26. Lima, R. M., Salcedo, R. L., Barbosa, D., 2006. SIMOP: Efficient reactive distillation optimization using stochastic optimizers. *Chem. Eng. Sci.* 61 (5), 1718–1739.
27. Chvatal, V., 1983. *Linear Programming* (Series of books in the mathematical sciences). W. H. Freeman.
28. Charnes, A., Cooper, W., Mellon, B., 1952. Blending aviation gasolines—a study in programming interdependent activities in an integrated oil company. *Econometrica* 20 (2), 135–159.
29. Choi, S., Ko, J., Manousiouthakis, V., 1999. A stochastic approach to global optimization of chemical processes. *Comput. Chem. Eng.* 23 (9), 1351–1356.
30. Manousiouthakis, V., Sourlas, D., 1992. A global optimization approach to rationally constrained rational programming. *Chem. Eng. Commun.* 115, 127–147.

31. Floudas, C., Ciric, A., 1989. Strategies for overcoming uncertainties in heat-exchanger network synthesis. *Comput. Chem. Eng.* 13 (10), 1133–1152.
32. Mendez, C., Myers, J., Roberts, S., Logdson, J., Vaia, A., Grossmann, I., 2005. MINLP model for synthesis of paraxylene separation processes based on crystallization technology. In: Puigjaner, L. (Ed.), *European Symposium on Computer Aided Process Engineering*, Vol. 15. Elsevier.
33. Lima, R. M., Grossmann, I. E., 2009. Optimal synthesis of p-xylene separation processes based on crystallization technology. *AIChE J.* 55 (2), 354–373.
34. Morari, M., Grossmann, I. (Eds.), 1991. *Chemical Engineering Optimization Models with GAMS.* In: *CACHE of process design case studies,* Vol. 6. CACHE.
35. Conejo, A., Castillo, E., Minguez, R., Garcia-Bertrand, R., 2006. *Decomposition Techniques in Mathematical Programming: Engineering and Science Applications.* Springer.

14 Use of GAMS for Optimal Process Synthesis and Operation

Antonio Sánchez, Mariano Martín

CONTENTS

14.1 INTRODUCTION

The idea behind process synthesis and design is the selection of the topology, the flowsheet, and the operating conditions to transform a set of raw materials into products [1]. For a long time, the objective was to design the process for the minimum cost (or maximum profit). However, environmental and social concerns have transformed this problem into a multi-objective one so that we account not only for economic issues but also for environmental ones such as greenhouse gas emissions and freshwater consumption, whose impact cannot easily be addressed via economic evaluations alone as well as social indicators such as jobs generated. We can distinguish two major methods: the use of heuristics and physical know-how, such as hierarchical decomposition [2] and pinch analysis [3], or the use of superstructure optimization [4,5]. The more recent approach consists of combining both, where we first develop a superstructure of alternative technologies for the various steps of the transformation of raw materials to products [6], we model the different technologies using different approaches such as first principles, mass and energy balances, chemical equilibria, or surrogate models based on rules of thumb, experimental correlations, design of experiments, reactor kinetics [7] formulating a problem whose solution yields the optimal production process. This type of formulation involves discrete and continue decisions for the selection of technologies and operating conditions giving rise to mixed-integer non-linear programming problem (MINLP) or a generalized disjunctive programming (GDP) model.

From a mathematical stand point the type of problems we usually deal with in process design are LP, MILP and most typically NLP and MINLP. The main characteristics of the different types of problems and solvers that can be used are briefly described in Chapter 13.

14.2 HINTS FOR MODELING PROCESSES IN GAMS

We provide some tips that can be useful to model a process in GAMS. We start with "Sets" that are typically used to represent groups of items, in order that we may apply common equations to all of them. We could have as sets the units within a process, the chemical species involved or the streams that link the units. For example, if J is the set of components and S the set of streams of the flowsheet such as:

$$J = \{CO, H_2, H_2O, CO_2\}$$

$$S = \{1,2,3,4,5\}$$

in GAMS we model them as:

```
Set
J /CO, H2, H2O, CO2/;
S /1*5/;
```

A major challenge in equation-based modeling is the representation of the thermodynamics. We need to provide heat capacities, formation heats, and other items within a mathematical formulation. We can make use of "table" and "parameter" as function of the sets of chemical species involved, so that we have the necessary properties of the species available for our calculations. For instance, we can have the information for molecular weights:

$MW = \{28,2,18,44\}$ which, when represented in, GAMS:

```
Parameter
      MW(J)
                /CO     28
                 H2     2
                H2O     18
                CO2     44/;
```

For single constant values such as conversions, rules or thumb, yields, we can define "scalar". For instance:

$$Conversion = 0.85$$

```
Scalar
      Conversion /0.85/;
```

We define "variables", for example the flows from one unit to another, temperatures and pressures, the size of equipment, its cost or other rules or thumb or experimental data to predict the performance of the units such as conversions, removal rates, etc. Using the default command "variable", the variables are considered to be free (no sign defined for them) and continuum. Special types such as "positive", "integer" or "binary" ones must specified using these words before the command "variable" in GAMS. Variables can also be a function of the sets. To illustrate we can have a variable to define the flow of a component J in all streams in the flowsheet. We can use the set to define all the flows in our flowsheet. We define in GAMS as follows:

```
Variable
        F(J,S);
```

Thus, we can define positive variables, binaries, integer or free variables. The objective function must be defined in GAMS as a free variable.

Next, we define the equations in GAMS. We use the command "equation" and provide a name per equation or set of equations. The advantage of using sets to define equations is that they can be written for an entire set. For instance, an equation to represent a stage with no change in composition between stream 1 and stream 2 would be as follows:

```
MB(J)..   F(J,'1')=E= F(J,'2 ')
```

Finally, we are ready to include the solver statement. We declare the equations that formulate the problem, provide the options for the solver to be used and write the solve statement defining the type of problem at hand (LP, MIP, MILP, MINLP, NLP).

14.3 EXAMPLES

14.3.1 Process Operation

14.3.1.1 SO_2 Catalytic Converter

Consider a converter consisting of four catalytic beds for the oxidation of SO_2 to SO_3. Such a reaction is highly exothermic, and the conversion is limited by the equilibrium of the reaction [8].

$$
\begin{aligned}
&SO_2 + \frac{1}{2}O_2 \Leftrightarrow SO_3 \\
&K_p = \frac{P_{SO_3}}{P_{SO_2} \cdot P_{O_2}^{0.5}} \\
&\log_{10}(K_p) = \frac{4956}{T} - 4.678
\end{aligned}
\tag{14.1}
$$

The flow needs to be cooled at the outlet of each bed if a high final conversion is to be achieved. In this example, we optimize the inlet temperatures of each bed to maximize the final conversion of the reactor. The SO_2 entering the reactor is 10% of the total inlet feed, the pressure is assumed to be 1 atm and the temperature 600 K.

TABLE 14.1
Feed Conditions to the Converter

	$S \rightarrow SO_2$		$SO_2 \rightarrow SO_3$	
	Initial	**Final**	**Initial**	**Equilibrium**
SO_2	–	a	a	a·(1–x)
N_2	0.79	0.79	0.79	0.79
O_2	0.21	0.21–a	0.21–a	0.21–a–0.5·a·x
SO_3	–			a·x

We start with sulfur as initial raw material, which is burned with dry air and later oxidized. We need to dry the atmospheric air in order to avoid any corrosion due to early formation of sulfuric acid. For 1 kmol of air the mass balance to the sulfur combustion and the SO_3 formation is as given in Table 14.1, where "a" is the molar fraction of SO_2 in the feed to the converter. We first determine the equilibrium curve. The amount of moles at equilibrium is given by:

$$n_T = a - a \cdot x + 0.79 + 0.21 - a - 0.5 \cdot a \cdot x + a \cdot x = 1 - 0.5 \cdot a \cdot x \qquad (14.2)$$

The partial pressure of the different gases in the equilibrium is given as:

$$P_{SO_2} = \frac{a \cdot (1-x)}{1 - \frac{a \cdot x}{2}} P_T; \quad P_{SO_3} = \frac{a \cdot (x)}{1 - \frac{a \cdot x}{2}} P_T; \quad P_{O_2} = \frac{0.21 - a - \frac{a \cdot x}{2}}{1 - \frac{a \cdot x}{2}} P_T \qquad (14.3)$$

Thus the equilibrium constant (K_p) as function of the total conversion, is calculated as shown in eq. (14.4).

$$K_p = \frac{P_{SO_3}}{P_{SO_2} \cdot P_{O_2}^{0.5}} = \frac{x \cdot \left[1 - \frac{a \cdot x}{2}\right]^{0.5}}{(1-x)\left[0.21 - a - \frac{a \cdot x}{2}\right]^{0.5} P_T^{0.5}} \qquad (14.4)$$

We plot the conversion versus the temperature in Figure 14.1. We use EXCEL® and refer to Chapter 3 for further reference, although we could have used any other package. As the reaction progresses, the gases are heated up. A simplified energy balance assuming constant heat capacity and constant total molar flow rated for the first bed is given by eq. (14.5).

$$mol \cdot Cp_{mix}(T - T_o) = (1 - 0.5 \cdot a \cdot x) \cdot Cp_{mix}(T - T_o) = a \cdot x \cdot 23200 \qquad (14.5)$$

The error is negligible in the case of using constant Cp, due to the large amount of nitrogen, and small for the assumption of constant molar flow rate as we will see

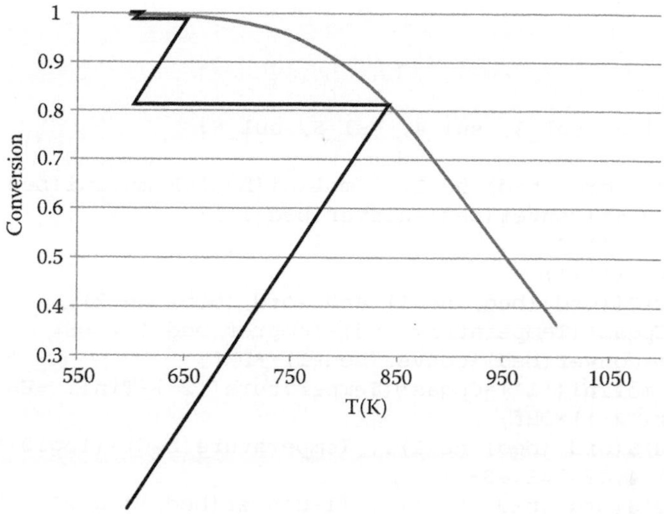

FIGURE 14.1 Optimal operation of the four-bed converter.

in the results. As we can see in Figure 14.1, the conversion is limited by the equilibrium. The energy balance given by eq. (14.5) represents a curve in Figure 14.1 that intersects at the equilibrium at the end of each bed; see Figure 14.1.

To write the model in GAMS®, we define as a set the number of streams considered, from the feed to the exit of the fourth bed. As scalars, we denote the heat of reaction, ΔH_f, the heat capacity of the gas mixture, the fraction of SO_2 in the feed to the reactor and the total pressure. The variables, temperatures, initial moles, conversion, equilibrium constant and the cool down temperature, are defined as arrays depending on the set of streams. Next, we write the equations that formulate the problem and we provide some bounds for the variables. Finally, we optimize the conversion obtained after the fourth bed.

```
Set
bed beds /1*5/

Scalar
DHf /23200/
Cpgas  /7.9/
SO2ini   /0.1/
Ptotal /1/
Tini /600/;

Positive Variables
Temperature(bed)
kp(bed)
molini(bed)
conver(bed)
Tempref(bed)       ;
```

```
Variable
z;

Equations
sul_1, sul_2, sul_3, sul_4, sul_5, sul_6;

sul_1(bed)$(ord (bed) ne 1).. molini(bed)=E=molini(bed-1)-
  SO2ini*0.5*(conver(bed)-conver(bed-1));
conver.fx('1')=0;
molini.fx('1')=1;
sul_2(bed)$((ord (bed) ne 1) and (ord (bed) ne 2)).. molini
  (bed)*Cpgas*(Temperature(bed)-Tempref(bed-1)) =E=
  SO2ini*(conver(bed)-conver(bed-1))*DHf;
sul_3.. molini('1')*Cpgas*(Temperature('2')-Tini) =E= SO2ini*
  (conver('2'))*DHf;
sul_4(bed)$(ord (bed) ne 1).. Temperature(bed)*(log10(kp(bed)+
  0.0001)+4.678)=E=4956;
sul_5(bed)$(ord (bed) ne 1).. (1-conver(bed))*(0.21-SO2ini-
  0.5*SO2ini*conver(bed))**0.5*Ptotal**0.5*kp(bed)-
conver(bed)*(1-0.5*SO2ini*conver(bed))**0.5=E=0;
Tempref.fx('1')=600;
Tempref.LO(bed)=600;
Tempref.UP(bed)=700;
Temperature.fx('1')=600;
Temperature.LO(bed)=600;
Temperature.UP(bed)=900;
sul_6.. z=E= conver('5');

Model sulfuricreact /ALL/;

option NLP = conopt3;

Solve  sulfuricreact Using NLP Maximizing Z;
```

In a new screen the results file, .lst, opens. To the left we can access the model statistics, solution report, equations and variables solved. We can click on each one to see different characteristics of the model (e.g. "Model Statistics", "Solution Report", "SolEQU" to provide the size of the model, information on the use of the solver and the equations, respectively).

We see that model status results in 2 Locally Optimal. CONOPT is a local solver and thus the solution is not guaranteed to be global in non-linear problems. Finally we can also click on "SolVAR" to see the optimal values of the variables in the model.

QUESTIONS

- Evaluate the effect of the cost of cooling on the optimum conversion and the cooling temperatures.
- Evaluate the effect of the fraction of SO_2 and that of the total operating pressure on the optimal temperatures for different final conversions.

14.3.1.2 Ammonia Reactor

Ammonia is one of the most important chemicals. Its production is about 157 Mt of ammonia (2010) [9–11]. Ammonia uses are diverse but the main one is as raw material in the nitrogen-based fertilizers. Nowadays, Haber-Bosch process is almost the unique technological alternative in use. A mixture of hydrogen and nitrogen (approximately 3:1) is employed like fed stream in the ammonia synthesis loop. Two alternatives have been considered for the reactor: tubular or multi-bed. Tubular is suitable for small capacities. In the common big facilities, multi-bed reactor is the most selected option. To cool down between beds in these multi-bed reactors, direct and indirect cooling have been considered.

In this example, a two-bed reactor with direct cooling is considered. The flowsheet is shown in Figure 14.2.

The flows, temperature, etc. are define as "variables" and the component data such as heat capacity or vapor pressure as "table." The variables regarding to stream are defined between the leaving unit and the inlet unit. For this, we use the command "Alias" to duplicate the same set and, then, we create the matrix "Arc" to determine the possible combination between the units according to the flowsheet.

```
Set
J 'Components' /N2,H2,NH3/
unit 'unit' /Src1*Src3,sep1,HX1,Reac1,Reac2,Mix1,HX2,sep2/
mar 'heat capacity equation' /1*4/
cnt 'Antoine equation' /A, B, C/ ;
Alias(unit, unit1)

Set
Arc(unit,unit1) stream matrix;
Arc(unit, unit1)=No;
Arc('Src1','sep1')=Yes;
Arc('sep1','HX1')=Yes;
Arc('HX1','Reac1')=Yes;
Arc('Reac1','Mix1')=Yes;
Arc('sep1','Mix1')=Yes;
```

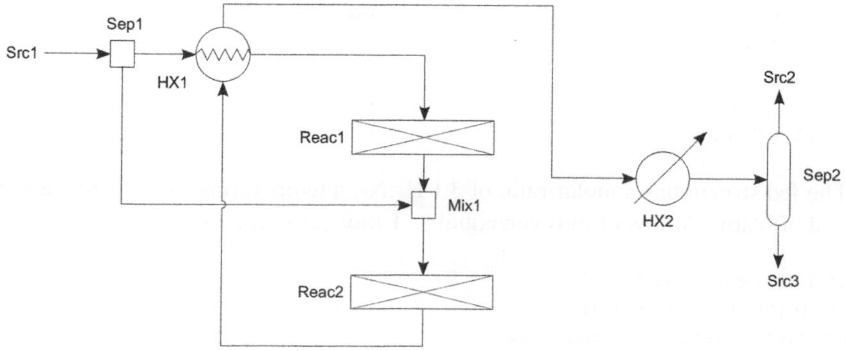

FIGURE 14.2 Flowsheet of ammonia reaction and separation.

```
Arc('Mix1','Reac2')=Yes;
Arc('Reac2','HX1')=Yes;
Arc('HX1','HX2')=Yes;
Arc('HX2','sep2')=Yes;
Arc('sep2','Src2')=Yes;
Arc('sep2','Src3')=Yes;

Table
*Heat capacity of gases
*Cp° = A + B*t + C*t2 + D*t3 where Cp0 (J/mol K) and t(T(K))
   cp_v(J,mar)
            1          2              3            4
     N2     31.150     -1.3570E-02    2.6796E-05   -1.168E-08
     H2     27.143     9.2738E-03     -1.381E-05   7.6451E-09
     NH3    27.315     2.3831E-02     1.7074E-05   -1.185E-08;
Table
*Vapor pressure
*lnP=A-(B/(C+T)) with P(mmHg) and T(K)
  Ant(J,cnt)
              A          B          C
    N2     14.9542    588.72     -6.60
    H2     13.6333    164.90      3.19
    NH3    16.9481    2132.5     -32.98;
Positive variable
T(unit,unit1) 'Temperature'
Kp1_1 'Equilibrium constant 1° bed reactor'
Kp1_2 'Equilibrium constant 2° bed reactor'
P(J,unit,unit1) 'Partial Pressure'
Xreac1 'Conversion 1° bed reactor'
Xreac2 'Conversion 2° bed reactor'
n(J,unit,unit1) 'mol per second for each component'
nT(unit,unit1)  'total flow'
QHX2 'Heat in HX2'
ySep2(J) 'Gas compositions in ammonia condensation'
xSep2(J) 'Liquid compositions in ammonia condensation'
Ksep2(J) 'G-L equilibirum constant';
Variable
Q(J,unit,unit1) 'specific enthalpy per component';
Xreac1.LO=0;
Xreac1.UP=1;
Xreac2.LO=0;
Xreac2.UP=1;
```

The fed stream has a molar ratio of 3:1 H_2:N_2, a temperature of 25°C and a pressure of 300 atm. A flow of nitrogen equal to 1 mol per second is assumed.

```
*Feed Stream (Src1)
T.LO(unit,unit1)=260;
n.fx('N2','Src1','sep1')=1;
n.fx('NH3','Src1','sep1')=0;
T.fx('Src1','sep1')=298;
```

```
equations
f_1,f_2;
f_1.. n('H2','Src1','sep1')=E=3*n('N2','Src1','sep1');
f_2.. nT('Src1','sep1')=E=sum(J,n(J,'Src1','sep1'));
```

The pressure is assumed constant in all units. The separator divides the flow into two streams, one to feed the first reactor bed (through heat exchanger HX1) and the other one to cool down the stream from the first bed reactor in the mixer Mix1. The composition is the same in both streams.

```
*Separator (Sep1)
equations
sep_1,sep_2,sep_3,sep_4,sep_5,sep_6,sep_7;
sep_1(J).. n(J,'Src1','sep1')=E=n(J,'sep1','HX1')+n(J,'sep1',
  'mix1');
sep_2.. nT('sep1','HX1')=E=sum(J,n(J,'sep1','HX1'));
sep_3.. nT('sep1','mix1')=E=sum(J,n(J,'sep1','mix1'));
sep_4.. T('Src1','sep1')=E=T('sep1','HX1');
sep_5.. T('Src1','sep1')=E=T('sep1','mix1');
sep_6(J).. n(J,'Src1','sep1')*nT('sep1','HX1')=E=n(J,'sep1',
  'HX1')*nT('Src1','sep1');
sep_7(J).. n(J,'Src1','sep1')*nT('sep1','Mix1')=E=n(J,'sep1',
  'Mix1')*nT('Src1','sep1');
```

In the exchanger HX1, the stream to feed the first bed reactor is heated up using the flow that leaves the second reaction bed. The inlet temperature of the first bed is fixed to 400ºC.

```
*Heat exchanger 1 (HX1)
equations
hx1_1,hx1_2,hx1_3,hx1_4,hx1_5,hx1_6,hx1_7,hx1_8;
hx1_1(J).. n(J,'sep1','HX1')=E=n(J,'HX1','Reac1');
hx1_2.. nT('HX1','Reac1')=E=sum(J,n(J,'HX1','Reac1'));
hx1_3(J).. Q(J,'HX1','Reac1')=E=(cp_v(J,'1')*(T('HX1','Reac1')
  -Tref))+((1/2)*cp_v(J,'2')*(T('HX1','Reac1')**2-Tref**2))+
  ((1/3)*cp_v(J,'3')*(T('HX1','Reac1')**3-Tref**3))+((1/4)
  *cp_v(J,'4')*(T('HX1','Reac1')**4-Tref**4));
hx1_4(J).. Q(J,'sep1','HX1')=E=(cp_v(J,'1')*(T('sep1','HX1')
  -Tref))+((1/2)*cp_v(J,'2')*(T('sep1','HX1')**2-Tref**2))+
  ((1/3)*cp_v(J,'3')*(T('sep1','HX1')**3-Tref**3))+((1/4)
  *cp_v(J,'4')*(T('sep1','HX1')**4-Tref**4));
hx1_5(J).. n(J,'Reac2','HX1')=E=n(J,'HX1','HX2');
hx1_6.. nT('HX1','HX2')=E=sum(J,n(J,'HX1','HX2'));
hx1_7(J).. Q(J,'HX1','HX2')=E=(cp_v(J,'1')*(T('HX1','HX2')
  -Tref))+((1/2)*cp_v(J,'2')*(T('HX1','HX2')**2-Tref**2))+
  ((1/3)*cp_v(J,'3')*(T('HX1','HX2')**3-Tref**3))+((1/4)
  *cp_v(J,'4')*(T('HX1','HX2')**4-Tref**4));
hx1_8.. sum(J,n(J,'sep1','HX1')*Q(J,'sep1','HX1'))+sum(J,n(J,
  'Reac2','HX1')*Q(J,'Reac2','HX1'))=E= sum(J,n(J,'HX1',
  'Reac1')*Q(J,'HX1','Reac1'))+sum(J,n(J,'HX1','HX2')
  *Q(J,'HX1','HX2'));
```

Both reaction beds are modeled like a chemical equilibrium reactor. The equilibrium constant as a function of the temperature is given in the equations eq. (14.6) and eq. (14.7):

$$P_{NH_3} = \frac{n_{NH_3}}{n_T} P_T; P_{N_2} = \frac{n_{N_2}}{n_T} P_T; P_{H_2} = \frac{n_{H_2}}{n_T} P_T;$$
$$K_p = \frac{P_{NH_3}}{P_{N_2}^{\ 0.5} \cdot P_{H_2}^{\ 1.5}} = \frac{n_T}{P_T} \frac{n_{NH_3}}{n_{N_2}^{\ 0.5} \cdot n_{H_2}^{\ 1.5}} \tag{14.6}$$

$$\log_{10}(K_p) = \frac{2250.322}{T} - 0.85430 - 1.51049 \log_{10} T - 2.58987x10^{-4} T + 1.48961 \cdot 10^{-7} T^2 \tag{14.7}$$

The mass balances in the reactor are as follow:

$$\begin{aligned} n_{NH_3}\Big|_{out} &= n_{NH_3}\Big|_{in} + 2X n_{N_2}\Big|_{in} \\ n_{N_2}\Big|_{out} &= n_{N_2}\Big|_{in}(1-X) \\ n_{NH_2}\Big|_{out} &= n_{H_2}\Big|_{in} - 3X n_{N_2}\Big|_{in} \end{aligned} \tag{14.8}$$

The outlet temperature is calculated based on the energy balance considering adiabatic conditions. The heat of reaction is taken from [9] with a value of 45720 kJ/kmol NH_3. The two reactors are modeled following the same procedure.

```
*Reaction Bed 1 (Reac1)
T.fx('HX1','Reac1')=673;
equations
eq1,eq2,eq3,eq4,eq5,eq6,eq7,eq8,eq9;
eq1.. Kp1_1=E=10**((2250.322/T('Reac1','Mix1'))-0.85430-1.51049
  *log10(T('Reac1','Mix1'))-2.58987e-4*T('Reac1','Mix1')+
  1.48961e-7*T('Reac1','Mix1')**2);
eq2..
Kp1_1*((P('N2','Reac1','Mix1')**0.5*P('H2','Reac1','Mix1')
  **1.5))=E=P('NH3','Reac1','Mix1');
eq3(J).. P(J,'Reac1','Mix1')*(nT('Reac1','Mix1'))=E=n(J,
  'Reac1','Mix1')*Ptotal;
eq4.. n('NH3','Reac1','Mix1')=E=n('NH3','HX1','Reac1')+
  2*n('N2','HX1','Reac1')*Xreac1;
eq5.. n('N2','Reac1','Mix1')=E=n('N2','HX1','Reac1')*(1-Xreac1);
eq6.. n('H2','Reac1','Mix1')=E=n('H2','HX1','Reac1')-3*n('N2',
  'HX1','Reac1')*Xreac1;
eq7.. nT('Reac1','Mix1')=E=sum(J,n(J,'Reac1','Mix1'));
eq8(J).. Q(J,'Reac1','Mix1')=E=(cp_v(J,'1')*(T('Reac1',
  'Mix1')-Tref))+((1/2)*cp_v(J,'2')*(T('Reac1','Mix1')**2-Tref
```

```
  **2))+((1/3)*cp_v(J,'3')*(T('Reac1','Mix1')**3-Tref**3))+
  ((1/4)*cp_v(J,'4')*(T('Reac1','Mix1')**4-Tref**4));
eq9.. sum(J,n(J,'Reac1','Mix1')*Q(J,'Reac1','Mix1'))=E=
        sum(J,n(J,'HX1','Reac1')*Q(J,'HX1','Reac1'))+
          (n('NH3','Reac1','Mix1')-n('NH3','HX1','Reac1'))
          *45720;
```

After the first bed reactor, the gas stream is directly cooled down using a fraction of the fed stream in the mixer Mix1. The outlet temperature is limited to 400°C as minimum temperature due to below this temperature, kinetic in the reactor bed could be really low. The mixer is assumed adiabatic.

```
*Mixer (Mix1)
equations
eq1_1,eq1_2,eq1_3,eq1_4,eq1_5;
eq1_1(J).. n(J,'sep1','Mix1')+n(J,'Reac1','Mix1')=E=n(J,
  'Mix1','Reac2');
eq1_2.. nT('Mix1','Reac2')=E=sum(J,n(J,'Mix1','Reac2'));
eq1_3(J).. Q(J,'sep1','Mix1')=E=(cp_v(J,'1')*(T('sep1','Mix1')
  -Tref))+((1/2)*cp_v(J,'2')*(T('sep1','Mix1')**2-Tref**2))+
  ((1/3)*cp_v(J,'3')*(T('sep1','Mix1')**3-Tref**3))+((1/4)
  *cp_v(J,'4')*(T('sep1','Mix1')**4-Tref**4));
eq1_4(J).. Q(J,'Mix1','Reac2')=E=(cp_v(J,'1')*(T('Mix1','R
  eac2')-Tref))+((1/2)*cp_v(J,'2')*(T('Mix1','Reac2')**2-Tref*
  *2))+((1/3)*cp_v(J,'3')*(T('Mix1','Reac2')**3-Tref**3))+
  ((1/4)*cp_v(J,'4')*(T('Mix1','Reac2')**4-Tref**4));
eq1_5.. sum(J,n(J,'Reac1','Mix1')*Q(J,'Reac1','Mix1'))
        +sum(J,n(J,'sep1','Mix1')*Q(J,'sep1','Mix1'))=E=
        sum(J,n(J,'Mix1','Reac2')*Q(J,'Mix1','Reac2'));
```

The ammonia from the gas stream is separated through condensation. For this purpose, the gas stream is cooled down in two steps. One in HX1, to transfer the heat to the inlet stream, and other one, in HX2. In this unit, the temperature is reduced to –13°C, using a refrigerant.

```
*Heat exchanger (HX2)
T.fx('HX2','sep2')=260;
equations
eq3_1,eq3_2,eq3_3,eq3_4;
eq3_1(J).. n(J,'HX1','HX2')=E=n(J,'HX2','sep2');
eq3_2.. nT('HX2','sep2')=E=sum(J,n(J,'HX2','sep2'));
eq3_3(J).. Q(J,'HX2','sep2')=E=(cp_v(J,'1')*(T('HX2','sep2')-
  Tref))+((1/2)*cp_v(J,'2')*(T('HX2','sep2')**2-Tref**2))+
  ((1/3)*cp_v(J,'3')*(T('HX2','sep2')**3-Tref**3))+((1/4)
  *cp_v(J,'4')*(T('HX2','sep2')**4-Tref**4));
eq3_4.. sum(J,n(J,'HX1','HX2')*Q(J,'HX1','HX2'))=E=
        sum(J,n(J,'HX2','sep2')*Q(J,'HX2','sep2'))+QHX2;
```

TABLE 14.2
Antoine Parameters for the Involved Components [12]

	A	B	C
N_2	14.9542	588.72	–6.60
H_2	13.6333	164.90	3.19
NH_3	16.9481	2132.5	–32.98

To calculate the fraction that condenses, Gas-Liquid equilibrium relations have been used.

$$\begin{aligned} &F \cdot z_J = L \cdot x_J + V \cdot y_J && \forall J \\ &y_J = K_J \cdot x_J && \forall J \\ &\sum_{J=1}^{n} y_J = 1 && \\ &\sum_{J=1}^{n} x_J = 1 && \\ &K^J = \frac{P_{vap}{}^J}{P_{total}} && \forall J \end{aligned} \tag{14.9}$$

The vapor pressure is calculated using the Antoine correlation for the hydrogen, nitrogen and ammonia, see Table 14.2 [11].

$$\ln(P(mmHg)) = A - \frac{B}{C + T(K)} \tag{14.10}$$

```
*Separator (Sep2)
equations
eq4_1,eq4_2,eq4_3,eq4_4,eq4_5,eq4_6,eq4_7,eq4_8,eq4_9;
eq4_1(J).. Ksep2(J)*Ptotal*760=E=(exp(Ant(J,'A')-(Ant(J,'B')/
   (T('HX2','sep2')+Ant(J,'C')))));
eq4_2(J).. ySep2(J)=E=Ksep2(J)*xSep2(J);
eq4_3.. sum(J,ySep2(J))=E=1;
eq4_4.. sum(J,xSep2(J))=E=1;
eq4_5(J).. n(J,'HX2','sep2')=E=n(J,'sep2','Src2')+n(J,'sep2',
   'Src3');
eq4_6.. nT('sep2','Src2')=E=sum(J,n(J,'sep2','Src2'));
eq4_7.. nT('sep2','Src3')=E=sum(J,n(J,'sep2','Src3'));
eq4_8(J).. ySep2(J)*nT('sep2','Src2')=E=n(J,'sep2','Src2');
eq4_9(J).. xSep2(J)*nT('sep2','Src3')=E=n(J,'sep2','Src3');
```

These equations provide a simple approach to the phenomena involved in the condensation of the gases. Due to this, the results obtained present a lack of accuracy with respect to the actual behavior. In this case, using the G-L equilibrium equations, a large amount of gases go along with ammonia in the liquid stream. If, for example, the reader simulates these same conditions in a commercial simulator (like ASPEN; see Chapter 12), the obtained ammonia has a purity about 99% molar, with only few gases leaving with the liquid stream. A solution to solve this issue is used surrogate models based on experimental data or rigorous simulations [12].

The selected objective function is to maximize the ammonia production in the reactor (two beds). The results show, for example, the optimal split in the separator sep1, to obtain the maximum conversion in the reactor or the inlet and outlet temperature in each reactor bed.

```
variable z;
equations obj;
obj.. z=E=n('NH3','Reac2','HX1');
```

The results obtained show that the inlet flow is equal to 4 mol per second (3:1 H_2:N_2). In the separator, 2.657 mol per second go to the first bed reactor and 1.343 mol per second go to the mix1. The final temperatures in the beds are 864.916 K and 812.014 K respectively. In both reaction bed the inlet temperature is equal to 673 K.

QUESTIONS

- In general, equilibrium conditions are not reached in the reactor. Model this same example fixing a final reactor temperature lower than the equilibrium one.
- Due to the low conversion in the reactor, limited by equilibrium, the gases separate in the ammonia condensation are recycled to the reactor. Try to close the recycled loop.

14.3.1.3 Steam Reforming of Natural Gas

Reforming of natural gas is a common process for the production of hydrogen, where the yield is a function of the equilibrium established among the different chemical species and depends on the operating pressure, due to the difference in the number of moles in the stoichiometry, and on the temperature, since the reforming reaction is endothermic. The stoichiometry and the equilibrium constants are taken from the literature [14].

The problem would be to optimize the operating pressure and temperature of the furnace taking into account that part of the natural gas, is needed to provide the energy to maintain the reaction isothermal and in based on the use of shale gas for liquid fuels production [15]. Figure 14.3 shows the flowsheet. We start by defining the sets for the units, the streams and the chemicals.

```
Set
*         Define units
          unit    units
         / Furnace,  Compres, Prod, Src1*Src2 , Spl  /

*         Define components
          J        components
          /Wa, CO2, CO, H2, CH4/
```

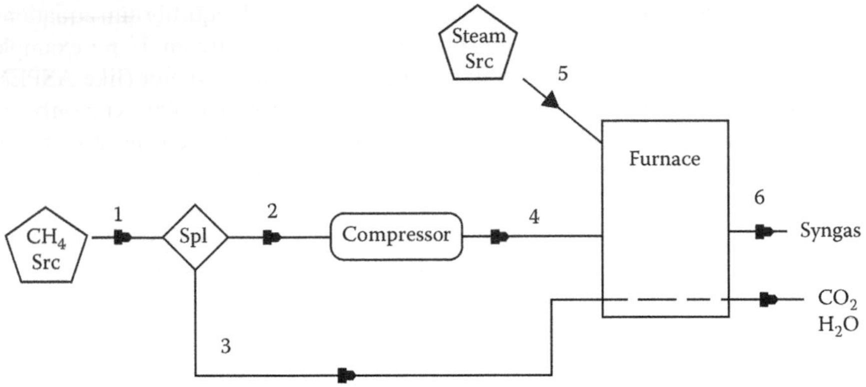

FIGURE 14.3 Steam reforming.

```
        liquid(J)
        /Wa/
        gases(J)
        /Wa, CO2, CO, H2, CH4/
        S      streams
        /1*6/
*       running variable for cp vapor
        mar /1*4/;
```

Next, we define as parameters, tables and scalars the general data for the process such as heat capacities (the coefficients for the polynomial correlation), molecular weights, heat of formation and the boiling and critical temperature for water [11]. We can compute the heat of vaporization as a function of temperature using Watson formula [16]. We also define a small parameter, espl, to avoid dividing by 0 in the equations.

```
Parameters
*       individual liquid/solid heat capacity of a
          component
        c_p_ind(J)
        /Wa              4.18
         CO2             0
         CO              0
         H2              0
         CH4             0/
*       standard heat of vaporization: kJ/kg
        dH_vap_0(liquid)
       /Wa               2254.62     /
*       critical temperature: C
        Tc(liquid)
       /Wa              374.15 /
*       boiling point temperature: C
        Tb(liquid)
       /Wa              100 /
*       Molecular weight: g/mol
        MW(J)
```

```
          /Wa              18
          CO2              44
          CO               28
          H2               2
          CH4              16/
     *Heat of formation of the components in gas form    (kJ/kg)
         dH_f (J)
          /Wa              -13444.44444
          CO2              -8946.590909
          CO               -3950.713286
          H2               0
          CH4              -4678.75/                 ;
      Table          c_p_v(J,mar)
               1               2                3               4
Wa     32.243        1.92E-03        1.06E-05          -3.60E-09
CO2    19.795        7.34E-02        -5.60E-05         1.72E-08
CO     30.869        -1.29E-02       2.79E-05          -1.27E-08
H2     27.133        92.738E-04      -1.381E-05        76.451E-10
CH4    19.251        5.21E-02        1.20E-05          -1.13E-08;

          Scalar   n_watson        exponent in Watson
            correlation /0.38/;
          Scalars
          *Define temperatures in C
                 T_amb       ambient temperature /20/ ;
          scalar   eps1    small number to avoid div. by zero /
            1.e-04 / ;
```

We define the variables to represent the flows, mass fractions, temperatures. Next, we define the outlet pressure of the compressor, the energy as heat or work required or needed at each unit and lower and upper bounds on the variables.

```
Positive Variables
*       streams and mass fractions:
        F(s)          total flow in stream 's' in kg s^-1
        fc(J,s)       individual component flow in each stream 's'
        x(J,s)        mass fraction of comp J in stream
*       heat of evaporation
        dH_v(liquid,s)       heat of vap. (KJ per kg)
*       temperatures in C
        T(s)          temperature of stream in C ;

Variables
*       power
        W(unit)  power consumption of a process unit in kW
          (efficiency included)
*       heat
        Q(unit)       heat produced or consumed in unit
          (efficiency included)   ;
```

```
*Define global bounds and fix specific variables
*Upper bound on mass fractions
x.UP(J,s)=1;
*Upper bound on total flow of all streams
F.UP(s)=600;
*Upper limit on component flow in all streams
fc.UP(J,s)=600;
*Specifying heat and power consumption of certain units
Q.Fx('Src1')=0;
Q.Fx('Src2')=0;
Q.Fx('Compres')=0;
Q.Fx('Prod')=0;
W.Fx('Furnace')=0;
W.Fx('Src1')=0;
W.Fx('Src2')=0;
W.Fx('Prod')=0;
*Temperature settings
*global temperature bounds - bounds get redefined for specific
  streams
T.LO(s)=20;
T.UP(s)=1000;
```

We also define the input streams in terms of temperature and composition. We use the conditional "$" so that we fix the components in stream 1 to be 0 except for methane in the feedstock, and similarly for the steam, we fix all the components to be 0 except for water. T is given in °C.

```
*Specifying temperatures
*Src
fc.fx(J,'1')$(ord (J) ne 5)=0;
x.fx('CH4','1')=1;
T.Fx('1')=  T_amb ;
*Src
x.fx('Wa','5')=1;
fc.fx(J,'5')$(ord (J) ne 1)=0;
T.Fx('5')=  233 ;
```

The total flow in all streams is equal to the sum of the flows of the individual components:

```
*Global relationships
Equations
         Rel_1,Rel_2;

Rel_1(J,s)..   fc(J,s) =E= F(s)*x(J,s);

Rel_2(s).. Sum(J,fc(J,s)) =E= F(s);
```

We start modeling the process by defining extra variables that help us calculate the enthalpy of products and reactants, the equilibrium constants and the operating pressure of the furnace (which is the same as the outlet pressure of the compressor). PCI is the inferior combustion power of the methane.

```
*                                         ***************
*                                         Rerforming Nat Gas
*                                         **************
Variables
 Q_prod, Q_reac;

Positive Variables
P_refor, K_met, K_WGS;

Scalar
PCI_gas        In kJ per kg               /39900/
nu_c                             /1/ ;

Equations
Refor_1, Refor_2, Refor_3, Refor_4, Refor_5, Refor_6, Refor_7,
  Refor_8, Refor_9, Refor_10, Refor_11, Refor_12, Refor_13,
  Refor_14, Refor_15;
```

14.3.1.3.1 Balance to the Splitter and Compressor

A general mass balance for a splitter is formulated where methane is the only component. We define the components as 0. For this we use the "$" symbol and the command ord(J) so that we specify the zero components within the set J. We can use "ne", non-equal, or "eq", equal to, to define the elements involved in the calculation. We can also use the command card(J) to determine the size of the set if needed.

```
Refor_1..    fc('CH4','1') =E= fc('CH4','2')+fc('CH4','3');
fc.fx(J,'2')$(ord (J) ne 5)=0;
fc.fx(J,'3')$(ord (J) ne 5)=0;
*inicial feedstock
fc.fx('CH4','1')= 4;
```

For the power consumed by the compressor, we assume polytropic behavior and 100% efficiency in this example. MW_gas is the average molar weight, T is the operating temperature, η_c the isentropic efficiency, P_{reform}, the reforming pressure, P_{inlet}, the initial pressure, F the total flow in (kg/s) and W the power consumed by the compressor.

$$\mathrm{W}(Compres)=(F)*\frac{8.314*1.4*(\mathrm{T}+273)}{((\mathrm{MW\_gas})*(1.4-1))}\frac{1}{\eta_c}\left(\left(\frac{P_{reform}}{P_{inlet}}\right)^{0.4/1.4}-1\right); \tag{14.11}$$

```
Refor_2.. W('Compres') =E= F('2')*(8.313*1.4*(T_amb+273))
  *nu_c* ((P_refor/1)**((0.4/1.4))-1)/((MW('CH4'))*(1.4-1));

P_refor.LO=2;
P_refor.UP=10;
```

And the temperature:

$$T=\left((T_{in}+273)+\frac{1}{\eta_c}(T_{in}+273)*\left(\left(\frac{P_{refor}}{P_{inlet}}\right)^{0.4/1.4}-1\right)\right)-273 \tag{14.12}$$

```
Refor_3..     T('4')=E= (T_amb+273)+nu_c*(T_amb+273)*((P_refor/1)
  **((0.4/1.4)) -1)-273;
```

14.3.1.3.2 Energy Balance for the Reformer

The energy to be provided to the furnace is calculated from an energy balance to the unit as:

$$Q(Furnance)=\sum_j \Delta H_f(j)\big|_{T(6)}\Big|_{(6)} -\sum_j \Delta H_f(j)\big|_{T(5)}\Big|_{(5)} -\sum_j \Delta H_f(j)\big|_{T(4)}\Big|_{(4)} \tag{14.13}$$

To compute this energy we use the two variables defined above as Q_prod and Q_react for the sake of simplicity.

```
Refor_4..          Q_prod =E= sum(J,fc(J,'6')*(dH_f(J)+(1/
  MW(J))*(c_p_v(J,'1')*(T('6')-25)+
(1/2)*c_p_v(J,'2')*((T('6')+273)**2-(25+273)**2)+
(1/3)*c_p_v(J,'3')*((T('6')+273)**3-(25+273)**3)+
(1/4)*c_p_v(J,'4')*((T('6')+273)**4-(25+273)**4))));
T.LO('6')=600;
T.UP('6')=800;
Refor_5..          Q_reac =E=    sum(J,fc(J,'4')*(dH_f(J)+(1/
  MW(J))*(c_p_v(J,'1')*(T('4')-25)+
(1/2)*c_p_v(J,'2')*((T('4')+273)**2-(25+273)**2)+
(1/3)*c_p_v(J,'3')*((T('4')+273)**3-(25+273)**3)+
(1/4)*c_p_v(J,'4')*((T('4')+273)**4-(25+273)**4))))+

sum(J,fc(J,'5')*(dH_f(J)+(1/MW(J))*(c_p_v(J,'1')*(T('5')-25)+
(1/2)*c_p_v(J,'2')*((T('5')+273)**2-(25+273)**2)+
(1/3)*c_p_v(J,'3')*((T('5')+273)**3-(25+273)**3)+
(1/4)*c_p_v(J,'4')*((T('5')+273)**4-(25+273)**4))));

Refor_6..          Q('Furnance')=E=   (Q_prod-Q_reac);
```

Part of the natural gas is computed as that which provides the energy required to maintain the temperature at the reformer as:

```
Refor_7..          fc('CH4','3') =E= Q('Furnace') /PCI_gas;

Refor_8..          fc('CH4','2')  =E= fc('CH4','4') ;
fc.fx(J,'4')$(ord (J) ne 5)=0;
```

14.3.1.3.3 Mass Balance for the Reforming Process

In the furnace, a mass balance must hold. Due to the equilibrium, we consider atomic mass balance for the hydrogen, carbon and oxygen based on the stoichiometry of the two reactions. Next, we define the equilibrium constants for the two reactions, and we also use the correlations from the literature where, as in previous examples, Pi is the partial pressure of component i given in atm while T is in K.

$$\begin{aligned} & mol_{CH_4}\Big|_{in} = mol_{CH_4} + mol_{CO} + mol_{CO_2}\Big|_{out} \\ & 4{\cdot}mol_{CH_4} + 2{\cdot}mol_{H_2O}\Big|_{in} = 4{\cdot}mol_{CH_4} + 2{\cdot}mol_{H_2} + 2{\cdot}mol_{H_2O}\Big|_{out} \\ & mol_{H_2O}\Big|_{in} = mol_{H_2O} + mol_{CO} + 2{\cdot}mol_{CO_2}\Big|_{out} \end{aligned} \tag{14.14}$$

$$CH_4 + H_2O \leftrightarrow CO + 3H_2 \qquad kp = 10^{\left[-\frac{11650}{T}+13{,}076\right]} = \frac{P_{H_2}{}^3 \cdot P_{CO}}{P_{CH_4} \cdot P_{H_2O}} \tag{14.15}$$

$$CO_{(g)} + H_2O_{(g)} \leftrightarrow CO_{2(g)} + H_{2(g)} \qquad kp = 10^{\left[\frac{1910}{T}-1{,}784\right]} = \frac{P_{H_2} \cdot P_{CO_2}}{P_{CO} \cdot P_{H_2O}} \tag{14.16}$$

```
*Carbon Balance
Refor_9..          fc('CH4','4')/MW('CH4') =E=  fc('CH4','6')/
  MW('CH4')   +fc('CO','6')/MW('CO') + fc('CO2','6')/MW('CO2');

*Hydrogen balance
Refor_10..          4*fc('CH4','4')/MW('CH4')  + 2*fc('Wa','5')/
  MW('Wa') =E=   4*fc('CH4','6')/MW('CH4')   +2*fc('H2','6')/
  MW('H2') + 2*fc('Wa','6')/MW('Wa');

*Oxygen balance
Refor_11..          fc('Wa','5')/MW('Wa') =E=  fc('CO','6')/
  MW('CO')   +2*fc('CO2','6')/MW('CO2') + fc('Wa','6')/MW('Wa');

Refor_12..       K_met =E=  10**(-11650/(T('6')+273) +13.076)    ;

Refor_13..       K_WGS   =E= 10**(1910/(T('6')+273)-1.784);

Refor_14..         (fc('CH4','6')/MW('CH4'))      *(fc('Wa','6')/
  MW('Wa'))   *(  fc('CO2','6')/MW('CO2')+fc('CH4','6')/
  MW('CH4')+fc('Wa','6')/MW('Wa')+(fc('CO','6')/
  MW('CO'))+fc('H2','6')/MW('H2')  )**2* K_met
  =E=((fc('CO','6')/MW('CO'))*(fc('H2','6')/
  MW('H2'))**3)*(P_refor+eps1)**2;
Refor_15..      (fc('CO2','6')/MW('CO2'))*(fc('H2','6')/
  MW('H2')) =E= K_WGS*(fc('CO','6')/MW('CO'))*(fc('Wa','6')/
  MW('Wa'));
```

Finally, we write the objective function and the solve statement. We want to optimize the hydrogen production taking into account the price for electricity and cost of the steam added. Thus we have:

```
Equations
         Obj;
Variables
         Z;
Scalar
P_elect /1.67e-5/
P_steam /0.019/;

Obj..     Z =E=fc('H2','6')*1.6-W('Compres')* P_elect
  -fc('Wa','5')* P_steam;

Model Natgas /ALL/;
option NLP = snopt;
Solve Natgas Using NLP Maximizing Z;
```

It turns out that the optimal operating temperature of the reactor is 750°C and the pressure is 2 atm. We use almost 30% of the natural gas to provide energy for the process.

QUESTIONS

- Evaluate the effect of the cost of steam and electricity on the solution. Make use of the loop command.
- Model an autoreforming process for glycerol.

14.3.2 Technology Selection: Process Design

We consider a simple flowsheet for the production of a chemical C following the reaction.

$$A+B \rightarrow C \tag{14.17}$$

We consider three synthetic paths I, II and III with increasing conversions and operating and fixed costs. Once C is produced, we need to separate the product from the unreacted species. For the separation, we considered a flash separation, cheap but with a low recovery ratio, and a distillation column, that allows 99% recovery of C but more expensive. The flowsheet is given as Figure 14.4, where the streams are given numbers and the reactors are denoted by names.

The data for the reactor operation and costs are given in Table 14.3, and the information for the separation equipment is given in Table 14.4.

TABLE 14.3
Reactors Data

Reactor	Conversion	Cost
I	0.5	5000*(flow(C,7))^0.6
II	0.65	5000*(flow(C,8))^0.7
III	0.9	10000*(flow(C,9))^0.7

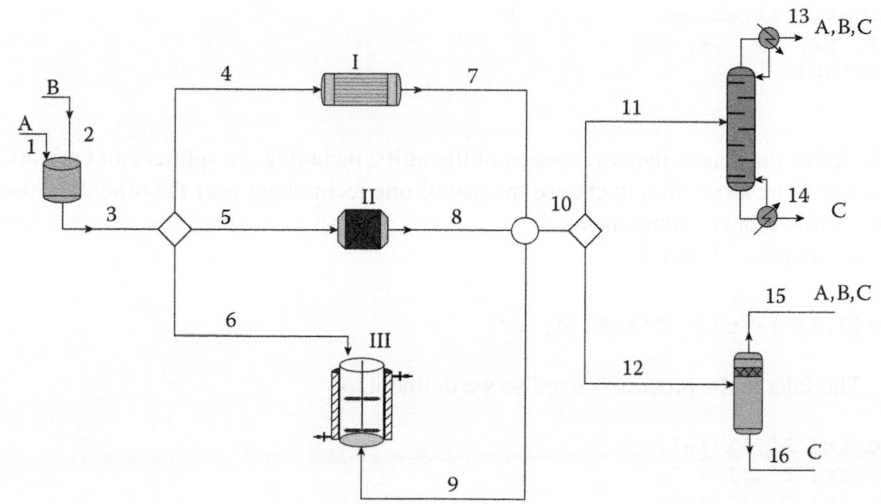

FIGURE 14.4 Flowsheet for the production of C.

We define the sets of components and streams.

```
set
J /A,B,C/
S         /1*16/
```

Next, we define the scalars to include the conversions and separation efficiencies:

```
scalar
      conver1 /0.5/
      conver2 /0.65/
      conver3  /0.9/
      sep_col /0.99/
      sep_flash  /0.8/
      priceC /75000/;
```

Followed by the variables including the flowrates and the costs, the decision variables and the objective variable:

```
Positive variables
flow(J,S), costI, costII, costIII, costcol, costflash  ;
```

TABLE 14.4
Separation Equipment Data

Separation	Separation	Cost
Column	0.99	10000 + flow(C,14)*2000
Flash	0.6	2000 + flow(C,16)*1000;

```
Binary variables
y1, y2, y3, y4, y5;
variable
z
```

We define the mass balances for each of the units, including the splitters and mixers. We use binary variables to choose the use of one technology over the other. We use molar flows for the components.

Thus for $J = \{A, B, C\}$

```
Fc(J,'1')+Fc(J,'2')=FC(J,'3')
```

The inlet to the process is fixed so we define it as:

```
Fc.fx('A','1')=1;
Fc.fx('B','1')=0;
Fc.fx('C','1')=0;
```

And:

```
Fc.fx('A','2')=0;
Fc.fx('B','2')=1;
Fc.fx('C','2')=0;
```

For the selection of a reactor, we define binary variables y1, y2 and y3 so that only one is used:

$$y1 + y2 + y3 = 1 \quad (14.18)$$

The flow of the components is 0 if a reactor is not selected and equal to Fc(J,3) otherwise. This is modeled used a big M constraint [16] where the M is the total maximum flow, 1 in this example:

$$S = \{4,5,6\} \, y = \{y1, y2, y3\}$$

```
flow(J,S)=L=M*(y);
```

Based on the stoichiometry we model the output of each of the reactors. For example for reactor I:

```
costI =E= 2000*(flow('C','7')+0.01)**0.6;
```

In the above equation, since we have an exponent, in order to avoid a base value of 0 we include a small parameter or value.

```
flow('C','7')=E=flow('A','4')*converl ;
flow('A','7')=E=flow('A','4')*(1-converl) ;
flow('B','7')=E=flow('B','4')*(1-converl) ;
```

The mixer after the three alternative reactors is modeled as:

```
flow(J,'7')+flow(J,'8')+flow(J,'9')=E=flow(J,'10');
```

We designate y4 and y5 as the binary variables involved in the decision regarding the separation process. Alternatively, we can define only one and the decision will be made between y and (1–y). S = {11,12} y = {y4,y5}

```
flow(J,S)=L=1*(y);
flow(J,'12')+flow(J,'11')=E=flow(J,'10');
y4+y5=E=1;
```

Next, the two separation technologies are modeled similarly. For instance, for the distillation column. First the mass balance based on the separation:

```
flow('C','14')=E= sep_col*flow('C','11');
flow.fx('A','14')=0;
flow.fx('B','14')=0;
flow('A','13')=E=flow('A','11');
flow('B','13')=E=flow('B','11');
flow('C','13')=E=(1-sep_col)*flow('C','11');
```

Subsequently we model the cost functions. In order for the fixed term to be 0 in case the alternative is not selected, we use the following syntax [16]. We use the associated binary variable so that if it takes the value of 1, the fix cost exists while for the variable cost, it is enough to use the flow of the product since if it take a value of 0 then the variable cost is also 0.

```
costcol =E=10000*y4+flow('C','14')*2000;
```

Finally, we define the objective function given by a simplified production cost involving the product obtained minus the cost of operation of the separation stages and the reactors.

```
z=E=  flow('C','14')*priceC+flow('C','16')*priceC- costcol
  - costflash-costI-costII-costIII;
```

We write the solve statement. In this case the problem is an MINLP and thus we need to ask that an MINLP solver is used. DICOPT is used by default and a local solver such as CONOPT.

```
Model flowsheet /ALL/;
option NLP = conopt;
Solve  flowsheetd Using MINLP Maximizing Z;
```

The complete code is as follows:

```
set

J /A,B,C/
S         /1*16/
```

```
react /I,II,III/
sep /col,flash/;

Positive variables
flow(J,S)
costI
costII
costIII
costcol
costflash  ;
scalar
conver1 /0.5/
conver2 /0.65/
conver3  /0.9/
sep_col /0.99/
sep_flash  /0.6/

priceC /75000/;

Binary variables
y1, y2, y3, y4, y5;

variable
z;
Equations
mix_1, spl1_1, spl1_2, spl1_3, spl1_4,  spl1_5,
react1_1, react1_2, react1_3, react1_4,
react2_1, react2_2, react2_3, react2_4,
react3_1, react3_2, react3_3, react3_4,
mix2_1, spl2_1, spl2_2, spl2_3,  spl2_4,
col_1, col_2, col_3, col_4, col_5,
flash_1, flash_2, flash_3, flash_4,  flash_5,
obj;

flow.fx('A','1')=1;
flow.fx('B','1')=0;
flow.fx('C','1')=0;

flow.fx('A','2')=0;
flow.fx('B','2')=1;
flow.fx('C','2')=0;

mix_1(J)..  flow(J,'3')=E=flow(J,'1')+flow(J,'2');
spl1_1(J)..  flow(J,'4')=L=1*(y1);
spl1_2(J)..  flow(J,'5')=L=1*(y2);
spl1_3(J)..  flow(J,'6')=L=1*(y3);
spl1_4(J)..  flow(J,'3') =E=flow(J,'4')+  flow(J,'5')+
  flow(J,'6');
spl1_5..  y1+y2+y3=E=1;
react1_1..   costI =E= 5000*(flow('C','7')+0.001)**0.6;
react1_2..    flow('C','7')=E=flow('A','4')*conver1 ;
react1_3..    flow('A','7')=E=flow('A','4')*(1-conver1) ;
```

```
react1_4..   flow('B','7')=E=flow('B','4')*(1-conver1) ;
react2_1..  costII =E= 5000*(flow('C','8')+0.001)**0.7;
react2_2..   flow('C','8')=E=flow('A','5')*conver2 ;
react2_3..   flow('A','8')=E=flow('A','5')*(1-conver2) ;
react2_4..   flow('B','8')=E=flow('B','5')*(1-conver2) ;
react3_1..  costIII =E= 10000*(flow('C','9')+0.001)**0.7;
react3_2..   flow('C','9')=E=flow('A','6')*conver3 ;
react3_3..   flow('A','9')=E=flow('A','6')*(1-conver3) ;
react3_4..   flow('B','9')=E=flow('B','6')*(1-conver3) ;
mix2_1(J).. flow(J,'7')+flow(J,'8')+flow(J,'9')=E=flow
  (J,'10');
spl2_1(J).. flow(J,'11')=L=1*(y4);
spl2_2(J).. flow(J,'12')=L=1*(y5);
spl2_3(J).. flow(J,'12')+flow(J,'11')=E=flow(J,'10');
spl2_4.. y4+y5=E=1;
col_1..    flow('C','14')=E= sep_col*flow('C','11');

flow.fx('A','14')=0;
flow.fx('B','14')=0;
col_2..    flow('A','13')=E=flow('A','11');
col_3..    flow('B','13')=E=flow('B','11');
col_4..   flow('C','13')=E=(1-sep_col)*flow('C','11');
col_5.. costcol =E=10000*y4+flow('C','14')*2000;
flash_1..    flow('C','16')=E= sep_flash*flow('C','12');
flow.fx('A','16')=0;
flow.fx('B','16')=0;
flash_2..    flow('A','15')=E=flow('A','12');
flash_3..    flow('B','15')=E=flow('B','12');
flash_4..    flow('C','15')=E=(1-sep_flash)*flow('C','12');
flash_5.. costflash =E= 2000*y5+flow('C','16')*1000;

obj..  z=E=  flow('C','14')*priceC+flow('C','16')*priceC-
  costcol - costflash-costI-costII-costIII;

Model flowsheet /ALL/;
option NLP = conopt;
Solve  flowsheetd Using MINLP Maximizing Z;
```

The solution obtained is y2 = 1 y4 = 1 and the objective function is 33113.9. Thus we use reactor II and the distillation column.

QUESTION

- We have commented on the fact that we use a local solver. Consider the use of a global solver such as Baron and evaluate the solution.

14.3.3 Air Separation Unit: Tray-by-Tray Solution

Air Separation Unit (ASU) based on cryogenic distillation is a useful technology to separate air into its components, especially suitable for large nitrogen production capacities. In the MATLAB chapter of this book, the reader can find a description and simulation of this Linde Double Column using the McCabe Thiele method. In Figure 4.12, a scheme of the tower is showed. In this case, a tray-by-tray modeling is used to determine the optimal number of trays and/or optimal reflux ratio in this kind of columns. To do this, binary variables are employed combined with the MESH (mass, equilibrium, summation and enthalpy) equations following the methodology proposed by Viswanathan and Grossmann [17]. In this example, our attention is focused on the high-pressure column.

A feed stream of 100 moles/s of air is assumed. A binary system is considered with a 79%n of nitrogen and a 21%n of oxygen. The inlet pressure is fixed to 5 atm.

```
*Feed Data
Scalar
       Ffeed 'total feed (mol)' /100/
       Pfeed 'pressure in high pressure column (mmHg)' /3800/;
Set
       J 'Components' /N2, O2/
       cnt 'Antoine Parameters' /A, B, C/;
Parameters
       zfi(J)'Feed compositions'
       /N2 0.79
       O2 0.21/;
```

First of all, the bubble and dew point have been calculated to select an inlet temperature within this range. The equations to calculate the bubble and dew point are:

$$\text{Bubble point} \rightarrow \sum_{J=1}^{n} K_J z_J = 1$$
$$\text{Dew point} \rightarrow \sum_{J=1}^{n} \frac{z_J}{K_J} = 1 \tag{14.19}$$

Where z_i is the feed composition and K_i is the gas-liquid equilibrium constant computed as follows:

$$K_J = \frac{P_{vap}^J}{P_{total}} \tag{14.20}$$

The value of vapor pressure is calculated with the Antoine equations for each component [11]:

$$\ln(\mathrm{P}_{vap}^{\mathrm{N2}}(\mathrm{mmHg})) = 14.9542 - \frac{588.72}{T(K) - 6.6}$$
$$\ln(\mathrm{P}_{vap}^{\mathrm{O2}}(\mathrm{mmHg})) = 15.4075 - \frac{734.55}{T(K) - 6.45} \tag{14.21}$$

```
*Feed Data
Table
      Ant(J,cnt) 'Antoine correlation'
      *Vapor pressure
      *lnP=A-(B/(C+T)) with P(mmHg) and T(K)
                A          B          C
         N2   14.9542    -588.72    -6.6
         O2   15.4075    -734.55    -6.45
;
*Bubble point sum(Kx)=1
Positive variable
      Tbubble;
Tbubble.LO=50;
equations
      bp_1;
bp_1.. sum(J,((exp(Ant(J,'A')+(Ant(J,'B')/(Ant(J,'C')+
  Tbubble))))/Pfeed)*zfi(J))=E=1;
*Solutions: Tbubble equal to 96.399

*Dew Point (y/K)=1
Positive variable
      Tdew;
Tdew.LO=50;
equations
      bd_1;
bd_1.. sum(J,zfi(J)/((exp(Ant(J,'A')+(Ant(J,'B')/
  (Ant(J,'C')+Tdew))))/Pfeed))=E=1;
*Solutions: Tdew equal to 99.205
```

In this case, instead of using a fixed feed temperature, an inlet liquid fraction is assumed to be equal to 0.8. To determine the liquid and vapor flows and compositions, the following equilibrium equations have been used:

$$
\begin{aligned}
&F \cdot z_f^J = L_f \cdot x_{fJ} + V_f \cdot y_{fJ} \\
&y_{fJ} = K_J \cdot x_{fJ} \\
&\sum_{J=1}^{n} y_{fJ} = 1 \\
&\sum_{J=1}^{n} x_{fJ} = 1 \\
&\frac{L_f}{L_f + V_f} = 0.8
\end{aligned}
\tag{14.22}
$$

```
*Liquid-Vapour fraction
Positive variable
      Lf
      Vf
      xf(J)
      yf(J);
```

```
Scalar
      fliquid /0.8/;

Positive variable
      Tfeed;
Tfeed.LO=50;
equations
eqf_1,eqf_2,eqf_3,eqf_4,eqf_5;
eqf_1(J).. Ffeed*zfi(J)=E=Lf*xf(J)+Vf*yf(J);
eqf_2(J).. yf(J)=E=((exp(Ant(J,'A')+(Ant(J,'B')/(Ant(J,'C')+
  Tfeed))))/Pfeed)*xf(J);
eqf_3.. sum(J,xf(J))=E=1;
eqf_4.. sum(J,yf(J))=E=1;
eqf_5.. Lf=E=fliquid*(Lf+Vf);
```

With the equilibrium data is easy to compute the enthalpy of the stream as shown as follows:

$$h_{feed} = \sum_{J=1}^{n} F \cdot z_f^J \int_{T_{ref}}^{T} C_p^J dT - \sum_{J=1}^{n} L_f \cdot x_{fJ} \cdot \left|Q_{latente}^J\right| \tag{14.23}$$

Therefore, the feed is characterized.

```
*Enthalpy in the feed stream
Scalar Tref 'Reference temperature for enthalpy calculations'
  /298/;
Set
*for heat capacity
mar /1*4/;
Table
*Heat capacity of gases
*equation Cp° = A + B*t + C*t2 + D*t3 where Cp0 (J/mol K) and
  t(T(K))
   cp_v(J,mar)
             1           2              3             4
     N2    31.150    -1.3570E-02    2.6796E-05    -1.168E-08
     O2    28.106    -3.6800E-06    1.7459E-05    -1.065E-08;
Parameter
Qlatent(J)  'latent heat (J/mol)'
 /N2   6824
  O2   5581/;
Variable
hfeed;
equations
entf_1;
entf_1.. hfeed=E=sum(J,Ffeed*zfi(J)*(cp_v(J,'1')*(Tfeed-
  Tref))+((1/2)*cp_v(J,'2')*(Tfeed**2-Tref**2))+((1/3)
  *cp_v(J,'3')*(Tfeed**3-Tref**3))+((1/4)*cp_v(J,'4')*
  (Tfeed**4-Tref**4)))-sum(J,Lf*xf(J)*Qlatent(J));
```

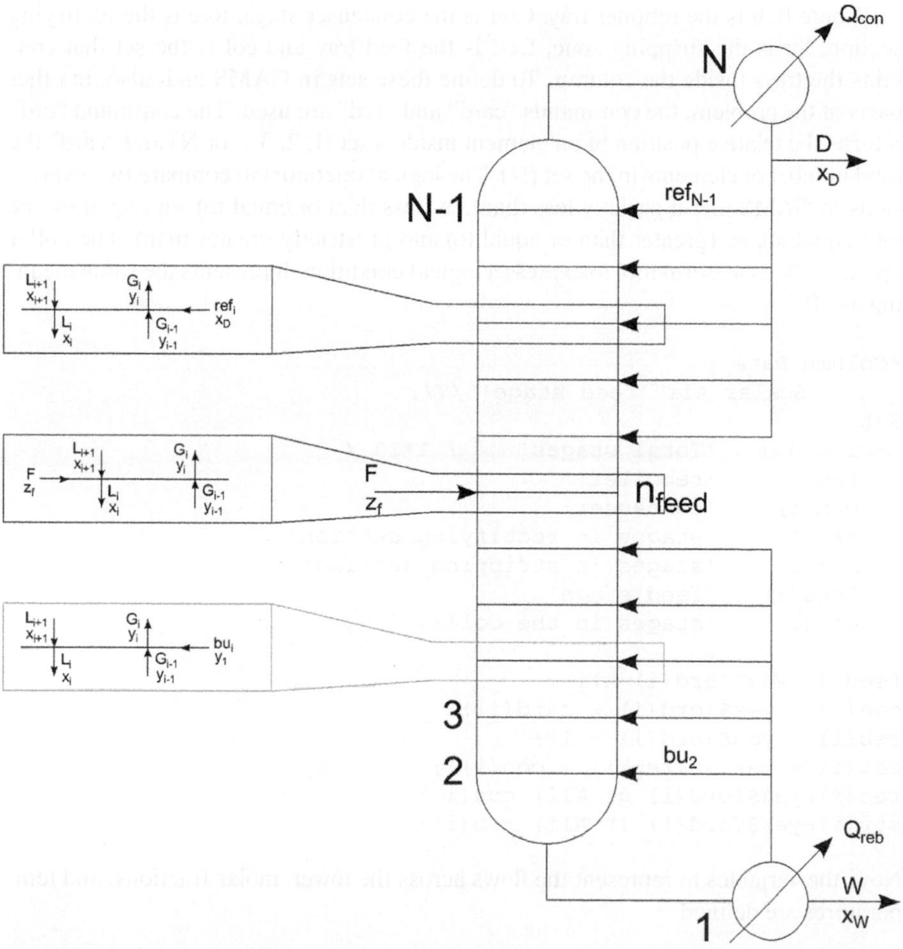

FIGURE 14.5 Scheme of the tower.

The next step is to describe the tower (Figure 14.5). In this example, the optimal number of trays will be selected, however, it is necessary to establish a set of trays to allow the problem to select which trays are in use or not. In this problem, the reboiler is the first tray and the condenser the last one. Thirteen stages (N) are fixed as maximum in the tower, and the feed stage $\left(n_{feed}\right)$ is located in the tray number 7. The following set are defined to help to solve the tower balances:

$$\begin{aligned}
&\text{Total stages} = \left\{1,2,\ldots N\right\}\\
&\text{Reb} = \left\{1\right\}\\
&\text{Con} = \left\{N\right\}\\
&\text{Rec} = \left\{n_{feed}+1,\ldots,N-1\right\}\\
&\text{Str} = \left\{2,\ldots,n_{feed}-1\right\}\\
&\text{feed} = \left\{n_{feed}\right\}\\
&\text{col} = \left\{2,\ldots,N-1\right\}
\end{aligned} \tag{14.24}$$

Where Reb is the reboiler tray, Con is the condenser stage, Rec is the rectifying section, Str is the stripping zone, feed is the feed tray and col is the set that contains the trays inside the column. To define these sets in GAMS and, also, in other parts of the problem, the commands "card" and "ord" are used. The command "ord" returns the relative position of an element inside a set (1, 2, 3… or N) and "card" the total number of elements in the set (N). The logical operators to compare two expressions in GAMS are: lt (strictly less than), le (less than or equal to), eq (equal to), ne (not equal to), ge (greater than or equal to) and gt (strictly greater than). The dollar operator "$" is a useful tool to express a logical condition. It presents the same meaning as "If."

```
*Column Data
      Scalar Ali 'feed stage' /7/;
Set
   i          'Total stages'    / 1*13 /
   reb(i)     'reboiler'
   con(i)     'condenser'
   rec(i)     'stages in rectifying section'
   str(i)     'stages in stripping section'
   feed(i)    'feed stage'
   col(i)     'stages in the col';

feed(i)=yes$(ord(i)=Ali);
con(i) = yes$(ord(i) = card(i));
reb(i) = yes$(ord(i) = 1);
col(i) = yes - (reb(i) + con(i));
rec(i)=yes$(ord(i) gt Ali)-con(i);
str(i)=yes$(ord(i) lt Ali)-reb(i);
```

Now, the variables to represent the flows across the tower, molar fractions, and temperatures are defined.

```
Positive variables
      x(J,i) 'liquid mole fraction of J component in i tray'
      y(J,i) 'vapour mole fraction of J component in i tray'
      L (i)  'Liquid flow rate in i tray (mol)'
      G(i)   'Vapour flow rate in i tray(mol)'
      T(i)   'Temperature in i tray'
      D      'Distillate flow (mol)'
      xD(J)  'Distillate composition'
      TD     'Distillate temperature'
      W      'Bottom flow(mol)'
      xW(J)  'Bottom composition'
      TW     'Bottom temperature'
      K(J,i) 'equilibrium relationship'
      ref(i) 'reflux flows'
      bu(i)  'reboiler flows';
x.LO(J,i)=0;
x.UP(J,i)=1.01;
y.LO(J,i)=0;
```

```
y.UP(J,i)=1.01;
xD.LO(J)=0;
xD.UP(J)=1.01;
xW.LO(J)=0;
xW.UP(J)=1.01;
T.LO(i)=50;
TD.LO=50;
TW.LO=50;
Variable
       hl(i)   'liquid entalphy in i tray'
       hg(i)   'vapour entalphy in i tray'
       hD 'distillate enthalpy'
       hW 'bottom enthalpy'
       Qcon 'condenser heat'
       Qreb 'reboiler heat';
```

The following equations describe the tower behavior. In this case study, a total condenser and a partial reboiler are considered. Pressure drop is negligible.

Mass balances inside the tower:

$$
\begin{aligned}
&G_{i-1}y_{i-1}^{J} = Dx_D^J + \left(\sum_{i=1}^{N} ref_i\right)x_D^J \quad i \in Con\ \forall J\\
&G_{i-1}y_{i-1}^{J} + L_{i+1}x_{i+1}^{J} + ref_i x_D^J = G_i y_i^J + L_i x_i^J \quad i \in \mathrm{Re}c\ \forall J\\
&Fz_f^J + G_{i-1}y_{i-1}^{J} + L_{i+1}x_{i+1}^{J} = G_i y_i^J + L_i x_i^J \quad i \in feed\ \forall J\\
&G_{i-1}y_{i-1}^{J} + L_{i+1}x_{i+1}^{J} + bu_i y_1^J = G_i y_i^J + L_i x_{i+1}^J \quad i \in Str\ \forall J\\
&L_{i+1}x_{i+1}^{J} = \left(\sum_{i=1}^{N} bu_i\right)y_i^J + Wx_W^J \quad i \in \mathrm{Re}b\ \forall J
\end{aligned}
\tag{14.25}
$$

G_i denotes the molar gas flow for each tray, L_i the molar liquid flow, y_i^J the gas molar fraction and x_i^J the liquid molar fraction. The amount of reflux is denoted as ref_i and the amount of reboiler vapor as bu_i. Finally, D and x_D^J represent the total molar flow and the molar fraction in the distillate, and W and x_W^J the total molar flow and the molar fraction in the bottom stream.

```
*condenser mass balance per component
     eq1_1(J,i)$con(i).. G(i-1)*y(J,i-1)=E=D*xD(J)+sumref*xD(J);
*balances in the rectifier section
     eq1_3(J,i)$rec(i).. G(i-1)*y(J,i-1)+L(i+1)*x(J,i+1)+ref(i)
       *xD(J)=E=G(i)*y(J,i)+L(i)*x(J,i);
*Feed tray balances
     eq1_8(J,i)$feed(i).. Ffeed*zfi(J)+G(i-1)*y(J,i-1)+L(i+1)
       *x(J,i+1)=E=G(i)*y(J,i)+L(i)*x(J,i);
*balances in the stripping section
     eq1_9(J,i)$str(i).. G(i-1)*y(J,i-1)+L(i+1)*x(J,i+1)+bu(i)*y
       (J,'1')=E=G(i)*y(J,i)+L(i)*x(J,i);
*reboiler mass balances per condenser
     eq1_10(J,i)$reb(i).. L(i+1)*x(J,i+1)=E=sumbu*y(J,i)+W*xW(J);
```

Summation of composition:

$$\sum_{J=1}^{n} y_i^J = 1 \ \forall i$$
$$\sum_{J=1}^{n} x_i^J = 1 \ \forall i$$
$$\sum_{J=1}^{n} x_D^J = 1 \tag{14.26}$$
$$\sum_{J=1}^{n} x_W^J = 1$$

```
*Summation of molar fraction
      eq1_4(i)$((ord(i) ne card(i)) and (ord(i) ne 1))..
        sum(J,x(J,i))=E=1;
      eq1_5(i)$((ord(i) ne card(i))).. sum(J,y(J,i))=E=1;
      eq1_6.. sum(J,xD(J))=E=1;
      eq1_7.. sum(J,xW(J))=E=1;
```

Equilibrium equations:

$$K_i^J = \frac{P_{vap_i^J}}{P_{total}} \ \forall i \ \forall J \tag{14.27}$$
$$y_i^J = K_i^J x_i^J \ \forall i \ \forall J$$

Where the vapor pressure is calculated using the Antoine equations.

```
*equilibrium relationship for all trays in the column
      eq1_17(J,i)$col(i).. K(J,i)=E=((exp(Ant(J,'A')+
        (Ant(J,'B')/(Ant(J,'C')+T(i)))))/Pfeed);
      eq1_18(J,i)$col(i).. y(J,i)=E=K(J,i)*x(J,i);
*equilibrium relationship in the partial reboiler
      eq1_19(J,i)$reb(i).. K(J,i)=E=((exp(Ant(J,'A')+
        (Ant(J,'B')/(Ant(J,'C')+T(i)))))/Pfeed);
      eq1_20(J,i)$reb(i).. y(J,i)=E=K(J,i)*xW(J);
```

Enthalpy balances:

$$G_{i-1}hg_{i-1} = Dh_D + \left(\sum_{i=1}^{N} ref_i\right)h_D + Q_{con} \quad i \in Con$$
$$G_{i-1}hg_{i-1} + L_{i+1}hl_{i+1} + ref_i h_D = G_i hg_i + L_i hl_i \quad i \in \mathrm{Re}c$$
$$Fh_f + G_{i-1}hg_{i-1} + L_{i+1}hl_{i+1} = G_i hg_i + L_i hl_i \quad i \in feed \tag{14.28}$$
$$G_{i-1}hg_{i-1} + L_{i+1}hl_{i+1} + bu_i hg_1 = G_i hg_i + L_i hl_{i+1} \quad i \in Str$$
$$L_{i+1}hl_{i+1} + Q_{reb} = \left(\sum_{i=1}^{N} bu_i\right)hg_i + Wh_W \quad i \in \mathrm{Re}b$$

Where hg_i is the specific gas enthalpy, hl_i is the specific liquid enthalpy, h_D is the distillate specific enthalpy and h_W is the bottom specific enthalpy.

```
*entalphy column
eq1_21(i)$col(i).. hg(i)=E=sum(J,y(J,i)*(cp_v(J,'1')*(T(i)-Tre
  f))+((1/2)*cp_v(J,'2')*(T(i)**2-Tref**2))+((1/3)*cp_v(J,'3')
  *(T(i)**3-Tref**3))+((1/4)*cp_v(J,'4')*(T(i)**4-Tref**4)));
eq1_22(i)$col(i).. hl(i)=E=sum(J,x(J,i)*(cp_v(J,'1')*(T(i)-Tre
  f))+((1/2)*cp_v(J,'2')*(T(i)**2-Tref**2))+((1/3)*cp_v(J,'3')
  *(T(i)**3-Tref**3))+((1/4)*cp_v(J,'4')*(T(i)**4-Tref**4)))
  -sum(J,x(J,i)*Qlatent(J));
*entalphy of condenser
eq1_24.. hD=E=sum(J,xD(J)*(cp_v(J,'1')*(TD-Tref))+((1/2)*cp_v
  (J,'2')*(TD**2-Tref**2))+((1/3)*cp_v(J,'3')*(TD**3-Tref**3))+
  ((1/4)*cp_v(J,'4')*(TD**4-Tref**4)))-sum(J,xD(J)*Qlatent(J));
eq1_25(i)$con(i).. T(i)=E=TD;
*enthalpy of reboiler
eq1_26(i)$reb(i).. hg(i)=E=sum(J,y(J,i)*(cp_v(J,'1')*(T(i)-Tre
  f))+((1/2)*cp_v(J,'2')*(T(i)**2-Tref**2))+((1/3)*cp_v(J,'3')
  *(T(i)**3-Tref**3))+((1/4)*cp_v(J,'4')*(T(i)**4-Tref**4)));
eq1_27.. hW=E=sum(J,xW(J)*(cp_v(J,'1')*(TW-Tref))+((1/2)
  *cp_v(J,'2')*(TW**2-Tref**2))+((1/3)*cp_v(J,'3')*(TW**3-Tref
  **3))+((1/4)*cp_v(J,'4')*(TW**4-Tref**4)))-sum(J,xW(J)
  *Qlatent(J));
eq1_28(i)$reb(i).. T(i)=E=TW;
*enthalpy balance condenser
eq1_29(i)$con(i).. G(i-1)*hg(i-1)=E=D*hD+sumref*hD+Qcon;
*enthalpy balance in rectifier zone
eq1_30(i)$rec(i).. G(i-1)*hg(i-1)+L(i+1)*hl(i+1)+ref(i)*hD=E=G
  (i)*hg(i)+L(i)*hl(i);
*enthalpy balance in stripping section
eq1_32(i)$str(i).. G(i-1)*hg(i-1)+L(i+1)*hl(i+1)+bu(i)*hg('1')
  =E=G(i)*hg(i)+L(i)*hl(i);
*enthalpy balance in reboiler
eq1_33(i)$reb(i).. L(i+1)*hl(i+1)+Qreb=E=sumbu*hg(i)+W*hW;
```

Bubble temperature in the distillate stream:

$$\sum_{J=1}^{n} K_i^J x_D^J = 1 \tag{14.29}$$

```
*Distillate Temperature (bubble temperature)
eq1_34.. sum(J,((exp(Ant(J,'A')+(Ant(J,'B')/(Ant(J,'C')+TD))))/
  Pfeed)*xD(J))=E=1;
```

Now, a set of binary variables (n_i) is defined to indicate whenever a tray (i) is selected to achieve the desired objective (final compositions, in general). The variable is only effective with the trays inside the column because the condenser and the reboiler always exist. The idea is to select a number of trays to achieve the desired objective, so that, above the last tray inside the column (in which enter the reflux) and below the first tray (in which reboiled vapor is fed) no trays will be selected. Since no reflux above the last tray in the column is allowed, no mass and heat transfer take place in these stages and the compositions, flows, and temperatures are kept constant. The same reasoning applies in the stripping section below the first tray.

```
Binary variable
z(i);
```

A series of constrains are added to ensure that these conditions is satisfied into the tower.

To satisfy that if a tray is selected in the rectifier section, below it, all trays in the section must be selected. The same reasoning is followed in the stripping section.

$$
\begin{aligned}
z_i - z_{i-1} &\le 0 \quad i \in \mathrm{Rec} \\
z_i - z_{i+1} &\le 0 \quad i \in Str
\end{aligned}
\tag{14.30}
$$

```
nw_1(i)$rec(i).. z(i)-z(i-1)=L=0;
nw_2(i)$str(i).. z(i)-z(i+1)=L=0;
```

Monotonicity of temperatures. The maximum temperature must be in the reboiler and decreases across the column.

$$
\begin{aligned}
T_i - T_{i-1} &\le 0 \quad i \in \mathrm{Rec}\ \mathrm{i} \in \mathrm{Con} \\
T_i - T_{i+1} &\ge 0 \quad i \in Str\ i \in \mathrm{Re}b
\end{aligned}
\tag{14.31}
$$

```
nw_3(i)$con(i).. T(i)-T(i-1)=L=0;
nw_4(i)$rec(i).. T(i)-T(i-1)=L=0;
nw_5(i)$str(i).. T(i)-T(i+1)=G=0;
nw_6(i)$reb(i).. T(i)-T(i+1)=G=0;
```

Constrains to ensure that the reflux and the boiled vapor enter only in the last selected stage in the rectifier section and in the first one in the stripping section.

$$
\begin{aligned}
ref_i &\le f_{\max}(z_i - z_{i+1}) \quad i \in Rec \\
bu_i &\le f_{\max}(z_i - z_{i-1}) \quad i \in Str
\end{aligned}
\tag{14.32}
$$

Where $f_{\max}$ is an upper bound for the flows (a value big enough). In this case of study, this value has been fixed to 500.

```
nw_7(i)$rec(i).. ref(i)=L=fmax*(z(i)-z(i+1));
nw_8(i)$str(i).. bu(i)=L=fmax*(z(i)-z(i-1));
```

In the current example, a distillate composition is assumed to be 99%n in nitrogen and bottom composition equal to 60%n in nitrogen.

```
*Final desired compositions
xD.fx('N2')=0.99;
xW.fx('N2')=0.60;
```

The objective function is a minimization and it is defined as the sum between the reflux ratio (R_{ratio}) and the total number of trays.

$$obj = R_{ratio} + \sum_{i=1}^{N} n_i \tag{14.33}$$

Where the reflux ratio is computed as following:

$$R_{ratio} = \frac{\sum_{i=1}^{N} ref_i}{D} \tag{14.34}$$

The problem is an MINLP. We use the MINLP solver DICOPT. Due to the complexity of the problem, a multistart optimization has been employed with CONOPT as preferential NLP solver and CPLEX as MIP solver.

```
Positive variable
Rratio;
*Reflux Ratio
equations
mn_1;
mn_1.. Rratio*D=E=sum(i,ref(i));
variable tt;
equations obj;
obj.. tt=E=Rratio+sum(i,z(i));
```

The solution shows that seven trays are necessary—the condenser, the reboiler and five stages inside the column (the feed stage, one in the stripping section, and three in the rectifier zone). The reflux ratio is 1.247.

If the objective function is modified, different results can be found. For example, if the objective function is to find the minimum reflux ratio, all trays are selected, as expected, and a reflux ratio of 0.727 is taken. Instead, if the objective is only to minimize the number of trays, the optimization results show that related to only five trays are necessary, like in the first cases.

There is a direct relationship between the number of trays and the investment cost while the reflux ratio determines the operating cost. So, a trade-off exists between these two variables. Traditionally, the rules of thumb recommend an economically optimal reflux ratio about 1.2 times the minimum reflux ratio [18]. Representing the number of trays versus reflux ratio, it is possible to check if this rule is appropriate in the system of study. Figure 14.6 shows the figure for the nitrogen-oxygen system. The minimum reflux corresponds with the maximum number of stages. At the opposite end, the minimum number of stages determines the maximum reflux ratio. In the graph, the optimal reflux ratio is in the section where the curve begins to be smooth and trends to achieve a constant value in the number of trays. In this case, a reflux ratio about 0.85 is appropriate, in agreement with the rule of thumb.

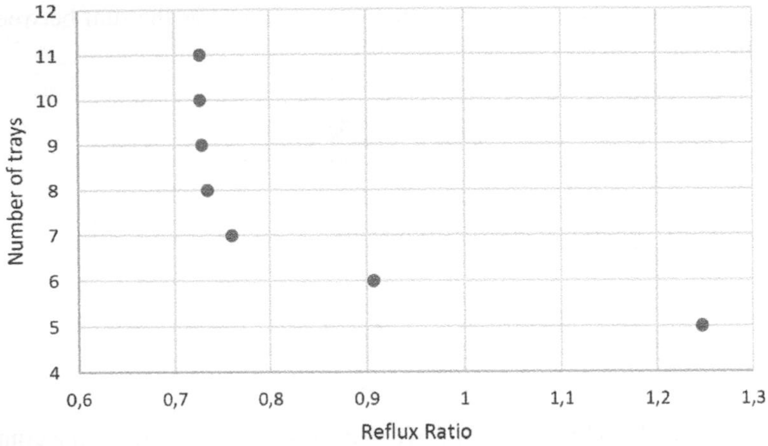

FIGURE 14.6 Number of trays vs. reflux ratio for high-pressure air distillation column.

QUESTIONS

- Evaluate the influence of top and bottom compositions on the total number of trays.
- The McCabe-Thiele method considers liquid and gas flux constant in each section inside the tower. According with this tray-by-tray simulation, is it a good approximation?

14.4 FURTHER READING: SIMULTANEOUS OPTIMIZATION AND TOPOLOGY DESIGN

As this book serves as an introduction, we do not go into more complex examples whose formulation and solution procedure are beyond the scope of this work. For those interested in the topic, we recommend specialized literature on systematic process design and optimization [16, 19]. Furthermore, the literature is rich in papers on the design of different facilities of industrial interest such as hydrocarbon separation [20], HDA production process optimization [4], distillation columns sequence design, heat exchanger networks, power plants and IGCC modeling and optimization [16], and recently, the design of biofuel production processes [20]. The main advantage of this approach is the possibility of evaluating a large number of technologies to conceptually synthesize the best process determining the operating conditions, for additional examples we refer the reader to [21–31], where process and product design is a new trend. This topic is also explored in Chapter 20.

REFERENCES

1. Rudd, D.; Powers, G.; Siirola, J. (1973) *Process Synthesis*. Prentice Hall, Englewood Cliffs, NJ.
2. Douglas, J. (1988) *Conceptual Design of Chemical Processes*. McGraw-Hill, New York.

3. Linnhoff, B. (1993) Pinch analysis—A state-of-the-art overview. *Chem. Eng. Res. Des.*, 71(a5), 503–522.
4. Kocis, G.R.; Grossmann, I.E. (1989) A modelling and decomposition strategy for the minlp optimization of process flowsheets. *Comput. Chem. Eng.*, 13(7), 797–819.
5. Yee, T.; Grossmann, I. E. (1990) Simultaneous-optimization models for heat integration. 2. Heat-exchanger network synthesis. *Comput. Chem. Eng.*, 13(10), 1165–1184.
6. Grossmann, I.E.; Caballero, J.; Yeomans, H. (1999) Mathematical programming approaches to the synthesis of chemical process systems. *Korean J. Chem. Eng.*, 16(4), 407–426.
7. Martín, M.; Grossmann, I.E. (2012) BIOptPACK: A library of models for optimization of biofuel production processes. *Computer Aid. Chem. Eng.*, 30, 16–20.
8. Ullmann, F., ed. (1998b) *Ullmann's Encyclopedia of Industrial Chemistry.* Wiley-VCH, Weinheim, Federal Republic of Germany. Sulfur Dioxide: Hermann Müller, Lurgi Metallurgie GmbH, Frankfurt/Main, Federal Republic of Germany.
9. Ullmann, F., ed. (1998a) *Ullmann's Encyclopedia of Industrial Chemistry.* Wiley-VCH, Weinheim, Federal Republic of Germany. Ammonia: Max Appl, Dannstadt-Schauernheim, Federal Republic of Germany.
10. Martín, M. (2016) *Industrial Chemical Process Analysis and Design.* 1st edn. Elsevier, Netherlands.
11. Tock, L.; Maréchal, F.; Perrenoud, M. (2015) Thermo-economic evaluation of the ammonia production. *Can. J. Chem. Eng.*, 93, 356–362.
12. Sinnot, R.K. (1999) *Coulson and Richardson's Chemical Engineering*, Vol 6: Chemical Engineering Design. Butterworth-Heinemann, Singapore.
13. Sánchez, A.; Martín, M. (2018) Optimal renewable production of ammonia from water and air. *J. Clean. Prod.*, 178, 325–342.
14. Ullmann, F., ed. (1998a) *Ullmann's Encyclopedia of Industrial Chemistry.* Wiley-VCH, Weinheim, Federal Republic of Germany. Methanol: Eckhard Fiedler,; Georg Grossmann, D. Burkhard Kersebohm, Günther Weiss, Claus Witte, BASF Aktiengesellschaft, Ludwigshafen, Federal Republic of Germany.
15. Martín, M; Grossmann I.E. (2013) Optimal use of hybrid feedstock, switchgrass and shale gas for the simultaneous production of hydrogen and liquid fuels. *Energy, 55 (15)*, 378–91.
16. Biegler, L.; Grossmann, I.E., Westerberg, A (1997) *Systematic Methods of Process Design.* Prentice Hall, Upper Saddle River, NJ.
17. Viswanathan, J.; Grossmann, I.E. (1990) A combined penalty function and outer–approximation method for MINLP optimization. *Comput. Chem. Eng.*, 14(7), 769–782.
18. Walas, S.M. (1990) *Chemical Process Equipment: Selection and Design.* Butterworth–Heinemann Series in Chemical Engineering. Butterworth–Heinemann Limited, Newton, MA.
19. Morari, M., Grossmann, I.E. (1986) Chemical engineering optimization models with GAMS. In: *CACHE Process Design Case Study*, Vol. 6: CACHE.
20. Lee, S.; Logsdon, J.S.; Foral, M.J.; Grossmann. I.E. (2003) Superstructure optimization of the olefin separation process. In: *Proceedings of ESCAPE-13.* Lappeeranta, Finland.
21. Martín, M.; Grossmann, I.E., (2013) On the systematic synthesis of sustainable biorefineries *Ind. Eng. Chem. Res.*, 9(52), 3044–3064.
22. Martín, M.; Grossmann, I.E. (2011) Energy optimization of lignocellulosic bioethanol production via gasification. *AIChE J.*, 57(12), 3408–3428.
23. Duran, M.A.; Grossmann, I.E. (1986) Simultaneous optimization and heat integration of chemical processes. *AIChE J.*, 32, 123–138.
24. Martín, M.; Grossmann, I.E. (2012) Simultaneous optimization and heat integration for biodiesel production from cooking oil and algae. *Ind. Eng. Chem Res.*, 51(23), 7998–8013.

25. Severson, K.; Martín, M.; Grossmann, I.E. (2013) Process optimization bioDiesel production using bioethanol. *AICHE J.* DOI: 10.1002/aic.13865.
26. Kravanja, Z.; Grossmann, I.E. (1990) PROSYN—an MINLP process synthesizer, *Comput. Chem. Eng.*, 13(12), 1363–1378.
27. Cucek, L.; Martín, M.; Kravanja, Z.; Grossmann, I.E. (2011) Integration of process technologies for the simultaneous production of fuel ethanol and food from corn grain and stover. *Comput. Chem. Eng.*, 35(8), 1547–1557.
28. Ahmetovic, E.; Martín, M.; Grossmann, I.E. (2010) Optimization of water consumption in process industry: Corn-based ethanol case study. *Ind. Eng. Chem Res.*, 49(17), 7972–7982.
29. Yang L.; Grossmann, I.E. (2013) Water targeting models for simultaneous flowsheet optimization. *Ind. Eng. Chem. Res.*, 9(52), 3209–3224.
30. Martín, M., Grossmann, I.E., (2013) Optimal engineered algae composition for the integrated simultaneous production of bioethanol and biodiesel. *AIChE J.*, 59(8), 2872–2883.
31. Guerras, L.S., Martín, M (2019) Optimal gas treatment and coal blending for reduced emissions in power plants: a case study in Northwest Spain. *Energy,* 169, 739–749.

15 AIMMS for Scheduling of Chemical Plants

Edwin Zondervan, Martijn A.H. van Elzakker

CONTENTS

15.1 INTRODUCTION

In its most general form, scheduling is a *decision-making* activity within the manufacturing and service industries. A scheduler tries to allocate *resources* efficiently to *processing tasks* over a given time. Each task requires a certain amount of resources at a specified time instant, often denoted as *processing time*. The resources may be processing equipment in a chemical plant, but they could be just as well runways at airports or crews at construction sites. Tasks may be the operations performed in the chemical facility, but they could also be takeoffs and landings at an airport or the activities in a construction project.

Generally, the efficient allocation of resources to tasks involves a certain objective, for example, to minimize the production *makespan* (time required to finish all tasks), to maximize the plant throughput, to maximize profit or to minimize costs. The allocation of resources to tasks is usually limited by the availability of the resources. For example, a reactor can only produce one product at the time. Overall: scheduling decisions may include the allocation of resources to tasks, the sequencing of tasks allocated to the same resource and the overall timing [1].

Scheduling is similar to *Planning*. However, scheduling normally involves short-term decisions (from days to weeks) while planning activities entail a longer time horizon (months to even years).

Once decisions have been made at the planning level, the optimized planning is sent to the scheduling level, which normally fills in the shorter time. In turn scheduling will optimize the short-term decisions, which ultimately will be passed on to the so-called *control level*, operating at the minutes to hours time scale.

Nevertheless, the focus in this chapter is on scheduling of (chemical) processing plants. While many chemicals are produced in large-scale *continuous processes*, many products are still produced *batch wise*. This is especially common in facilities that produce low volume products. These plants are often made flexible to changing consumption markets and can produce different products in the same equipment (*multiproduct*). These multiproduct production facilities can be found especially in the production of fine chemicals, pharmaceuticals, foods and to a certain extent polymers [2].

For these types of facilities production needs to be scheduled, i.e. the order in which the products will be produced and the time allocation to each of them must be decided. It is finally noted that the possibilities in the production scheduling strongly depend on the design of the plant and that this can have major economic consequences [3].

In the following sections we will introduce the reader to the basics of batch scheduling and the formulation of so-called state-task networks. In addition, we will provide a flavour of the mathematical formulation of scheduling problems and with a simple example we will familiarize the reader with the implementation of such problems into an appropriate software.

15.2 BASICS OF BATCH SCHEDULING

15.2.1 Single Product versus Multiproduct

Pharmaceuticals and specialty chemicals are often produced in dedicated equipment that is operated batch wise. In Figure 15.1 an example of a batch process for the production of a *single product* C is given. There are four processing units that are operated in batch mode; a reactor, a static mixer, a belt filter, and a dryer.

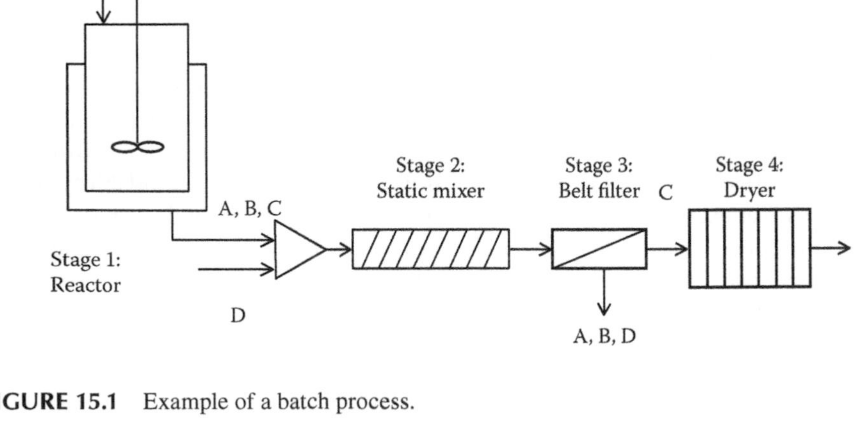

FIGURE 15.1 Example of a batch process.

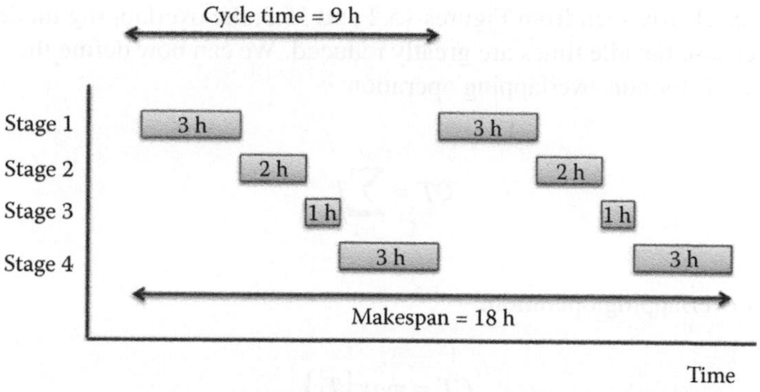

FIGURE 15.2 Gantt chart for the plant of Figure 1 (without transfer times).

In this setup a single product C is produced according to the following *recipe*:

1. Mix the raw materials A and B. Heat the reactor to 75°C and react for 3 hours to form product C.
2. Mix the reaction mixture with solvent D for 2 hours at room temperature.
3. Filtrate to recover product C for one hour.
4. Dry the product in a dryer for 3 hours at 50°C.

The activities at each processing step can be represented by a *Gantt chart*, as shown in Figure 15.2. It is noted that the Gantt chart could also include the times required for filling and emptying the equipment.

Figure 15.2 further shows us that in this case the batch or *lot* is produced in a *non-overlapping* way; a batch is produced after the previous one has completed; two batches cannot be manufactured parallel. It is of course also possible to simultaneously process the batches. The *idle times* are in such case eliminated as much as possible. Figure 15.3 shows the Gantt chart of an *overlapping* mode of operation.

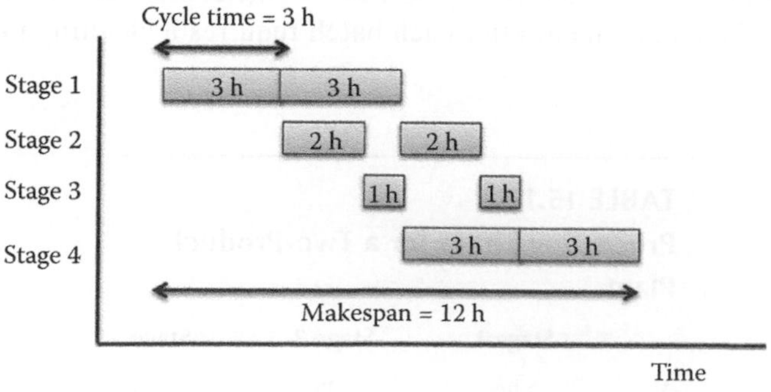

FIGURE 15.3 Gantt chart with overlapping mode of operation.

As can be clearly seen from Figures 15.2 and 15.3, the overlapping mode is more efficient because the idle times are greatly reduced. We can now define the so-called cycle time, CT, for non-overlapping operation:

$$CT = \sum_{j=1}^{M} T_j \tag{15.1}$$

And for overlapping operation:

$$CT = max\{T_j\} \tag{15.2}$$

Where T_j is the processing time in stage j. For the given example; the cycle time for non-overlapping operation is 9 hours, while the cycle time for overlapping operation is only 3 hours. It is further noted that the *makespan* corresponds to the total time needed to produce a given number of batches. For the non-overlapping operation the makespan is 18 hours, while for the overlapping operation the makespan is only 12 hours.

In a *multiproduct plant*, production campaigns are setup to manufacture a fixed number of batches for the various products. Take as an example the production of three batches of products A and B, produced in a plant with three stages. The processing times are given in Table 15.1.

The three batches for product A and B can basically be produced in two different ways, namely in so-called *single product campaigns* (SPC)—where all batches of one product are produced before switching to the second one—or in *mixed-product campaigns* (MPC) where batches are produced according to a chosen switching sequence. In Figure 15.4 the two production campaigns are compared.

For the example sketched above the mixed-product campaign seems to be more efficient, as the cycle time is only 18 hours, as compared to the single product campaign, which has a cycle time of 20 hours. It is noted here that MPCs are not necessarily more efficient than SPCs as cleanup time or product changeovers could influence the cycle time significantly. What would be the cycle times, if the equipment after each batch requires a cleaning procedure of one hour?

TABLE 15.1
Processing Times for a Two-Product Plant

	Stage 1	Stage 2	Stage 3
A	5 hrs	1 hr	1 hr
B	1 hrs	2 hrs	2 hrs

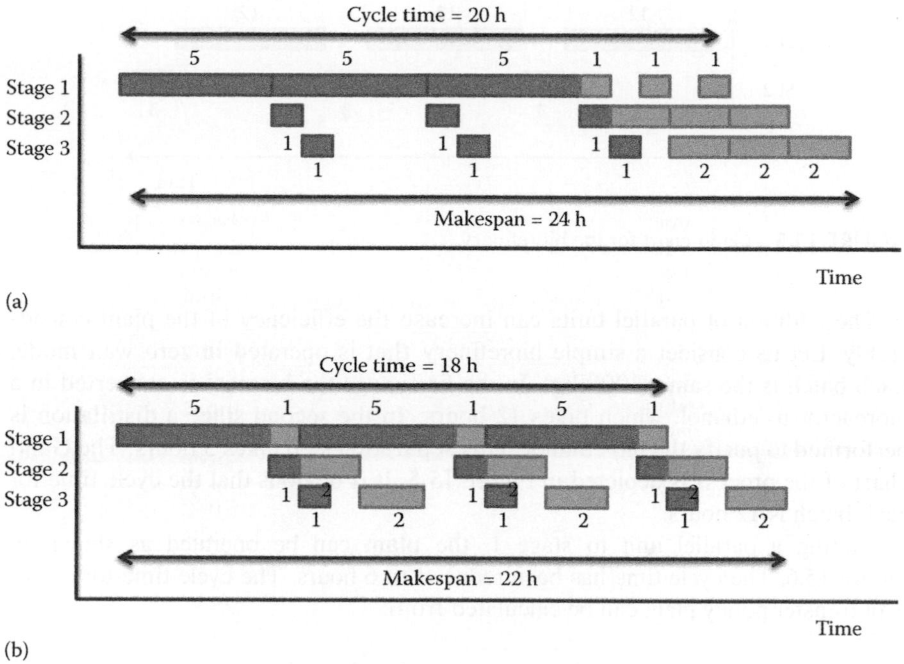

FIGURE 15.4 Schedules for single and mixed-product campaigns. Light bars for product A, dark bars for product B.

15.2.2 Flowshop versus Jobshop

The multiproduct plant example given in Section 15.2.1 is sometimes also referred to as a *flowshop plant*; in which all products go through all stages, following the same order of operations. If a plant produces very similar products, for example different polymer grades, the plant approaches a flowshop. However, there are also plants where not all products require all stages, or follow the same processing sequence, these plants are called *jobshop plants* or *multipurpose plants.*

15.2.3 Transfer Policies, Parallel Units, Storage, and Inventory

The so-called *zero wait transfer policy* is used when a batch at any stage is immediately transferred to the next stage because there is no intermediate storage vessel available or when it cannot be kept in the current vessel. This policy is extremely restrictive. The other extremum is the *unlimited intermediate storage* policy, where a batch can be stored without any capacity limit in a storage vessel. Lastly there is a transfer option called *no-intermediate storage,* which allows the batch to be kept inside the vessel. Normally the zero wait transfer requires the longest cycle time. In practice plants normally have a mixture of the three transfer policies.

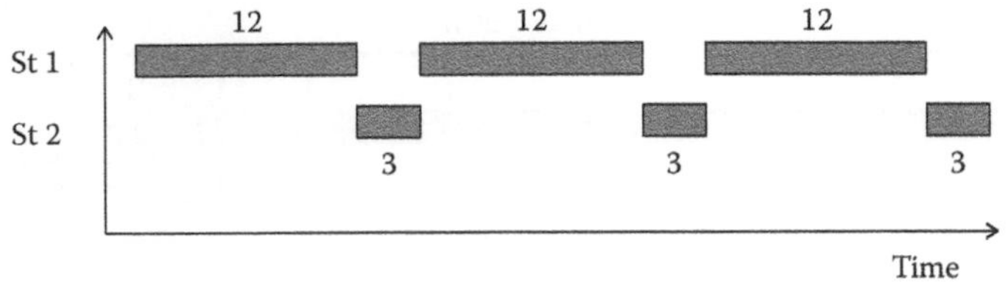

FIGURE 15.5 Gantt chart for the biorefinery.

The addition of parallel units can increase the efficiency of the plant considerably. Let us consider a simple biorefinery that is operated in zero wait mode. Each batch is the same (1000 kg). In the first stage the biomass is converted in a bioreactor to ethanol, which takes 12 hours. In the second stage a distillation is performed to purify the bio ethanol. This separation step takes 3 hours. The Gantt chart of the process is depicted in Figure 15.5. It is obvious that the cycle time for each batch is 12 hours.

Adding a parallel unit to stage 1, the plant can be operated as shown in Figure 15.6. The cycle time has been reduced to 6 hours. The cycle time for a zero wait transfer policy plant can be calculated from:

$$CT = max\left\{T_{ij} / NP_j\right\} \tag{15.3}$$

Where NP_j is the number of parallel units with $j = 1...M$.

If a large number of batches has to be processed, then the batch size could be reduced to 500 kg because the cycle time has been halved. Another option to increase the efficiency of the plant is by introducing intermediate storage between the stages. This basically allows the two stages to operate independently from each other with their own cycle times and batch sizes.

Inventory plays an important role in the selection of the production cycle. The main trade off is between inventory and cleanups. The shorter the production cycle, the less inventory we need since the products are available more frequently. However, the number of transitions will increase. If the production cycle is longer, the number of transitions decreases. However, the inventories will increase, as the products are produced less frequently.

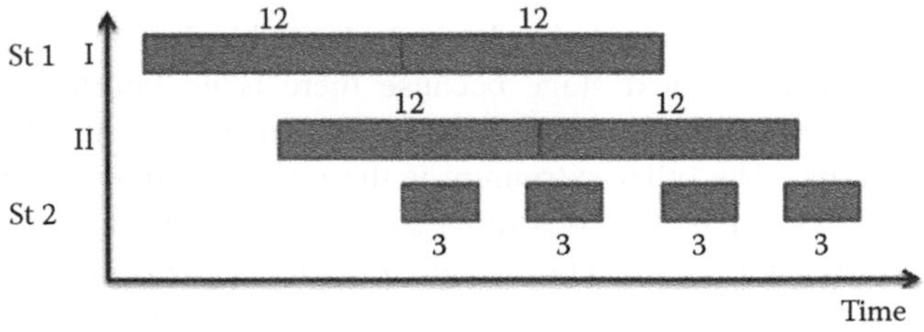

FIGURE 15.6 Plant with parallel units in the bioreactor.

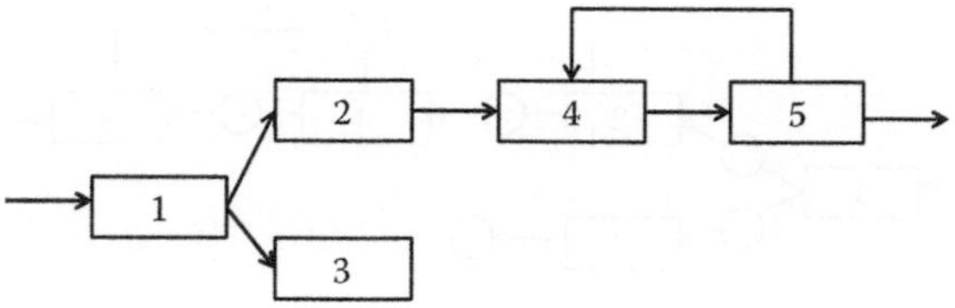

FIGURE 15.7 Recipe network representation of a chemical process.

15.3 STATE-TASK NETWORKS

In the beginning of the 1990s the so-called *recipe networks* were introduced [4]. These networks look very similar to process flowsheets of continuous plants. However, these networks are intended to describe the process and not so much a plant. In Figure 15.7 an example of a recipe network is given. Basically, each block in the sequence represents a task that should be executed.

Recipe networks are certainly useful for serial processing structures, but they can also lead to ambiguities, when they are applied to more complex networks [5]. From Figure 15.7, it is not clear whether task 1 will produce *two* different products or whether it has only *one* type of product, which will be shared between task 2 and 3. Or as a second example: it is not clear whether task 4 requires *two* different feedstocks (from task 2 and 5) or whether it only needs one type of feedstock by either task 2 *or* task 5. Both interpretations are plausible.

To deal with the ambiguities in recipe networks Kondili [5] introduced the concept of the *State-Task Network* (*STN*) representation. The main difference that the STN has with the recipe network is that it contains two types of nodes; *tasks* and *states*. The state nodes represent the feeds, intermediate and final products, while the task nodes represent the processing operation that transform materials in one form to another. The state nodes are represented by circles, and the tasks by rectangles.

In Figures 15.8 and 15.9 two STN's are shown which both correspond to the recipe network of Figure 15.7.

Figure 15.8 shows clearly that task 1 has only one product, which is then shared by tasks 2 and 3. In addition, task 4 only requires one feedstock, which is produced both by task 2 and 5.

Figure 15.9 shows that task 1 actually has two different products, forming the inputs to tasks 2 and 3 respectively. Furthermore it shows that task 4 also has two different feedstocks, produced by task 2 and task 5.

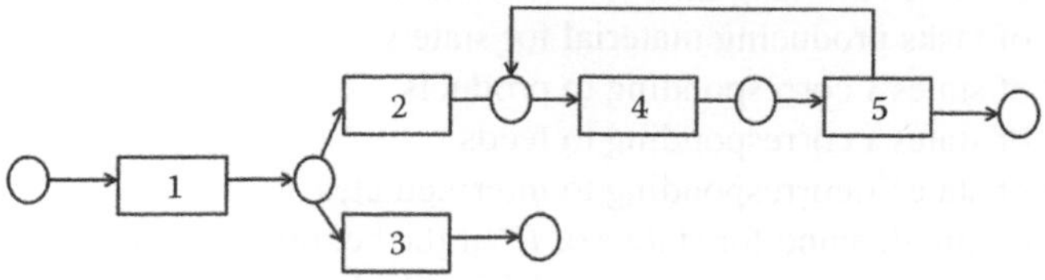

FIGURE 15.8 STN representation of a chemical process.

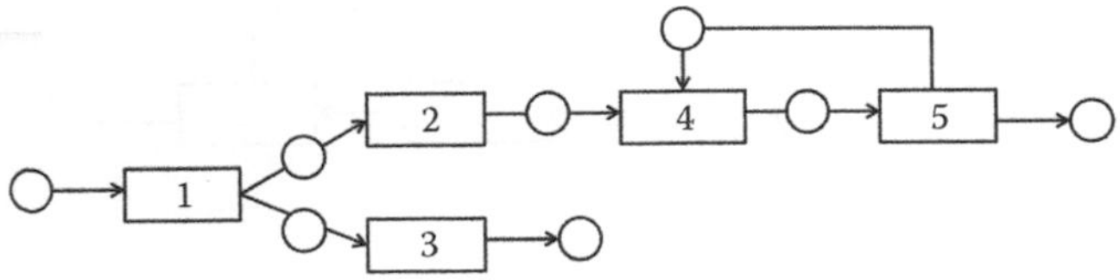

FIGURE 15.9 STN representation of a chemical process.

15.4 MATHEMATICAL FORMULATIONS OF SCHEDULING PROBLEMS

From the STN representation we can go to a mathematical formulation of the batch scheduling problem. This class of optimization models are called *Mixed Integer Linear Programs* (MILP's).

The MILP model that we will describe in this chapter is based on the one by Kondili [5]. For this model, the assignment of equipment to tasks does not need to be fixed, the batch size may be variable and can be handled with the possibility of mixing and splitting and different storage and transfer policies are allowed, as well as the limitation in the resources.

The MILP model will be formulated according to a provided STN, given one or more feeds, demands and deadlines. Now we will have to decide the timing of the operations, the assignments of the equipment to the operations and the flows of the materials through the network. The objective is to maximize a given profit function. We will further work with a uniform, discretized time domain, where h time periods of equal size are considered. Task i is defined by:

S_i = Set of states inputs to task i
S^*_i = Set of states outputs of task i
$\rho_{i,s}$ = Proportion input to task i from state s
$\rho^*_{i,s}$ = Proportion output of task i for state s
P_i = processing time for task i
K_i = Set of units j capable of processing task i

The last parameter K relates the process equipment units to the STN. The state s is defined by:

T_s = Set of tasks receiving material from state s
T^*_s = Set of tasks producing material for state s
IP = Set of states s corresponding to products
IF = Set of states s corresponding to feeds
II = Set of states s corresponding to intermediates
$d_{s,t}$ = Minimum demand for state s of IP at the beginning of period t
$r_{s,t}$ = Maximum purchase for state s of IF at the beginning of period t
C_s = Maximum storage of state s

The unit j may be capable of performing one or more tasks and is characterized by:

V_j = Maximum capacity
I_j = Set of tasks i for which equipment j can be used.

As for the variables, we will require binary and continuous variables:

$W_{i,j,t}$ = 1 if unit j starts processing task i at the beginning of period t
$B_{i,j,t}$ = Amount of material starts task i in unit j at beginning of period t
$S_{s,t}$ = Amount of material stored in state s at beginning of period t
$U_{u,t}$ = Demand of utility u over time interval t
$R_{s,t}$ = Purchases of state s at beginning of period t
$D_{s,t}$ = Sales of state s at beginning of period t

With all variables and parameters in place we can start with the formulation of the constraints of the MILP model. We will start with constraining the assignment of equipment j to tasks i over the various time periods t.

$$\sum_{i \in I_j} \sum_{t^*=t}^{t-p_i-1} W_{i,j,t^*} \le 1, \quad \forall j,t \tag{15.4}$$

If $W_{i,j,t} = 1$, unit j cannot be assigned to tasks other than i during the interval $[t - p_i + 1, t]$. The capacity limits for the units and storage tanks can be expressed as:

$$0 \le B_{i,j,t} \le V_j W_{i,j,t} \quad \forall i,t,j \in K_i \tag{15.5}$$

$$0 \le S_{s,t} \le C_s, \quad \forall s,t \tag{15.6}$$

The mass balances for each state and time are given as:

$$S_{s,t-1} + \sum_{i \in T^*_s} \rho^*_{i,s} \sum_{j \in K_i} B_{i,j,t-pi} + R_{s,t} = S_{s,t} - 1 + \sum_{i \in T_s} \rho_{i,s} \sum_{j \in K_i} B_{i,j,t} + D_{s,t} \quad \forall s,t \tag{15.7}$$

Which is the initial, plus the amounts produced and purchased must be equal to the hold-up plus the amount consumed and sold. Furthermore the following bounds on the purchases and sales apply:

$$D_{s,t} \ge d_{s,t}, \quad s \in IP \tag{15.8}$$

$$R_{s,t} \le r_{s,t}, \quad s \in IF \tag{15.9}$$

For the maximum amount of utility that is available we can introduce the following constraint, where α and β are the appropriate cost factors:

$$U_{u,t} = \sum_{i}\sum_{j\in K_i}\sum_{\theta=1}^{p_i-1}(\alpha_{u,i}W_{i,j,t-\theta} + \beta_{u,i}B_{i,j,t-\theta}), \quad \forall u,t \tag{15.10}$$

$$0 \le U_{u,t} \le U_u^{max} \tag{15.11}$$

Lastly, the objective function can be defined as: sales − purchases + final inventory − utilities:

$$Z = \sum_{s}\sum_{t=1}^{h} C_{s,t}^{D} D_{s,t} - \sum_{s}\sum_{t=1}^{h} C_{s,t}^{R} R_{s,t} + \sum_{s} C_{s,h=1} S_{s,h+1} - \sum_{u}\sum_{t=1}^{h} C_{u,t} U_{u,t} \tag{15.12}$$

Where ${C^D}_{s,t}$, ${C^R}_{s,t}$, $C_{s,h+l}$ and $C_{u,t}$ are the appropriate cost coefficient. It is possible to include a zero wait policy by adding constraints that specify that task i^* follows task i:

$$\sum_{j\in K_i} W_{i,j,t} = \sum_{j\in K^*_i} W_{i^*,j,t+pi} \tag{15.13}$$

The objective function in Eq. 15.12 subject to constraints Eqs 15.4–15.11 and Eq. 15.13 is an MILP problem that can be solved within reasonable computational time, if the number of time intervals is not too large.

15.5 EXAMPLE: SCHEDULING OF AN ICE CREAM FACTORY

In the following section, a small scheduling problem is formulated and solved for the so-called fast moving consumer goods industry (FMCG). Although the example is not a typical chemical process, the food industry is of great interest to the process engineering community. The example concerns a small ice-cream factory. The problem is formulated in a way very similar to the MILP formulation introduced in the previous section. The interested reader is referred to [6] in which a scheduling problem at industrial scale for the FMCG industry is solved.

For this example, consider a small ice-cream factory containing two production stages. On the first stage the ingredients are mixed and the ice cream is frozen. On the second stage the ice cream is packed. Between mixing and packing the products are stored in intermediate storage with an unlimited capacity. The factory contains one mixing line and two packing lines, as is shown in Figure 15.10.

Four varieties of ice cream are produced in this factory. The mixing line can produce all four products. Products A and B are packed on packing line 1 and products C and D are packed on packing line 2. The mixing rate is 3000 kg/hr and the packing rate is 1000 kg/hr. When switching from one product to another, a 4-hour

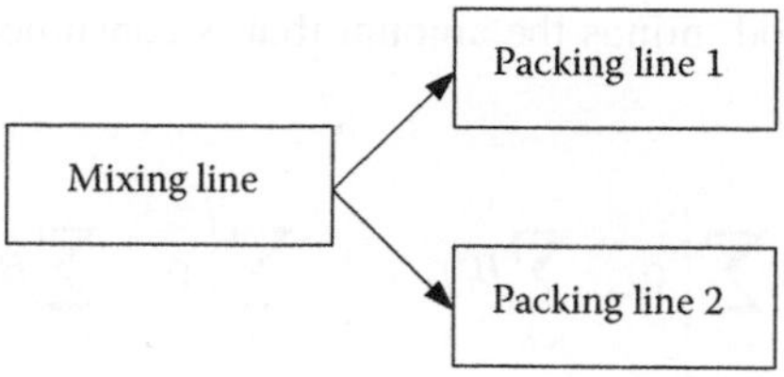

FIGURE 15.10 Schematic overview of the ice cream production process.

changeover is required to clean the mixing/packing line. The ice cream is produced in batches of 12,000 kg. The demand, which is given in Table 15.2, must be met at the end of the week. The objective is to minimize the makespan.

We can use the following model to optimize the production schedule for this ice cream factory. First, we ensure that at most one task can be active on any unit at any time.

$$\sum_{i \in I_j} \sum_{t^*=t}^{t+p_i-1} W_{i,j,t^*} \leq 1, \quad \forall j,t \tag{15.14}$$

Secondly, different tasks on the same unit cannot start before the end of the current task plus the changeover time:

$$\sum_{(i' \neq i) \in I_j} \sum_{t'=t}^{t+P_i+\tau_j-1} W_{i',j,t'} \leq M\left(1-W_{i,j,t}\right) \quad \forall i \in I_j, j, t \tag{15.15}$$

If task i is active on unit j in period t, then the amount of material undergoing this task is equal to the batch size:

$$B_{i,j,t} = BatchSize \cdot W_{i,j,t} \quad \forall i, j, t \tag{15.16}$$

The amount of material in state s at the end of period t is equal to the amount in the previous period, plus the amount produced by the tasks that are completed at the

TABLE 15.2
Demand for the Different Products

	Demand
Product A	36,000 kg
Product B	48,000 kg
Product C	60,000 kg
Product D	36,000 kg

end of the current period, minus the amount that is consumed in the tasks starting in the current period.

$$S_{s,t} = S_{s,t-1} + \sum_{i \in T_s^+} \left(\rho_{i,s} \cdot \sum_{j \in K_i} B_{i,j,t-P_i} \right) - \sum_{i \in T_s^-} \left(\rho_{i,s}^* \cdot \sum_{j \in K_i} B_{i,j,t} \right) \quad \forall s,t \qquad (15.17)$$

The amount of material in state s should be equal to the demand at the end of the scheduling horizon.

$$S_{s,\text{last}(t)} = D_s \quad \forall s \qquad (15.18)$$

The makespan is equal to the latest end time of all tasks.

$$MS \geq t \cdot W_{i,j,t} + P_i - 1 \quad \forall i \in I_j, j, t \qquad (15.19)$$

This MILP model can be used to optimize the weekly production schedule. The resulting schedule will have a makespan of 104 hours. It should be noted that several different schedules with a 104-hour makespan could be obtained. One example of such a schedule is in Figure 15.11.

However, in practice the schedule should satisfy another constraint. After the mixing stage, the ice cream must remain in intermediate storage for several hours before it can be packed to allow the product to age. The minimum required ageing time depends on the product. Products A and D have an ageing time of 4 hours, product B of 8 hours, and product C of 12 hours.

We can consider this ageing time by modifying Eq. 15.18. To ensure that an intermediate product cannot be packed before it is aged, the amount of material in state s is only increased after the production and ageing time of this material in state s.

$$S_{s,t} = S_{s,t-1} + \sum_{i \in T_s^+} \left(\rho_{i,s} \cdot \sum_{j \in K_i} B_{i,j,t-P_i-AgeT_i} \right) - \sum_{i \in T_s^-} \left(\rho_{i,s}^* \cdot \sum_{j \in K_i} B_{i,j,t} \right) \quad \forall s,t \qquad (15.20)$$

Using the modified model, a schedule with a makespan of 112 hours is obtained. A Gantt chart of this schedule is given below, see Figure 15.12. This time products A and D are produced first because their shorter ageing time allows the packing to start earlier.

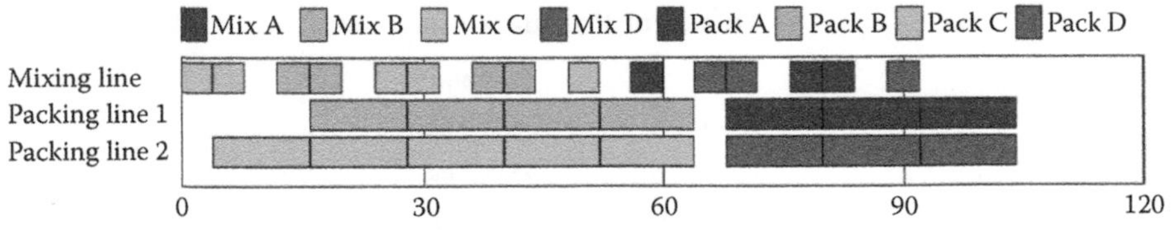

FIGURE 15.11 Ice cream production Gantt chart.

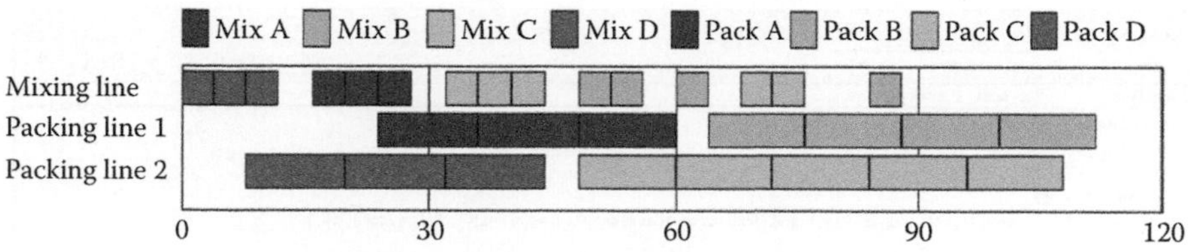

FIGURE 15.12 Ice cream production Gantt chart for the modified model.

15.6 IMPLEMENTATION

The above formulated MILP model was coded in the software package AIMMS®. The model contains 1058 constraints and 721 variables of which 240 are binary. In Section 15.7 the modeling code can be found.

AIMMS® offers an all-round development environment for the creation of high-performance decision support and advanced planning applications to optimize strategic operations. It allows organizations to rapidly improve the quality, service, profitability, and responsiveness of their operations. The AIMMS® development environment possesses a unique combination of advanced features and design tools, such as the graphical model explorer, which allow you to build and maintain complex decision support applications and advanced planning systems in a fraction of the time required by conventional programming tools.

As a beginning user of AIMMS® it is advised to visit the website of AIMMS® under http://aimms.com/downloads/aimms. The reader can find the software package for which a free trial license can be obtained.

A one hour instruction to the basics of AIMMS® can be obtained at http://aimms.com/aimms/download/manuals/aimms_tutorial_beginner.pdf in different languages.

After installation and start up of AIMMS® a *new project* can be created. The Main AIMMS® interface opens; see Figure 15.13. In the so-called *Model explorer*

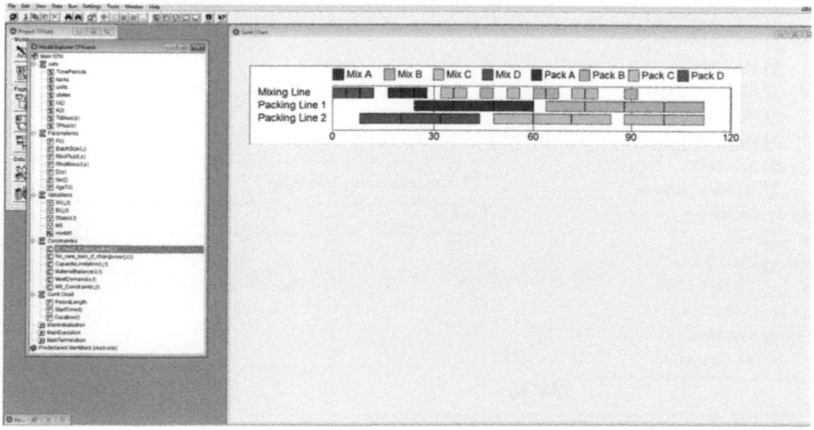

FIGURE 15.13 Screenshot of the main AIMMS® user interface.

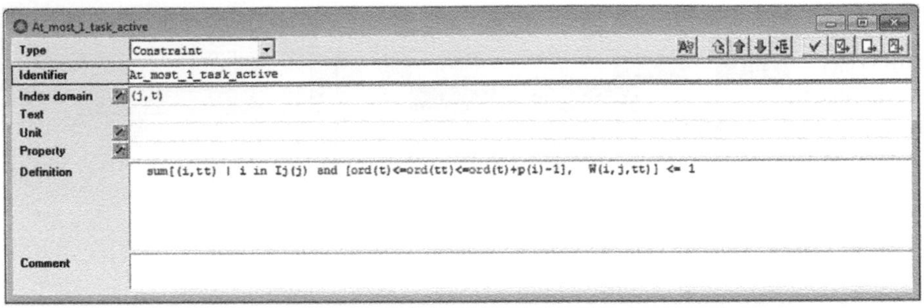

FIGURE 15.14 Screenshot of an identifier block for a constraint.

a small *declaration icon* appears that can be used to insert new *types of identifiers*, such as constraints, parameters and data.

In Figure 15.14 an example of a constraint type of identifier is shown. Under the header *definition* the constraint can be inserted.

AIMMS® uses a predefined solver configuration. However, if the user wishes to change the settings under the folder *Settings\Project Options* specific preferences for solver and options can be modified. Figure 15.15 shows a screenshot of the AIMMS® options folder.

In the model explorer it is also possible to define the preferred output, for example as *Gantt chart*, or *Table*. Finally, to run the model the user selects from

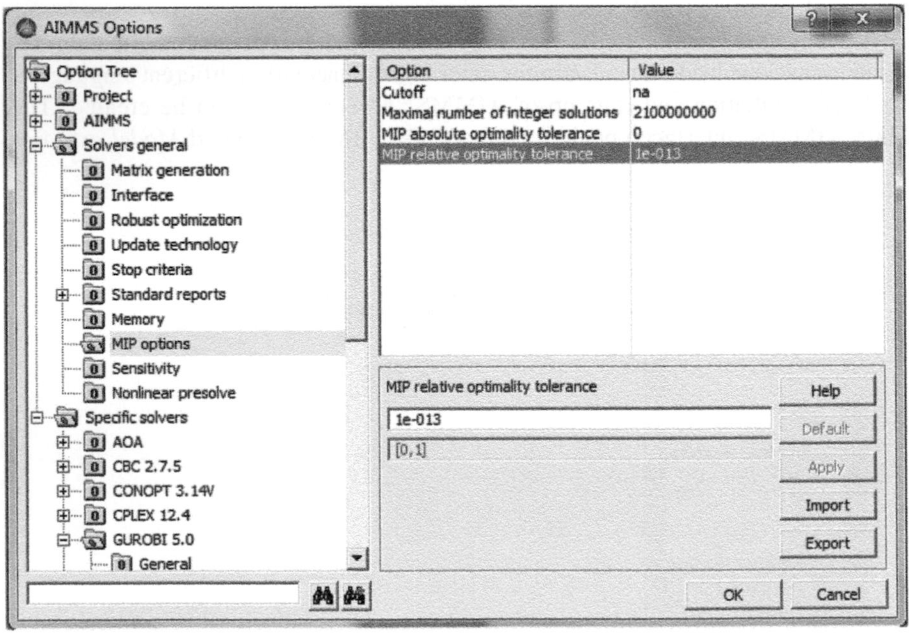

FIGURE 15.15 Screenshot of the AIMMS Options with the solver settings.

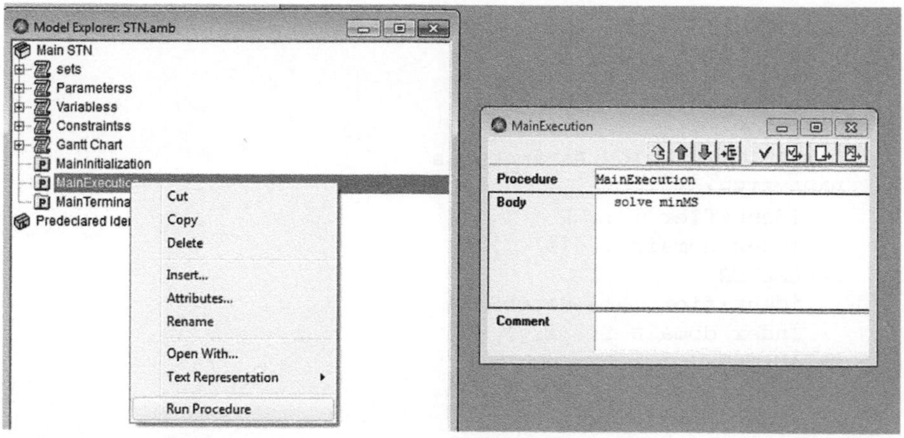

FIGURE 15.16 Screenshot for running the AIMMS® model.

the Model explorer the *Main Execution* folder and selects *Run Procedure* (see Figure 15.16).

15.7 AIMMS® CODE OF THE SCHEDULING MODEL

```
MAIN MODEL Main_STN
  DECLARATION SECTION sets
    SET:
       identifier    :  TimePeriods
       indices       :  t, tt
       definition    :  elementrange(001,030,1,"Period ") ;
    SET:
       identifier    :  tasks
       indices       :  i, ii ;
    SET:
       identifier    :  units
       index         :  j ;
    SET:
       identifier    :  states
       index         :  s ;
    SET:
       identifier    :  IJ
       index domain  :  (j)
       subset of     :  tasks ;
    SET:
       identifier    :  K
       index domain  :  (i)
       subset of     :  units ;
    SET:
       identifier    :  TMinus
       index domain  :  s
       subset of     :  tasks ;
```

```
  SET:
     identifier   :  TPlus
     index domain :  s
     subset of    :  tasks ;
ENDSECTION  ;
DECLARATION SECTION Parameterss
  PARAMETER:
     identifier   :  P
     index domain :  (i) ;
  PARAMETER:
     identifier   :  BatchSize
     index domain :  (i,j) ;
  PARAMETER:
     identifier   :  RhoPlus
     index domain :  (i,s) ;
  PARAMETER:
     identifier   :  RhoMinus
     index domain :  (i,s) ;
  PARAMETER:
     identifier   :  D
     index domain :  (s) ;
  PARAMETER:
     identifier   :  tau
     index domain :  (j) ;
  PARAMETER:
     identifier   :  AgeT
     index domain :  (i) ;
ENDSECTION  ;
DECLARATION SECTION Variabless
  VARIABLE:
     identifier   :  W
     index domain :  (i,j,t) | i in IJ(j)
     range        :  binary ;
  VARIABLE:
     identifier   :  B
     index domain :  (i,j,t)| i in IJ(j)
     range        :  nonnegative ;
  VARIABLE:
     identifier   :  State
     index domain :  (s,t)
     range        :  nonnegative ;
  VARIABLE:
     identifier   :  MS
     range        :  free ;
  MATHEMATICAL PROGRAM:
     identifier   :  minMS
     objective    :  MS
     direction    :  minimize
     constraints  :  AllConstraints
     variables    :  AllVariables
     type         :  Automatic ;
```

```
 ENDSECTION  ;
 DECLARATION SECTION Constraintss
   CONSTRAINT:
      identifier    :  At_most_1_task_active
      index domain :  (j,t)
      definition    :  sum[(i,tt) | i in Ij(j) and [ord(t)<=or
        d(tt)<=ord(t)+p(i)-1],  W(i,j,tt)] <= 1 ;
   CONSTRAINT:
      identifier    :  No_new_task_if_changeover
      index domain :  (j,t,i) | i in Ij(j)
      definition    :  sum[ii | ii in Ij(j) and ord(ii)
        <>ord(i), sum[tt | (ord(tt)>= ord(t)) and (ord(tt) <=
        ord(t) + P(i)+tau(j)-1), W(ii,j,tt)]] <= 20*(1 -
        W(i,j,t)) ;
CONSTRAINT:
      identifier    :  CapacityLimitation
      index domain :  (i,j,t)
      definition    :  B(i,j,t) = BatchSize(i,j) * W(i,j,t) ;
   CONSTRAINT:
      identifier    :  MaterialBalance
      index domain :  (s,t)
      definition    :  State(s,t) = State(s,t-1)
                       + sum[i | i in TPlus(s), RhoPlus(i,s) *
                          sum[j | j in K(i), B(i,j,t-P(i)-
                          AgeT(i))]]
                       - sum[i | i in TMinus(s), RhoMinus(i,s)
                          * sum[j | j in K(i), B(i,j,t)]] ;
   CONSTRAINT:
      identifier    :  MeetDemand
      index domain :  (s,t) | ord(t)=30
      definition    :  State(s,t) = D(s) ;
   CONSTRAINT:
      identifier    :  MS_Constraint
      index domain :  (i,j,t)| i in IJ(j)
      definition    :  MS >= 4*(ord(t)* W(i,j,t) + P(i)-1) ;
 ENDSECTION  ;
 DECLARATION SECTION Gantt_Chart
   PARAMETER:
      identifier    :  PeriodLength ;
   PARAMETER:
      identifier    :  StartTime
      index domain :  (t)
      definition    :  Periodlength*(ord(t)-1) ;
   PARAMETER:
      identifier    :  Duration
      index domain :  i
      definition    :  Periodlength*P(i) ;
 ENDSECTION  ;
 PROCEDURE
   identifier :  MainInitialization
 ENDPROCEDURE  ;
```

```
  PROCEDURE
    identifier :  MainExecution
    body       :
      solve minMS
  ENDPROCEDURE  ;
  PROCEDURE
    identifier :  MainTermination
    body       :
      return DataManagementExit();
  ENDPROCEDURE  ;
ENDMODEL Main_STN ;
```

15.8 FURTHER READING

This chapter was intended to give the reader an introduction to batch scheduling. It has not discussed scheduling of multipurpose batch processes. A complete MILP model has been reported by Ierapetritou and Floudas [7]. In Floudas and Lin [8] an overview of the developments in scheduling of multiproduct/multipurpose batch and continuous processes is presented, with a particular focus on the time representation of the resulting models.

REFERENCES

1. Mendez, C.A., Cerda, J., Grossmann, I.E., Harjunkoski, I., Fahl, M. (2006) State-of-the-art review of optimization methods for short term scheduling of batch processes, *Comp. Chem. Eng.*, 30, 913–946.
2. Rippin, D.W.T. (1983) Simulation of single- and multiproduct batch chemical plants for optimal design and operation, *Comp. Chem. Eng.*, 7(3), 137–156.
3. Biegler, L.T., Grossmann, I.E., Westerberg, A.W. (1997) *Systematic methods of chemical process design*, Prentice Hall.
4. Reklaitis, G.V. (1991) Perspectives of scheduling and planning of process operations. PSE '91, Montebello, Canada.
5. Kondili, E., Pantelides, C.C., Sargent, R.W.H. (1993) A general algorithm for short-term scheduling of batch operations-I. MILP formulation, *Comp. Chem. Eng.*, 17 (2), 211–227.
6. van Elzakker, M.A.H., Zondervan, E., Raikar, N.B., Grossmann, I.E., Bongers, P.M.M., (2012) Scheduling in the FMCG industry: an industrial case study, *Ind. Eng. Chem. Res.*, 51(22), 7800–7815.
7. Ierapetritou, M.G., Floudas, C.A. (1998) Effective continuous-time formulation for short-term scheduling. 1. Multipurpose batch processes, *Ind. Eng. Chem. Res.*, 37, 4341–4359.
8. Floudas, A.F., Lin, X. (2004) Continuous-time versus discrete-time approaches for scheduling of chemical processes: a review, *Comp. Chem. Eng.*, 28, 2109–2129.

16 Plant Location: Supply Chain Management

Gonzalo Guillén-Gosálbez, Fengqi You

CONTENTS

16.1 INTRODUCTION

In this chapter we will address the problem of plant location. That is, where should a chemical facility be placed, considering cost and demand data. This problem is quite common in the field of chemical engineering, and lies in the interface between chemical engineering itself and operations research. We will provide a basic general formulation and then give some additional references to other similar models.

The plant location problem is related to the concept of Supply Chain Management (SCM) (see Figure 16.1), which appeared in the early 1990s, and which has recently raised a lot of interest since it can affect decisively the profitability of a company. SCM looks for the integration of a plant with its suppliers and its customers to be managed as a whole, and the co-ordination of all the input/output flows (materials, information and finances) so that products are produced and distributed at the right quantities, to the right locations, and at the right time [1-4]. The main objective is to achieve suitable economic results, together with the desired consumer satisfaction levels.

The SCM problem may be considered at different levels, depending on the strategic, tactical and operational variables involved in decision-making. Therefore, a

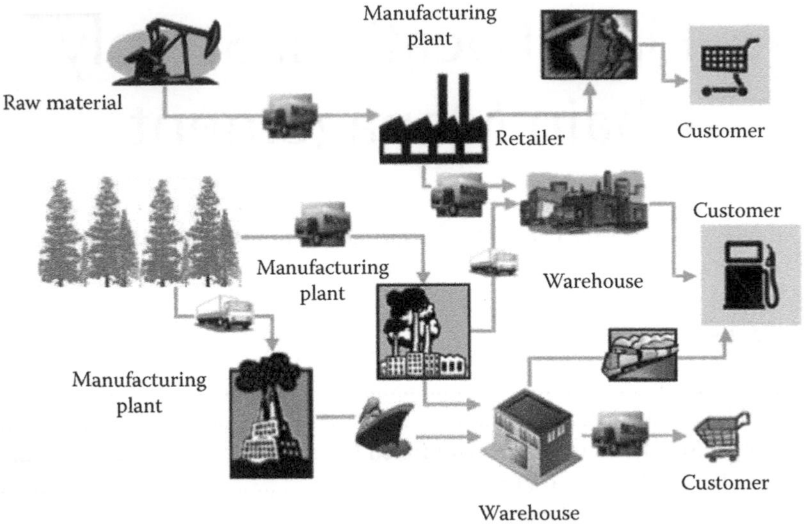

FIGURE 16.1 Generic supply chain.

large spectrum of a firm's strategic, tactical and operational activities are encompassed by SCM:

- The strategic level concerns those decisions that will have a long-lasting effect on the firm. It is focused on SC design, and entails determining the optimal configuration for an entire SC network. Hence, at this level we deal with the plant location problem, which must define the optimal location of a facility taking into account information such as raw materials availability, final demand, distribution channels, and economic data.
- The tactical level encompasses long-/medium-term management decisions, which are typically updated at a rate ranging between once every quarter and once every year. These include overall purchasing and production decisions, inventory policies, and transport strategies.
- The operational level refers to day-to-day decisions such as scheduling, lead-time quotations, routing, and lorry loading.

In this chapter we will study the strategic SCM level and introduce a general mathematical formulation for optimizing plant location decisions considering different information. We will first formally define the problem of interest, and then describe the mathematical formulation in detail. We end the chapter with some notes and recommendations for further readings.

16.2 PROBLEM STATEMENT

We will focus in this chapter on the SC design problem, which has as objective to determine the optimal configuration of a production-distribution network according to a predefined performance indicator and considering a set of physical constraints. Without

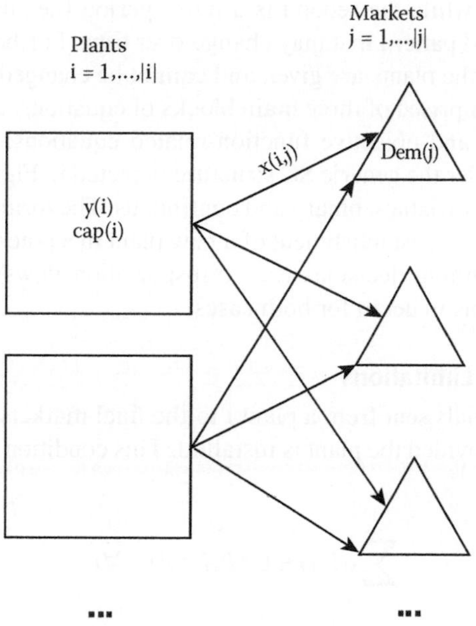

FIGURE 16.2 Two-echelon SC taken as reference in the MILP models.

loss of generality, we will consider a generic two-echelon SC (production-market) whose configuration is to be determined with the goal at maximizing the economic performance. Decisions to be made are of two types: structural (location of the plants), and planning (production rates, and material flows between plants and markets).

The structure of the two-echelon SC taken as reference in this work is depicted in Figure 16.2. This network includes the following elements:

- A set I of potential plant locations where products are manufactured.
- A set J of existing markets where products are available to customers.

Final products are manufactured in the i (potential) plants and then sent to the j (existing) final markets where they become available to final customers. The demand of the final markets is given, along with the transportation and investment costs. The problem consists then on deciding where to open the new plants, and from which plants we should supply the final markets (i.e., transportation flows). In the next section we describe in detail a mathematical formulation that tackles this problem.

16.3 SIMPLE STEADY-STATE MODEL WITH FIXED CAPACITIES

16.3.1 Mathematical Formulation

The most common approach to address the SC design problem is to formulate a large-scale mixed-integer program (MIP) that captures the relevant features associated with the network. In this chapter we present two MIPs to address the plant location problem. The first considers steady-state conditions and provides a snapshot of the SC operation, while the second is a multi-period (i.e., dynamic) model that accounts for a demand pattern that may change over time. Furthermore, we consider that the capacities of the plants are given and cannot be changed.

Both MIPs are comprised of three main blocks of equations: capacity limitations, demand satisfaction, and objective function-related equations. To formulate these constraints, we consider the generic SC structure depicted in Figure 16.2. The model contains two types of variables: binary and continuous. The former are used to model the SC configuration (i.e., establishment of a new plant in a potential location), while the latter denote planning decisions (i.e., transportation flows). We describe next each of these equations in detail for both cases.

16.3.1.1 Capacity Limitations

The amount of materials sent from a plant i to the final markets must be lower than the plant capacity, provided the plant is installed. This condition is enforced with the following constraint:

$$\sum_{j} x(i,j) \leq CAP(i)y(i) \quad \forall i \tag{16.1}$$

Where y(i) is a binary variable that is one if the facility is opened at location i, and 0 otherwise. Hence, if the model decides to open the plant, then the binary variable will be one, and the amount produced and sent to the markets will be forced to be lower than the plant capacity (parameter CAP(i)). If the plant is not opened, then the binary variable will be zero, and the flows of materials between the plant and the markets (denoted by the continuous variable x(i,j)) will be all set to zero as well.

16.3.1.2 Demand Satisfaction

The following constraint ensures that the amount of materials sent from all the plants to every market (continuous variable x(i,j)) must be greater or equal to the demand in that market (parameter DEM(j)) (i.e., the demand of each final markets is fully covered).

$$\sum_{i} x(i,j) \geq DEM(j) \quad \forall j \tag{16.2}$$

16.3.1.3 Objective Function

In general, we can use different objectives functions to drive the optimization, such as economic (e.g. cost, profit, net present value, etc.), environmental (e.g. global warming, water footprint, etc.), and social indicators. Without loss of generality, we consider here that the SC design is carried out considering only the economic performance of the network, which is quantified through the total cost. This indicator is determined via the following equation:

$$TC = \sum_{j}\sum_{j} x(i,j)trc(i,j) + \sum_{i} y(i)invc(i) \tag{16.3}$$

As seen, the total cost (denoted by the continuous viable TC), accounts for the transportation cost (obtained from the transportation flows, continuous variable x(i,j), and unitary transportation cost, parameter trc(i,j)), and the investment cost. The investment cost is determined from the network topology (i.e., binary variable y(i)) and the capital investment associated with each potential location (parameter inv(i)). Note that if the binary variable y(i) takes a value of one, then the investment cost of the plant at location i will be accounted for in the equation, while if it is zero, it will be omitted.

16.3.2 Solution Procedure

Finally, the overall model can be expressed in compact form as follows:

$$(M1) \quad \min \quad TC = \sum_{j}\sum_{j} x(i,j)trc(i,j) + \sum_{i} y(i)invc(i)$$

$$s.t. \quad Eqs.1-3$$

$$x(i,j) \in \Re; y(i) \in \{0,1\}$$

The task of the optimization algorithm is to identify the values of the decision variables that minimize the objective function. The model contains two types of variables: binary and continuous. In addition, all the constraints are linear. Hence, we are dealing with a mixed-integer linear program (MILP) that can be solved by the methods already discussed in the previous chapter.

Note that the size of the problem grows with the number of potential plant locations and markets. Hence, for very large instances, the traditional branch-and-cut method implemented in standard MIP solvers might lead to prohibitive CPU times. In this case, we can resort to specific decomposition algorithms that exploit the problem structure and which are not described in detail here due to space limitations.

EXERCISE 16.1

The task here is to identify the optimal structure of a SC that will cover the demand of three final markets (j1, j2 and j3) with demands equal to 325, 300 and 275, respectively. Two potential locations are considered for the plants, i1 (capacity of 250 and investment cost of 20) and i2 (capacity of 1200 and investment cost of 140). The transportation costs are shown in Table 16.1.

TABLE 16.1
Transportation Cost between Plants and Markets

Plants/Markets	j1	j2	j3
i1	2.5	1.7	1.8
i2	2.5	1.8	1.4

Using the MILP formulation defined by (M1), we find a SC structure with 2 plants, the one in i1 sends 250 to j2, while i2 sends 325 to j1, 50 to j2, and 275 to j3, with a total cost of 1872.5 monetary units.

```
Sets

i plants / i1*i2 /
j markets / j1*j3 / ;

Parameters

cap(i) capacity of plant i
/ i1 250
i2 1200 /

invc(i) investment cost of plant i
/ i1 20
i2 140 /

dem(j) demand at market j
/ j1 325
j2 300
j3 275 / ;

Table trc(i,j) unitary transportation cost
        j1      j2      j3
i1      2.5     1.7     1.8
i2      2.5     1.8     1.4 ;

Variables

x(i,j) amount sent from plant i to market j
y(i) binary variable 1 if plant is opened at location i 0
  otherwise
tc total cost ;

Positive Variable x ;
Binary variable y ;

Equations

cost total cost
capacity(i) capacity limitations at plant i
demand(j) satisfy demand at market j ;

cost .. tc =e= sum((i,j), trc(i,j)*x(i,j)) +
  sum(i,invc(i)*y(i)) ;
capacity(i) .. sum(j, x(i,j)) =l= cap(i)*y(i) ;
demand(j) .. sum(i, x(i,j)) =g= dem(j) ;
Model SC_design_1 /all/ ;
```

```
option MIP = CPLEX;
option optcr = 0;

Solve SC_design_1 using mip minimizing tc ;
```

16.4 MULTI-PERIOD OPTIMIZATION MODEL WITH CAPACITY EXPANSIONS

16.4.1 Mathematical Formulation

The model presented before assumes steady-state conditions and provides a snapshot of the SC operation. In this section we describe a more general formulation that considers a varying demand over time. The model includes several time periods in each of which it is possible to take decisions regarding the SC structure. Hence, the capacity of the network can be modified over time in order to follow a predefined demand pattern, and we must find the optimal timing of capacity investments. In addition, we consider that the capacities of the plants are not fixed, but rather optimized along with the plants locations. Similarly, as in the previous case, the model comprises three main blocks of equations that are next described in detail.

16.4.1.1 Capacity Limitations

Eq. (1) is now defined for every time period as follows:

$$\sum_{j} x(i,j,t) \leq CAP(i,t) \quad \forall i,t \tag{16.4}$$

Where x(i,j,t) denotes the flow of materials between plant i and market j in period t. As oppose to the previous formulation, here CAP(i,t) is a continuous variable (not a fixed parameter) that denotes the capacity of plant i in period t. This variable is determine using the following equation:

$$CAP(i,t) = CAP(i,t-1) + ECAP(i,t) \quad \forall i,t \tag{16.5}$$

That is, the capacity of plant i in period t is obtained from the capacity in the previous period plus the expansion in capacity executed in the present period. Here ECAP(i,t) is a continuous variable denoting the expansion in capacity executed in period t. This expansion in capacity is constrained within lower and upper limits as follows:

$$ECAP^{LO}(i)y(i,t) \leq ECAP(i,t) \leq ECAP^{UP}(i)y(i,t) \quad \forall i,t \tag{16.6}$$

where $ECAP^{LO}(i)$ and $ECAP^{UP}(i)$ are given parameters representing lower and upper bounds imposed on the capacity expansions, while binary variable y(i,t) takes a value of one if plant i is expanded in capacity in period t, and it is zero otherwise.

16.4.1.2 Demand Satisfaction

The demand satisfaction constraint, which is equivalent to Eq. (16.2), can be expressed as follows:

$$\sum_{i} x(i,j,t) \geq DEM(j,t) \quad \forall j,t \tag{16.7}$$

Where DEM(j,t) represents now the demand of market j in period t.

16.4.1.3 Objective Function

Finally, the objective function accounts for the operation costs over the entire time horizon plus the investment cost, as follows:

$$TC = \sum_{j}\sum_{j}\sum_{t} x(i,j,t)trc(i,j,t) + \sum_{i}\sum_{t} \alpha(i,t)y(i,t) + \sum_{i}\sum_{t} \beta(i,t)ECAP(i,t) \tag{16.8}$$

The first term of Eq. (16.8) accounts for the transportation costs. Note that the unitary transportation cost parameter (trc(i,j,t)) is now defined for every time period t. The second and third terms determine the capital investment from the capacity expansions. We assume that the capital investment is a linear function of the capacity expansions, where alpha(i,t) and beta(i,t) are the fixed and variable capital investment parameters, respectively. For simplicity, we have not considered the temporal value of money, but note that this could be easily accounted for in the model by introducing the interest rate.

16.4.2 Solution Procedure

The multi-period model can be expressed in compact form as follows:

$$\begin{aligned}
(M2) \quad \min \quad & TC = \sum_{j}\sum_{j}\sum_{t} x(i,j,t)trc(i,j,t) + \sum_{i}\sum_{t} \alpha(i,t)y(i,t) \\
& + \sum_{i}\sum_{t} \beta(i,t)ECAP(i,t) \\
s.t. \quad & Eqs.4-7 \\
& x(i,j,t), CAP(i,t), ECAP(i,t) \in \Re; y(i,t) \in \{0,1\}
\end{aligned}$$

Again, the model contains both binary and continuous variables and only linear constraints. Hence, it is an MILP that can be solved by standard MIP solvers, in a similar manner as commented before.

EXERCISE 16.2

In this task we determine the optimal structure of a SC over a time horizon of five time periods. We consider three markets (j1, j2 and j3) with an initial demand of 325, 300 and 275, respectively, that grows by 10% every time period. Two potential locations are considered for the plants, i1 (alpha equal to 350 and beta equal to 3) and i2 (alpha equal to 600 and beta equal to 6). The transportation costs are the same as in the previous case (shown in Table 16.1). The cost parameter (alpha, beta and the unitary transportation cost are increased by 5% every tie period).

Using the MILP formulation defined by (M2), we find a SC structure that opens a single plant at location i1 in the first time period with a capacity of 1000. This plant is expanded in capacity in the second time period by 317.69 units, reaching a total capacity of 1317.69 in period 2 and subsequent time periods. The total cost of this design is 16271.402 monetary units.

```
Sets

i plants / i1*i2 /
j markets / j1*j3 /
t time periods /t1*t5/ ;

Parameters

alpha_ini(i) initial fixed investment cost parameter of
  plant i
/ i1 350
i2 600 /

beta_ini(i) initial variable investment cost parameter of
  plant i
/ i1 3
i2 6 /

dem_ini(j) intial demand at market j
/ j1 325
j2 300
j3 275 / ;

Table trc_ini(i,j) intial unitary transportation cost
         j1      j2      j3
i1       2.5     1.7     1.8
i2       2.5     1.8     1.4 ;

Parameters
alpha(i,t) fixed investment cost parameter of plant i at
  period t
beta(i,t) variable investment cost parameter of plant i at
  period t
dem(j,t) demand at market j at period t
trc(i,j,t) unitary transportation cost at period t
inc annual increment in the costs
inc_d annual increment in the demand
ecap_lo(i) lower bound on the capacity expansion of plant i
ecap_up(i) upper bound on the capacity expansion of plant i;
```

```
inc = 0.02;
inc_d = 0.1;
ecap_lo(i)=0;
ecap_up(i)=1e3;

alpha(i,t)=alpha_ini(i)*(1+inc)**(ord(t)-1);
beta(i,t)=beta_ini(i)*(1+inc)**(ord(t)-1);
trc(i,j,t)=trc_ini(i,j)*(1+inc)**(ord(t)-1);
dem(j,t)=dem_ini(j)*(1+inc_d)**(ord(t)-1);

Variables

x(i,j,t) amount sent from plant i to market j
y(i,t) binary variable 1 if plant is opened at location i 0
  otherwise
cap(i,t) capacity of plant i at time period t
ecap(i,t) expansion in capacity of plant i at time period t
tc total cost ;

Positive Variable x,cap,ecap ;
Binary variable y ;

Equations

cost total cost
capacity1(i,t) capacity limitations at plant i
capacity2(i,t) capacity limitations at plant i
capacity3(i,t) capacity limitations at plant i
capacity4(i,t) capacity limitations at plant i
capacity5(i,t) capacity limitations at plant i
demand(j,t) satisfy demand at market j ;

cost .. tc =e= sum((i,j,t), trc(i,j,t)*x(i,j,t)) +
  sum((i,t),alpha(i,t)*y(i,t)) +
  sum((i,t),beta(i,t)*ecap(i,t)) ;

capacity1(i,t) .. sum(j, x(i,j,t)) =l= cap(i,t) ;
capacity2(i,t) $ (ord(t) eq 1) .. cap(i,t) =e= ecap(i,t) ;
capacity3(i,t) $ (ord(t) gt 1) .. cap(i,t) =e= cap(i,t-1) +
  ecap(i,t) ;

capacity4(i,t) .. ecap(i,t) =l= ecap_up(i)*y(i,t) ;
capacity5(i,t) .. ecap(i,t) =g= ecap_lo(i)*y(i,t) ;
demand(j,t) .. sum(i, x(i,j,t)) =g= dem(j,t) ;
Model SC_design_2 /all/ ;

option MIP = CPLEX;
option optcr = 0;

Solve SC_design_2 using mip minimizing tc ;
```

16.5 FINAL REMARKS AND FURTHER READINGS

There are many MILP models based on the ones presented in this chapter. A general review of mathematical models and decomposition strategies for SCM problems can be found in the work by Mula et al. [5]. A more specific review devoted to process and chemical industries can be found in the work by Grossmann [6–7].

These optimization modeling tools for plant location and supply chain network design have been widely used by industry to tackle complex process operations and design problems. For example, Norton and Grossmann [8] extended the multi-period optimization model to network design and production planning of a multi-product chemical process network at Dow Chemical. Terrazas-Moreno and co-workers [9–10] recently investigated the optimal design of integrated chemical production site at Dow Chemical based on a similar model introduced in this section.

REFERENCES

1. Simchi-Levi, D., Kamisky, P., Simchi-Levi, E. (2002). *Designing and managing the supply chain. Concepts, strategies and case Studies*. Irwin McGraw-Hill.
2. Shah, N. (2005) Process industry supply chains: Advances and challenges. *Comp. Chem. Eng.*, 29, 1225–1235.
3. Papageorgiou, L. G. (2009) Supply chain optimisation for the process industries: Advances and opportunities. *Comp. Chem. Eng.*, 33, 1931–1938.
4. Wassick, J. M., Agarwal, A., Akiya, N., Ferrio, J., Bury, S., You, F. Q. (2012) Addressing the operational challenges in the development, manufacture, and supply of advanced materials and performance products. *Comp. Chem. Eng.*, 47, 157–169.
5. Mula, J., Peidro, D., Díaz-Madroñero, M., Vicens, E. (2010) Mathematical programming models for supply chain production and transport planning. *Eur. J. Oper. Res.*, 204(3), 377–390.
6. Grossmann, I. (2005) Enterprise-wide optimization: A new frontier in process systems engineering. *AIChE J.* 51(7), 846–1857.
7. Grossmann, I. E. (2012) Advances in mathematical programming models for enterprise-wide optimization. *Comp. Chem. Eng.*, 47, 2–18.
8. Norton, L.C., Grossmann. I.E. (1994) Strategic planning model for complete process flexibility. *Ind. Eng. Chem. Res.*, 33, 1, 69–76.
9. Terrazas-Moreno, S., Grossmann, I.E., Wassick, J.M., Bury, S.J. (2010) Optimal design of reliable integrated chemical production sites. *Comp. Chem. Eng.*, 34, 12, 1919–1936.
10. Terrazas-Moreno, S., Grossmann, I.E., Wassick, J.M., Bury, S.J., Akiya, N. (2012) An efficient method for optimal design of large-scale integrated chemical production sites with endogenous uncertainty. *Comp. Chem. Eng.*, 37, 89–103.

17 Dynamic Optimization in Process Systems

María Soledad Díaz, Lorenz T. Biegler

CONTENTS

17.1 INTRODUCTION

Dynamic models describe many operations and processes that take place in several disciplines, including chemical engineering, economics, ecological engineering, management of communications services and aeronautics, among others. Many processes and applications in the chemical industry are intrinsically dynamic. Such processes include the operation of batch and semibatch reactors, intensively used for the production of specialty chemicals, pharmaceutical and high-value products, and polymers. For continuous processes, dynamic optimization is used in the design of distributed systems, such as plug flow reactors and packed distillation columns; as well as in the determination of optimal trajectories in the transition between operating conditions and in handling load changes. For model building of dynamic systems and model validation with experimental data, parameter estimation also requires dynamic optimization. Moreover, process control problems in chemical engineering as an example of online applications require dynamic optimization, especially in the case of multivariable systems that are nonlinear with input and output constraints. In particular, nonlinear model predictive control and dynamic real-time optimization are recent online applications of dynamic optimization.

Mathematical models describing dynamic optimization problems involve large sets of partial differential algebraic equations, with constraints on control and state variables, leading to infinite-dimensional problems. The development of robust numerical strategies, together with the increasing computational capacity has paved the way to the formulation and solution of dynamic optimization problems within key applications in chemical engineering. Therefore, dynamic optimization has become an important tool in current industrial operations and decision-making processes.

This chapter provides a general description of dynamic optimization problems and available numerical methods. These methods can be broadly classified as indirect or variational approaches and direct approaches, which can be further divided into sequential and simultaneous. In direct methods, the problem is discretized and the infinite-dimensional nature of the dynamic optimization problem is transformed into a finite-dimensional problem. Available software for both approaches is mentioned and briefly described. Finally, two typical examples in process and ecological engineering are presented. The first problem is the dynamic optimization between two operation states in a continuous stirred tank, which is solved with sequential and simultaneous strategies in gPROMS [1], and IPOPT [2] within AMPL [3], respectively. The objective is to minimize the transient between both steady states. Numerical results are presented for increasing discretization degree, with comparison of number of variables in the nonlinear problem and computational time. Different objective function weights are also explored. The second example is a parameter estimation problem for a water quality model that includes phosphorus cycle through phytoplankton, phosphate and organic phosphorus dynamics. The model is solved with a simultaneous approach with IPOPT [2] in GAMS [4].

17.2 PROBLEM DESCRIPTION

17.2.1 General Problem Formulation

A constrained optimization problem subject to a differential algebraic equation (DAE) system, with or without inequality constraints is referred to as a dynamic optimization problem or optimal control problem. This problem can be posed as follows with the DAEs (17.2–17.3) in semi explicit form:

$$min\ \Phi\left(z\left(t_f\right)\right) \tag{17.1}$$

Subject to:

$$\frac{dz}{dt} = f\left(z(t), y(t), u(t), p\right) \tag{17.2}$$

$$h\left(z(t), y(t), u(t), p\right) = 0 \tag{17.3}$$

$$g\left(z(t), y(t), u(t), p\right) \le 0 \tag{17.4}$$

$$h_P\left(z\left(t_f\right)\right)=0 \tag{17.5}$$

$$g_P\left(z\left(t_f\right)\right)\leq 0 \tag{17.6}$$

$$z\left(t_0\right)=z^o \tag{17.7}$$

$$z^{LB}\leq z(t)\leq z^{UB} \tag{17.8}$$

$$y^{LB}\leq y(t)\leq y^{UB} \tag{17.9}$$

$$u^{LB}\leq u(t)\leq u^{UB} \tag{17.10}$$

$$p^{LB}\leq p\leq p^{UB} \tag{17.11}$$

In this formulation, $t\in\left[0,t_f\right]$ is the independent variable, which represents time in most chemical engineering applications, but it can also stand for the spatial coordinate.

The DAE system (17.2–17.7) represents the system *model*. Differential equations are derived from the application of conservation principles to fundamental quantities: the rate of accumulation of a quantity within the boundaries of a system is the difference between the rate at which this quantity enters the system and the rate at which it comes out, plus the rate of its net internal production. In chemical process systems, the fundamental quantities that are being conserved are mass, energy and momentum, and conservation laws are expressed as balances on these quantities.

Algebraic equations 17.3 correspond to constitutive equations, which are generally based on physical and chemical laws. They include basic definitions of mass, energy and momentum in terms of physical properties, like density, temperature, etc.; thermodynamic equations, through equations of state and chemical and phase equilibria; transport rate equations, such as Fick's law for mass transfer, Fourier's law for heat conduction, Newton's law of viscosity for momentum transfer; chemical kinetic expressions; hydraulic equations; etc.

State differential and algebraic variables are represented by $z(t)$ and $y(t)$, respectively. The differential variables usually correspond to the fundamental quantities that are conserved, which typically include component holdups, internal energy, in chemical engineering examples. Algebraic variables are generally related to differential variables and correspond to physical and chemical properties, reaction rates, thermodynamic properties, etc.

The degrees of freedom in problem 17.1–17.11 are represented by $u(t)$ and p. The first vector, $u(t)$, depends on t and represents control or optimization variables. In chemical engineering examples, they may correspond to profiles for process stream flowrates, utility fluid flowrates, etc. Vector p corresponds to parameters, degrees of freedom that do not depend on t, which can represent the reactor volume in a batch reactor design problem or any parameter in kinetic expressions within a parameter estimation problem.

The scalar objective function Φ is calculated at the end of the integration horizon. This representation is general enough to also represent a variety of alternative situations. If the objective is a function of several state differential and algebraic variables, a new algebraic equation is added to the model, so as to define the new algebraic variable as the objective function. If the objective function Φ includes an integral term over the independent variable *t:*

$$\Phi = \int_{o}^{tf} \Psi(z(t), y(t), v(t), p)\, dt \tag{17.12}$$

it can be re-written to scalar form by defining a new differential variable z_0 replacing the integral term in the objective function and adding an additional differential equation in which the RHS is the integral term with initial condition $z_0 = 0$, as follows:

$$\Phi = z_0\left(t_f\right) \tag{17.13}$$

$$\frac{dz_0}{dt} = \Psi(z(t), y(t), v(t), p) \tag{17.14}$$

$$z_0\left(0\right) = 0 \tag{17.15}$$

Lower and upper bounds on variables are stated by *LB* and *UB*.

Point constraints correspond to conditions that must be fulfilled at a certain point in the integration horizon; they can be either equalities ($\boldsymbol{h}_p$) or inequalities ($\boldsymbol{g}_p$). They are referred to as end-point constraints if they are imposed at the end of the horizon t_f (a typical example in chemical engineering is a desired product concentration at the end of the operation in a batch process). If these additional conditions are imposed at any other point in the horizon, they are called interior-point constraints.

Path constraints $\boldsymbol{g}$ represent conditions that must be fulfilled throughout the entire integration horizon. These inequality constraints augment the algebraic equations $\boldsymbol{h}$.

For the sake of clarity, the mathematical description 17.1–17.11 assumes that the system behaviour is defined in terms of DAEs, in a semi-explicit form, where $y(t)$ can be implicitly eliminated through the algebraic variables $\boldsymbol{h}$. On the other hand, mixed lumped and distributed systems, described by general integral, partial differential and algebraic equations in time and one or more space dimensions, can be appropriately transformed into this form for the application of the numerical solution strategies described below.

Notes

1. The dynamic optimization problem has an infinite-dimensional nature and the solution strategies discussed below transform it into a finite-dimensional problem.
2. With nonlinear DAEs Problem 17.1–17.11 is nonconvex and may have multiple optima, so any solution we determine to the dynamic optimization problem is locally optimal.

17.2.2 Index of a DAE System

Initial value problems with ordinary differential equations have well defined conditions (based on Lipschitz continuity of the time derivatives) that guarantee unique solutions. Conditions for unique solutions of DAEs (17.2–17.3) are less well defined. One way to guarantee existence and uniqueness of DAE solutions is to confirm that the DAE can be converted (at least implicitly) to an initial value ODE. A general analysis of these DAE properties can be found in [5] and is beyond the scope of this chapter. On the other hand, for a workable analysis one needs to ensure a regularity condition on the DAE characterized by its index.

Practically speaking, the index of a DAE system is an integer that represents the minimum number of differentiations of (at least part of) the DAE system (with respect to the independent variable), that reduce the DAE to a pure ordinary differential equation (ODE) system for the original algebraic and differential variables. Based on this definition, pure ODE systems are index 0. For index 1 systems, differentiating algebraic equations (17.3) with *fixed* values of $u(t)$ and p, becomes:

$$(\partial h / \partial z)^T f(z, y, u, p) + (\partial h / \partial y)^T dy / dt + dh / dt = 0, \tag{17.16}$$

and if $\partial h / \partial z$ is nonsingular we can directly write a pure ODE system for $z(t)$ and $y(t)$. Index-1 systems can be solved as easily as ODEs because initial conditions are only needed for the differential variables, $z(0)$, and the algebraic variables $y(t)$ are implicit functions of $z(t)$ through (17.3). Consequently, initial values for all the differential variables may be specified freely, and we can use similar numerical methods to those for solving ODEs. However, for high-index DAEs, the algebraic equations usually lead to restrictions on the differential variables. In this case the differential variables are not independent of each other and an arbitrary specification of their initial conditions will cause ODE-type numerical methods to fail. An additional difficulty with high index DAE systems is that the usual numerical algorithms are generally incapable of controlling the error of integration, and this often leads to failure or spurious solutions.

High-index problems generally arise in chemical engineering due to the application of simplifying assumptions when modeling, such as deleting negligible dynamic terms as vapor holdup in the mass balance around a stage within a distillation column, internal energy holdup in the stage energy balance, or accumulations in reactive systems that satisfy quasi-steady state assumptions. High index DAEs may also be due to poor selection of input variables. Most high-index problems share the following features: a) differential variables do not evolve independently of each other (so they cannot have arbitrary initial values); b) algebraic variables $y(t)$ cannot be solved by algebraic equations (17.3), i.e., given differential and control variable values, high-index DAE problems cannot be reduced to ODEs by using algebraic equations to eliminate algebraic variables directly; additional manipulation and differentiation may be required. Several algorithms have been proposed for high-index problem reduction; one of these is implemented within gPROMS [1].

It is usually better to develop or reformulate the DAE process model so that it is index-1, rather than rely on algorithms to reduce the index automatically, as automated index reduction techniques can be computationally expensive. To develop index 1 models, we attempt to avoid certain simplifying assumptions or variable specifications.

Treatment of high-index problems and index reduction are beyond the scope of this chapter. Instead, detailed discussion on DAE index concepts and index reduction is provided in [6] and several examples can be found in [7].

17.3 SOLUTION STRATEGIES FOR DYNAMIC OPTIMIZATION PROBLEMS

Dynamic optimization problems, also referred to as optimal control problems, are generally nonlinear and most of them do not have analytical solutions. There are two main numerical approaches for solving dynamic optimization problems: indirect and direct methods.

17.3.1 Indirect Methods

Indirect or variational approaches are based on Pontryagin's Maximum Principle [8], in which the first-order optimality conditions are derived by applying calculus of variations. For problems without inequality constraints, the optimality conditions can be written as a set of differential algebraic equations and solved as a two-point boundary value problem. If there are inequality path constraints, additional optimality conditions are required and the determination of entry and exit points for active constraints along the integration horizon renders a combinatorial problem, which is generally hard to solve. There are several developments and implementations of indirect methods, including [9] and [10].

17.3.2 Direct Methods

Direct approaches transform problem (17.1–17.11) into a nonlinear programming (NLP) problem. In these methods, either the control or both control and state variables are approximated by appropriate functions, generally polynomials or piecewise constant parameterizations. The coefficients of the function approximations are treated as optimization variables, $x \varepsilon R^n$ and the problem is "transcribed" into a nonlinear programming problem of the form:

$$min\ F(x) \tag{17.17}$$

Subject to:

$$h(x) = 0$$

$$g(x) \le 0$$

$$x^{LB} \le x \le x^{UB}$$

The number of variables, n, and constraints in the nonlinear optimization problem depends on the type of direct method, ranging from quite small in the so-called control vector parameterization approaches to very large in the simultaneous approaches.

Direct methods can be broadly classified into control vector parameterization (CVP) and simultaneous, which are described in more detail in the following section.

17.3.2.1 Control Vector Parameterization

In these approaches, control variables are approximated by piecewise-constant, piecewise linear or, in general, polynomial functions, over a specified number of control intervals. As shown in Figure 17.1, an NLP is formulated at the outer level, with coefficients of these polynomials as optimization variables. With fixed profiles for control variables, there are no degrees of freedom and the DAE system can be

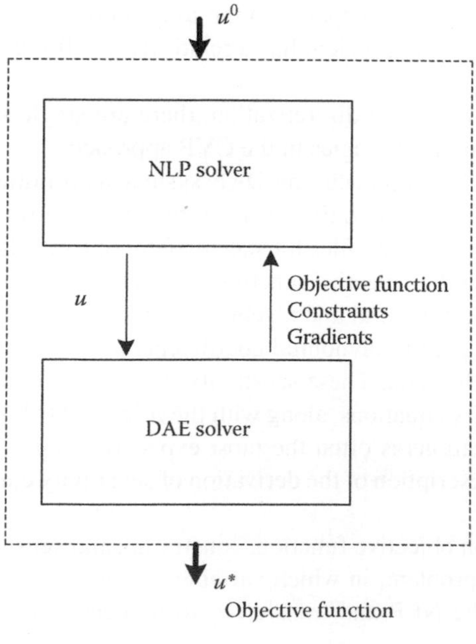

FIGURE 17.1 Scheme of control vector parameterization approach.

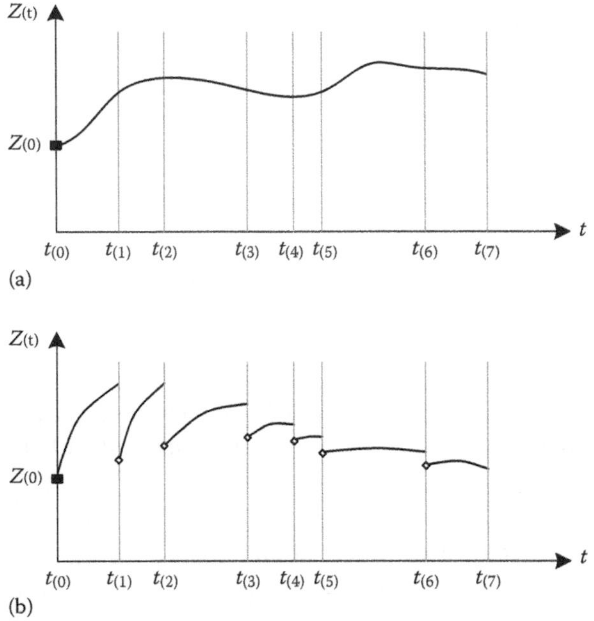

FIGURE 17.2 Control vector parameterization: (a) Single shooting (or sequential); (b) Multiple shooting.

solved in an inner level with any commercial DAE solver. As the number of control variables is usually small as compared to the total number of variables, the resulting nonlinear programming problem has a relatively small number of optimization variables.

Depending on the degree of discretization, there are single shooting (or sequential) and multiple shooting strategies in the CVP approach.

In the single shooting approach, the DAE system with initial values at $t = 0$, is integrated along the entire integration horizon, for fixed control variable profiles, as shown in Figure 17.2a. Thus, profiles for state differential and algebraic variables are determined along with the objective function value and constraints. At this step, the partial derivatives of differential and algebraic variables with respect to the optimization variables (e.g., p and polynomial coefficients for $u(t)$) have to be determined along the integration horizon. These sensitivity profiles are obtained by integrating the so-called sensitivity equations, along with the original DAE system. The solution of the sensitivity equations is often the most expensive step in the single shooting approach. Detailed description of the derivation of sensitivity equations can be found in [7], [11], and [12].

The information on objective function, constraints and sensitivities is transferred to the external NLP problem, in which variables p and polynomial coefficients for $u(t)$ are updated in the NLP and transferred to proceed with the next inner DAE system solution.

In multiple shooting, the integration horizon is divided into time intervals, with the control variables approximated by polynomials in each control interval and differential variables assigned initial values at the beginning of each interval. The DAE system is solved separately within each control interval, as shown in Figure 17.2b. Profiles for partial derivatives with respect to the optimization variables, as well as the initial conditions of the state variables in each time interval, are obtained through integration of the sensitivity equations. These state and sensitivity profiles are solved independently over each time interval and can even be computed in parallel. Additional equations are included in the NLP to enforce continuity of state variables at the time interval boundaries.

In principle, the single shooting approach is more appropriate for large DAE models (potentially involving several hundreds or thousands of differential variables), but with relatively few control variables and control intervals. On the other hand, multiple shooting should normally be preferred for problems which may require more time intervals, but with relatively few differential variables.

Moreover, multiple shooting is an intermediate strategy between sequential and simultaneous approaches. It exploits the structure of problem 17.1–17.11 and, as a result of its construction, it can handle open-loop unstable systems efficiently, which is an important advantage over single shooting. On the other hand, multiple shooting employs similar solver components as single shooting. Both DAE solver and sensitivity components can be handled by commercial DAE software, and SQP codes (such as [13], [14]) are applied for the optimization step. The complexity of multiple shooting is often favorable for problems with large integration horizons. More information on multiple shooting can be found in [15] and [16].

Control vector parameterization strategies are relatively easy to construct and to apply as they can make use of robust commercial DAE solvers as DASSL [17], DASOLV[11], DAEPACK[12] as well as NLP solvers, such as NPSOL [18], SNOPT[13], [14], IPOPT [2]. Both single and multiple shooting methods are implemented within the control vector parameterization environment in gPROMS® [1].

17.3.2.2 Simultaneous Approaches

In simultaneous methods, both control and state differential and algebraic variables are approximated in each integration interval by piecewise polynomials, whose coefficients are the optimization variables for the resulting large-scale NLP. In this way the differential-algebraic equations are discretized to algebraic equations along the integration horizon and there is no longer an embedded DAE problem. For the discretization, an efficient method such as orthogonal collocation on finite elements is widely used for the solution of DAE problems.

With collocation on finite elements, the state and control variable profiles are approximated by a family of polynomials on a time interval mesh defined by breakpoints ($t_0 < t_1 < ... < t_N = t_f$). These profile polynomials can be represented as power series, sums of orthogonal polynomials or as Lagrange interpolation polynomials.

Here, we use the following monomial basis representation for the differential profiles, which is popular for Runge-Kutta discretizations:

$$z(t) = z_{i-1} + h_i \sum_{q=1}^{K} \Omega_q\left(\frac{t - t_{i-1}}{h_i}\right)\frac{dz}{dt_{i,q}}, \quad z_i = z_{i-1} + h_i \sum_{q=1}^{K} \Omega_q(1)\frac{dz}{dt_{i,q}} \tag{17.18}$$

where z_{i-1} is the value of the differential variable at the beginning of element i, h_i is the length of element i, $dz/dt_{i,q}$ is the value of its first derivative in element i at the collocation point q, and Ω_q is a polynomial of degree K, satisfying $\Omega_q(0) = 0$ for $q = 1\ldots,K$; $\Omega'_q(\rho_r) = \delta_{r,q}$ for q, $r = 1,\ldots,K$. Here $\rho_r \in [0, 1]$ is the location of the rth collocation point within each element, which relates to $t_{ir} = t_{i-1} + h_i\rho_r$. Note that z_0, at the beginning of the first element is specified by the initial conditions for the DAE, and continuity is maintained between element i and $i + 1$ through the definition of z_i. Figure 17.3 shows the state variable approximation with piecewise cubic polynomials ($K = 3$).

Similarly, we define the control and algebraic variable profiles as:

$$u(t) = \sum_{q=1}^{K} \psi_q\left(\frac{t - t_{i-1}}{h_i}\right) u_{i,q}, \quad y_i = \sum_{q=1}^{K} \psi_q\left(\frac{t - t_{i-1}}{h_i}\right) y_{i,q} \tag{17.19}$$

where $y_{i,q}$ and $u_{i,q}$ represent the values of the algebraic and control variables, respectively, in element i at collocation point q. Here ψ_q is the Lagrange polynomial of degree K-1 satisfying $\psi_q(\rho_r) = \delta_{r,q}$, q, $r = 1,\ldots,K$. Substituting these polynomials into the differential algebraic equations leads to the following algebraic equations:

$$z_{i,q} = z_{i-1} + h_i \sum_{q=1}^{K} \Omega_q(\rho_q)\frac{dz}{dt_{i,q}}, \quad z_i = z_{i-1} + h_i \sum_{q=1}^{K} \Omega_q(1)\frac{dz}{dt_{i,q}},$$

$$\frac{dz}{dt_{i,q}} = f(z_{i,q}, y_{i,q}, u_{i,q}, p) \tag{17.20}$$

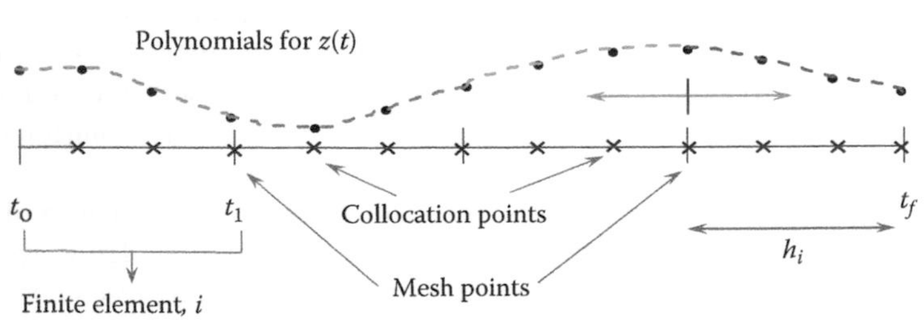

FIGURE 17.3 Orthogonal collocation over finite elements with piecewise cubic polynomials.

With this representation, the general dynamic optimization problem can be rewritten as the following NLP:

$$
\begin{aligned}
& Min\ \Phi(z_f) \\
& z_{i,q} = z_{i-1} + h_i \sum_{q=1}^{K} \Omega_q(\rho_q) \frac{dz}{dt_{i,q}},\ i=1,\ldots N,\ q=1,\ldots K \\
& \frac{dz}{dt_{i,q}} = f(z_{i,q}, y_{i,q}, u_{i,q}, p),\ i=1,\ldots N,\ q=1,\ldots K \\
& z_i = z_{i-1} + h_i \sum_{q=1}^{K} \Omega_q(1) \frac{dz}{dt_{i,q}},\ i=1,\ldots N-1 \\
& z_0 = z(0),\ z_f = z_{N-1} + h_N \sum_{q=1}^{K} \Omega_q(1) \frac{dz}{dt_{N,q}} \\
& g(z_{i,q}, y_{i,q}, u_{i,q}, p) \le 0,\ h(z_{i,q}, y_{i,q}, u_{i,q}, p) = 0 \\
& g(z_f) \le 0,\ h(z_f) = 0 \\
& z^{LB} \le z_{iq} \le z^{UB} \\
& u^{LB} \le u_{iq} \le u^{UB} \\
& y^{LB} \le y_{iq} \le y^{UB} \\
& p^{LB} \le p \le p^{UB}
\end{aligned}
\tag{17.21}
$$

Orthogonal collocation on finite elements is a special type of implicit Runge–Kutta method, and consequently, it inherits the strong stability and high order properties of these methods. The collocation points $\rho_r \in [0, 1]$ are chosen as shifted roots of Gauss-Legendre or Radau orthogonal polynomials. Radau points are preferred because they allow constraints to be set at the end of each element and they stabilize the system more efficiently if high index path constraints are present. Gauss–Legendre and Radau collocation methods are order 2K and 2K-1, respectively, are algebraically stable, and have no stability limitations on the integration step, even for stiff problems. The simultaneous approach has a number of advantages for dynamic optimization:

- Control variables are discretized at the same level as the state variables, and the Karush Kuhn Tucker (KKT) conditions of the simultaneous NLP are consistent with the optimality conditions of the discretized variational problem.
- Under mild conditions, convergence properties to the optimal control solution can be shown. Moreover, convergence rates have been derived that relate NLP solutions to the true solutions of the infinite dimensional optimal control problem.
- As with multiple shooting approaches, simultaneous approaches can deal with instabilities that occur for a range of inputs. Because they can be seen

as extensions of robust boundary value solvers, they are able to "pin down" unstable modes (or increasing modes in the forward direction). This property has important advantages for problems that include transitions to unstable points and optimization of systems with limit cycles and bifurcations.

On the other hand, the resulting NLP in simultaneous approaches is generally large and it may appear that solving the nonlinear optimization problem cannot be easier than solving the boundary-value problem in the indirect approach. However, the NLP is easier to solve, particularly in a direct collocation method, because the NLP is sparse and many well-known software programs exist (IPOPT [2], KNITRO [19], LOQO [20]) that exploit exact first and second derivative information and take advantage of sparse structures in the constraint Jacobians and Lagrange Hessian of the NLP. Additional details on simultaneous methods for dynamic optimization are given in [7].

17.4 DYNAMIC OPTIMIZATION EXAMPLES

17.4.1 Continuous-Stirred Tank Reactor (CSTR): Dynamic Optimization between Steady-States

A typical example of dynamic optimization in chemical engineering is the change between steady states in a continuous-stirred tank reactor (CSTR) in which the irreversible reaction A→B takes place ([21], [22], see Figure 17.4). The reaction is first order, exothermic and follows Arrhenius rate law. The reactor is equipped with a cooling jacket with refrigerant fluid at constant temperature T_w. To develop model equations, we formulate mass and energy balances.

Assumptions

Perfect mixing (C_A is the concentration in the outlet stream)
A and B are in dilute solution (constant density)

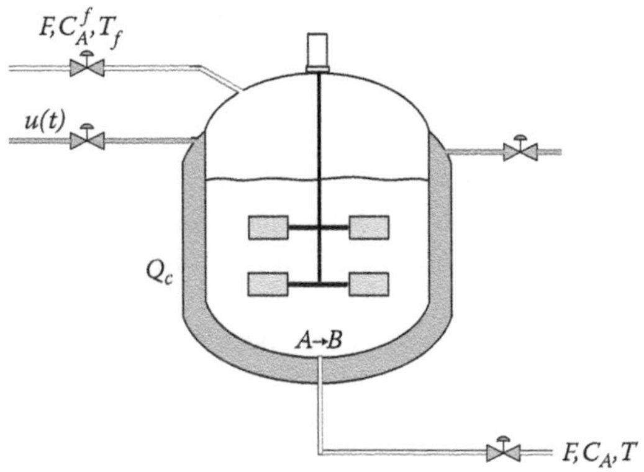

FIGURE 17.4 CSTR reactor with cooling jacket.

The mass balance for component A gives:

$$\frac{d(VC_A)}{dt} = FC_A^f - FC_A - kVC_A \tag{17.22}$$

where V, the reactor volume, F, feed flowrate and C_A^f, feed concentration, are assumed constant; k is the reaction rate constant and C_A corresponds to the concentration of A within the tank.

Defining $\theta = {}^{V}\!/_{F}$ and $C^* = {}^{C_A}\!/_{C_A^f}$, the mass balance can be re-written as:

$$\frac{dC^*}{dt} = \left(\frac{1-C^*}{\theta}\right) - kC^* \tag{17.23}$$

The reaction rate is described by Arrhenius law:

$$\frac{dC^*}{dt} = \left(\frac{1-C^*}{\theta}\right) - k_o exp\left(-{}^{E_a}\!/_{RT}\right)C^*. \tag{17.24}$$

An energy balance within the reactor is as follows:

$$\rho C_p \frac{d(VT)}{dt} = \rho C_p FT_f - \rho C_p FT + (-\Delta H)kVC_A - UA(T - T_w) \tag{17.25}$$

where the last term in the RHS corresponds to heat transfer rate to the cooling fluid, ρ and C_p are density and specific heat capacity in the fluid within the tank, respectively; $(-\Delta H)$ is the heat of reaction and T_w is the cooling fluid temperature. Assuming that the heat transfer coefficient is a linear function of the coolant flowrate, u, we can re-write the equation as follows:

$$\frac{dT}{dt} = \frac{FT_f}{V} - \frac{FT}{V} + \frac{(-\Delta H)kC_A}{\rho c_p} - \propto u(T - T_w) \tag{17.26}$$

by considering $\alpha u = \frac{UA}{\rho C_p V}$. Additionally defining $\beta = {}^{-\Delta H}\!/_{\rho C_p}$, $N = \frac{E_a}{R\beta C_A^f}$, $T^* = \frac{T}{\beta C_A^f}$, $T_w^* = \frac{T_w}{\beta C_A^f}$, $T_f^* = \frac{T_f}{\beta C_A^f}$, mass and energy balances can be re-written as follows:

$$\frac{dC^*}{dt} = \left(\frac{1-C^*}{\theta}\right) - k_o exp\left(-{}^{N}\!/_{T^*}\right)C^* \tag{17.27}$$

$$\frac{dT^*}{dt} = \frac{\left(T_f^* - T^*\right)}{\theta} + k_o exp\left(-{}^{N}\!/_{T^*}\right)C^* - \alpha u\left(T^* - T_w^*\right) \tag{17.28}$$

$$c^*(t=0)=c_0^* \tag{17.29}$$

$$T^*(t=0)=T_0^* \tag{17.30}$$

The dynamic optimization problem is formulated with the following objective function to minimize the transient between steady states:

$$min\int_0^{t_f}\left\{\omega_1\left(T^*-T_{sp}\right)^2+\omega_2\left(C^*-C_{sp}\right)^2+\omega_3\left(u-u_{des}\right)^2\right\}dt \tag{17.31}$$

where T_{sp}, C_{sp} and u_{des} correspond to desired values at the final steady-state for temperature, concentration and coolant flowrate, respectively, and ω_1, ω_2, ω_3 are weighting factors. The control variable is the coolant flowrate $u(t)$.

Values for the constant coefficients, desired values for states in the new steady state and weights are shown in Tables 17.1 and 17.2 [21].

Initial conditions for the model are:

$$C_0^*=0.7930 \tag{17.32}$$

$$T_0^*=0.1367 \tag{17.33}$$

$$u(0)=390\ (\text{L/min}) \tag{17.34}$$

Additional inequality constraints are bounds on state and control variables:

$$0\le C^*(t)\le 1.5 \tag{17.35}$$

TABLE 17.1
Scalar Parameters in CSTR Problem

Parameter	Value	Definition (Units)
T_w	0.380	Dimensionless cooling water temperature
T_f	0.395	Dimensionless feed temperature
N	5	$N=\frac{E_a}{R\beta C_A^f}$ (K)
α	1.95×10^{-4}	Coefficient for heat transfer (1/L)
θ	20	Reactor residence time (min)
k_0	300	Arrhenius pre-exponential factor (1/min)
β	100	$\beta=-\Delta H/\rho C_p$ (K.L/mol)

TABLE 17.2
Objective Function Parameters in CSTR Problem

Parameter	Value	Definition (units)
T_{sp}	0.7766	Dimensionless temperature value at final steady-state
C_{sp}	0.0944	Dimensionless conc. value at final steady-state
u_{des}	340	Coolant flowrate (control) at final steady-state (L/min)
ω_1	1×10^6	Weight coefficient for temperature
ω_2	2×10^3	Weight coefficient for concentration
ω_3	1×10^{-3}	Weight coefficient for control variable

$$0 \le T^*(t) \le 1.5 \tag{17.36}$$

$$0 \le u(t) \le 600 \tag{17.37}$$

The dynamic optimization problem 17.27–17.37 is solved with a control vector parameterization (sequential) approach in gPROMS® [1] and with a simultaneous (orthogonal collocation on finite elements) approach in AMPL [3], using IPOPT [2], as the NLP solver. As a first step, the problem is solved with increasing number of intervals (finite elements, in the simultaneous approach). Numerical results are reported in Figures 17.5–17.11 and Table 17.3. The control variable for the dynamic optimization problem is refrigerant flowrate, which is set to zero at initial time and changed to the new set point value. Figure 17.5 shows control variable profiles; it can be seen that even though the profiles have the same trend in all cases, better approximations are obtained for the finer discretization. This is achieved at higher computational cost, especially in the sequential approach, as shown in Table 17.3. We recall that the sequential strategy proceeds by solving an external NLP in which the optimization variables correspond to coefficients of the polynomials that approximate the control profiles in each time interval (piecewise constant, in this case) and an embedded DAE system that is integrated along the time horizon in each iteration of the NLP problem. Table 17.1

TABLE 17.3
Computational Time and NLP Size for Increasing Number of Control Intervals (or Finite Elements)

	Sequential approach		Simultaneous approach	
Control intervals	Number NLP variables	CPU time [s]	Number NLP variables	CPU time [s]
9	9	0,750	81	0,046
18	18	2,141	162	0,087
36	36	4,921	324	0,091
90	90	9,281	810	0,094

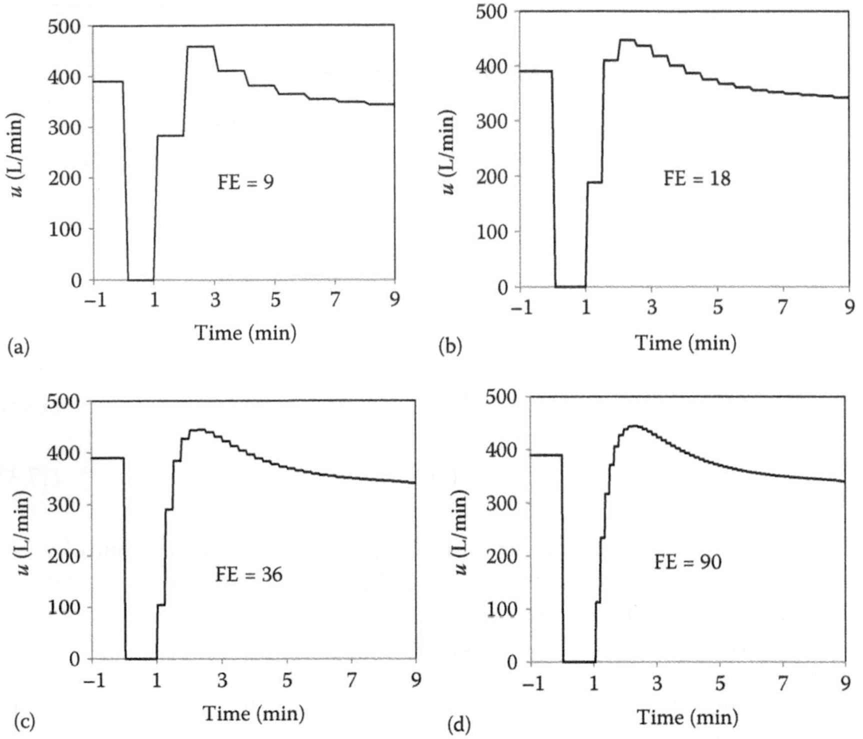

FIGURE 17.5 Control variable profiles for simultaneous approach with increasing number of finite elements: (a) 9 FE; (b) 18 FE; (c) 36 FE; (d) 90 FE.

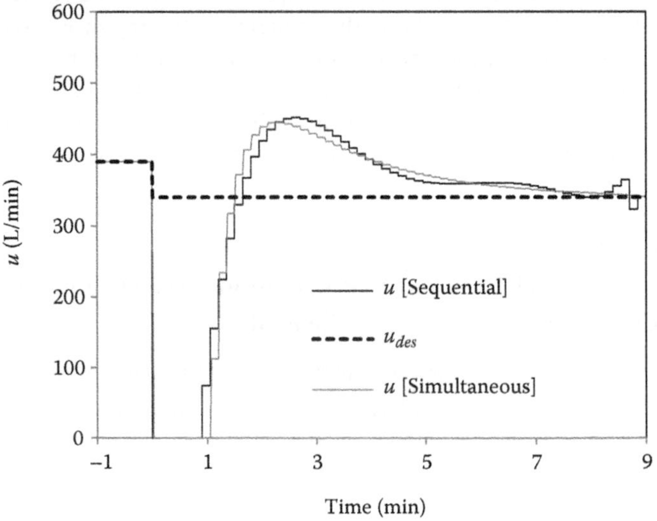

FIGURE 17.6 Optimal control profiles determined with both simultaneous and sequential approaches, 90 time intervals.

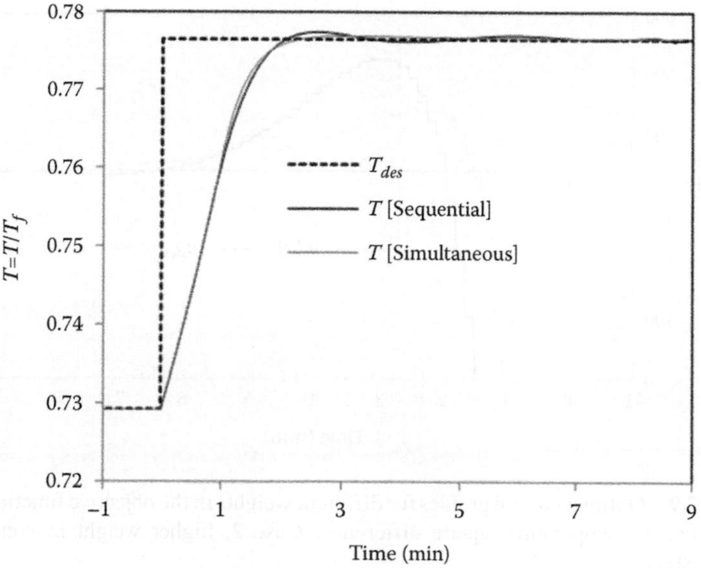

FIGURE 17.7 Optimal temperature profiles determined with both simultaneous and sequential approaches, 90 time intervals.

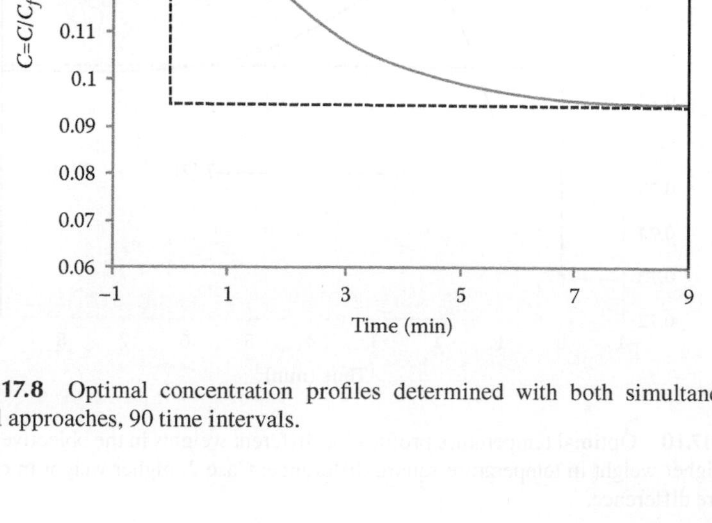

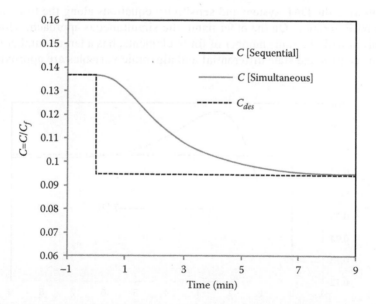

FIGURE 17.8 Optimal concentration profiles determined with both simultaneous and sequential approaches, 90 time intervals.

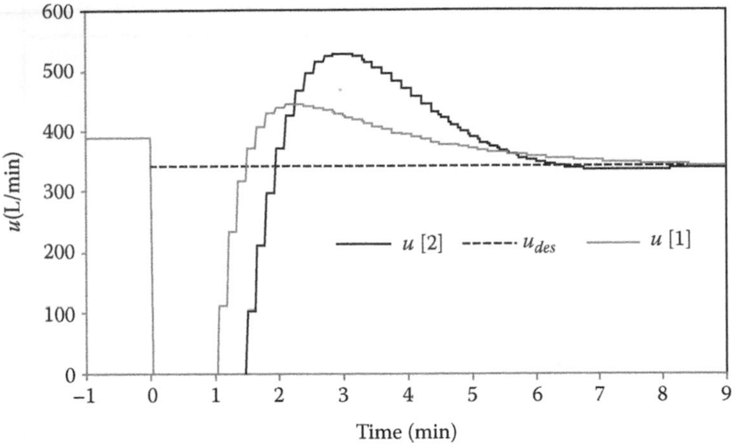

FIGURE 17.9 Optimal control profiles for different weights in the objective function: Case 1, higher weight in temperature square difference; Case 2, higher weight in concentration square difference.

shows that computational time in CVP increases from 0.75 CPU s, for 9 time intervals to 9.28 CPU s, for 60 time intervals in the sequential approach. The number of NLP iterations ranges from 11 to 16, with most of the time consumed by integrating the DAE system and sensitivity equations along the time horizon at each NLP iteration. On the other hand, the simultaneous approach, which has been solved with the same number of finite elements, has a larger number of variables, as control and state differential and algebraic variables are approximated

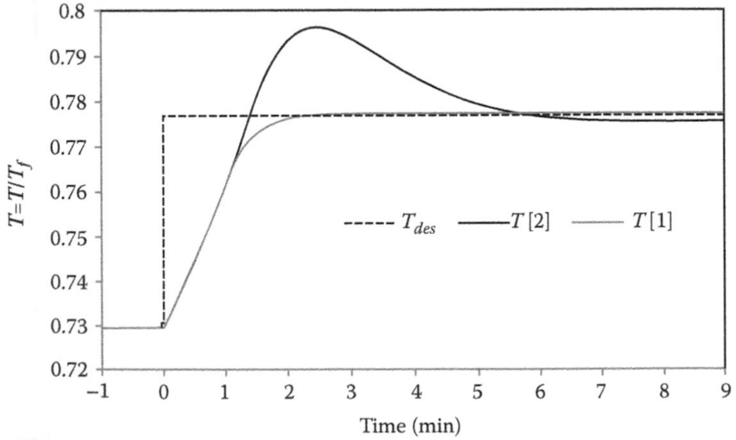

FIGURE 17.10 Optimal temperature profiles for different weights in the objective function: Case 1, higher weight in temperature square difference; Case 2, higher weight in concentration square difference.

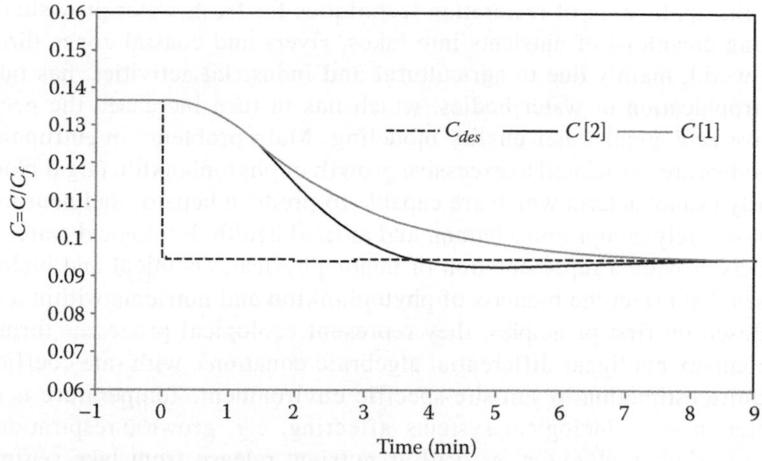

FIGURE 17.11 Optimal concentration profiles for different weights in the objective function: Case 1, higher weight in temperature square difference; Case 2, higher weight in concentration square difference.

with cubic polynomials over finite elements. However, the lack of an embedded DAE system to solve and the use of an NLP solver that takes advantage of the sparse structure of the resulting system of nonlinear equations renders very low CPU times, between 0.046 (9 finite elements) and 0.094 seconds (90 finite elements), in all cases.

Figures 17.6 to 17.8 show profiles for the control and state variables, as obtained with both the sequential and simultaneous approach. For the simultaneous approach we have imposed a constant control variable within each finite element so that profiles are comparable to those obtained with the sequential approach. It can be seen that numerical results are equivalent in both approaches.

Finally, Figures 17.9 to 17.11 show a comparison between optimal control and state variable profiles, as function of the objective function weights. In the first case, we have run the problem with $\omega_1 = 1\times10^6$ and $\omega_2 = 2\times10^3$; while in the second case, $\omega_1 = 2\times10^3$ and $\omega_2 = 1\times10^6$. It can be seen that in the first case, with higher weight on the temperature square difference, renders a much tighter temperature profile to the new steady state value, while it takes longer to reach the desired final concentration in the outlet stream. More details on the solution to this problem, including the collocation equations coded in AMPL, can be found on http://www.mcs.anl.gov/~vzavala/dynopt.html.

17.4.2 Parameter Estimation in a Water Quality Model

In this case, we present a small ecological water quality model, which has been formulated as a simplification of the model proposed by [23]. Ecological models are of interest in water management because eutrophicated water bodies require the application of restoration techniques for fresh water provision. The increasing download of nutrients into lakes, rivers and coastal zones throughout the world, mainly due to agricultural and industrial activities, has intensified eutrophication of water bodies, which has in turn increased the need for predictive ecological water quality modeling. Main problems in eutrophicated water bodies are associated to excessive growth of phytoplankton (algal blooms), especially cyanobacteria which are capable to produce hepato- and neurotoxins that can severely compromise human and animal health. Ecological water quality models provide a representation of major physical, chemical and biological processes that affect the biomass of phytoplankton and nutrients within a water body. Based on first principles, they represent ecological processes through a set of complex nonlinear differential algebraic equations, with rate coefficients that require estimation to suit site-specific environment. Temperature is a key parameter in most biological systems affecting, e.g. growth, respiration and death rate of phytoplankton, as well as nutrient release from lake sediments. The use of mathematical models becomes necessary for careful planning of restoration actions.

The present model is based on mass balances for three main components representing phytoplankton (*A*), nutrients (*B*) and organic phosphorus (*C*), which can describe phosphorus cycle within a water body as follows. Phytoplankton biomass is produced by the photosynthesis reaction, consuming nutrients (in lakes and reservoirs the limiting nutrient is phosphorus), dissolved carbon dioxide, with solar radiation and adequate temperature. Upon death, phytoplankton biomass increases the pool of organic phosphorus, which is in turn converted to phosphate by mineralization bacteria. The model has several kinetic parameters that have to be calibrated based on collected data from the specific reservoir under study.

Mass balances for components *A*, *B* and *C* are formulated as follows:

$$\frac{dC_A}{dt} = \frac{C_{A_{IN}}}{\varphi_{IN}} - \frac{C_A}{\varphi_{OUT}} + \mu_{max}\left(\theta^{T-Tref}\right)\left(\frac{C_B}{C_B + Kp}\right)C_A - r_d C_A - r_{max}\left(\frac{C_A}{C_A + Kg}\right)C_Z \tag{17.38}$$

$$\frac{dC_B}{dt} = \frac{C_{B_{IN}}}{\varphi_{IN}} - \frac{C_B}{\varphi_{OUT}} + k_m C_C - \mu_{max}\left(\theta^{T-Tref}\right)\left(\frac{C_B}{C_B + Kp}\right)C_C - r_m C_B \tag{17.39}$$

$$\frac{dC_C}{dt} = \frac{C_{C_{IN}}}{\varphi_{IN}} - \frac{C_C}{\varphi_{OUT}} + \alpha_{PC} r_d C_A - k_m C_C \tag{17.40}$$

$$T = A_1 t^3 + A_2 t^2 + A_3 t + A_4 \tag{17.41}$$

where φ_{IN} and φ_{OUT} correspond to the ratio between volume and inlet and outlet streams flowrates, respectively, which are constant throughout the entire time

TABLE 17.4
Experimental Data for Phytoplankton (*A*), Phosphate (*B*) and Organic Phosphorus (*C*) Concentrations, for Water Quality Model

Time [days]	*A*[mgC/l]	*B*[mgPO$_4$/l]	*C*[mgOP/l]
3	0.342	0.200	0.072
13	0.840	0.148	0.087
23	1.250	0.155	0.136
33	1.185	0.219	0.135
43	0.725	0.220	0.090
53	0.273	0.180	0.035
63	0.065	0.099	0.009
73	0.009	0.040	0.001
83	0.001	0.019	1.00E-04
93	2.90E-04	0.007	5.10E-06

horizon of 93 days; $C_{A_{IN}}, C_{B_{IN}}, C_{C_{IN}}$ *and* C_A, C_B, C_C are inlet and outlet concentrations for components *A*, *B* and *C*, see Table 17.4. The third term in Eqn. (17.37) stands for phytoplankton biomass (*A*) production, which is first order in *A* and follows the Monod law in nutrients (*B*) concentration, it is modified by a factor depending on temperature and μ_{max} is the maximum growth rate. The fourth term in Eqn. (17.37) stands for phytoplankton natural death and the fifth term represents phytoplankton consumption through grazing by zooplankton (*Z*), with constant zooplankton concentration (C_Z).

Equation (17.38), the mass balance for phosphate (*B*), includes a term that takes into account phosphate production from organic phosphorus (*C*) as a first order mineralization reaction (third term) in which k_m is the mineralization rate constant. Consumption is due to phytoplankton growth and uptake by macrophytes.

The mass balance for organic phosphorus (*C*) is formulated in Equation (17.40), where it can be seen that organic phosporus is produced by phytoplankton death (third term) and consumed by mineralization to phosphate (fourth term).

Water temperature profile (*T*) is represented as a polynomial in time, with coefficients A_1, A_2, A_3 and A_4 in algebraic Eqn. (17.41) and initial conditions:

$$C_A(0) = 0.2495 \text{ mgCarbon/L} \tag{17.42}$$

$$C_B(0) = 0.3700 \text{ mgPO}_4\text{/L} \tag{17.43}$$

$$C_C(0) = 0.5801 \text{ mgOP/L} \tag{17.44}$$

For parameter estimation, the scalar objective function is the summation of weighted least squares, where weights (W_l) correspond to the inverse of the variance of the error of experimental concentration data, as follows:

$$min\ \phi = \sum_{l\in\{A,B,C\}} \sum_{i=1}^{N} \frac{\left(C_l^M(t_i) - C_l(t_i)\right)^2}{W_l} \tag{17.45}$$

Here N is the total number of sampling times and $C_l^M(t_i)$ and $C_l(t_i)$ correspond to measured and predicted values for state differential variables at sampling times t_i, respectively. The predicted values are provided by the corresponding Lagrange piecewise polynomials over finite elements. The parameters to be estimated, which are the time independent optimization variables in this problem, are μ_{max} (maximum growth rate for A), k_m (organic phosphorus mineralization rate [1/day]) and K_P (half-saturation constant for B uptake [mg/L]). Table 17.5 shows scalar parameters for problem (17.38–17.45).

The dynamic parameter estimation problem is fully discretized by Radau collocation over finite elements and it is formulated in GAMS [4]. We have solved the problem with ten and thirty finite elements, respectively. State variable profiles are shown in Figures 17.12 and 17.13. It can be seen that the finer discretization gives better profiles for state variables with 574 algebraic variables in the NLP (211 variables for 10 FE) and the same in CPU time (0.032 s). The use of 10 finite elements is not enough to adjust model predictions to experimental data. Final values for model estimated parameters are as follows:

$$\mu_{max} = 3.097\,[1/\text{day}]$$
$$k_m = 0.888\,[1/\text{day}]$$
$$K_P = 0.004\,[\text{mg/L}]$$

TABLE 17.5
Scalar Parameters for Water Quality Model

Parameter	Value	Nomenclature
A_1	0.1492E-05	Polynomial Coefficient (°C/day³)
A_2	-0.4053E-03	Polynomial Coefficient (°C/day²)
A_3	-0.5593E-01	Polynomial Coefficient (°C/day)
A_4	5.326	Polynomial Coefficient (°C)
α_{PC}	0.137	Phosphorus to carbon ratio in phytoplankton biomass (mgP/mgCarbon)
r_d	0.730	Mortality rate for phytoplankton (1/day)
θ	1.090	Temperature adjustment factor (-)
$C_{AIN}, C_{BIN}, C_{CIN}$	0.000	Concentration in inlet stream for components A, B and C (mgCarbon/L)
r_{max}	0.500	Maximum grazing rate of zooplankton over phytoplankton (1/day)
r_m	0.100	Phosphate uptake rate by macrophyte (1/day)
K_g	15.00	Half-saturation constant for grazing (mgCarbon/L)
C_Z	0.100	Zooplankton concentration (mgCarbon/L)

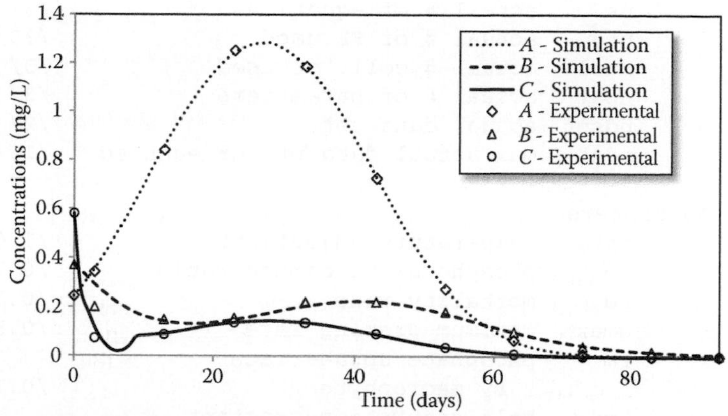

FIGURE 17.12 Predicted and experimental data for phytoplankton (*A*), phosphate (*B*) and organic phosphorus (*C*) concentrations. Number of finite elements = 10.

The GAMS code for the water quality model is as follows:

```
* Sets definition
sets  k        equation # (max 10)                    /k1*k3/
      i        finite elements # (max 20)             /i1*i30/
      j        collocation coeff. #                   /j1*j3/
      col      # possible coll pt                     /c1*c3/
      nd       # of data points per eq                /d1*d10/
      np       # of parameter                         /p1*p3/   ;

alias          (k,kp)
               (j,jp)
               (nd,ndp,nds)
               (i,ip) ;
```

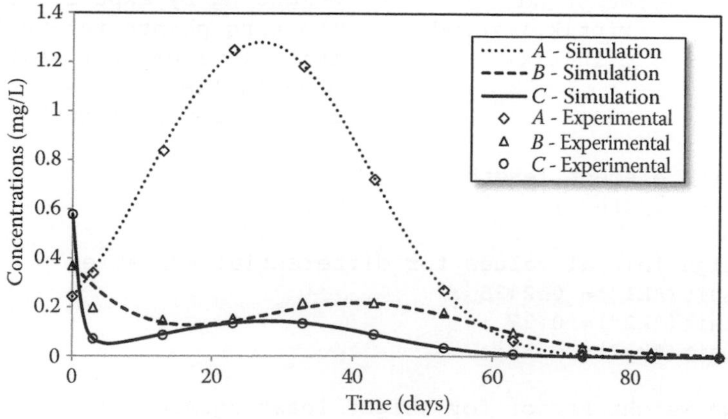

FIGURE 17.13 Predicted and experimental data for phytoplankton (*A*), phosphate (*B*) and organic phosphorus (*C*) concentrations. Number of finite elements = 30.

```
scalars      nk     actual # of equations                  /3/
             nfe    actual # of FE used                    /30/
             ncol   actual # coll. pt used                 /3/
             npar   actual # of parameters                 /3/
             ndset  actual data set                        /2/
             ndat   max actual data pt for each eq         /10/;

* Model Parameters
scalars      tita   temperature adjustment                 /1.09/
             apc    phosphorus to carbon ratio             /0.137/
             rd     mortality rate                         /0.73/
             gmax   maximum grazing rate                   /0.5/
             rm     phosphate uptake rate
                      by macrophyte                        /0.1/
             Kg     half-saturation constant
                      for grazing                          /15/
             zoo    zooplankton concentration              /0.1/
             tf     final time                             /93/;

set    sdata(k,nd)        actual dim of data points ;
       sdata(k,nd)      = yes $ ( (ord(k) le nk) $ (ord(nd) le
         ndat) ) ;

set    snedat(k,nd,i)    actual location of data on finite
         elements

set    sxint(k,i,jp,nd) actual dim of xcol starting pts ;

parameter    tau(jp)              tau at specified ncol
             tglob(k,i,jp)        tau in normalized t domain
             tglobf(k,i,jp)       tau in global t domain for
                                    functions
             wvar(k,kp)           weighting by var for lsfobj
             wexpt(nd)            weighting by expt for lsfobj
             xint(k,i,jp,nd)      starting points for xcol
             c_init(k)            initial values for state
                                    variables
             h(i)                 finite element length  ;

*   Finite element length
    h(i)= 1./nfe  ;

*   Assign initial values for differential variables
    c_init("k1")= 0.2435 ;
    c_init("k2")= 0.37 ;
    c_init("k3")= 0.58 ;

* Assign weight factor for weight least square obj func
    wvar(k,k)  = 1 ;
    wexpt(nd)  = 1 ;
```

```
Table  Acol(j,jp)  collocation matrix for three collocation
  points

              j1                  j2                  j3
 j1  0.19681547722366   0.39442431473909   0.37640306270047
 j2  -0.06553542585020  0.29207341166523   0.51248582618842
 j3  0.02377097434822   -0.04154875212600  0.11111111111111;

table          dx_aux(k,nd)     the time domain of the data

         d1          d2          d3          d4          d5
k1        3          13          23          33          43
k2        3          13          23          33          43
k3        3          13          23          33          43

 +       d6          d7          d8          d9          d10
k1       53          63          73          83          93
k2       53          63          73          83          93
k3       53          63          73          83          93   ;

parameter      dx(k,nd)     the normalized time domain of the
  data ;

dx(k,nd) $ sdata(k,nd) = dx_aux(k,nd)/sum(ndp $ (ord(ndp) eq
  ndat), dx_aux(k,ndp)) ;

table            dy(k,nd)      experimental data

        d1          d2          d3          d4          d5
k1      0.342       0.840       1.250       1.185       0.725
k2      0.200       0.148       0.155       0.219       0.220
k3      0.072       0.087       0.136       0.135       0.090

 +      d6          d7          d8          d9          d10
k1      0.273       0.065       0.009       0.001       2.90e-4
k2      0.180       0.099       0.040       0.019       0.007
k3      0.035       0.009       0.001       1.0e-4      5.1e-6 ;

* Assign tau
   tau("j1")= 0.15505102572168 ;
   tau("j2")= 0.64494897427832 ;
   tau("j3")= 1.;

* Calculate tau in overall t domain
   tglob(k,i,jp)   = sum (ip $ (ord(ip) lt ord(i)), h(ip)) +
     tau(jp)*h(i) ;
   tglobf(k,i,jp) = sum(ndp $ (ord(ndp) eq ndat), dx_
     aux(k,ndp)) *
   (sum (ip $ (ord(ip) lt ord(i)), h(ip)) +
     tau(jp)*h(i)) ;
```

```
variables      yres(k,nd)    residual term (dy-y) at dx
               xcol(k,i,jp)  collocation coefficients
               par(np)       parameters to be estimated (degrees
                               of freedom)
               lsfobj        least square objective function
               cdot(k,i,jp)  first derivatives of state
                               variables   ;

equations      eyres(k,nd)      equation of residual term (y-dy)
                                  at data points
                  elsobj      least squares objective function
                  eqCAdot(i,j)  state first order derivatives
                    for A concentration
                  eqCBdot(i,j)  state first order derivatives
                    for B concentration
                  eqCCdot(i,j)  state first order derivatives
                    for C concentration
                  colCA(i,j)    collocation equations for comp A
                    concentration
                  colCB(i,j)    collocation equations for comp B
                    concentration
                  colCC(i,j)    collocation equations for comp C
                    concentration
                  colCAi(j)    collocation equations for A conc
                    (initial element)
                  colCBi(j)    collocation equations for B conc
                    (initial element)
                  colCCi(j)    collocation equations for C conc
                    (initial element) ;

* Define set snedat
snedat(k,nd,i) $    ( (ord(k) le nk) $ (ord(nd) le ndat)
                      $(dx(k,nd) ge h(i)*(ord(i)-1)) $
                      (dx(k,nd) lt h(i)*(ord(i)) )) = yes ;

snedat(k,nd,i) $    ( (ord(k) le nk) $ (ord(nd) le ndat)
  $(ord(i) eq nfe) $ (dx(k,nd) eq h(i)*(ord(i))) ) = yes ;

* Equations
* Residuals for objective function

eyres(k,nd) $ sdata(k,nd) ..

    yres(k,nd) =e=  ( dy(k,nd) - sum(i $ snedat(k,nd,i), sum(j
                      $ (ord(j) le ncol) , xcol(k,i,j)*
                      prod(jp $ (ord(jp) ne ord(j) and ord(jp)
                      le ncol), ((dx(k,nd) - tglob(k,i,jp))/
                      (tglob(k,i,j) - tglob(k,i,jp)))))));

* State first order derivatives
```

```
eqCAdot(i,j)..  cdot("k1",i,j) =e= ((par('p1') *
  tita**((0.1492592e-5*tglobf("k1",i,j)**3 -
  0.4053333e-3*tglobf("k1",i,j)**2 -
    0.5592667e-1*tglobf("k1",i,j) + 5.326) - 20)*(
    xcol("k2",i,j)/ (xcol("k2",i,j) + par('p3')) ) - rd )
    *xcol("k1",i,j) - (gmax * (xcol("k1",i,j)/
    (xcol("k1",i,j)+kg)) * zoo)) ;

eqCBdot(i,j)..  cdot("k2",i,j) =e= (par('p2') * xcol("k3",i,j)
   - par('P1')*tita**((0.1492592593e- 5*tglobf("k2",i,j)**3
   - 0.4053333333e-3*tglobf("k2",i,j)**2
   -0.5592666667e-1*tglobf("k2",i,j) + 5.326) - 20 ) *
     xcol("k2",i,j) * xcol("k3",i,j) / (xcol("k2",i,j) +
       par('p3')) - rm*xcol("k2",i,j) ) ;

eqCCdot(i,j)..  cdot("k3",i,j) =e= rd*apc*xcol("k1",i,j)
   - par('p2')*xcol("k3",i,j);

* Collocation equations

colCA(i,j) $ ((ord(i) ne 1)).. xcol("k1",i,j) =e= xcol("k1",
  i-1,"j3") + tf*h(i)*sum(jp, Acol(jp,j)*cdot("k1",i,jp) );

colCB(i,j) $ ((ord(i) ne 1)).. xcol("k2",i,j) =e= xcol("k2",
  i-1,"j3") + tf*h(i)*sum(jp, Acol(jp,j)*cdot("k2",i,jp) );

colCC(i,j) $ ((ord(i) ne 1)).. xcol("k3",i,j) =e= xcol("k3",
  i-1,"j3") + tf*h(i)*sum(jp, Acol(jp,j)*cdot("k3",i,jp) );

*Initial element

colCAi(j)..    xcol("k1","i1",j) =e=    c_init("k1") +
  tf*h("i1")*sum(jp, Acol(jp,j)*cdot("k1","i1",jp) );

colCBi(j)..    xcol("k2","i1",j) =e=    c_init("k2") +
  tf*h("i1")*sum(jp, Acol(jp,j)*cdot("k2","i1",jp) );

colCCi(j)..    xcol("k3","i1",j) =e=    c_init("k3") +
  tf*h("i1")*sum(jp, Acol(jp,j)*cdot("k3","i1",jp) );

* Weighted least square objective function

elsobj..       lsfobj =e= sum( nd , wexpt(nd)
  *sum((k,kp),(yres(k,nd)*wvar(k,kp)*yres(kp,nd)) ) ) ;

* Bounds on variables
    yres.lo(k,nd) $ sdata(k,nd)   = -100 ;
    yres.up(k,nd) $ sdata(k,nd)   =  100 ;
    xcol.lo(k,i,jp)               = -1e-2 ;
    xcol.up(k,i,jp)               = 10 ;
    par.lo('p1')                  = 1 ;
```

```
    par.up('p1')                = 5 ;
    par.lo('p2')                = 0.1 ;
    par.up('p2')                = 2.0 ;
    par.lo('p3')                = 4e-3 ;
    par.up('p3')                = 1e-1 ;

* Starting points for collocation coefficients
*  Define set sxint
    sxint(k,i,jp,nd)     $ ((ord(nd) eq 1) $ (tglob(k,i,jp) ge
      0.) $ (tglob(k,i,jp) lt dx(k,nd)) ) = yes ;

    sxint(k,i,jp,nd)    $ ((ord(nd) le ndat) $ (tglob(k,i,jp)
      ge dx(k,nd)) $ (tglob(k,i,jp) lt dx(k,nd+1)) ) = yes ;

    sxint(k,i,jp,nd)     $ ((ord(nd) eq ndat) $ (ord(i) eq nfe)
      $ (tglob(k,i,jp) ge dx(k,nd))$ (tglob(k,i,jp) le
      h(i)*(ord(i))) ) = yes ;

*  Define parameter xint
    xint(k,i,jp,nd) $ sxint(k,i,jp,nd) = ( sum( nds $ ( (ord
      (nds) ge ord(nd)) $(ord(nds) le (ord(nd)+2)) ) ,
    dy(k,nds) *prod(ndp $ ( (ord(ndp) ne ord(nds))$ (ord(ndp)
      ge ord(nd))$ (ord(ndp) le (ord(nd)+2)) ) ,
    (tglob(k,i,jp)-dx(k,ndp))/(dx(k,nds)-dx(k,ndp))) ) ) $ (
      ((ord(nd)+1) le ndat) $ ((ord(nd)+2) le ndat) )
    +( sum( nds $ ( (ord(nds) ge (ord(nd)-1)) $ (ord(nds) le
      (ord(nd)+1)) ) , dy(k,nds) * prod(ndp $ ( (ord(ndp)
    ne ord(nds)) $ (ord(ndp) ge (ord(nd)-1)) $ (ord(ndp) le
      (ord(nd)+1)) ) , (tglob(k,i,jp)-dx(k,ndp))/
    (dx(k,nds)-dx(k,ndp))) ) ) $ ( ((ord(nd)+1) le ndat) $
      ((ord(nd)+2) gt ndat) ) + ( sum( nds $ ( (ord(nds) ge
    (ord(nd)-2)) $(ord(nds) le ord(nd)) ) , dy(k,nds) *
      prod(ndp $ ( (ord(ndp) ne ord(nds)) $ (ord(ndp)
    ge (ord(nd)-2)) $ (ord(ndp) le ord(nd)) ) ,
      (tglob(k,i,jp)-dx(k,ndp))/ (dx(k,nds)-dx(k,ndp))) ) )
    $ ( ((ord(nd)+1) gt ndat) $ ((ord(nd)+2) gt ndat) ) ;

* Assign initial xcol
    xcol.l(k,i,jp) = sum(nd, xint(k,i,jp,nd) ) ;

* Assign initial yres
    yres.l(k,nd) =  dy(k,nd) - sum(i $ snedat(k,nd,i), sum(j $
      (ord(j) le ncol) , xcol.l(k,i,j)*prod(jp $ (ord(jp) ne
      ord(j) and ord(jp) le ncol), ((dx(k,nd) -
      tglob(k,i,jp))/(tglob(k,i,j) - tglob(k,i,jp))) ) ));

    cdot.l("k1",i,j) = ((par.l('p1') * tita**((0.1492592593e-
      5*tglobf("k1",i,j)**3
     - 0.4053333333e-3*tglobf("k1",i,j)**2 - 0.5592666667e-1*
       tglobf("k1",i,j) +
```

```
    5.326) - 20) * ( xcol.l("k2",i,j) / (xcol.l("k2",i,j) +
      par.l('P3')) ) - rd )
    *xcol.l("k1",i,j) - (gmax * (xcol.l("k1",i,j)/
       (xcol.l("k1",i,j)+kg)) * zoo)) ;

   cdot.l("k2",i,j) = (par.l('p2') * xcol.l("k3",i,j) -
     par.l('p1') * tita**(
    (0.1492592593e-5*tglobf("k2",i,j)**3 - 0.4053333333e-
      3*tglobf("k2",i,j)**2 -
    0.5592666667e-1*tglobf("k2",i,j) + 5.326) - 20 ) *
      xcol.l("k2",i,j) *
    xcol.l("k3",i,j) /(xcol.l("k2",i,j) + par.l('P3')) -
      apc*xcol.l("k2",i,j) ) ;

    cdot.l("k3",i,j) = apc*xcol.l("k1",i,j) - par.l('P2') *
      xcol.l("k3",i,j);

* Assign initial parameters
    par.l(np) = 0.5*par.lo(np) + 0.5*par.up(np) ;

* Assign initial value for the objective function
    lsfobj.l = sum( nd, wexpt(nd) *sum( (k,kp) ,
      (yres.l(k,nd)*wvar(k,kp)*yres.l(kp,nd)) ) ) ;

    model water /all/ ;

    option nlp = CoinIpopt ;

    option limcol = 1000 ;
    option limrow = 1000 ;

    solve water using nlp minimizing lsfobj ;
```

17.5 CONCLUSIONS

In this chapter, we have given a general overview of dynamic optimization problems and applications in chemical engineering and other disciplines such as ecological engineering. Dynamic optimization is used both in offline and online applications. In the case of online applications, computations are time-limited, enforcing the need for efficient optimization algorithms and solvers, particularly for large-scale systems that render NLPs with thousands of variables. A brief description of current numerical solution methods has been presented, as well as their advantages and difficulties. We have mainly focused on control vector parameterization and simultaneous approaches and described the main environments in which these methods can be implemented, as well as the most efficient NLP and DAE algorithms and solvers. The concept of index of a DAE system has been introduced and methods for index reduction have been described and cited.

Finally, we present two examples in which parameter estimation and optimal transition between different operating conditions are solved. Finally, a comparison of numerical results, solution time and number of variables for the resulting NLPs are provided.

REFERENCES

1. g-PROMS, PSEnterprise (2011), User guide. http://www.psenterprise.com.
2. Waechter, A., Biegler, L.T. (2006) On the Implementation of an Interior Point Filter Line Search Algorithm for Large-Scale Nonlinear Programming, *Math. Program.*, 106, 1, 25–57. (http://projects.coin-or.org/Ipopt).
3. Fourer R., Gay D. M. and Kernighan B. W. (2012) *AMPL: A Modelling Language for Mathematical Programming.* 2nd edition, Duxbury Press, Cole Publishing Company.
4. Brooke, Kendrick, Meeraus and Raman (2012) *GAMS: A User's Guide.* GAMS Development Cooperation, Washington DC.
5. Kunkel, P., V. Mehrmann (2006) *Differential-Algebraic Equations, Analysis and Numerical Solution.* EMS Publishing House, Zurich, Switzerland.
6. Ascher, U.M., L.R. Petzold (1998) *Computer Methods for Ordinary Differential Equations and Differential-Algebraic Equations.* SIAM, Philadelphia, PA.
7. Biegler, L. T. (2010) *Nonlinear Programming: Concepts, Algorithms and Applications to Chemical Processes.* SIAM, Philadelphia, PA.
8. Pontryagin, V., Y. Boltyanskii, R. Gamkrelidzeand E. Mishchenko (1962) *The Mathematical Theory of Optimal Processes.* Interscience Publishers, New York.
9. vonStryk, O., R. Bulirsch (1992) Direct and Indirect Methods for Trajectory Optimization, *Ann. Oper. Res.*, 37, 357–373.
10. Oberle, H.J., W. Grimm (2001) *BNDSCO: A Program for the Numerical Solution of Optimal Control Problems.* Inst. für Angewandte Mathematik, Hamburg, Germany.
11. Jarvis R.B., C.C. Pantelides (1992) *DASOLV – A Differential-Algebraic Equation Solver.* Technical Report, Centre for Process Systems Engineering, Imperial College, London, United Kingdom.
12. Tolsma, J., P. Barton (2000) DAEPACK: An Open Modeling Environment for Legacy Models, *Ind. Eng. Chem. Res.*, 39, 1826–1839.
13. Gill, P. E.; W. Murray, M.A. Saunders (2005) SNOPT: An SQP Algorithm for Large-Scale Constrained Optimization, *SIAM Rev.*, 47, 99–131.
14. Gill, P.E., W. Murray, M.A. Saunders (1996) *User's Guide for SNOPT: A FORTRAN Package for Large-Scale Nonlinear Programming.* Technical Report SOL96-0, Department of Mathematics, University of California, San Diego, CA.
15. Bock, H.G. (1978) Numerical Solution of Nonlinear Multipoint Boundary Value Problems with Applications to Optimal Control, *Zeitschriftfuer Angewandte Mathematik und Mechanik*, 58, 407.
16. Leineweber, D. B., I. Bauer, H.G. Bock, J.P. Schloeder (2003) An Efficient Multiple Shooting Based Reduced SQP Strategy for Large-Scale Dynamic Process Optimization, *Comp. Chem. Eng.*, 27, 157–166.
17. Petzold, L. R. (1982) *A Description of DASSL: A Differential/Algebraic System Solver.* Technical Report SAND 82-8637, Sandia National Laboratory, Livermore, CA.
18. Gill, P. E., W. Murray, M. A. Saunders, M. H. Wright (1986) *User's Guide for NPSOL. (Version 4.0): A Fortran Package for Nonlinear Programming.* Report SOL 86-2, Department of Operations Research, Stanford University, Stanford, CA.
19. Byrd, R., J. Nocedal, R. Waltz (2006) KNITRO: An integrated package for non-linear optimization. In: G. Di Pillo and M. Roma, editors, *Large-Scale Nonlinear Optimization*, 35–60. Springer, Berlin, Germany.

20. Benson, H.Y., R.J. Vanderbei (1998) *Using LOQO to Solve Second-Order Cone Programming Problems.* Technical Report SOR-98-09, Department of Operations Research and Financial Engineering, Princeton University, Princeton, NJ.
21. Hicks, G.A., W.H. Ray (1971) Approximation Methods for Optimal Control Synthesis, *Can. J. Chem. Eng.*, 49, 522–528.
22. Flores Tlacuahuac, A., S. Terrazas Moreno, L.T. Biegler (2008) On Global Optimization of Highly Nonlinear Dynamic Systems, *Ind. Eng. Chem. Res.*, 47(8), 2643–2655.
23. Estrada V., E. Parodi, M.S. Diaz (2009) Determination of Biogeochemical Parameters in Eutrophication Models as Large Scale Dynamic Parameter Estimation Problems, *Comp. Chem. Eng.*, 33, 1760–1769.

18 Optimization in Chemical and Biological Engineering using Julia

Jordan Jalving, Victor M. Zavala

CONTENTS

18.1 INTRODUCTION

Scientific computing environments are essential for training engineers on how to construct and solve optimization problems. Matlab is (by far) the most widely used environment in engineering education and practice. Reasons for this are Matlab's interpreted language (which makes it user-friendly) and the fact that it provides a flexible environment to construct complex computing workflows that merge diverse computing tasks (e.g., data processing, matrix manipulations, dynamic simulations, plotting/visualization). Unfortunately, algebraic modeling capabilities in Matlab are limited and this has hindered its ability to easily express complex optimization problems. Specifically, algebraic modeling is key to facilitate compact representations of optimization problems and to compute derivative information. On the other hand, environments such as GAMS [1], AMPL [2], and gPROMS [3] provide scalable and easy-to-use algebraic modeling capabilities but they use self-contained environments that are difficult to embed in computational workflows. Moreover, these environments use proprietary languages that cannot be easily extended by users. Open-source programming languages such as Python [4] and Julia [5] provide user-friendly syntax and, since the languages are open, they are also extensible. Extensibility has facilitated the recent development of powerful algebraic modeling capabilities such as Pyomo and JuMP [6, 7] and of other specialized extensions [8, 9, 10]. Compact syntax and flexibility provided by Python and Julia can greatly facilitate engineering education and the construction of sophisticated computing workflows.

In this chapter we discuss capabilities available in Julia that facilitate the construction and solution of optimization problems that arise in chemical and biological engineering applications. We begin by providing examples that highlight basic syntax and features and we then discuss specific applications that arise from design, control, and estimation. The content is tutorial in nature and is targeted towards undergraduate/graduate students and practitioners.

18.2 SYNTAX AND BASIC FEATURES

We begin our exposition by presenting a Julia script that solves the simple optimization problem:

$$\min \varphi = x_1 + x_2^2 \tag{18.1a}$$

$$\text{s.t.}\, x_1 + x_2 = p \tag{18.1b}$$

$$x_1/x_2 = 1 \tag{18.1c}$$

$$x_1 \geq 0, x_2 \geq 0. \tag{18.1d}$$

Here, p is a parameter that is set to two. This problem is a nonlinear program and has a single unique solution $x_1 = x_2 = 1$ with optimal objective value $\phi = 2$. Note that the nonlinear constraint (18.1c) can also be written in linear form as $x_1 = x_2$ (but we leave this in nonlinear form for illustration purposes). The Julia script that solves problem (18.1) using Julia is presented in Figure 18.1.

The above script contains all the basic elements that are needed to solve an optimization problem using the algebraic modeling language JuMP. We first load the necessary packages; in this case, we require the algebraic modeling language JuMP that provides syntax to express the problem in algebraic form and we also require Ipopt, which is a numerical solver for nonlinear optimization problems that is interfaced to JuMP. We then proceed by defining a JuMP modeling *object* that we call m (but any name is possible); this object will be populated using different attributes that define the problem (variables, constraints, objective, solver, and so on). The numerical solver to be used to solve the model object is Ipopt (this is specified directly in the definition of the object). Here, we can also communicate algorithmic options to Ipopt (e.g., the maximum number of iterations is 100 and the solution tolerance is 1×10^{-8}). We then proceed to add the *variable* attributes to the model object and note that we can embed variable bounds directly in the definition. The syntax start=1 indicates that the variable will be initialized at a value of one (if this is not specified, JuMP or Ipopt can take an arbitrary value within the bounds specified). Needless to say, initializing variables is very important in nonlinear problems (e.g., the constraint $x_1/x_2 = 1$ cannot be evaluated if x_2 is initialized at $x_2 = 0$). We then proceed to define the *constraint* attributes; here, we note that nonlinear constraints are defined using NLconstraint (if they are linear then one uses constraint) and note that equality constraints require a double equal sign ==. Finally, we define the *objective* attribute and note that NLobjective is used

```
# load necessary packages
using JuMP, Ipopt

# define JuMP object model
m = Model(solver=IpoptSolver(max_iter=100,tol=1e-8))

# define parameter data
p = 2

# define variables
@variable(m,x1>=0)
@variable(m,x2>=0,start=1)

# define constraints
@constraint(m, x1+x2 == p)
@NLconstraint(m, x1/x2 == 1)

#define objective function
@NLobjective(m, Min, x1+x2^2)

# call Ipopt to solve model
status=solve(m)
# display results
println("\nstatus = ",status)
println("x1 = ",getvalue(x1))
println("x2 = ",getvalue(x2))
println("ϕ = ",getobjectivevalue(m))
```

FIGURE 18.1 Julia implementation of problem (18.1).

because the objective is nonlinear (objective would be used if it is linear). We also note that the syntax Min indicates that the objective is to be minimized (use Max if the objective is to be maximized). The model object now has all the attributes needed for it to be solved. Consequently, we proceed to solve the model by using solve(m); this command calls the model processing capabilities of JuMP and the numerical solver Ipopt. Model processing tasks include the automatic computation of derivatives (first and second order) for the objective and constraints. Derivative information is essential to achieve efficient performance of solvers such as Ipopt. Notably, the user does not have to provide derivatives to the solver (this is an important benefit of using algebraic modeling languages). Once the problem has been solved, we use the Julia print command println to display the solution in the command line. Here, note that the syntax getvalue(x1) is needed to extract the actual numerical value of the JuMP variable attribute x_1. This syntax is necessary because x_1 is a complex variable object (as opposed to p, which is an actual scalar value). The objective value that is output from the solver is displayed using the command getobjectivevalue(m). If the getvalue commands are called before calling solve(m) then the value reported will be the initial guess of the variables (this is a useful way of verifying that the variables are properly initialized).

We highlight that, by default, all variables are treated as continuous. In our example, if we want to enforce that variable x_1 is binary (i.e., can only take a value of

TABLE 18.1
Examples of Optimization Solvers Available in Julia

Solver	Open-Source	Julia Package	LP	MILP	NLP	MINLP
GLPK	Yes	GLPKMathProgInterface.jl	x	x		
Gurobi	No	Gurobi.jl	x	x		
Ipopt	Yes	Ipopt.jl	x		x	
SCIP	Yes	SCIP.jl		x		x
BARON	No	BARON.jl			x	x

zero or one) we use the syntax @variable(m,x1,Bin), and if we want it to be any positive integer (i.e., 0,1,2,...) we use @variable(m,x1>=0,Int). To handle such problems, the solver to be used needs to be able to handle integer variables (solvers such as Gurobi and SCIP can be used with JuMP). Table 18.1 provides an overview of some open-source and commercial solvers available in Julia to handle different problems. Linear programs (LP) contain all continuous variables and linear constraints, mixed-integer linear programs (MILP) introduce integer variables, nonlinear programs (NLP) introduce nonlinear constraints, and mixed-integer nonlinear programs (MINLP) contain both nonlinear constraints and integer variables. In this chapter we focus exclusively on nonlinear optimization problems (NLP) with continuous variables (which can be solved with Ipopt). The capabilities discussed are sufficient to help the reader handle problems with integer variables as well.

The output of the Julia script in Figure 18.1 is shown in Figure 18.2. This output contains information from Ipopt that reports the number of variables and constraints, iteration history, number of iterations, solution time and final status (e.g., optimal solution, infeasible solution, maximum number of iterations reached). From the output of the solver we see that the problem has two variables and two equality constraints (i.e., the problem does not have any degrees of freedom). We also see that the solver converges to the solution in one iteration. Finally we see the output of lines 25–27 of the Julia script, which print the solution and optimal objective value to the command line.

A key capability of algebraic modeling packages is the notion of *index sets*. Set notation enables the expression of models using compact expressions that resemble mathematical representations (as written in paper). For instance, in the above example, the variables x_1 and x_2 can also be defined as x_k, $k \in \mathrm{K}$ with set $\mathrm{K} = \{1,2\}$. By using index sets, one can also express operations in compact mathematical form. For instance, the constraint $x_1 + x_2 = 2$ can be expressed as $\sum_{k \in \mathrm{K}} x_k = 2$. Set notation also enables the definition of string indexes. For instance, in JuMP, we can define a set K={"myvar1","myvar2"}, in which case the variables will have names x_{myvar1} and x_{myvar2}. The Julia script shown in Figure 18.3 expresses the optimization problem (18.1) using set notation. Here, we highlight that the definition of the variables and access to their values simplifies because one can express the entire variable array directly (as opposed to element by element). In this Julia script we also observe the use of syntax for specifying a sum over the elements of the set (line 14). We also highlight that Julia and JuMP accept unicode (such as

```
******************************************************************************
This program contains Ipopt, a library for large-scale nonlinear optimization.
 Ipopt is released as open source code under the Eclipse Public License (EPL).
         For more information visit http://projects.coin-or.org/Ipopt
******************************************************************************

This is Ipopt version 3.12.1, running with linear solver mumps.
NOTE: Other linear solvers might be more efficient (see Ipopt documentation).

Number of nonzeros in equality constraint Jacobian...: 4
Number of nonzeros in inequality constraint Jacobian.: 0
Number of nonzeros in Lagrangian Hessian.............: 4

Total number of variables............................: 2
                     variables with only lower bounds: 2
                variables with lower and upper bounds: 0
                     variables with only upper bounds: 0
Total number of equality constraints.................: 2
Total number of inequality constraints...............: 0
        inequality constraints with only lower bounds: 0
   inequality constraints with lower and upper bounds: 0
        inequality constraints with only upper bounds: 0

iter objective inf_pr inf_du lg(mu) ||d|| lg(rg) alpha_du alpha_pr ls
   0 1.0100000e+00 9.90e-01 1.00e+00 -1.0 0.00e+00 - 0.00e+00 0.00e+00 0
   1 2.0000000e+00 2.22e-16 9.70e+01 -1.7 9.90e-01 - 1.01e-02 1.00e+00h 1

Number of Iterations....: 1

Number of objective function evaluations = 2
Number of objective gradient evaluations = 2
Number of equality constraint evaluations = 2
Number of inequality constraint evaluations = 0
Number of equality constraint Jacobian evaluations = 2
Number of inequality constraint Jacobian evaluations = 0
Number of Lagrangian Hessian evaluations = 1
Total CPU secs in IPOPT (w/o function evaluations) = 0.216
Total CPU secs in NLP function evaluations = 0.052

EXIT: Optimal Solution Found.

status = Optimal
x1 = 1.0
x2 = 0.9999999999999999
φ = 1.9999999999999998
```

FIGURE 18.2 Output of Julia script for solving example problem (18.1).

Greek letters) to create strings and to name arrays, parameters, objects, and variables. This feature can facilitate readability and consistency with mathematical representations of the problem (line 26).

Imagine now that you would like to solve problem (18.1) for different values of the parameter p in the domain [1,10] (to perform sensitivity analysis). To do so, we can implement a function that creates a JuMP model for a given value of p and then we can execute a loop that runs these models over a given set of points in [1,10]. The Julia script that accomplishes this task is given in Figure 18.4. Some additional features of Julia are highlighted in this script. First, note that the definition of the function follows a similar syntax to that of Matlab and that the output of the function is a JuMP model itself (abstract objects can be easily manipulated in Julia). Note also that syntax to create and operate with arrays and plots is similar to that of Matlab.

```
# load necessary packages
using JuMP, Ipopt

# define JuMP object model
m = Model(solver=IpoptSolver())

# define set
K=["myvar1","myvar2"]

# define variables
@variable(m,x[K]>=0,start=1)

# define constraints
@constraint(m, sum(x[k] for k in K) == 2)
@NLconstraint(m, x["myvar1"]/x["myvar2"] == 1)

#define objective function
@NLobjective(m, Min, x["myvar1"]+x["myvar2"]^2)

# call Ipopt to solve model
status=solve(m)

# display results
println("status = ",status)
println("x = ",getvalue(x))
println("ϕ = ",getobjectivevalue(m))
```

FIGURE 18.3 Julia implementation of example problem (18.1) using set indexes.

Here, plotting is performed using the PyPlot package but other packages such as Plots and Gadfly are available. The script generates a plot that is saved to the file plot.pdf. This plot is also shown in Figure 18.4.

Julia scripts can be executed on the public web portal http://juliabox.com without having to install Julia in a local machine. Juliabox also enables the creation of Jupyter *notebooks*, which are editable files that embed code, documentation, and output. Moreover, the Juliabox portal has preinstalled copies of common packages such as JuMP, Ipopt, and Pyplot (and many others). All these capabilities allow users to quickly gain access to Julia tools. In Figure 18.4 we present a screenshot of a Jupyter notebook that implements Julia code to perform sensitivity analysis for problem (18.1). Detailed information on syntax and capabilities of Julia and JuMP can be found at https://julialang.org and https://github.com/JuliaOpt/JuMP.jl. All scripts discussed in this chapter are available in the form of Jupyter notebooks at https://github.com/zavalab/JuliaBox/tree/master/CBE_Chapter.

18.2.1 Chemical and Biological Engineering Examples

Having established some basic syntax and features, we now proceed to illustrate how to use Julia to solve problems that arise in chemical and biological engineering. These will reveal some common patterns and introduce additional features.

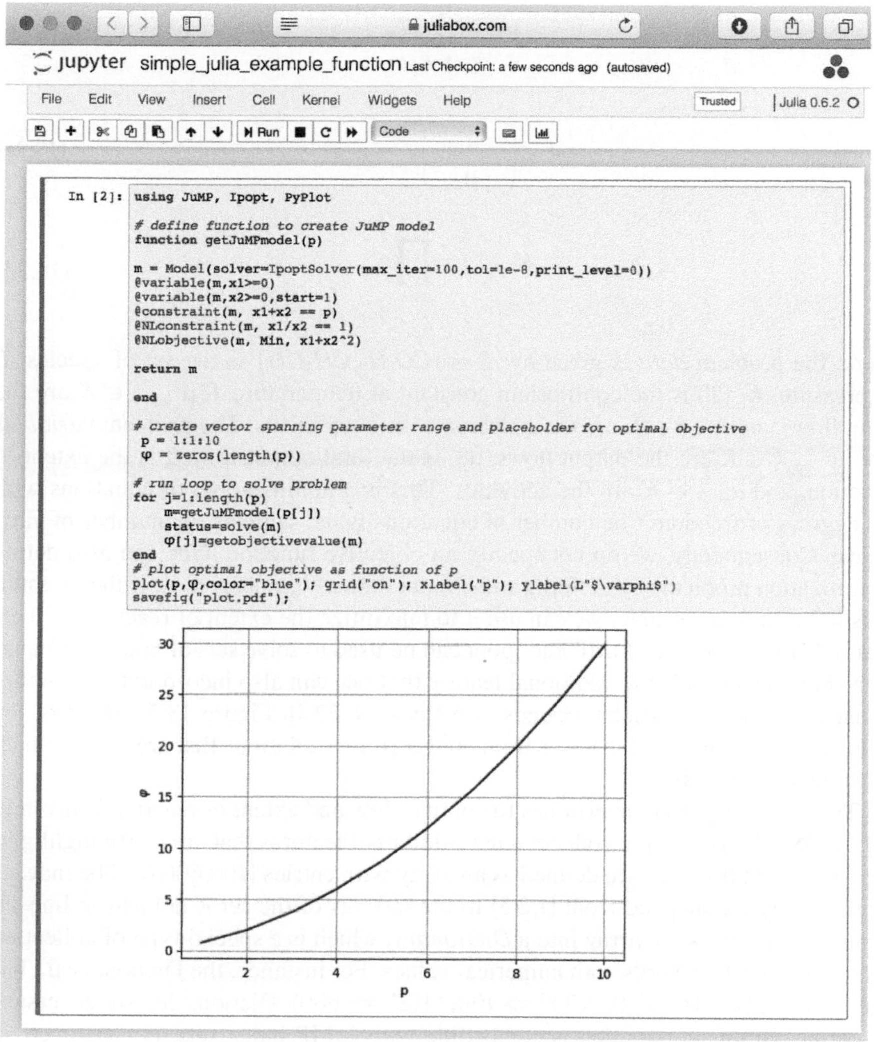

FIGURE 18.4 Jupyter notebook for sensitivity analysis example (executed in http://juliabox.com).

18.2.1.1 Gibbs Reactor

Given an inlet flow stream, temperature, and pressure, we want to determine the output flow (and composition) of a reactor in which the reversible reaction $CO + 2H_2 \leftrightarrow CH_3OH$ takes place. The reaction is assumed to take place in the gas phase, under ideal conditions, and reach equilibrium.

The composition of the output flow can be obtained by solving the following set of equations:

$$\mu_{out}^k = \mu_{in}^k + \gamma^k \xi, \; k \in K \tag{18.2a}$$

$$\mu_{tot} = \sum_{k \in K} \mu_{out}^{k} \tag{18.2b}$$

$$a^k = \left(P \frac{\mu_{out}^{k}}{\mu_{tot}} \right)^{\gamma^k}, k \in K \tag{18.2c}$$

$$K_{eq}(T) = \prod_{k \in K} a^k . \tag{18.2d}$$

Here, the problem *data* is given by: $K = \{CO, H_2, CH_3OH\}$ is the set of species, P is pressure, $K_{eq}(T)$ is the equilibrium constant at temperature T, μ^k_{in}, $k \in K$ are the inlet flows, and γ^k, $k \in K$ are the stoichiometric coefficients. The problem *variables* are: μ^k_{out}, $k \in K$ are the output flows, μ_{tot} is the total output flow, ξ is the extent of reaction, and a_k, $k \in K$ are the activities. This is a nonlinear set of equations with no degrees of freedom (the number of equations is the same as the number of variables). Consequently, we do not specify an objective function. One can also define optimization problems by defining some input data as variables (e.g., if the pressure P is left as a free variable, we can use it to maximize the extent of reaction ξ). This example highlights that JuMP and Ipopt can be used to solve sets of nonlinear equations but they provide the additional feature that one can also incorporate inequality constraints such as variable bounds (see lines 19–22 in Figure 18.5). Bounds can help prevent the numerical solver from converging to solutions that are non-physical (e.g., negative flows).

The Julia script that determines the output flow and extent of reaction is given in Figure 18.5. This script introduces a few additional features that are worth highlighting. The input flows μ_{in} are defined as an array with entries [100,600,0]. The indexes of this array are mapped from [1,2,3] to the indexes of the set K defined in line 11. Line 11 transforms the array into a *Dictionary*, which is a special type of collection that associates key words with numerical values. For instance, the Dictionary μ_{in} has the form ("CH3OH" => 0, "CO" => 100, "H2" => 600). Dictionaries are necessary to define parameter/data that is compatible with JuMP index sets. For instance, in line 25, we reference the entries of the inlet flow μ_{in} using the indexes of the set K. In the script we also note that index sets enable compact representation of not only variables but also of constraints. For instance, line 25 defines the index constraint (18.2a). In the Julia script we have also reformulated the problem by factoring out the pressure from the equilibrium condition (this is done in order to improve scaling of the constraint and numerical behavior). The solution of the problem is $\xi = 76.7$ and $\mu_{out} = (23.3, 446.61, 76.69)$.

18.2.1.2 Stream Mixing (Blending)

We now consider an optimal mixing (blending) problem that arises in the context of wastewater treatment. Consider that you have a set of wastewaster streams W and a set of treatment facilities T. We are given as data a set of wastewater input flows f_{in}^{w}, $w \in W$ with concentrations C_{in}^{w}, $w \in W$ and maximum processing capacities

```
using JuMP, Ipopt
# define JuMP model object
m =Model(solver=IpoptSolver())
# define sets
K = ["CO","H2","CH3OH"] # set of components
# define problem data
μ_in=[100,600,0] # inlet molar flows (kmol/hr)
μ_in=Dict(zip(K,μ_in)) # Create a dictionary with keywords
K and values μ in
P=150 # pressure (bar)
T=300+273.15 # temperature (K)
Keq=10^(-12.275+4938/T) # equilibrium constant
γ=[-1,-2,1] # stoichiometric coefficients (-)
γ=Dict(zip(K,γ)) # rename indexes

# define variables
@variable(m, μ_out[K]>=0) # outlet molar flows (kmol/hr)
@variable(m, μ_tot>=0) # total outlet flow (kmol/hr)
@variable(m, a[K]>=0) # component activities (-)
@variable(m, ξ>=0) # extent of reaction (kmol/hr)

# define constraints
@constraint(m, [k in K], μ_out[k] == μ_in[k]+γ[k]*ξ) #
conservation
@constraint(m, μ_tot == sum(μ_out[k] for k in K)) # total
flow
@NLconstraint(m, [k in K], a[k] == (μ_out[k]/μ_tot)^γ[k]) #
activities
@NLconstraint(m, a["CO"]*a["H2"]*a["CH3OH"] == (P^2)*Keq)
# equilibrium condition

# solve and display solution
solve(m)
println("ξ=",getvalue(ξ))
println("μ_out=",getvalue(μ_out))
```

FIGURE 18.5 Julia script for solving the Gibbs equilibrium problem (18.2).

for the facilities (in terms of total flow f^t_{max}, $t \in T$ and concentration $C_{max}{}^t$, $t \in T$). We are also given processing costs for the facilities α^t, $t \in T$. The goal is to find an optimal strategy to mix the input streams in order to satisfy the facility capacity constraints and to minimize the total processing cost. This stream mixing problem can be expressed in mathematical form as:

$$\min \sum_{t \in T} \alpha_t f^t_{tot} \tag{18.3a}$$

$$\text{s.t. } f^w_{in} = \sum_{t \in T} f^{w,t}, w \in W \tag{18.3b}$$

$$f_{tot}^{t} = \sum_{w \in W} f^{w,t}, \; t \in T \quad (\lambda_t) \tag{18.3c}$$

$$C_{tot} f_{tot}^{t} = \sum_{w \in W} C_{in}^{w} f^{w,t}, \; t \in T \tag{18.3d}$$

$$0 \le C_{tot}^{t} \le C_{max}^{t}, \; t \in T \tag{18.3e}$$

$$0 \le f_{tot}^{t} \le f_{max}^{t}, \; t \in T. \tag{18.3f}$$

Here, $f^{w,t}$ is the stream allocation from stream w to facility t, f_{tot}^{t} is the total input stream to facility t, and C_{tot}^{t} is the input concentration to facility t. In this problem, we are also interested in determining the Lagrange multipliers λ_t (dual variable or shadow price) of the conservation constraint (18.3c). The Lagrange multiplier λ_t can be interpreted as the flow price at facility t (increment in the optimal cost associated with adding one more unit of flow to facility t). The Julia script that implements this problem for a setting with three streams and two facilities is given in Figure 18.6. For this script, we have used the syntax @variables on line 23 as another way to define the variables in the problem. In line 29 note that we have specified the name λ to the conservation constraint. This name is used to access the values of its Lagrange multipliers in line 39. From the solution of this problem we obtain that the Lagrange multipliers have a value of 110 USD/gpm at both facilities. In this problem we also see an example in which variables are defined over multiple sets using a more compact syntax.

In JuMP, a variable can be indexed over an arbitrary number of sets.

We note that the options of Ipopt include printlevel=0 (this option suppresses the output from the solver). As a result, the only outputs from the script are the results of the print statements (lines 36–39), which are shown in Figure 18.7. Here, we can see that JuMP displays the optimal value of the variables using the index set notation and we also see the optimal value of the Lagrange multipliers.

18.2.1.3 Controller Tuning

Consider now the problem of finding tuning parameters for a proportional-integral-derivative (PID) controller so that a given dynamic system is capable of tracking different set-point changes as closely as possible and with minimum control effort.

This problem can be cast as the nonlinear program:

$$\min = \sum_{s \in S} \sum_{t \in T} \alpha_x \varepsilon_{t,s}^2 + \alpha_u u_{t,s}^2 \tag{18.4a}$$

$$\text{s.t.} \frac{1}{\tau} \frac{\left(x_{t+1,s} - x_{t,s}\right)}{h} + x_{t,s} = K u_{t+1,s}, \; t \in T_m, \; s \in S \tag{18.4b}$$

$$u_{t,s} = K_c \varepsilon_{t,s} + \tau_I I_{t,s} + \tau_D \frac{\left(\varepsilon_{t+1,s} - \varepsilon_{t,s}\right)}{h}, \; t \in T_m, \; s \in S \tag{18.4c}$$

```
using JuMP, Ipopt

# define JuMP model object
m = Model(solver = IpoptSolver(tol = 1e-6, max_iter = 200,
print_level = 0));

# define sets
W = ["w1","w2","w3"]; # waste streams
T = ["tA","tB"]; # waste treatment facilities

# define data
fin = [10;20;10]; # waste flows (gpm)
fin = Dict(zip(W,fin));
Cin = [0.1; 0.5; 0.25]; # waste concentration (kg/gpm)
Cin = Dict(zip(W,Cin));
fmax = [20; 30]; # max waste flow capacity (gpm)
fmax = Dict(zip(T,fmax));
Cmax = [0.4; 0.3]; # max waste concentration capacity (kg/gpm)
Cmax = Dict(zip(T,Cmax));
α = [50; 10] # processing cost (usd/gpm)
α = Dict(zip(T,α ));

# define optimization problem
@variables m begin
f[W,T]>=0 # waste to treatment flow
ftot[T]>=0 # treatment inlet flow
Ctot[T]>=0 # treatment inlet concentration
end
@constraint(m, [w in W], fin[w] == sum(f[w,t] for t in T))
@constraint(m,λ[t in T], ftot[t] == sum(f[w,t] for w in W))
@constraint(m, [t in T], Ctot[t]*ftot[t] ==
sum(f[w,t]*Cin[w] for w in W))
@constraint(m, [t in T], Ctot[t] <= Cmax[t])
@constraint(m, [t in T], ftot[t] <= fmax[t])
@objective(m, Min, sum(α[t]*ftot[t] for t in T))

# get solution
println(getvalue(ftot))
println(getvalue(Ctot))
println(getvalue(f))
println(getdual(λ))
```

FIGURE 18.6 Julia script for solving mixing problem (18.3).

$$\varepsilon_{t,s} = \left(x_{t,s} - \bar{x}_s\right),\ t \in T,\ s \in S \tag{18.4d}$$

$$\frac{\left(I_{t+1,s} - I_{t,s}\right)}{h} = \varepsilon_{t,s},\ t \in T_m, s \in S \tag{18.4e}$$

$$x_{1,s} = 0,\ I_{1,s} = 0,\ s \in S. \tag{18.4f}$$

```
ftot: 1 dimensions:
  [tA] = 14.999996004541718
  [tB] = 25.000003995458282
Ctot: 1 dimensions:
  [tA] = 0.4000000099848527
  [tB] =0.30000000999091847
f: 2 dimensions:
  [w1,:]
   [w1,tA] = 2.0466180073504843
   [w1,tB] = 7.953381992649517
 [w2,:]
   [w2,tA] = 10.227969006227072
 [w2,tB] = 9.772030993772928
 [w3,:]
   [w3,tA] = 2.725408990964162
   [w3,tB] = 7.274591009035837
 __anon__: 1 dimensions:
 [tA] = -110.00000391568517
 [tB] = -110.00000395740864
```

FIGURE 18.7 Output of Julia script for solving mixing problem (18.3).

Here, $T = \{1,\ldots,N\}$ and $T_m = \{1,\ldots,N-1\}$ are the sets of times (with mesh size h), $S = \{1,\ldots,N_S\}$ is the set of scenarios, $\bar{x}_s$ is the desired set-point in scenario s, $x_{t,s}$ is the system dynamic trajectory, $u_{t,s}$ is the control trajectory, $\varepsilon_{t,s}$ is the error trajectory, and $I_{t,s}$ is the integral (accumulated) error trajectory. The goal is to find the optimal design parameters K_c,τ_I,τ_D for the control law (18.4c) that minimize the accumulated squared error over time and the control effort for all scenarios. α_x and α_u are tunable optimization parameters that can be used to obtain solutions that minimize the squared error versus the control effort respectively. The Julia script that solves a specific instance of the tuning problem is given in Figure 18.8.

```
using Ipopt, JuMP, PyPlot, Distributions
# set of times
N=100; Tf=10; h=Tf/N; T=1:N; Tm=1:N-1;
# set time vector
time=zeros(N); for t=1:N time[t] = h*(t-1); end
# set of scenarios
NS=3; S=1:NS;
# problem data
K = 1.0; x0 = 0.0; (τ) = 1.0;
# generate random set-point scenarios
srand(0) #generate a fixed random seed. This will ensure
   the script always produces the same random numbers
μ = 0; σ = 2; d = Normal(μ,σ) #Create a normal
   distribution with mean μ and variance σ
```

FIGURE 18.8 Julia script for solving controller tuning problem (18.4).

```
xsp = rand(d,NS); #Create a random vector of setpoints
using the distribution d
# Create JuMP model object
m = Model(solver=IpoptSolver())

# variables (states and inputs)
@variable(m,-2.5<=x[T,S]<=2.5) #physical limits of the
state
@variable(m,-2.0<=u[T,S]<=2.0) #physical control limits
@variable(m,ε[T,S])
@variable(m, I[T,S])
@variable(m, C[T,S])
# variables (controller design)
@variable(m, -10<= Kc <=10)
@variable(m, 0<= τI <=100)
@variable(m, 0<= τD <=1000)
# constraints
@constraint(m, [t in Tm,s in S],(1/(τ))*(x[t+1,s]-x[t,s])/h
+ x[t+1,s]== K*u[t+1,s]);
@constraint(m, [t in Tm,s in S], u[t,s] == Kc*[t,s]+
τI*I[t,s] +τD*(ε[t+1,s]-[t,s])/h);
@constraint(m, [t in Tm,s in S], (I[t+1,s]-I[t,s])/h ==
[t,s]);
@constraint(m, [t in T ,s in S], [t,s]==(xsp[s]-x[t,s]));
@constraint(m, [s in S], x[1,s] == 0);
@constraint(m, [s in S], I[1,s] == 0);
#With αx = 100, αu = 0.01, the optimization problem will
seek parameters that minimize the error trajectory over
control actions.
@constraint(m, [t in T, s in S], C[t,s]==(100*[t,s]^2 +
0.01*u[t,s]^2));

# objective function
@objective(m, Min, sum(C[t,s] for t in T,s in S));

# solve problem
solve(m)
# display results
println(getvalue(Kc)); println(getvalue(τI)); println
(getvalue(τD))

# plot responses
x=zeros(NS,N)
for s in 1:NS; for j=1:N; x[s,j]=getvalue(getindex(m,:x)
[j,s]) ; end; end
plot(T, x[1,:]); plot(T, x[2,:]); plot(T, x[3,:]);
xlabel("Time"); ylabel("x(t)"); grid("on");
savefig("PID.pdf")
```

FIGURE 18.8 *(Continued)*

In the objective function of this example, we see an instance of using a summation over multiple indexed sets. We also see how to interrogate variable entries of a model object using the commands getindex and getvalue on line 51. On line 51 we loop over the scenarios (s = 1:NS) and time horizon (j = 1:N) and retreive the variable x from the model using getindex(m,:x) and retrieve its value at the indices [j,s] using getvalue(getindex(m,:x)[j,s]). In this example we also see how to use a time discretization scheme to covert a differential equation into a set of algebraic equations (in the example we use an implicit Euler scheme). For instance, in the above example we assume that the dynamical system has the form:

$$\frac{1}{\tau}\frac{dx}{dt}+x(t)=Ku(t),\ x(0)=0. \tag{18.5}$$

This differential equation can be converted using an implicit Euler scheme into the set of algebraic equations of the form:

$$\frac{1}{\tau}\frac{\left(x_{t+1}-x_t\right)}{h}+x_t=Ku_{t+1},\ t\in T_m. \tag{18.6}$$

The optimal tracking trajectories for the system ($x_{t,s}$) are shown in Figure 18.9. Here, we can see that the trajectories are smooth, indicating that the discretization scheme is accurate (if not accurate, one needs to decrease the mesh size h). The optimal values for the controller parameters are $K_c = 1.56$, $\tau_I = 1.46$, $\tau_D = 0.34$ (one can compare the performance of the PID controller using these optimal parameters against parameters obtained with heuristic tuning rules such as Ziegler-Nichols). The Julia

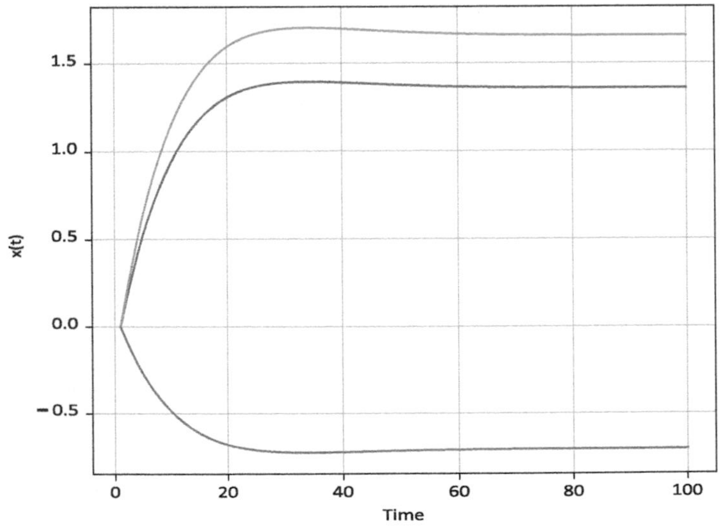

FIGURE 18.9 Optimal tracking trajectories obtained from the solution of controller tuning problem (18.4).

script provided also shows how to generate random samples for a given distribution (this is done using the package Distributions).

Generating random scenarios is an important aspect of the PID tuning problem because it expresses the uncertainty to the optimization problem. In our case, the random parameter is the set-point and we assume this to be Gaussian with a mean of zero and a standard deviation of two.

18.2.1.4 Parameter Estimation

We now show how to use Julia to perform parameter estimation. Specifically, we consider the estimation of parameters for a dynamical model that describes competitive and cooperative effects of a microbial community with two species (this behavior is captured using a Lotka-Volterra model). The estimation problem takes the form:

$$\min \sum_{i \in S} \sum_{k \in T_{exp}} \left(y_{i,t_k^{exp}} - y_{i,k}^{exp} \right)^2 \tag{18.7a}$$

$$\text{s.t. } y_{i,k+1} = y_{i,k} + (t_{k+1} - t_k) \left(r_i + \sum_{j \in S} a_{i,j} \cdot y_{j,k+1} \right) y_{i,k+1}, \; i \in S, k \in T_m \tag{18.7b}$$

$$0 \le r_i \le 10, \; i \in S \tag{18.7c}$$

$$\left| a_{i,j} \right| \le 3, \; i \in S, \; j \in S \tag{18.7d}$$

$$0 \le y_{i,k} \le 1, \; i \in S, \; k \in T. \tag{18.7e}$$

Here, $S = \{1,2\}$ is the set of species, $T = \{1...N\}$ and $T_m = \{1...N-1\}$ are the set of times of the discretized dynamical model, and T_{exp} is the set of experimental times (at which measurements are taken). The objective is to find the growth r_i and interaction $a_{i,j}$ parameters that best match the experimental data $y_{i,k}^{exp}$. The real parameter values are $r_1 = 0.1$, $r_2 = 0.2$, $a_{1,1} = -1$, $a_{1,2} = 0.5$, $a_{2,1} = -1$, and $a_{2,2} = -0.5$. In this problem we note that the initial conditions of the differential equation are not specified and are also estimated from the data. The Julia script that solves this problem is shown in Figure 18.10. Here, we use an implicit Euler discretization scheme and we use the plotting package Plots instead of Pyplot. From the script output in Figure 18.11 we can see that the real parameters are inferred with high precision. The Ipopt output also tells us that the estimated parameters are unique. This can be determined from the lg(rg) column, which shows a value of zero at the solution in iteration 16 (this indicates that the reduced Hessian matrix is positive definite at the solution [11]). From the Ipopt output we also observe that the problem has 208 variables and 200 equality constraints, indicating that we have 8 degrees of freedom. These degrees of freedom correspond to the number of parameters (six) and the number of initial conditions (two). The match of the experimental data obtained is shown in Figure 18.12.

```
using JuMP, Ipopt, Plots
# Data
t = linspace(0,24,101) # set of simulation times
t_exp = 1:10:101 # set of experimental times
# experimental observations
y_exp=[0.0106 0.0120 0.0170 0.0202 0.025 0.031 0.039
0.058 0.077 0.083 0.111
0.0091 0.0147 0.0245 0.0358 0.045 0.081 0.092 0.133 0.153
0.160 0.168]
# create JuMP model object
m=Model(solver=IpoptSolver())

# define variables
@variable(m, 0 <= r[i in 1:2] <=10, start = 1) # growth
rates
@variable(m,-3 <= a[i in 1:2, k in 1:2] <= 3, start =1) #
interaction parameters
@variable(m, 0 <= y[i in 1:2, k in 1:length(t)] <= 1,
start = 1) # abundance

# define model
@NLconstraint(m, [i in 1:2,k in 1:length(t)-1],
y[i,k+1] == y[i,k]+( r[i] + sum( a[i,j] *y[j,k+1] for j
in 1:2 )) *y[i,k+1]*(t[k+1]-t[k]))

# define objective
@objective(m, Min, sum((y[i,t_exp[k]]-y_exp[i,k])^2 for i
in 1:2 for k in 1:length(t_exp)))

solve(m)

# Obtain the solution
r_sol = getvalue(r)
a_sol = getvalue(a)
y_sol = getvalue(y)
p=plot(t,y_sol',xlabel="Time (hr)", ylabel="Abundance",
label=["Species 1" "Species 2"],
color =[:blue :red], legend=:topleft)
plot!(p,t[t_exp],y_exp',label=["Species 1 (exp)" "Species
2 (exp)"],
seriestype=:scatter, color =[:blue :red])
```

FIGURE 18.10 Julia script for parameter estimation problem of microbial community (18.7).

18.3 CONCLUDING REMARKS

In this chapter we have presented basic capabilities for solving optimization problems in Julia. Julia is a user-friendly language for scientific computing that provides access to a wide range of tools for algebraic modeling and optimization. Such tools facilitate the creation of optimization models and enable embedding such models in complex computational workflows that might involve statistics, data processing, and

```
This is Ipopt version 3.12.1, running with linear solver mumps.
NOTE: Other linear solvers might be more efficient (see Ipopt documentation).

Total number of variables............................:      208
                     variables with only lower bounds:        0
                variables with lower and upper bounds:      208
                     variables with only upper bounds:        0
Total number of equality constraints.................:      200
Total number of inequality constraints...............:        0
        inequality constraints with only lower bounds:        0
   inequality constraints with lower and upper bounds:        0
        inequality constraints with only upper bounds:        0

iter    objective    inf_pr   inf_du lg(mu)  ||d||  lg(rg) alpha_du alpha_pr  ls
   0  1.8937115e+01 7.08e-01 1.72e+00  -1.0 0.00e+00    -  0.00e+00 0.00e+00   0
   1  1.5033829e+01 1.22e-01 2.43e+00  -1.0 9.80e-01    -  3.98e-01 1.00e+00f  1
   2  1.0004476e+01 3.34e-02 6.60e-01  -1.0 2.32e-01    -  6.82e-01 1.00e+00f  1
   3  6.1546928e+00 1.14e-02 1.89e-01  -1.0 1.78e-01   0.0 1.00e+00 1.00e+00f  1
   4  4.4221248e+00 2.79e-03 9.41e-02  -2.5 1.47e-01  -0.5 7.64e-01 1.00e+00f  1
   5  1.5627923e+00 3.98e-03 2.50e-02  -2.5 2.03e-01  -1.0 8.30e-01 1.00e+00f  1
   6  2.1729348e-01 3.93e-03 4.59e-02  -3.8 7.92e-01    -  7.65e-01 1.00e+00f  1
   7  4.9899697e-02 1.44e-03 8.24e-03  -3.8 3.99e-01    -  8.62e-01 1.00e+00h  1
   8  1.3293523e-02 7.55e-04 5.91e-03  -3.8 1.35e+00    -  4.93e-01 1.00e+00h  1
   9  7.3892863e-03 4.89e-04 1.37e-03  -3.8 1.85e+00    -  1.00e+00 7.86e-01h  1
iter    objective    inf_pr   inf_du lg(mu)  ||d||  lg(rg) alpha_du alpha_pr  ls
  10  1.2008549e-03 1.39e-04 1.59e-03  -5.7 2.32e-01    -  7.06e-01 1.00e+00h  1
  11  5.1485793e-04 2.05e-04 2.62e-03  -5.7 1.44e+00    -  2.61e-01 1.00e+00h  1
  12  3.5586500e-04 2.37e-04 9.24e-04  -5.7 1.98e+00    -  6.98e-01 1.00e+00h  1
  13  3.3924029e-04 2.96e-05 6.01e-05  -5.7 4.08e-01    -  1.00e+00 1.00e+00h  1
  14  3.3660968e-04 3.70e-06 5.79e-06  -5.7 2.76e-01    -  1.00e+00 1.00e+00h  1
  15  3.3431811e-04 5.50e-07 7.23e-06  -8.6 3.65e-02    -  9.63e-01 1.00e+00h  1
  16  3.3428472e-04 2.74e-09 4.83e-09  -8.6 8.32e-03    -  1.00e+00 1.00e+00h  1

Number of Iterations....: 16

EXIT: Optimal Solution Found.
r_1 = 0.09952955486059546, a_11 = -1.3562024340101704, a_12 = 0.6729280251178505
r_2 = 0.21316884757385257, a_21 = -1.4447177885295344, a_22 = -0.373465951193943
```

FIGURE 18.11 Output of Julia script for parameter estimation of microbial community model.

plotting/visualization tools. Julia is an ideal tool for prototyping and teaching new optimization concepts to engineers and scientists. This is currently being routinely used to teach optimization to senior undergraduate students at UW-Madison. In our experience, the compact and intuitive syntax of Julia drastically lowers the learning barrier for users with limited expertise in optimization.

ACKNOWLEDGEMENTS

V.M. Zavala would like to thank the students of the undergraduate process design course at UW-Madison (CBE 450) for providing feedback on how to best use Julia in the classroom. We also acknowledge funding from the National Science Foundation under award CBET-1748516.

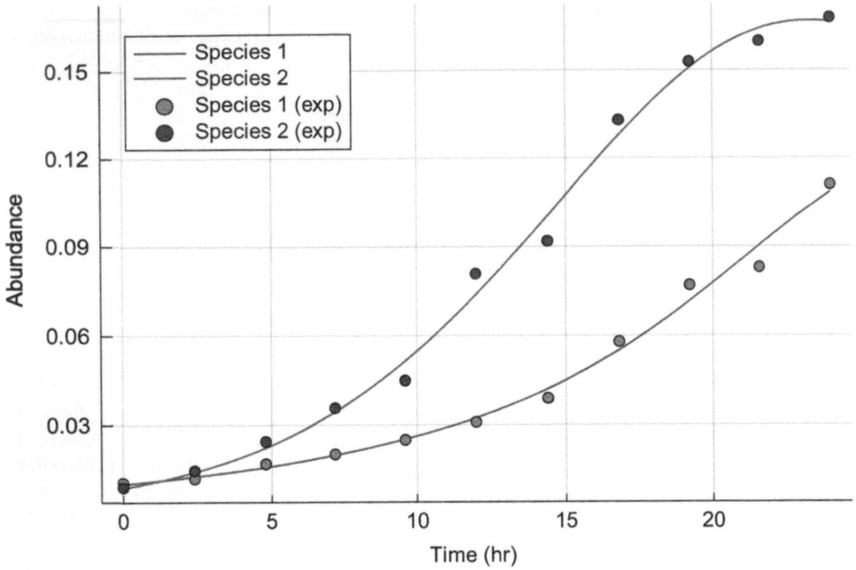

FIGURE 18.12 Experimental and model trajectories for microbial community obtained from solution of (18.7).

REFERENCES

1. Anthony Brooke, David Kendrick, Alexander Meeraus, Ramesh Raman, and U America. The general algebraic modeling system. *GAMS Development Corporation*, Washington DC, 1050, 1998.
2. Robert Fourer, David M Gay, and Brian W Kernighan. A modeling language for mathematical programming. *Manag Sci*, 36(5):519–554, 1990.
3. Paul Inigo Barton and CC Pantelides. gPROM—a combined discrete/continuous modelling environment for chemical processing systems. *Simulation Series*, 25: 25–25, 1993.
4. Michel F Sanner et al. Python: a programming language for software integration and development. *J Mol Graph Model*, 17(1):57–61, 1999.
5. Miles Lubin and Iain Dunning. Computing in operations research using julia. *INFORMS J. Comput.*, 27(2):238–248, 2015.
6. William E Hart, Carl D Laird, Jean-Paul Watson, David L Woodruff, Gabriel A Hackebeil, Bethany L Nicholson, and John D Siirola. *Pyomo-optimization Modeling in Python*, vol 67, Springer, 2012.
7. Iain Dunning, Joey Huchette, and Miles Lubin. Jump: A modeling language for mathematical optimization. *SIAM Review*, 59(2):295–320, 2017.
8. Bethany Nicholson, John D Siirola, Jean-Paul Watson, Victor M Zavala, and Lorenz T Biegler. pyomo. dae: a modeling and automatic discretization framework for optimization with differential and algebraic equations. *Math Program Comput*, 10(2):187–223, 2018.
9. Federico Lozano Santamaría and Jorge M Gomez. Framework in pyomo for the assessment and implementation of (as) nmpc controllers. *Comput Chem Eng*, 92:93–111, 2016.

10. Jordan Jalving, Shrirang Abhyankar, Kibaek Kim, Mark Hereld, and Victor M Zavala. A graph based computational framework for simulation and optimisation of coupled infrastructure networks. *IET Generation, Transmission & Distribution*, 11(12):3163–3176, 2017.
11. V.M. Zavala. Computational Strategies for the Operation of Large-Scale Chemical Processes. Carnegie Mellon University (2008).

19 Fast Deployment of Optimization Applications Using SolverStudio

Qi Zhang, W. Alex Marvin

CONTENTS

19.1 INTRODUCTION

Although mathematical optimization has proven to be highly effective in solving a large variety of challenging engineering problems, only very few chemical engineers make active use of optimization modeling tools. This is in part due to the lack of optimization courses in the chemical engineering curriculum, but can also be attributed to the relatively large effort required to implement user-friendly optimization applications. Here, we emphasize user-friendliness because, in practice, acceptance by the end-user is key to the successful deployment of a software application. In this chapter, we introduce SolverStudio [1], a tool that allows fast deployment of small- to medium-scale optimization applications and thus helps lower the threshold for initiating optimization projects in practical settings.

SolverStudio is an add-in for Microsoft Excel that allows the user to embed and solve optimization problems in Excel. A large set of optimization modeling languages are supported, including AMPL, GAMS, JuMP, PuLP, and Pyomo. Therefore, you can continue using your modeling language of choice while leveraging Excel's data manipulation and visualization capabilities. The code for the optimization model can be hidden, leaving a clean user interface for the end-user who is not going to make changes to the model.

SolverStudio is free, simple to use, and can be very effective because most people are already familiar with Excel. As a spreadsheet-based tool, Excel is popular especially among engineers because of its intuitive calculation and plotting functionalities. As a result, SolverStudio is very easy to learn for optimization application developers and convenient for application users because they will be able to familiarize themselves quickly with the user interface.

Needless to say, SolverStudio implementations come with a computational overhead. Therefore, it is only well-suited for applications of moderate computational intensity with data packages that are not overly large. However, even for projects involving large-scale optimization, SolverStudio can be useful in the prototyping phase when the model must be refined in an iterative process in discussion with the intended users. For this model development process, it is helpful to have an early implementation with decent data visualization. Once the model is finalized, one can switch to a different modeling environment and start working on a computationally more efficient implementation.

In this chapter, we first demonstrate the use of SolverStudio with an illustrative example. In this example, we solve a multilevel lot-sizing problem, a classical problem in production planning, while introducing the main features of the tool. We then present a real-world industrial optimization application developed in SolverStudio at BASF and provide insights on how the tool is used, and could be used more broadly, in the chemical industry. We close the chapter with some final remarks.

19.2 DEMONSTRATION

The features of SolverStudio are best explained with an illustrative example. In the following, we state the problem chosen for this example, present the model formulation, and then provide step-by-step instructions for the implementation of the optimization application in Excel using SolverStudio.

19.2.1 PROBLEM STATEMENT

We consider the multilevel lot-sizing problem (MLLSP)—which is an extension of the classical lot-sizing problem—in relation to the case with multiple items and multiple production stages [2]. The MLLSP represents an important and widely used class of production planning problems as it can be used to model a large variety of production structures. In these production systems, two or more items are produced, and at least one item is required as an input of another. Such a production structure can be represented by a directed graph such as the one shown in Figure 19.1. Each node represents an item, and the arcs indicate which items are required to produce each item. The number on each arc indicates how many units of the preceding item are required to produce one unit of the subsequent item.

Figure 19.1 shows the given production structure for our specific MLLSP that will be used throughout this chapter. Item 1 is the desired product, which can be directly produced from Items 2 and 3. Item 2 is produced from Items 3 and 4, while Item 3 is directly produced from Item 5. Given demand for Item 1, production capacities, initial inventory levels, and production and inventory holding costs, the objective is

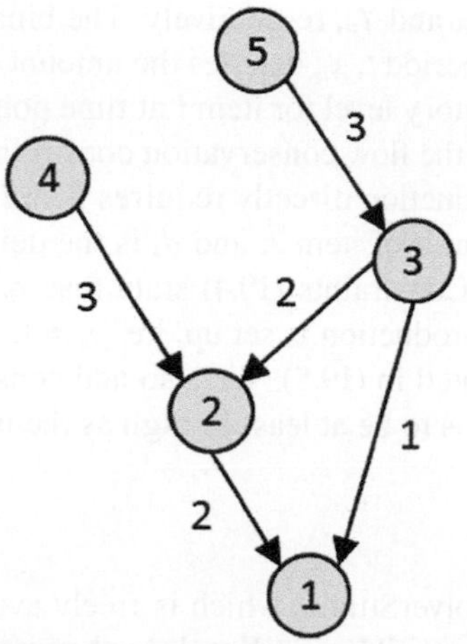

FIGURE 19.1 Given production structure represented by a directed graph.

to optimize the production plan for a planning horizon of 12 time periods such that the total operating cost is minimized.

19.2.2 Model Formulation

The MLLSP can be formulated as the following mixed-integer linear program (MILP):

$$\min \sum_{i \in I} \sum_{t \in T} \left(\alpha_{it} x_{it} + \beta_{it} s_{it} + \gamma_{it} y_{it} \right) \tag{19.1}$$

$$\text{s.t.} \quad s_{i,t-1} + x_{it} = \sum_{j \in A_i} \rho_{ij} x_{jt} + s_{it} \quad \forall i \in I \setminus \{1\},\ t \in T \tag{19.2}$$

$$s_{1,t-1} + x_{1t} = d_t + s_{1t} \quad \forall t \in T \tag{19.3}$$

$$x_{it} \le C_{it} y_{it} \quad \forall i \in I, t \in T \tag{19.4}$$

$$s_{i0} = s_i^0 \quad \forall i \in I \tag{19.5}$$

$$s_{i|T|} \ge s_i^0 \quad \forall i \in I \tag{19.6}$$

$$s_{it} \ge 0, x_{it} \ge 0, y_{it} \in \{0,1\} \quad \forall i \in I, t \in T \tag{19.7}$$

where $I = \{1,2,\ldots,m\}$ is the set of m items, and $T = \{1,2,\ldots,n\}$ is the set of n time periods. The unit production costs, unit inventory holding costs, and fixed setup costs are denoted by α_{it}, β_{it}, and γ_{it}, respectively. The binary variable y_{it} equals 1 if item i is produced in time period t, x_{it} denotes the amount of item i produced in time period t, and s_{it} is the inventory level for item i at time point t, which by definition is the end of time period t. In the flow conservation constraints (19.2)-(19.7), A_i denotes the set of items whose production directly requires item i, ρ_{ij} is the amount of item i required to produce one unit of item j, and d_t is the demand for the final product (Item 1) in time period t. Constraints (19.4) state that x_{it} is bounded above by the production capacity C_{it} if production is set up, i.e. $y_{it} = 1$. The initial inventory level for item i is set to s_i^0, as stated in (19.5). We also add constraints (19.6), which force the terminal inventory levels to be at least as high as the initial ones.

19.2.3 Implementation

Once you have installed SolverStudio, which is freely available at *solverstudio.org*, you can find it under the *Data* ribbon in Excel, as shown in Figure 19.2.

We first input all data into Excel and make them available for the optimization model using SolverStudio. Figure 19.3 shows the data for the two-index ρ-parameter. By clicking on *Edit Data*, we access SolverStudio's data items editor in which one can specify the data used in the optimization model. As shown in Figure 19.3, a data item is defined by its name, the cell range in which the data is written, and possibly index ranges if the data is not a scalar. In a similar fashion, we fill the worksheet with all input data as well as define the cells in which the optimization results will be displayed. The result is shown in Figure 19.4, and the data items are defined according to the list shown in Figure 19.5.

Two useful features of SolverStudio are its *Show/Hide Data* and *Show Data in Color* functionalities, which show (or hide) and color-code explicitly the data items and their corresponding cell and index ranges on the worksheet. Figure 19.6 shows what our worksheet looks like if the data items are displayed. These features allow a quick visual check of whether the data items are defined correctly.

Now, with all the required data made available to the optimization engine of SolverStudio, we can write our model. As mentioned in the Introduction, you can choose among a large set of optimization modeling languages, making it easy to migrate any existing code you may have. In this example, we use JuMP [3], which is a Julia package for algebraic modeling of mathematical programs. The code is displayed when clicking on *Show Model*. Figure 19.7 shows our code for the implementation of the MILP model presented in Section 19.2.2. Note that no model parameters are defined and no data are input here. Instead, the command

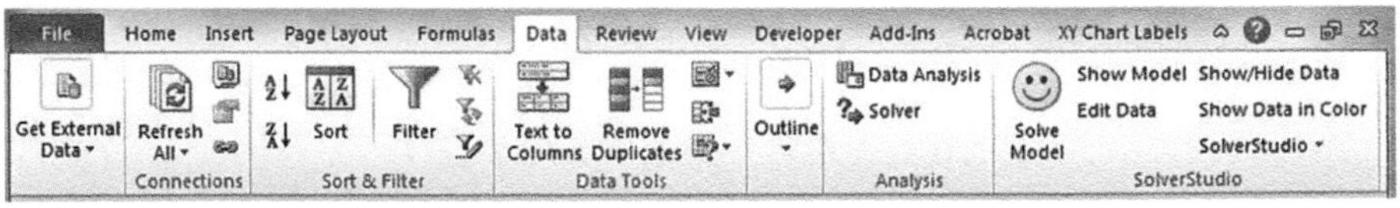

FIGURE 19.2 SolverStudio add-in under the *Data* ribbon in Excel.

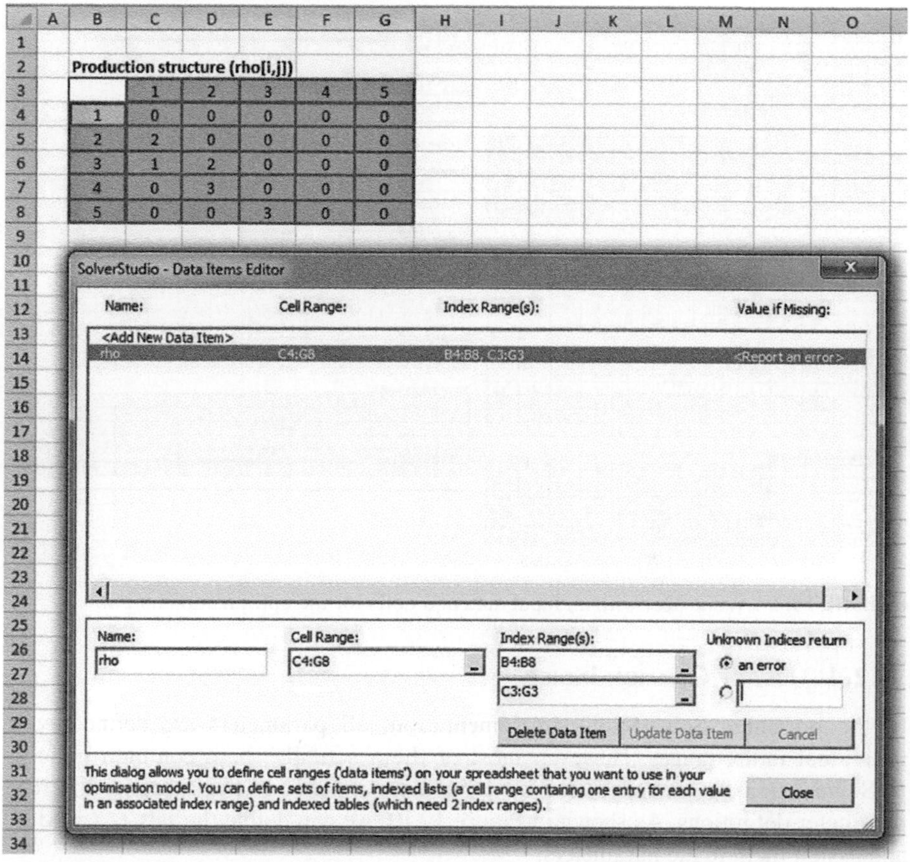

FIGURE 19.3 Make data available to the optimization model using SolverStudio's Data Items Editor.

include("SolverStudio.jl") in the second line of the code executes *SolverStudio.jl*, which uses the data items defined in SolverStudio and converts them into data that can be directly used in Julia under their given names.

Solve Model executes the code, and the progress is shown in the model output window. After the model is solved, the results are written to the output data items defined in SolverStudio. As a result, the corresponding cells on the worksheet are populated with the new data as shown in Figure 19.8.

Once the data from the solution of the optimization model are transferred to Excel, they can be readily used for further data manipulation and visualization. The rich plotting functionalities in Excel allow us to quickly create a decent user interface, which could take the form of a dashboard as the one shown in Figure 19.9, where the model is hidden from the end-user. Note that although each worksheet can only have one embedded model, the data of that model can be used across multiple worksheets.

Production structure (rho[i,j])

	1	2	3	4	5
1	0	0	0	0	0
2	2	0	0	0	0
3	1	2	0	0	0
4	0	3	0	0	0
5	0	0	3	0	0

Production capacities (C[i,t])

	1	2	3	4	5	6	7	8	9	10	11	12
1	160	160	160	160	160	160	160	0	0	160	160	160
2	320	320	320	320	320	320	320	320	320	320	320	320
3	950	950	950	950	950	950	950	950	950	950	950	950
4	1000	1000	1000	1000	1000	1000	1000	1000	1000	1000	1000	1000
5	2500	2500	2500	2500	2500	2500	2500	2500	2500	2500	2500	2500

Unit production costs (alpha[i,t])

	1	2	3	4	5	6	7	8	9	10	11	12
1	50	50	50	55	50	50	50	50	50	50	50	50
2	45	45	45	50	45	45	45	45	45	45	45	45
3	55	55	55	60	55	55	55	55	55	55	55	55
4	30	30	30	35	30	30	30	30	30	30	30	30
5	55	55	55	55	55	55	55	55	55	55	55	55

Unit inventory holding costs (beta[i,t])

	1	2	3	4	5	6	7	8	9	10	11	12
1	4	4	4	4	4	4	4	4	4	4	4	4
2	3	3	3	3	3	3	3	3	3	3	3	3
3	2	2	2	2	2	2	2	2	2	2	2	2
4	1	1	1	1	1	1	1	1	1	1	1	1
5	2	2	2	2	2	2	2	2	2	2	2	2

Fixed setup costs (gamma[i,t])

	1	2	3	4	5	6	7	8	9	10	11	12
1	6000	6000	6000	7000	7000	6000	6000	6000	6000	6000	6000	6000
2	9000	9000	9000	10000	10000	9000	9000	9000	9000	9000	9000	9000
3	10000	10000	10000	11000	11000	10000	10000	10000	10000	10000	10000	10000
4	8000	8000	8000	9000	9000	8000	8000	8000	8000	8000	8000	8000
5	10000	10000	10000	11000	11000	10000	10000	10000	10000	10000	10000	10000

Demand (d[t])

1	2	3	4	5	6	7	8	9	10	11	12
100	120	120	130	140	140	120	90	80	100	110	120

Initial inventory levels (s0[i])

1	2	3	4	5
50	50	70	100	100

Total operating cost:

Production setups (y[i,t])

	1	2	3	4	5	6	7	8	9	10	11	12
1												
2												
3												
4												
5												

Production amounts (x[i,t])

	1	2	3	4	5	6	7	8	9	10	11	12
1												
2												
3												
4												
5												

Inventory levels (s[i,t])

	0	1	2	3	4	5	6	7	8	9	10	11	12
1													
2													
3													
4													
5													

FIGURE 19.4 Worksheet with all input data and cells for the optimization output.

19.2.4 Use of Common Indices

In the presented SolverStudio implementation, all parameters are defined with individual index ranges. However, most of them actually share common indices. In SolverStudio, index ranges can be defined separately and then used in different parameter definitions. As shown in Figure 19.10, we can define the sets I, J, and T, and assign them to the parameters.

The index ranges can then also be used in the model code, allowing a more generic implementation in which the cardinalities of the sets change automatically with the given data. In our case, this can be implemented in Julia/JuMP as shown in Figure 19.11.

Name:	Cell Range:	Index Range(s):
<Add New Data Item>		
alpha	C20:N24	B20:B24, C19:N19
beta	C28:N32	B28:B32, C27:N27
C	C12:N16	B12:B16, C11:N11
d	P4:AA4	P3:AA3
gamma	C36:N40	B36:B40, C35:N35
rho	C4:G8	B4:B8, C3:G3
s	Q31:AC35	P31:P35, Q30:AC30
s0	P8:T8	P7:T7
x	Q23:AB27	P23:P27, Q22:AB22
y	Q15:AB19	P15:P19, Q14:AB14
z	S11	

FIGURE 19.5 List of data items that are used in the optimization model.

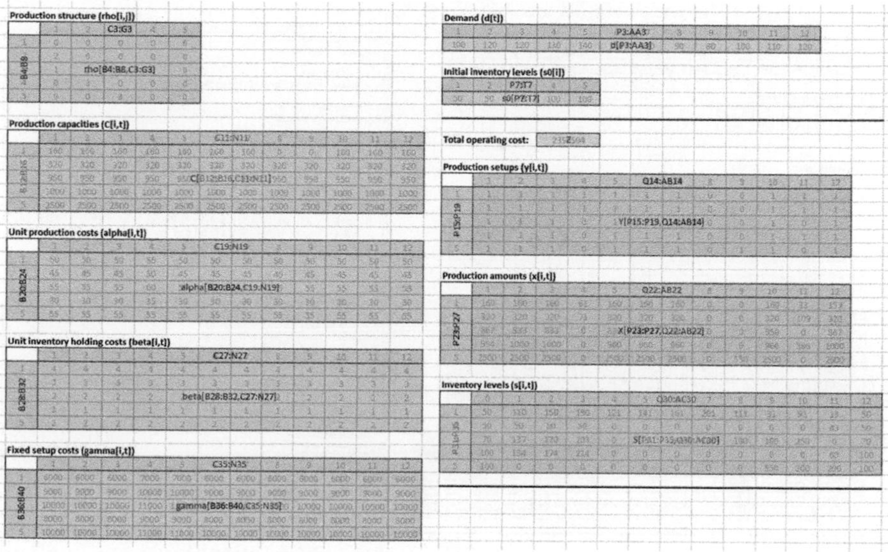

FIGURE 19.6 Worksheet if data items are shown.

SolverStudio ©Andrew Mason

File Edit Language Julia

```
using JuMP, CPLEX
include("SolverStudio.jl")

# Create model.
m = Model(solver=CplexSolver(CPX_PARAM_EPGAP=0, CPX_PARAM_TILIM=300))

# Variables:
@variables m begin
    s[i=1:5, t=0:12] >= 0
    x[i=1:5, t=1:12] >= 0
    y[i=1:5, t=1:12], Bin
end

# Constraints:
@constraints m begin
    c1[i=2:5, t=1:12], s[i,t-1] + x[i,t] == sum(rho[i,j]*x[j,t] for j=1:5) + s[i,t]
    c2[t=1:12], s[1,t-1] + x[1,t] == d[t] + s[1,t]
    c3[i=1:5, t=1:12], x[i,t] <= C[i,t]*y[i,t]
    c4[i=1:5], s[i,0] == s0[i]
    c5[i=1:5], s[i,12] >= s0[i]
end

# Objective function:
@objective(m, Min, sum(alpha[i,t]*x[i,t] + beta[i,t]*s[i,t] + gamma[i,t]*y[i,t] for i=1:5, t=1:12))

# Solve model.
solve(m)
S = getvalue(s)
X = getvalue(x)
Y = getvalue(y)
Z = getobjectivevalue(m)
```

Model Output

FIGURE 19.7 Julia/JuMP implementation of the MILP model.

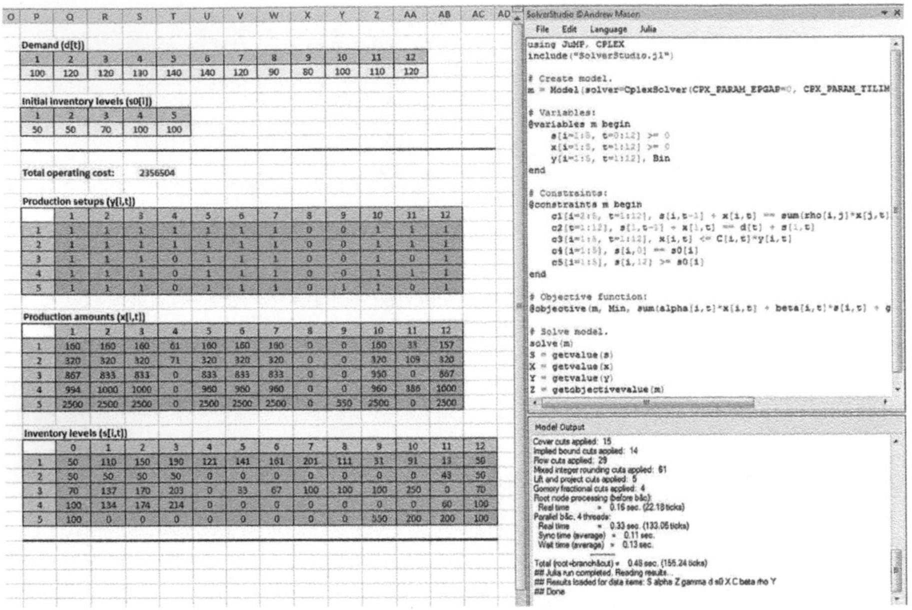

FIGURE 19.8 Result of the successful solution of the optimization model.

19.3 INDUSTRIAL APPLICATION

One industrial application of SolverStudio at BASF is the Advanced Tactical Planner (ATP), shown in Figure 19.12. It was initially developed in 2015 as a support tool for the Sales & Operational Planning (S&OP) process of business units and production plants, which is typically a monthly process to align the many job functions (e.g. procurement, operations, supply chain) on a shared tactical plan for the next year. ATP provides necessary transparency and quantitative rigor to S&OP. It was

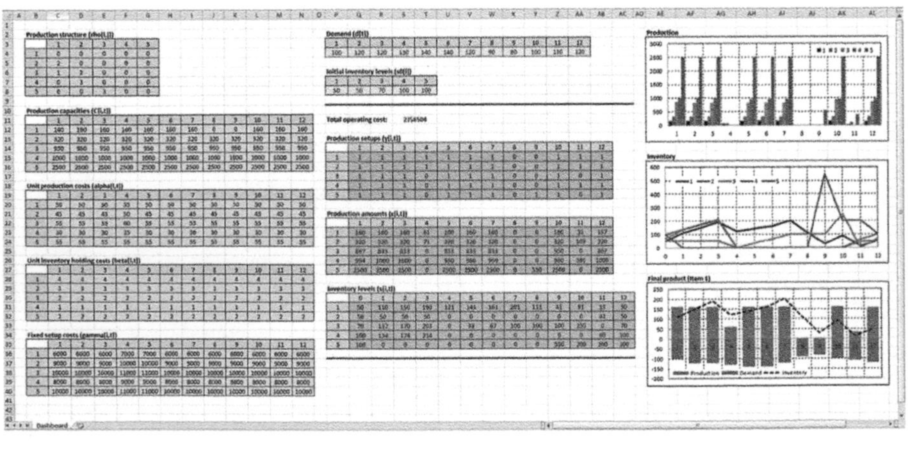

FIGURE 19.9 Dashboard for the end-user (not showing the model). (See color insert.)

Name:	Cell Range:	Index Range(s):
<Add New Data Item>		
alpha	C20:N24	I, T
beta	C28:N32	I, T
C	C12:N16	I, T
d	P4:AA4	T
gamma	C36:N40	I, T
I	B4:B8	
J	C3:G3	
rho	C4:G8	I, J
S	Q31:AC35	P31:P35, Q30:AC30
s0	P8:T8	I
T	P3:AA3	
X	Q23:AB27	I, T
Y	Q15:AB19	I, T
Z	S11	

FIGURE 19.10 Data item list indicating common indices.

a major upgrade from previous manual tools that could not guarantee feasibility of plans and did not attempt to find optimal plans.

ATP involves one MILP model on the Model worksheet that is written in IronPython, formulated with the PuLP module, and solved with either the Gurobi or CBC solver. There are multiple other worksheets for documentation, input data,

```
using JuMP, CPLEX
include("SolverStudio.jl")

# Create model.
m = Model(solver=CplexSolver(CPX_PARAM_EPGAP=0, CPX_PARAM_TILIM=300))

# Set cardinalities:
II = length(I)
JJ = length(J)
TT = length(T)

# Variables:
@variables m begin
    s[i=1:II, t=0:TT] >= 0
    x[i=1:II, t=1:TT] >= 0
    y[i=1:II, t=1:TT], Bin
end

# Constraints:
@constraints m begin
    c1[i=2:II, t=1:TT], s[i,t-1] + x[i,t] == sum(rho[i,j]*x[j,t] for j=1:JJ) + s[i,t]
    c2[t=1:TT], s[1,t-1] + x[1,t] == d[t] + s[1,t]
    c3[i=1:II, t=1:TT], x[i,t] <= C[i,t]*y[i,t]
    c4[i=1:II], s[i,0] == s0[i]
    c5[i=1:II], s[i,TT] >= s0[i]
end

# Objective function:
@objective(m, Min, sum(alpha[i,t]*x[i,t] + beta[i,t]*s[i,t] + gamma[i,t]*y[i,t] for i=1:II, t=1:TT))

# Solve model.
solve(m)
S = getvalue(s)
X = getvalue(x)
Y = getvalue(y)
Z = getobjectivevalue(m)
```

FIGURE 19.11 Julia/JuMP model with set definitions from data.

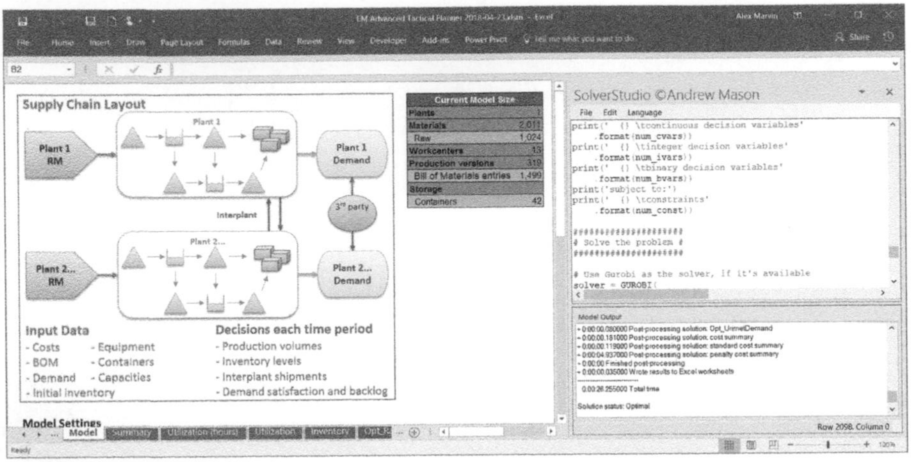

FIGURE 19.12 Advanced Tactical Planner (ATP). (See color insert.)

optimization results, and data visualization dashboards. Status updates, warnings, and errors are printed to the Model Output window of SolverStudio. A few VBA macros are used to sync data with company databases, pre- and post-process data tables, and run the optimization model.

SolverStudio is used in ATP for all the previously mentioned advantages, but also due to some reasons unique to industry. Firstly, the installation of an Excel add-in does not require admin rights, so no specialized IT request is required. Secondly, it comes with compiled IronPython with PuLP, and a few solvers (e.g. IPOPT, CBC), so users can be solving optimization models within minutes.

19.4 CONCLUDING REMARKS

In this chapter, we have introduced SolverStudio, a tool that allows the user to embed and solve optimization problems in Excel. In a short tutorial, we have shown how to use the tool to define data sets in Excel and directly use them in an optimization model. As most people are familiar with Excel, it serves as an easy-to-use platform for creating user interfaces. Hence, by bringing optimization models into Excel, SolverStudio allows the fast deployment of light-weight optimization applications and helps increase the acceptance of optimization solutions in practice. Its use in real-world industrial applications is further demonstrated in an example from BASF.

REFERENCES

1. A. J. Mason, "SolverStudio: A new tool for better optimisation and simulation modelling in Excel," *INFORMS Transactions on Education*, vol. 14, no. 1, pp. 45–52, 2013.
2. Y. Pochet and L. A. Wolsey, *Production planning by mixed integer programming*, New York: Springer Series in Operations Research and Financial Engineering, 2006.
3. M. Lubin and I. Dunning, "Computing in Operations Research Using Julia," *INFORMS Journal on Computing*, vol. 27, no. 2, pp. 238–248, 2015.

20 Use of Lingo for Product Design

Salvador I. Pérez-Uresti, Lidia S. Guerras, Mariano Martín, and Arturo Jiménez Gutiérrez

CONTENTS

20.1 INTRODUCTION

For decades, industry understood product design more related to marketing that engineering. However, the complexities in the thermodynamics of mixtures, new chemicals based on molecular design have attracted the attention of chemical engineers over the last two decades [1]. As a result of the interest, the chemical industry moved towards the manufacture and sale of high added-value materials marketed on performance for a novel paradigm, the product-centered approach rather than the process centered. Three types of products are typically defined [2]:

- Speciality chemicals which provide a specific benefit. Pharmaceuticals are the obvious example. These range from molecules to mixtures.
- Products whose microstructure—rather than molecular structure—creates value. Examples include paint and ice cream.
- The third category of chemical products are devices which affect chemical change. An example is the blood oxygenator used in open-heart surgery.

The integration of product design with process design is a straightforward next step so that the interactions between both are evaluated simultaneously and the issues resolved. For instance, if a mixture with the appropriate composition is to be produced, it must go through a process and mixing, mass transfer, heat transfer and pressure drop are some of the process constraints that link the final product with the technical feasibility to produce it [3]. Furthermore, this problem is also a multi-scale one from molecules to supply chain. The aim of the efforts is to win the customers' appreciation by providing products with desirable properties [4].

From the initial attempts, based on experience, heuristics and following an iterative procedure, a systematic way is needed to reduce costs and time to market. Textbooks have included two different approaches to deal with this problem—a heuristic approach, [2] and lately, a mathematical optimization approach [5].

1. The heuristic procedure for chemical product design [1] consists of the following stages:
   a. Needs. What needs should cover the product?
   b. Ideas. What different products could fill this need?
   c. Selection. Which ideas are the most promising?
   d. Manufacture. How these products are produced?
2. Mathematical optimization approach [5].

Typical problems such as the pooling problem to evaluate detergent mixtures [3] or fuels mixtures, where the problem is commonly known as blending, have been evaluated in the literature. Furthermore, crude oil or biomass composition shows another type of problem, which is related to the optimal (bio) refinery operation for the maximum profit [6]. On the other side, molecular design has also been a wide area of research targeting molecules with appropriate properties [1].

From a mathematical stand point the type of problems we usually deal with in product design are LP, MILP and most typically NLP and MINLP. The main characteristics of the different types of problems and solvers that can be used are briefly described in Chapter 13.

20.2 HINTS FOR MODELING IN LINGO

Lingo allows us to write the model in terms of sets. A section in Lingo is defined by writing its name followed by a colon, and it ends by writing END plus the name of the section [7]. To define a set, we first write its elements between slashes, //; then we write the name of each variable associated with it. This is represented as follows:

```
SETS:

Product/apple, pear, orange, banana /: fixed_cost;
Cities /London, Paris, Madrid/: Location;
ENDSETS
```

The DATA section is used to assign a fixed value for a variable:

```
DATA:
fixed_cost=4 3 4.5;
ENDDATA
```

In this case, the first value of the variable *fixed cost* 4, is assigned to the first element of set *Product*, apple.

We also can feed data as a table. This can be done as follows:

```
SETS:
Table(Product,cities):transport_cost;
ENDSETS
```

Where `Table(Product,cities)`indicates that the set *Table* includes all pairs formed by each element of *Product* and *cities* sets.

```
DATA:
transport_cost= 5  3  3  7
                8  9  7  3
                9  3  4  4 ;
ENDDATA
```

Once sets and data sections have been defined, we can write the equations in terms of sets. Consider the following equation:

$$\text{selling\_price}_{i,j} = \text{Fixed\_cost}_i + \text{Transport\_cost}_{i,j}$$

This equation is used to determine the selling price of fruit, i, which is function of its supplying place, j. It can be written in Lingo as flows:

```
SETS:
Table(Product,cities):transport_cost,selling_price;
ENDSETS
@for(Product(i):@for(cities(j):selling_price(i,j)=
  Fixed_cost(i)+Transport_cost(i,j)));
```

In this example, we need first to define the dimension of the variable `selling _ price` in the SETS section, then we use the operator @for (…) to calculate the selling price for each product, i, which is supplied from site, j. In general, we write inside @for (…) the name of the set followed by a variable, in our example, *i* and *j*, which is an element of the set written before, then we write the expression we want to calculate. The operator @for (…) repeats this task for every, i and j.

In addition, Lingo has other mathematical functions and logical operator, as shown in Table 20.1.

TABLE 20.1
Mathematical Functions and Logical Operator

Lingo's Operators	Description
@SUM(SET_Name(i):)	It sums up an expression for every i
@BND(L,X,U)	It limits the variable X to greater to L and less than to U
@BIN(X)	Define the Variable X to have a binary integer value
@PROD(SET_NAME(i):)	It multiplies an expression for every i
#NE#	Non-equal
#GE#	Greater or equal than
#GT#	Greater than
#LE#	Lower or equal than
#LT#	Lower than
#AND#	And
#OR#	Or

20.3 EXAMPLES

20.3.1 Food Industry: Brewing Company

A brewing company produces three kinds of beers: lager, ale and an economic edition. Table 20.2 shows the required amount of feedstock and adjuncts cereals used for their preparation. The planning manager wants to know how the optimal planning production of each kind of beer should be in order to maximize the profit of company.

To determine the optimal planning production, we need to calculate the number of barrels produced of each kind of beer and the amount of feedstock used in their preparation. This last is given by the next equations:

$$Total\_Cereal_j = \sum_i cereal\_content_{i,j} * \text{Barrel}_i \quad \forall \text{ j} \tag{20.1}$$

$$Total\_cereal_j \le Max_j \tag{20.2}$$

TABLE 20.2
Feedstock Used in Beer Production

	Beer, i			
Feedstock, j(Cereals)	Lager	Ale	Economic edition	Feedstock in warehouse, MAX_j (kg)
Corn	5	1.5	2	400
Wheat	1	4	3	300
Barley	3	7	4	500
Hop	0.4	0.4	0.4	100
Rye	0	0	8	100
Operating cost, Op_Cost_i (\$/barrel)	1.85	5	2.5	
Selling price, SP_i (\$/barrel)	14	17	15	

where $cereal\_content_{i,j}$ is the amount of cereal, j, used to produce beer, i. On the other hand, the operating cost and incomes are calculated as follows:

$$Incomes = \sum_i SP_i * Barrel_i \tag{20.3}$$

$$Production\_cost = \sum_i OP\_cost_i * Barrel_i \tag{20.4}$$

where SP_i and OP_cost_i are the selling price and operating cost of beer, i, respectively. We define the objective function in terms of profit of the company which is determined by the following equation:

$$Profit = Incomes - Production\_cost \tag{20.5}$$

To code this problem in Lingo, we first define two SETS, *cereals and beer*, and the variables related to them. In our case we define the variables `Total _ cereal, barrel` to calculate the total amount of cereal used and the number of barrels produced, respectively.

```
Cereals/Corn Wheat Barley hop rye/: Total_cereal;
Beer /lager ale economic/: beer_price, barrel,op_cost;
```

We also define the variables `beer _ price, and op _ cost` to feed the data in Table 20.2. Additionally, we can construct a Table by using the variable `Cereals _ content` to feed feedstock information. Then, we define its dimension by using sets *Cereals and Beer* as follows:

```
Table_content(beer,cereals):Cereals_content;
```

To write Equation 20.1, we can use the following expression:

```
@for(cereals(j):Total_cereal=@sum(beer(i):cereals_content
  (i,j)*barrel(i)));
```

In this case, we use operator @for to compute the total amount of each cereal for beer production. Whereas, the profit can be calculated by the followings expressions coded in Lingo:

```
Income=@sum(beer(i):(beer_price(i)*barrel(i)));
Production_cost=@sum(beer(i):(op_cost(i)*barrel(i)));
profit=Income-Production_cost;
```

In this case, we use the operator @sum to calculate the incomes for the sale of beer, i, and to determine the production cost, as well. It is worth mentioning, that the amount of each cereals to be used in beer production is restricted by the amount which remain in the warehouse, this last can be written as follows:

```
Total_cereal(1)<=400;
Total_cereal(2)<=300;
Total_cereal(3)<=500;
Total_cereal(4)<=100;
Total_cereal(5)<=100;
```

The entire model coded in Lingo is shown below.

```
Max=profit;
sets:
Cereals/Corn Wheat Barley hop rye/:Total_cereal;
beer/lager ale special/:beer_price,barrel,op_cost;
Table_content(beer,cereals):Cereals_content;
endsets
data:
cereals_content= 5     1  3  0.4 0
                 1.5   4  7  0.4 0
                 2     3  4  0.4 8;
beer_price= 14 17 15;
op_cost=1.85 5 2.5;
enddata
@for(cereals(j):Total_cereal=@sum(beer(i):
  cereals_content(i,j)*barrel(i)));
Total_cereal(1)<=400;
Total_cereal(2)<=300;
Total_cereal(3)<=500;
Total_cereal(4)<=100;
Total_cereal(5)<=100;
Income=@sum(beer(i):(beer_price(i)*barrel(i)));
Production_cost=@sum(beer(i):(op_cost(i)*barrel(i)));
profit=Income-Production_cost;
```

Lingo generates a document .lgr, which contains the results. It consists of three sections. The first section provides information about model size, 11 variables and 14 constraints, and type (Class), it is an LP, and the optimal value of the objective function, 1375.676.

The second section includes two columns see below, *value* and *reduced cost.* The value column shows the optimal value of each variable, while the reduced cost column shows the Karush-Kuhn-Tucker multipliers of each variable, which can be interpreted as the amount that the objective function would be affected if the variable value was increased or decreased.

In the same solution report, we can observe a third section, which consists of two columns, *Slack or Surplus* and *Dual price. Slack or Surplus* column indicates how close the constraints are satisfied as an equality. Whereas, *Dual price* column shows the Karush-Kuhn-Tucker multipliers of the constraints. It indicates the amount that the objective function would improve as the right-hand side of the constraint is increased by one unit. In general terms, they can be interpreted as sensitivity coefficients.

TABLE 20.3
Composition, Raw Material Cost and Composition Products

	Raw Materials							Products	
	Wheat flakes	Oat Bran	Toasted almond	Nuts	Toasted hazelnut	Raisins	Sugar	New Cereals	Current Cereals
Carbohydrates	60	45	10	8	6	50	99.8	47.60	44.33
Sugar	4	0	4	2	4	1	99.8	5	5.1
Water	5	6.5	2.23	3.8	2.52	24	0.2	10	4.9
Minerals	1.1	1.5	2	1	1.6	1.1	0	1.254	1.3
Proteins	9	15	19	16.3	14	1.9	0	9.13	12
Fat	3	7	41	45	59	0.5	0	11	15
Saturated fats	0.7	1.2	3.5	5.5	3.9	0.15	0	1.3	1.74
Price (€/kg)	*3*	*16*	*25*	*22*	*19*	*20*	*1*		

```
Row     Slack or Surplus        Dual Price
1        1375.676               1.000000
2           0.000000           -1.608197
3           0.000000            0.000000
4           0.000000           -1.369672
5           0.000000            0.000000
6           0.000000           -0.4756148
7           0.000000            1.608197
8          51.024590            0.000000
9           0.000000            1.369672
10         54.672130            0.000000
11          0.000000            0.475614
12          0.000000            1.000000
13          0.000000           -1.000000
14          0.000000            1.000000
```

The optimal planning production was determined to be 64 barrels of lager beer, 37 barrels of ale beer, and 13 barrels of special edition. In this way, the maximum profit was calculated to be $ 1,375.676.

EXERCISE

1. Design a new breakfast cereal with reduced sugars and fat, developing a diet problem [8.9] using the same ingredients by minimizing the price, see Table 20.3 for the data.
2. Evaluate the economics between the current and the new product.

The solution can be found in the CRC webpage of the book.

20.3.2 Refinery Operation

A company in Mexico wants to increase its gasoline production. Half of the crude oil production in Mexico corresponds to Olmeca crude, which is considered a heavy

TABLE 20.4
Data of the Raw Materials and Products Obtained from Crude Oil Processing

Product/Composition	Ural	WTI	Olmeca	Price (€/Barrel)	Maximum Capability (Barrels/day)
Gasoline	75	87	37	90	78000
Kerosene	11	8	10	30	7000
Fuel Oil	10	4	21	47	13600
Residues	4	1	32	5	
Processing cost (€/day)	0.9	0.4	0.6		
Price(€/Barrel)	20	50	80		

crude oil due to its high content of asphaltenes and residues. This fact reduces its efficiency to produce gasoline and other products. Thus, the company has considered importing light crude oil. Two options have been proposed: 1) Ural crude, which is considered as an intermediate light crude oil; and 2) the West Texas Intermediate (WTI) crude, which is considered an extra-light crude oil. Currently, the federal law in Mexico restricts crude oil importations to 75,000 barrels/day. Taking data shown in Table 20.4, find the optimal portion of each crude to be processed and the optimal distribution of products generated that maximize the profit of the company.

To solve this problem we define the objective function as the summation of the incomes minus operating and feedstock cost.

$$Profit = Incomes - Feedstock - Op\_cost \tag{20.6}$$

Assuming that the income in the company is given by the sale of the products generated,

$$Incomes = \sum_{i} SP_i * x_i \tag{20.7}$$

where SP_i is the selling price of product, i, and x_i is the amount of the product, i. x_i which depends on the crude used for its production,

$$x_i = \sum_{j} Yield_{i,j} * C_j \quad \forall \text{i} \tag{20.8}$$

where C_j is the amount of crude, j, processed by the refinery. Additionally, the feedstock and operating costs are computed as follows:

$$Feedstock = \sum_{i} CP_i * C_j \tag{20.9}$$

$$Op\_cost = \sum_j PC_j * C_j \tag{20.10}$$

Moreover, the usage of Ural and WTI are restricted to be as much as 75,000 barrels/day, which is expressed by the following equation:

$$\sum_{\substack{j \\ j \neq Olmeca}} C_j \leq 75{,}000 \tag{20.11}$$

In the same way, the production of each commodity is also restricted by the maximum capability.

$$x_i \leq Max\_Cap_i \quad \forall \text{i} \tag{20.12}$$

To code this problem in Lingo, we define two sets, *Product and Crude*, and the variables related with them. We define variable *C* to calculate the amount of crude to be processed and *X* to calculate the amount generated of products such as gasoline, kerosene and so on.

```
Product /gas keros fuel resi/:price,x;
Crude /Ural WTI Olmeca/: c,P_crude, operation;
```

We define another set to feed yield data in Table 20.2:

```
production(Product,Crude):Yield;
```

Then, we write Equations 20.6–20.10 as follows:

```
Incomes=@sum(Product(i):price(i)*x(i));
Feedstock=@sum(Crude(i):p_crude(i)*c(i));
Op_Cost=@sum(Crude(i):operation(i)*c(i));
@for(Product(i):x(i)=@sum(production(i,j):yield(i,j)*c(j)));
Profit=Incomes-Op_Cost-Feedstock;
```

Finally, Equation 20.11 which restricts the usage of Ural and WTI is written as follows:

```
@sum(Crude(i)|i#NE#3:C(i))<=75000;
```

This expression indicates that the summation of elements, i, of set Crude, such that, i is different to 3 (equals to Olmeca crude) has to be lower than 75,000.

We can use a very helpful tool in Lingo called "Display model" to see our summarized model. We select from menu bar Solver/Generate/Display model and we get:

```
MODEL:
 [_1] MAX= PROFIT;
 [_2] INCOMES - 90 * X_GAS - 30 * X_KEROS - 47 * X_FUEL - 5 *
   X_RESI = 0;
```

```
[_3] FEEDSTOCK - 50 * C_URAL - 80 * C_WTI - 20 * C_OLMECA = 0;
[_4] OP_COST - 0.6 * C_URAL - 0.4 * C_WTI - 0.9 * C_OLMECA = 0;
[_5] - 0.75 * C_URAL - 0.87 * C_WTI - 0.37 * C_OLMECA
  + X_GAS = 0;
[_6] - 0.11 * C_URAL - 0.08 * C_WTI - 0.1 * C_OLMECA
  + X_KEROS = 0;
[_7] - 0.1 * C_URAL - 0.04 * C_WTI - 0.21 * C_OLMECA
  + X_FUEL = 0;
[_8] - 0.04 * C_URAL - 0.01 * C_WTI - 0.32 * C_OLMECA +
  X_RESI = 0;
[_9] PROFIT - INCOMES + FEEDSTOCK + OP_COST = 0;
[_10] C_URAL + C_WTI <= 75000;
[_11] X_GAS <= 78000;
[_12] X_KEROS <= 7000;
[_13] X_FUEL <= 13600;
[_14] C_URAL >= 0;
[_15] C_WTI >= 0;
[_16] C_OLMECA >= 0;
END
```

The following is the solution:

```
Variable        Value          Reduced Cost
PROFIT          1843           0.000000
INCOMES         3538315.       0.000000
FEEDSTOCK       1635115.       0.000000
OP_COST         59725.19       0.000000
PRICE( GAS)     90.00000       0.000000
PRICE( KEROS)   30.00000       0.000000
PRICE( FUEL)    47.00000       0.000000
PRICE( RESI)    5.000000       0.000000
X( GAS)         28780.15       0.000000
X( KEROS)       7000.000       0.000000
X( FUEL)        13600.00       0.000000
X( RESI)        19780.15       0.000000
C( URAL)        8396.947       0.000000
C( WTI)         0.000000       14.91107
C( OLMECA)       60763.36      0.000000
P_CRUDE( URAL) 50.00000       0.000000
P_CRUDE( WTI)  80.00000       0.000000
P_CRUDE( OLMECA) 20.00000     0.000000
OPERATION( URAL) 0.6000000     0.000000
OPERATION( WTI) 0.4000000    0.000000
OPERATION( OLMECA) 0.9000000  0.000000
YIELD( GAS, URAL)  0.7500000  0.000000
YIELD( GAS, WTI)  0.8700000   0.000000
YIELD( GAS, OLMECA) 0.3700000 0.000000
YIELD( KEROS, URAL) 0.1100000  0.000000
YIELD( KEROS, WTI) 0.8000000E-01  0.000000
YIELD( KEROS, OLMECA) 0.1000000  0.000000
YIELD( FUEL, URAL) 0.1000000    0.000000
```

```
YIELD( FUEL, WTI)    0.4000000E-01 0.000000
YIELD( FUEL, OLMECA) 0.2100000   0.000000
YIELD( RESI, URAL)   0.4000000E-01 0.000000
YIELD( RESI, WTI)    0.1000000E-01 0.000000
YIELD( RESI, OLMECA) 0.3200000    0.000000
```

EXERCISE

Crude oil contains sulfur whose concentration in the final product is limited by regulations. Assume that Olmeca, Ural and WTI has 10, 3, and 0.5% of sulfur per volume respectively.

1. Determine the sulfur content by maximizing the profit of the company, assuming no restrictions in sulfur content.
2. Determine the profit value by minimizing the sulfur content.

The results can be found in the CRC press webside.

20.3.3 Metabolic Engineering: Bacteria Engineering

A company has a process to generate the products P_1, P_2, and P_3 that is carried out using a bacteria that follows the metabolic process shown in the Figure 20.1.

The company would like to increase the production of product, P_3 and it is necessary to determine the maximum amount that can be generated by using the substrates S_1, S_2 and S_3. The maximum availability of substrates is restricted to be less than 100, 175, and 200, respectively.

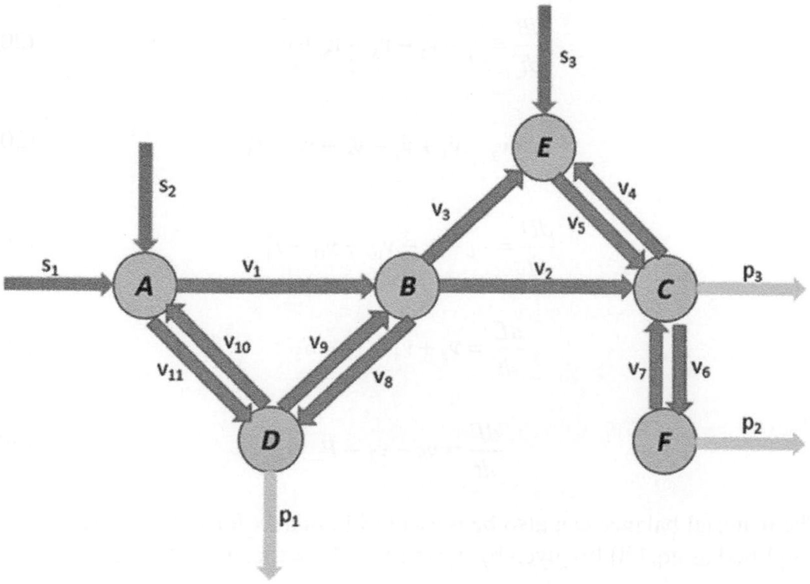

FIGURE 20.1 Metabolic process.

Furthermore, the company has collected experimental data that relates the fluxes. These relationships are as follows:

- Flux one (v_1) is equal to the sum of the flow of substrates S_1 and S_2 minus flow 3 (v_3).
- Flux two (v_2) is the same as flow one (v_1).
- Flux three (v_3) is twice flow 4 (v_4)
- Flux four (v_4) is equal to half the flow of substrate one (S_1).
- Flux five (v_5) is similar to the sum of flows four and six (v_4 and v_6).
- Flux six (v_6) is the same as flow eight (v_8).
- Flux seven (v_7) is equal to the sum of flows one, two and three (v_1, v_2 and v_3).
- Flux eight (v_8) is the same as the flow of substrate one (S_1).
- Flux nine (v_9) is the same as flow ten (v_{10}).
- Flux ten (v_{10}) is equal to the sum of the flow of products one and two (P_1 and P_2).
- Flux eleven (v_{11}) is the same as flow ten (v_{10}).

The flux of product, P_1, equals the summation of the products, P_2 and, P_3. Whereas, the flowrate of the second product, P_2, is twice as high as the flow obtained from the third product, P_3. Finally, the third product depends on the flow seven, v_7, and v_{10} as Equation 20.36 shows.

The first stage is to perform the material balance to the metabolites:

$$\frac{dA}{dt} = -v_1 + v_{10} - v_{11} + S_1 + S_2 \tag{20.13}$$

$$\frac{dB}{dt} = v_1 - v_2 - v_3 - v_8 + v_9 \tag{20.14}$$

$$\frac{dC}{dt} = v_2 - v_4 + v_5 - v_6 + v_7 - P_3 \tag{20.15}$$

$$\frac{dD}{dt} = v_8 - v_9 - v_{10} + v_{11} - P_1 \tag{20.16}$$

$$\frac{dE}{dt} = v_3 + v_4 - v_5 + S_3 \tag{20.17}$$

$$\frac{dF}{dt} = v_6 - v_7 - P_2 \tag{20.18}$$

The material balance can also be presented in matrix form. In steady state it can be simplified as eq. (20.19) given by the matrix of coefficients S by the flow vector v:

$$\overline{T} = \overline{0} = S \cdot \overline{v} \tag{20.19}$$

Based on to the experimental data, the fluxes have the following constraints:

$$v_1 = S_1 + S_2 - v_3 \tag{20.20}$$

$$v_2 = v_1 \tag{20.21}$$

$$v_3 = 2 \cdot v_4 \tag{20.22}$$

$$v_4 = \frac{S_1}{2} \tag{20.23}$$

$$v_5 = v_4 + v_6 \tag{20.24}$$

$$v_6 = v_8 \tag{20.25}$$

$$v_7 = v_1 + v_2 + v_3 \tag{20.26}$$

$$v_8 = v_9 \tag{20.27}$$

$$v_9 = v_{10} \tag{20.28}$$

$$v_{10} = P_1 + P_2 \tag{20.29}$$

$$v_{11} = v_{10} \tag{20.30}$$

The substrates constraints are:

$$0 \le S_1 \le 100 \tag{20.31}$$

$$0 \le S_2 \le 175 \tag{20.32}$$

$$0 \le S_3 \le 200 \tag{20.33}$$

Finally, the products constraints are as follows:

$$P_1 = P_2 + P_3 \tag{20.34}$$

$$P_2 = 2 \cdot P_3 \tag{20.35}$$

$$P_3 = v_{10} - v_7 \tag{20.36}$$

We solve this problem in a matrix form, to code this problem in Lingo the first stage we define six SETS. Three are to define rows and columns:

```
Rate /rA,rB,rC,rD,rE,rF/;
Reaction/1..17/;
Column/1/;
```

Next, we define the matrixes:

```
Matrix1(Rate,Column):T;
Matrix2(Rate,reaction):S;
Matrix3(Reaction,column):v;
```

Subsequently, the matrix of coefficients S is added as it was explained above. The steady state implies that the matrix T always takes null values.

In this case, we use the command @MTXMUL. The @MTXMUL function multiplies matrix S (6, 17) and matrix v (17,1) to generate matrix T (6,1) as the product.

```
Calc:
T=@MTXMUL(S,v);
endcalc
```

Finally, the constraints in vectorial form are introduced in Lingo and we maximize the product three, called v(17,1) in Lingo. The maximum amount to product three (P_3) is 112.5.

```
v(1,1)=v(12,1)+v(13,1)
  -v(3,1);
v(2,1)=v(1,1);
v(3,1)=2*v(4,1);
v(4,1)=v(12,1)/2;
v(5,1)=v(4,1)+v(6,1);
v(6,1)=v(8,1);
v(7,1)=v(1,1)+v(2,1)
  +v(3,1);
v(8,1)=v(12,1);
v(9,1)=v(10,1);
v(10,1)=v(15,1)+v(16,1);
v(11,1)=v(10,1);
v(12,1)=100;
v(13,1)=175;
v(14,1)=200;
v(15,1)=v(16,1)+v(17,1);
v(16,1)=2*v(17,1);
v(17,1)=v(10,1)-v(7,1);
Max=v(17,1);
```

Variable	Value	Reduced Cost
V(1, 1)	175.0000	0.000000
V(2, 1)	175.0000	0.000000
V(3, 1)	100.0000	0.000000
V(4, 1)	50.00000	0.000000
V(5, 1)	150.0000	0.000000
V(6, 1)	100.0000	0.000000
V(7, 1)	450.0000	0.000000
V(8, 1)	100.0000	0.000000
V(9, 1)	562.5000	0.000000
V(10, 1)	562.5000	0.000000
V(11, 1)	562.5000	0.000000
V(12, 1)	100.0000	0.000000
V(13, 1)	175.0000	0.000000
V(14, 1)	200.0000	0.000000
V(15, 1)	337.5000	0.000000
V(16, 1	225.0000	0.000000
V(17, 1)	112.5000	0.000000

REFERENCES

1. Gani, R., & Ng, K.M. (2015) Product design–molecules, devices, functional products, and formulated products. *Comput. Chem. Eng.*, 81, 70–79.
2. Cussler, E.L., & Moggridge, G.D., (2012) *Chemical product design.* Cambridge University Press Cambridge, U.K.
3. Martín, M., Martínez, (2013) A Methodology for simultaneous process and product design in the consumer products industry: The case study of the laundry business. *Chem. Eng. Res. Des.* (91) 795–809.
4. Seider, W.D., Lewin, D.R., Seader, J.D., Widagdo, S., Gani, R., & Ng, K.M. (2017). *Product and Process Design Principles: Synthesis, Analysis and Evaluation.* Wiley, New York.
5. Martín, M.M., Eden, M.R., & Chemmangattuvalappil, N.G. (Eds.). (2016) *Tools for Chemical Product Design: From Consumer Products to Biomedicine.* (Vol. 39) Elsevier, Netherlands.
6. Martín, M., Grossmann, I.E., (2013) Optimal engineered algae composition for the integrated simultaneous production of bioethanol and biodiesel. *AIChE J.* 59 (8) 2872–2883.
7. Lindo, T. (2017) Lingo: The Modeling Languange and Optimizer, 973. Retrieved from: http://www.lindo.com/downloads/PDF/LINGO.pdf
8. Fourer, R., Gay, D.M., Kernighan, B.W. (2003) Diet and other input models: minimizing costs. Retrieved from: https://ampl.com/BOOK/CHAPTERS/05-tut2.pdf
9. Hretcanu, C.E., Hretcanu, C.I. (2010) A linear programming model for a diet problem. Jounal Food and Environment safety of the Suceava University. Food Engeneering. Retrieved from: http://www.fia.usv.ro/fiajournal/index.php/FENS/article/viewFile/394/392

Index

C

E

H

I

J

K

L

Q

R

S

T

U

V

W

Printed and bound by CPI Group (UK) Ltd, Croydon, CR0 4YY
21/10/2024
01777044-0017